Geometrical and Physical Optics

GEOMETRICAL AND PHYSICAL OPTICS

R. S. Longhurst
Ph.D., A.R.C.S., D.I.C., F.Inst.P.

Senior Lecturer in Physics,
Chelsea College, London

Third Edition

Longman
London and New York

LONGMAN GROUP LIMITED
Longman House, Burnt Mill, Harlow,
Essex CM20 2JE, England
Associated companies throughout the world.

Published in the United States of America
by Longman Inc., New York

First published 1957
Second edition 1967
Third edition 1973
Reprinted 1976, 1981, 1984, 1986

ISBN 0-582-44099-8

Produced by Longman Group (FE) Ltd
Printed in Hong Kong

PREFACE TO THIRD EDITION

It is again a pleasure to record my gratitude to those who have criticised the text constructively. Since the book is intended primarily for students, detailed criticisms of particular portions of the text are as valuable as general criticisms of the overall plan. Thus many passages have been re-written with the aim of making the text easier to follow. Again bearing in mind the needs of students, I have responded to the request that I should give more than the bare solutions to the examples. I see little merit in disguising some of the standard bookwork as worked examples and, although a few worked examples have been included in the main text where it seemed that they reinforced the discussion usefully or (as in §2–27) where it seemed useful to illustrate methods of attacking a particular class of problem, the main change in this connection has been to supply outline solutions, together with comments, for the collection of examples at the back of the book.

My impression is that in optics the literature is becoming more mathematical but, bearing in mind the needs of undergraduates, I have not made this edition significantly more mathematical than earlier editions. There is one important exception: I have added several sections that require a knowledge of elementary matrix algebra. There has been something of a revolution in elementary mathematics teaching, and most students now encounter matrices fairly early in their careers—many at school. For their benefit I have included accounts of the use of matrix methods in paraxial geometrical optics and in the theory of multilayers. More important, perhaps, I have added a discussion of the Jones vectors and matrices, the coherency matrix, and Stokes parameters and Mueller matrices; associated with these additions, there is now included a discussion of the Poincaré sphere. The reader who is not familiar with matrix algebra could very quickly learn enough to follow these sections but can take comfort from the fact that other sections of the book do not depend upon them. There are several sections in which there has been some expansion of the discussions involving Fourier methods but these do not involve an increase in the initial mathematical knowledge demanded from the reader. For example, there is now a more detailed discussion of the analysis of finite wave trains but this merely involves the application of results that were included in earlier editions. The Fourier transform relationship associated with Fraunhofer diffraction patterns is discussed in a little more detail and there is now a section on spatial filtering, but again no significant increase in mathematical prowess is required.

Other major changes, reflecting current interest and progress, include an expansion of the discussion of coherence, with special reference to lasers. The discussion of the elementary theory and applications of holography has been expanded considerably, the section on fibre optics has been expanded, and a brief account has been given of the flying spot microscope. Recent developments in spectroscopic instruments and techniques have made it desirable to expand the section on Fourier transform spectroscopy and to add brief accounts of multislit and grille spectrometers, the Mock interferometer, and the spherical Fabry-Perot; it has been necessary to revise completely the comparison of the various classes of spectrometers. Major modifications have been made to the chapter on photometry: much of the discussion in earlier editions on visual methods is now of historical interest only and has been deleted in order to make room for an expansion of the discussion of physical photometry and for the inclusion of a discussion of the photometric properties of high resolution spectrometers. Many minor modifications and additions have been made in order to bring the book as a whole more up to date.

The need to include new material has made it necessary to curtail severely the accounts of polarimetry and of the early determinations of c. Other abbreviations and omissions include elementary material relating to the simple prism spectrometer since this is probably more appropriate to a laboratory manual.

The question of terminology and units has again caused me much anguish (and, in photometry, some anger). I am delighted to find that many other authors of books on optics refuse to be bullied and I have retained the mixed Gaussian system in electromagnetism. It is true that one has the irritation caused by some quantities being measured in esu and others in emu but I see little advantage in changing to the Giorgi system which is equally old-fashioned, having been conceived over 70 years ago when it was reasonable to plan a system that was suitable for those who accepted the 'luminiferous aether' as a reality, and also, in my view, has the disadvantage of making the transition from classical to modern physics more difficult. Apart from this I have used SI units and recommended terms and symbols as far as seems reasonable (the vintage that permits cm and is reconciled to heretical angstroms and cm^{-1}). I do accept the desirability of uniformity in the use of agreed symbols for various quantities and my main transgression is the Poynting vector; I have continued to use S for surface since A for area invites confusion with amplitude and I feel that the retention of bold π in this book is less liable to cause confusion. Further, an impecunious student is unlikely to vote to pay extra for the resetting of the type that would be required to have the full stops removed from, say, cm. throughout those sections of the book that have not otherwise been changed; to do this would be pointless pedantry.

I still believe that in a textbook such as this a list of books and review articles in English is more useful for most readers than a long list of references to original papers; anyone wishing to locate the original paper by a worker named in the text simply has to consult the Author Index of Physics Abstracts for the year quoted (or, in some cases, the following year).

Finally, I wish to record my gratitude to my wife who has again borne the preparation of a new edition with fortitude—and given a great deal of help throughout in preparing the typescript and reading the proofs.

R. S. LONGHURST

Chelsea College, London.
May 1972.

PREFACE TO FIRST EDITION

THIS book is intended primarily for students reading for an honours degree in physics, but it is hoped that it will be of use to those taking general or pass degrees and also to those who intend to specialize in optics at the post-graduate level. Thus, on the one hand I have tried to give a sufficiently detailed account of the subject to make the book useful to the more advanced student, and on the other hand I have borne in mind the fact that the mathematical equipment of some readers might be somewhat limited. For example, although an account is given of Kirchhoff's formulation of Huygens' principle, the chapters on diffraction and diffraction gratings are developed as far as possible along elementary lines. Most of the more mathematical parts of the book fall into separate sections and can be omitted by the less advanced student.

Since this is a students' textbook rather than a review, I have appended some suggestions for further reading instead of giving references to original papers. I feel this is the proper course since the undergraduate is always desperately short of time. The suggestions include more advanced books and some recent reviews; all are in English.

The book contains a number of photographs, but it has been necessary to group these together to form four plates. Since it is not possible to locate each photograph alongside the section that refers to it, I have thought it best to put the plates together near the front of the book, between pages 16 and 17.

It is a pleasure to record my gratitude to many friends for their help at various stages in the preparation of this book. In particular, I wish to thank Dr. L. J. Freeman, who read most of the original draft of the manuscript and with whom I have had many helpful discussions. I also wish to thank Dr. G. W. Series, who read and commented on the entire manuscript, Dr. H. H. Hopkins, who made numerous suggestions in connection with Chapters XIV–XVI, Dr. M. P. James, who read and commented on Chapters XVII and XVIII, and Dr. H. R. Paneth, who read and commented on Chapter XIX. I also wish to thank my wife for her help in preparing the typescript and reading the proofs.

I acknowledge with thanks permission to reproduce copyright material as follows:

The Director, N.P.L., the Council of the Royal Society, and the authors for Figs 9–13 and 9–14 from a paper by Sears and Barrell; the Oxford University Press for a quotation in Chapter IX from Tolansky's *Multiple Beam Interferometry of Surfaces and Films*; Constable and Co. Ltd. for Fig. 18–11 from Walsh's *Photometry*; Longmans, Green and Co. Ltd. for Fig. 5–25 from Watson's *Physics*; and Professor Partington for Figs 22–32 and 22–36 from *Advanced Treatise on Physical Chemistry*, Volume IV.

Chelsea Polytechnic   R. S. LONGHURST
February 1957

CONTENTS

CHAPTER I

INTRODUCTION AND ELEMENTARY THEORY

		page
1–1.	Rays, Waves, and Particles	1
1–2.	Light as a Wave Motion—Progressive Waves	1
1–3.	Simple Harmonic Waves	2
1–4.	Refractive Index and Optical Path	3
1–5.	Huygens' Principle	3
1–6.	Reflection of a Plane Wave at a Plane Surface	4
1–7.	Refraction of a Plane Wave at a Plane Surface	5
1–8.	Rays of Light	5
1–9.	Fermat's Principle	6
1–10.	Image Formation by an Optical System	7
1–11.	Refraction at a Spherical Surface—Notation	8
1–12.	Method of Deriving Formulae	10
1–13.	Refraction at a Spherical Surface	10
1–14.	The Paraxial Region	11
1–15.	Thin Lens in Air	11
1–16.	Smallest Separation of Object and Real Image	12
1–17.	Two Thin Lenses in Contact	14
1–18.	Refraction of a Wave at a Spherical Surface	14
1–19.	Reflection at a Spherical Surface	15

CHAPTER II

THE GENERAL GEOMETRICAL THEORY OF OPTICAL SYSTEMS

2–1.	Principal Planes and Focal Planes	17
2–2.	Location of Objects and Images	18
2–3.	Location of Conjugate Ray. General Case	19
2–4.	Transverse and Angular Magnifications. The Helmholtz Equation	20
2–5.	The Nodal Points	21
2–6.	Longitudinal or Axial Magnification	22
2–7.	Reference of Conjugate Distances to Principal Planes	22
2–8.	Reference to Nodal Planes	23
2–9.	Reference to any Pair of Conjugate Planes	24
2–10.	Helmholtz Equation—Infinite Conjugate	24
2–11.	Telescopic Systems	25
2–12.	The Combination of Two Systems	26
2–13.	Use of Powers and Reduced Distances	29
2–14.	Two Systems in Air	30
2–15.	Paraxial Refraction at a Spherical Surface	30
2–16.	The Thin Lens	32
2–17.	The Thick Lens	33
2–18.	The Spherical Lens	35

page

2–19. Examples of Principal Planes of a Thick Lens in Air   36
2–20. Two Separated Thin Lenses in Air   36
2–21. The Deviation Method for Tracing Paraxial Rays   38
2–22. Location of Cardinal Points of Complicated Systems   39
2–23. Application of the Deviation Method to Two Systems   40
2–24. Use of Matrix Notation for Paraxial Ray Tracing. The System
          Matrix   41
2–25. Image Formation. The Location of the Cardinal Planes   43
2–26. The Conjugate Distances Measured from the Principal Planes   46
2–27. Numerical Example of the Use of the Various Techniques for
          Solving Thick Lens Problems   47

CHAPTER III

THE PARAXIAL THEORY OF OPTICAL INSTRUMENTS

3–1.   Stops and Diaphragms   52
3–2.   The Human Eye   54
3–3.   Magnifying Power of a Telescope   55
3–4.   Simple and Telephoto Camera Lenses   57
3–5.   Projection Systems and Enlargers   58
3–6.   The Simple Microscope or Hand Magnifier   59
3–7.   The Principle of the Compound Microscope   60
3–8.   Chromatic Aberration   61
3–9.   The Thin Achromatic Doublet   62
3–10. Two Separated Thin Lenses of the Same Material   64
3–11. Huygens and Ramsden Eyepieces   65
3–12. The Optical Systems of Telescopes and Microscopes   66
3–13. Terrestrial Telescopes   68
3–14. Reflecting Systems   69

CHAPTER IV

THE DETERMINATION OF THE PARAXIAL CONSTANTS
OF OPTICAL SYSTEMS

I. SIMPLE METHODS FOR MIRRORS AND THIN LENSES

4–1.   Neutralization Test for the Power of a Thin Lens   71
4–2.   Location of Object and Image   71
4–3.   Simple Object and Image Coincidence Methods   71
4–4.   Refractive Index of a Lens   72

II. ACCURATE MEASUREMENT OF RADII OF CURVATURE

4–5.   The Spherometer   74
4–6.   Microscope Method for Short Radii   75
4–7.   The Knife-Edge Method for Concave Surfaces   75
4–8.   Shallow Surfaces (Convex and Concave)   76
4–9.   Newton's Rings   76

III. LOCATION OF THE CARDINAL POINTS OF A THICK LENS OR LENS SYSTEM

4–10. Collimator and Plane Mirror Methods for Principal Foci   76
4–11. Location of Principal Planes   77
4–12. The Nodal Slide   77

page

IV. INDIRECT DETERMINATION OF FOCAL LENGTH
4–13. Magnification Methods 79
4–14. The Goniometer 80
4–15. Magnifying Power of a Telescope 81
4–16. Magnifying Power of a Microscope 81

CHAPTER V

PRISMATIC INSTRUMENTS FOR REFRACTOMETRY AND SPECTROSCOPY

5–1. Refraction Through a Prism 82
5–2. Formation of a Spectrum 83
5–3. Dispersion of a Prism 84
5–4. Thin Prisms. Combinations of Prisms 85
5–5. The Simple Table Prism Spectrometer 86
5–6. Adjustment of a Spectrometer 87
5–7. Measurement of Refracting Angle and Angle of Minimum Deviation 88
5–8. Use of Hollow Prism for Liquids 89
5–9. Critical Angle Methods Using a Prism Spectrometer 89
5–10. The Auto-collimator 91
5–11. Wollaston's Critical Angle Method 92
5–12. The Pulfrich Refractometer 93
5–13. The Abbe Refractometer 93
5–14. The Hilger-Chance Refractometer 94
5–15. Refractive Index of Solids of Irregular Shape 95
5–16. The Simple Spectrograph 96
5–17. The Constant Deviation Spectrograph 97
5–18. Other Spectrographs and Monochromators. Spectrophotometers 97

CHAPTER VI

THE WAVE THEORY OF LIGHT

6–1. The General Equation of Wave Motion and the Principle of Superposition 100
6–2. Graphical Method of Adding Disturbances of the Same Frequency 103
6–3. The Use of Complex Quantities in Wave Theory 103
6–4. Coherent, Incoherent, and Partially Coherent Disturbances 105
6–5. Spatial Coherence with an Extended Quasi-monochromatic Source 109
6–6. The Coherence Properties of Lasers 111
6–7. Fourier Series 114
6–8. Extension of the Range. The Fourier Integral 116
6–9. Wave Trains of Finite Length 118
6–10. Group Velocity 122
6–11. Passage of a Pulse through a Diffraction Grating 124
6–12. Passage of a Pulse through a Prism 125
6–13. Talbot's Bands 125
6–14. Stationary Waves 128
6–15. The Doppler Effect 128
6–16. Malus' Experiment. Malus' Law 130
6–17. Brewster's Law 132
6–18. Superposition of Beams of Polarized Light 133
6–19. Polarized, Unpolarized and Partially Polarized Light 136

CHAPTER VII

INTERFERENCE OF TWO BEAMS. DIVISION OF WAVEFRONT

                                                                    page
7–1. Classification of Interference and Diffraction Phenomena   137
7–2. Young's Experiment   137
7–3. Elementary Discussion of Young's Experiment Using Slits   138
7–4. Fresnel's Biprism   140
7–5. Fresnel's Double Mirror   142
7–6. Lloyd's Mirror   142
7–7. Other Arrangements   143
7–8. Achromatic Fringes with Lloyd's Mirror   144
7–9. General Theory of Achromatic Fringes   145
7–10. Introduction of a Thin Transparent Plate   145

CHAPTER VIII

INTERFERENCE OF TWO BEAMS. DIVISION OF AMPLITUDE

8–1. Parallel-sided Plates   147
8–2. Interference in Very Thin Films   150
8–3. The Wedge Film   150
8–4. Newton's Rings   152
8–5. Non-reflecting Films   154
8–6. Further Applications of Localized Fringes   155
8–7. The Michelson Interferometer   156
8–8. Localized Fringes. White Light   158
8–9. Adjustment of the Michelson Interferometer   158
8–10. Visibility of the Fringes. Structure and Width of Spectral Lines 159
8–11. Fourier Transform Spectroscopy   161
8–12. The Twyman and Green Interferometer   166
8–13. The Jamin Interferometer   168
8–14. The Mach-Zehnder Interferometer   170
8–15. The Wavefront Shearing Interferometer   171
8–16. Gauge Measuring Interferometers   172

CHAPTER IX

MULTIPLE BEAM INTERFEROMETRY

9–1. Multiple Beam Fringes of Equal Inclination   175
9–2. The Fabry-Perot Interferometer and Etalon   179
9–3. Resolving Power of Fabry-Perot Instruments   180
9–4. Comparison of Wavelengths. Method of Coincidences. Hyper-
        fine Structure   184
9–5. Method of Exact Fractions   185
9–6. Measurement of the Fractions   187
9–7. Brewster's Fringes   189
9–8. Central Spot Scanning   192
9–9. The Spherical Fabry-Perot Interferometer   195
9–10. Comparison of Wavelengths with Mechanical Standards of
        Length. Method of Sears and Barrell   196
9–11. Summary of Results of All Determinations. The New Metre   200
9–12. Two Fabry-Perot Instruments in Series   203
9–13. Method of Edser and Butler for Calibration of a Spectroscope 203

page

9–14. The Lummer-Gehrcke Interferometer 204
9–15. Multiple Beam Fringes of Equal Thickness 205
9–16. Fringes of Equal Chromatic Order 208
9–17. Use of Highly Reflecting Dielectric Films. Interference Filters 210
9–18. Summary of Types of Fringes 213

CHAPTER X

HUYGENS' PRINCIPLE AND
THE SCALAR THEORY OF DIFFRACTION

10–1. Early Geometrical Picture 215
10–2. Fresnel's Extension 215
10–3. Propagation of a Spherical Wave 216
10–4. Defects of Fresnel's Theory 219
10–5. Circular Aperture. Circular Obstacle. Irregular Obstacle 220
10–6. Kirchhoff's Formulation of Huygens' Principle 221
10–7. Application to Spherical Waves 223
10–8. Kirchhoff's Diffraction Formula. St. Venant's Principle 224
10–9. Classification of Diffraction Patterns 225
10–10. Fresnel and Fraunhofer Patterns for a Single Slit 227
10–11. Fresnel and Fraunhofer Patterns for a Double Slit 229
10–12. Babinet's Principle 230
10–13. Young's Edge-Wave Interpretation of Diffraction 231

CHAPTER XI

FRAUNHOFER DIFFRACTION PATTERNS

11–1. Rectangular Aperture and Point Source. Location of the Minima 232
11–2. Intensity Distribution. Use of the Graphical Method 234
11–3. Use of the Kirchhoff Integral 237
11–4. Slit Aperture and Slit Source 238
11–5. Chromatic Resolving Power of a Prism Spectroscope. The Rayleigh Criterion 239
11–6. Limit of Resolution with a Slit Aperture 241
11–7. Circular Aperture and Point Source. Qualitative Discussion 241
11–8. Derivation of the Airy Pattern 242
11–9. Young's Eriometer 245
11–10. The Double Slit 246
11–11. Derivation of the Double Slit Pattern from Kirchhoff's Integral 249
11–12. Double Source and Source of Finite Width 249
11–13. Michelson's Stellar Interferometer 253
11–14. The Rayleigh Refractometer 255
11–15. Further Discussion of Fraunhofer Patterns 257

CHAPTER XII

DIFFRACTION GRATINGS

12–1. Gratings with Narrow Apertures 261
12–2. Intensity Distribution near Principal Maxima 264
12–3. Oblique Incidence. Minimum Deviation. Reflection Gratings 265
12–4. Grating Spectra 266
12–5. Dispersion. Overlapping of Orders 267

page

12–6.   Resolving Power   269
12–7.   Intensity Distribution in the Pattern for Narrow Slits   271
12–8.   Wide Slits and Other Diffracting Elements   273
12–9.   Blazed Gratings. Echelons, Echelettes, and Echelles   276
12–10. The Concave Grating   281
12–11. The Theory of the Concave Grating   282
12–12. Mountings for Gratings   285
12–13. Multislit and Grille Spectrometers   291
12–14. Mock Interferometry   292
12–15. The Production of Gratings   294
12–16. Irregularities in Gratings. Ghosts   297
12–17. Preliminary Review of the Various Classes of Spectrometers   298
12–18. Two- and Three-dimensional Gratings   302
12–19. Crossed Gratings—Moire Fringes   304

CHAPTER XIII

FRESNEL DIFFRACTION PATTERNS

13–1.   Elementary Discussion of the Division of a Plane Wavefront into
        Half-period Annular Zones   306
13–2.   Circular Obstacle and Circular Aperture   307
13–3.   The Zone Plate   309
13–4.   Rectangular Aperture. Cornu's Spiral and Fresnel's Integrals   311
13–5.   Derivation of Fresnel's Integrals from Kirchhoff's Formula   314
13–6.   The Use of a Slit Source   315
13–7.   Diffraction at a Straight Edge   316
13–8.   Single Slit   318
13–9.   Opaque Strip   319
13–10. Double Slit   320

CHAPTER XIV

THE PHYSICAL THEORY OF IMAGE FORMATION IN OPTICAL INSTRUMENTS

14–1.   Imagery by Perfect Optical Systems   321
14–2.   Optical Paths Along Neighbouring Rays   321
14–3.   The Sine Relation and the Sine Condition   322
14–4.   Herschel's Condition   324
14–5.   Limit of Resolution. The Rayleigh Criterion   326
14–6.   Limit of Resolution of a Telescope Objective   327
14–7.   Limit of Resolution of a Microscope Objective. (Incoherent
        Object Points)   328
14–8.   Necessary Magnifying Power of a Microscope   329
14–9.   Depth of Focus. The Rayleigh Quarter-wave Limit   331
14–10. Imagery by a Microscope of Non Self-luminous Objects   333
14–11. Abbe Theory of the Microscope   334
14–12. Derivation of Limit of Resolution from Abbe Principle   338
14–13. Dark-ground, Phase-contrast, and Interference Microscopy   340
14–14. Experimental Arrangements   345

page

14–15.   The Flying-spot Microscope   352
14–16.   Image Formation with Coherent Illumination. Spatial Filtering   353
14–17.   Image Formation with Incoherent Illumination. The Optical
       Transfer Function   354
14–18.   Wavefront Reconstruction (Holography). Gabor's Original
       Method   359
14–19.   The Basis of Modern Holography   362
14–20.   Applications of Holographic Image Formation   367
14–21.   The Foucault Test. Schlieren Systems   369
14–22.   Fibre Optics   370

CHAPTER XV

THE MONOCHROMATIC ABERRATIONS

15–1.   Introduction. Ray and Wavefront Aberration   376
15–2.   Choice of Reference Surface   377
15–3.   Longitudinal Shift of Focus   378
15–4.   Transverse Shift of Focus   378
15–5.   The Aberration Function   379
15–6.   Spherical Aberration   382
15–7.   Coma   384
15–8.   Astigmatism   385
15–9.   Field Curvature   386
15–10.   Distortion   387
15–11.   Coma and the Sine Condition   388
15–12.   Tangent Condition for the Elimination of Distortion   388
15–13.   Astigmatic Correction in the Presence of Petzval Field Curvature 389
15–14.   Permissible Aberration   390

CHAPTER XVI

THE OCCURRENCE AND CONTROL OF ABERRATIONS
IN OPTICAL SYSTEMS

16–1.   Factors Influencing the Monochromatic Aberrations   392
16–2.   Variation of Coma with Stop Position   394
16–3.   Variation of Distortion with Stop Position   394
16–4.   The Aberrations of a Single Spherical Refracting Surface. The
       Aplanatic Points   395
16–5.   The Aberrations of a Single Thin Lens   399
16–6.   The Achromatic Doublet   402
16–7.   The Astigmatism of a Concave Mirror   406
16–8.   The Design of Small Objectives   407
16–9.   Microscope Objectives   408
16–10.   Substage Condensers   411
16–11.   Photographic Lenses   411
16–12.   Eyepieces   418
16–13.   Reflecting Systems   419
16–14.   Zoom Lenses   421
16–15.   Computerized lens design   422

CHAPTER XVII

VISUAL OPTICS

page
17–1.   General Structure of the Eye   424
17–2.   The Optical System of a Normal Unaccommodated Eye   425
17–3.   Defects in the Optical System. Spectacles   427
17–4.   The Retina   428
17–5.   Adaptation   428
17–6.   Spectral Sensitivity of the Eye   430
17–7.   Visual Acuity   432
17–8.   The Stiles-Crawford Effect   433
17–9.   Colour Vision. Additive Mixtures   433
17–10. The Trichromatic System of Colour Measurement   435
17–11. The Colour of Non Self-luminous Objects   438
17–12. Subtractive Colour Mixture   439
17–13. Stereoscopic Effect   439
17–14. Stereoscopic Range. Binocular Telescopes   442
17–15. Binocular Microscopes   443

CHAPTER XVIII

RADIOMETRY, PHOTOMETRY, AND THE RADIOMETRIC AND PHOTOMETRIC PROPERTIES OF IMAGE-FORMING AND SPECTROSCOPIC INSTRUMENTS

18–1.   Fundamental Photometric and Radiometric Quantities   445
18–2.   Uniformly Diffusing Surfaces   448
18–3.   Brightness of a Diffuse Radiator   449
18–4.   Total Flux Radiated by a Uniformly Diffusing Surface Element   449
18–5.   The Luminance and Illumination of the Image Formed by an
        Optical System   450
18–6.   Observation with Visual Instruments   452
18–7.   The Maxwellian View   454
18–8.   Radiometric and Photometric Units. Absolute measurements   455
18–9.   The Visual Photometer Head and Bench   458
18–10. Flicker Photometers   460
18–11. Physical Photometry, Radiometry, Spectrophotometry and
        Spectroradiometry   461
18–12. The Radiometric Properties of High Resolution Spectroscopic
        Instruments   464
18–13. Directional Properties of Sources   470
18–14. The Integrating Sphere   470
18–15. Transmission and Reflection Factors   472
18–16. Illumination Photometers   473
18–17. Optical Pyrometers   474

CHAPTER XIX

THE ELECTROMAGNETIC THEORY OF LIGHT

19–1.   Vector Notation   475
19–2.   Maxwell's Equations. The Displacement Current   475
19–3.   Propagation of an Electromagnetic Disturbance in Isotropic
        Dielectrics   476
19–4.   Plane Waves   477
19–5.   The Poynting Vector   479
19–6.   Hertz's Experiments   481

page

19–7. Electrical Polarization in Anisotropic Media   482
19–8. Plane Waves in Anisotropic Media   484
19–9. Wave Surface Expanding from a Point Source   486
19–10. Uniaxial Media   489

CHAPTER XX

ABSORPTION, EMISSION, DISPERSION, AND SCATTERING

20–1. General and Selective Absorption   491
20–2. Lambert's Law   492
20–3. Dichromatism   494
20–4. The Scattering of Light by Small Particles   494
20–5. The Quantum Theory of Absorption, Emission and Scattering. The Laser   495
20–6. The Electromagnetic Theory of Absorption and Scattering   497
20–7. Normal Dispersion. Cauchy's Formula   498
20–8. Anomalous Dispersion. Sellmeier's Formula   500
20–9. Experimental Demonstration of Anomalous Dispersion   502
20–10. Propagation of Light in an Absorbing Medium   503
20–11. The Theory of Dispersion in Dielectrics   504
20–12. Regions Remote from Absorption Bands   506
20–13. Regions Close to Absorption Bands   507
20–14. Connection Between Dispersion and Molecular Scattering   508
20–15. The Propagation of Light in Metals   509
20–16. The Dispersion of Metals   510

CHAPTER XXI

REFLECTION AND REFRACTION

21–1. Boundary Conditions   512
21–2. Reflection and Refraction by Transparent Isotropic Dielectrics   512
21–3. Normal Incidence   516
21–4. Discussion of Results   516
21–5. Total Internal Reflection   519
21–6. Disturbance in the Second Medium   521
21–7. Reflection by an Absorbing Medium   522
21–8. The Reflected Disturbance   524
21–9. The Refracted Disturbance   526
21–10. Stationary Light Waves   527
21–11. Thin Dielectric Films and Multilayers   529

CHAPTER XXII

THE PROPAGATION OF LIGHT IN ANISOTROPIC MEDIA AND THE ANALYSIS OF POLARIZED LIGHT

22–1. Waves in Anisotropic Media   533
22–2. Double Refraction in Uniaxial Crystals   534
22–3. Plane Waves at Oblique Incidence   536
22–4. Observation of Double Refraction in Calcite   537
22–5. The Nicol Prism   539
22–6. Two Nicols in Series   540
22–7. Other Polarizing Prisms   541
22–8. Polarization by Selective Absorption (Dichroism). Sheet-type Polarizers   542
22–9. Ray and Wave Velocity for Biaxial Crystals   544

page

22–10. Internal Conical Refraction   546
22–11. External Conical Refraction   547
22–12. Passage of Light through Crystal Plates   548
22–13. Quarter- and Half-wave Plates   550
22–14. Use of a Quarter-wave Plate in the Analysis of Polarized Light   552
22–15. The Babinet and Soleil Compensators   553
22–16. Analysis of Elliptically Polarized Light using a Babinet Com-
        pensator   553
22–17. Passage of Collimated White Light through a Crystal Plate   555
22–18. Passage of Convergent Plane Polarized Pencils through Crystal
        Plates   557
22–19. The Polarizing Microscope   561
22–20. Jones Vectors and Matrices   562
22–21. The Coherency Matrix   565
22–22. Stokes Vectors and Mueller Matrices   567
22–23. The Poincaré Sphere   570
22–24. The Kerr Electro-optical Effect   572
22–25. The Voigt and Cotton-Mouton Effects   573
22–26. The Photo-elastic Effect   573

CHAPTER XXIII

OPTICAL ACTIVITY

23–1.   Optically Active Crystals   574
23–2.   Rotatory Dispersion   574
23–3.   Fresnel's Theory   575
23–4.   Wave Surface for Quartz   577
23–5.   The Cornu Prism   579
23–6.   Optical Activity in Liquids   580
23–7.   Polarimeters   581
23–8.   Half-shadow End-point Indicators   582
23–9.   Compensating Saccharimeters   585
23–10. Circular Dichroism. The Cotton Effect   586
23–11. The Faraday Effect   587

CHAPTER XXIV

THE VELOCITY OF LIGHT

24–1.   Introduction   588
24–2.   Römer's Method   588
24–3.   The Aberration of Light   589
24–4.   Early Terrestrial Methods   589
24–5.   The Use of a Kerr Cell Optical Shutter   590
24–6.   The Position in 1941   592
24–7.   Use of Radar, Cavity Resonators, and Micro-wave Inter-
        ferometry and Spectroscopy   593
24–8.   Recent Optical Experiments and Summary   596
24–9.   The Velocity of Light in Material Media   601
24–10. The Velocity of Light in a Moving Medium   602
24–11. The Michelson-Morley Experiment   603

        APPENDIX. Some Suggestions for Further Reading   605
        MISCELLANEOUS EXAMPLES   613
        OUTLINES OF SOLUTIONS WITH COMMENTS   626
        INDEX   651

CHAPTER I

INTRODUCTION AND ELEMENTARY THEORY

1-1. Rays, Waves, and Particles

THE energy emitted by a source of light may be considered to be ejected as a stream of particles or *photons*, as a continuous stream of energy along a *ray*, or as a *wave motion*. The idea of discrete packets of energy is mainly of use when considering the origin of spectra, and the interaction of light with matter as in the photo-electric effect. It will be found that as far as the *propagation* of light is concerned it is convenient to suppose that a periodic wave motion is propagated from a source. It is then found that in certain cases the conception of streams of energy or rays provides a useful basis for calculating the approximate effect of the waves.

It is easily seen that the conception of energy streaming out from the source along the rays can be used to explain the rectilinear propagation of light and the formation of shadows. If, however, one attempts to isolate a ray by means of a series of small holes as in Fig. 1-1, it is found that the smaller the holes are made the less like a ray does the light behave. This phenomenon is readily explained in terms of the wave theory of light, and illustrates the fact that in some cases the ray conception breaks down.

Fig. 1-1. Attempt to isolate a ray.

1-2. Light as a Wave Motion—Progressive Waves

In order to consider energy to be transmitted as a wave motion, one usually requires a medium whose particles may oscillate. An oscillating particle then exerts a force on its neighbour, causing the latter to oscillate. In this way the motion is transmitted from one particle to the next and the energy is conveyed through the medium. Since light can be propagated in a vacuum the existence of an all-pervading medium was postulated. This was known as the *aether*, and was assumed to pervade all space and to provide the means whereby the light waves were propagated. For the present purpose it is quite unnecessary to assume anything about the nature of this medium. However, to explain certain properties of light it is necessary to assume that the actual "particle displacements" which occur as the disturbance passes a point in space are perpendicular to the direction of propagation, i.e. light waves are transverse waves. For example, in *plane polarized* light all the "particles" execute linear S.H.M. Later theory replaces the oscillating particles by an oscillatory electromagnetic field, but discussion of this is deferred to later chapters. At this stage it is convenient to think of oscillating particles. Let ϕ denote the transverse displacement of the particles set in motion by the passage of the wave. It is easy to see that any function of the type $\phi = f(ct - x)$ represents a disturbance moving in the positive x-direction with velocity c. For example, at time t_1, ϕ will have a certain value at $x=x_1$, and at a later time t_2 it will have the same value at x_2 if $ct_1-x_1=ct_2-x_2$. Hence the disturbance travels a distance (x_2-x_1) in time (t_2-t_1) and the velocity is $(x_2-x_1)/(t_2-t_1)=c$. Similarly, $\phi=f(ct+x)$ represents a disturbance in the negative x-direction.

1-3. Simple Harmonic Waves

A point source in an isotropic medium will emit spherical waves and, assuming that the particles execute linear S.H.M., the above considerations indicate that one may write, for the expanding spherical waves,

$$\phi=\frac{a}{r}\sin k(ct-r), \qquad \ldots \ldots \quad (1\text{-}1)$$

where ϕ is the displacement of the particles distance r from the source at time t, and a is the amplitude at unit distance from the source. k is a constant. If, $\omega=kc$ one has

$$\phi=\frac{a}{r}\sin(\omega t-kr). \qquad \ldots \ldots \quad (1\text{-}2)$$

Since $\sin\theta$ is a periodic function of period 2π (i.e. $\sin(\theta\pm 2\pi)=\sin\theta$) it is seen, on putting $r=$constant, that at any particular point in space the particles execute S.H.M. of period $T=2\pi/\omega$ and of amplitude a/r. Now, for any progressive wave, the *intensity* is the rate of flow of energy across unit area perpendicular to the direction of propagation of the waves and, since the energy of a vibrating particle is proportional to the square of the amplitude, it will be seen that the intensity falls off as $1/r^2$. This gives the necessary agreement with the well established inverse square law for radiant energy, and is the reason for making the amplitude proportional to $1/r$. Equation (1-2) also shows that although all the particles oscillate with the same frequency, those at various distances from the source are at different stages in their oscillations. They are said to have different *phases*. The phase of any particle is given by the quantity $(\omega t-kr)$ so that at any instant the phase difference between particles at distances r and $r+d$ is kd. Since $\sin\theta$ has a period of 2π the particles will be in similar phases of their vibrations if $kd=2\pi$, i.e. if $d=\lambda$ where $\lambda=2\pi/k$. λ is called the *wavelength*. Now the particles whose distances from the source differ by λ have a phase difference of 2π, so that those whose distances differ by d have, by direct proportion, a phase difference $2\pi(d/\lambda)$. This is also evident if one puts $k=2\pi/\lambda$ so that $[\omega t-(2\pi/\lambda)r]$ specifies the phase. Referred to the phase of the source, the phase of a particle is then $(2\pi/\lambda)r$, and if one has particles at r and $r+d$ their phase difference is $(2\pi/\lambda)d$.

A continuous surface that is the locus of points where the particles are in phase is called a *wavefront*; any sphere centred on a point source is obviously a wavefront. Consider now the disturbance travelling outwards from the source. The quantity $k(ct-r)$ specifies the phase and the particle at r_2 at time t_2 will have the same phase as the particle at r_1 had at time t_1 if $ct_1-r_1=ct_2-r_2$. The particular phase may be said to have been propagated a distance (r_2-r_1) in time (t_2-t_1) and the velocity, $(r_2-r_1)/(t_2-t_1)=c$, may be referred to as the phase velocity. The wavefront associated with any particular phase may therefore be said to advance with velocity c, so that it advances a distance λ in time T. It is frequently convenient to look upon the advance of a wavefront as the advance of the initial disturbance from the source. Although this is usually done, it is fundamentally unsound since equation (1–1) implies a wave train of infinite extent.

The above discussion was concerned with a spherical wave. Another important type is the plane wave in which the wavefront is always a plane perpendicular to the direction of propagation. A simple harmonic plane wave travelling in the positive x-direction is represented by $\phi = a \sin k(ct-x)$. The phase is independent of y and z—it is constant over any plane perpendicular to the direction of propagation.

1–4. Refractive Index and Optical Path

It is found that the velocity of monochromatic light in a material medium is different from that in a vacuum. If the velocities are v and c respectively, the ratio c/v is called the *refractive index* and in this book is denoted by n. If f is the frequency of the vibrations, it follows from above that $f\lambda = c$. Now if light passes from one medium to another, the frequency of the separate oscillating particles remains unchanged, so that the wavelength alters; if λ and λ_m are the wavelengths in a vacuum and material medium respectively, $\lambda/\lambda_m = n$. From § 1–3 it follows that, referred to the source, the phase at a distance r from a source in a material medium is $(2\pi/\lambda_m)r$, which may be written $(2\pi/\lambda)rn$. If light reaches a point after passing through several different media, the phase at that point will be given by $(2\pi/\lambda)\Sigma rn$. A geometrical distance multiplied by the refractive index of the space in which it is measured is referred to as an *optical path*. It will be seen that in a given time a disturbance always traverses the same optical path. Frequently one is concerned with light which reaches a point or points after passing along different paths, and in this case it will be seen that the phase difference between the disturbances arriving by different routes is given by

$$\text{Phase difference} = (2\pi/\lambda)(\text{optical path difference}). \quad . \quad . \quad (1\text{–}3)$$

Sometimes several disturbances arrive at a point with different phases and are superposed. The way in which one calculates the total disturbance will be discussed in Chapter VI.

1–5. Huygens' Principle

It was stated above that the mechanism by which a disturbance can be propagated is that a vibrating particle exerts a force on its neighbour, causing it to vibrate. Thus every vibrating particle may be said to behave as a point source. According to *Huygens' Principle* it is supposed that every point on a wavefront acts as a point source in this way and becomes a source of so-called secondary wavelets. The subsequent disturbance is then found as a combination of all the secondary wavelets. This accounts at once for the failure of the principle of rectilinear propagation through small apertures, and explains why one fails to isolate a ray by using a succession of pinholes as in Fig. 1–1. When, for example, a plane wave impinges on a screen with a small hole, the element of wavefront at the hole acts as a point source, so that a spherical wave emerges on the far side of the screen (see Fig. 1–2). In practice most of the energy remains within

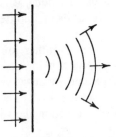

Fig. 1–2. Passage of light through a small aperture.

a fairly small cone with the hole as apex. In order to explain the approximately rectilinear propagation through large apertures and the formation of shadows, it is necessary to postulate that the secondary wavelets produce an appreciable effect on their forward envelope only. Consider a plane wave passing through a large aperture (see Fig. 1–3). When the wavefront lies in the plane of the aperture each point may be taken as a source. After a time t the secondary wavefronts have travelled a distance ct and the resultant disturbance is the forward envelope of the wavelets as shown. Subsequent positions of the wavefront may be found in a similar manner, and the result indicates that the disturbance passes through the aperture as a broad beam of light travelling in a direction perpendicular to the wavefront.

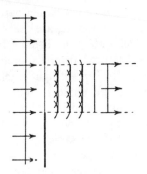

Fig. 1–3. Passage of a plane wave through a large aperture.

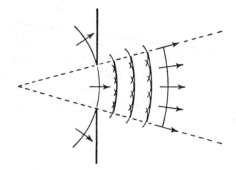

Fig. 1–4. Passage of a spherical wave through a large aperture.

The passage of a spherical wave through an aperture can be explained in an exactly similar manner, and, in this case also, the disturbance is propagated along the normals to the wavefront (see Fig. 1–4). It is obvious that the way in which light passes through a wide aperture can also be explained on the assumption that the light consists of streams of energy or·rays. For isotropic media the rays in this interpretation of the phenomenon are always normal to the waves in the previous interpretation.

1–6. Reflection of a Plane Wave at a Plane Surface

In Fig. 1–5 let ABC be a plane wave incident on a reflecting surface XY. As each element of the wave reaches the surface, it can be regarded as a source of secondary wavelets. Some elements of the wave will, of course, reach the mirror before others, and in order to find the subsequent path of the wave, one must allow for the fact that some points on the mirror act as sources before others. In particular, A_1 acts as a source when C is at C_1, so that when C gets to C_2 the secondary wavelet from A_1 has expanded to a radius $A_1A_2 = C_1C_2$. Similarly,

B reaches B_2 before C reaches C_2, and the secondary wavelet from B_2 has a radius $B_2B_3 = (C_1C_2 - B_1B_2)$ when C reaches C_2. Treating all parts of the wave in this way and taking the envelope of these secondary wavelets, one obtains $A_2B_3C_2$ as the reflected position of the wave. The wave normals before and after reflection are in the same plane as, and equally inclined to, the normal to the mirror at the point of incidence, i.e. $I = I'$.

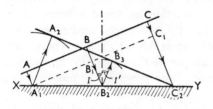

Fig. 1–5. Reflection of a plane wave at a plane surface.

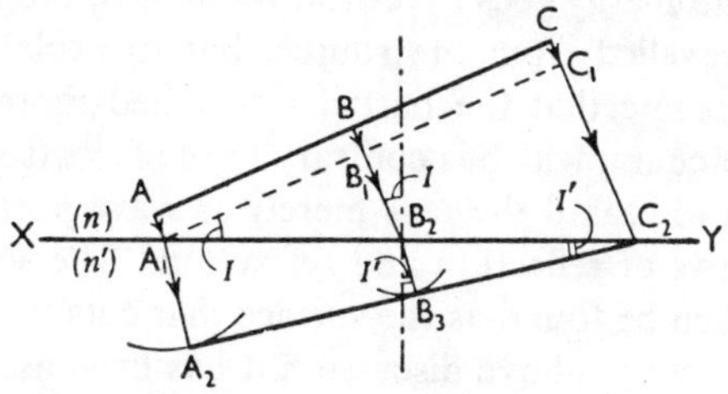

Fig. 1–6. Refraction of a plane wave at a plane surface

1–7. Refraction of a Plane Wave at a Plane Surface

In Fig. 1–6, suppose XY is a plane interface separating media whose refractive indices are n and n'. As before, one regards points A_1, B_2, C_2, as secondary sources. In this case the disturbance passes into the second medium where the velocity of light is different. The ratio of the velocities is given by $(v/v') = (n'/n)$. Consequently, while the disturbance travels from C_1 to C_2, the secondary wavelet from A_1 expands to a radius $A_1A_2 = C_1C_2(v'/v) = C_1C_2(n/n')$. Similarly, when C is at C_2 the secondary wavelet from B_2 has a radius $B_2B_3 = (C_1C_2 - B_1B_2)(n/n')$. The envelope $A_2B_3C_2$ consequently gives the refracted position of the wave.

From the figure,
$$\frac{\sin I}{\sin I'} = \frac{C_1C_2/A_1C_2}{A_1A_2/A_1C_2} = \frac{v}{v'} = \frac{n'}{n},$$

i.e.
$$n \sin I = n' \sin I'. \quad \ldots \ldots \quad (1\text{–}4)$$

This relates the directions of the wave normals before and after refraction. As before, the two wave normals and the normal to the refracting surface lie in the same plane. These laws of refraction were first given by Snell. It will be seen that the larger is the angle of incidence I the greater is the deviation produced by refraction.

The reflection and refraction of plane and spherical waves at spherical surfaces may be treated in the same way.

1–8. Rays of Light

In the case of reflection and refraction one again has the energy propagated along the wave normals. Consequently, if the rays are defined to be wave normals, one can say that the light behaves as a stream of energy travelling along the rays which are directed outwards from the source and obey the classical laws of reflection and refraction given above. This is very convenient in practice since the laws governing the ray paths are so simple. However, there are cases when the assumption that the energy travels along rays leads to erroneous results. One example has already been discussed: the passage of light through a small hole. Another example of the failure of the ray conception is the case of a converging spherical wave. Here all the rays pass through the centre of curvature of the spherical wavefront, indicating an infinite energy density at that point. The simple wave theory as discussed above also fails to explain what happens in this case. On the other hand, there is no doubt that there is a concentration of energy near the focus of the wave, so that the rays do give some idea of where the energy goes. Newton thought of the rays as the paths along which particles travelled from the source, but to explain refraction he found it necessary to assume that the particles travelled more rapidly in material media than in a vacuum, which is contrary to the facts (see Chapter XXIV). The best procedure is to regard the rays merely as wave normals, where the rays obey the classical laws of reflection and refraction. The shape of the wavefront at any time can then be found as the surface that cuts the rays orthogonally.

In the above discussion it has been assumed that the media are isotropic. In anisotropic media the energy is not propagated in the direction of the wave normals—even when one is far removed from a focus or a screen. The rays are then taken to be the lines of energy flow and are not wave normals.

1-9. Fermat's Principle

The laws governing the behaviour of the rays can be summarized in one fundamental law known as *Fermat's principle*. This states that the *ray path along which a disturbance travels from one point to another is such that the time taken is at a stationary value.* If c is the velocity of light in a vacuum, the velocity in a

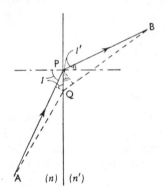

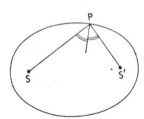

Fig. 1–7. Snell's law and Fermat's   Fig. 1–8. Focal property of
           principle.   an ellipse.

medium of refractive index n is, by definition, c/n; in this medium the time taken for a disturbance to travel a distance l is nl/c. If the light passes through a number of media, the total time taken is $(1/c)\Sigma nl$ (or $(1/c)\int ndl$ if the refractive index varies continuously). Now Σnl is the total optical path, so that Fermat's principle states that the path of a ray is such that the *optical path* is at a stationary value. This principle is obviously in agreement with the fact that the rays are straight lines in a homogeneous isotropic medium. It is found that it also agrees with the classical laws of reflection and refraction. Consider, for example, refraction at a plane surface separating media whose refractive indices are n and

n' (see Fig. 1–7). Of the possible routes from A to B that give a straight path in each medium, it is easily proved that the path APB given by Snell's law is also the route that gives a stationary optical path length. Thus let Q be a point on the refracting surface close to P. The optical path difference [APB]−[AQB] is nPQ sin $I−n'$PQ sin I'. (It is usual to employ square brackets to denote optical path lengths.) Now, from Snell's law, n sin $I−n'$ sin $I'=0$, so that the optical path difference is zero. That is, [APB] is at a stationary value when Snell's law is obeyed. Reflection at a plane surface can be dealt with in a similar manner.

The fact that the optical path is sometimes a maximum and sometimes a minimum can be demonstrated by using a well known property of ellipses. In Fig. 1–8, if S and S′ are the foci of an ellipse, it can be shown that SP+PS′ remains constant as P moves round the circumference, and SP and S′P make

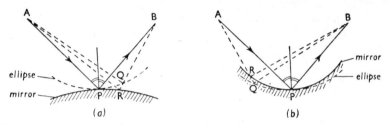

Fig. 1–9. Fermat's principle. (a) Ray path a minimum. (b) Ray path a maximum.

equal angles with the normal at P. Referring now to Fig. 1–9(a) or (b), the ray path along which light travels from A to B via the mirror is, according to the classical law of reflection, that for which the angles of incidence and reflection are equal. The point of incidence P must be the point where the mirror touches an ellipse whose foci are A and B since AP and PB are then equally inclined to the normal. Also, it has been seen that (AP+BP)=(AQ+QB) if Q is another point on the ellipse. It follows that (AR+RB) is greater than (AP+PB) in (a), and less than (AP+PB) in (b), so that the ray path gives a minimum optical path in (a) but a maximum optical path in (b).

1–10. Image Formation by an Optical System

Consider the case of a single convex lens forming a real image (see Fig. 1–10). It can be shown that, within certain limits, points A and B exist such that the

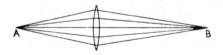

Fig. 1–10. Image formation by a lens.

optical path from A to B is the same by any of the possible routes through the lens. Consequently, all the rays from A that enter the lens will intersect at B so that, corresponding to point sources on the left-hand side of the lens, there are concentrations of energy on the right-hand side. Since all the rays from A through the lens pass through B, the latter point is located if one can find where any *two* rays intersect. This is the interpretation of image formation in terms of rays but, as remarked earlier, one must ultimately treat this ray picture as an approximation since the energy does not flow exactly along the rays.

In terms of waves one may say that all the disturbances from A arrive in phase at B, i.e. a spherical wave from the point source A emerges from the lens as a spherical wave converging to B. A study of the ray paths is, however, of considerable importance in discussing the formation of images by optical systems, and, if necessary, the shapes of the wavefronts can be deduced from knowledge of the ray paths.

1-11. Refraction at a Spherical Surface—Notation

The reader is warned that, in places, the notation in this book appears at first sight to be rather cumbersome. It is certainly not the easiest convention to

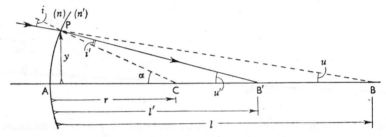

Fig. 1–11. Refraction at a spherical surface.

describe, but it has the merit of completely universal applicability; when fully understood, its advantages become obvious. It is unfortunate that some of the conventions which may be employed very conveniently in elementary optics become almost unusable in the more advanced stages and are quite different from those employed in industry and commerce.

In Fig. 1–11, a ray directed towards B is refracted so as to pass through B' where B and B' lie on the axis of the refracting surface AP, centre C, and radius r, separating media whose refractive indices are n and n'. The ray intersection distances l and l', and the radius r, are measured from A, the pole of the surface. i and i' are the angles of incidence and refraction; u and u' are the angles between the ray and the optical axis before and after refraction; y is the height of the point of incidence, P, above the axis; and α is the angle subtended at C by the arc AP. For an object or image of finite size perpendicular to the axis, the object and image heights measured from the axis are h and h'. These symbols are also used to denote the distance off-axis of extra-axial point objects or images.

In Fig. 1–11 the undashed symbols refer to the ray before refraction (i.e. to the object) and the dashed symbols refer to the ray after refraction (i.e. to the image), but this is not universally the case. When a ray strikes a refracting surface it is deviated, but the undeviated continuation of the incident ray is referred to as a virtual ray in the second medium. Similarly, the refracted ray, produced backwards, is a virtual ray in the first medium. Thus any ray can be said to exist on both sides of the surface, being real on one side and virtual on the other. This concept is now extended and one says that the refracting surface divides space into two parts which are known as the *left-* and *right-hand spaces.*

One thinks of both spaces extending to infinity in both directions, the former being real on the left-hand side of the surface and virtual on the right-hand side, whilst the reverse is true of the right-hand space. One now says that *symbols associated with quantities or positions in the left-hand space are undashed and those in the right-hand space are dashed.* If the light travels from left to right, the incident ray is real on the left-hand side and is therefore in the left-hand space. The refracted ray is real on the right-hand side and is in the right-hand space. Similarly, if the light travels from right to left the incident ray is in the right-hand space and the refracted ray is in the left-hand space. It will be seen that the convention can be stated in the following alternative form. If the light travels from left to right, the undashed symbols refer to the ray before refraction (i.e. to the object) and the dashed to the ray after refraction (i.e. to the image). On

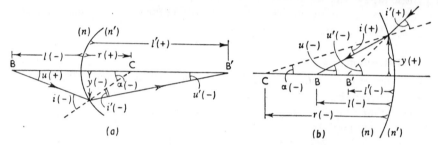

Fig. 1–12. Refraction at a spherical surface.

the other hand, if the light travels from right to left, the dashed quantities refer to the object and the undashed to the image. (Note that although it has not yet been proved that B′ is actually the image of B, this will be deduced shortly and, in the meantime, it is convenient to speak of object and image.) Usually the light will be taken as travelling from left to right so that the dashed quantities refer to the image, but it is frequently convenient to discuss an instrument with the light travelling, first one way, and then the other.

The sign convention for distances is the same as that used in co-ordinate geometry. Consequently, l and l' (measured from the pole of the surface) are positive if measured to the right, and y, h, and h' (measured from the axis) are positive if measured upwards.

The sign convention for angles is now chosen in such a way that it is possible to draw a diagram in which r, l, l', y, h, h', α, u, u', i, and i', are positive. One can then recall the notation and sign convention by remembering this diagram. Fig. 1–11 is an *all-positive* diagram.

To determine the sign of u or u', rotate about B or B′ through the angle concerned from the ray to the axis. A clockwise rotation defines a negative angle, and an anti-clockwise rotation a positive angle. The sign of α is similarly defined by a rotation from radius to axis. The sign of i or i' is determined by rotating about the point of incidence through the angle from ray to radius, when a clockwise rotation now defines a positive angle and an anti-clockwise rotation indicates a negative angle (see Fig. 1–12). The angles u, u', i, i', and α, are always acute.

This notation and sign convention has the advantage that if the direction of the light is reversed (i.e. object and image interchanged), the same symbols represent the same quantities and positions without change of sign, and the same equations apply.

1-12. Method of Deriving Formulae

When it is necessary to derive from first principles a formula relating to some phenomenon, the procedure is to draw a diagram illustrating any special case, treat all the distances and angles as positive, and, from the geometry of the diagram, deduce an equation relating the quantities in which one is interested. When the final equation is obtained, it is necessary to change the sign in front of all the quantities that are negative in the figure. The result is the algebraic equation which can be applied to any possible case. It is never necessary to consider the signs of quantities that appear during the derivation but are absent from the final equation.

1-13. Refraction at a Spherical Surface

The simplest procedure is to deduce the relation between l and l' from the all-positive diagram but, to illustrate the method, Fig. 1–12(a) will be used.

It is assumed that the rays are at all times close to the optical axis and never inclined to it at large angles. That is to say, y is always small and the angles may be equated to their sines or tangents. In addition, the sagitta in the refracting surface may be neglected. Snell's law then becomes $ni=n'i'$.

Now, from Fig. 1–12(a),

$$i=\alpha+u \quad \text{and} \quad i'=\alpha-u'$$

and, writing the angles as their tangents, one has

$$n\left(\frac{y}{r}+\frac{y}{l}\right)=n'\left(\frac{y}{r}-\frac{y}{l'}\right)$$

or

$$\frac{n'}{l'}+\frac{n}{l}=\frac{n'-n}{r}.$$

In the figure, however, l is negative so that, inserting a negative sign in front of l, the algebraic equation becomes

$$\frac{n'}{l'}-\frac{n}{l}=\frac{n'-n}{r}.$$

Note also that u', i, i', and α, are negative so that, algebraically, $\alpha=u+i=u'+i'$. If $l=f$ when $l'=\infty$, and $l'=f'$ when $l=\infty$, f and f' are called the *first and second focal lengths*. In addition, it is frequently convenient to put $(n'/f')=-(n/f)=F$, where F is called the *power*. Consequently, one has

$$\frac{n'}{l'}-\frac{n}{l}=\frac{n'}{f'}=-\frac{n}{f}=\frac{n'-n}{r}=F. \quad \cdots \cdots \quad (1\text{–}5)$$

Alternatively, one may write

$$\frac{f'}{l'}+\frac{f}{l}=1. \quad \cdots \cdots \cdots \quad (1\text{–}6)$$

1–14. The Paraxial Region

In the above discussion it was assumed that the height of incidence, y, is small so that all the rays are close to the optical axis. Such rays are known as *paraxial rays*, and one speaks of the region immediately surrounding the axis as the *paraxial region*. That is to say, within the paraxial region the angles u, u', α, i, and i', may be equated to their sines or tangents. It will be observed that equation (1–5), relating l and l', does not involve y. Consequently, within the paraxial region, any ray directed through B before refraction will pass through B' after refraction. Therefore B' is the image of B, and may be located by considering any one ray. It is easy to see that within the paraxial region the image of a small but finite object perpendicular to the axis is itself perpendicular to the axis. For example, in Fig. 1–13, suppose the axial point B is imaged at B' by

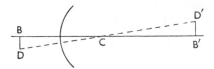

Fig. 1–13. Imagery in the paraxial region.

the spherical refracting surface centred at C. For a spherical surface any line through the centre may be taken as the optical axis since there is nothing to distinguish one from another. Take a new axis DCD' where DC=BC and D'C=B'C. Then it is obvious that, for the new axis, D' is the axial image of D. If BD and B'D' are small, they may be assumed to be perpendicular to the original axis, the latter being the image of the former. It will be seen that for different object sizes the image sizes are in direct proportion. In addition, since there is symmetry about the axis BCB', a plane object perpendicular to the axis at B will give a geometrically similar image perpendicular to the axis at B'.

It is a common practice to use small letters to denote paraxial quantities and capital letters to denote quantities associated with rays incident at finite distances from the axis. Consequently, a ray lying outside the paraxial region will be inclined to the axis at an angle U (or U') and will intersect the axis at a distance L (or L') from the pole of the surface, the height of incidence being Y, and the angle of incidence I.

It is assumed that the reader is familiar with the simple properties of mirrors and lenses, especially the graphical constructions that can be used to locate the image. Nevertheless, in order to familiarize the student with the notation, the following paragraphs are included.

1–15. Thin Lens in Air

If one refracting surface is followed by another so that the separation between them may be neglected, one has a thin lens. The image formed by the first surface becomes the object for the second and one may write $l_2 = l_1'$. If one assumes that the thin lens is of refractive index n and is in air, one has

$$n_1 = n_2' = 1,$$
$$n_1' = n_2 = n.$$

Double application of equation (1–5) now gives

$$\frac{n}{l_1'} - \frac{1}{l_1} = \frac{n-1}{r_1}$$

and

$$\frac{1}{l_2'} - \frac{n}{l_1'} = \frac{1-n}{r_2}.$$

Now l_1 and l_2' are the object and image distances from the thin lens, and it is convenient to write them simply as l and l' respectively. Adding the last two equations, one then obtains

$$\frac{1}{l'} - \frac{1}{l} = (n-1)\left[\frac{1}{r_1} - \frac{1}{r_2}\right]$$

and again one writes

$$l = f \quad \text{if} \quad l' = \infty,$$
$$l' = f' \quad \text{if} \quad l = \infty,$$

so that

$$\frac{1}{l'} - \frac{1}{l} = \frac{1}{f'} = -\frac{1}{f} = F = (n-1)\left[\frac{1}{r_1} - \frac{1}{r_2}\right]. \quad \cdots \quad (1\text{–}7)$$

Thus the power of a thin lens is the algebraic sum of the powers of the separate surfaces. Since the reciprocals of the radii of curvature of refracting surfaces often appear, it is convenient to put $1/r = R$, where R is the curvature of a surface.

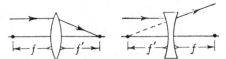

Fig. 1–14. The two focal lengths of a lens.

It will be seen that the second focal length of a converging lens is positive (and the first negative), whilst for a diverging lens the reverse is true (see Fig. 1–14). The two focal lengths of a thin lens in air are, of course, numerically equal. Frequently one refers simply to "the focal length" of a lens; it will be assumed that this always refers to the second focal length so that the focal length of a converging lens is positive, whilst for a diverging lens it is negative. The powers of these two lenses will be seen to be positive and-negative respectively, which is in agreement with the convention adopted in ophthalmic optics. The power of a thin lens is often expressed in dioptres where the power in dioptres is the reciprocal of the focal length in metres. Thus a lens of focal length $+25$ cm. is a converging lens whose power is $+4$ dioptres.

1–16. Smallest Separation of Object and Real Image

In Fig. 1–15(a) the converging lens L forms a real image h' of the object h. It will now be shown that the smallest possible separation of h and h' is $4f'$.

One has

$$\frac{1}{l'} - \frac{1}{l} = \frac{1}{f'}. \quad \cdots \cdots \cdots \quad (1\text{–}8)$$

If the image is real, h and h' must lie on opposite sides of the lens as shown. That is, l is negative, and the separation of object and image is $l'-l$.

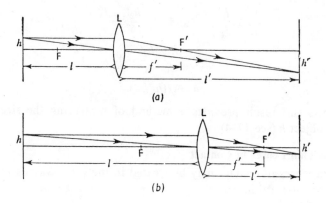

Fig. 1–15. Formation of a real image.

(1–8) gives
$$l'-l=-\frac{ll'}{f'}. \qquad \cdots \cdots \cdots \quad (1\text{–}9)$$

At a maximum or a minimum $d(l'-l)/dl=0$. Using (1–9), this gives

$$-\frac{1}{f'}\left(l'+l\frac{dl'}{dl}\right)=0$$

or
$$\frac{dl'}{dl}=-\frac{l'}{l}. \qquad \cdots \cdots \cdots \quad (1\text{–}10)$$

But
$$\frac{d(l'-l)}{dl}=\frac{dl'}{dl}-1,$$

therefore at a max. or min.
$$\frac{dl'}{dl}=1.$$

Hence, from (1–10),
$$l'=-l.$$

Now, either l or l' (but not both) may be infinite. Therefore this condition corresponds to a *minimum* separation.

Putting $l'=-l$ in (1–8), one obtains

$$l'=2f',$$

$$\therefore \quad (l'-l)=4f'.$$

This shows that the object and image have their smallest separation when the lens is midway between them. For larger separations there are two possible positions of the lens. From the symmetry of the problem it is obvious that the object distance for one position is numerically equal to the image distance for the other; one has $l_2=-l_1'$ and $l_2'=-l_1$ (see Fig. 1–15). Now, it can be assumed that a ray which passes through the centre of a thin lens is undeviated. If h_1' and h_2' are the image sizes for the two positions of the lens one has

$$\frac{h_1'}{h} = \frac{l_1'}{l_1} \quad \text{and} \quad \frac{h_2'}{h} = \frac{l_2'}{l_2} = \frac{l_1}{l_1'},$$

$$\therefore \frac{h_1'}{h} \cdot \frac{h_2'}{h} = 1$$

or

$$h = \sqrt{(h_1'h_2')}. \quad \cdots \cdots \cdots \quad (1\text{--}11)$$

Sometimes this result provides a method of measuring the size of an inaccessible object h (see §7–4).

1–17. Two Thin Lenses in Contact

Two thin lenses in contact may be treated in the same way as two surfaces combined to form a thin lens.

One has, if the lenses are in air,

$$\frac{1}{l_1'} - \frac{1}{l_1} = \frac{1}{f_1'}$$

and

$$\frac{1}{l_2'} - \frac{1}{l_2} = \frac{1}{f_2'}.$$

Putting $l_1'=l_2$, $l_1=l$, $l_2'=l'$, as before, and adding the above equations, one obtains

$$\frac{1}{l'} - \frac{1}{l} = \frac{1}{f_1'} + \frac{1}{f_2'} = \frac{1}{f'} = -\frac{1}{f}, \quad \cdots \cdots \quad (1\text{--}12)$$

where f and f' have their usual meanings.

Thus the power of the combination is the sum of the powers of the separate lenses.

1–18. Refraction of a Wave at a Spherical Surface

It is interesting to deduce the equation relating l' to l for a single refracting surface by a consideration of the progress of the wavefront.

In Fig. 1–16 QAP is a refracting surface, centre C, radius r, separating media

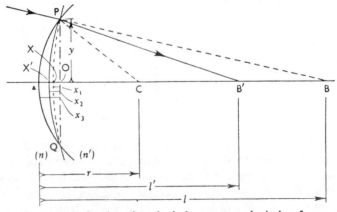

Fig. 1–16. Refraction of a spherical wave at a spherical surface

whose refractive indices are n and n'. Let QXP represent the position that a spherical wavefront would have in the absence of the refracting surface, and QX′P the position it does in fact have at the same instant. The centre of the wave QXP is the virtual object point B. Similarly, B′ is the real image point. Drop the perpendicular PO and put PO$=y$, OX$=x_1$, OX′$=x_2$, OA$=x_3$. $n(x_3-x_1)$ is the optical distance which the axial disturbance would travel during the time in which it actually travels $n'(x_3-x_2)$. Now a disturbance always travels the same optical distance in the same time (§ 1–4),

$$\therefore \quad n(x_3-x_1)=n'(x_3-x_2).$$

If y is small, the geometry of the figure gives

$$x_1=\frac{y^2}{2l}, \qquad x_2=\frac{y^2}{2l'}, \qquad x_3=\frac{y^2}{2r},$$

$$\therefore \quad n\left(\frac{y^2}{2r}-\frac{y^2}{2l}\right)=n'\left(\frac{y^2}{2r}-\frac{y^2}{2l'}\right),$$

$$\therefore \quad \frac{n'}{l'}-\frac{n}{l}=\frac{n'-n}{r},$$

which is the result obtained by the ray method using Snell's law.

1–19. Reflection at a Spherical Surface

In the case of reflection one may have a real object and a real image on the same side of the mirror. Consequently, one has both object and image distances real on the same side, i.e. both are measured in the same space. This, at first sight, is an embarrassment to the present notation since it would indicate either that both object and image distances are denoted by l or that both are l'. It is soon found, however, that there is no cause for alarm since the equation relating l and l' is such that it makes no difference which is which. One can, if one so desires, distinguish between object and image distances by using the

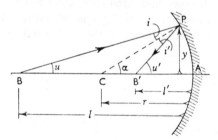

Fig. 1–17. Reflection at a spherical surface.

alternative statement of the convention. That is to say, l denotes the object distance if the light travels from left to right initially. From Fig. 1–17,

$$i=\alpha-u=\frac{y}{r}-\frac{y}{l},$$

$$i'=u'-\alpha=\frac{y}{l'}-\frac{y}{r}.$$

For reflection $i=i'$ and one has

$$\frac{y}{r}-\frac{y}{l}=\frac{y}{l'}-\frac{y}{r}$$

or

$$\frac{1}{l'}+\frac{1}{l}=\frac{2}{r}=\frac{1}{f'}. \qquad \cdot \quad \cdot \quad \cdot \quad \cdot \quad \cdot \quad \cdot \quad (1\text{–}13)$$

This is the true algebraic equation since all the quantities are negative and changing all the signs leaves the equation unchanged.

It is seen at once that it makes no difference which symbol—l or l'—one uses to denote the object or image.

Another important point also emerges. In Fig. 1-17 i is negative and i' is positive so that the law of reflection, written algebraically, becomes $i = -i'$. This follows as a special case of Snell's law of refraction on putting $n' = -n$. In fact, if one puts $n' = -n$ in the equation governing refraction at a spherical surface, the equation for reflection emerges at once. One has

$$\frac{n'}{l'} - \frac{n}{l} = \frac{n'-n}{r}$$

and, putting $n' = -n$, one obtains immediately

$$\frac{1}{l'} + \frac{1}{l} = \frac{2}{r}.$$

One finds that this is always the case and any equation associated with refraction may be converted into the equation governing the corresponding phenomenon in reflection by putting $n' = -n$. This is very useful since it means that it is never necessary to discuss reflecting surfaces explicitly.

From the original definition of refractive index it is seen that a negative index implies a negative velocity. That is to say, negative distance is traversed in positive time. Now in co-ordinate geometry a negative distance is simply a distance measured from right to left, so that putting $n' = -n$ indicates that the direction of the light is reversed. One can, in fact, consistently use negative refractive indices if the light travels from right to left, but there is no advantage in doing so.

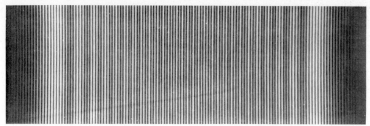

(a) Fringes obtained with a Fresnel Biprism.

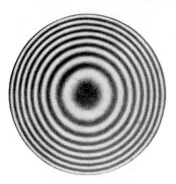

(b) Newton's rings by reflection. (Similar in appearance to the fringes of equal inclination obtained by reflection with a plane parallel plate or with a Michelson interferometer.)

(c) Newton's rings by transmission. (Similar in appearance to the fringes of equal inclination obtained by transmission with a plane parallel plate.)

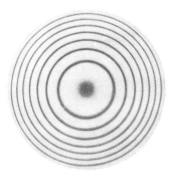

(d) Multiple beam fringes of equal inclination viewed by reflection. (Similar in appearance to multiple beam Newton's rings by reflection.)

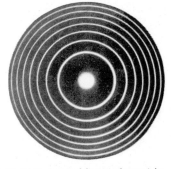

(e) Multiple beam fringes of equal inclination by transmission (Fabry-Perot fringes). (Similar in appearance to multiple beam Newton's rings by transmission.)

(f) Wedge fringes by reflection with unsilvered surfaces.

(g) Multiple beam Fizeau fringes between an optical flat and an aluminized mica sheet.

PLATE I. INTERFERENCE FRINGES

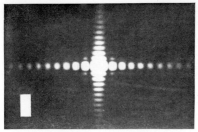

(a) Point source and rectangular aperture. The orientation of the aperture is shown by the white rectangle.

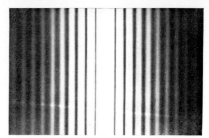

(b) Slit source and slit aperture of same width as rectangular aperture.

(c) Slit source and single narrow slit aperture.

(d) Slit source and two narrow slit apertures.

(e) Slit source and three narrow slit apertures.

(f) Slit source and four narrow slit apertures.

(g) Slit source and five narrow slit apertures.

(h) Slit source and six narrow slit apertures.

In (c)–(h) inclusive the centre of the pattern was over-exposed and the outer maxima intensified during processing.

PLATE II. FRAUNHOFER DIFFRACTION PATTERNS

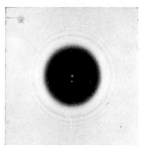

(a) Point source and circular obstacle.

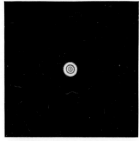

(b) Point source and small circular aperture.

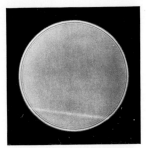

(c) Point source and large circular aperture.

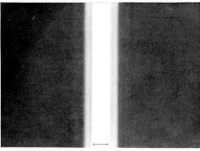

(d) Slit source and very narrow slit aperture.

(e) Slit source and slightly wider slit aperture.

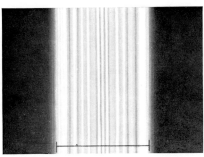

(f) Slit source and very wide slit aperture.

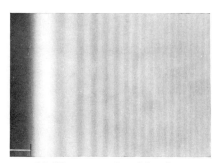

(g) Slit source and straight edge.

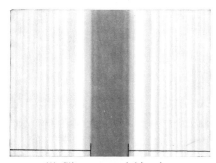

(h) Slit source and thin wire.

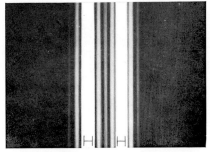

(i) Slit source and two narrow slit apertures.

In (d)–(i) inclusive the positions of the geometrical shadows of the diffracting screens are indicated at the bottom of the patterns.

PLATE III. FRESNEL DIFFRACTION PATTERNS

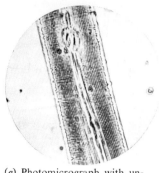

(a) Photomicrograph with un-obstructed objective.

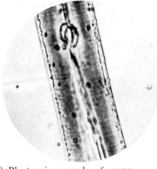

(b) Photomicrograph of same object, using reduced objective aperture.

(c) Airy disc (centre greatly over-exposed).

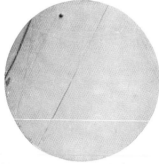

(d) Photomicrograph of sheet of mica—ordinary illumination.

(e) Same sheet of mica with phase contrast.

(f) Image of a point source with about $\frac{1}{2}$ wavelength of coma.

(g) Mica sheet with dark-ground illumination.

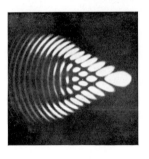

(h) Image of point source with about 5 wavelengths of coma.

(i) Focal line with about 1 wavelength of astigmatism.

(j) Focal line with about $2\frac{1}{4}$ wavelengths of astigmatism.

(k) Disc of least confusion with about $2\frac{1}{4}$ wavelengths of astigmatism.

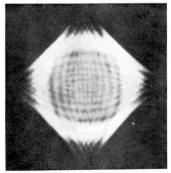

(l) Disc of least confusion with about $6\frac{1}{4}$ wavelengths of astigmatism.

PLATE IV. IMAGES IN OPTICAL INSTRUMENTS

CHAPTER II

THE GENERAL GEOMETRICAL THEORY
OF OPTICAL SYSTEMS

IN terms of geometrical (ray) optics, it was shown in Chapter I that, within the paraxial region, a single spherical refracting (or reflecting) surface forms a point image of a point object. Moreover, if one has a small line object perpendicular to the axis, the image is also linear and perpendicular to the axis, and it was shown that for any plane object perpendicular to the axis a geometrically similar plane image is formed perpendicular to the axis. There is said to be a *collinear* relationship between objects and images in the left- and right-hand spaces; the image is said to be *conjugate* to the object, and a refracted ray is conjugate to the incident ray. Now the image, as formed by one surface, can be the object for a second surface and so on. Consequently, for a system of spherical refracting surfaces, there is a collinear transformation from the object space to the final image space. It is found that the general features of such image formation are the same for all optical systems. It is also found that these features can be deduced without referring to laws of optics such as the laws of refraction. That is, it is not necessary to discuss the way in which the rays traverse the system—one need discuss only their orientation in the final image space. All that is necessary is the assumption that any plane object perpendicular to the axis gives a geometrically similar image perpendicular to the axis. Actually Maxwell showed that it is sufficient to assume that this occurs for *two* pairs of conjugate planes, since it can then be shown that it must occur for all positions of the object and image planes.

In this chapter the general geometrical properties of such a "perfect" image-forming system will be discussed. In particular, it will be shown that if one combines two systems for which certain reference planes are known, the properties of the combination can be found. Now, since within the paraxial region a simple refracting surface gives geometrically perfect imagery, it is possible to deduce the paraxial properties of the combinations of surfaces forming thin and thick lenses and lens systems.

In the first fourteen sections the discussion is of a general nature and the conjugate points, lines, and planes, are not necessarily optical objects and images. Nevertheless one has its application to optics in view and it is often convenient to speak of "rays entering the system" instead of referring vaguely to "lines in the left-hand space", etc.

2-1. Principal Planes and Focal Planes

As mentioned above, the initial assumption is that by means of some un-specified mechanism plane objects perpendicular to an axis in one space give geometrically similar plane images perpendicular to an axis in another space. For simplicity it will be assumed that the axis is common to the two spaces—it is referred to as the axis of the system. Since there is geometrical similarity between plane objects and images perpendicular to the axis, there is a certain magnification associated with each pair of conjugate planes. This *transverse magnification* is denoted by m. The planes for which $m = +1$ are called the *principal planes* (or *unit planes*). These planes intersect the axis in the principal points P and P' (see Fig. 2-1). If H and H' are conjugate points in the principal planes, H'P' = HP, by definition. The plane in the left-hand space that is perpendicular to the axis and is conjugate to a similar plane at infinity in the right-hand space is called the *first focal plane*. Similarly, the *second focal plane* is the plane in the right-hand space that is conjugate to the plane at infinity in the left-hand space. The first and second focal planes intersect the axis at the points F and F', which are called the *first* and *second principal foci*. The distances PF and P'F' are called the *first* and *second focal lengths* of the system, and are denoted by f and f'.

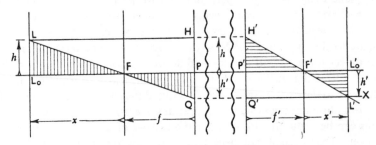

Fig. 2-1. Image formation by a perfect system.

It is convenient to write $f'/f = -n'/n$ where, for the purposes of the general theory, n and n' can be regarded simply as constants associated with the left- and right-hand spaces respectively. Actually they become identified with the refractive indices of these spaces when the theory is applied to an optical system. In general the ratio n'/n is more important than the absolute values n and n'.

Consider a line LH, parallel to the axis, which intersects the first principal plane at H. If HL is produced to the left, it intersects the axis at infinity, and F' is conjugate to this point of intersection. Also H' is conjugate to H, so that the line that is conjugate to LH passes through H' and F'. Thus, if the incident ray is parallel to the axis, the emergent ray passes through the appropriate principal focus, and vice versa.

If the paths of the rays LH and H'F' are known, one can quickly determine the point of intersection, H', and so find the second principal plane. This simple fact is of considerable practical importance. Obviously the position of the first principal plane may be determined in a similar manner.

2–2. Location of Objects and Images

Suppose it is necessary to determine the plane that is conjugate to the given plane L_0L (Fig. 2-1). As pointed out above, H'F' is conjugate to LH. Produce LF to Q. Q' is conjugate to Q where P'Q' = PQ. The line conjugate to LFQ must, therefore, pass through Q'. In addition, the emergent ray Q'X must be parallel to the axis since the incident ray passes through F. Since L lies on both LH and LFQ it follows that L' must lie on H'F' and Q'X, so that L' is the point of intersection of H'F' and Q'X. One may say that $L_0'L'$ is the image of L_0L.

Put $FL_0=x,\quad F'L'_0=x',\quad L_0L=h,\quad L_0'L'=h'.$

One has, also, $P'Q'=PQ=h'$

and $P'H'=PH=h.$

The triangles LL_0F, QPF (shaded vertically) are similar. Hence

$$\frac{h'}{h}=\frac{f}{x}$$

Also, the triangles $L'L_0'F'$, $H'P'F'$ (shaded horizontally) are similar. Hence

$$\frac{h'}{h}=\frac{x'}{f'}.$$

In the figure x, f, and h', are negative so that, algebraically, one has

$$\frac{h'}{h}=-\frac{x'}{f'}=-\frac{f}{x}. \quad \cdot \quad \cdot \quad \cdot \quad \cdot \quad \cdot \quad \cdot \quad \cdot \quad (2\text{–}1)$$

Hence $$xx'=ff'. \quad \cdot \quad \cdot \quad \cdot \quad \cdot \quad \cdot \quad \cdot \quad \cdot \quad \cdot \quad (2\text{–}2)$$

This is known as Newton's equation.

2–3. Location of Conjugate Ray. General Case

Suppose that the principal and focal planes of a system are given, and that it is necessary to determine the path of the ray conjugate to the ray AB, incident

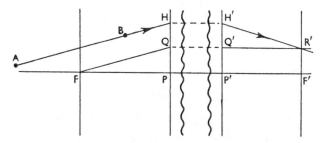

Fig. 2–2. Location of a conjugate ray.

in the left-hand space (see Fig. 2–2). Produce AB to intersect the first principal plane in H. Then the conjugate ray must emerge through H', where $P'H'=PH$. Draw FQ parallel to ABH, and let it intersect the first principal plane in Q. Then Q'R' must be conjugate to FQ, where $P'Q'=PQ$ and Q'R' is parallel to the axis. Now ABH and FQ intersect at infinity, and the point which is conjugate to this point of intersection must lie in the second focal plane. It follows that if any two incident rays are parallel, their emergent rays intersect in the appropriate focal plane. In this case, the emergent rays which are conjugate to ABH and FQ intersect at R' in the second focal plane. This determines the path of the ray H'R' which is conjugate to AB.

2–4. Transverse and Angular Magnifications. The Helmholtz Equation

In Fig. 2–3 L_0L and $L_0'L'$ are conjugate elements with heights h and h', as shown. If a ray through L_0 at angle u emerges through L_0' at angle u', it is convenient to refer to $\tan u'/\tan u$ as the *angular magnification* for the conjugate planes L_0Y, $L_0'Y'$. Helmholtz's equation relates this to the transverse linear magnification h'/h associated with the same pair of conjugate planes.

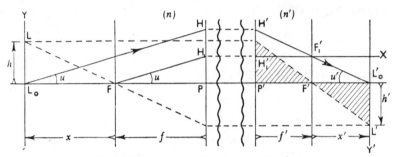

Fig. 2–3. Angular and lateral magnification.

From the similarity of the shaded triangles, it has already been shown that, algebraically,

$$\frac{h'}{h} = \frac{-x'}{f'} . \quad \ldots \ldots \ldots \quad (2\text{–}3)$$

The ray FH_1, which is parallel to L_0H, emerges along $H_1'F_1'X$, which is parallel to the axis, and intersects $H'L_0'$ at F_1' in the second focal plane.

In the figure, $\qquad\qquad\qquad PH_1 = f \tan u$

and $\qquad\qquad\qquad P'H_1' = F'F_1' = x' \tan u'$.

Since $PH_1 = P'H_1'$, and f and u are negative in the figure, one has, algebraically,

$$f \tan u = x' \tan u', \quad \ldots \ldots \ldots \quad (2\text{–}4)$$

$$\therefore \quad \frac{\tan u'}{\tan u} = \frac{f}{x'}. \quad \ldots \ldots \ldots \quad (2\text{–}5)$$

From equations (2–3) and (2–5),

$$\left(\frac{h'}{h}\right)\left(\frac{\tan u'}{\tan u}\right) = -\frac{f}{f'} = +\frac{n}{n'} \quad \ldots \ldots \quad (2\text{–}6)$$

which relates the angular and transverse linear magnifications for a pair of conjugate planes. This is usually written in the form

$$nh \tan u = n'h' \tan u' \quad \ldots \ldots \ldots \quad (2\text{–}7)$$

and is known as Helmholtz's equation. For small angles one has

$$nhu = n'h'u'. \quad \ldots \ldots \ldots \quad (2\text{–}8)$$

This is often called Lagrange's equation. It will be seen that $H = nh \tan u = n'h' \tan u'$ is an invariant for the system (and, incidentally, for a series of systems).

Now an optical system forms the image of an off-axis object point by means of the cone of rays that enters the entrance pupil from the point (see Fig. 2–4). The semi-angle of this cone is a "generalized u" and the constant H may be regarded as a constant characterizing both the aperture (u) and the distance off-axis (h). As one increases the aperture and field the aberrations increase, and in discussing the aberrations it is found that H is a convenient measure of u and h.

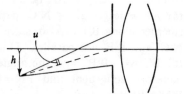

Fig. 2–4. Cone of light entering a lens system.

2–5. The Nodal Points

If N and N′ are two conjugate axial points such that a ray through N′ is conjugate to a parallel ray through N, then N and N′ are called the *nodal points*. Thus for rays passing through the nodal points one has $u'=u$. The planes perpendicular to the axis at N and N′ are known as the *nodal planes* For imagery in these planes it follows from equation (2–6) that

$$\frac{h'}{h} = \frac{-f}{f'}. \qquad \cdots \cdots \cdots \quad (2\text{–}9)$$

The position of the second nodal point follows at once from equation (2–4) on putting $u=u'$. One obtains

$$\left.\begin{array}{l} x'_{N'}=f. \\ x_{N} =f'. \end{array}\right\} \qquad \cdots \cdots \cdots \quad (2\text{–}10)$$

Similarly,

The principal points, principal foci, and nodal points, are sometimes referred to as the *cardinal* points of the system. The six cardinal points for a typical case are illustrated in Fig. 2–5.

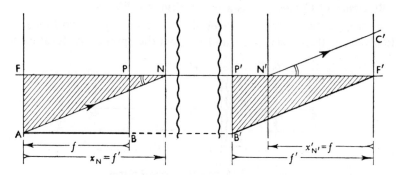

Fig. 2–5. Location of the nodal points.

If the refractive indices in the left- and right-hand spaces are equal, equations (2–10) show that the nodal and principal planes coincide. This fact also follows from (2–6) since, on putting $n=n'$, one has $h'/h=1$ when $u'=u$. That is, the nodal planes become the planes of unit magnification which, by definition, are the principal planes. The positions of the nodal planes can be deduced very quickly from first principles. In Fig. 2–5 take a point A in the first focal plane. Then the ray AN emerges along N'C' parallel to AN. The ray AB parallel to the axis emerges along B'F'. Also, since AN and AB intersect in the first focal plane, N'C' and B'F' are parallel. Thus, in the shaded right-angled triangles AFN, B'P'F', AF=B'P', and the angles at N and F' are equal. The triangles are therefore congruent and FN=P'F'=f'. Similarly, it may be shown that F'N'=PF=f.

2–6. Longitudinal or Axial Magnification

Consider conjugate points on the axis which are specified by their distances, x and x', from the focal planes. One has $xx'=ff'$. For small movements along the axis dx' will be conjugate to dx where $xdx'+x'dx=0$.

$$\therefore \quad \frac{dx'}{dx}=-\frac{x'}{x}.$$

From equations (2–1)

$$\frac{h'}{h}=-\frac{f}{x}=-\frac{x'}{f'},$$

whence

$$\frac{x'}{x}=\left(\frac{h'}{h}\right)^2\cdot\frac{f'}{f}$$

so that

$$\frac{dx'}{dx}=-m^2\cdot\frac{f'}{f}=+m^2\cdot\frac{n'}{n}. \quad \ldots \ldots \quad (2\text{–}11)$$

This expression gives the magnification for small objects and images along the axis. It follows that while conjugate figures in planes perpendicular to the axis are geometrically similar, conjugate three-dimensional figures are not similar since the transverse and longitudinal magnifications are different.

2–7. Reference of Conjugate Distances to Principal Planes

It was seen to be convenient, theoretically, to specify the positions of conjugate planes by their distances, x and x', from the appropriate focal planes. If

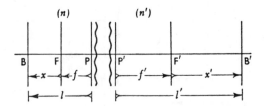

Fig. 2–6. Reference to principal planes.

this is done, the relation between the conjugate distances is very simple (equation 2–2). It is frequently more convenient in practice to measure the conjugate distances from the principal planes. In this case they are denoted by l and l'. The conjugate planes may also be specified by their distances, z and z', from the nodal planes, or their distances, l_R and l'_R, from any other pair of conjugate planes. First, the conjugate distances will be referred to the principal planes. In Fig. 2–6 B and B' are any pair of conjugate planes. From the figure one has

$$l'=f'+x'$$

and

$$l=f+x.$$

These are also the algebraic equations, so that the values $x'=l'-f'$, and $x=l-f$, may be substituted into Newton's equation.

One has

$$(l'-f')(l-f)=ff'$$

or

$$ll'-l'f-lf'=0$$

or

$$\frac{f}{l}+\frac{f'}{l'}=1. \qquad \cdots \cdots \qquad (2\text{–}12)$$

Alternatively, since $nf'=-n'f$,

$$\frac{n'}{l'}-\frac{n}{l}=\frac{n'}{f'}=-\frac{n}{f}=F \text{ (say)}, \qquad \cdots \cdots \qquad (2\text{–}13)$$

where F is called the *power* of the system. If the indices of the initial and final spaces are equal, $n'=n$ and (2–13) reduces to

$$\frac{1}{l'}-\frac{1}{l}=\frac{1}{f'}. \qquad \cdots \cdots \cdots \qquad (2\text{–}14)$$

2–8. Reference to Nodal Planes

From Fig. 2–7 it follows that

$$z=x+f'$$

and

$$z'=x'+f.$$

But x, z, and f, are negative so that, algebraically,

$$z=x-f',$$
$$z'=x'-f.$$

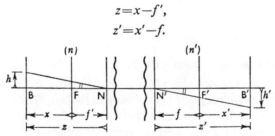

Fig. 2–7. Reference to nodal planes.

Then, from Newton's equation,

$$(z+f')(z'+f)=ff'$$

whence

$$\frac{f'}{z}+\frac{f}{z'}=-1. \qquad \cdots \cdots \qquad (2\text{–}15)$$

Hence, putting $nf'=-n'f$, one has

$$\frac{n}{z'}-\frac{n'}{z}=\frac{n'}{f'}=-\frac{n}{f}=F \qquad \cdots \cdots \qquad (2\text{–}16)$$

which reduces, as it should, to equation (2–14) when one puts $n=n'$, the condition that the nodal planes should coincide with the principal planes.

It is obvious, from Fig. 2–7, that the transverse magnification for the planes B and B' is given by $m=z'/z$ which becomes $m=l'/l$ if $n'=n$.

2–9. Reference to Any Pair of Conjugate Planes

In Fig. 2–8, let R and R' be the pair of conjugate planes chosen as reference planes. Let B and B' be any other pair of conjugate planes. Let m_R and m be

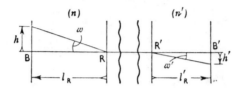

Fig. 2–8. Reference to a pair of conjugate planes.

the transverse magnifications associated with the pairs of planes RR' and BB' respectively.

Applying the Helmholtz equation (2–6) *to the planes* R, R',

$$n \tan \omega = n' m_R \tan \omega'.$$

From the figure, $h=l_R \tan \omega = -l_R \tan \omega$ algebraically,

$$h'=l'_R \tan \omega' = -l'_R \tan \omega' \text{ algebraically,}$$

whence $\dfrac{nh}{l_R} = n' m_R \dfrac{h'}{l'_R}$ algebraically

or, since $\dfrac{h'}{h} = m,$

$$\frac{l'_R}{l_R} = \frac{n'}{n} m_R m. \qquad \ldots \ldots \ldots \text{(2–17)}$$

2–10. Helmholtz Equation—Infinite Conjugate

For a finite image, $h \to \infty$ and $u \to 0$ as $l \to \infty$. $nh \tan u$ then becomes indeterminate. Suppose the object plane is at infinity to the left. Then the second focal plane is the image plane. Consider a ray in the left-hand space parallel to the axis at height y (see Fig. 2–9). This converges at an angle u' in the right-hand space to pass through F'. Consider also a ray inclined at an angle ω to the axis. This and the parallel ray through F may be considered as having come from an off-axis object point at infinity which is specified by the angle ω which it subtends at the system. On emerging from the system, these two rays intersect in the second focal plane. Thus h' is determined, since the ray incident through F emerges parallel to the axis.

It follows immediately from the figure that

$$y=f' \tan u' \qquad \ldots \ldots \ldots \text{(2–18)}$$

and $h'=f \tan \omega \qquad \ldots \ldots \ldots \text{(2–19)}$

and these are also the algebraic equations. Equation (2–19) is important since it shows where, in the second focal plane, one will find the image of an extra-axial point at infinity; it gives the size of the image in terms of the angular size of the object.

From (2–18) and (2–19), it follows that

$$n'h' \tan u' = n'f \tan \omega . \frac{y}{f'} = -ny \tan \omega.$$

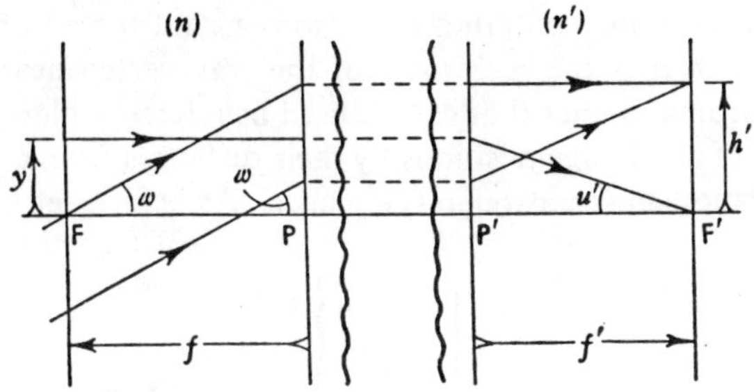

Fig. 2–9. Helmholtz equation with infinite conjugate.

Therefore if the first conjugate is infinite one may write the Helmholtz equation in the form

$$H = -ny \tan \omega = n'h' \tan u'. \qquad \ldots \quad (2\text{–}20)$$

For telescopic systems both conjugates can be infinite and one has

$$H = -ny \tan \omega = -n'y' \tan \omega'. \qquad \ldots \quad (2\text{–}21)$$

2–11. Telescopic Systems

The discussions so far have assumed that the cardinal points lie at finite distances from the system. In what is known as a *telescopic* system the principal foci are at infinity and the focal lengths f, f' are infinite. In this case one cannot specify the positions of conjugate planes by means of their distances from the cardinal points but only by means of their distances from another pair of conjugate planes. It should be noted here that not all the systems colloquially referred to as telescopes are true telescopic systems. For example, a "telescope" adjusted for observation of a relatively near object is not truly telescopic. It should be noted also that, although one usually thinks of a truly telescopic system as being used to form an image of an object at infinity, such a system also images near objects.

In Fig. 2–10, suppose AB is a ray entering a telescopic system from an axial object point at infinity so that AB is parallel to the axis. In general, the emergent ray must pass through the second principal focus. If the second focal plane is at infinity, the ray CD which is conjugate to AB must also be parallel to the axis although it is not usually at the same height as AB. CD may, in fact, be below the axis. Consider the object L_0L of height h; the image of L_0 must be on the axis and the image of L must be on the line CD. Consequently, the image height

h' will be independent of the location of the object. Thus the magnification, $m = h'/h$, is the same for all pairs of conjugate planes L_0L, L'_0L'. This magnification may be referred to as the magnification of the telescope. The Helmholtz equation (2–6) gives

$$\frac{h'}{h} = -\frac{f}{f'}\frac{\tan u}{\tan u'}.$$

If f and f' each tend to infinity while the ratio f'/f remains constant (and equal to $-n'/n$), then the ratio $\tan u'/\tan u$ is also constant for all object and image positions. If $n' = n$ it is the reciprocal of the transverse linear magnification. There are no cardinal planes that can be used as reference planes and one must specify the object and image positions by their distances from a convenient pair of conjugate planes chosen as reference planes. As before, let l_R and l'_R be the

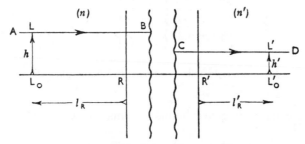

Fig. 2–10. Telescopic system.

distances of object and image from the reference planes. Since, for a telescopic system, the magnification is the same for all pairs of conjugate planes, one can put $m_R = m$ in equation (2–17). Consequently, for a telescopic system one has

$$\frac{l'_R}{l_R} = \left(\frac{n'}{n}\right) m^2. \qquad \ldots \ldots \quad (2\text{–}22)$$

This equation also follows from equation (2–11) since, in the present case, m is constant and one may take finite increments along the axis in place of dx and dx'.

2–12. The Combination of Two Systems

In general terms, suppose a transformation from one space to another is effected by a system. The geometry of the transformation can be evolved if the cardinal points of the system are known. It may happen that a second system now effects a transformation from the second to a third space. It is obvious that one can look upon the two systems as being equivalent to a single system which effects a transformation from the first to the third space. In terms of optics the image formed by one optical system becomes the object for a second system and the two systems act as one—forming the final image from the original object. The problem now is to find the cardinal points of this combined system, given the cardinal points of the two separate systems and the position of the second system relative to the first.

It will be assumed that there is a common axis of symmetry. In this case the separation of the two systems is specified by the separation of their "adjacent" principal foci, $F_1'F_2=g$, or by the separation of their "adjacent" principal planes, $P_1'P_2=d$. The sign convention for g is that g is positive when measured to the right from F_1' to F_2. Similarly, d is positive when measured to the right from P_1' to P_2.

From Fig. 2–11, $d=f_1'+g+f_2$, but f_2 is negative; one has, algebraically,

$$d=f_1'+g-f_2. \quad \ldots \ldots \ldots \quad (2\text{-}23)$$

Consider a ray in the first space parallel to the axis and at height y. The conjugate ray in the second space is AF_1' which is determined at once since it must emerge from the point A, at height y, and pass through F_1'. Let this ray

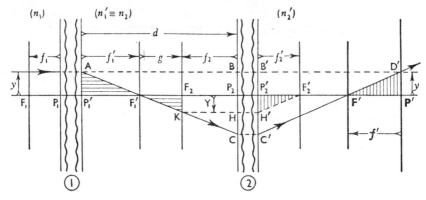

Fig. 2–11. Combination of two systems.

intersect the first focal plane and the first principal plane of the second system in K and C respectively. It follows at once that the conjugate ray in the third space must pass through C'. To find the direction of the final ray, consider a line KH parallel to the axis. This is obviously conjugate to $H'F_2'$. The ray C'D' conjugate to $AF_1'KC$ is now located since it must be parallel to $H'F_2'$. The second principal plane P'D' of the complete system is determined by the point D', where the ray C'D' in the third space intersects the path of the initial ray parallel to the axis. The second principal focus F' of the complete system is also determined; it is the point where the final ray C'D' intersects the axis. The second focal length of the combination is $f'=P'F'$ which, in the figure, is negative. It follows that the complete system as illustrated in Fig. 2–11 is a diverging system; it is interesting to note that it represents the system of the compound microscope (see §3–7).

In the figure the large triangle ABC and the smaller triangles $AP_1'F_1'$, KF_2F_1' (shaded horizontally) are similar. It follows that

$$\frac{y}{Y}=\frac{f_1'}{g} \quad \ldots \ldots \ldots \quad (2\text{-}24)$$

and

$$\frac{BC}{Y}=\frac{d}{g}. \quad \ldots \ldots \ldots \quad (2\text{-}25)$$

Also, the large triangle B'C'D', and the smaller triangles $P_2'H'F_2'$, P'D'F' (shaded vertically) are similar.

$$\therefore \quad \frac{y}{Y} = \frac{f'}{f_2'} \qquad \cdots \cdots \cdots \quad (2\text{–}26)$$

and

$$\frac{B'C'}{Y} = \frac{B'D'}{f_2'}. \qquad \cdots \cdots \cdots \quad (2\text{–}27)$$

From (2–24) and (2–26)

$$\frac{f_1'}{g} = \frac{f'}{f_2'}, \qquad \cdots \cdots \cdots \quad (2\text{–}28)$$

and from (2–25) and (2–27), since B'C'=BC and B'D'=$P_2'P'$,

$$\frac{d}{g} = \frac{P_2'P'}{f_2'}. \qquad \cdots \cdots \cdots \quad (2\text{–}29)$$

Now, in the figure, f_1', f_2', g, d, and $P_2'P'$, are positive but f' is negative. Thus (2–28) and (2–29) give, algebraically,

$$f' = -\frac{f_1'f_2'}{g} \qquad \cdots \cdots \cdots \quad (2\text{–}30)$$

and

$$P_2'P' = \frac{f_2'd}{g} = -\frac{df'}{f_1'}. \qquad \cdots \cdots \cdots \quad (2\text{–}31)$$

Similarly it may be shown that

$$f = \frac{f_1f_2}{g} \qquad \cdots \cdots \cdots \quad (2\text{–}32)$$

and

$$P_1P = \frac{f_1d}{g} = \frac{df}{f_2}. \qquad \cdots \cdots \cdots \quad (2\text{–}33)$$

(P and F are not shown in the figure.)

Using equations (2–30) to (2–33) it is possible to find the cardinal points of the combined system in terms of the constants of the constituent parts. The combination is equivalent to a single system whose cardinal points are P, P', F, and F', and the quantities f and f' are usually referred to as the *equivalent focal lengths* (E.F.L.s). In terms of optics: if a thin lens of the same E.F.L. as a complete optical system is placed at the first principal plane, the imagery produced is identical to that of the system except for a bodily lateral shift of magnitude PP'. The thin lens is known as the *equivalent lens*. (See §2–17 and Fig. 2–16.)

One special case is worthy of recognition at once: if $g=0$, f, f', P_1P, and $P_2'P'$, are infinite, i.e. the system is telescopic.

2-13. Use of Powers and Reduced Distances

The powers F, F_1, and F_2, of the combination and the separate systems are respectively given by

$$\left.\begin{aligned}F&=-\frac{n_1}{f}=\frac{n_2'}{f'},\\[4pt]F_1&=-\frac{n_1}{f_1}=\frac{n_1'}{f_1'},\\[4pt]F_2&=-\frac{n_1'}{f_2}=\frac{n_2'}{f_2'}.\end{aligned}\right\} \quad \cdots \cdots \quad (2\text{–}34)$$

Any distance divided by the refractive index of the space in which it is measured is called a *reduced distance*, and is denoted by the use of a "bar". For example, the reduced separation is $\bar{d}=d/n_1'$. A *reduced angle*, on the other hand, is the product of an angle and the appropriate refractive index: $n'u'=\bar{u}'$. Using this notation one has, from (2–31),

$$P_2'P'=-\frac{n_2'\bar{d}F_1}{F}, \quad \cdots \cdots \quad (2\text{–}35)$$

and similarly, from (2–33),

$$P_1P=\frac{n_1\bar{d}F_2}{F}. \quad \cdots \cdots \quad (2\text{–}36)$$

Also, from (2–30) and (2–23),

$$\frac{n_2'}{F}=-\frac{\dfrac{n_1'}{F_1}\cdot\dfrac{n_2'}{F_2}}{\bar{d}-\dfrac{n_1'}{F_2}-\dfrac{n_1'}{F_1}}=-\frac{n_2'}{F_1F_2\dfrac{d}{n_1'}-F_1-F_2}$$

or

$$F=F_1+F_2-\bar{d}F_1F_2. \quad \cdots \cdots \quad (2\text{–}37)$$

Examination candidates should note that (2–35), (2–36), and (2–37) are the most generally useful equations for solving problems involving two systems. (See the worked example in §2–27.) Sometimes it is useful to have expressions for P_1F and $P_2'F'$. In Fig. 2–11 one has

$$P_2'F'=P_2'P'-P'F'$$

or, since $P'F'$ is negative, one has, algebraically,

$$P_2'F'=P_2'P'+f'$$

$$=f'-\frac{\bar{d}f'}{f_1'}=f'(1-\bar{d}/f_1'), \quad \cdots \cdots \quad (2\text{–}38)$$

or

$$P_2'F'=\frac{n_2'(1-\bar{d}F_1)}{F}. \quad \cdots \cdots \quad (2\text{–}39)$$

Similarly

$$P_1F=-\frac{n_1(1-\bar{d}F_2)}{F}. \quad \cdots \cdots \quad (2\text{–}40)$$

P_1F and $P_2'F'$ are sometimes called the *front and back focal lengths* respectively (F.F.L. and B.F.L.). $-n_1/P_1F$ $[=F/(1-dF_2)]$, and $n_2'/P_2'F'$ $[=F/(1-dF_1)]$, are known as the *front and back vertex powers*.

It should be noted that the proof given in §2–12 breaks down if one of the systems has zero power (infinite focal length). g and either f_2 and f_2' or f_1 and f_1' are then infinite and equations (2–30) and (2–32) become indeterminate. However, the proof is valid as this condition is approached, and equations (2–35) to (2–40) inclusive are still applicable in the limiting case. The alternative proof given in §2–23 does not break down and gives the same results.

2–14. Two Systems in Air

For two optical systems in air one has $n_1=n_1'=n_2'=1$ and the general equation (2–37) becomes

$$\frac{1}{f'}=\frac{1}{f_1'}+\frac{1}{f_2'}-\frac{d}{f_1'f_2'}, \qquad \dots \dots \quad (2\text{–}41)$$

i.e.
$$f'=\frac{f_1'f_2'}{f_1'+f_2'-d}.$$

The positions of the principal planes are given by

$$P_1P=\frac{df_1'}{f_1'+f_2'-d}$$

and
$$P_2'P'=-\frac{df_2'}{f_1'+f_2'-d}, \qquad \left.\begin{array}{c} \\ \\ \end{array}\right\} \quad \dots \dots \quad (2\text{–}42)$$

and the front and back focal lengths are

$$P_1F=-\frac{f_1'(f_2'-d)}{f_1'+f_2'-d}$$

and
$$P_2'F'=\frac{f_2'(f_1'-d)}{f_1'+f_2'-d}. \qquad \left.\begin{array}{c} \\ \\ \end{array}\right\} \quad \dots \dots \quad (2\text{–}43)$$

2–15. Paraxial Refraction at a Spherical Surface

At the beginning of this chapter it was pointed out that within the paraxial region there is a collinear relationship between the object and image spaces of a spherical refracting surface. Thus, within the paraxial region, such a surface is an image-forming system of the type which has been under discussion above. Now it was shown in Chapter I that, if one specifies the positions of the conjugate planes by their distances l and l' from the pole of the surface, one has

$$\frac{n'}{l'}-\frac{n}{l}=\frac{n'-n}{r}=-\frac{n}{f}=\frac{n'}{f'} \qquad \dots \dots \quad (2\text{–}44)$$

with the usual notation, where f and f' are the values of l and l' when l' and l respectively are infinite. If one compares this equation with equation (2–13), where l and l' were measured from the principal points, it will be seen that a single refracting surface is a system whose principal points coincide with the pole of the surface. The paraxial approximation is such that the sagitta in the refracting surface is negligible, and in this sense one may say that the principal *planes* coincide with the refracting *surface*. The constants n and n' in the general theory can now definitely be identified with the refractive indices of the media separated by the refracting surface. The focal lengths and the power of a single surface are given by

$$f=-\frac{nr}{n'-n}, \qquad f'=\frac{n'r}{n'-n}, \qquad F=\frac{n'-n}{r}. \qquad . \quad . \quad (2\text{–}45)$$

A ray through the centre of curvature of the surface will be incident normally at the surface and will therefore pass through undeviated. By definition, therefore, both the nodal points of a single refracting surface coincide with the centre of curvature. Happily, this result also follows analytically. The second nodal point, for example, is distance l' from the pole of the surface where, as always, $l'=f'+x'$. Thus, for this point one has

$$l'=\frac{n'r}{n'-n}-\frac{nr}{n'-n}=r.$$

The cardinal points of a single refracting surface are shown in Fig. 2–12, for $n=1$ and $n'=1{\cdot}5$.

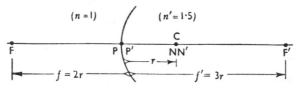

Fig. 2–12. Cardinal points of a single spherical refracting surface.

In the first fourteen sections of this chapter a number of results have been deduced for a system giving geometrically perfect imagery. Within the paraxial region these results can now be applied to a single spherical refracting surface. For example, the equations giving the cardinal points of a combination of systems can now be applied to find the cardinal points of combinations of refracting surfaces. The most important applications are to the combination of two surfaces to form a thin lens or a thick lens, and to the combination of two separated thin lenses. Before considering these applications, the paraxial form of the Helmholtz equation will be deduced from first principles for a refracting surface. This serves as further evidence that the results deduced in the general analysis do, in fact, apply to the paraxial region of an optical system.

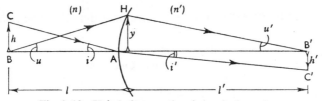

Fig. 2–13. Helmholtz equation for a single surface.

In Fig. 2–13 the small object BC, of height h, is imaged at B′C′ by the refracting surface AH separating media whose refractive indices are n and n'. The ray from B at angle u is incident at height y and is refracted to B′ at angle u'. The ray CA is refracted along AC′ and the angles of incidence and refraction are as indicated. The paraxial form of Snell's law of refraction gives

$$ni = n'i'$$

or, from the figure,
$$n(h/l) = n'(h'/l'),$$

$$\therefore \quad \frac{nh}{n'h'} = \frac{l}{l'}. \quad \cdots \cdots \quad (2\text{--}46)$$

Again from the figure one has

$$y = lu = l'u',$$

$$\therefore \quad \frac{l}{l'} = \frac{u'}{u}. \quad \cdots \cdots \quad (2\text{--}47)$$

From (2–46) and (2–47) one has, at once,

$$nhu = n'h'u'$$

which is the paraxial form of the Helmholtz equation.

2–16. The Thin Lens

When two refracting surfaces are combined to form a lens the positions of the cardinal points of the lens are given by the general equations of §§2–12 and 2–13. If the lens is thin, the separation of the surfaces, i.e. of the adjacent principal points, is negligible. Thus one puts $d=0$. Equation (2–37) then shows that the power of the lens is the sum of the powers of the separate surfaces. One has

$$F = F_1 + F_2 = \frac{n_1' - n_1}{r_1} + \frac{n_2' - n_1'}{r_2}. \quad \cdots \cdots \quad (2\text{--}48)$$

Equations (2–35) and (2–36) show that the principal planes of the lens coincide with the lens itself. The relation between the conjugate distances l and l' is given by the usual equation

$$\frac{n'}{l'} - \frac{n}{l} = F, \quad \cdots \cdots \quad (2\text{--}49)$$

where F is given by (2–48) and n and n' are now the refractive indices of the initial and final media.

The case of a thin lens in air follows if one puts $n_1 = n_2' = 1$ and $n_1' = n$, the refractive index of the material of the lens. One has

$$F = \frac{n-1}{r_1} + \frac{1-n}{r_2},$$

i.e.
$$\frac{1}{l'} - \frac{1}{l} = (n-1)\left(\frac{1}{r_1} - \frac{1}{r_2}\right). \quad \cdots \cdots \quad (2\text{--}50)$$

The principal planes again coincide with the lens, and since the initial and final media are the same the nodal planes also coincide with the lens (see §2–5).

It is convenient to point out here that the physical positions of the principal foci of a system have no special significance in the spaces to which they do not belong. For example, for a diverging lens the principal foci are virtual. The second principal focus is, as always, in the right-hand space; here it is virtual and lies on the left-hand side of the lens. A real object placed at this point lies in the *left*-hand space and its image is not at infinity; it lies between the object and the lens (see Fig. 2–14). Similarly, a virtual object at a

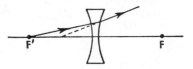
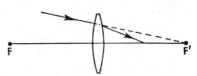

Fig. 2–14. Principal foci of lenses.

principal focus of a convex lens is in a space different from that of the focus and, again, the image is not at infinity (Fig. 2–14).

2–17. The Thick Lens

Consider the thick lens consisting of the combination of surfaces whose radii are r_1 and r_2 and whose poles A and B are distance d apart. Equations (2–35), (2–36), and (2–37), give the power and the positions of the principal planes. One has

$$
\left.
\begin{array}{l}
AP = \dfrac{n_1 d F_2}{n_1' F} \\[2ex]
BP' = -\dfrac{n_2' d F_1}{n_1' F},
\end{array}
\right\} \qquad \cdots \cdots \quad (2\text{–}51)
$$

and

where

$$
F_1 = \frac{n_1' - n_1}{r_1}
$$

and

$$
F_2 = \frac{n_2' - n_1'}{r_2}
$$

and

$$
F = \frac{n_1' - n_1}{r_1} + \frac{n_2' - n_1'}{r_2} - \frac{d(n_1' - n_1)(n_2' - n_1')}{n_1' r_1 r_2}. \quad \cdots \quad (2\text{–}52)
$$

For a thick lens of index n, thickness d, in air one has $F_1 = (n-1)/r_1$, $F_2 = (1-n)/r_2$, and the expression for F reduces to

$$
F = (n-1)\left(\frac{1}{r_1} - \frac{1}{r_2}\right) + \frac{d(n-1)^2}{n r_1 r_2} \quad \cdots \cdots \quad (2\text{–}53)
$$

$$
= \frac{n(n-1)(r_2 - r_1) + d(n-1)^2}{n r_1 r_2}. \quad \cdots \cdots \quad (2\text{–}54)
$$

In equation (2–53) the first term is the power of the thin lens that has the same radii of curvature. If one imagines the surfaces of this thin lens being pulled apart the power is modified by the amount given by the second term.

The positions of the principal points of a thick lens in air follow at once from equations (2–51). One has

$$AP = \frac{dF_2}{nF}$$

and

$$BP' = -\frac{dF_1}{nF},$$

$$\qquad \qquad (2\text{–}55)$$

whence

$$AP = -\frac{dr_1}{n(r_2-r_1)+d(n-1)},$$

$$BP' = -\frac{dr_2}{n(r_2-r_1)+d(n-1)}.$$

$$\qquad \qquad (2\text{–}56)$$

Frequently $n(r_2-r_1)$ is considerably larger than $d(n-1)$ and the latter can be neglected in comparison with the former. Thus, for a moderately thick lens, one has

$$AP = -\frac{dr_1}{n(r_2-r_1)},$$

$$BP' = -\frac{dr_2}{n(r_2-r_1)}.$$

$$\qquad \qquad (2\text{–}57)$$

Since the initial and final media are the same, the nodal and principal planes coincide (Fig. 2–15). For a biconvex or biconcave lens r_1 and r_2 have opposite

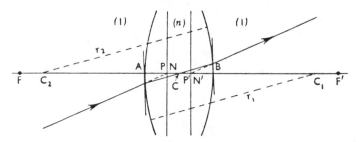

Fig. 2–15. Thick lens.

signs, and for many meniscus lenses the radii will be widely different. The result is that $|r_2-r_1|$ is often fairly large and (2–57) sufficiently accurate.

If one has a ray that is directed towards the first nodal point, it emerges from the direction of N′ and the incident and emergent rays are parallel. Consequently, the tangents to the surfaces at the points of incidence and emergence must be parallel, and it may be said that the ray passes through the lens as though through a plane-parallel plate. The axial point C, through which such a ray passes in the middle space, is known as the *optical centre*. Since it must remain in the paraxial region, the ray must, in fact, make a small angle with the optical axis. The tangents to the surfaces at the points of incidence and emergence are therefore almost normal to the axis. Now the points P and P′ are the images of C as formed by the first and second refracting surfaces respectively, and since the lens behaves as a parallel-sided plate for the ray illustrated, it follows that $AC = nAP$ and $BC = nBP'$.

$$\therefore \quad \frac{AC}{BC} = \frac{AP}{BP'} = \frac{r_1}{r_2}. \qquad \ldots \ldots (2\text{--}58)$$

It will be seen that, in general, the optical centre is not the physical centre of the lens. In fact, C is at the pole of the curved surface if the lens is plano, and lies outside the lens if the latter is meniscus.

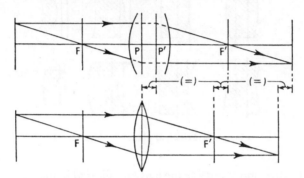

Fig. 2–16. Imagery by a thick lens.

The general paraxial image-forming properties of a thick lens or lens system are completely determined by the positions of the cardinal planes. If these are known one can, for example, find the image of any object without making a detailed study of the passage of the image-forming rays through the system. That is, it is not necessary to consider the refractions of the rays at the various surfaces. The geometrical construction for finding the image is as given in §2–2 and is the same as the simple method used for a thin lens. This is illustrated by Fig. 2–16 which shows the construction for a thick lens and the equivalent thin lens. The figure confirms the statement, made in §2–12, that the latter gives the same image except for a displacement PP'.

2–18. The Spherical Lens

Consider the case of a lens consisting simply of a sphere of index n in air. It is at once obvious that the nodal and principal points must coincide with the centre. This also follows from the equations given above. (Note that, since r_2 is negative, one has $d = r_1 - r_2$ or, since $r_2 = -r_1$, $d = 2r_1$.) The power of the lens follows from equation (2–54). One has

$$F = \frac{-n(n-1)2r + 2r(n-1)^2}{-nr^2}$$

$$= \frac{2(n-1)}{nr}$$

or
$$\text{E.F.L.} = \frac{nr}{2(n-1)}. \qquad \ldots \ldots (2\text{--}59)$$

2-19. Examples of Principal Planes of a Thick Lens in Air

Fig. 2-17 illustrates the positions of the principal planes of a *moderately* thick lens in air. If the lens is equiconvex or equiconcave and is not very thick, one may put $F_1=F_2=F/2$. Taking $n=1\cdot5$ one has, from equations (2-55), $AP=-BP'=d/3$. For a plano-convex or plano-concave lens where $r_2=\infty$, $F_2=0$ and equations (2-55) give $AP=0$ and $BP'=-2d/3$. For meniscus lenses

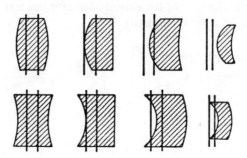

Fig. 2-17. Principal planes of thick lenses.

the principal planes move outside the lens as indicated in the figure. It can easily be shown from equations (2-57) that if $n=1\cdot5$ the second principal plane reaches the first surface when $r_2=3r_1$. In the figure $|F|$ is approximately the same for all the lenses. It should be noted that in order to apply the approximate equations (2-57) to a steep meniscus lens, the lens must be relatively thin.

2-20. Two Separated Thin Lenses in Air

Suppose one has two thin lenses A and B whose focal lengths are f_1' and f_2' respectively, the lenses being separated by a distance d. The results of §2-14 will apply so that one has

i.e.

$$\left.\begin{array}{c} \dfrac{1}{f'}=\dfrac{1}{f_1'}+\dfrac{1}{f_2'}-\dfrac{d}{f_1'f_2'}, \\[2mm] f'=\text{E.F.L.}=\dfrac{f_1'f_2'}{f_1'+f_2'-d}. \end{array}\right\} \quad \ldots \ldots \quad (2\text{-}60)$$

$$\left.\begin{array}{c} AP=\dfrac{df_1'}{f_1'+f_2'-d}, \\[2mm] BP'=-\dfrac{df_2'}{f_1'+f_2'-d}, \end{array}\right\} \quad \ldots \ldots \quad (2\text{-}61)$$

with corresponding equations for the front and back focal lengths. Since the system is in air, the nodal points coincide with the principal points.

It should be noted that for two convex lenses a short distance apart, the cardinal planes have positions that are similar to those for a biconvex thick lens except that in the present case the principal planes are "crossed". That is to say, P' actually lies to the left of P [see Fig. 2-18(a)].

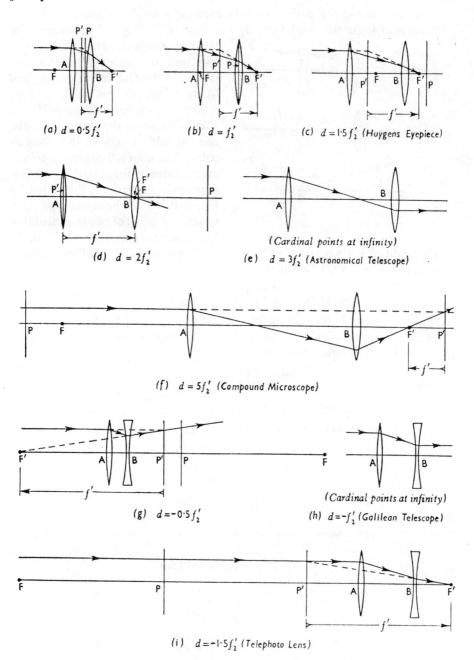

(a) $d = 0{\cdot}5f_2'$

(b) $d = f_2'$

(c) $d = 1{\cdot}5f_2'$ (Huygens Eyepiece)

(d) $d = 2f_2'$

(e) $d = 3f_2'$ (Astronomical Telescope)

(Cardinal points at infinity)

(f) $d = 5f_2'$ (Compound Microscope)

(g) $d = -0{\cdot}5f_2'$

(h) $d = -f_2'$ (Galilean Telescope)

(Cardinal points at infinity)

(i) $d = -1{\cdot}5f_2'$ (Telephoto Lens)

Fig. 2–18. Cardinal points of a system of two separated thin lenses. In (a)–(f) $f_1' = 2f_2'$, and in (g)–(i) $f_1' = -2f_2'$.

Fig. 2–18 shows the positions of the cardinal points for a number of combinations of separated thin lenses. In each case $|f_1'|=2|f_2'|$. (c) represents the

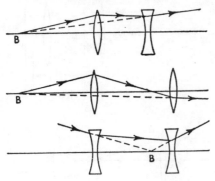

Huygens' eyepiece, (e) the astronomical telescope, (f) the compound microscope, (h) the Galilean telescope, and (i) the telephoto lens. Each of these is discussed further in Chapter III.

In addition to the cardinal points, one should mention the *Bravais* points. These are self-conjugate points; their existence for a system of two separated thin lenses is illustrated by Fig. 2–19. It can be shown that for a system consisting of two separated thin lenses, Bravais points exist if $d^2>4f_1'f_2'$.'

Fig. 2–19. Bravais points.

This condition is always satisfied if one lens is converging and the other diverging because this makes $f_1'f_2'$ negative.

2–21. The Deviation Method for Tracing Paraxial Rays

Figs 2–20(a) and (b) illustrate the passage of a paraxial ray through a complete optical system and through a simple refracting surface respectively; the discussion applies equally well to either.

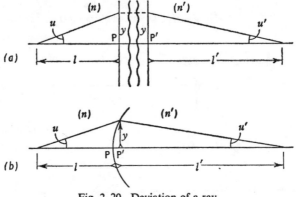

Fig. 2–20. Deviation of a ray.

One has

$$\frac{n'}{l'} - \frac{n}{l} = F,$$

$$\therefore \quad \frac{n'y}{l'} - \frac{ny}{l} = Fy,$$

$$\therefore \quad n'u' - nu = Fy. \quad \cdots \cdots \quad (2\text{–}62)$$

If u and y are known, the path of the incident ray is specified completely. Using (2–62) it is then possible to find the path of the emergent ray. For a system in air ($n'=n=1$), Fy is simply the deviation of the ray ($\because \delta=u'-u$ algebraically), and it will be seen that the angle of deviation of a paraxial ray is independent of its initial direction. The most useful application of (2–62) is the tracing of a paraxial ray through a system of refracting surfaces. The path of the incident ray is specified by u_1 and y_1 and (2–62) gives the path after refraction at the first surface. In order to apply the equation to the second surface it is necessary to find y_2, the height of incidence. It will be seen, from Fig. 2–21, that

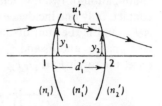

$$y_2 = y_1 - d_1' u_1'. \quad . \quad . \quad (2\text{–}63)$$

Remembering that $u_2 \equiv u_1'$, one can now apply (2–62) to the second refraction. By repeated alternate application of (2–62) and (2–63) one can

Fig. 2–21. Incidence heights at adjacent surfaces.

find the path of the ray through any number of refracting surfaces (or through any number of complete systems). (See §2–27.)

2–22. Location of Cardinal Points of Complicated Systems

For any optical system, it has been shown (§2–1) that if one has a ray entering parallel to the axis, the emergent ray passes through the appropriate principal focus, and the initial and final rays (produced if necessary) intersect in the corresponding principal plane. For complicated systems this provides the

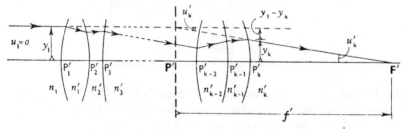

Fig. 2–22. Cardinal points of complicated systems.

easiest method of calculating the positions of the cardinal planes. Suppose one has a system of k refracting surfaces. Starting with $u_1 = 0$, and using any convenient value of y_1 (usually $y_1 = 1$) one can use (2–62) and (2–63) to find the path of the ray through the system. These equations are applied alternately until y_k and u_k' are found. Then, from Fig. 2–22, one has

$$P_k' P' = -\frac{y_1 - y_k}{u_k'} \quad . \quad . \quad . \quad . \quad . \quad (2\text{–}64)$$

and

$$f' = \frac{y_1}{u_k'} \quad . \quad . \quad . \quad . \quad . \quad . \quad (2\text{–}65)$$

so that the positions of P′ and F′ can be found (§2–27). In order to find P and F for the complete system it is necessary to trace the path of a ray entering parallel to the axis from the right. If the indices of the initial and final media are different, the nodal planes can be found from the relations $x'_N = f$, $x_N = f'$.

The above equations, which apply to a number of complete systems as well as to a number of refracting surfaces, provide an alternative method of deriving the equations governing the combination of two optical systems.

2-23. Application of the Deviation Method to Two Systems

Before discussing the general problem, one special case should be noted: the combination of two thin lenses in contact in air. Since the lenses are thin and in contact, only one incidence height (y) is involved. Obviously the total deviation is Fy, and is the sum of the deviations F_1y and F_2y, F being the power of the combination and F_1 and F_2 the powers of the separate lenses. Thus one has, at once, the result $F=F_1+F_2$.

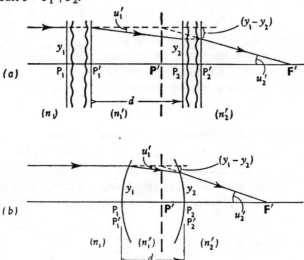

Fig. 2-23. Combination of two systems.

Figs 2-23(a) and (b) illustrate the combination of two complete systems and two refracting surfaces respectively. Again the discussion applies equally well to either. If the incident ray is parallel to the axis one has $u_1=0$.

Equation (2-62) gives

$$n_1'u_1'=F_1y_1. \qquad \cdots \cdots \quad (2\text{-}66)$$

For the second surface one has, since $n_2u_2\equiv n_1'u_1'$,

$$n_2'u_2'-n_1'u_1'=F_2y_2. \qquad \cdots \cdots \quad (2\text{-}67)$$

On adding these equations one has

$$n_2'u_2'=F_1y_1+F_2y_2. \qquad \cdots \cdots \quad (2\text{-}68)$$

Now the incident ray is parallel to the axis. Hence the incidence height in the first principal plane of the combination must be y_1 and one has, for the complete system,

$$n_2'u_2'=Fy_1. \qquad \cdots \cdots \quad (2\text{-}69)$$

(2-68) and (2-69) give

$$Fy_1=F_1y_1+F_2y_2. \qquad \cdots \cdots \quad (2\text{-}70)$$

Now from the figure

$$y_2=y_1-du_1', \quad \text{algebraically,}$$

or, using (2-66),

$$y_2=y_1-\frac{dF_1y_1}{n_1'}. \qquad \cdots \cdots \quad (2\text{-}71)$$

Therefore, from (2–70),

$$Fy_1 = F_1 y_1 + F_2 y_1 - \frac{F_2 dF_1 y_1}{n_1'}$$

or
$$F = F_1 + F_2 - dF_1 F_2. \quad . \quad . \quad . \quad . \quad . \quad (2\text{–}72)$$

The figure also gives

$$P_2'P' = -\frac{y_1 - y_2}{u_2'}, \quad \text{algebraically.}$$

Substituting from (2–71) for $y_1 - y_2$, and from (2–69) for u_2', one has

$$P_2'P' = -\frac{n_2' dF_1}{F}. \quad . \quad . \quad . \quad . \quad (2\text{–}73)$$

Similarly, by tracing a ray incident parallel to the axis from the right, it can be shown that

$$P_1 P = \frac{n_1 dF_2}{F}. \quad . \quad . \quad . \quad . \quad . \quad (2\text{–}74)$$

Equations (2–72), (2–73), and (2–74), are identical to those derived in §2–12 but the present proof does not break down when F_1 or F_2 is zero.

2–24. Use of Matrix Notation for Paraxial Ray Tracing. The System Matrix

In this and the following sections there is presented an alternative approach to the problem of tracing paraxial rays through optical systems, finding the image of a given object, and finding the positions of the cardinal planes. The method is presented for the benefit of those who are familiar with matrix algebra and who prefer an algebraic to a geometrical approach; it is particularly advantageous if one is dealing with a complex system. It should be noted that the paraxial approximation is made throughout whereas in the more abstract general discussions in §§2–1 to 2–14 this limitation was not imposed.

Consider a ray travelling from left to right through a system of refracting surfaces. For the jth surface, equation (2–62), written in terms of reduced quantities (as defined in §2–13), is

$$\bar{u}_j' - \bar{u}_j = F_j y_j \quad . \quad . \quad . \quad . \quad . \quad (2\text{–}75)$$

(2–63) gives

$$y_{j+1} = y_j - \bar{d}_j \bar{u}_j' \quad . \quad . \quad . \quad . \quad . \quad (2\text{–}76)$$

The first equation gives the change of direction caused by refraction at the jth surface and the second gives the change of incidence height resulting from the passage of the ray across the space between the jth surface and the $(j+1)$th surface. One refers to a *refraction* equation and a *transfer* or *translation* equation. A complete set of refraction and translation equations would give the effect on *both direction and incidence height* of *both refraction at* a surface and *translation between* one surface and the next. Obviously refraction leaves y unaffected and translation leaves u unaffected, but it is convenient to include the trivial equations that express these facts and write, for the complete set of equations,

1. Refraction:
$$\bar{u}_j' = 1.\bar{u}_j + F_j.y_j \quad \ldots \ldots \quad (2\text{–}77)$$

$$y_j' = 0.\bar{u}_j + 1.y_j \quad \ldots \ldots \quad (2\text{–}78)$$

2. Translation:
$$\bar{u}_{j+1} = 1.\bar{u}_j' + 0.y_j' \quad \ldots \ldots \quad (2\text{–}79)$$

$$y_{j+1} = -\bar{d}_j'.\bar{u}_j' + 1.y_j'. \quad \ldots \ldots \quad (2\text{–}80)$$

These equations are written with the superfluous unit and zero multipliers in order to make more obvious the fact that the equations written in matrix form are:

1. Refraction:
$$\begin{bmatrix} \bar{u}_j' \\ y_j' \end{bmatrix} = \begin{bmatrix} 1 & F_j \\ 0 & 1 \end{bmatrix} \begin{bmatrix} \bar{u}_j \\ y_j \end{bmatrix}. \quad \ldots \ldots \quad (2\text{–}81)$$

2. Translation:
$$\begin{bmatrix} \bar{u}_{j+1} \\ y_{j+1} \end{bmatrix} = \begin{bmatrix} 1 & 0 \\ -\bar{d}_j' & 1 \end{bmatrix} \begin{bmatrix} \bar{u}_j' \\ y_j' \end{bmatrix}. \quad \ldots \ldots \quad (2\text{–}82)$$

One writes
$$\begin{bmatrix} \bar{u}_j' \\ y_j' \end{bmatrix} = \mathbf{R}_j \begin{bmatrix} \bar{u}_j \\ y_j \end{bmatrix} \quad \ldots \ldots \quad (2\text{–}83)$$

and
$$\begin{bmatrix} \bar{u}_{j+1} \\ y_{j+1} \end{bmatrix} = {}_j\mathbf{T}_{j+1} \begin{bmatrix} \bar{u}_j' \\ y_j' \end{bmatrix} \quad \ldots \ldots \quad (2\text{–}84)$$

where $\mathbf{R}_j$ is called the *refraction matrix* for surface j and $_j\mathbf{T}_{j+1}$ is the *translation matrix* between surfaces j and $j+1$.

If one has a lens consisting of two refracting surfaces the direction and height of emergence from the second surface is related to the direction and height of incidence at the first surface by

$$\begin{bmatrix} \bar{u}_2' \\ y_2' \end{bmatrix} = \mathbf{R}_2 \begin{bmatrix} \bar{u}_2 \\ y_2 \end{bmatrix}$$

$$= \mathbf{R}_2 \cdot {}_1\mathbf{T}_2 \begin{bmatrix} \bar{u}_1' \\ y_1' \end{bmatrix}$$

$$= \mathbf{R}_2 \cdot {}_1\mathbf{T}_2 \cdot \mathbf{R}_1 \begin{bmatrix} \bar{u}_1 \\ y_1 \end{bmatrix} \quad \ldots \ldots \quad (2\text{–}85)$$

In general, for a system of k refracting surfaces,

$$\begin{bmatrix} \bar{u}_k' \\ y_k' \end{bmatrix} = \mathbf{R}_k \cdot {}_{k-1}\mathbf{T}_k \cdot \mathbf{R}_{k-1} \quad \ldots \ldots \quad {}_1\mathbf{T}_2 \cdot \mathbf{R}_1 \begin{bmatrix} \bar{u}_1 \\ y_1 \end{bmatrix} \quad (2\text{–}86)$$

It should be noted that since, in matrix multiplication, $\mathbf{AB} \neq \mathbf{BA}$ it is essential that the matrices be written in the order given.

One writes
$$\begin{bmatrix} \bar{u}_k' \\ y_k' \end{bmatrix} = \mathbf{M} \begin{bmatrix} \bar{u}_1 \\ y_1 \end{bmatrix} \quad \ldots \ldots \quad (2\text{–}87)$$

where $\mathbf{M}$ is called the *system matrix*; since it is the product of a number of 2×2 matrices, it is itself a 2×2 matrix.

Put
$$\mathbf{M} = \begin{bmatrix} M_{11} & M_{12} \\ M_{21} & M_{22} \end{bmatrix} \quad \cdots \cdots \quad (2\text{–}88)$$

The elements of $\mathbf{M}$ are definite constants of the system. It was Gauss who first defined the image-forming properties of a system by means of these constants and, in consequence, they are called the *Gaussian Constants* of the system. Since all refraction and translation matrices have unit determinant values, the system matrix will also have a determinant value of unity, i.e. one always has $M_{11}M_{22} - M_{12}M_{21} = 1$, and this is a useful check on the arithmetic when the matrix of a system has been calculated numerically. (See §2–27.)

[Equations (2–75) and (2–76), which provide the starting point for the development of the present theory, apply to a thin lens, to a thick lens, or to a complete lens system, as well as to a single refracting surface (see §2–21). In general, j denotes the complete lens or system; F_j is simply the power of system j; the incidence height y_j is measured in *each* principal plane of system j; and d_j' is the distance $P_j' P_{j+1}$. "Refraction" and Translation matrices for complete systems can then be defined exactly as above. In fact, the method is most frequently applied with j denoting a refracting surface, and in what follows this will be assumed to be the case, but it is worth noting that the theory applies far more generally.]

2–25. Image Formation. The Location of the Cardinal Planes

In the above discussion $j, j+1$, etc., denote refracting surfaces. In general they can also denote any planes perpendicular to the axis provided there is no refracting surface between planes j and $j+1$. Obviously, for a surface that is not a refracting surface one simply has $n' = n$, giving $F = 0$ and $u' = u$.

Suppose, now, that one has a system of k refracting surfaces and that planes B and B′ lie in the left- and right-hand spaces of this system (see Fig. 2–24). In

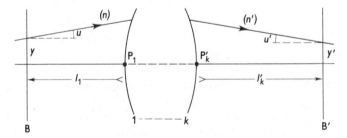

Fig. 2–24. Imagery by a complete system.

the figure B and B′ are shown as real planes outside, and on the left- and right-hand *sides* of, the system. In general they can lie anywhere in their proper optical *spaces*. I.e., they can be virtual and located anywhere along the axis in the figure. The refractive indices of the left- and right-hand spaces are n and n'. The distance *to* B *from* surface 1 is l_1 (negative in the figure) and B′ is the positive distance l_k' from surface k. A ray is defined by (u, y) at B and (u', y') at B′. The system matrix $\mathbf{M}$ transforms from surface 1 to surface k. There will also be a 2×2 matrix $\mathbf{N}$ which transforms from plane B to plane B′. One has

$$\mathbf{N} = \mathbf{B'MB}, \quad \ldots \ldots \ldots \ldots \quad (2\text{-}89)$$

where $\mathbf{B}$ is the (translation) matrix that transforms from plane B to the first surface of the system, and $\mathbf{B'}$ transforms from the last surface of the system to plane B'.

Let

$$\mathbf{N} = \begin{bmatrix} N_{11} & N_{12} \\ N_{21} & N_{22} \end{bmatrix} \quad \ldots \ldots \ldots \quad (2\text{-}90)$$

It will be seen at once that det $\mathbf{N} = 1$. One has

$$\begin{bmatrix} \bar{u}' \\ y' \end{bmatrix} = \begin{bmatrix} N_{11} & N_{12} \\ N_{21} & N_{22} \end{bmatrix} \begin{bmatrix} \bar{u} \\ y \end{bmatrix} \quad \ldots \ldots \ldots \quad (2\text{-}91)$$

$$= \begin{bmatrix} 1 & 0 \\ -l'_k & 1 \end{bmatrix} \begin{bmatrix} M_{11} & M_{12} \\ M_{21} & M_{22} \end{bmatrix} \begin{bmatrix} 1 & 0 \\ l_1 & 1 \end{bmatrix} \begin{bmatrix} \bar{u} \\ y \end{bmatrix} \quad \ldots \quad (2\text{-}92)$$

(noting that in the general form of a translation matrix as given in (2–82) d'_j is positive and measured from left to right).

Referring now to the matrix $\mathbf{N}$, some special cases will be noted. From (2–91) and Fig. 2–24 one sees:

1. If $N_{11} = 0$, $\bar{u}' = N_{12}y$. I.e. the direction of the emergent ray is determined by the height of intersection of the incident ray at plane B. That is to say, all rays that enter the system from the same point in plane B emerge from the system in the same direction—B is the first focal plane of the system.
2. If $N_{12} = 0$, $\bar{u}' = N_{11}\bar{u}$. I.e. the direction of the emergent ray is determined solely by the direction of the incident ray. Thus if a collimated beam enters the system, the emergent beam is collimated—the system is telescopic.
3. If $N_{21} = 0$, $y' = N_{22}y$. I.e. all the rays that pass through the point in plane B specified by y pass through the point in plane B' specified by y'—B and B' are conjugate planes.
4. If $N_{22} = 0$, $y' = N_{21}\bar{u}$. I.e. the ray intersection point in plane B' is determined by the angle u at which the ray enters the system. That is, if parallel rays enter the system the emergent rays come to a focus on plane B'—B' is the second focal plane of the system.

Consider in more detail the case where B and B' are conjugate planes. On carrying out the matrix multiplication in (2–92), one obtains

$$\begin{bmatrix} N_{11} & N_{12} \\ N_{21} & N_{22} \end{bmatrix} = \begin{bmatrix} (M_{11} + M_{12}l_1) & M_{12} \\ -l'_k(M_{11} + M_{12}l_1) \cdot - M_{21} + M_{22}l_1 & (-l'_k M_{12} + M_{22}) \end{bmatrix} . \quad (2\text{-}93)$$

and in the case being considered $N_{21} = 0$. This gives

$$M_{12}l_1 l'_k + M_{11}l'_k - M_{22}l_1 - M_{21} = 0 \quad \ldots \ldots \quad (2\text{-}94)$$

or

$$l'_k = \frac{M_{22}l_1 + M_{21}}{M_{12}l_1 + M_{11}} \quad \text{and} \quad l_1 = \frac{M_{21} - M_{11}l'_k}{M_{12}l'_k - M_{22}} \quad \ldots \quad (2\text{-}95)$$

These equations give the relationship between the object and image distances measured from the first and last surfaces of the system. Since N_{21} is zero, it follows at once from (2–91) and (2–93) that the transverse linear magnification for the conjugate planes is given by

$$m = \frac{y'}{y} = N_{22} = M_{22} - M_{12}l'_k \quad \ldots \ldots \text{(2–96)}$$

Hence one now has

$$N = \begin{bmatrix} (M_{11}+M_{12}l_1) & M_{12} \\ 0 & m \end{bmatrix} \quad \ldots \ldots \text{(2–97)}$$

and, since det $N = 1$,

$$\frac{1}{m} = M_{11} + M_{12}l_1 \quad \ldots \ldots \ldots \ldots \text{(2–98)}$$

so that for conjugate planes

$$N = \begin{bmatrix} 1/m & M_{12} \\ 0 & m \end{bmatrix} \quad \ldots \ldots \ldots \text{(2–99)}$$

For the axial points of the conjugate planes $y' = y = 0$, and putting (2–99) into (2–91) gives $\bar{u}' = (1/m)\bar{u}$.

I.e.

$$m\left(\frac{u'}{u}\right) = \frac{n}{n'} \quad \ldots \ldots \ldots \text{(2–100)}$$

which is the paraxial form of the Helmholtz equation (2–6).

If one sets $m = +1$, the conjugate planes are, by definition, the principal planes. In terms of the usual notation (2–96) and (2–98) give the positions of the principal planes:

$$(l'_k)_{P'} = \frac{P'_k P'}{n'} = \frac{M_{22}-1}{M_{12}} \quad \ldots \ldots \text{(2–101)}$$

$$(l_1)_P = \frac{P_1 P}{n} = \frac{1-M_{11}}{M_{12}} \quad \ldots \ldots \text{(2–102)}$$

If the conjugate planes are the nodal planes, $u' = u$ by definition and (2–100) gives $m = n/n'$. (2–96) and (2–98) then give the positions of the nodal planes:

$$(l'_k)_{N'} = \frac{M_{22}-n/n'}{M_{12}} \quad \ldots \ldots \ldots \text{(2–103)}$$

$$(l_1)_N = \frac{n'/n-M_{11}}{M_{12}} \quad \ldots \ldots \ldots \text{(2–104)}$$

These results show at once that if $n' = n$ the nodal planes coincide with the principal planes.

The position of the first focal plane is obtained by setting $N_{11} = 0$. (2–93) gives

$$(l_1)_F = \frac{P_1 F}{n} = -\frac{M_{11}}{M_{12}} \quad \ldots \ldots \text{(2–105)}$$

and the first focal length is obtained from

$$f = PF = PP_1 + P_1F = -P_1P + P_1F$$

whence

$$\frac{f}{n} = \frac{M_{11}-1}{M_{12}} - \frac{M_{11}}{M_{12}} = -\frac{1}{M_{12}} \quad . \quad . \quad (2\text{--}106)$$

Similarly, the position of the second focal plane is obtained by setting $N_{22}=0$. (2–93) gives

$$(l'_k)_{F'} = \frac{P'_kF'}{n'} = \frac{M_{22}}{M_{12}}. \quad . \quad . \quad . \quad . \quad . \quad (2\text{--}107)$$

and the second focal length is obtained from

$$f' = P'F' = P'P'_k + P'_kF' = -P'_kP' + P'_kF'$$

whence

$$\frac{f'}{n'} = \frac{1-M_{22}}{M_{12}} + \frac{M_{22}}{M_{12}} = \frac{1}{M_{12}}. \quad . \quad . \quad (2\text{--}108)$$

(2–106) and (2–108) show that

$$M_{12} = \frac{n'}{f'} = -\frac{n}{f} = F, \text{ the power of the system.} \quad . \quad (2\text{--}109)$$

Equations (2–101), (2–102), (2–103), (2–104), (2–105), and (2–107) enable one to find the positions of the cardinal planes of any system of refracting surfaces from the system matrix, which in turn can be found from the powers and reduced separations of the surfaces, using (2–86) and the expressions for the refraction and translation matrices given by (2–81) and (2–82). This method of locating the cardinal planes is particularly useful for complicated systems. Once the elements of the system matrix have been found there is often no need to find the cardinal planes since equations (2–95) and (2–96) enable one to find the position of the image plane and the magnification directly from the object distance measured from the first surface of the system (see §2–27 and no. 7 of the Miscellaneous Examples on Chapters I–V.

2–26. The Conjugate Distances Measured from the Principal Planes

As an exercise, the matrix method will be used to find the conjugate distance equation when the distances are measured from the principal planes. The matrix N for a pair of conjugate planes B,B' is, from (2–93),

$$N = \begin{bmatrix} (M_{11}+M_{12}l_1) & M_{12} \\ 0 & (-l'_kM_{12}+M_{22}) \end{bmatrix}. \quad . \quad (2\text{--}110)$$

As before, the object and image distances measured from the principal planes are denoted by l and l'. One has

$$l_1 = P_1B = P_1P + PB = P_1P + l$$
$$l'_k = P'_kB' = P'_kP' + P'B' = P'_kP' + l'$$

(2–101) and (2–102) give

$$l_1 = \frac{1 - M_{11}}{M_{12}} + l$$

$$l'_k = \frac{M_{22} - 1}{M_{12}} + l'$$

$$\therefore \quad N = \begin{bmatrix} (1 + M_{12}l) & M_{12} \\ 0 & (1 - M_{12}l') \end{bmatrix} \quad \cdots \quad (2\text{–}111)$$

Since the determinant value of N is always unity, one has

$$(1 + M_{12}l)(1 - M_{12}l') = 1$$

Noting, (2–109), that M_{12} is the power of the system, and that $l' = l'/n'$ and $l = l/n$, this yields the result

$$\frac{n'}{l'} - \frac{n}{l} = \frac{n'}{f'} = -\frac{n}{f} = F$$

which is the result obtained using the purely geometrical treatment [equation (2–13)].

2–27. Numerical Example of the Use of the Various Techniques for Solving Thick Lens Problems

Three methods of solving thick lens problems are considered here:

(a) By treating the thick lens as the combination of two systems, each of which is simply a refracting surface, and quoting the general equations (2–35), (2–36), (2–37), to obtain the power of the lens [and hence the focal lengths from (2–34)], and the positions of the principal planes, and by finding the nodal planes from (2–10). Usually object and image distances are then measured from the cardinal planes that are most convenient for the problem in hand; the use of the nodal planes yields the magnification.

(b) By direct application of what has here been called the deviation method, using (2–62), (2–63), (2–64), (2–65), followed by (2–10) to find the nodal planes. Object and image positions can again be referred to the cardinal planes or they can be found independently by using (2–62) and (2–63). One can trace a ray from an axial or an extra-axial object point; the former locates the image plane and the latter yields the magnification.

(c) By using the matrix method, which is based on the deviation method and can be used to find the cardinal planes and hence relate the object and image positions as in the other methods, or can be used to relate object and image positions and to find the magnification without finding the cardinal planes explicitly.

In what follows, these three methods will be illustrated by using them to solve a particular thick lens problem. The problem is: A converging meniscus lens has radii of curvature 10 cm and 20 cm and has axial thickness 5 cm. It is made of glass of refractive index 1·650 and has water (index 1·330) on its concave side and air on its convex side. Find the positions of the cardinal planes. If a real

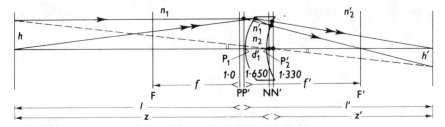

Fig. 2–25. The Cardinal planes of the lens.

object plane is 50 cm from the convex surface, find the position of the image plane, and find the magnification for this pair of planes.

The lens specification is:

$$n_1 = 1\cdot000 = n$$

$$
\begin{aligned}
r_1 &= +10 \\
d_1' &= 5 \qquad\qquad n_1' \equiv n_2 = 1\cdot650, \qquad\qquad d_1' = d_1'/n_1' = 3\cdot0303 \\
r_2 &= +20
\end{aligned}
$$

$$n_2' = 1\cdot330 = n'$$

(Throughout, distances are in cm and powers in cm^{-1}).

P_1 and P_1' are at the pole of surface 1; P_2 and P_2' are at the pole of surface 2. Each method requires the values of the powers of the individual surfaces. Using (2–45),

$$F_1 = \frac{n_1' - n_1}{r_1} = +0\cdot0650 \qquad F_2 = \frac{n_2' - n_2}{r_2} = -0\cdot0160$$

Method (a)

The power of the complete lens is given by (2–37):

$$F = +0\cdot0650 - 0\cdot0160 + 3\cdot0303 \times 0\cdot0650 \times 0\cdot0160 = +0\cdot052152$$

whence, from (2–34), $f = -1/F = -19\cdot175$, $f' = +n'/F = +25\cdot502$. (2–35) and (2–36) give

$$P_1P = +\frac{1 \times 3\cdot0303 \times (-0\cdot0160)}{0\cdot052152} = -0\cdot9297$$

$$P_2'P' = -\frac{1\cdot330 \times 3\cdot0303 \times 0\cdot0650}{0\cdot052152} = -5\cdot023$$

(N.B. The results in §2–17 are not applicable because the present lens does not have air on both sides.)

The actual positions of the principal foci can be obtained directly from (2–39) and (2–40) or simply from f and f' and the positions of P and P'. One obtains

$P_1F = -20\cdot105$: F is 20·105 to the left of the convex surface.
$P_2'F' = +20\cdot479$: F' is 20·479 to the right of the concave surface.

(2–10) gives $FN = f'$ and $F'N' = f$, whence

N is $25{\cdot}502-20{\cdot}105=5{\cdot}397$ to the right of the convex surface, i.e. $0{\cdot}397$ to the right of the concave surface.

N' is $20{\cdot}479-19{\cdot}175=1{\cdot}304$ to the right of the concave surface.

(A useful check is that $NN'=PP'$.)

If one specifies the position of the object plane by means of its distance l from the first principal plane of the complete lens, P, one has $l=-(50-0{\cdot}9297)=-49{\cdot}070$ and (2–13) gives $l'=+41{\cdot}859$. Hence the image plane is $41{\cdot}859-5{\cdot}023=36{\cdot}836$ to the right of the concave surface. In fact, since one requires the magnification, it is easier to use the conjugate distances from the nodal planes. One has $z=-(50+5{\cdot}397)=-55{\cdot}397$, and (2–16) gives $z'=+35{\cdot}532$. This yields the same position for the image plane $(35{\cdot}532+1{\cdot}304=36{\cdot}836)$ and at once gives the magnification, $m=z'/z=-0{\cdot}6414$. The positions of the cardinal planes and of the object and image planes are shown in the figure.

Method (b)

The positions of P' and F' are found by tracing a ray incident from the left parallel to the axis, i.e. $u_1=0$. The ray can have any height. For convenience take $y_1=1$. (2–62) gives

$$1{\cdot}650\,u_1'=0{\cdot}0650 \qquad \text{or} \qquad u_1'=+0{\cdot}039394$$

(2–63) gives $\quad y_2=1-5\times0{\cdot}039394=1-0{\cdot}19697=+0{\cdot}80303$

Since $n_2u_2=n_1'u_1'$, (2–62) gives

$$1{\cdot}330\,u_2'=0{\cdot}0650-0{\cdot}160\times0{\cdot}80303, \qquad \text{or} \qquad u_2'=+0{\cdot}039212$$

(2–64) gives $\quad P_2'P'=-0{\cdot}19697/0{\cdot}039212=-5{\cdot}023$

(2–65) gives $\quad f'=1/0{\cdot}039212=25{\cdot}502$

$P_2'F'$ is given by y_2/u_2'; one obtains

$$P_2'F'=0{\cdot}80303/0{\cdot}039212=20{\cdot}479$$

By taking a ray parallel to the axis from the right $(u_2'=0)$ at, say, $y_2=1$, P_1P, P_1F, and f can be found. As in method (a) the actual positions of the principal foci can be found from f and f' and the positions of P and P'. The nodal planes, also, are found as in method (a), i.e. using (2–10). The position of the image plane and the value of the magnification can be found as in method (a) but it is interesting to see how the deviation (i.e. ray tracing) method enables one to find the image plane and the magnification without finding the cardinal planes. To find the image plane one takes a ray from an axial point 50 cm to the left of the lens. One is free to choose any convenient value of y_1. Take $y_1=1$ so that $u_1=-1/50=-0{\cdot}0200$. (2–62) gives

$$1{\cdot}650\,u_1'=0{\cdot}0650-0{\cdot}0200=+0{\cdot}0450,$$

whence $\qquad u_1'=+0{\cdot}027273$

(2–63) gives $\quad y_2=1-5\times0{\cdot}027273=+0{\cdot}863635$

(2–62) gives $1\cdot330\ u_2' = -0\cdot0160 \times 0\cdot863635 + 0\cdot4500,$

whence $u_2' = +0\cdot023445$

$\therefore\quad l_2' = 0\cdot863635/0\cdot023445 = +36\cdot837$

To find the magnification one can take *any* ray from a height h in the object plane and find where it intersects the image plane. In practice one might wish to find the paths of principal rays (see §3–1) but these are not defined in the present problem and for economy of effort it is convenient to take the ray that was used initially to find P′ and f'. This corresponds to $h=1$. The ray emerged with $y_2 = 0\cdot80303$ and $u_2' = +0\cdot039212$. In the image plane h' is negative with

$$y_2 - h' = l_2' u_2',$$

whence
$$h' = -(36\cdot837 \times 0\cdot039212 - 0\cdot080303)$$
$$= -0\cdot64142$$
$$\therefore\quad m = h'/h = h'/1 = -0\cdot6414$$

Method (c)

From (2–81) the refraction matrices are

$$\mathbf{R}_1 = \begin{bmatrix} 1 & 0\cdot0650 \\ 0 & 1 \end{bmatrix} \quad \text{and} \quad \mathbf{R}_2 = \begin{bmatrix} 1 & -0\cdot0160 \\ 0 & 1 \end{bmatrix}$$

and the translation matrix is, from (2–82),

$$_1\mathbf{T}_2 = \begin{bmatrix} 1 & 0 \\ -3\cdot0303 & 1 \end{bmatrix}$$

Hence, from (2–85), the system matrix is given by

$$\mathbf{M} = \mathbf{R}_2\,_1\mathbf{T}_2\,\mathbf{R}_2 = \begin{bmatrix} 1 & -0\cdot0160 \\ 0 & 1 \end{bmatrix}\begin{bmatrix} 1 & 0 \\ -3\cdot0303 & 1 \end{bmatrix}\begin{bmatrix} 1 & 0\cdot0650 \\ 0 & 1 \end{bmatrix}$$

which gives

$$\mathbf{M} = \begin{bmatrix} 1\cdot048485 & 0\cdot052152 \\ -3\cdot0303 & 0\cdot80303 \end{bmatrix}$$

N.B. It is a useful check on arithmetic that $M_{11}M_{22} - M_{12}M_{21} = 1$. (2–101) and (2–102) give

$$P_1P = (1 - 1\cdot048485)/0\cdot052152 = -0\cdot9297$$

and $P_2'P' = 1\cdot330(0\cdot80303 - 1)/0\cdot052152 = -5\cdot023$

(2–103) and (2–104) give

$$(l_1)_N = (1\cdot330 - 1\cdot048485)/0\cdot052152 = +5\cdot398$$

$$(l_2')_{N'} = (1\cdot330 \times 0\cdot80303 - 1)/0\cdot052152 = +1\cdot304$$

(2–105) and (2–107) give

$$(l_1)_F = -1\cdot048485/0\cdot052152 = -20\cdot104$$

$$(l_2')_{\text{F}'} = +(1 \cdot 330 \times 0 \cdot 80303)/0 \cdot 052152 = +20 \cdot 479$$

(2–106) and (2–108) give

$$f = -1/0 \cdot 052152 = -19 \cdot 175 \quad \text{and} \quad f' = +1 \cdot 330/0 \cdot 052152 = +25 \cdot 502$$

(2–95) gives

$$l_2' = +1 \cdot 330 \, \frac{0 \cdot 80303 \times (-50) - 3 \cdot 0303}{0 \cdot 052152 \times (-50) + 1 \cdot 048485}$$

$$= +1 \cdot 330 \times 27 \cdot 696 = +36 \cdot 836.$$

(2–96) gives

$$m = 0 \cdot 80303 - 0 \cdot 052152 \times 27 \cdot 696$$
$$= -0 \cdot 6414.$$

As in method (a), the positions of the nodal and focal planes could be obtained from the positions of P and P′, using the values of f and f', and the position of the image plane and the value of the magnification could be obtained by using the conjugate distance relation in terms of z and z'. However, once the system matrix has been obtained all the results follow in a straightforward manner as indicated above.

If one requires just the positions of the cardinal planes for a combination of two systems, method (a) is probably the easiest. If one wishes to find the position of the image of an object plane, without finding the cardinal planes method (b) is probably the easiest. If one wishes to find the image planes and the magnifications for *several* object planes, particularly if one does not wish to know the positions of the cardinal planes, method (c) is the easiest and the advantage becomes more marked the more complicated the system. Method (a) can be used for systems consisting of more than two components by combining systems in pairs, e.g. (1) with (2) followed by (1 + 2) with (3) etc.; clearly this is clumsy. For those who are not familiar with matrix multiplication, method (b) is to be recommended for systems consisting of more than two components.

CHAPTER III

THE PARAXIAL THEORY OF OPTICAL INSTRUMENTS

IN this chapter an account is given of the paraxial theory* of the chief optical instruments such as the camera lens, the telescope, and the microscope. The object of the discussion is to show the principle of each instrument; a more detailed account of the lenses employed is given in Chapter XVI, after the monochromatic aberrations have been discussed. Chromatic aberration and the achromatic combination of lenses are included in this chapter, however, because this subject forms part of the paraxial theory. In the chapters on interference and diffraction, which precede the discussion of aberrations (and in many chapters which follow it), the lenses and optical instruments that are employed are assumed to be perfect. That is to say, it is assumed that the systems have the same properties outside the paraxial region as they have inside it. For example, the optical path from an object point to the image point is the same through all zones of the lens. That is to say, a spherical wave emerges and disturbances from the object arrive exactly in phase at the image, even if the lens has large aperture. In particular, if a plane wave enters a perfect lens a perfect spherical wave emerges and the disturbances arrive in phase at the image point in the focal plane.

3-1. Stops and Diaphragms

It was seen in Chapter II that, in terms of geometrical optics, a system gives perfect imagery within the paraxial region. This means that the imagery is perfect for all object distances. Later it will be shown that it is impossible to secure perfect imagery in this sense with image-forming pencils of large cross-section and objects of finite size. Usually, however, it is found that one can obtain very good imagery for one position of the object plane, but it is always necessary to place some restriction on the size of the object that can be imaged, and on the cross-section of the image-forming pencils from each object point. The necessary restrictions can be introduced by the use of two diaphragms.

Consider the case of a simple lens forming an image of a distant object [Fig. 3-1(a)]. The image of an axial object point can be of good quality if the angle u' is not excessive—one must limit the aperture of the system. This is done by means of the diaphragm A, known as the *aperture stop*. It will be obvious that the aperture stop controls the light-gathering power; it will be seen later that it also limits the resolving power. In designing a lens system, the position of the aperture stop is of paramount importance since, as is seen from the figure, it controls the part of the lens through which oblique image-forming pencils pass.

The presence of the stop A will not prevent the passage of image-forming

---

* The contributions made by Gauss to the theory of perfect optical systems were so extensive that the term *Gaussian* is frequently regarded as a synonym of *paraxial*.

pencils inclined to the axis at a large angle ω, but it is necessary to do this in order to limit the image size if good images alone are required. It is done by means of a diaphragm known as the *field stop* F which coincides with the image plane. That is to say, the field stop is conjugate to the object. The field stop does nothing to improve the image quality—it merely ensures that only the good images are utilized. In the case of a simple system as shown in Fig. 3–1(a), the use of a receiving screen of limited extent is equivalent to the use of a field stop. For example, in the case of a photographic lens, the film or plate holder acts as the field stop.

It will be seen that the aperture and field stops together control the value of the Helmholtz invariant, H, for the system.

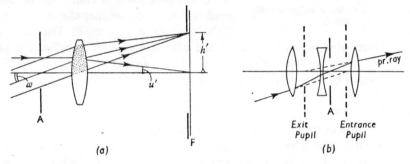

Fig. 3–1. (a) Function of aperture and field stops. (b) Positions of entrance and exit pupils.

The aperture stop may be anywhere in the optical system or outside it. The image of the aperture stop formed in the object space, i.e. formed by that part of the optical system which is nearer the object is known as the *entrance pupil* [Fig. 3–1(b)], and the image of A formed in the image space by that part of the system which is on the image side is known as the *exit pupil*. The pupils therefore lie in conjugate planes of the complete optical system. Either or both of the pupils may be real or virtual images of the aperture stop, and the exit pupil can be either to the left or to the right of the entrance pupil.

In any image-forming pencil of rays that passes through a system there will be one ray which passes through the centre of the aperture stop. This is known as the *principal* (or *chief*) *ray* of the pencil. It will be seen that in the object space the principal ray is directed towards the axial point of the entrance pupil, and that in the image space it emerges from the centre of the exit pupil. Quantities that refer to the principal ray are distinguished by the suffix *pr*.

If the centre of the aperture stop is at a principal focus, one is said to have a *telecentric* system. If the entrance pupil is at infinity, the system is telecentric on the side of the object, and if the exit pupil is at infinity, it is telecentric on the image side. It will be seen that a system cannot be telecentric on the object side if the object is at infinity since A would be located at the plane where the image-forming pencils are focused, and would be unable to control their diameters.

Although, ideally, only two stops are necessary in an optical system, one frequently finds other diaphragms introduced in order to obstruct stray light. In addition, some of the components of the optical system may have inadequate apertures. Fig. 3–2 shows an effect encountered frequently in camera lenses. For image-forming pencils at large angles, the edge of the lens obstructs some of the rays. Often *vignetting* of this kind is introduced in order to remove rays

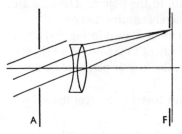

Fig. 3–2. Vignetting of image-forming pencils.

that, owing to aberrations, would spoil the image quality. In a complex system there may be several diaphragms in addition to the nominal aperture and field stops. Consider the image of each stop as formed by that part of the optical system that lies on the object side of it. The stop image that subtends the smallest angle at the centre of the entrance pupil is called the *entrance window* since it determines the field of view if the aperture stop is small. The image of the entrance window, as formed by the complete system, is called the *exit window*. In the ideal case the entrance window is simply the image of the field stop as formed by that part of the optical system that precedes it.

3–2. The Human Eye

The structure of the human eye is discussed in Chapter XVII. It is sufficient here to regard the eye as a lens system that focuses the image on to a light-sensitive screen, the *retina*. There is an aperture stop, the *iris*, whose diameter varies involuntarily as the intensity of the incident light varies. The entrance pupil is very close to the first principal point. There is no field stop but the resolving power of the retina falls off very rapidly from the centre, so that,

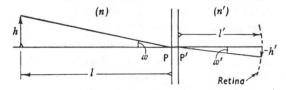

Fig. 3–3. Action of the eye.

although the total field of view is very large, the field of distinct vision is very small. To "look" at an object, the eye is rotated in its socket until the image falls on the central part of the retina where the resolving power is good. The apparent size of the object is determined by the size of the retinal image. Suppose Fig. 3–3 represents, diagrammatically, the formation of an image on the retina. Suppose the object and image (h and h') subtend angles ω and ω' at the first and second principal points of the lens system of the eye.

Helmholtz's equation, applied to imagery in the principal planes, gives

$$n\omega = n'\omega'.$$

In addition one has

$$h' = -l'\omega' = -l'(n/n')\omega \qquad \cdots \cdots \quad (3\text{–}1)$$

so that the apparent size of the object is proportional to the angle that the object subtends at the first principal point.

The power of the lens system may be varied by the observer when it is required to view objects at different distances. This process is known as *accommodation*. During accommodation, the distance l' of the retina from the second principal point of the optical system remains almost constant. In consequence, *the apparent size of an object is proportional to the angle it subtends at the first principal point of the optical system, irrespective of the state of accommodation.*

3–3. Magnifying Power of a Telescope

In a telescopic system, rays that enter the system parallel leave it parallel. Such a system has infinite focal length, and if the object is at infinity the image

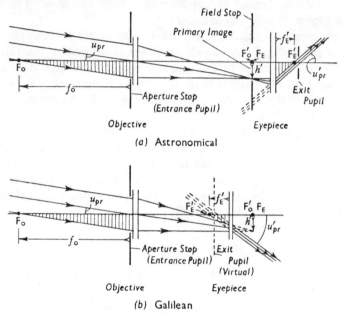

(a) Astronomical

(b) Galilean

Fig. 3–4. Telescopic systems.

is also at infinity. The main use of such systems is the observation of distant objects, but the student should not forget that such a system forms a finite image of a finite object.

A telescopic system can result from the combination of two converging systems, or one converging and one diverging system, such that $g=0$. In the latter case the focal length of the diverging system must be numerically smaller than that of the converging system. An example of the first type is the *astronomical* telescope which, in its simplest form, consists of two single convex lenses. The *Galilean* telescope consists of one convex and one concave lens. The two systems are illustrated in Fig. 3–4. In each case, the lens nearer the object is termed the *objective* and that nearer the observer, the *eyepiece*. Usually the objective itself constitutes the aperture stop and also, therefore, the entrance pupil. It follows that the exit pupil is the image of the objective formed by the eyepiece.

In the astronomical telescope the exit pupil is real and one may place the entrance pupil of the observer's eye at that point (sometimes known as the *eye-point*). In this case the whole field of view is visible at once. The distance from the last surface of the system to the eye-point is called the *eye clearance*. The field stop coincides with the second focal plane of the objective and the first focal plane of the eyepiece, where the objective forms the so-called *primary image* of the object.

In the case of the Galilean telescope the exit pupil is virtual and, since there is no real primary image, there can be no field stop. In consequence, the field of view has no sharp boundary and usually the whole field cannot be seen at once. When the eye is placed just behind the eyepiece, the field of view of such a system is eventually limited by the diameter of the objective. The following discussion applies equally well to either system.

If the telescope is adjusted by the observer to be truly telescopic, the final image is at infinity and the *magnifying power* is defined as the ratio of the apparent size of the image to the apparent size of the object viewed without the telescope. This, as explained in the last section, is the same as the ratio of the angle subtended at the first principal point of the eye by the image, to the angle subtended by the object.

In Fig. 3–4 rays are shown coming from the object at infinity and entering the observer's eye after passing through the telescope. The magnifying power is evidently u'_{pr}/u_{pr} since all the incident and emergent rays corresponding to a single object and image point are parallel, and the object subtends an angle u_{pr} at the eye as well as at the entrance pupil. The figure shows the usual case where the objective itself is the entrance pupil.

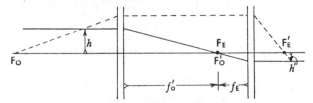

Fig. 3–5. Magnification of a telescope.

If h' is the size of the primary image, it can be seen from the shaded triangles that $h'=f_{0}u_{pr}=f'_{E}u'_{pr}$. Hence the magnifying power is given by

$$M=\frac{u'_{pr}}{u_{pr}}=-\frac{f'_{0}}{f'_{E}} \quad \cdot \quad \cdot \quad \cdot \quad \cdot \quad \cdot \quad (3\text{–}2)$$

since $-f_{0}=f'_{0}$.

It has been shown that the linear magnification is constant for all object and image positions in the case of a telescopic system. The linear magnification is the inverse of the angular magnification so that, for any pair of conjugates,

$$\frac{h'}{h}=-\frac{f'_{E}}{f'_{0}}. \quad \cdot \quad \cdot \quad \cdot \quad \cdot \quad \cdot \quad (3\text{–}3)$$

This is seen to be the case for a finite object and image from Fig. 3–5. One may then ask how it is that if the object and image both tend to infinity while $(h'/h)=-(f'_{E}/f'_{0})$ and is constant, the ratio of the apparent sizes can be f'_{0}/f'_{E}.

The answer lies in the fact that, as shown by equation (2–22) (p. 26), the ratio of the image and object distances measured from any conjugate reference planes is $(l'_R/l_R)=(h'/h)^2$ if $n'=n$. If l'_R and l_R approach infinity they can be measured from any fixed point in the system without error. One can say that the image and object are distances l'_R and l_R from the observer's eye. Obviously the angular dimensions of the image and object are h'/l'_R and h/l_R and consequently $M=(h'/l'_R)\div(h/l_R)=h/h'$. Thus, when the object and image both tend to infinity, the image does so more slowly, so that although the actual linear magnification or ratio of image to object size is h'/h, the angles subtended at the observer by the image and object are in the ratio h/h'.

3–4. Simple and Telephoto Camera Lenses

Consider the case of a camera lens used to produce an image of a distant object. Equation (2–19) (p. 24) gives the size of the image in terms of the angle

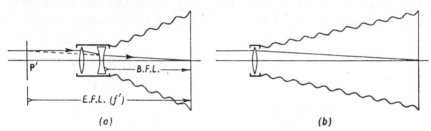

$$(a) \qquad\qquad (b)$$

Fig. 3–6. Action of a telephoto lens.

subtended at the camera by the object. It follows that to obtain a large scale photograph one requires a lens having a long focal length.

Suppose the separation of the lenses of a Galilean telescope is increased so that g is positive. The image of a distant object will now be formed at a finite distance. The position of the second principal plane of the system is determined by the intersection of a ray parallel to the axis in the object space with its conjugate ray in the final image space (produced if necessary, as it is in this case)—see Fig. 3–6(a). The second principal plane P' is seen to lie some distance to the left of the system so that the equivalent focal length f' is quite large. As a result, the image is as large as it would be if formed by a single convex lens of focal length f'. In this latter case, however, the camera would have to be of length f', whereas in the former case it could be somewhat shorter. (Compare Figs. 3–6(a) and (b).) For this reason the combination of a convex and concave lens system is frequently employed. Such a combination is known as a *telephoto* lens. The ratio of the equivalent focal length to the back focal length is sometimes referred to as the *telephoto magnification*.

The view of a scene which is obtained by a camera lens is that of an eye placed at the first nodal point; that is to say, N is the centre of perspective. Thus the centre of perspective in the image space is N'. Consequently, the photograph is seen in proper perspective only if viewed from a distance equal to that of the plate or film from N'. This is known as the *viewing distance* and, for distant objects, it is simply the focal length of the camera lens. Since the focal length is usually less than D_v, the least distance of distinct vision, one cannot use the correct viewing distance unless the photograph is viewed through a convex lens. One can relax one's accommodation and employ a lens having the same focal length as the camera lens, the photograph being placed in the focal plane. Alternatively one can enlarge the photograph until the viewing distance, which increases in proportion, is equal to or greater than D_v.

With a telephoto lens, the problem is complicated by the fact that the combination of image size and field of view resembles that of a close-up photograph while the perspective corresponds to a camera at the larger distance actually employed; it looks wrong. The perspective would be correct if the proper viewing distance were employed (i.e. f' in most cases) but the advantage of the large image size would then be lost.

3–5. Projection Systems and Enlargers

Figs 3–7(a) and (b) show, diagrammatically, two systems for projecting the image of a transparency on to a screen. In the first, which is used for fairly large

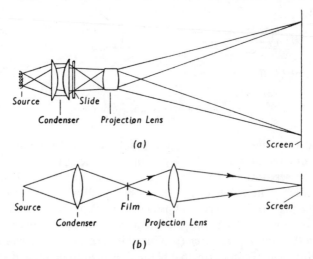

Fig. 3–7. Projection systems. (a) Lantern slides. (b) Ciné film.

slides and a moderately distant screen, the *condenser* images the source in the entrance pupil of the *projection* lens. It will be seen that the conditions of *Maxwellian view* obtain (see §18–7) and, consequently, the screen is evenly illuminated in the absence of the slide. The action of the projection lens resembles that of a camera lens in reverse, although the angular field may be smaller; *Petzval* or *Triplet* type lenses are frequently used (see Chapter XVI). The correction of the condenser is less critical and the system shown is common in the less expensive instruments. Photographic enlargers have a construction which is similar to this projection system; in the cheaper ones the condenser is replaced by a ground glass diffusing screen.

The system shown in Fig. 3–7(b) represents that used in the Kinematograph projector—actually a reflecting condenser is usually employed. The source is an arc and the image is sufficiently large to cover a 35 mm. film. By employing a large aperture condenser and projection lens, one can secure greater illumination on the screen with this system. (If, in the first system, the entrance pupil of the projection lens is as large as the image of the source, an increase in aperture does not give a brighter image. Also, the aperture of the condenser controls the field of view only.)

In the *Episcope*, which is used to project the image of an opaque picture, the major problem is shortage of light. The picture to be projected is illuminated as strongly as possible but, since this light is scattered in all directions, only a small proportion of the total light enters the projection lens. The latter must, therefore, have the largest possible aperture; frequently this is accomplished at the expense of the image quality.

3–6. The Simple Microscope or Hand Magnifier

Consider a single convex lens system used as a hand magnifier. The observer may adjust the final image to be at infinity or at some finite distance not less than the least distance of distinct vision. The magnifying power, M, of such a system is the ratio of the apparent size of the image when using the lens to the apparent size of the object when it is in the best position for observation by the naked eye. That is to say, M is the ratio of the angle subtended at the eye by the final image to the angle subtended by the object when at the least distance of distinct vision, D_v.

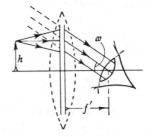

Fig. 3–8. Simple microscope.

Consider first the case where the final image is formed at infinity. The object, h, is therefore in the focal plane of the lens. From Fig. 3–8 it is seen that the angle subtended at the eye by the final image at infinity is $\omega = h/f'$.

$$\therefore \quad M = \frac{h}{f'} \div \frac{h}{D_v} = \frac{D_v}{f'}. \quad \quad \ldots \ldots \quad (3\text{–}4)$$

Suppose, now, that the observer moves the lens closer to the object, so that a magnified virtual image h' is formed by the lens (Fig. 3–9). Suppose object and image are distances l and l' from the lens which may be taken as thin. (If the lens is not thin, l and l' must, of course, be measured from the first and second principal planes.) Since the lens is in air one has

$$\frac{1}{l'} - \frac{1}{l} = \frac{1}{f'}$$

and since the nodal planes coincide with the lens

$$\frac{h'}{h} = \frac{l'}{l} = 1 - \frac{l'}{f'}.$$

If the eye is close to the lens and the observer adjusts the image to be at the least distance of distinct vision, one may put $l' = -D_v$, the negative sign being necessary since l' is negative. In this case

$$M = \frac{h'}{D_v} \div \frac{h}{D_v} = \frac{h'}{h},$$

i.e. $$M = 1 + \frac{D_v}{f'}. \quad \ldots \ldots \quad (3\text{-}5)$$

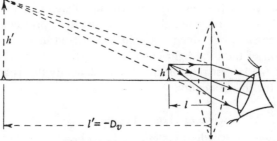

Fig. 3-9. Simple microscope.

It is seen that there is a gain in magnification if the image is brought up as close as possible to the eye. If the magnifying power is 10 with the infinity adjustment, for example, the gain in magnifying power is 10%. One may put $D_v = 10''$ so that for the infinity adjustment M is equal to ten times the reciprocal of the focal length in inches. Alternatively, one may take $D_v = 25$ cm. and it will be seen that M is one-quarter of the power of the lens in dioptres.

3-7. The Principle of the Compound Microscope

For very large magnifications the simple magnifying lens is impracticable and a highly magnified image must be produced in two stages. This is done in the *compound microscope* illustrated in Fig. 3-10. The instrument consists of two

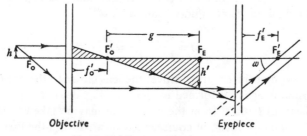

Fig. 3-10. Compound microscope.

converging systems whose adjacent principal foci are a considerable positive distance apart. The objective, or object glass, produces a real image h' of the object h and this image is further magnified by the eyepiece.

Consider the case where the final image is formed at infinity. The path of the image-forming rays is indicated in the figure. It will be seen that the angle under which the final image is viewed is $\omega = h'/f'_E$. From the figure, since the shaded triangles are similar,

$$\frac{h'}{h} = -\frac{g}{f'_0}. \quad \ldots \ldots \ldots \quad (3\text{-}6)$$

Consequently the angle subtended at the eye by the final image is $hg/f'_o f'_E$ and the magnifying power is given by

$$M = -\frac{hg}{f'_o f'_E} \div \frac{h}{D_v},$$

$$\therefore \quad M = -\frac{gD_v}{f'_o f'_E} = -\left(\frac{g}{f'_o}\right)\left(\frac{D_v}{f'_E}\right). \quad \cdots \quad (3\text{-}7)$$

This is the *primary magnification*, produced by the objective, multiplied by the *power of the eyepiece*, the eyepiece being looked upon as a simple microscope used to examine the primary image formed by the objective. The negative sign expresses the fact that the image is inverted. (The eyepiece may be considered to act in this way in the case of the astronomical telescope also.) The distance g is called the *optical tube length* of the microscope, and is usually 160 mm.

Although, as stated above, a simple microscope consisting of a single lens is unsuitable for high magnifications, the compound microscope may be considered analytically as a single system of focal length f'. It is just that a compound instrument must be used in order to obtain a sufficiently short focal length. If the final image is formed at infinity, the magnifying power of such a system is given by $M = D_v/f'$. But from equation (2–30) (p. 28) it follows that $f' = -f'_o f'_E/g$ so that

$$M = -\frac{D_v g}{f'_o f'_E}.$$

Many continental writers put $M = \left(\frac{D_v}{f'_o}\right)\left(\frac{g}{f'_E}\right)$ and call D_v/f'_o and g/f'_E the objective and eyepiece magnifications respectively.

The field stop of a compound microscope is located in the plane of the primary image and the aperture stop is usually in the region of the objective. The exit pupil, therefore, is beyond the eyepiece, a distance slightly greater than f'_E from the second principal plane of the latter. As for a telescope, the eye pupil should be placed as close as possible to the exit pupil of the instrument.

3–8. Chromatic Aberration

The instruments described above are usually used with a source of "white" light such as daylight or an electric lamp. Such light may be looked upon as a mixture of light of many wavelengths. So far, however, the theory has taken no account of the way in which the refractive index of a medium varies with the wavelength of the light. For transparent substances the refractive index decreases with increasing wavelength (see §20–7, Fig. 20–4, p. 499) and a thin lens, for example, will have a shorter focal length for the shorter wavelengths. Consider a point source on the axis of an optical system. In general, the light of each wavelength present in the source will form a separate axial image and there will be no "white" image point. This defect is referred to as *longitudinal chromatic aberration*. Similarly, light of each wavelength will form an image of a finite object, and each image will have a different size—there is said to be a *chromatic difference of magnification*. Thus longitudinal chromatic aberration causes the images to be at different distances along the axis, and chromatic difference of magnification causes the images to have different sizes. These two chromatic aberrations seriously impair the image quality, and one must investigate ways in which their effects may be eliminated or, at any rate, reduced. It should be noted that a purely reflecting system will exhibit no chromatic errors of any kind, and, in view of this fact, reflecting systems start off with an inherent advantage over refracting systems.

Consider the refractive index of a medium at two wavelengths B and R. These letters are chosen for the following reason: in the visible spectrum, the blue end corresponds to the shorter wavelengths and the red end to the longer wavelengths. Thus, while the two wavelengths considered here need not correspond to blue and red light, it is convenient to use the letters B and R to serve as a reminder of which is the shorter of the two. In the same way it is convenient to denote an intermediate wavelength by Y (corresponding to yellow light). With the usual notation, the focal lengths of a thin lens for these three wavelengths are given by

$$\frac{1}{f_B{}'} = (n_B - 1)R, \qquad \frac{1}{f_Y{}'} = (n_Y - 1)R, \qquad \frac{1}{f_R{}'} = (n_R - 1)R,$$

where

$$R = \frac{1}{r_1} - \frac{1}{r_2}.$$

Hence

$$\frac{1}{f_B{}'} - \frac{1}{f_R{}'} = (n_B - n_R)R$$

or

$$\frac{f_R{}' - f_B{}'}{f_B{}' f_R{}'} = (n_B - n_R)R.$$

If $f_B{}'$, $f_Y{}'$, and $f_R{}'$, are not widely different, one may write $f_Y{}'^2$ for $f_B{}' f_R{}'$, so that one has

$$\frac{f_R{}' - f_B{}'}{f_Y{}'} = (n_B - n_R)R f_Y{}' = \frac{(n_B - n_R)}{n_Y - 1}. \qquad \ldots \quad (3\text{-}8)$$

The quantity $\dfrac{n_B - n_R}{n_Y - 1}$ is called the *dispersive power* of the medium between wavelengths B and R. Its reciprocal is denoted by V and sometimes is referred to as the *constringence*. It will be seen from (3-8) that the dispersive power of the material of a single thin lens measures the chromatic difference of focal length as a fraction of the mean focal length $f_Y{}'$.

It will be seen below that if it is possible to make lenses of materials which have different dispersive powers, it is possible for the combination of two thin lenses in contact to have the same focal length for two wavelengths. Newton thought that all media had the same dispersive power and therefore thought such a combination was impossible.

3-9. The Thin Achromatic Doublet

Consider two thin lenses in contact. Denote quantities that refer to the separate lenses by suffixes 1 and 2. The powers of the combination for wavelengths B and R are given by

$$F_B = (n_{1B} - 1)R_1 + (n_{2B} - 1)R_2,$$
$$F_R = (n_{1R} - 1)R_1 + (n_{2R} - 1)R_2.$$

If $F_B = F_R$ the combination is said to be *achromatized* for these two wavelengths. Thus for an achromatic combination

$$R_1(n_{1B} - n_{1R}) + R_2(n_{2B} - n_{2R}) = 0,$$

$$\therefore \quad \frac{R_1(n_{1Y} - 1)}{V_1} + \frac{R_2(n_{2Y} - 1)}{V_2} = 0$$

or

$$\frac{F_{1Y}}{V_1} + \frac{F_{2Y}}{V_2} = 0. \quad \cdots \cdots \cdots \quad (3\text{–}9)$$

But

$$F_{1Y} + F_{2Y} = F_Y, \quad \cdots \cdots \cdots \quad (3\text{–}10)$$

where F_Y is the power of the combination for the intermediate wavelength Y.

If V_1 and V_2 are different, these two equations may be satisfied. (If $V_1 = V_2$, then the equations can be satisfied only if $F_Y = 0$.)

It follows that

$$\left. \begin{aligned} F_{1Y} &= F_Y \left(\frac{V_1}{V_1 - V_2} \right), \\ F_{2Y} &= -F_Y \left(\frac{V_2}{V_1 - V_2} \right). \end{aligned} \right\} \quad \cdots \cdots \quad (3\text{–}11)$$

If it is desired to make a combination of given power F_Y using lenses whose materials are characterized by V_1 and V_2, then equations (3–11) give the necessary powers of the components. It must be pointed out at once that the combination is not completely free from chromatic aberration but is achromatic for the two wavelengths B and R which are used in calculating V_1 and V_2. The focal lengths for other wavelengths will be different, and this residual chromatic error is referred to as the *secondary spectrum*.

In the earlier days there were only two main glass types: *crown* and *flint*. A typical pair had refractive indices and V-numbers of about 1·52 and 60, and 1·63 and 35, where the refractive indices are for the Helium d ($\lambda 5876$) line and the V-numbers involve the hydrogen C ($\lambda 6563$) and F ($\lambda 4861$) lines for the "red" and "blue". Using these glasses for a doublet of unit power achromatized for the C and F lines, one has

$$F_{1d} = +2\cdot4,$$
$$F_{2d} = -1\cdot4.$$

A doublet achromatized in this way for wavelengths in the red and blue regions has approximately constant focal length over an appreciable range of wavelengths in the middle of the visible spectrum where the eye is most sensitive (see Fig. 16–15, p. 403). There is now a wide range of glass types, the importance of which will be seen in Chapter XVI.

Roughly speaking, glasses which have low dispersions (i.e. high V-numbers) are referred to as crowns and those with high dispersions as flints. However, an achromatic combination may consist of two nominal "crowns" or two "flints" as long as the dispersive powers are *different*. In such a case it is usual to refer to the component having the smaller dispersion as the crown and the other as the flint. It follows from equations (3–11) that for an achromatic doublet of positive power the crown component is of positive, and the flint of negative power. If the adjacent surfaces are given equal curvatures, the components may be cemented together. Many telescopes, low power microscopes, and cheap photographic objectives, consist of doublets (see §§16–8 to 16–11).

3–10. Two Separated Thin Lenses of the Same Material

It is possible to achromatize a system consisting of two separated thin lenses, even when the lenses are made from the same glass. If the lenses are in air the focal length of the combination is given by

$$\frac{1}{f'} = \frac{1}{f_1'} + \frac{1}{f_2'} - \frac{d}{f_1'f_2'}$$
$$= (n-1)R_1 + (n-1)R_2 - d(n-1)^2 R_1 R_2$$

$$\therefore \quad \frac{\partial(1/f')}{\partial n} = R_1 + R_2 - 2d(n-1)R_1 R_2$$

$$= 0 \text{ if } f' \text{ is to be stationary.}$$

$$\therefore \quad R_1 + R_2 = 2d(n-1)R_1 R_2$$

$$\therefore \quad R_1(n-1) + R_2(n-1) = 2d(n-1)R_1(n-1)R_2$$

or
$$\frac{1}{f_1'} + \frac{1}{f_2'} = \frac{f_1' + f_2'}{f_1'f_2'} = \frac{2d}{f_1'f_2'}$$

$$\therefore \quad d = \frac{f_1' + f_2'}{2}. \qquad \cdots \cdots \quad (3\text{–}12)$$

If (3–12) is satisfied the focal length of the combination is stationary, but this does not mean that the system can be made entirely free from chromatic error. f_1' and f_2' are different for different wavelengths and hence the lenses can be given the correct separation for only one wavelength; it is only at this wavelength that the focal length of the combination is rendered stationary, and there is a secondary spectrum similar to that described in §16–6.

In view of the discussion of the simple microscope in §3–6 it might be thought that by using the above system one obtains a simple microscope or eyepiece whose magnifying power is achromatized. However, (3–4) holds provided the object is in the focal plane, and the magnifying power would be achromatized only if both f' and the *actual position* of the focal plane were achromatized. But the actual position of the focal plane is not achromatized.

When an eyepiece is used in a compound instrument such as an astronomical telescope, the image-forming rays are obviously restricted to those that pass through the objective (which is the aperture stop and entrance pupil). Only these pencils can enter the eye to enable it to view the primary image with the eyepiece. Now, when an image-forming pencil enters the observer's eye, the direction in which the image appears is determined by the direction of the principal ray of the pencil. The image quality depends on the surrounding rays. Suppose, in the present problem, the principal rays of the pencils are parallel to the axis when entering the eyepiece. Then, if the latter is achromatized for focal length, the angle of emergence of the principal rays is achromatized ($u' = y/f'$). This corresponds to an entrance pupil at infinity, so that (3–12) is the condition for achromatism of the magnifying power. For eyepieces used under different conditions the achromatism condition differs from (3–12).

In practice, many instruments are quite long and there is an approach to the conditions described above. The chromatic difference of magnification introduced by the eyepiece is then small if it consists of two thin lenses separated by half the sum of their focal lengths. It may be noted that for a thin achromatic doublet one may neglect the chromatic differences in the cardinal points and, in particular, for an objective consisting of a thin achromatic doublet which is also the aperture stop, one can say that the image position is achromatized and no chromatic difference of magnification is introduced.

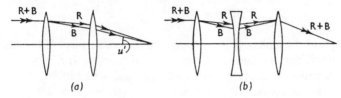

(a)   (b)

Fig. 3–11. Achromatism.

Suppose a system consisting of two separated thin lenses is free from both chromatic aberrations for "red" and "blue" light. Then both "red" and "blue" rays must intersect the axis at the same point; in addition, if the initial and final media are non-dispersing, Helmholtz's equation shows that in order to secure freedom from chromatic difference of magnification they must emerge at equal angles u'. This means that the final paths of the "red" and "blue" rays must coincide, and the two rays must emerge from the second lens at the same height. It is at once evident from Fig. 3–11(a) that *this can occur only if each thin lens is itself achromatic.*

In the case of three separated thin lenses, however, the emergent "red" and "blue" rays may coincide even if the separate lenses are not achromatic. The paths in the system are as shown in Fig. 3–11(b). In this case there is a diverging lens between two converging lenses, the separations of the lenses being determined by the dispersive powers involved.

3–11. Huygens and Ramsden Eyepieces

Since many telescopes and microscopes are quite long, it follows from what has been said above that the chromatic difference of magnification introduced by the eyepiece is small if it consists of two thin lenses separated by half the sum of their focal lengths. Two simple eyepieces in common use employ this principle. They are the *Huygens* and *Ramsden* eyepieces, and each consists of two plano-convex lenses arranged so as to minimize the aberrations. Fig. 3–12(a) shows a Huygens eyepiece, it being

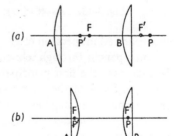

Fig. 3–12. Cardinal points of eyepieces. (a) Huygens. (b) Ramsden.

assumed that the objective lies to the left. That is, the plane surfaces of the eyepiece lenses are towards the eye. The lens nearer the eye, which is known as the *eye lens*, usually has a focal length about half that of the other component,

Suppose that $f_1'=2f_2'=2$. Then, for achromatism of the focal length, $d=3/2$. The focal length of the combination is given by

$$f'=\frac{f_1'f_2'}{f_1'+f_2'-d}=\frac{4}{3}$$

and for the positions of the principal planes one has

$$\text{AP}=\frac{df_1'}{f_1'+f_2'-d}=2,$$

$$\text{BP}'=-\frac{df_2'}{f_1'+f_2'-d}=-1.$$

In the Ramsden eyepiece [Fig. 3-12(b)] the convex surfaces of the components face each other, and the lenses have equal focal lengths. If $f_1'=f_2'=1$, the condition for achromatism gives $d=1$ and one has, in addition, $f'=1$, $\text{AP}=1$, $\text{BP}'=-1$. The principal and focal points of each eyepiece are shown in the figures.

3-12. The Optical Systems of Telescopes and Microscopes

A simple achromatic doublet serves as an objective for a wide range of telescopes—from astronomical to small laboratory instruments such as those

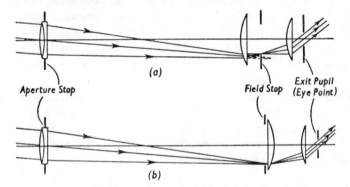

Fig. 3-13. Passage of rays through eyepieces. (a) Huygens. (b) Ramsden.

incorporated in spectrometers. Figs 3-13(a) and (b) show the paths of the image-forming pencils through telescopes employing Huygens and Ramsden eyepieces. In each case the first principal focus of the eyepiece coincides with the second principal focus of the objective. It will be seen that with a Huygens eyepiece the intermediate image is formed between the eyepiece lenses, and that with the Ramsden it coincides with the first lens. Since with a Ramsden eyepiece each image-forming pencil is focused on the first lens, all the rays in a given pencil have the same incidence height and, therefore, are deviated through the same angle. (Deviation=Fy.) That is to say, the first lens does not affect the position of the image points. However, the presence of the lens is important because, without it, the pencils would pass more eccentrically through the eye lens and the latter would need to have a larger diameter. A lens that coincides with an intermediate image is known as a *field lens*. In the Huygens eyepiece, the first lens is a considerable distance from the intermediate image and affects the image. However, it does behave as a field lens in the sense that it directs the pencils through the inner zones of the eye lens, and for this reason it is usually referred to as the field lens.

It will be seen that the field lens is in focus when one makes observations through a Ramsden eyepiece. This is a nuisance because any dust that settles on the lens, or any scratches that are made on its surface, appear in focus also. To overcome this it is customary to reduce the separation between the lenses— the focal plane then lies outside the eyepiece. The chromatic correction is less good if the separation of the lenses is reduced but this disadvantage is outweighed by the advantage of having an external focal plane. The freedom from chromatic difference of magnification makes the Huygens eyepiece preferable for visual observations when there is no eyepiece scale.

If a telescope is to be used as a measuring instrument, a graduated scale is placed in the plane in which the intermediate image is formed. With a Huygens eyepiece both lenses are used to view the object but only the eye lens would be used to view the eyepiece scale. Now, although the complete instrument may give a good final image of the object, the eye lens alone does not give a good image of the scale. This may be expressed in another way: the scale is used to measure the size of the intermediate image and, when a Huygens eyepiece is used, this image is of poor quality and cannot be the subject of an accurate measurement. It will be seen that a Huygens eyepiece is not satisfactory if an eyepiece scale is employed. With a Ramsden eyepiece the scale is viewed through the complete eyepiece and can be imaged more clearly. In order to be able to see the scale and the object in focus together

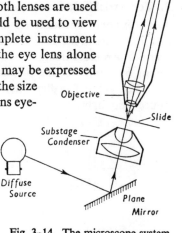

Fig. 3–14. The microscope system.

it is obviously necessary for the intermediate image to be of good quality; image and scale are then viewed with a corrected eyepiece. For more accurate work the simple eyepiece scale is sometimes replaced by a fine wire which can be moved across the image plane by means of a micrometer screw. Such an arrangement is known as a *micrometer eyepiece*. If the telescope carries a simple cross-wire and no eyepiece scale (as in a spectrometer telescope, for example) one requires good definition in the centre of the field of view only. The Ramsden eyepiece is certainly the better.

A Huygens eyepiece is usually employed in microscopes but the objective may be very complex (see §§16–9 and 16–12). High-power microscopes must be equipped with an additional lens system in order to concentrate light on to the object. When the object is not opaque, the extra system is placed on the far side of the slide, co-axial with the objective and eyepiece; it is referred to as the *substage condenser*. Fig. 3–14 represents the system of a typical microscope.

As with telescopes, a scale can be placed in the plane of the intermediate image.

This is satisfactory if the object to be measured can be contained within the field of view of the microscope. Again one should use a Ramsden eyepiece, but one often finds a Huygens eyepiece employed. If the object more than fills the field of view of the instrument, its diameter may be measured by moving the whole microscope and bringing the extremities of the object to the centre of the field in turn. Only a simple cross-wire is required in the focal plane of the eyepiece and a Huygens type is frequently employed because the cross-wire, which is easily damaged, is then enclosed within the eyepiece. The movement of the microscope is measured with the aid of either a vernier or a micrometer screw arrangement. It must be remembered that an eyepiece scale simply gives a measure of the size of the intermediate image and, in order to find the size of the object, one must determine the primary magnification of the instrument. That is to say, one must calibrate the eyepiece scale by viewing a mm. or divided mm. scale; this gives the number of divisions of the eyepiece scale that correspond to 1 mm. in the object plane.

3–13. Terrestrial Telescopes

Most telescopes encountered in physics laboratories are astronomical telescopes. These are of little use in everyday terrestrial observations since the image is inverted. An erect image may be obtained by employing an additional lens system or by using erecting prisms. The action of an *erecting eyepiece* is illustrated in Fig. 3–15. The diaphragm D is conjugate to the aperture stop and helps to reduce stray light.

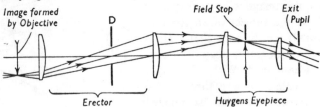

Fig. 3–15. An erecting eyepiece.

The use of an additional pair of convex lenses in the erecting system has certain very serious disadvantages associated with the aberrations of the instrument (see Chapter XVI) and, in addition, the telescope is made considerably longer.

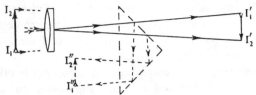

Fig. 3–16. Action of erecting prisms.

Fig. 3–16 represents the path of the principal rays through a telescope objective forming the inverted image $I_1'I_2'$ of an object I_1I_2. Suppose a right-angled isosceles prism is interposed as shown. The paths of the principal rays will be changed as indicated but the image points $I_1''I_2''$ will still lie on the respective principal rays. It will be seen that the prism has introduced an inversion so that the image is rendered erect. The original image $I_1'I_2'$ formed by the lens is also

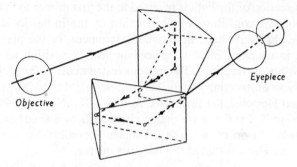

Fig. 3–17. Arrangement of erecting prisms.

reversed left to right, and this latter reversal is not removed by the prism. If, however, a second prism is introduced as indicated in Fig. 3–17, the necessary reversal is secured without interfering with the inversion produced by the first prism. In addition, the final direction of the light is the same as the original. Prism systems of this and similar types are employed in prismatic binocular telescopes and possess distinct advantages over erecting eyepieces.

3–14. Reflecting Systems

Newton supposed that all materials had the same dispersive power and concluded that it was impossible to remove the chromatic aberration of a refracting system. Now the law of reflection does not involve the refractive index, so that a purely reflecting system can give no chromatic aberration. Consequently,

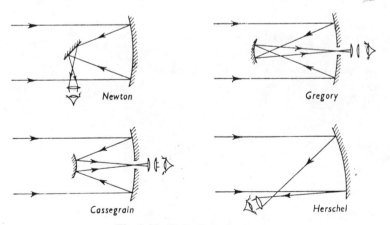

Fig. 3–18. Reflecting telescopes.

many reflecting systems were devised. With the discovery of the possibility of an achromatic lens, interest in reflecting systems decreased, and attention was again focused on refracting systems. Recently, however, there has been a renewed interest in reflecting instruments; some of the later developments are described in §16–13. In the present section a brief description will be given of some of the older reflecting instruments. The chief difficulty encountered is the fact that the direction of the light is reversed at the first mirror so that subsequent optical components constitute an obstruction of the initial incident rays. In particular, the observer's head in a visual instrument, or the plate holder in a photographic instrument, must be placed in such a position as to cause the minimum possible obstruction. The various instruments differ in their methods of reducing these obstructing effects. The telescopes due to Newton, Gregory, Cassegrain, and Herschel, are illustrated in Fig. 3–18. Newton's and Herschel's telescopes are similar in the sense that each consists of a single concave mirror plus an eyepiece. In order to make the image accessible Newton employed a plane mirror and Herschel tilted the concave mirror.

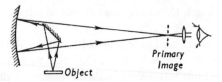

Fig. 3–19. Reflecting microscope.

Newton also suggested the use of a concave mirror as a microscope objective. Amici used an objective which resembled a Newtonian telescope objective in reverse; he viewed the magnified primary image with an eyepiece in the usual way (see Fig. 3–19).

Most of the earlier reflectors were of speculum metal (an alloy of 67% copper and 33% tin) but current practice is to use glass surfaces coated with a suitable metal film.

CHAPTER IV

THE DETERMINATION OF THE PARAXIAL CONSTANTS OF OPTICAL SYSTEMS

I. SIMPLE METHODS FOR MIRRORS AND THIN LENSES

4–1. Neutralization Test for the Power of a Thin Lens

IT was seen in Chapters I and II that the power of two thin lenses in contact is simply the sum of the powers of the separate lenses. Consequently, if two thin lenses having equal and opposite powers are placed in contact, the combination has zero power. A spectacle trial case as used by ophthalmic opticians contains a set of thin lenses whose powers vary from $\pm\frac{1}{4}$ to about ± 12 dioptres. Consequently, an approximate value for the power of an unknown lens can be obtained by selecting a lens from the trial case which, when placed in contact with the unknown lens, gives zero power.

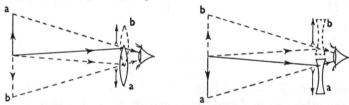

Fig. 4–1. The displacement test for the power of a lens.

In order to decide whether the combination has positive, negative, or zero power, it is moved up and down immediately in front of the eye whilst viewing a distant object. It will be observed that the object appears to move. If the lens has negative power (i.e. is a diverging lens) the movement is in the same direction as the movement of the lens, whereas for a positive lens the two movements are in opposite directions. For a combination of zero power there is no apparent movement of the object. The reason for this movement lies in the prismatic effect of a decentred lens and is illustrated in Fig. 4–1. When the power of the combination is numerically small, increased sensitivity may be obtained by holding the lenses farther away from the eye.

4–2. Location of Object and Image

The lens or mirror can be used to form an image of some convenient object, the object and image distances measured, and equations (1–13) or (1–7) used to calculate the focal length. The image of a pin (either real or virtual) may be located by a no-parallax adjustment. For converging systems a real image may be received on a screen. It is assumed that the reader is familiar with these elementary methods.

4–3. Simple Object and Image Coincidence Methods

The radius of curvature of a concave mirror is easily found by using the fact that the object and image coincide when the object is at the centre of

71

curvature. It will be seen from Fig. 4–2 that a similar adjustment is possible using:

(a) A concave mirror.
(b) A plane mirror and a convex lens.
(c) A convex mirror and lens.
(d) A concave mirror and lens.
(e) A convex and a concave lens with a plane mirror.

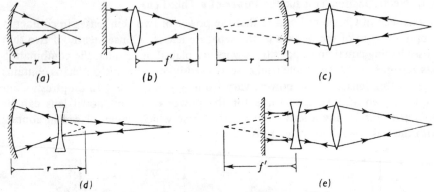

(a)   (b)   (c)

(d)   (e)

Fig. 4–2. Object and image coincidence methods.

It is obvious that if the focal length of one component in (c), (d), or (e), is known, the other can be found. It should be noted that in (c) the radius of curvature of the mirror must be shorter, and in (d) longer, than the focal length of the lens. If in (e) the lenses are in contact, the method yields the sum of their powers.

4–4. Refractive Index of a Lens

Equation (1–7) shows that if the radii of curvature and the focal length of a lens are known the refractive index may be calculated. Any of the above methods may be used to determine the focal length. The radii of curvature may then be determined fairly easily. Consider the case of a double convex lens [Fig. 4–3(a)]. Suppose rays from the point source B are incident normally on the second surface. The rays then emerge without refraction at that surface and the virtual image point B′ will coincide with the centre of curvature of the second surface C_2. For a thin lens the image distance is given by $l'=r_2$ so that, from equation (1–7),

$$\frac{1}{r_2} - \frac{1}{l} = \frac{1}{f'}, \qquad \cdots \cdots \cdots \quad (4\text{–}1)$$

where, in the figure, r_2 and l are negative. It is easy to decide when the rays do, in fact, emerge along the normal to the second surface. As always, some light is internally reflected and under these circumstances it must retrace its path to B so that a faint image will be seen coincident with the object. Thus, in practice, the object distance is varied until this image is observed and equation

(4–1) applied. It should be noted that when using the plane mirror method to find the focal length of a convex lens one is liable to confuse the image formed by internal reflection with that formed after reflection by the mirror. The latter will, of course, disappear if the mirror is removed.

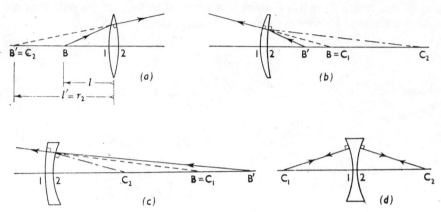

Fig. 4–3. Determination of radii of curvature.

Similar conditions obtain in the case of meniscus lenses. For each meniscus lens there is one surface, and for a double concave lens there are two surfaces, which can be treated as separate concave mirrors [Fig. 4–3(b–d)].

Example

A plane mirror is placed behind the convex surface of a converging meniscus lens. A small object is placed on the other side of the lens close to the axis. It is found that an image is formed alongside the object for the following distances of the object from the lens: 10 cm, 7·5 cm, 2·3 cm. If the plane mirror is removed, the first of these images disappears. Find the refractive index of the lens.

Assume that the mirror is to the left of the lens, which resembles the lens shown in Fig. 4–3(b). The image that disappears yields $f' = +10$ cm. The image at 7·5 cm corresponds to reflection at the concave surface, i.e. is at C_2, so that $r_2 = +7·5$ cm. The image at 2·3 cm corresponds to reflection at normal incidence at the convex surface, i.e. is at B': for the light that is transmitted, it corresponds to a real object at B' giving a virtual image at $B = C_1$. The situation is similar to that described by equation (4–1) except that in the present case one has $l' = +2·30, f' = 10·0$, and $r_1 = l$, i.e.

$$\frac{1}{2·30} - \frac{1}{r_1} = \frac{1}{10·0}$$

which gives $r_1 = +2·987.$

Equation (1–7) can now be used, with $f' = +10·0$, $r_1 = +2·987$, $r_2 = +7·50$. One obtains $n = 1·496$.

II. ACCURATE MEASUREMENT OF RADII OF CURVATURE

4-5. The Spherometer

In Fig. 4–4(a) AB $(=2\varrho)$ is the diameter of a "small circle" on the surface of the sphere, centre C, radius r which it is required to find. If the pole P of the sphere is distance h above the plane of the small circle

$$(r-h)^2+\varrho^2=r^2,$$

whence
$$r=\frac{\varrho^2+h^2}{2h}. \qquad \ldots \ldots \ldots \quad (4\text{--}2)$$

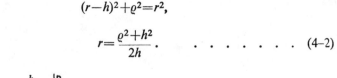
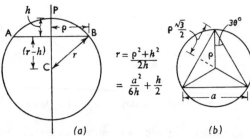

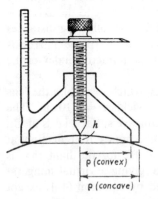

Fig. 4–4. The spherometer formulae.

In the ring spherometer shown in Fig. 4–5 an accurately ground metal ring of known radius ϱ fits on to the spherical surface to be measured. The distance h is measured with a micrometer screw or a spring-loaded plunger whose movement is recorded on a calibrated scale. h is found as the difference between the readings of the screw or plunger scale when the instrument rests on a plane surface and on the surface to be measured. The method applies equally well to convex or concave surfaces which employ the inner and outer radii of the ring respectively. In a common form of spherometer the metal ring is replaced by the tips of three legs located at the vertices of an equilateral triangle. The micrometer screw then moves along the axis of the circumcircle of the triangle and the radius ϱ now refers to this circumcircle. If a is the distance between adjacent legs of the spherometer it is easily shown that $a=\varrho\sqrt{3}$ [see Fig. 4–4(b)].

Fig. 4–5. The ring spherometer.

$$\therefore \quad r=\frac{a^2}{6h}+\frac{h}{2}. \qquad \ldots \ldots \ldots \quad (4\text{--}3)$$

If the tips of the legs are ground to a point, the same value of a is used for convex or concave surfaces.

4–6. Microscope Method for Short Radii

For radii of curvature not larger than about 3–4 cm. the best method is probably that illustrated by Fig. 4–6. Light from the source enters the Gauss eyepiece (see § 5–5) as shown, and it will be seen from the figure that for either a convex or a concave mirror there are two positions of the mirror which give an image of the cross-wires coincident with the cross-wires themselves. The distance between these positions gives the radius of curvature of the mirror in each case.

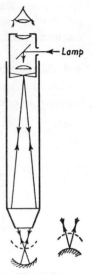

4–7. The Knife-Edge Method for Concave Surfaces

This method is particularly useful for shallow surfaces where the quantity h in the spherometer method is very small and therefore difficult to measure with sufficient accuracy. The arrangement is illustrated in Fig. 4–7(*a*) and consists of a knife edge which can move in the plane of the blade and in a direction perpendicular to the edge, an illuminated pinhole being located in the plane traced out by the knife edge. Suppose the principal axis of the mirror passes approximately midway between the knife edge and the pinhole. If the latter is closer to the mirror than the centre of curvature its image will lie beyond the knife edge as in (*b*). The opposite con-

Fig. 4–6.
Measurement
of short radii.

ditions are shown in (*c*). If the eye is placed just behind the knife edge so as to receive the light from the mirror, the latter will appear to be evenly illuminated. If the knife edge is now moved across the line of sight, a shadow will be seen to cross the mirror from A to B for the condition shown in (*b*),

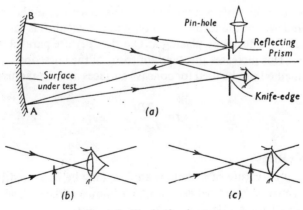

Fig. 4–7. The knife-edge test.

or from B to A for the condition shown in (*c*). If, however, the centre of curvature lies exactly in the plane of the pinhole and the knife edge, the whole aperture of the mirror will suddenly darken as the knife edge crosses the image of the pinhole. In this way the centre of curvature can be located very accurately.

The whole aperture of the mirror will not darken at exactly the same instant if the mirror is not truly spherical, and the shadow pattern across the mirror yields information concerning the state of perfection of the sphericity and optical polish of the surface. This method of examining the form of a mirror surface is discussed further in § 14–21; it was first used by Foucault in 1859.

4–8. Shallow Surfaces (Convex and Concave)

Consider first the case of a convex surface. A rectangular aperture of width *l* is placed in contact with the surface to be measured (see Fig. 4–8). The eye (or

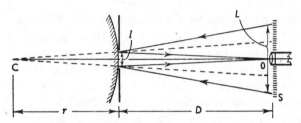

Fig. 4–8. Measurement of shallow surfaces.

the objective of an observing telescope) is placed at O and the scale S viewed by reflection at the surface. Suppose the length *L* of the scale is visible. Then the extreme rays are as shown and, from similar triangles if *D* is large,

$$\frac{l}{r} = \frac{L/2}{r+D},$$

whence

$$r = \frac{2Dl}{L-2l}. \qquad \cdots \cdots \quad (4\text{--}4)$$

The aperture *l* and the scale S cannot be in focus together, so that if the latter is clear the boundaries of the former are not, and the length *L* is rather uncertain. This effect can be reduced by placing at O a slit aperture parallel to the rulings of the scale S. This increases the depth of focus of the telescope.

The same method can be used for concave surfaces and the radius is then given by

$$r = \frac{2Dl}{L+2l}. \qquad \cdots \cdots \quad (4\text{--}5)$$

4–9. Newton's Rings

The radius of curvature of a shallow surface may be measured by observing the interference rings found between it and a known surface. The method will be discussed in Chapter VIII.

III. LOCATION OF THE CARDINAL POINTS OF A THICK LENS OR LENS SYSTEM

4–10. Collimator and Plane Mirror Methods for Principal Foci

By placing a small aperture or pinhole at the principal focus of a well-corrected convex lens, accurately parallel rays are produced. Such a device is called a *collimator*, and the emergent beam is said to be *collimated*. If such parallel rays enter a system parallel to the axis, they will emerge so as to pass through the principal focus which can then be located. Alternatively, the method of so-called *auto-collimation* may be employed which, in principle, is the plane mirror method already referred to in §4–3. In the case of a lens system that is not thin this merely serves to locate the principal focus since the focal length is not simply the distance from the lens to the focus. The knife-edge method may be used to locate the principal focus, but the sudden darkening of the aperture may be upset by the presence of aberrations in the system.

4–11. Location of Principal Planes

Since the principal planes usually lie within the lens system, a real object cannot be placed in one of the planes and a real image received on a screen in the other. By using an auxiliary lens, however, it is possible to project the image

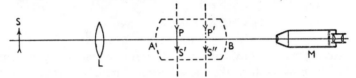

Fig. 4–9. Location of principal planes.

of a scale into any plane within the lens system under test. The image in the conjugate plane can be viewed by means of a low-power microscope. The principal planes may then be located as the planes that give unit magnification. The experimental arrangement is illustrated in Fig. 4–9.

The illuminated object S is imaged at S′ by the auxiliary lens L and the image observed by the low-power microscope M, which contains an eyepiece scale. Suppose the primary image of S′ in the microscope covers *n* divisions of the eyepiece scale. The lens AB is now interposed as shown and the positions of both AB and M adjusted until the image of S′ again covers *n* divisions of the eyepiece scale. When this adjustment has been made S′ must lie in the first principal plane of the lens AB and its image S″ in the second principal plane. The microscope is then focused on S″, so that the distance moved by the microscope gives the separation PP′ of the principal planes. (If the principal planes are "crossed", M will have to be moved to the left.) If M is now moved so that dust on the surface B is in focus, the movement of M will give the distance BP′.

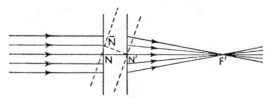

Fig. 4–10. The nodal points.

4–12. The Nodal Slide

Consider rays parallel to the axis entering a lens system whose nodal points are N and N′ (Fig. 4–10). Suppose the lens is rotated about N′ so that N rotates to N̄. The incident rays now constitute an oblique pencil but the emergent rays must still pass through a single image point. Now, the ray incident through N̄ must emerge along N'F' parallel to its direction of incidence. Hence the image point must still lie on N'F', i.e. it does not move laterally. Therefore if a collimated beam enters a lens system, the second nodal point can be located as the point about which rotation of the lens system produces no lateral shift of the image point.

The nodal slide consists of a lens holder in which the lens can be moved along its own axis, the holder itself being capable of rotation about a vertical axis

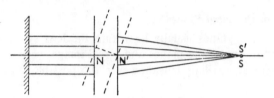

Fig. 4-11. Auto-collimation method.

which intersects the optical axis of the lens system. By mounting the lens in the nodal slide it is possible to make the adjustment indicated above. Instead of employing a separate collimator, the method of auto-collimation can be employed. A plan of the arrangement is shown in Fig. 4-11. The position of the source S is arranged so that its image S' lies alongside and the position of the lens in the nodal slide is adjusted until rotation about the vertical axis produces no lateral movement of S'. The vertical axis of rotation then passes through the second nodal point N'.

In principle, the same method may be used to find the radius of curvature of a mirror, in which case the nodal points coincide with the centre of curvature. The method is useful for convex mirrors and in that case a real object gives a virtual image which may be observed with a low-power microscope. If the centre of curvature of the mirror lies on the axis of rotation of the nodal slide, the image will obviously remain stationary [see Fig. 4-12(a)]. The mirror is then

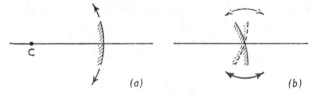

Fig. 4-12. Use of nodal slide for mirrors.

moved along the slide parallel to its principal axis until dust particles at the pole of the mirror remain stationary during rotation about the vertical axis [see Fig. 4-12(b)]. The distance moved by the mirror gives the radius of curvature.

IV. INDIRECT DETERMINATION OF FOCAL LENGTH

If the principal planes are known, a simple determination of the object and image distances l and l' will enable the focal length of a system to be calculated. Similarly, if the focal planes are known for a system in air, the focal length may be found from a single determination of the object and image distances x and x'. For a system in air Newton's equation becomes

$$xx'=-f'^2.$$

4–13. Magnification Methods

For a lens system in air

$$\frac{1}{l'}-\frac{1}{l}=\frac{1}{f'}$$

and since the nodal and principal planes coincide the magnification m is simply given by l'/l.

Hence

$$1-m=\frac{l'}{f'}$$

or

$$l'=f'-f'm. \quad \cdots \cdots \quad (4\text{–}6)$$

If the magnification is found for two image positions, one has

$$l_1'=f'-f'm_1,$$
$$l_2'=f'-f'm_2,$$
$$\therefore \quad l_1'-l_2'=f'(m_2-m_1).$$

Hence one needs only the movement of the image plane, and it is not necessary to know the positions of the principal planes. Alternatively, if l' is plotted against m, f' may be found from the slope of the resulting straight line or the intercept on the l' axis. Actually l' cannot be measured unless the position of the second principal plane is known. However, if a scale is fixed along the axis of the lens and C is the reading of the second principal plane, then the scale reading S at the image position will be $(l'+C)$. Equation (4–6) then becomes

$$S-C=f'-f'm$$

or

$$S=-f'm+f'+C, \quad \cdots \cdots \quad (4\text{–}7)$$

so that *if the lens is not moved* a graph of S against m is also a straight line of slope $-f'$; the position of the second principal plane is not needed in order to find f', but it can be found from the intercept if f' has been found from the slope.

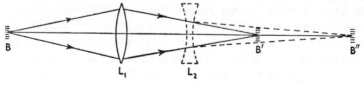

Fig. 4–13. Use of virtual object.

In order to apply this method to a diverging system it is necessary to employ a virtual object as indicated in Fig. 4–13. The object B is imaged at B' by the auxiliary lens L_1, and the diameter of the image measured. The lens under test (L_2) is then interposed as shown so that B' becomes a virtual object. Provided $L_2B'<f_2$ a real image B'' is formed and the magnification may be determined

To change the magnification L_2 must be moved so that in equation (4–7) C is no longer constant. This difficulty is most easily overcome by rewriting the equation in terms of l and m. The movement of L_2 then gives the change in object distance directly.

4–14. The Goniometer

If a scale is placed in the focal plane of a lens system, the image is at infinity and the rays from each object point emerge as a collimated beam. If the object point is a small distance h from the axis the collimated beam will be inclined to the axis at an angle $\omega = h/f$ (see equation (2–19) and Fig. 2–9), and if this angle

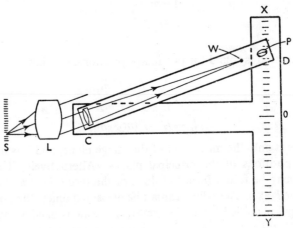

Fig. 4–14. The Searle goniometer.

can be measured for a given value of h, f can be deduced. More generally, for any two object points distance h apart in the focal plane, the emergent beams of parallel rays have an angular separation ω given by $\omega = h/f$ if ω is small. The goniometer devised by G. F. C. Searle is a simple instrument for measuring this angle, and a plan of the instrument is shown in Fig. 4–14. It consists of an achromatic objective at whose principal focus a fine vertical wire W is located. The lens and wire are mounted on a wooden base CD capable of rotation about a pivot located beneath the lens. The angle of rotation is measured by recording the linear motion of a pointer P over the scale XY fixed at a known distance from the pivot. CD is adjusted so that the pointer P lies on the zero of the scale XY. The lens L under test is then placed as shown and the scale S adjusted to be in the focal plane by observing when there is no parallax between its image and W. The reading of S is taken in this position and CD rotated until a second point of S is imaged at W. The reading on the scale XY determines the angle ω, and the difference between the readings of the scale S gives the object height h. The focal length of the lens L may then be found. For a lens of moderate size one can mount the lens on the prism table and use the telescope and circular scale of a simple table spectrometer of the form illustrated by Fig. 5–5.

4–15. Magnifying Power of a Telescope

The most obvious method for finding the magnifying power of a telescope is that of direct comparison. The object viewed directly with one eye is compared with the image viewed through the telescope with the other eye. Many observers have difficulty in making this measurement, and for small terrestrial telescopes the following method is better. It was seen in §§2–11 and 3–3 that a telescopic system forms a finite image of a finite object and that the transverse magnification is the same for all object positions and is equal to the reciprocal of the magnifying power if the indices in the object and image spaces are equal. The magnifying power may therefore be found by finding the transverse magnification for any convenient object. Usually it is convenient to place a scale in front of the objective (which is usually the entrance pupil) and measure the image by means of a low-power microscope. Even if the telescope is focused for viewing a finite object (i.e. is not truly telescopic), the above method is applicable if the object is located in the entrance pupil. This is because the magnifying power can be taken as the angular magnification for the pupil planes; it is the inverse of the transverse magnification for these conjugates if $n=n'$.

4–16. Magnifying Power of a Microscope

The magnifying power of a microscope can be found by measuring the focal lengths of the objective and the eyepiece and applying equation (3–7). It is

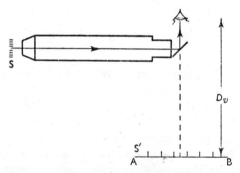

Fig. 4–15. Magnifying power of a microscope.

possible to measure M directly by the method illustrated by Fig. 4–15. A transparent scale S is observed through the microscope via the semi-reflecting plate. (A cover glass forms a suitable reflector.) Assuming that the image is brought to the least distance of distinct vision, a sheet of paper AB is placed a similar distance from the eye. The image S' as seen on AB is then traced with a pencil and the sizes of S' and S compared to give M.

CHAPTER V

PRISMATIC INSTRUMENTS FOR REFRACTOMETRY AND SPECTROSCOPY

5–1. Refraction Through a Prism

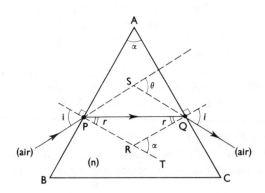

Fig. 5–1. Symmetrical path of a ray through a prism.

The majority of refracting prisms are either in fact (as discussed here—Fig. 5–1), or in effect (as in §5–17), right prisms on triangular bases. In Fig. 5–1 only the *refracting faces* AB and AC are of significance in the present discussion; the face BC can be replaced by a surface of any shape, and the nature of the base is equally unimportant in principle. The refracting faces need not intersect physically, but the line of intersection of their planes is called the *refracting edge* and the angle between them—the *refracting angle* or simply the *angle of the prism*—is denoted by α. A plane section of a prism perpendicular to the refracting edge is called a *principal section*. Since the principal section is perpendicular to both refracting faces, it follows from Snell's law that a ray incident in a principal section remains in that section during its passage through the prism. Fig. 5–1 shows a ray in a principal section.

The deviation of a ray, θ, depends on the angle of incidence, i, and simply from the symmetry of the system one can say at once that the deviation must reach either a maximum or a minimum when the angles of incidence and emergence are equal. If the ray is rotated from the symmetrical position, either the angle of incidence or the angle of emergence increases, and since the angle of deviation at a single surface increases as the angle of incidence is increased (and does so at an increasing rate), it follows that the deviation is a minimum for the ray that passes through the prism symmetrically.

In Fig. 5–1 $\angle APR = \angle AQR = 90°$ so that APRQ is a cyclic quadrilateral.

82

Hence $\angle\,\mathrm{TRQ}=\alpha$ and, since it is the exterior angle of $\triangle\mathrm{PQR}$, one has

$$r=\tfrac{1}{2}\alpha \qquad\qquad (5\text{–}1)$$

θ is the exterior angle of $\triangle\mathrm{SPQ}$. Hence $\angle\,\mathrm{QPS}=\angle\,\mathrm{PQS}=\tfrac{1}{2}\theta$

$$\therefore\quad i=r+\tfrac{1}{2}\theta=\tfrac{1}{2}\alpha+\tfrac{1}{2}\theta. \quad\ldots\ \ldots\ \ldots \quad (5\text{–}2)$$

Now Snell's law gives

$$\sin i=n\sin r. \qquad\ldots\ \ldots\ \ldots\ .(5\text{–}3)$$

Hence for minimum deviation one has

$$n=\frac{\sin\tfrac{1}{2}(\alpha+\theta)}{\sin\tfrac{1}{2}\alpha}. \qquad\ldots\ \ldots\ \ldots\ \ldots \quad (5\text{–}4)$$

It should be noted that the maximum possible value of r is the critical angle, and there is then grazing incidence and grazing emergence. Hence if the angle of a prism is greater than twice the critical angle, it is impossible for rays to be transmitted through the prism.

5–2. Formation of a Spectrum

Since the refractive index of a prism varies with the wavelength of the light, light incident on a prism in a given direction is deviated through an angle which depends upon the wavelength. Hence a prism can be used to separate the various wavelengths emitted by a source. Fig. 5–2 shows a suitable optical

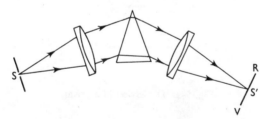

Fig. 5–2. Production of spectrum of good quality.

system. The source is located behind the narrow slit S which is perpendicular to the principal section of the prism, i.e. to the plane of the figure. In terms of geometrical optics a pure spectrum will be formed if the beams traversing the prism are collimated. (If a converging or a diverging beam traverses the prism the constituent rays are deviated through different angles and are not brought to a sharp focus.) The lenses are achromatic doublets, which not only reduces the chromatic aberration but also gives an equally important correction of the monochromatic aberrations (see §16–6). It will be seen that different wavelengths form images of the slit at different positions along VR. If the source emits a number of discrete wavelengths, each will produce a discrete image of S and a *line spectrum* is formed; if the source emits a continuous distribution of wavelengths over a certain range, the corresponding *continuous spectrum* is formed. V (for violet) and R (for red) denote the short and long wavelength ends respectively of the spectrum. With the system shown there is no geometrical optical requirement to work at minimum deviation although it could be advantageous to do so (see §11–5).

Rays from the off-axis parts of the slit do not traverse the prism in a principal section. The deviation for these rays is greater than for those in a principal section and this causes the image of the slit to be curved. It is concave in the direction of increased deviation, i.e. towards the short wavelength end of the spectrum.

5–3. Dispersion of a Prism

For a particular angle of incidence, consider the change in θ consequent upon a small change in the wavelength λ. The quantity $d\theta/d\lambda$ is known as the *angular dispersion* of the prism. Now, one may write

$$\frac{d\theta}{d\lambda} = \frac{d\theta}{dn} \cdot \frac{dn}{d\lambda}.$$

The first term on the right-hand side depends on the geometry of the system (i.e. on the angle of incidence and the angle of the prism) and the second is a property of the material of the prism. $d\theta/dn$ is most easily found by the following method due to Rayleigh.

Consider a collimated beam entering a prism. Suppose that the beam includes wavelengths λ and $\lambda+\delta\lambda$ and that these are deviated through angles θ and $\theta+\delta\theta$. In Fig. 5–3, AB represents the incident plane wave for each wavelength

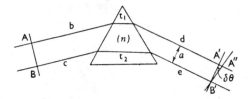

Fig. 5–3. Dispersion of a prism.

and A'B' the emergent wave for wavelength λ. If $\delta\theta$ is small, the path taken by the extreme rays of the beam between the planes AB and A'B' is the same for both wavelengths (to the second order approximation) and the emergent wave-front of $\lambda+\delta\lambda$ is A"B'. Consequently, one may equate the optical paths [A to A'] and [B to B'] for λ and [A to A"] and [B to B'] for $\lambda+\delta\lambda$. If the cross-section of the emergent beam is a, A'A" $\fallingdotseq a\delta\theta$, and if the edges of the beam have geometrical paths t_1 and t_2 in the prism and the other distances are as indicated in the figure, one has

$$b+nt_1+d=c+nt_2+e$$

and $$b+(n+\delta n)t_1+d+a\delta\theta=c+(n+\delta n)t_2+e.$$

By subtraction, this yields

$$\delta nt_1+a\delta\theta=\delta nt_2$$

or
$$\frac{d\theta}{dn} = \frac{t_2 - t_1}{a} \qquad \cdots \cdots \cdots (5\text{–}5)$$

If the beam fills an isosceles prism at minimum deviation, this reduces to

$$\frac{d\theta}{dn} = \frac{t}{a},$$

where t is the length of the base of the prism. Hence the angular dispersion is given by

$$\frac{d\theta}{d\lambda} = \frac{t}{a} \cdot \frac{dn}{d\lambda}. \qquad \cdots \cdots \cdots (5\text{–}6)$$

According to Cauchy's formula, the refractive index of most optical materials can be represented by

$$n = A + \frac{B}{\lambda^2},$$

where A and B are numerical constants (see § 20–7). In this case,

$$\frac{dn}{d\lambda} = -\frac{2B}{\lambda^3}$$

and
$$\frac{d\theta}{d\lambda} = -\frac{2t}{a} \cdot \frac{B}{\lambda^3}. \qquad \cdots \cdots \cdots (5\text{–}7)$$

Thus the angular dispersion varies through the spectrum, being greater at the blue (shorter wavelength) end. The negative sign expresses the fact that θ decreases as λ increases.

In general, the dispersion reaches a minimum value for a particular angle of incidence but it should be noted that this does not occur at the position of minimum deviation.

5–4. Thin Prisms. Combinations of Prisms

In Fig. 5–1, suppose the ray does not pass through the prism symmetrically. With an obvious notation one has, in general,

$$\alpha = r_1 + r_2$$

and
$$\theta = (i_1 - r_1) + (i_2 - r_2)$$

If, now, the angles are small, Snell's law gives

$$i_1 = nr_1 \qquad \text{and} \qquad i_2 = nr_2$$

$$\therefore \quad \theta = (n-1)r_1 + (n-1)r_2$$

i.e.
$$\theta = (n-1)\alpha. \qquad \cdots \cdots \cdots (5\text{–}8)$$

Hence for a thin prism the angle of deviation is independent of the angle of incidence provided the latter is small. The angle of deviation is also small under these circumstances and (5–4) reduces to (5–8) immediately. In addition, it follows at once that the angular dispersion of a thin prism is given by

$$\frac{d\theta}{d\lambda}=\alpha\frac{dn}{d\lambda}.$$

Consider the refraction of rays of light of wavelengths λ_B and λ_R which may be widely different. Let n_B and n_R be the refractive indices for these wavelengths and let the two rays enter the prism along the same path. The angular separation of the emergent rays is given [from (5–8)] by

$$\theta_B-\theta_R=(n_B-n_R)\alpha.$$

The deviation for an intermediate wavelength λ_Y is given by

$$\theta_Y=(n_Y-1)\alpha.$$

It will be seen that the ratio of the dispersion to the mean deviation gives the dispersive power of the material of the prism.

For a number of prisms in series, the total dispersion and total deviation are given by $\Sigma(n_B-n_R)\alpha$ and $\Sigma(n_Y-1)\alpha$ respectively. It is possible to combine prisms so that either of these quantities is zero. When applying these expressions to a series of prisms, one must regard inverted prisms as prisms with negative angles. For example, in Fig. 5–4 the middle prism has a negative angle.

Fig. 5–4. Direct vision spectroscope.

If one has a pair of prisms such that

$$(n_{1B}-n_{1R})\alpha_1=(n_{2B}-n_{2R})\alpha_2$$

the combination is achromatic and rays whose wavelengths are λ_B and λ_R are equally deviated.

On the other hand, if the ray of mean wavelength λ_Y is to be undeviated one must have

$$(n_{1Y}-1)\alpha_1=(n_{2Y}-1)\alpha_2.$$

In this case there will, in general, be some dispersion. This is the principle of the direct-vision spectroscope. In this instrument, which is usually pocket-sized, the so-called *Amici prism system* often contains two crown and one flint, or three crown and two flint prisms cemented together so that the mean wavelength of the visible spectrum is undeviated (see Fig. 5–4). (It must be remembered that the prisms in this case are not thin.) The light from the slit S is collimated by the lens L and the virtual spectrum viewed by the unaccommodated eye. The spectrum is of good quality since the beams traversing the prism system are collimated.

5–5. The Simple Table Prism Spectrometer

The optical system of the spectrometer is similar to the one already described for producing a spectrum of good quality, the spectrum being viewed by means

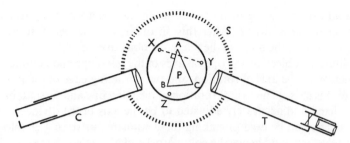

Fig. 5–5. The spectrometer.

of an eyepiece. A plan view of the complete optical system is illustrated in Fig. 5–5. C is a collimator consisting of a slit at the principal focus of an achromatic objective. T is a telescope consisting of an achromatic objective and a Ramsden eyepiece which carries cross-wires in its focal plane. The prism P rests on a table, and can be levelled by means of three screws X, Y, and Z. The telescope and prism table can rotate about the axis of the divided circle S. Usually both are fitted with clamps and fine adjustment screws. As the telescope is swung round, the slit image for each wavelength is brought into coincidence with the cross-wires.

A telescope equipped with a modified eyepiece known as a *Gauss eyepiece* can be useful in adjusting the optical system of a spectrometer initially and also for making certain measurements subsequently. The device usually consists of a Ramsden eyepiece to which has been added a thin parallel-sided plate of glass (see Fig. 5–6). Light enters through an aperture in the eyepiece tube and is reflected down the axis of the telescope. Suppose that the cross-wires are at the foci of the objective and eyepiece. An eye placed in the usual viewing position will see the cross-wires illuminated, and if a plane reflecting surface is located beyond the telescope objective and is perpendicular to the

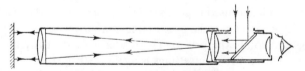

Fig. 5–6. The Gauss eyepiece.

optical axis, an image of the cross-wires will be seen alongside the cross-wires themselves. Such an eyepiece enables the telescope to be focused and to be placed with its axis perpendicular to a plane surface since it is only when these adjustments have been made that the image of the cross-wires is in focus and correctly positioned in the field of view.

5–6. Adjustment of a Spectrometer

To obtain a good spectrum and to be able to apply the simple theory for refraction by a prism as given above, it is essential for light from the centre of the slit to be collimated and for the rays to lie in principal sections of the prism.

The first adjustment to be made before using the spectrometer is to focus the eyepiece on the cross-wires. After this, in order, it is necessary to adjust the cross-wires to be at the focus of the telescope objective, the slit to be at the focus of the collimator objective, the optical axes of the telescope and collimator to be perpendicular to the axis of rotation, and the refracting edge of the prism to be parallel to this axis. (The axis of rotation of the telescope and prism table is permanently fixed and accurately adjusted to be the axis of the divided circle.) A Gauss eyepiece can be used in making these adjustments since it enables one to focus the telescope (and hence the collimator) and also to set the axis of the telescope (and hence the collimator) normal to the surface of the prism. For full details of this and alternative methods of adjusting a spectrometer the reader is referred to a laboratory manual. (The details were given in earlier editions of the present book but have now been omitted in order to make room for the addition of material of more fundamental importance.) With regard to the adjustment of the prism by means of the levelling screws XYZ (Fig. 5–5), it is important to note that it is possible that the base of the prism is not exactly perpendicular to the refracting edge, i.e. is not a principal section. The adjustments level the prism, not the prism table, and once they have been made the prism must not be moved relative to the table during the course of an experiment.

5–7. Measurement of Refracting Angle and Angle of Minimum Deviation

There are several methods of using a spectrometer to measure the refracting angle of a prism:

1. The telescope is clamped with its axis inclined to the axis of the collimator—in principle any angle can be used but roughly 90° is convenient. The prism table is then rotated until light from the collimator is reflected into the telescope from each refracting face of the prism in turn, the slit image being set on the telescope cross-wires on each occasion. The reader should satisfy himself that the angle turned through by the prism is then $(180° \pm \alpha)$.

2. The Gauss eyepiece is used to set the telescope axis normal to each refracting face in turn. One can clamp the prism table and move the telescope, or clamp the telescope and rotate the prism. The reader should satisfy himself that in each case the rotation between settings is again $(180° \pm \alpha)$.

3. The prism table is rotated until the refracting edge of the prism faces the collimator and light from the collimator is reflected from both refracting faces simultaneously. The prism table is clamped and the telescope rotated to receive light from each refracting face in turn. In each case the cross-wires are set on the slit image; the angle turned through by the telescope is 2α. The reader should satisfy himself that this is correct, and also that it is not necessary for the angles of incidence on the refracting faces to be equal.

The angle of minimum deviation is measured in a fairly obvious manner by setting the telescope cross-wires on the slit image when the prism has been rotated to give minimum deviation. This is repeated with the prism rotated to deviate the light in the opposite direction. The angle between the two telescope positions is obviously 2θ. This procedure avoids the need for a "straight through" setting with the prism removed. (As pointed out at the end of §5–6, one should not move the prism relative to the table after the prism has been levelled.)

When θ and α have been measured, the refractive index of the prism material can be calculated from equation (5–4). With a first-class spectrometer θ and α can be measured to about one second and n found to about ± 0.00001.

5–8. Use of Hollow Prism for Liquids

The spectrometer may be used to determined the refractive index of a liquid by employing a hollow prism whose faces are parallel-sided plates of glass. The walls of such a vessel have no effect on the angle of deviation, and the angle of the liquid prism is the same as that of the vessel. Hence α and θ can be measured in the usual way.

The refractive index of a liquid varies rapidly with temperature and there is usually no point in attempting an accuracy better than 0.0001 in the refractive index unless special care is given to the temperature control.

5–9. Critical Angle Methods Using a Prism Spectrometer

In Fig. 5–7 ABC is the principal section of a prism whose face BC is a ground but unpolished surface. This face is illuminated with monochromatic light, and each point acts as a source sending rays into the prism in all directions. In particular, there will be rays which, after reflection at AB (partial or total, according to the angle of incidence), emerge from AC and enter the telescope T. In the primary image plane of T, each point along the line in the plane of the figure (assumed horizontal) is the focus of rays that emerge from the prism at the same angle. Now if ϕ is positive as shown, larger angles of incidence at face AB correspond to smaller angles of emergence from face AC. Hence, if θ is the critical angle and corresponds to an angle of emergence ϕ, rays emerging at angles less than ϕ are more intense than those emerging at angles greater than ϕ since the latter are only partially reflected at face AB. Thus, along the horizontal line in the field of view of T, the intensity will show a sudden decrease

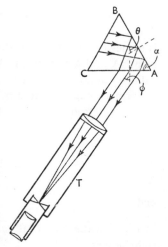

Fig. 5–7. Critical angle method.

at the point corresponding to the angle of emergence ϕ; the right-hand side of the field seen through the eyepiece will be the brighter. Since the field of view of the telescope is small, the inclination of rays to the principal section may be ignored and it will be seen that a similar argument can be applied to all horizontal lines in the field of T. Consequently, the whole field of view is divided by a vertical demarcation line on the right-hand side of which the field is brighter.

To measure ϕ, the prism is mounted on a spectrometer whose telescope T is adjusted to bring the demarcation line on to the cross-wires. The telescope is then swung round until, using the Gauss eyepiece, it is set with its axis perpendicular to face AC. The angle turned through by the telescope is obviously ϕ.

An alternative method is to position an extended monochromatic source beyond B so that rays fall on face BA in the direction B to A at grazing and near-grazing incidence (see Fig. 5–8). Care must be taken to shield the source from the diffuse surface BC which must remain dark. The rays falling on face BA at grazing incidence emerge from face AC at angle ϕ, and those incident on BA at angles of incidence less than 90° emerge from AC at angles greater than ϕ. No light emerges at angles less than ϕ. Hence, using the telescope as before, the demarcation line in the field of view will now separate an illuminated area of the field from an area that is completely dark. With the arrangement shown in the figure, it is the right-hand side of the field that is dark. Using this method it is obviously easier to locate the demarcation line and set the cross-wires on it. If the refracting angle is measured in the usual way, the refractive index of the prism can be calculated using the equation (5–10) deduced below.

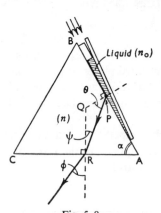

Fig. 5–8.
Critical angle method for a liquid.

The same method may be employed to determine the refractive index of a liquid of which only a small quantity is available provided the index of the prism is somewhat greater than that of the liquid. A drop of the liquid is placed on face BA and a thin glass plate placed over it. The latter is held in place by capillary action. If the second method for determining ϕ is used, it is advisable to form a wedge of liquid by means of a small piece of glass held in position by capillary action (see Fig. 5–8). In this case, θ is the critical angle for glass to liquid.

If the refractive indices of the prism and liquid are respectively n and n_0 one has

$$n_0 = n \sin \theta \qquad . \; . \; . \; . \; . \; . \quad (5\text{–}9)$$

and

$$n \sin \psi = \sin \phi. \qquad . \; . \; . \; . \; . \; . \quad (5\text{–}10)$$

In Fig. 5–8 the quadrilateral APQR is cyclic since the angles at P and R are right angles.

$$\therefore \quad \angle PQR = \pi - \alpha.$$

Hence, from the triangle PQR,

$$\pi - \alpha + \theta + \psi = \pi,$$

$$\therefore \quad \alpha = \psi + \theta.$$

Hence, from (5–9),

$$n_0 = n \sin (\alpha - \psi)$$

$$= n(\sin \alpha \cos \psi - \cos \alpha \sin \psi)$$

or, from (5–10),

$$n_0 = \sin \alpha \sqrt{n^2 - \sin^2 \phi} - \cos \alpha \sin \phi. \qquad . \; . \; . \quad (5\text{–}11)$$

If there is no liquid on face AB, $n_o=1$ and one has

$$(1+\cos \alpha \sin \phi)^2=\sin^2 \alpha(n^2-\sin^2 \phi)$$

which reduces to

$$n^2=1+\left(\frac{\sin \phi+\cos \alpha}{\sin \alpha}\right)^2. \quad \cdot \quad \cdot \quad \cdot \quad \cdot \quad \cdot \quad (5\text{--}12)$$

If ϕ is measured with the prism alone, (5–12) gives its refractive index. If ϕ is then measured with the liquid on face AB, (5–11) gives the refractive index of the liquid.

5–10. The Auto-collimator

In this instrument, the telescope and collimator are combined into a single unit. The lower half of the primary image plane of the telescope T is occupied

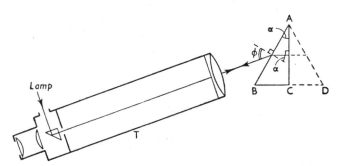

Fig. 5–9. Auto-collimator.

by a vertical slit illuminated by means of a reflecting prism (see Fig. 5–9). The prism ABC, whose refractive index is required, is placed on the prism table, and the telescope or prism table is rotated until the image of the slit is observed with the telescope eyepiece to lie vertically above the slit itself. Light is then falling normally on face AC and, from the figure, it will be seen that the prism can be regarded as half the prism ABD of refracting angle 2α. Since the ray path in prism ABC corresponds to symmetrical refraction through ABD, the position of minimum deviation is used automatically in this method. It will be seen that

$$n=\frac{\sin \phi}{\sin \alpha},$$

where ϕ is the angle between the incident/emergent ray and the normal to face AB. This method cannot be used for prisms where α is greater than the critical angle since the maximum value of $\sin \phi$ is unity (see §5–1). Frequently 30° prisms are used in refractive index determinations and the spectrometer manufactured by Ross gives n to an accuracy of 0·00001, the angles being measured to one second of arc. A calibrated fine adjustment on the prism table facilitates the measurement of the variation of ϕ with wavelength.

The above discussion refers to the use of the auto-collimator for the measurement of the refractive index of a prism. In fact it is used for the measurement of the angle between two plane reflecting surfaces, or for the alignment of a surface, in a wide variety of contexts.

5–11. Wollaston's Critical Angle Method

The equations given in §5–9 apply to any prism with refracting angle α. The special case where α is a right angle is worthy of note. Wollaston employed two adjacent faces of a right-angled glass block (often a cube). Using a very simple arrangement, it is possible to find the refractive index of the glass block or of a liquid to about 0·1%. In Fig. 5–10 ABCD is a right-angled glass block resting

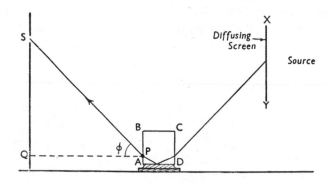

Fig. 5–10. Wollaston's method.

on a film of liquid. XY is a diffusing screen (tissue paper is satisfactory) illuminated by any convenient source. The principle is the same as in the first method described in §5–9. Rays from the screen are either partially or totally reflected at the face AD, and the ray incident on this face at the critical angle emerges at angle ϕ. Consequently, on looking into the prism from S (usually with the naked eye), one sees a demarcation line where the brightness of the field changes suddenly. The figure illustrates a very simple method of measuring ϕ. A horizontal scratch is made on face AB at P and viewed through a horizontal slit at S. The cube (mounted on a square of black material such as ebonite) is then moved along the bench until the demarcation line appears to coincide with the scratch at P. The path of the critical ray is then PS and the angle ϕ is given by $\tan \phi = SQ/PQ$. If the experiment is performed carefully it is possible to detect the variation of ϕ with wavelength. When a source of white light is used a coloured demarcation line is seen. From equation (5–9), putting $\alpha=90°$, the refractive index of the liquid is given by

$$n_o = \sqrt{n^2 - \sin^2 \phi}. \qquad \cdots \cdots \quad (5\text{--}13)$$

Alternatively, if the index of the liquid is known, that of the glass is given by

$$n = \sqrt{n_o^2 + \sin^2 \phi}. \qquad \cdots \cdots \quad (5\text{--}14)$$

5–12. The Pulfrich Refractometer

The Pulfrich refractometer frequently employs a right-angled prism as above, but is fitted with a telescope to view the demarcation line. The arrangement is shown diagrammatically in Fig. 5–11(a). T represents the telescope which receives the light emerging from the block of known index n. The instrument can be used to determine the refractive index n_o of a liquid or of a solid body provided $n_o < n$. In the case of an unknown solid, at least one surface must be polished flat. To measure the index of a liquid it is simply placed on the top surface of the block; to measure the index of a solid the plane surface of the specimen is placed in contact with the top surface of the block. Frequently, in the latter case, a liquid of index n (or somewhat higher) is placed between the

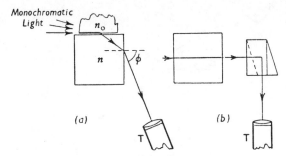

Fig. 5–11. Pulfrich refractometer.

surfaces to reduce the effect of imperfect polish. The theory is as given in the preceding section and the results can be found from equation (5–13). Usually the instrument is supplied with tables to give n_o in terms of ϕ; alternatively, the arm carrying the telescope can read n_o directly. The telescope is usually of the auto-collimating type so that, when necessary, ϕ can be measured by setting the telescope normal to the vertical face of the block before the cross-wires are brought on to the demarcation line. In most instruments a reflector is fitted outside the objective of the telescope so that the axis of the latter can be made horizontal. In Fig. 5–11(a) this corresponds to the telescope being perpendicular to the paper; the view from above is then of the form shown in (b). The reflector and telescope can move round a vertical circular scale which indicates the values of ϕ or n_o. When the instrument is to be used solely for liquids, a short length of wide bore glass tube is cemented on to the top surface of the block to form a cell.

In some of the more recent instruments the 90° prism is replaced by one of 60°. In the Bellingham and Stanley instrument a 60° prism is interchangeable with a V-block to convert the refractometer into an instrument of the Hilger-Chance type (see §5–14).

5–13. The Abbe Refractometer

The principle of this instrument is the same as that of the previous critical angle methods. Fig. 5–12 shows the arrangement employed for the measurement of the refractive index of a liquid. P and Q are prisms of high refractive index, and the liquid under test is placed between them. P corresponds to the prism in the total reflection method employing a spectrometer (§5–9). The prism Q and the mirror M simply provide a convenient method of passing light from the liquid into P. (In the dipping refractometer Q is absent and P is dipped into the liquid.) If the top face of Q is not a diffusing surface, an extended diffuse source of light must be used. Light that strikes P at grazing incidence emerges at angle ϕ and the usual demarcation line is viewed by means of the telescope T. As in §5–9, measurement of ϕ would enable the refractive index of the liquid to be calculated from equation (5–11) if α and the refractive index of P are known. Usually T is carried on an arm attached to a scale which is calibrated to read

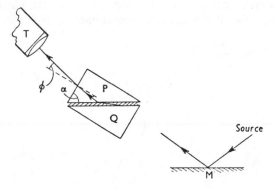

Fig. 5–12. Abbe refractometer.

directly the index of the liquid. When measuring the refractive index of a solid, Q is removed and a polished face of the solid is placed in contact with P, a small film of high index liquid being interposed as in the Pulfrich refractometer. If white light enters the instrument, rays entering P at grazing incidence will emerge at an angle ϕ which depends upon the wavelength. Consequently, the demarcation line has coloured edges.

In order to enable the instrument to be used with white light, it is usual to place an Abbe compensator before the objective of the telescope. This consists of a pair of Amici direct-vision prisms in series (see §5–4); these can be rotated by equal and opposite amounts about the axis of T. In this way, the dispersion of the combination can be varied until it is sufficient to recombine the rays emerging from P. The demarcation line is then achromatized. Since the compensator is arranged to give no deviation for sodium light, it is for this light that the instrument gives the refractive index of the specimen.

5–14. The Hilger-Chance Refractometer

In the critical angle refractometers the eyepiece cross-wires have to be set on a demarcation line separating bright and dark areas. It is found that the unsymmetrical nature of this setting is responsible for a reduction in the accuracy that an observer can obtain. Further disadvantages of the critical angle methods are: (1) the refractive index of the glass prism must be greater than that of the specimen under test; (2) the measured index corresponds to a surface layer and may differ from that of the main bulk of the specimen. On the other hand, the spectrometer methods require two highly polished surfaces on the specimen. The Hilger-Chance instrument was developed in order to avoid these difficulties. A side elevation of the arrangement is shown in Fig. 5–13(*a*). The composite glass block B is made by combining two prisms as indicated. The components are united by heat treatment and cannot be separated by any of the usual solvents. The angle between the faces of the V-block is exactly 90°, and the specimen under test is positioned in the groove as indicated. Monochromatic light from the slit S is collimated by C and passes through the V-block and specimen which together resemble an Amici prism system. The direction of the emergent light obviously depends upon the refractive index of the specimen and is measured by means of the telescope T. As in the Pulfrich refractometer, the telescope is fitted with a reflector and has its axis horizontal.

It will be seen that two perpendicular plane surfaces must be produced on the

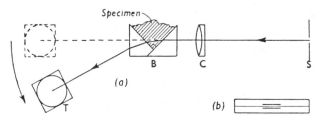

Fig. 5–13. Hilger-Chance refractometer.

specimen, but it is not necessary for these to be of the highest quality since a film of liquid can be placed between the faces of the specimen and the V-block. Since the method is not a critical angle method, the refractive index of the specimen can be greater than or less than that of the V-block. Consequently, the latter can be made of a durable glass of moderate index. In addition, it is obvious that the direction of the emergent beam does not depend solely upon the refractive index of the surface layers. In designing the instrument, special attention was given to the method of setting. The eyepiece carries two short parallel lines in its focal plane, and the setting is made on a fine line which runs along the middle of the wide slit S; the appearance of the field of view is shown in Fig. 5–13(*b*). The symmetrical nature of this setting, and the care taken to facilitate the reading of the divided circle, are important factors contributing to the accuracy of the instrument. The refractive index of a specimen of glass can be found to an accuracy of 0·00001. The instrument can be used for liquids if the V-block is provided with side plates to form a trough; 2–3 c.c. of liquid are required.

5–15. Refractive Index of Solids of Irregular Shape

The methods described above for the measurement of the refractive index of a solid require the production of at least one polished surface on the solid. This can be avoided as follows: If a transparent solid of irregular shape is immersed in a liquid of the same refractive index, there is no change in index at the boundary of the solid and it becomes invisible. The method consists simply of finding a liquid or mixture of liquids for which the outline of the solid disappears. The index of the liquid can then be measured by one of the methods described above. Frequently, the liquids with suitable refractive indices are volatile, and the different rates of evaporation cause the refractive index of the mixture to change while it is being transferred to the refractometer. This difficulty can be overcome by employing a hollow prism and a spectrometer as described in § 5–8, or by using the Hilger-Chance refractometer. It is then possible to mix the liquids in the vessel containing the irregular solid whilst viewing the slit in the usual way. The slit image will become clear when the refractive index of the liquid is equal to that of the solid, and when this stage is reached the telescope setting is made. If necessary, the evaporation of the liquids can be compensated while making the setting. The refractive indices of most glasses can be matched by a mixture of readily available liquids. These include carbon disulphide (1·632), benzene (1·504), nitrobenzene (1·553), α-monobromnaphthalene (1·658).

5–16. The Simple Spectrograph

Frequently, it is necessary to know the wavelengths present in the light emitted by a source. The spectrum can, of course, be examined visually by

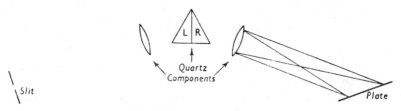

Fig. 5–14. Quartz spectrograph.

means of a spectrometer, but it is sometimes desirable to obtain a permanent record of the spectrum. To do this the telescope is replaced by a camera focused for parallel light, the spectrum being in focus on the plate. If it is necessary to photograph the spectrum in the ultra-violet or infra-red as well as in the visible region, it is obviously necessary to employ lenses and a prism of materials which transmit the wavelengths concerned. A commonly used substance is quartz. Now quartz is birefringent, and it is necessary for the optical axis to be parallel to the direction of the light in the prism. Again, since quartz is also optically active the prism must be made in two halves, one of right-handed and the other of left-handed quartz (see Chapter XXIII). Furthermore, a quartz lens exhibits considerable longitudinal chromatic aberration and, if the spectrum is to be in focus over a wide range of wavelengths, the photographic plate must be inclined to the axis of the camera lens. A typical arrangement is shown in Fig. 5–14.

Quartz is transparent from about $0 \cdot 19\mu$ to about 4μ. For the farther U.V. one can use prisms cut from *vacuum-grown* crystals of calcium fluoride (fluorite) and lithium fluoride down to $0 \cdot 125\mu$ and $0 \cdot 105\mu$ respectively. Substances useful for infra-red spectroscopy include lithium fluoride (up to 6μ), fluorite (up to 9μ), sodium chloride (rock salt) (up to 17μ), silver chloride (up to 25μ), potassium bromide (up to 30μ), and thallium bromide-iodide—42–58 % (up to 40μ). Beyond about 40μ crystalline quartz again becomes transparent and, apart from a strong narrow absorption band at 78μ, remains useful into the far I.R. (to beyond 1000μ). Fused quartz has also been used beyond 200μ. Also, alkali halides become transparent again in the far I.R. It should be noted that some of these substances are very soluble in water and must be protected from the atmosphere. To reduce the number of components made from these rather inconvenient materials, it is usual to employ mirrors instead of lenses.

5-17. The Constant Deviation Spectroscope

In the ordinary spectrometer the telescope is moved in order to view different parts of the spectrum. In the constant deviation instrument the collimator and telescope are fixed at right angles and the prism is rotated. The general arrangement of the instrument is shown in Fig. 5–15. The action of the prism is most easily understood by regarding it as a combination of three parts as indicated in the figure. The slit image for one particular wavelength will coincide with the cross-wires of the telescope, and the path of the principal ray for this wavelength is as shown. It will be seen that the angles of incidence and emergence are equal and opposite, so that the two refractions cause no total deviation. The presence of the reflecting surface, however, causes the two dispersions to be in the same sense.

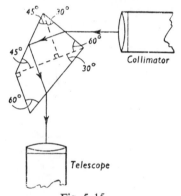

Fig. 5–15.
Constant deviation spectroscope.

The prism can be rotated to bring the slit image for any wavelength on to the cross-wires when, for that wavelength, the action of the prism is equivalent to that of a 60° prism at minimum deviation. Usually the prism table is calibrated to enable the wavelength to be read off directly. Frequently a camera attachment replaces the eyepiece and the spectrum, or a region of it, can be photographed. Also, by placing a second slit at the focus of the telescope objective, a narrow region of the spectrum can be isolated. The emergent light is then almost monochromatic and the instrument is referred to as a *monochromator*.

5-18. Other Spectrographs and Monochromators. Spectrophotometers

For a given prism the angular dispersion is fixed, so that the linear dispersion on the photographic plate (that is, the linear separation of slit images for neighbouring wavelengths) will be proportional to the focal length of the camera

Fig. 5–16(a). Littrow spectrograph.

lens. The *Littrow spectrograph* is a convenient arrangement which employs a long focus lens without being unduly cumbersome [see Fig. 5–16]. The light is doubly dispersed by P and the mirror is tilted about a horizontal axis so that the light returns above or below the small prism. Alternatively, the small prism can be moved horizontally a short distance away from the axis of the lens. One can achieve a saving of light (through the elimination of two refractions) at the expense of a loss of dispersion by using a rear-silvered prism as in Fig. 5–9. The dispersion of the original system can be increased by using two (or more) prisms in series.

In some spectral regions it is more convenient to replace the lens with a mirror which has the additional advantage of freedom from chromatic aberration. It is unavoidable that the mirror is used off-axis, and for good definition an off-axis parabola should be used. Some systems use two spherical mirrors—one

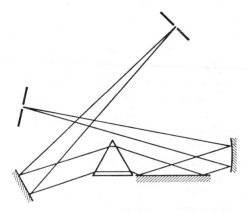

Fig. 5–17. System used by Czerny and Plettig.

for collimating and the other for focusing. To produce sharp spectral lines it is then essential to use the two mirrors off-axis by the same amount in opposite directions. To a first approximation this eliminates unsymmetrical aberration (i.e. coma—see Chapters XV and XVI), but since the correction will apply only to the spectral line at the appropriate distance from the axis of the focusing mirror, the arrangement is more appropriate for a monochromator than for a spectrograph. As was pointed out in §5–2, a prism gives curved spectral lines. In consequence, it is necessary for the exit slit of a monochromator to be curved. There is some difficulty in arranging the layout of the system in such a way that the mirrors are used as close as possible to the axis and the various components do not get in each other's way. Fig. 5–17 shows the system due to Czerny and Plettig (1930). (When a plane mirror is parallel to the base of a prism at minimum deviation as shown, the light leaving the plane mirror is parallel to that entering the prism.) At first sight the mirrors appear to be off-axis in the same direction, but the plane mirror introduces the necessary reversal.

A photographic plate has spatial discrimination and by using a spectrograph one can record a wide spectrum with a single exposure. This is a satisfactory way of recording the wavelengths present in a spectrum but is less satisfactory if accurate measurements of intensity are required. In a *spectrophotometer* or *spectroradiometer* it is better to use a receiver such as a photo-cell or thermopile, and since these detectors do not have spatial discrimination (but have short response times), it is necessary for the detector to scan across the spectrum, recording the intensity at each wavelength as it proceeds. Usually, this is done by using a monochromator in which successive wavelengths are directed to a stationary exit slit behind which the detector is mounted. (See also §18–11.)

CHAPTER VI

THE WAVE THEORY OF LIGHT

Sections 7 to 15 can be omitted on first reading

6–1. The General Equation of Wave Motion and the Principle of Superposition

IT was pointed out in Chapter I that for many purposes the light disturbance at any point can be represented by a single scalar quantity ϕ. It was assumed that the variations of ϕ are propagated in the form of a wave motion, so that for a plane wave of infinite width propagated in the positive x-direction one can write:

$$\phi = a \sin k(ct-x) = a \sin (\omega t - kx) \ . \ . \ . \ . \ . \ (6\text{--}1)$$

and for a spherical wave expanding from the origin

$$\phi = \frac{a}{r} \sin k(ct-r) = \frac{a}{r} \sin (\omega t - kr). \ . \ . \ . \ (6\text{--}2)$$

For a sound wave the particle displacements are always in the direction of propagation and the disturbance at any point is specified completely simply by specifying the magnitude of this particle displacement (or by specifying the excess pressure, condensation, etc.). That is to say, a single scalar quantity adequately specifies the disturbance in a sound wave. In the case of light it is found that a single scalar quantity is not completely adequate. In terms of the old elastic solid theory of the propagation of light, it is found to be necessary to assume that the light disturbance consists of vibrations perpendicular to the direction of propagation. It is then necessary to specify the direction of these vibrations. That is to say, the quantity specifying the light disturbance at any point is a vector quantity. In terms of the electromagnetic theory of light, the vibrating particle is replaced by oscillatory electric and magnetic fields. For example, a transverse linear oscillation of a particle becomes an electric field along a given line (the magnitude varying sinusoidally), plus an associated magnetic field.

An investigation into the propagation of sound waves leads to the general result that the velocity potential satisfies the equation

$$\frac{\partial^2 \phi}{\partial x^2} + \frac{\partial^2 \phi}{\partial y^2} + \frac{\partial^2 \phi}{\partial z^2} = \frac{1}{c^2} \frac{\partial^2 \phi}{\partial t^2}. \ . \ . \ . \ . \ (6\text{--}3)$$

This is known as the *general equation of wave motion*. For plane waves of infinite width propagated in the x-direction, ϕ is a function of x only. One has

$$\frac{\partial^2 \phi}{\partial x^2} = \frac{1}{c^2} \frac{\partial^2 \phi}{\partial t^2} \ . \ . \ . \ . \ . \ . \ (6\text{--}4)$$

and, by direct substitution, it is easily proved that (6–1) is a solution.

For spherical waves symmetrical about the origin (6–3) becomes

$$\frac{\partial^2(r\phi)}{\partial r^2} = \frac{1}{c^2} \frac{\partial^2(r\phi)}{\partial t^2} \ . \ . \ . \ . \ . \ . \ (6\text{--}5)$$

and, by comparison with (6–4) and (6–1), it will be seen that (6–2) is a solution of (6–5). Classical electromagnetic theory shows that the electric and magnetic field vectors also satisfy the general equation of wave motion; in a scalar theory of light the light disturbance must satisfy the same fundamental equation. When thinking in terms of the scalar wave theory, it is not necessary to specify the nature of the scalar quantity which varies—it is simply a quantity which varies sinusoidally and is such that the square of its amplitude is proportional to the energy flux density, i.e. to what in this context is called the *intensity* of the light. (In radiometry and photometry the terminology is somewhat different and intensity is used in a different way—see §18–1.)

Frequently it is necessary to find the resultant disturbance at a point when a number of disturbances arrive simultaneously. According to the *Principle of Superposition*, the resultant disturbance is simply the sum of the separate disturbances. It should be noted that since the general equation of wave motion is a linear equation, the sum of any number of solutions is also a solution. If this were not so the principle of superposition would need to be modified. In fact, when the light intensity is extremely high the linear equation of wave motion is not satisfied and the principle of superposition does not apply. It is only in recent years that very high power *lasers* have enabled these conditions of *non-linear optics* to be produced in the laboratory. The intensities required for these non-linear effects to be observed are never produced in the laboratory with the classical non-laser sources (referred to as *thermal sources*) and are not produced with the lower power lasers used in most of the traditional types of optical experiments. For this reason it is assumed in the discussion that follows that these very high intensities are not involved, so that the linear equation of wave motion is satisfied and the principle of superposition can be applied. An important result of the principle of superposition is that when beams of light intersect, the propagation of each beam is completely unaffected by the others. Thus after one beam has crossed another it proceeds exactly as it would have done if the other had not been present. Consider, for example, the disturbance at a point where two disturbances of the same frequency arrive with different phases. The separate disturbances can be represented by

$$\left.\begin{aligned}\phi_1 &= a_1 \sin(\omega t - \delta_1), \\ \phi_2 &= a_2 \sin(\omega t - \delta_2),\end{aligned}\right\} \qquad \cdots \cdots \quad (6\text{–}6)$$

where δ_1 and δ_2 are sometimes called the phase constants. The resultant disturbance is given by

$$\phi = a_1 \sin(\omega t - \delta_1) + a_2 \sin(\omega t - \delta_2)$$
$$= \sin \omega t (a_1 \cos \delta_1 + a_2 \cos \delta_2) - \cos \omega t (a_1 \sin \delta_1 + a_2 \sin \delta_2).$$

It is convenient to call the first of these brackets $A \cos \delta$ and the second $A \sin \delta$. That is, one puts

$$\left.\begin{aligned}A \sin \delta &= a_1 \sin \delta_1 + a_2 \sin \delta_2, \\ A \cos \delta &= a_1 \cos \delta_1 + a_2 \cos \delta_2.\end{aligned}\right\} \qquad \cdots \quad (6\text{–}7)$$

The expression for ϕ then reduces to

$$\phi = A \sin(\omega t - \delta), \qquad \cdots \cdots \cdots \quad (6\text{–}8)$$

where $$A^2 = a_1^2 + a_2^2 + 2a_1a_2 \cos{(\delta_2 - \delta_1)} \quad \cdots \quad (6\text{--}9)$$

and $$\tan{\delta} = \frac{a_1 \sin{\delta_1} + a_2 \sin{\delta_2}}{a_1 \cos{\delta_1} + a_2 \cos{\delta_2}}. \quad \cdots \quad (6\text{--}10)$$

If $a_1 = a_2$ one has

$$\left.\begin{array}{l} A^2 = 2a^2[1 + \cos{(\delta_2 - \delta_1)}] \\ \quad = 4a^2 \cos^2{\left(\dfrac{\delta_2 - \delta_1}{2}\right)}. \end{array}\right\} \quad \cdots \quad (6\text{--}11)$$

Thus the resultant intensity at the point (A^2) is not simply the sum of the intensities due to the separate disturbances $(a_1^2 + a_2^2)$. The two disturbances are said to *interfere*. In particular, if $a_1 = a_2$ and $(\delta_2 - \delta_1) = \pi$ they interfere destructively to give zero intensity. At first sight this would seem to violate the principle of conservation of energy, but closer inspection shows that whenever there are points where destructive interference causes the resultant intensity to be less than the sum of the separate intensities, there are also other points where constructive interference causes the resultant intensity to be greater than the sum of the separate intensities.

Frequently one is concerned with disturbances that travel from a source to a point via two different routes. They therefore arrive with different phases owing to the different optical paths traversed. Equation (6–9) shows that the superposed disturbances interfere constructively if the phase difference $(\delta_2 - \delta_1)$ is an even multiple of π, and destructively if it is an odd multiple of π. Now the phase difference is related to the difference between the optical paths traversed by the two disturbances by the relation

phase difference$= (2\pi/\lambda)$(optical path difference).

Hence there is constructive interference if the path difference is a multiple of λ, and destructive interference if it is an odd multiple of $\lambda/2$. The analysis can be extended in an obvious manner to the superposition of any number of disturbances.

It must be remembered that the discussion has been concerned with the superposition of disturbances each of which is completely specified by a single scalar quantity. This corresponds to the superposition of simple harmonic oscillations that are along the same straight line. For disturbances that are not along the same straight line it would be necessary to specify the directions of the various oscillations.

The above results give the resultant at any point when two simple harmonic scalar disturbances are superposed there. If one has two progressive waves of the same frequency propagated along the same line, the phase difference is constant at all points along the line, and the resultant disturbance is given by (6–8). For waves of the same amplitude the resultant amplitude is $2a \cos{\frac{1}{2}(\delta_2 - \delta_1)}$. where $(\delta_2 - \delta_1)$ is the phase difference. If the waves are propagated in the positive x-direction and one is distance Δ ahead of the other, one may write

$$\phi_1 = a \sin{(\omega t - kx)},$$
$$\phi_2 = a \sin{[\omega t - k(x + \Delta)]}$$

and it is easily shown that the resultant is given by

$$\phi = 2a \cos \frac{k\Delta}{2} \sin \left[\omega t - k \left(x + \frac{\Delta}{2} \right) \right]. \quad \ldots \quad (6\text{--}12)$$

If $\Delta = 0$, λ, 2λ, . . ., the crests and troughs of one wave coincide with those of the other and there is maximum constructive interference to give a wave of amplitude $2a$. If $\Delta = \lambda/2$, $3\lambda/2$, $5\lambda/2$, . . ., the crests of one wave coincide with the troughs of the other and there is complete destructive interference.

6–2. Graphical Method of Adding Disturbances of the Same Frequency

In Fig. 6–1 the disturbances ϕ_1 and ϕ_2 of the previous section are represented by the lines OP and PQ. These have lengths proportional to a_1 and a_2 and are inclined to the x-axis at angles δ_1 and δ_2 respectively. These two lines form the adjacent sides of a parallelogram whose diagonal is inclined to the x-axis at angle δ and is of length A where A and δ are given by equations (6–9) and (6–10). The value of A follows immediately from the triangle OPQ and it is easily seen that the components of A along and perpendicular to the x-axis are given by $A \cos \delta = a_1 \cos \delta_1 + a_2 \cos \delta_2$ and $A \sin \delta = a_1 \sin \delta_1 + a_2 \sin \delta_2$ respectively. That is to say, if the amplitudes and phases of two disturbances of the same frequency are represented by the lengths and directions of two vectors, the amplitude and phase of the resultant is represented by the length and direction of the vector sum. To find the effect of superposing a third disturbance one simply adds the corresponding vector QR. OR then represents the resultant. It is obvious that by proceeding in this way one can use a vector polygon to find the resultant of any number of superposed disturbances. The resultant intensity depends upon the amplitudes and relative phases of the

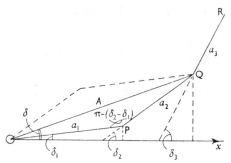

Fig. 6–1.
Graphical method of adding disturbances.

superposed disturbances and varies from zero, when the polygon is a closed figure, up to a maximum of $(\Sigma a)^2$, when all the disturbances are in phase and the polygon becomes a straight line.

It is worthy of note that if a resultant is built up of a number of superposed disturbances and the phase of each component is changed by the same amount, the phase of the resultant is changed by the same amount, there being no change in amplitude. The change is simply represented by a rotation of the complete vector polygon.

6–3. The Use of Complex Quantities in Wave Theory

The complex quantity e^{ix} has real and imaginary parts given by

$$e^{ix} = \cos x + i \sin x.$$

Hence $a \cos (\omega t - \delta)$ is the real part of the complex quantity $\phi = ae^{i(\omega t - \delta)}$ and, if it is understood that only its real part has physical significance, the latter may be used to represent the disturbance defined by the former. This representation is possible because on adding a number of complex quantities one adds the real and imaginary parts separately. That is to say, the real part of the sum is equal to the sum of the real parts. Therefore to find the resultant of a number of superposed disturbances one can add the corresponding complex quantities and take the real part of the sum. In wave theory such a summation frequently takes the form of an integral which can be evaluated more easily if the complex representation is used. Another advantage of using the complex representation is that the complex quantity that is used to represent a disturbance may be split into its space and time parts to give

$$\phi = ae^{-i\delta}e^{i\omega t}.$$

It is seen to consist of a *complex amplitude*

$$\psi = ae^{-i\delta}$$

and a harmonic time factor

$$e^{i\omega t}.$$

The modulus and argument[1] of the complex amplitude give the amplitude and phase of the disturbance represented. When adding a number of disturbances of the same frequency, $e^{i\omega t}$ is a common factor and the sum of the complex amplitudes gives the complex amplitude of the resultant, the modulus and argument respectively giving the amplitude and phase of the resultant disturbance. Frequently one is interested only in the resultant intensity and it is not then necessary to know both the modulus and the argument. The resultant intensity a^2 is obviously given by the product of ψ and its conjugate complex quantity ψ^* and this is the usual method of finding the intensity when an expression for ψ has been obtained.

To illustrate the procedure, consider the superposition of two simple harmonic disturbances. One has

$$a_1 \cos(\omega t - \delta_1) + a_2 \cos(\omega t - \delta_2)$$
$$= (a_1 \cos \delta_1 + a_2 \cos \delta_2) \cos \omega t + (a_1 \sin \delta_1 + a_2 \sin \delta_2) \sin \omega t$$
$$= A \cos \delta \cos \omega t + A \sin \delta \sin \omega t$$
$$= A \cos(\omega t - \delta), \quad \ldots \ldots \ldots \ldots \ldots \quad (6\text{--}13)$$

where

$$A^2 = (a_1 \cos \delta_1 + a_2 \cos \delta_2)^2 + (a_1 \sin \delta_1 + a_2 \sin \delta_2)^2$$
$$= a_1{}^2 + a_2{}^2 + 2a_1 a_2 \cos(\delta_2 - \delta_1)$$

and

$$\tan \delta = \frac{a_1 \sin \delta_1 + a_2 \sin \delta_2}{a_1 \cos \delta_1 + a_2 \cos \delta_2}.$$

In terms of the complex representation, the complex amplitude of the resultant is given by

$$\psi = a_1 e^{-i\delta_1} + a_2 e^{-i\delta_2} \quad \ldots \ldots \ldots \quad (6\text{--}14)$$

which reduces to

$$\psi = Ae^{-i\delta} = C - iS,$$

---

* The modulus $|z|$ of the complex quantity $z = C + iS$ is defined as $\sqrt{C^2 + S^2}$ and the argument θ is given by $\tan \theta = S/C$. If $z = ae^{ix}$, $|z| = a$ and $\arg z = x$. The complex conjugate of z is obtained by substituting $-i$ for i in the expression for z and is written $\bar{z}$. Thus, if $z = C + iS$ $\bar{z} = C - iS$ and $z\bar{z} = C^2 + S^2 = |z|^2$.

where $C = A \cos \delta$ and $S = A \sin \delta$ and A and δ are given by the above expressions. It will be seen that (6–13) gives the real part of $\psi e^{i\omega t}$.

Alternatively, if only the resultant intensity is required, one can write at once

$$\text{Intensity} = \psi\psi^* = (a_1 e^{-i\delta_1} + a_2 e^{-i\delta_2})(a_1 e^{+i\delta_1} + a_2 e^{+i\delta_2})$$
$$= a_1^2 + a_2^2 + a_1 a_2 (e^{i(\delta_2 - \delta_1)} + e^{-i(\delta_2 - \delta_1)})$$
$$= a_1^2 + a_2^2 + 2a_1 a_2 \cos(\delta_2 - \delta_1). \qquad (6\text{–}15)$$

The disturbance $a \sin(\omega t - \delta)$ may be represented by the same complex quantity as $a \cos(\omega t - \delta)$ if it is understood that throughout the analysis it is the imaginary part which represents the physical disturbance (see §6–1).

6–4. Coherent, Incoherent, and Partially Coherent Disturbances

It would be logical for this section to be placed after the sections dealing with Fourier theory since the results of the Fourier theory are quoted in two places. On the other hand, the ideas discussed in this section are important and can be appreciated without a full understanding of the more mathematical sections; it would be a pity if those whose mathematics is weak were to omit this section because the order of presentation implied that a mastery of the mathematics was a prerequisite.

The discussion in this section applies to thermal, i.e. non-laser, sources and is entirely in terms of classical physics (see also §6–6).

An equation of the form $\phi = a \sin k(ct - x)$ represents a continuous train of waves stretching from $x = -\infty$ to $x = +\infty$ and proceeding in the positive x-direction with speed c. That is, if one plots ϕ against x for a given instant of time, one

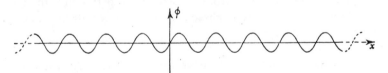

Fig. 6-2. Sine wave.

has a sine wave stretching from $-\infty$ to $+\infty$ (see Fig. 6–2). The variation with time of the disturbance at any point x is given by imagining the profile of the wave to move along with speed c; it is given by the movement of the point of intersection of the ordinate at x with the sine curve as the latter moves past the former. The disturbance is sinusoidal and continues for ever.

Suppose that at a given point one has two such disturbances of the same frequency superposed. They will not, in general, be in phase, but *the phase difference between them is constant* and the resultant intensity is given by equation (6–9).

Now, as will be seen later, a source of light does not emit a continuous train of waves but can be said to emit a succession of wave trains of finite length, there being no fixed phase relation between successive wave trains. That is, the disturbance emitted by a source exhibits random abrupt variations of phase; these changes occur at the rate of 10^8 or more per second. When the disturbances from two independent sources are superposed, the instantaneous intensity is still given by equation (6-9) but cos $(\delta_2-\delta_1)$ will vary rapidly in the range between -1 and $+1$. Now the eye and most existing physical receptors cannot appreciate such rapid variations in intensity and the effective intensity is given by the average value of A^2 over a period of time which includes many variations. If the phase changes are random the mean value of cos $(\delta_2-\delta_1)$ will be zero and the mean value of A^2 is $a_1{}^2+a_2{}^2$. If there is a constant phase relation between two or more disturbances they are said to be *coherent* and if there is no fixed phase relation they are *incoherent*. When disturbances are added and the time average of the resulting intensity is different from the sum of the separate intensities the disturbances are said to interfere. That is to say, coherent disturbances interfere and incoherent disturbances do not. It will be seen that to add coherent disturbances one adds the complex amplitudes, and to add incoherent disturbances one adds the intensities. Since every source exhibits random changes of phase, disturbances from different sources (or different points of an extended source) are incoherent.

Suppose light from a single point source is divided into two parts (by a half-silvered mirror for example) and that after traversing slightly different optical path lengths these disturbances are superposed. The disturbances superposed at any instant will have originated in the source at only slightly different times so that the abrupt changes of phase which occur in one of the disturbances will occur almost simultaneously in the other. Consequently, the phase difference between them is constant and equal to $(2\pi/\lambda)$(optical path difference). The disturbances are coherent and can interfere. On the other hand, if disturbances traverse widely different optical path lengths before being brought to the same point, they must have been emitted by the source at widely different times, and light belonging to one wave train from the source is superposed with light from another. There is then no fixed phase relation between them—they are incoherent and do not interfere.

Between the two extreme examples quoted in the above paragraph, there is the possibility that the superposed disturbances are *partially coherent*. Thus, suppose that the path difference is increased from zero until it is such that the abrupt changes in one disturbance occur about half-way between those of the other. Then for half the time the superposed disturbances are derived from the same wave train and for half the time they belong to different wave trains. When averaging over a relatively large period of time one could say that the disturbances are half coherent and half incoherent, i.e. they are partially coherent. Thinking in terms of propagation of wave trains of finite length, one may say that in an interference experiment in which light is divided into two beams, each finite wave train is divided into two similar wave trains. If these are superposed after traversing similar optical path lengths, they overlap along their entire length and interference effects can be observed. If they traverse widely different optical paths before passing along the same path, the whole of one wave train will have passed by before the other has arrived, and obviously they cannot interfere. On the average there is no interference between wave trains that have not been derived from the same original wave train so that no interference effects will be observable. Between these two extremes it may happen that if they have traversed moderately different optical path lengths, the wave trains partially overlap and the overlapping portions interfere. With such an arrangement it is only for part of the time that one has overlapping wave trains derived from the same original one. The average intensity over a relatively large period of time shows the effect of interference but the effect is less marked than when the wave trains overlap along their entire length.

For any source the average length of a wave train is called the *coherence length*, and the time taken by light to travel this distance (i.e. the interval of time during which the mean wave train is emitted) is called the *coherence time*.

The above discussions show that the transition from coherence to incoherence is a gradual one, and explain why the interference effects which can be observed when disturbances from the same source are superposed become less distinct as one increases the differences between the optical paths traversed by the disturbances. As the path difference is increased the disturbances may be said to be coherent over a smaller fraction of the total time and the time average of the intensity gradually changes from $[a_1{}^2 + a_2{}^2 + 2a_1a_2 \cos (2\pi/\lambda)(\text{path difference})]$ to $(a_1{}^2 + a_2{}^2)$.

The above discussion shows that a strictly simple harmonic wave would extend to infinity and hence two such strictly monochromatic disturbances would necessarily be completely coherent since neither would exhibit the random phase changes that have been referred to. However, it is obvious that one can never have strictly monochromatic light. On the other hand, light from a single line in a line spectrum can correspond to very long wave trains (a million or more wavelengths, or more than 50 cm. in the middle of the visible spectrum), and it will be shown in §6–9 that this is equivalent to a mixture of monochromatic disturbances, most of which have wavelengths close to the mean value. Such light is referred to as *quasi-monochromatic*, and may be described as light in which the spread of frequencies is sufficiently small for a single frequency to be associated with it for many purposes, whilst the non-monochromaticity and the associated finite lengths of the wave trains permit two beams that nominally have the same frequency to be incoherent. Consider, then, an extended source of quasi-monochromatic light, and suppose light from this source reaches two points P_1 and P_2 whose distances from any element of the source differ by much less than the coherence length of the light emitted by the element. The light arriving at P_1 and P_2 from a single source element will then be coherent. However, since disturbances from different elements of the source are incoherent, the *total* disturbances at P_1 and P_2 are partially coherent and there is a gradual transition from complete coherence as the size of the source is increased from infinitesimal to finite.

It will be seen from the discussions given above that for disturbances to be coherent they must have been emitted from the same source element at approximately the same time. Disturbances at different distances from a given source element are partially coherent owing to the finite coherence length which, as shown in §6–9, is related to the spectral impurity of the source. It will be seen in §§ 8–10 and 8–11 that this has been studied with the Michelson interferometer. Disturbances arriving at neighbouring points from an extended quasi-monochromatic source are also partially coherent as indicated above and this effect is studied in more detail in §6–5 and in §11–12 and is applied in the Michelson stellar interferometer (§11–13). One refers in these cases to *temporal* or *longitudinal* coherence, and to *spatial* or *lateral* (*transverse*) coherence respectively. So far, the effect of the finite extent of the source is neglected when the finite coherence length (spectral impurity) is discussed, and the spectral impurity is neglected when discussing a source of finite extent. The general problem involves light from an extended polychromatic source.

Suppose, then, that light from an extended polychromatic source reaches the point of observation via two different routes. If ϕ_1 and ϕ_2 denote the instantaneous complex light disturbances (sometimes called the *analytic signals*), one writes them as $\phi_1(t+\tau)$ and $\phi_2(t)$ where τ denotes the extra time taken for the light to traverse one of the routes. The instantaneous disturbance is

$$\phi_1(t+\tau)+\phi_2(t)$$

One writes

$$\text{"Instantaneous intensity"}=(\phi_1(t+\tau)+\phi_2(t))(\phi_1{}^*(t+\tau)+\phi_2{}^*(t))$$

but

$$\text{Observed intensity}=I=\langle(\phi_1(t+\tau)+\phi_2(t))(\phi_1{}^*(t+\tau)+\phi_2{}^*(t))\rangle \quad . \quad (6\text{–}16)$$

where the brackets $\langle \cdots \rangle$ denote time average taken over the exposure time of the detector.

$$\therefore \quad I=\langle\phi_1(t+\tau)\phi_1{}^*(t+\tau)\rangle+\langle\phi_2(t)\phi_2{}^*(t)\rangle$$
$$+\langle\phi_1(t+\tau)\phi_2{}^*(t)\rangle+\langle\phi_1{}^*(t+\tau)\phi_2(t)\rangle. \quad . \quad . \quad . \quad (6\text{–}17)$$

The first two terms are simply the intensities of the separate disturbances. The third term or its conjugate, the fourth term, measures the degree of correlation that exists between the two disturbances; it is called the *cross-correlation function* of ϕ_1 and ϕ_2 or the *mutual coherence* of the two light disturbances.

For a quasi-monochromatic source one may write

$$\phi_1=\psi_1 e^{i\omega t} \quad \text{and} \quad \phi_2=\psi_2 e^{i\omega t} \quad . \quad . \quad . \quad . \quad (6\text{–}18)$$

where ψ_1 is written as $a_1 e^{-i\delta_1}$ and ψ_2 as $a_2 e^{-i\delta_2}$, so that the correlation function is written as

$$\langle a_1 a_2 e^{i(\delta_2-\delta_1)}\rangle$$

or

$$\langle a_1 a_2 e^{-i\delta}\rangle \quad . \quad . \quad . \quad . \quad . \quad . \quad (6\text{–}19)$$

where δ is the instantaneous phase difference between the disturbances. In the discussion given below (§6–5) it is assumed that τ is much smaller than the coherence time so that one discusses the purely spatial coherence due to an extended quasi-monochromatic source.

Suppose, now, that a single beam from the source is amplitude-divided by a partially reflecting mirror and that the disturbances are recombined after a time delay τ has been introduced between them by sending them along unequal optical path lengths as indicated earlier (and as, for example, in a Michelson interferometer—see §8–7). The forms of the two light beams are the same so that $\phi_2(t)=\phi_1(t)$ and after the time delay has been introduced the correlation between the two beams is measured by

$$\langle \phi_1(t+\tau)\phi_1^*(t)\rangle. \quad \cdots \cdots \quad (6\text{-}20)$$

This is called the *auto-correlation function* or the *self-coherence* and measures the temporal or longitudinal coherence. If one actually has sharply limited wave trains, one can use the simple definition of the coherence time already given, and one can say that if τ is greater than the coherence time the disturbances are incoherent and the auto-correlation function is zero. In fact, the situation that occurs naturally usually involves a wave train which is not sharply limited (see §6–9) and, in consequence, a more sophisticated definition of coherence time is required; there is not then a well-defined value of τ for which the auto-correlation function becomes zero. Whatever (reasonable) definition of coherence time is adopted one can say that if τ is much smaller than the coherence time, the disturbances are coherent.

6–5. Spatial Coherence with an Extended Quasi-monochromatic Source

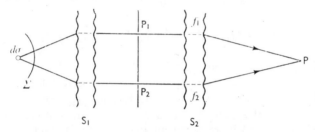

Fig. 6–3. Partially coherent disturbances.

In Fig. 6–3 light from the extended quasi-monochromatic source Σ passes through the optical system S_1, through the two pinholes P_1 and P_2, through the system S_2, and arrives at the point P. The disturbances at P_1 and P_2 and hence at P are partially coherent and one needs to be able to calculate the total intensity at P. The treatment given here and in §11–12 follows that given by Hopkins (1951). Consider the effect of a small source element $d\sigma$. This produces at P_1 and P_2 intensities proportional to $d\sigma$, i.e. amplitudes proportional to $\sqrt{d\sigma}$. Suppose, then, that one has complex amplitudes $u_1\sqrt{d\sigma}$ and $u_2\sqrt{d\sigma}$ at P_1 and P_2 respectively. Suppose disturbances of unit amplitude and zero phase at P_1 and P_2 produce complex amplitudes f_1 and f_2 respectively at P. Then light passing from $d\sigma$ to P via P_1 and P_2 gives complex amplitudes $u_1 f_1\sqrt{d\sigma}$ and $u_2 f_2\sqrt{d\sigma}$, and since these are coherent the resultant complex amplitude at P due to $d\sigma$ is $(u_1 f_1 + u_2 f_2)\sqrt{d\sigma}$, and the intensity, δI_p, is given by

$$\delta I_p = (u_1 f_1 + u_2 f_2)(u_1^* f_1^* + u_2^* f_2^*)d\sigma.$$

Now, for any two complex quantities, z_1 and z_2, $z_1 z_2^* + z_1^* z_2 = 2R\{z_1 z_2^*\}$, where R denotes "real part of".

$$\therefore \ \delta I_p = |u_1|^2 d\sigma |f_1|^2 + |u_2|^2 d\sigma |f_2|^2 + 2R\{u_1 u_2^* d\sigma f_1 f_2^*\}.$$

Since different source elements are incoherent, the total intensity at P due to the extended source Σ is given by

$$I_p = \int\int_\Sigma \delta I_p$$

$$= (\int_\Sigma \int |u_1|^2 d\sigma)|f_1|^2 + (\int_\Sigma \int |u_2|^2 d\sigma)|f_2|^2 + 2R\{(\int_\Sigma \int u_1 u_2{}^* d\sigma) f_1 f_2{}^*\}$$

Now $\int\int_\Sigma |u_1|^2 d\sigma (=I_1)$ and $\int\int_\Sigma |u_2|^2 d\sigma (=I_2)$ are the total intensities emerging from P_1 and P_2. One puts

$$I_{12} = \int_\Sigma \int u_1 u_2{}^* d\sigma \quad . \quad . \quad . \quad . \quad . \quad (6\text{--}21)$$

where I_{12} is called the *mutual intensity* and one puts

$$\gamma_{12} = \frac{I_{12}}{\sqrt{I_1 I_2}} \quad . \quad . \quad . \quad . \quad . \quad . \quad (6\text{--}22)$$

where γ_{12} is called the *phase coherence factor* or the *complex degree of coherence*. One has

$$I_p = I_1 |f_1|^2 + I_2 |f_2|^2 + 2\sqrt{I_1 I_2} R\{\gamma_{12} f_1 f_2{}^*\}. \quad . \quad . \quad (6\text{--}23)$$

One denotes $|\gamma_{12}|$ by V_{12} and arg γ_{12} by β_{12}, i.e. $\gamma_{12} = V_{12} e^{i\beta_{12}}$. One can put $f_1 = g_1 e^{-i\delta_1}$ and $f_2 = g_2 e^{-i\delta_2}$, where δ_1 and δ_2 are the phase delays corresponding to the optical paths from P_1 to P and P_2 to P. Then $|f_1|^2 = g_1{}^2$, $|f_2|^2 = g_2{}^2$, and $f_1 f_2{}^* = g_1 g_2 e^{i(\delta_2 - \delta_1)}$. If $a_1{}^2$ and $a_2{}^2$ are the total intensities at P due to light that has travelled via P_1 and P_2 respectively, then $a_1{}^2 = g_1{}^2 I_1$ and $a_2{}^2 = g_2{}^2 I_2$, and one has

$$I_p = a_1{}^2 + a_2{}^2 + V_{12} \cdot 2a_1 a_2 \cos(\delta_2 - \delta_1 + \beta_{12}). \quad . \quad . \quad (6\text{--}24)$$

It will be seen that when $V_{12} = 0$ the result corresponds to complete incoherence between the disturbances arriving at P, and that when $V_{12} = 1$ and $\beta_{12} = 0$ the result corresponds to complete coherence. In general, V_{12} measures the extent to which the interference term is involved and is reasonably called the *degree of coherence*. It will be seen that when partially coherent disturbances are added there is, in effect, an additional phase difference β_{12} between them, although it simply becomes either zero or π in an axially symmetrical arrangement. The degree of coherence between points illuminated by an extended source is calculated in §11–12.

It follows from the discussion of §11–12 that to illuminate a wide area coherently, using a quasi-monochromatic non-laser source, it is necessary to employ a pinhole source, with the result that one cannot obtain a high intensity. One of the many important properties of lasers is that they enable this difficulty to be overcome.

6–6. The Coherence Properties of Lasers

The fundamental physics involved in the production of light by a laser will not be discussed in detail here; the discussion is concerned with the coherence properties of the light that is emitted, rather than with why it has these properties. Moreover, the discussion is purely in terms of classical theory. This cannot deal with all aspects of the properties of laser light; it cannot, in fact, deal fully and quantitatively with the statistical fluctuations in the output from a laser, i.e. it cannot, in fact, deal fully with the coherence properties. Thus the discussion given here really only gives a qualitative indication of the differences between thermal and laser sources. It is only in quantum mechanical terms that a full discussion is possible. A very brief indication of the mechanism underlying the operation of a laser is given in §20–5. In the past lasers have been of two main types: pulsed lasers employing solids, of which the ruby laser is the classic example, and continuously operating gas lasers, of which the helium-neon laser is probably the most common example. Until recently one could say that the solid state lasers gave vastly greater powers, whilst the gas lasers gave much greater coherence, especially longitudinal coherence. However, a few years ago a new class of continuous gas laser emerged, of which the outstanding example was the carbon dioxide laser. This produced a continuous beam of very high power (several kilowatts) and also maintained the coherence properties of the earlier gas lasers. One should mention that liquid lasers have also been produced and considerable effort is now being put into their development.

In a laser the light is generated in an optical resonant cavity resembling a Fabry-Perot etalon (see §9–2) and in such a cavity many standing wave patterns or *modes* are possible. A gas laser can be made to oscillate in a single axial mode, yielding accurately spherical diverging waves or a narrow pencil (of the order of 1 cm. diameter) of accurately plane waves with the power concentrated into an extremely narrow spectral line. Obviously one can employ a suitable lens system to convert the output into spherical waves with any desired spread or plane waves with any desired width. Although the phase is completely uniform across the wavefront, i.e. one has perfect lateral coherence, the amplitude is non-uniform; the intensity distribution across the wave follows an approximately Gaussian curve with the maximum at the centre. It follows from diffraction theory that any plane wavefront of limited width must have an angular spread, and it is interesting to note here that the non-uniform wavefront just described actually exhibits a smaller angular spread than a uniform wave of the same width; the system is said to be *self-apodizing* (see §11–15).

As far as the coherence properties of a laser are concerned, it will be seen that with the aid of a suitable lens system the excellent coherence across the beam enables one to illuminate a wide area coherently. This does not, in fact, constitute a breakthrough *in principle* since if one illuminates a very small pinhole by means of a classical thermal source, the emergent wave can exhibit comparable transverse spatial coherence and one can, for example, produce a highly collimated beam of plane waves by placing the pinhole at the focus of a good lens. Moreover, the total power output from many gas lasers is actually much smaller than that obtainable from a laboratory thermal source; the important difference is that with the laser the power is concentrated into a highly directional beam whereas with a thermal source the light is emitted in all directions from a spatially incoherent region and, since one has to interpose the pinhole, one utilises a very small fraction of the total power.

Since the laser can illuminate a large area coherently and with high intensity it is a very suitable source for many classical optical experiments—in particular, interference and diffraction experiments involving division of wavefronts, such as those described in Chapters VII, XI, XII, XIII.

So far, the discussion has been concerned only with the lateral coherence obtained with a laser. More significant in principle is the fact that the light from a laser, particularly a continuous gas laser, has a very long coherence time—the longitudinal coherence is very high. As shown in §6–9 this means that the power is concentrated into a very narrow spectral line. There is no way in which a comparable temporal coherence (spectral purity) can be obtained with a thermal source. Thus, whilst the spatial coherence of laser light enables one to perform many traditional experiments more conveniently, the temporal coherence enables one to perform experiments that were literally impossible with thermal sources. The longer wave trains (corresponding to a narrower spectral line) enable one to perform interference experiments involving much larger path differences (see, for example, §8–10). The long coherence length combined with the excellent lateral coherence obtainable with high intensity enables one to extend the technique of holography to three-dimensional object fields (see §14–19). These two applications are really extensions (although large ones!) of experiments that were carried out with thermal sources.

A phenomenon that has never been observed with thermal sources is interference of light from two independent sources; as explained in §6–4 it is not observed with thermal sources even when they emit the same spectral line. This is because the widths of the spectral lines emitted by thermal sources correspond to coherence times of the order 10^{-8} sec (and often much less) and it would require a detector that could respond in this time if the interference effects were to be detected. However, with a good laser one can obtain coherence times of the order 10^{-2} sec so that one might suppose that the phase difference between the disturbances from two lasers can remain constant for a time considerably greater than the response time of many detectors and that, in consequence, interference fringes are easily observable. However, this is an over-simplification of the problem because the disturbances from two independent lasers, although "highly monochromatic", usually differ somewhat in absolute frequency. It can be shown that one should employ a detector whose response time is appreciably shorter than $1/\delta v$ where δv is the difference between the two laser frequencies; in practice the desired response time can be less than 1 μsec. Interference fringes produced by two independent laser beams were observed in 1963 by Magyar and Mandel. In such interference experiments there are random phase changes during times longer than the coherence times, so that the fringes suffer random displacements. Hence fringes observed during different very short exposures are in different random positions. It is worth noting that the δv referred to above is the beat frequency between the two lasers. The explicit observation of beats between independent sources is now possible—a phenomenon well-known in acoustics but almost impossible to observe in optics before the advent of the laser.

An application envisaged for the high intensity, highly directional, highly coherent beams of light produced by lasers is for long distance communications. It is possible, at least in principle, to replace the traditional radio-frequency wave by the almost monochromatic light as the carrier wave; in fact this has already been done at the "demonstration" level. The advantage of using light as a carrier wave is that the very high frequency could provide a communication channel which would have a very high information capacity—if suitable methods of modulation could be developed a very wide signal band-width (about 10^{11} cycles) could be used which would enable a single optical carrier wave to carry as much information as all the radio-communication channels now in existence. This would be suitable for direct application in space communication, but in the earth's atmosphere fog, and even turbulence, would be ruinous, and one would have to use enclosed light paths.

Finally, one should mention that the light intensities produced by lasers are sufficiently high to make possible the observation of phenomena hitherto unobservable—they can be sufficiently high for non-linear optics to be an experimental as well as a theoretical subject. It is the high lateral coherence which makes the beam from a laser highly directional, and this beam can be focused to produce an extremely high intensity. For example, pulsed lasers have produced peak powers of about 10^9 watts in a beam whose cross section area is less than 1 cm^2. At the focus of a 1 cm focal length lens this would give more than 10^{15} watts cm^{-2}, corresponding to an electric field of more than 10^9 volts cm^{-1}! At such power levels the reaction between light and a material medium, or between two beams of light, give a range of phenomena too wide even to mention here. At a more moderate level there is obviously a wide variety of possible applications for the sheer concentration of power that can be obtained with a laser.

In what has gone before, the emphasis has been on the advantages that derive from the high degree of coherence obtained with a laser. But there are also disadvantages. There are many optical systems in which light is reflected, refracted, diffracted, or scattered from the principal light paths. With thermal sources this stray lightly simply gives a low background illumination which reduces slightly the contrast in the intensity variations that are of prime interest. However, with laser sources the spatial and temporal coherence is so high that virtually all the stray light in the system is coherent so that many unwanted interference patterns obtrude. For example, it is likely that if a ground glass screen is used as a diffuser in a traditional system, the use of a laser instead of a thermal source will lead to the unwanted appearance of a complicated interference/diffraction pattern unless the light is, in effect, made spatially incoherent by oscillating the diffuser rapidly—an unwelcome complication. A similar effect is observed if a surface—such as this page—is illuminated with laser light. The light scattered from neighbouring elements gives complicated interference effects and the page has a grainy, speckly, appearance which makes it difficult to read since the graininess varies as one moves one's eyes and changes the various relevant phase differences. It will be appreciated, then, that the use of a laser often requires a traditional optical system to be redesigned in order to obtain the advantages of the high coherence and avoid the disadvantages.

6–7. Fourier Series*

In its simplest and most commonly quoted form, Fourier's theorem states that any periodic function can be expressed as the summation of a series of simple harmonic terms having frequencies that are multiples of that of the given function. (Actually the function must not have an infinite number of discontinuities during one period.) One refers to the function as being *analysed* into its simple harmonic components; the function is the *synthesis* of those components. To give some idea of what is involved, one simple example will be quoted. It can be shown that the rectangular wave form illustrated in Fig. 6–4(*a*) (which extends from $x = -\infty$ to $x = +\infty$) can be analysed into a series of sine waves, and can be expressed as the following summation:

$$y = \frac{4a}{\pi}\left(\sin x + \frac{1}{3}\sin 3x + \frac{1}{5}\sin 5x + \ldots\right).$$

Each of the first three terms is represented by a dotted curve in the figure and the sum of these three terms can be seen by inspection to be represented by the full curve. Fig. 6–4(*b*) shows the resultant of the first fifteen terms. It can be

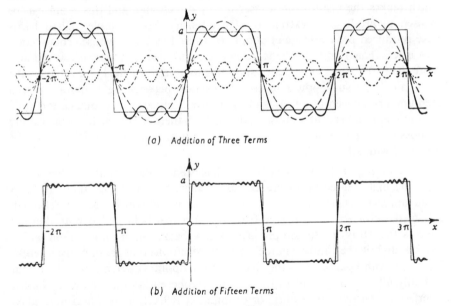

(*a*) Addition of Three Terms

(*b*) Addition of Fifteen Terms

Fig. 6–4. Analysis of a square wave.

seen that as more terms in the series are included in the summation, the resultant approaches the rectangular wave form more closely. In this example, the Fourier analysis of the function contains only sine terms which are initially in phase and which have frequencies which are odd multiples of that of the function represented. In general the series also contains cosine terms (which is equivalent to saying that it contains sine terms with various initial phases) and there are terms with every integral multiple of the original frequency. In general there is also a constant term.

* A rigorous discussion of Fourier series and Fourier Integrals is beyond the scope of this book. The present discussion merely serves to introduce the results that are required.

This way of analysing a periodic function into simple harmonic components can be extended to non-periodic functions in the following manner. Suppose that the non-periodic function $y=f(x)$ has the form represented by the full line in Fig. 6–5. Let $\phi(x)$ be the periodic function which is identical to $f(x)$ in the region $-\pi < x < +\pi$ but which repeats itself outside that region as indicated by the dotted curve in the figure. Then, according to Fourier's theorem as stated above, the function $\phi(x)$ can be analysed into a series of simple harmonic terms. It is obvious that within the range $-\pi < x < +\pi$ the series also represents $f(x)$. By proceeding in this way one can represent any non-periodic function over any desired range by an appropriate series of simple harmonic terms. Obviously

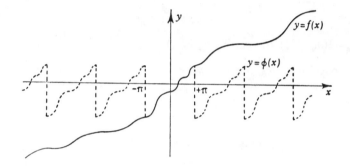

Fig. 6-5. Analysis of a non-periodic function over a limited range

these terms have frequencies which are multiples of that of $\phi(x)$. In particular one can say that in the range $-\pi$ to $+\pi$ the function $f(X)$ can be written

$$f(X)=a_0+\sum_{n=1}^{n=\infty} a_n \cos nX+\sum_{n=1}^{n=\infty} b_n \sin nX. \qquad . \ . \quad (6\text{–}25)$$

Now $\displaystyle\int_{-\pi}^{+\pi} \cos n_1 X \cos n_2 X dX=0$ if $n_1 \neq n_2$, and $=\pi$ if $n_1=n_2$. In addition,

$\displaystyle\int_{-\pi}^{+\pi} \cos n_1 X \sin n_2 X dX=0$ for all values of n_1 and n_2. Hence by integrating

each side of (6–25) between $-\pi$ and $+\pi$ one has

$$a_0=\frac{1}{2\pi}\int_{-\pi}^{+\pi} f(X)dX.$$

Similarly, by multiplying each side of (6–25) by $\cos nX$ and integrating, one has

$$a_n=\frac{1}{\pi}\int_{-\pi}^{+\pi} f(X) \cos nX \, dX$$

and, by multiplying each side of (6–25) by $\sin nX$ and integrating, one obtains

$$b_n = \frac{1}{\pi} \int_{-\pi}^{+\pi} f(X) \sin nX\, dX.$$

It is convenient to transform (6–25) into a summation of complex exponential terms. One has

$$\cos nX = \tfrac{1}{2}(e^{inX} + e^{-inX}) \qquad \text{and} \qquad \sin nX = \tfrac{i}{2}(e^{-inX} - e^{inX}).$$

$$\therefore \quad a_n \cos nX + b_n \sin nX = \frac{a_n}{2}e^{inX} + \frac{a_n}{2}e^{-inX} - \frac{ib_n}{2}e^{inX} + \frac{ib_n}{2}e^{-inX}$$

$$= c_n e^{inX} + c_{-n} e^{-inX},$$

where $c_n = \tfrac{1}{2}(a_n - ib_n)$ and $c_{-n} = \tfrac{1}{2}(a_n + ib_n)$.

Hence, putting $a_0 = c_0$, one can write

$$f(X) = \sum_{n=-\infty}^{n=+\infty} c_n e^{inX}, \qquad \cdots \cdots \quad (6\text{–}26)$$

where c_{-n} and c_n are conjugate complex quantities and

$$c_n = \frac{1}{2\pi} \int_{-\pi}^{+\pi} f(X) e^{-inX} dX. \qquad \cdots \cdots \quad (6\text{–}27)$$

6–8. Extension of the Range. The Fourier Integral

Equation (6–26) shows that over the range $-\pi < X < +\pi$ the non-periodic function $f(X)$ may be represented as the summation of an infinite series of simple harmonic terms whose amplitudes are given by (6–27). It will now be shown that the range can be extended as far as may be desired and that it can be made to extend from $-\infty$ to $+\infty$. In the latter case the function is expressed as a *continuous* series of simple harmonic terms whose frequencies have every possible value between $-\infty$ and $+\infty$.

To show that the range can be extended and that $f(x)$ can be represented over the range $-l < x < +l$, put $x = lX/\pi$. Then (6–26) and (6–27) give

$$f(x) = \sum_{n=-\infty}^{n=+\infty} c_n e^{\left(\frac{in\pi x}{l}\right)}, \qquad \cdots \cdots \quad (6\text{–}28)$$

where

$$c_n = \frac{1}{2l} \int_{-l}^{+l} f(x) e^{-\frac{in\pi x}{l}} dx. \qquad \cdots \cdots \quad (6\text{–}29)$$

If one introduces the dummy variable x' in place of x in (6–29) one can write

$$f(x) = \sum_{n=-\infty}^{n=+\infty} \frac{1}{2l} \int_{-l}^{+l} f(x') e^{\frac{in\pi}{l}(x-x')} dx' \qquad \cdots \quad (6\text{–}30)$$

and the summation on the right-hand side represents $f(x)$ in the range $-l < x < +l$. In (6–30) l may be made as large as one pleases and it can be shown that one can extend the range in this way to $\pm\infty$. Now the nth harmonic in the series has wavelength $\lambda_n = 2l/n$ and in terms of the wavelength constant $k = 2\pi/\lambda$ one has $k_n = n\pi/l$. Similarly $k_{n+1} = (n+1)\pi/l$. Hence $k_{n+1} - k_n = \pi/l = \Delta k$ (say). One may then write (6–30) in the form

$$f(x) = \frac{1}{2\pi} \sum_{n=-\infty}^{n=+\infty} \Delta k \int_{-l}^{+l} f(x') e^{i\{n\Delta k(x-x')\}} dx'. \qquad . \quad . \quad (6\text{–}31)$$

Now $n\Delta k$ is simply k_n and as $l \to \infty$ $\Delta k \to 0$. (6–31) then leads to the *Fourier Integral*:

$$f(x) = \frac{1}{2\pi} \int_{-\infty}^{+\infty} dk \int_{-\infty}^{+\infty} f(x') e^{ik(x-x')} dx'. \qquad . \quad . \quad . \quad (6\text{–}32)$$

Hence one can write

$$f(x) = \frac{1}{\sqrt{2\pi}} \int_{-\infty}^{+\infty} g(k) e^{ikx} dk, \qquad . \quad . \quad . \quad . \quad (6\text{–}33)$$

where

$$g(k) = \frac{1}{\sqrt{2\pi}} \int_{-\infty}^{+\infty} f(x) e^{-ikx} dx. \qquad . \quad . \quad . \quad . \quad (6\text{–}34)$$

This shows that the function $f(x)$ is represented by a continuous series of simple harmonic terms whose amplitudes are given by (6–34). When two functions f and g are related in this way, each is said to be the *Fourier Transform* of the other.

By writing (6–28) as $\displaystyle\sum_{n=-\infty}^{n=+\infty} c_{-n} e^{-\frac{in\pi x}{l}}$, one can reverse the signs in the exponents

of (6–33) and (6–34), but there must always be $+i$ in one and $-i$ in the other.

The factor $1/\sqrt{2\pi}$ can be replaced by $1/2\pi$ in (6–33) and omitted from (6–34). If one then puts $k' = k/2\pi$, $dk = 2\pi dk'$, and one has

$$f(x) = \int_{-\infty}^{+\infty} g(k') e^{2\pi ik'x} dk', \quad . \quad . \quad . \quad . \quad . \quad (6\text{–}35)$$

where

$$g(k') = \int_{-\infty}^{+\infty} f(x) e^{-2\pi ik'x} dx. \quad . \quad . \quad . \quad . \quad (6\text{–}36)$$

k' $(=1/\lambda)$ is now the number of cycles per unit of x. Thus, if one is thinking of a complicated motion as the superposition of simple harmonic motions, x is time, λ is period, and k' is frequency; $f(x)$ is the displacement as a function of time, and $g(k')$ is the amplitude of the simple harmonic component of frequency k'. If one is analysing a complicated variation of some quantity with distance, x is distance, λ is wavelength, and k' is known as *spatial frequency*.

Equation (6–35) can be written

$$f(x)=\int_0^\infty g(k')e^{2\pi ik'x}dk'+\int_0^\infty g(-k')e^{-2\pi ik'x}dk'$$

$$=\int_0^\infty [g(k')+g(-k')] \cos 2\pi k'xdk'+i\int_0^\infty [g(k')-g(-k')] \sin 2\pi k'xdk'.$$

Similarly, (6–36) gives

$$g(k')=\int_0^\infty [f(x)+f(-x)] \cos 2\pi k'xdx-i\int_0^\infty [f(x)-f(-x)] \sin 2\pi k'xdx,$$

and $g(-k')=\int_0^\infty [f(x)+f(-x)] \cos 2\pi k'xdx+i\int_0^\infty [f(x)-f(-x)] \sin 2\pi k'xdx.$

Suppose, now, that $f(x)$ is real and an even function, i.e. $f(-x)=f(x)$. Then $g(-k')=g(k')$ and one has

$$f(x)=2\int_0^\infty g(k') \cos 2\pi k'xdk' \quad . \ . \ . \ . \ . \quad (6\text{--}37)$$

and

$$g(k')=2\int_0^\infty f(x) \cos 2\pi k'xdx. \quad . \ . \ . \ . \ . \quad (6\text{--}38)$$

Each is referred to as the *Fourier cosine transform* of the other.

6–9. Wave Trains of Finite Length

The expression $\phi=a \cos k(ct-x)$ represents a simple harmonic disturbance of wavelength $2\pi/k$ travelling with speed c in the direction of x increasing. At any instant the wave extends from $x=-\infty$ to $x=+\infty$, and at any point whose position corresponds to a particular value of x the disturbance continues from $t=-\infty$ to $t=+\infty$. The wave can be represented by Fig. 6–6(a); at a given instant the disturbance in space is represented by having x as the abscissa, and the disturbance at a given position in space is represented by having t as the abscissa. The disturbance represented in (b) is *not* a simple harmonic wave since it follows the shape of a cosine curve over a certain range only, and is zero beyond that range. Strictly speaking, then, ϕ is a non-periodic function—it does not repeat itself to infinity. Consequently, it can be represented as the summation of a continuous distribution of simple harmonic terms by an expression of the form of (6–33) or (6–35), the relative amplitudes of the components being defined by (6–34) or (6–36). As an example of the use of these equations, it will now be shown that a sharply limited wave train of the form shown in Fig. 6–6(b) is the superposition of an infinite number of monochromatic waves whose relative amplitudes are given by a curve of the form shown in Fig. 6–6(c).

At the point $x=0$ the disturbance, as a function of time, can be written

$$\phi=f(t)=a \cos k_0ct=a \cos 2\pi v_0t \quad \text{for} \quad -\tfrac{1}{2}t_1\leqslant t\leqslant +\tfrac{1}{2}t_1$$

and $\phi=f(t)=0 \quad$ for $\quad |t|>\tfrac{1}{2}t_1$

where v_o is the frequency and t_1 is the coherence time as defined in §6–4. Since $f(t)$ is an even function of t one can, in fact, use (6–37) and (6–38). One has

$$f(t)=2\int_0^\infty g(v)\cos 2\pi vt dv \qquad . \quad . \quad . \quad . \quad (6\text{–}39)$$

where, from (6–38),

$$g(v)=2\int_0^\infty f(t)\cos 2\pi vt dt. \qquad . \quad . \quad . \quad . \quad (6\text{–}40)$$

That is to say, the disturbance is the superposition of a continuous spectrum of monochromatic disturbances, the amplitude of the component of frequency v being given by $g(v)$. Since $f(t)=0$ for $t>\frac{1}{2}t_1$, one has

$$g(v)=2\int_0^{t_1/2} a\cos 2\pi v_o t\cos 2\pi vt dt$$

$$=a\int_0^{t_1/2}\{\cos 2\pi(v-v_o)t+\cos 2\pi(v+v_o)t\}dt$$

$$=\frac{a\sin\pi(v-v_o)t_1}{2\pi(v-v_o)}+\frac{a\sin\pi(v+v_o)t_1}{2\pi(v+v_o)}. \qquad (6\text{–}41)$$

The first term gives a curve of the form shown in Fig. 6–6(c), and the second term gives a similar curve centred on $+v_o$. Now one need consider only positive frequencies, and if the coherence time t_1 is long (so that the wave train is long compared with the wavelength) the width of the central maximum is small enough to make the contribution of the second term negligible. The relative *intensities* of the simple harmonic (Fourier) components of a sharply limited wave train is given by the square of the first term, i.e. is of the form $\left(\dfrac{\sin\alpha}{\alpha}\right)^2$.

The function $\left(\dfrac{\sin\alpha}{\alpha}\right)$ is often referred to as the *sinc* function: written sinc α.

[Some writers use sinc α to denote $(\sin\pi\alpha)/\pi\alpha$.] A fairly accurate plot of this is given in Fig. 11–4. The important point to note is that most of the power is in the central maximum, which falls to zero when the frequency differs from v_o by $1/t_1$. That is to say, if a source were to emit sharply limited trains of waves the corresponding spectral line would have a profile (intensity distribution) whose full width is $2/t_1$ where t_1 is the coherence time; in other words, the longer the wave train the greater the spectral purity.

A source of light is unlikely to emit a train of waves that is sharply limited as discussed above; it is more likely to emit a train of waves whose amplitude increases gradually to a maximum and then decreases gradually as illustrated by Fig. 6–6(d). It can be shown that the spectral line profile (frequency distribution) is then of the form illustrated by Fig. 6–6(e). It is left as an exercise for the reader to verify that for a wave train which has a Gaussian envelope (i.e. is of the form $ae^{-t^2/b^2}\cos 2\pi v_o t$) the corresponding line profile also has the form of a Gaussian curve. (Hint: Although this is a symmetrical function it is probably easier to use equations (6–35) and (6–36). Write the cosine in terms of exponentials and manipulate the integrals until you can use the standard result $\int\limits_{-\infty}^{+\infty} e^{-mx^2}\,dx=\sqrt{\pi/m}$. One obtains, for the effective term, $g(v)=\frac{1}{2}ab\sqrt{\pi}\,e^{-\pi^2b^2(v-v_0)^2}$.) The reader should also verify that the Gaussian envelope falls to $1/e$ of its maximum value (a) when $t=\pm b$, and that the corresponding spectral line profile falls to $1/e$ of its maximum value ($\frac{1}{2}ab\sqrt{\pi}$) when the frequency differs from v_o by $\pm 1/\pi b$. Hence for a wave train with

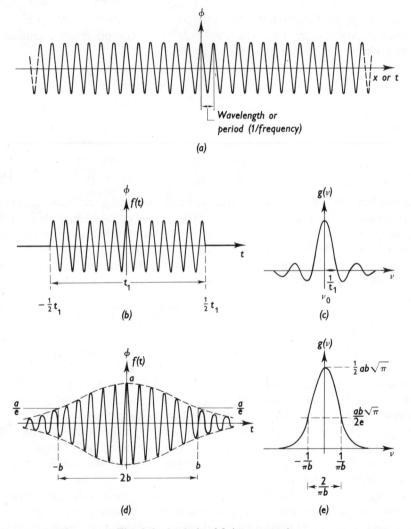

Fig. 6-6. Analysis of finite wave trains.

a Gaussian envelope one again has this important result: there is a reciprocal relationship between the length of the wave train and the width of the spectral line. Strictly speaking, both the envelope of the wave train and the intensity distribution curve tend to zero as t or $v\to\infty$ but one can define the length or width by taking the value of t or v when the curves fall to $1/e$ of their maxima. This convenient convention gives $2b$ for the coherence time and $2/\pi b$ for the spectral line width.

The results referred to above confirm quantitatively the qualitative discussion given in §6–4. In the context of experimental optics, the discussions show that if one is to obtain interference fringes between disturbances that have traversed widely different optical path lengths the wave groups must be long, i.e. long coherence lengths (or times) are required; in other words the light must be quasi-monochromatic with a small spectral line width. The line width (or, for more complicated cases, the line *structure*) of a nominally monochromatic (i.e. quasi-monochromatic) source can be deduced from observations of the way in which the visibility of the interference phenomena decrease as the optical path difference between the interfering disturbances is increased. (See §§8–10 and 8–11.)

With a source emitting what is known as a line spectrum one has, in crude terms, the emission of a number of discrete wavelengths; actually one has the emission of a number of sharp frequency distributions.

In a source emitting "white light" spectroscopic analysis shows that the energy is fairly evenly distributed over a wide range of wavelengths. From

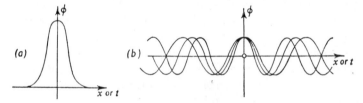

Fig. 6–7. A single pulse.

(6–33) it can be shown that the synthesis of such a distribution is a short *pulse* such as that represented in Fig. 6–7(a). In Fig. 6–6 this corresponds to the extreme case of the frequency distribution in (e) becoming very broad and the wave train (d) becoming very short. Fig. 6–7(b) illustrates how a continuous distribution of wavelengths gives a pulse. It is only at O that curves of every wavelength are in phase. As one moves away from this point the waves gradually get more and more out of phase to give destructive interference. With a number of discrete wavelengths the waves would begin to get in phase again later, but with a continuous distribution of wavelengths they never give constructive interference again. It can also be seen from this figure that when most of the energy is concentrated into a relatively small range of wavelengths one can move farther from O before one gets destructive interference and the resultant is a wave group of the form shown in Fig. 6–6. Fig. 6–7(b) can also be used to represent the way in which interference fringes for different wavelengths gradually get out of step away from the zero order fringe so that only the latter can be observed with white light. (See Chapter VII.)

In the past there have been many arguments on the question of whether white light "really consists" of a series of pulses or a mixture of disturbances of all wavelengths. It will be seen that the two are synonymous.

In many experiments it is found to be possible to obtain interference effects with white light by arranging that disturbances from the source are superposed after traversing almost equal optical path lengths. The necessity of having the paths almost equal can be explained by the fact that disturbances from the same pulse must overlap or by saying that the interference fringes for all wavelengths must coincide.

6–10. Group Velocity

In all the above discussions it has been assumed that simple harmonic waves are propagated with a speed c which is independent of the frequency—there is no dispersion. When this is so every component of a wave group travels at the same speed. This means that the whole group also travels at that speed and is completely unchanged during its passage through the medium. It can be shown that if there is dispersion the wave group constituting the disturbance of finite length travels at a speed which differs from that of any of the component simple harmonic waves. This phenomenon can be observed with the waves spreading across the surface when a stone is dropped into still water. A group of waves is observed to spread out across the water surface but, in the group, waves can be seen to be moving faster than the group itself. These waves appear at the rear of the group, travel through it, and die out at the front. It is found that for any type of wave motion the *group velocity* is less than the velocity of individual waves (the *wave velocity*) if the dispersion is such that the wave velocity increases with wavelength.

Suppose one has two superposed simple harmonic waves, with equal amplitudes and slightly different frequencies and velocities, propagated in the positive x-direction. The resultant is given by

$$\phi = a \cos \frac{2\pi}{\lambda}(ct-x) + a \cos \frac{2\pi}{\lambda+\delta\lambda}[(c+\delta c)t-x]$$

$$= 2a \left\{ \cos \pi \left[\left(\frac{c}{\lambda} + \frac{c+\delta c}{\lambda+\delta\lambda} \right) t - \left(\frac{1}{\lambda} + \frac{1}{\lambda+\delta\lambda} \right) x \right] \right.$$
$$\left. \times \cos \pi \left[\left(\frac{c}{\lambda} - \frac{c+\delta c}{\lambda+\delta\lambda} \right) t - \left(\frac{1}{\lambda} - \frac{1}{\lambda+\delta\lambda} \right) x \right] \right\}.$$

If $\delta\lambda \ll \lambda$ one can write λ^2 in place of $\lambda(\lambda+\delta\lambda)$ so that one has

$$\frac{1}{\lambda} - \frac{1}{\lambda+\delta\lambda} = \frac{\delta\lambda}{\lambda^2} \quad \text{and} \quad \frac{c}{\lambda} - \frac{c+\delta c}{\lambda+\delta\lambda} = \frac{c\delta\lambda - \lambda\delta c}{\lambda^2}$$

and

$$\phi = 2a \cos \frac{2\pi}{\lambda}(ct-x) \cos \frac{\pi\delta\lambda}{\lambda^2} \left(\frac{c\delta\lambda - \lambda\delta c}{\delta\lambda} t - x \right). \qquad . \quad (6\text{--}42)$$

At a given instant the variation of ϕ with x is of the form shown in Fig. 6–8. The wave shown as a full line corresponds to the first term in (6–42) and is modulated by the slowly varying envelope which corresponds to the second term.

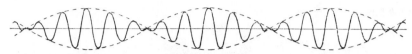

Fig. 6-8. Wave groups.

The first term defines a wave of the same speed and wavelength as the original waves but the envelope has wavelength $2\lambda^2/\delta\lambda$ and speed $(c\delta\lambda - \lambda\delta c)/\delta\lambda$. Thus the wave is divided up into groups which travel with the group velocity v given by

$$v = c - \lambda \frac{dc}{d\lambda}. \qquad \ldots \ldots \ldots \quad (6\text{-}43)$$

As the disturbance is propagated, the wave shown as a continuous line in Fig. 6-8 moves along with velocity c while the envelope moves with velocity v.

This example involves the superposition of only two waves. It has already been seen that if one has a continuous series of waves they form only one group and (6-43) shows that this has a unique group velocity only if the dispersion term $\lambda \, dc/d\lambda$ is constant. If the wave group is mainly confined to a narrow range of frequencies, the dispersion term is usually almost constant so that there is a unique group velocity. Equation (6-43) shows that the group velocity is smaller than the wave velocity if the latter increases with wavelength, is greater than the wave velocity if $dc/d\lambda$ is negative, and is equal to the wave velocity if there is no dispersion.

It is sometimes convenient to have alternative expressions for the group velocity. For example, if ω is the pulsatance and f is the frequency, one has

$$\omega = 2\pi f = \frac{2\pi c}{\lambda} = ck,$$

$$\therefore \quad \frac{d\omega}{dk} = c + k\frac{dc}{dk} = c - \lambda\frac{dc}{d\lambda},$$

i.e.

$$v = \frac{d\omega}{dk}. \qquad \ldots \ldots \ldots \quad (6\text{-}44)$$

Now $(d\omega/dk) = (d\omega/df)(df/d\lambda)(d\lambda/dk)$, where $d\omega/df = 2\pi$ and $dk/d\lambda = -2\pi/\lambda^2$.

$$\therefore \quad v = -\lambda^2 \frac{df}{d\lambda}. \qquad \ldots \ldots \ldots \quad (6\text{-}45)$$

If c_0 is the velocity of light in a vacuum the refractive index of the medium is defined by $n = c_0/c$. From (6-44) one has

$$\frac{1}{v} = \frac{dk}{d\omega} = \frac{d}{d\omega}\left(\frac{\omega}{c}\right) = \frac{1}{c} - \frac{\omega}{c^2}\frac{dc}{d\omega},$$

whence

$$\frac{1}{v} = \frac{1}{c} + \frac{\omega}{c_0}\frac{dn}{d\omega}. \qquad \ldots \ldots \ldots \quad (6\text{-}46)$$

In practice the dispersion of a material is often given in the form of a relation between the refractive index and the vacuum wavelength λ_0. (It should be noted that in (6-43) λ is the wavelength in the medium and is itself determined by the refractive index.) A manipulation of (6-46) yields a formula that is convenient in this situation.

One has

$$\frac{dn}{d\omega} = \frac{dn}{d\lambda_o} \frac{d\lambda_o}{d\omega}.$$

Now $\omega = k_o c_o$, or $\lambda_o = \dfrac{2\pi c_o}{\omega}$.

$$\therefore \quad \frac{d\lambda_o}{d\omega} = -\frac{2\pi c_o}{\omega^2} = -\frac{\lambda_o}{\omega}.$$

(6–46) then gives

$$\frac{1}{v} = \frac{1}{c} - \frac{\lambda_o}{c_o} \frac{dn}{d\lambda_o}. \quad \ldots \ldots \ldots \quad (6\text{–}47)$$

The use of this result is illustrated in the Miscellaneous examples—no. 1 in the collection on Chapters VI–IX.

6–11. Passage of a Pulse Through a Diffraction Grating

In this section (and in § 6–13) it is assumed that the reader is familiar with the elementary properties of diffraction gratings. These are discussed in detail in

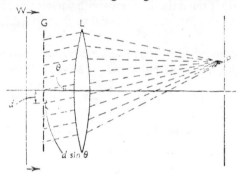

Fig. 6–9. Passage of a pulse through a grating.

Chapter XII where it is convenient to consider the incident light resolved into its simple harmonic components. It is convenient to discuss here the passage of a pulse through a grating. It has already been seen that a pulse can be analysed into waves whose frequencies are evenly distributed over a wide range and it is easy to see how a grating performs this analysis to give a continuous spectrum. In Fig. 6–9 a plane pulse W is incident on a grating G which consists of narrow slits. Thus a single pulse emerges from each slit and these pulses arrive at regular intervals at the point P in the focal plane of the lens L. There is thus a periodic disturbance at P. Referring only to the first order spectrum, the number of pulses arriving at P per second gives the frequency of the disturbance found at P, the wavelength being equal to the path difference from P to successive slits in the grating. This path difference is $d \sin \theta$, so that the wavelength of the disturbance found at P is given by $\lambda = d \sin \theta$, which agrees with the result obtained in § 12–1. The number of pulses involved gives the number of waves in the train at P. This is equal to the number of slits in the grating, i.e. to the resolving power of the grating in the first order (see § 12–6).

6–12. Passage of a Pulse Through a Prism

In Fig. 6–10 P is the centre of the spectral line for monochromatic light of wavelength λ. The optical path from A to A′ is then equal to that from B to B′. Thus disturbances travelling with the wave velocity throughout take equal times to travel from A to A′ and B to B′. If AB now represents a pulse entering the prism, it can be resolved into a broad spectrum of simple harmonic waves, but only those within a small range of wavelengths give an appreciable amplitude at P. Suppose that the group velocity for these waves in the prism is v. Now disturbances passing from A to A′ travel with the wave velocity throughout but

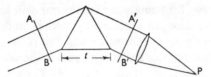

Fig. 6–10. Passage of a pulse through a prism.

the relevant disturbance from B to B′ travels at the relatively slow group velocity through a thickness t of the prism. This causes the time taken for the path t in the prism to be t/v instead of t/c. The time taken for the disturbance to travel from B to B′ therefore exceeds the time taken for the pulse to travel from A to A′ by an amount $(t/v - t/c)$ and the disturbance at P continues for this period of time. (The effect of dispersion in the lens is ignored.) Suppose that during this time N waves of frequency f pass. Then

$$N = \left(\frac{t}{v} - \frac{t}{c}\right) f.$$

Inserting the value of $1/v$ from (6–46) one has

$$N = \frac{tf\omega}{c_o} \frac{dn}{d\omega}$$

and if λ is now the wavelength in a vacuum one has

$$\omega = \frac{2\pi c_o}{\lambda} \quad \text{or} \quad \frac{d\omega}{d\lambda} = -\frac{2\pi c_o}{\lambda^2},$$

$$\therefore \quad N = -t\frac{dn}{d\lambda}. \quad \cdot \quad \cdot \quad \cdot \quad \cdot \quad \cdot \quad (6\text{–}48)$$

N is positive since, as has already been pointed out, $dn/d\lambda$ is negative when $v < c$. Again the number of waves in the wave train is equal to the resolving power. (See §11–5.)

6–13. Talbot's Bands

Suppose white light enters a prism spectroscope to form a continuous spectrum. In 1837 Talbot discovered that if a thin glass plate is placed across one half of the aperture of the prism as indicated in Fig. 6–11(a), a series of dark bands appears across the spectrum. It is found that the bands are most distinct for a certain thickness of the plate and that they cannot be observed at all if the plate has more than twice this thickness. It is also found that the same bands are formed whether the plate is placed before or after the prism or immediately in front of the observer's eye, provided that the disturbances that pass through the plate are those that pass through the thinner part of the prism. No bands are formed if the plate covers the other half of the prism aperture. Similar bands are observed in the continuous spectrum formed by a grating if the plate covers half the width of the grating as in Fig. 6–11(*b*), but again no bands are formed if the plate covers the other half of the aperture.

The formation of the bands can be explained quite simply by saying that the plate introduces a path difference between disturbances traversing the two halves of the aperture and that for some wavelengths one will have destructive interference. However, this does not explain why there is an optimum thickness for the plate nor why no bands are formed when the plate covers the other half of the aperture. These facts are easily explained if the white light which enters the prism or grating consists of a series of pulses. The arrangement involving a grating will be considered first. If there are N slits in the grating, each half of the

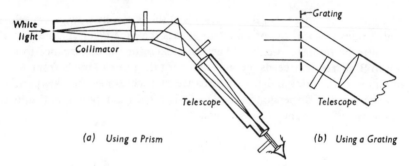

Fig. 6–11. Observation of Talbot's bands.

grating gives a wave train containing $N/2$ waves. It follows that two wave trains, each $N\lambda/2$ in length, pass through the appropriate point P in the focal plane of the telescope objective. If there is no plate in position, the wave train from the lower half of the grating arrives first at P and is immediately followed by the wave train from the upper half of the grating. Consequently, if the wave train from the lower half of the grating is retarded by the plate, it can overlap the other and the two can interfere. It is obvious that the wave trains cannot be made to overlap by retarding that from the upper half of the grating. This explains why the bands can be produced by interposing the plate across one half of the aperture but not the other. It is also obvious that the interference will be most marked when the thickness of the plate is just sufficient to make the wave trains overlap along the whole of their length, and that no bands will be produced if the retardation in the plate is increased until the wave trains are again prevented from overlapping. That is to say, the bands have maximum visibility when the plate produces a retardation of $N/2$ wavelengths and the visibility falls to zero if the retardation exceeds $N\lambda$.

A similar explanation is possible with the arrangement involving a prism. It was shown in §6–12 that a train of $-t\,dn/d\lambda$ waves arrives at a point in the spectrum and that the initial disturbance arrives from the apex of the prism. That is to say, the plate must be inserted as indicated by Fig. 6–11 in order to retard the first half of the wave train until it overlaps the second half. The optimum retardation is obviously $-\frac{1}{2}t\,dn/d\lambda$.

The formation of Talbot's bands has also been explained as follows. The plate and the space beside it can be looked upon as constituting an *echelon grating* of two elements. The properties of these gratings are discussed in Chapter XII and it is shown that one can observe only one principal maximum for those wavelengths for which the optical path difference introduced by the plate is an integral number of wavelengths. For other wavelengths two maxima are always contained beneath an envelope whose minima are only twice as far apart as the principal maxima. Suppose, then, that light is incident normally on an aperture of which half is occupied by a plate. Looking against the oncoming light, the

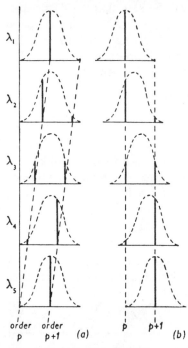

positions of the principal maxima for a number of slightly different wavelengths will be as indicated in Fig. 6–12(a) if $\lambda_1 > \lambda_2 > \lambda_3$, etc., and the plate covers the left-hand side of the aperture. If a prism is now inserted with its apex to the left, the light is deviated to the right and the deviations for $\lambda_1, \lambda_2, \lambda_3, \ldots$, are progressively larger. For a suitable combination of plate and prism the result is the superposition of the pth and $(p+1)$th orders as shown in Fig. 6–12(b). In the spectrum these give adjacent maxima between which there is a minimum. Throughout the banded spectrum, light with a narrow distribution of wavelengths collects in the maxima. If the prism had been reversed so that the plate covered the other half of its aperture, the principal maxima would not have been collected together in this way and the bands would not be observed. Since inadequate dispersion by the prism for a given plate thickness corresponds to excessive thickness of the plate for a given

order order (a)
p p+1

p p+1 (b)

Fig. 6–12. Talbot's bands with a prism.

prism, one can also explain how (for a given prism) it is possible for the plate to be too thick for the bands to be formed.

The envelopes shown in Fig. 6–12 represent the central maxima in the diffraction pattern for monochromatic disturbances when using half the aperture of the prism and no plate. If the full aperture is used the width of these maxima is halved and becomes equal to the separation of the Talbot's bands which were obtained with the plate in position across half the aperture. That is, the separation of the Talbot's bands which are obtained at a given point in the spectrum with a given spectroscope is equal to twice the distance from the centre of the maximum to the first minimum in the ordinary spectral line for one wavelength.

6–14. Stationary Waves

It was shown in §6–1 that when one superposes two simple harmonic waves of the same frequency travelling in the same direction the resultant is a similar wave. An interesting result is produced by superposing two waves of the same frequency travelling in opposite directions. The waves can be represented by

$$\phi_1 = a \sin k(ct-x),$$
$$\phi_2 = a \sin k(ct+x),$$

and the resultant is

$$\phi = 2a \cos kx \sin kct. \quad \ldots \ldots \quad (6\text{–}49)$$

This does not represent a progressive wave. That is to say, its profile does not move along; it is known as a *stationary* or *standing wave*. At all points along the x-direction the disturbance is simple harmonic with frequency $ck/2\pi$, the amplitude at any point being $2a \cos kx$. It will be seen that the disturbance has maximum amplitude at $kx = p\pi$ (p integral) but has zero amplitude at $kx = (p+\frac{1}{2})\pi$. The points where the disturbance is zero are called *nodes*, and those where the disturbance has maximum amplitude are called *antinodes*. The distance between adjacent nodes or adjacent antinodes is given by

$$k(x_2 - x_1) = p\pi - (p-1)\pi,$$

or

$$x_2 - x_1 = \frac{\lambda}{2}.$$

The occurrence of stationary waves on a finite stretched string is well known. The interference of light waves to give stationary waves has also been confirmed experimentally (see § 21–10). The phenomenon is more difficult to observe with light waves because the wavelength is very short.

6–15. The Doppler Effect

If there is relative motion between a source and a receiver, the frequency measured by an observer moving with the receiver differs from that measured by an observer moving with the source. If the source approaches the receiver the frequency observed by the receiver is higher than that for an observer at the source, and if the observer moves away from the source the apparent frequency is lower than that of the source. This explains the sudden fall in the apparent pitch of the whistle when a railway engine passes the observer. It will be shown below that the fractional change in frequency is given by the ratio of the velocity of the source relative to the observer, to the velocity of the waves. For light the velocity of the waves is very large and this ratio is usually very small.

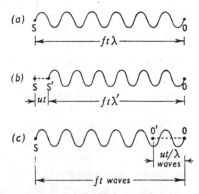

Fig. 6-13. The Doppler effect.

In Fig. 6–13(a) S is a stationary source of sound of frequency f and O is the stationary observer. Suppose that a disturbance passes from S to O in time t. Then ft waves are emitted during this time and occupy the space between S and O. If λ is the wavelength $SO = ft\lambda$. Suppose now that the source moves towards O with speed u. In time t the source moves a distance ut to S′ and ft waves are emitted. Since these waves now occupy the distance S′O the new wavelength is λ', where $ft\lambda' = S'O = SO - ut$,

i.e.
$$ft\lambda' = ft\lambda - ut,$$

$$\therefore \quad \lambda' = \lambda - \frac{u}{f}.$$

Now if f' is the frequency measured at O and c is the velocity of the waves through the air, one has $f'\lambda' = f\lambda = c$.

Hence
$$\lambda' = \lambda\left(1 - \frac{u}{c}\right), \qquad \ldots \ldots \quad (6\text{--}50)$$

or
$$f' = \frac{f}{1 - \dfrac{u}{c}}. \qquad \ldots \ldots \quad (6\text{--}51)$$

Suppose now that the source is stationary and that the observer moves towards the source with velocity u. During time t the observer moves from O a distance ut to O′ [Fig. 6–13(c)]. ft waves pass O during this time but the observer also receives the ut/λ waves which lie between O and O′. The total number of waves received is therefore $t(f + u/\lambda)$ and the apparent frequency is $f'' = f + u/\lambda$. Since $\lambda = c/f$ this gives

$$f'' = f\left(1 + \frac{u}{c}\right). \qquad \ldots \ldots \quad (6\text{--}52)$$

In the above discussion the waves are assumed to be propagated with definite velocity in a fixed medium with respect to which the motion of the source and the observer can be specified. According to the theory of relativity light waves are not propagated in a medium which can be said to be at absolute rest and with respect to which the absolute motion of a source or observer can be specified. The theory shows that the only movement which can be recognized is the *relative* motion of source and observer. It shows that there can be no distinction between the two cases discussed above and leads to the result

$$f' = f\sqrt{\frac{1 + \dfrac{v}{c}}{1 - \dfrac{v}{c}}}, \qquad \ldots \ldots \quad (6\text{--}53)$$

where c is the velocity of light and v is the relative velocity of source and observer towards one another.

The Doppler effect has been observed in light from the sun, the spectral lines from an approaching edge being of slightly shorter wavelength than those from a receding edge. In the laboratory the effect has been observed by using rotating mirrors to produce a virtual source in motion. Much more rapidly moving sources have been obtained in the laboratory by accelerating hydrogen ions and then neutralizing them without slowing them down appreciably. If the atoms are then excited and emit light, one can observe a Doppler shift in the wave-lengths of the spectral lines emitted. Owing to the fact that the atoms in the source are always in motion, the Doppler effect causes all spectral lines to cover a range of frequencies (see § 8–10).

One important application of the Doppler effect is in the discovery of double stars that are so close that they cannot be resolved with a telescope. If one has a bright star and a dark star rotating round one another the observed spectral lines exhibit a periodic shift about a mean position, moving to the shorter wave-lengths when the bright star is approaching the observer and to the longer wave-lengths when it is receding. If one has two stars of similar brightnesses rotating round one another the observed spectral lines become double and single periodically. They are double when the line joining the stars is perpendicular to the line of sight since one star is approaching and the other is receding. When one star is behind the other, they are both moving across the line of sight and the spectral lines are single.

6–16. Malus' Experiment. Malus' Law

In all the above discussions it has been assumed that the light disturbance at any point can be represented by a single scalar quantity. Malus' experiment is one of the simplest experiments which show the inadequacy of a scalar wave theory of light. The arrangement is shown in Fig. 6–14. P_1 and P_2 are glass plates on which light is incident at an angle of about 57°. (The significance of this particular angle of incidence will be seen later.) When the planes of the plates are parallel as indicated in the figure the light is reflected as shown. If, however, the plate P_2 is rotated about the ray 2 as axis, it is found that as the reflected ray 3 swings out of the plane of the figure its intensity decreases. The intensity of ray 3 falls to zero when P_2 is rotated 90°. The ray reappears as one continues to rotate P_2 and its intensity rises to a maxi-mum when the rotation of P_2 reaches 180°. Its intensity falls to zero again when the rotation reaches 270°,

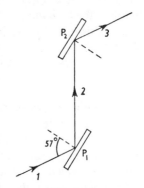

Fig. 6–14.
Malus' experiment.

and returns to its original maximum value at 360°. Thus the intensity of ray 3 is a maximum when the planes of incidence at P_1 and P_2 are parallel, and falls to zero when the planes of incidence are at right angles. If either of the angles of incidence is not 57° the intensity of ray 3 varies between maximum and mini-mum values but does not fall to zero. The reason for this will be seen later. One is led to the conclusion that under certain circumstances the intensity of a ray reflected from a glass plate varies as the incident ray is rotated relative to the plate about itself as axis. Obviously there is something unsymmetrical about such a ray. This behaviour cannot be explained in terms of a scalar wave theory. For example, the phenomenon is not observed with sound waves. In the case of sound the vibrations are known to be along the direction of propagation and can therefore be specified by a single scalar quantity. In order to explain the effect it is necessary to assume that light waves are transverse waves. For example, if the disturbance at any point consists of linear simple harmonic vibrations, these must be perpendicular to the direction of propagation or must have a component in that direction. To specify the disturbance completely, it is then necessary to specify the direction of the vibrations as well as the amplitude and frequency, i.e. one must use a vector. A beam of light obtained direct from the sun or from an ordinary filament lamp does not exhibit this non-symmetrical nature although it does show the effect after it has been reflected from a glass plate. Light showing this non-symmetrical property is said to be *polarized* and direct sunlight is *unpolarized*. In Malus' experiment the reflection from the plate P_1 causes ray 2 to be polarized and P_1 is referred to as the *polarizer*. The fact that the ray *is* polarized is shown by its behaviour at P_2. P_2 is called the *analyser*. The eye cannot differentiate between polarized and unpolarized light.

Malus' experiment can be explained if it is assumed that in the light reflected from P_1 the disturbances take the form of linear simple harmonic motions in a direction perpendicular to the direction of propagation. The fact that no effect is produced by changing the separation of P_1 and P_2 leads to the conclusion that these transverse vibrations remain in the same plane as the disturbance progresses. Such light is said to be *plane-polarized*. It can be assumed that when the angle of incidence is 57° the light reflected from a glass plate is plane polarized with the vibrations perpendicular to the plane of incidence, and that with this angle of incidence a plate cannot reflect light whose vibrations are *in* the plane of incidence. This accounts for the fact that ray 3 is absent when the planes of incidence at P_1 and P_2 are at right angles. An equally good assumption would be that for ray 2 the vibrations are parallel to the plane of incidence at P_1 and that there is no reflection if the vibrations are perpendicular to the plane of incidence. For the present purposes it does not matter which of these assumptions is made but the former will be adopted. The vibration direction then becomes the direction of the electric displacement in the electromagnetic theory, and it will be seen later that this is an advantage. One can therefore define the *plane of vibration* of plane polarized light as the plane of incidence at a glass plate at which, for an appropriate angle of incidence, there is no reflection. In the past, the term *plane of polarization* has been used for the plane of vibration as defined here or, more often, for the plane at right angles to it. The term will not be used in this book.

In Malus' experiment (still using angles of incidence of 57°) it is found that as P_2 is rotated about ray 2 as axis the intensity of ray 3 is proportional to $\cos^2 \theta$, where θ is the angle between the planes of incidence at P_1 and P_2. This is known as *Malus' law* and can be explained in the following way. It is assumed that the vibrations in ray 2 can be resolved into components in and perpendicular to the plane of incidence at P_2 and that the light corresponding to the component in the plane of incidence cannot be reflected from P_2 when the angle of incidence is 57°. On the other hand, light corresponding to the component perpendicular to the plane of incidence is reflected with a reflection factor which is constant for that particular angle of incidence. Now the vibrations in ray 2 are perpendicular to the plane of incidence at P_1 so that when the angle between the planes of incidence at P_1 and P_2 is θ a vibration of amplitude a in ray 2 has component $a \cos \theta$ perpendicular to the plane of incidence at P_2. This component can be reflected and ray 3 has amplitude $ra \cos \theta$, where r is a reflection factor. This means that as P_2 is rotated the intensity of ray 3 is proportional to $\cos^2 \theta$.

6–17. Brewster's Law

If direct sunlight or light from an ordinary filament lamp falls on a glass plate, it is found that the intensity of the reflected beam does not vary as the plate is rotated about an incident ray as axis—even if the angle of incidence is 57°. This implies that (at least when averaged over a finite period of time) the vibrations in the incident light have equal components in all directions and can be resolved into components of equal amplitude in any two mutually perpendicular directions. The light is unpolarized. (There is an alternative possibility as will be seen later.) A mixture of unpolarized light and plane polarized light is referred to as *partially plane polarized*. If it falls on a glass plate at 57°, the intensity of the reflected beam varies between a maximum and a minimum as the mirror is rotated about an incident ray as axis (instead of between a maximum and zero as it is does when the incident light is plane polarized).

It is found that when unpolarized light is incident on the surface of any transparent material, the reflected and refracted beams are partially plane polarized. For a particular angle of incidence known as the *polarizing angle*, the reflected beam becomes completely plane polarized with the plane of vibration perpendicular to the plane of incidence. (It will be seen in Chapter XXI that there are, in practice,

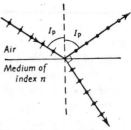

Fig. 6–15. Brewster's law.

small deviations from this condition, but they can be ignored in the present discussions.) Brewster discovered that when light is incident at the polarizing angle the plane polarized reflected beam is at right angles to the partially plane polarized refracted beam. Fig. 6–15 shows light incident at the polarizing angle I_p. The fact that in the reflected light the vibrations are perpendicular to the plane of incidence is denoted by the dots. The lines and dots on the incident and refracted rays denote that in these rays the vibrations have components both in and perpendicular to the plane of incidence. Snell's law of refraction gives

$$\frac{\sin I_p}{\sin I'_p} = n$$

and since the reflected and refracted rays are at right angles,

$$I'_p = \frac{\pi}{2} - I_p,$$

i.e. $$\tan I_p = n. \quad . \quad . \quad . \quad . \quad . \quad . \quad . \quad (6\text{–}54)$$

This is known as *Brewster's Law*. It is now possible to see the origin of the mysterious 57° which has been referred to frequently when dealing with glass plates. The refractive index of hard crown glass is about 1·52 and since $\tan^{-1}$ 1·52 is 56·7° it will be seen that 57° is an approximate value for the polarizing angle. When the refractive index is increased to 1·7 the polarizing angle increases to about 59·5°; i.e. it is not widely different for different glasses.

The polarization resulting from reflection and refraction can be explained by assuming that the reflection coefficient for light whose vibrations are in the plane of incidence is different from that for light whose vibrations are perpendicular to the plane of incidence and that for the former it falls to zero at the polarizing angle. This is confirmed experimentally. Unpolarized incident light can then be resolved into two components: one with vibrations in, and the other with them perpendicular to, the plane of incidence, and these are reflected and refracted in the same way as independent plane polarized beams. Thus the reflected light is plane polarized when the angle of incidence is equal to the polarizing angle, and partially plane polarized for other angles of incidence. If one has plane polarized incident light and the plane of vibration is inclined to the plane of incidence, it can be resolved into components in the same way. One can see at once that if the angle of incidence at either of the plates in Fig. 6–14 is not equal to the polarizing angle, the intensity of ray 3 will vary between a maximum and a minimum but will never fall to zero.

Now when unpolarized light is incident on a glass plate only half the energy is represented as having its vibrations perpendicular to the plane of incidence, and only about 15% of this is reflected at the polarizing angle. On the other hand, *all* the light whose vibrations are in the plane of incidence is refracted. If the light is passed through a series of parallel plates, some of the light with vibrations perpendicular to the plane of incidence is removed at each reflection and, apart from absorption losses, all the light polarized at right angles is transmitted. Thus the larger is the number of plates the more nearly is the transmitted light plane polarized, and very little of this light is lost. In a more detailed study it is, of course, necessary to take into account the multiple reflections between the various surfaces.

6–18. Superposition of Beams of Polarized Light

It is shown in Chapter XXII that if light is passed through a calcite crystal there are, in general, two emergent rays. Both these rays are plane polarized and their planes of vibration are at right angles. If the incident light is unpolarized the two emergent rays have equal intensities, and if the incident light is plane polarized the relative intensities of the emergent beams depends on the orientation of the plane of vibration of the incident light with respect to the crystal structure. These two plane polarized beams can be superposed after being made to traverse different optical paths. It is possible to rotate their planes of vibration before superposing them but the way in which this is done need not be discussed here. (See Chapters XXII and XXIII.) In this way it is possible to study the effects of superposing plane polarized disturbances under various conditions. The conditions under which it is possible to obtain interference fringes with the polarized disturbances are summarized in the *Fresnel-Arago laws*:

(1) Two plane polarized beams whose vibration directions are at right angles

cannot give interference fringes, i.e. they cannot give destructive interference.

(2) Two plane polarized beams whose vibration directions are the same interfere under the same conditions as two similar beams of unpolarized light, provided
they are originally derived from the same beam of plane polarized light or from
the same plane polarized component of unpolarized light.

(3) Two plane polarized beams which are derived from perpendicularly
polarized components of unpolarized light never produce interference fringes,
even if their planes of vibration have been rotated until they are the same.

Results 1 and 2 are what one might expect, since the two transverse vibrations
are obviously unable to interfere destructively when they are at right angles, and
are able to do so when they are in the same straight line unless something has
occurred to upset their coherence. For example, if the two beams had traversed
widely different optical paths, they would in any case be unable to interfere. The
third law given by Fresnel and Arago indicates that the perpendicularly polarized
components into which unpolarized light can be resolved must always be
regarded as being incoherent. This result will be discussed again later.

The first of the Fresnel-Arago laws does not imply that when one superposes
two perpendicularly polarized beams which are derived from the same plane
polarized beam the disturbances are not combined in any way; it simply says
that interference fringes cannot be produced. Consider now the way in which
the disturbances do combine. If the two beams are produced from a single plane
polarized beam by means of a calcite crystal as described above they will not, in
general, have the same amplitude. Let their amplitudes be denoted by a_x and
a_y. Suppose that, before being superposed, the beams traverse unequal optical
paths with the result that there is a phase difference δ between their vibrations.
The problem, then, is to find the resultant of two simple harmonic vibrations of
the same frequency at right angles. The vibrations can be assumed to occur in
the xy plane and if the vibrations in the y-direction lag behind those in the
x-direction with a phase difference δ, they can be represented by

$$\left. \begin{array}{l} x = a_x \sin \omega t \\ y = a_y \sin(\omega t - \delta) \end{array} \right\} \quad . \quad . \quad . \quad . \quad . \quad (6\text{--}55)$$

If these are components of the motion of a particle, the path of the resultant
motion is given by eliminating t from this pair of simultaneous equations. One
has

$$y = a_y \sin \omega t \cos \delta - a_y \cos \omega t \sin \delta$$

$$= a_y \frac{x}{a_x} \cos \delta - a_y \left(\sqrt{1 - \frac{x^2}{a_x^2}} \right) \sin \delta$$

or

$$\left(y - \frac{a_y}{a_x} x \cos \delta \right)^2 = a_y^2 \left(1 - \frac{x^2}{a_x^2} \right) \sin^2 \delta$$

which reduces to

$$\frac{x^2}{a_x{}^2} + \frac{y^2}{a_y{}^2} - \frac{2xy\cos\delta}{a_x a_y} = \sin^2\delta. \quad \ldots \quad (6\text{–}56)$$

The curve represented by this equation is, in general, an ellipse. It is plotted in Fig. 6–16 for various values of δ. In general, the angle θ between the coordinate axes and the axes of the ellipse is given by

$$\tan 2\theta = \frac{2a_x a_y \cos\delta}{a_y{}^2 - a_x{}^2}. \quad \ldots \quad \ldots \quad (6\text{–}57)$$

Fig. 6–16 can be said to represent the light disturbance in a beam of light that is approaching the observer (and is travelling in the positive z-direction for a right-handed co-ordinate system). The axes of the ellipse become parallel to the x and y axes when $\delta = \pi/2$, $3\pi/2$, $5\pi/2$, ..., and the curve reduces to a straight line when $\delta = 0$, π, 2π, ... By inspection of the original equation it is easy to see that the direction in which the particle traverses each path is as indicated.

Thus when $\delta = 0$, π, 2π, ..., the two superposed plane polarized beams give a plane polarized resultant. For other values of δ the resultant is an elliptic motion and one is said to have *elliptically polarized light*. If $a_x = a_y$ the resultant reduces to a circle when $\delta = \pi/2$, $3\pi/2$, ..., and one has *circularly polarized light*. It should be noted that although both circularly polarized and unpolarized light can be resolved into components of equal amplitude in any two mutually perpendicular directions, the components of the former are coherent and those of the latter must be considered to be incoherent. In the old elastic solid theory

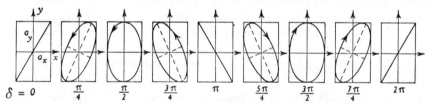

Fig. 6–16. The combination of two S.H.M.'s of the same frequency at right angles (y lags).

the disturbances in a light wave were thought of as the motion of a particle. This executed linear S.H.M. when the light was plane polarized, or traversed an ellipse or a circle for elliptically and circularly polarized light. In terms of the electromagnetic theory, the light disturbance consists of rapidly varying electric and magnetic fields. For example, in plane polarized light the electric displacement at a given point is always along the same straight line but its magnitude varies sinusoidally. That is to say, the tip of the arrow representing the electric displacement vector executes linear S.H.M. For elliptically and circularly polarized light the end-point of the vector traverses an ellipse and a circle respectively. Traditionally, the term *plane polarized* has been used in optics, but since one uses the terms *circularly* and *elliptically polarized*, one must concede that there is a good argument for referring to *linearly polarized* as is customary in microwave Physics.

The states of polarization shown in Fig. 6–16 are produced by the superposition of two components at right angles, and since there is no destructive interference each diagram represents light of the same intensity. Thus the total intensity is always equal to the sum of the intensities of the components, i.e. $a_x^2 + a_y^2$. It is in any case fairly obvious that this gives the intensity of the plane polarized resultant since it is simply the square of the amplitude. For elliptically polarized light the intensity is given by the sum of the squares of the semi-axes, and for circularly polarized light it is given by twice the square of the radius.

6.19. Polarized, Unpolarized, and Partially Polarized Light

It has already been seen that the vibrations in unpolarized light can be resolved into equal transverse linear components in any two mutually perpendicular directions, and that these components must be considered to be incoherent. Now when two disturbances are incoherent the phase difference between them is constantly changing in a random fashion. At any given instant the two components of unpolarized light correspond to a particular elliptical polarization. That is, unpolarized light can be regarded as elliptically polarized light in which the shape and orientation of the ellipse is constantly changing, circularly and plane polarized light being included as special cases. No matter what phase difference is introduced between the components of unpolarized light after they have been separated, the resultant will always be unpolarized light if the components are recombined.

It has already been shown that strictly monochromatic light is represented by an infinite train of waves. In general, the disturbance is represented by the resultant of two vectors in perpendicular planes and the variation of each of these vectors must be represented by an expression of simple harmonic form. Since each of these expressions must be valid from $-\infty$ to $+\infty$, there is a systematic phase relation between them. That is to say, a perfectly monochromatic beam must be completely polarized. Obviously the state of polarization varies as the phase difference between the components varies, but the polarization is complete.

It will be seen that one has the two extreme cases: unpolarized light, which can be resolved into incoherent components whose amplitudes are equal, and completely polarized light, which can be resolved into coherent components whose relative amplitudes and phases are determined by the state of polarization. Between these two extremes one has the general case of partially polarized light, which can be looked upon as having partially coherent components; alternatively, partially polarized light can be regarded as a mixture of polarized light and unpolarized light. The subject of partially polarized light is dealt with more fully in §§ 22–21 and 22–22.

CHAPTER VII

INTERFERENCE OF TWO BEAMS.
DIVISION OF WAVEFRONT

7-1. Classification of Interference and Diffraction Phenomena

As stated in Chapter I, each point on a wavefront may be regarded as a source of secondary waves. To account for the rectilinear propagation of light through large apertures it was assumed that these secondary waves produce an effect only on their forward envelope. However, careful investigation shows that the passage of light past an obstacle is not completely in accordance with the idea of rectilinear propagation: it is found that some light always appears within the geometrical shadow. The fact that the edge of a shadow is not sharp was first observed by Grimaldi (c. 1660). This complex phenomenon is known as diffraction. In order to explain the observed intensity patterns it is necessary to assume that the secondary waves interfere with one another according to the principle of superposition. This subject will be discussed in detail later (Chapter X), but it will be seen that every optical problem involving apertures is, strictly speaking, a problem in interference and diffraction.

One frequently has to deal with arrangements in which light from a source is separated into two or more beams by some suitable optical system, the separate beams being superposed subsequently. If the apertures of the system are sufficiently large, the amount of light involved in the departure from rectilinear propagation through the apertures is small, and the primary phenomenon is the subsequent interference of the separate beams of light. Thus in the majority of what are classed as interference phenomena, the behaviour of the light before the superposition of the beams is roughly in accordance with the principles of geometrical optics. The chief exception in this method of grouping is the case where two narrow slits are employed. These are simply regarded as sources of light and the diffraction mechanism is ignored.

The light may be divided into separate beams or waves by employing apertures and other optical components to separate neighbouring parts of the wavefront (division of wavefront) or by employing partial reflection to divide a single beam (division of amplitude). In the present chapter arrangements involving division of wavefront will be discussed. They always require the use of a point or line source.

The Rayleigh Refractometer and the Michelson Stellar Interferometer are sometimes included in discussions of two beam interference by division of wavefront but these are more conveniently discussed in Chapter XI. In the same way a diffraction grating could be regarded as a case of multiple division of the wavefront, especially when the apertures are narrow; in this case the chief phenomenon is associated with the interference between disturbances from different apertures. Diffraction gratings are discussed in Chapter XII.

7-2. Young's Experiment

One of the earliest experiments involving interference was that described by Thomas Young in 1807. Sunlight from a single pinhole P (Fig. 7-1) was allowed to fall on two pinholes P_1P_2. According to Huygens' principle P acts as a point source so that a spherical wave emerges and falls on P_1 and P_2. These two pinholes act as point sources emitting fresh spherical waves which spread out from P_1 and P_2 as indicated. In the space beyond P_1 and P_2 these secondary waves become superposed. Since P_1 and P_2 are illuminated by the same spherical wave, the superposed disturbances are coherent and the intensity at any point will depend on their relative phases. For example, assuming P_1 and P_2 are equidistant from P, the intensity will be a maximum for light of wavelength λ at any point X whose distances from P_1 and P_2 differ by $p\lambda$, where p is an integer. For a given value of p the locus of the point X is an hyperboloid of revolution having P_1 and P_2 as foci. (An hyperboloid may be defined as the surface generated by

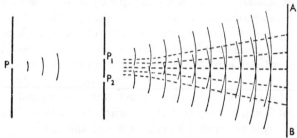

Fig. 7–1. Young's experiment.

a point which moves in such a way that the difference between its distances from two fixed points is a constant.) Other hyperboloidal surfaces of maximum intensity correspond to other values of p and similar surfaces of zero intensity will be generated by points which move in such a way that their distances from P_1 and P_2 differ by an odd multiple of $\lambda/2$. Since light passes through the pinholes in only one direction the complete pattern in space cannot be observed. Fig. 7–1 shows the intersection of the surfaces of maximum intensity with the plane PP_1P_2. As one recedes from P_1P_2 these hyperbolae quickly approximate to straight lines, and over the region near the middle of the screen AB the variations in intensity give a set of approximately linear fringes.

In the original experiments sunlight was used, so that the fringe systems for many wavelengths were superposed. In consequence, fringes could be observed near the centre of the field only (see below).

The amount of light that emerges from the pinhole P is very small and the fringes are rather faint and difficult to observe on a screen. However, they can be observed quite easily by looking at a small (preferably monochromatic) source and holding a double pinhole in front of the eye.

7–3. Elementary Discussion of Young's Experiment Using Slits

Fringes are more easily observed if monochromatic light is used and the pinholes replaced by parallel slits. The disturbances from the slits now have cylindrical envelopes and there will be symmetry about any plane perpendicular to the three slits. Consequently, it is usual to restrict the analysis to one of these planes. The interference pattern will consist of straight fringes parallel to the slits. The use of slit sources and apertures in diffraction experiments is discussed further in §§ 11–4, 13–6.

In Fig. 7–2 it will be assumed that the slits S_1 and S_2 are equidistant from S so that S_1 and S_2 act as coherent sources in the same phase; if they are of the same width, they will emit disturbances of equal amplitudes. These disturbances are superposed at P and since S_1P and S_2P are not widely different each disturbance may be assumed to have the same amplitude a at P. The path difference (S_2P-S_1P) causes the disturbances from S_1 and S_2 to be out of phase at P by an amount $\delta=(2\pi/\lambda)(S_2P-S_1P)=(2\pi/\lambda)(S_2A)$, where λ is the wavelength of the light concerned. Thus, one has, at P, two superposed disturbances of the same

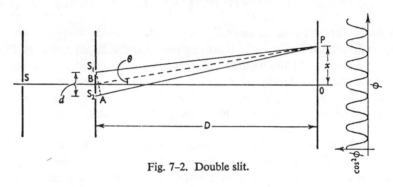

Fig. 7–2. Double slit.

frequency differing in phase by an amount δ. From equation (6–11) it will be seen that the resultant intensity at P is given by

$$I=2a^2(1+\cos\delta). \qquad \ldots \ldots \quad (7\text{–}1)$$

The intensity at P is therefore a maximum if $\cos\delta=+1$ and zero if $\cos\delta=-1$. That is, for a bright fringe $\delta=0$, 2π, 4π, 6π, ..., and for a dark fringe $\delta=\pi$, 3π, 5π, ...

From the figure, if $D\gg d$ or x, triangles BOP, S_1S_2A are similar so that

$$S_2A=\frac{dx}{D}.$$

Therefore for a bright fringe

$$\delta=\frac{2\pi}{\lambda}\cdot\frac{dx}{D}=0,\ 2\pi,\ 4\pi,\ 6\pi,\ \ldots=2p\pi,$$

and for a dark fringe

$$\delta=\frac{2\pi}{\lambda}\cdot\frac{dx}{D}=\pi,\ 3\pi,\ 5\pi,\ \ldots=(2p+1)\pi,$$

where p is an integer.

That is, for a bright fringe

$$x=p\frac{\lambda D}{d}, \qquad \ldots \ldots \quad (7\text{–}2)$$

and for a dark fringe

$$x=(p+\tfrac{1}{2})\frac{\lambda D}{d}, \qquad \ldots \ldots \quad (7\text{–}3)$$

p is called the *order* of the fringe. For example, the bright fringe which is at distance $x=2\lambda D/d$ from the axis is the second order bright fringe. As defined here p is an integer but, in general, the phase difference as a multiple of 2π is called the *order of interference* and is not necessarily integral.

When, as assumed above, $x\ll D$, the distance between successive maxima and minima is $\lambda D/d$ and is constant across the screen. Consequently, by measuring the fringe spacing, one can find λ. The intensity at any point is given [from (7–1)]

by
$$I=4a^2 \cos^2\left(\frac{\delta}{2}\right)$$
$$=4a^2 \cos^2\left(\frac{\pi d}{\lambda D}\cdot x\right). \qquad \ldots \ldots (7\text{–}4)$$

Such fringes are referred to as $\cos^2$ fringes. The intensity distribution in the fringes is illustrated in Fig. 7–2. It will be seen that the bright and dark fringes have equal widths. At the minima the intensity is zero and at the maxima it is $4a^2$ but the average intensity over the fringe system is $2a^2$, which is the sum of the intensities due to the disturbances from S_1 and S_2 considered separately. Hence, as always in interference phenomena, the principle of conservation of energy is not violated although a redistribution of the energy occurs.

If light of several wavelengths is present, each wavelength produces its own fringe system. For all wavelengths the central fringe is at O which is optically equidistant from S_1 and S_2. The next bright fringe is formed where the disturbances from S_1 and S_2 are 2π out of phase, i.e. where the path difference is λ. It will therefore be formed a distance x from O where x depends on λ [see equation (7–2)]. Thus for white light the central fringe is white whilst higher order maxima at first show coloured edges and then appear drawn out into spectra. Eventually, the maxima for various orders and different wavelengths overlap to give a white continuum where no fringes are observable.

If the primary slit S is wide each element of it gives fringes in different positions, and if S_1 and S_2 are wide corresponding elements act as separate slits and also give overlapping fringe systems. Thus to observe the above fringes clearly it is important to employ narrow slits. (See also §11–12.)

The fringes discussed above are formed by the superposition of two coherent cylindrical wave envelopes diverging from line sources a short distance apart. The same conditions may be produced in several ways by dividing a single expanding wavefront into two parts and causing the parts to become superposed.

7–4. Fresnel's Biprism

Although its function is that of a pair of thin prisms whose bases are cemented together, the biprism usually consists of a single isosceles prism whose apex A is just less than 180°. Light from a slit source is refracted through the prism as shown in Fig. 7–3 so that cylindrical wave envelopes diverge from S_1 and S_2. A convenient source is a slit illuminated by light from a gas discharge lamp. The waves become superposed in the region beyond the biprism to give fringes of the same type as in the modified form of Young's experiment. The virtual slit images S_1 and S_2 replace the two slits used in that experiment but the interference effects are the same. The effects are, of course, present only in the region XY where the waves overlap. Examination of the fringes shows an effect which is absent from the simple fringe system obtained in Young's experiment. Super-posed.on the main interference fringes there is a more gradual variation of intensity as shown in Plate I(a). The two cylindrical wave envelopes are of finite extent in this experiment, being limited on one side by the extreme edges of the biprism and on the other by the vertex A. The intensity variation that is super-posed on the interference pattern is caused by the limitation of the divergent waves by the vertex A and is characteristic of diffraction by a straight edge [see Chapter XIII and Plate III(g)]. If the biprism is distance D_1 from the slit whose

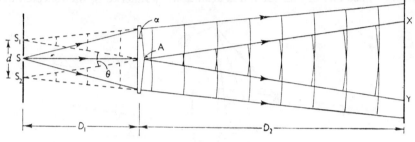

Fig. 7–3. Fresnel's biprism.

virtual images S_1S_2 are distance d apart, the fringe separation at distance D_2 from the biprism is given by

$$x = \frac{\lambda(D_1+D_2)}{d},$$

and, by measuring x, λ may be determined if d can be found. This can be done in several ways. If α is the refracting angle of each thin prism which effectively forms half of the biprism, and n is the refractive index, equation (5–8) gives

$$d = D_1\theta$$
$$= D_1 2(n-1)\alpha,$$

and n and α can be measured with a spectrometer.

Alternatively, the angle θ can be measured directly with a spectrometer by passing parallel light from the collimator through the biprism and measuring the angle θ between the two emergent pencils by means of the telescope. Another method, usually more convenient in practice, is to employ a convex lens to form two real images of the slit (i.e. images of S_1 and S_2) on a screen. The distance from S to the screen must be greater than four times the focal length of the lens. For a given position of the screen there are two possible positions for the lens. If the separations of the slit images for the two lens positions are d_1 and d_2, it can be shown that $d = \sqrt{d_1 d_2}$. (See §1–16.)

To observe the fringes and measure x, d_1, and d_2, it is usual to replace the screen by a micrometer eyepiece. It must be remembered that d_1 and d_2 must be measured with the same value of (D_1+D_2) and with the same value of D_1 as is used when x is measured. In performing the experiment it is necessary to ensure that the edge A of the biprism is accurately parallel to the slit S. A single point on S will give a complete set of approximately linear fringes parallel to the edge

A of the prism. If S is not parallel to A the fringes for a neighbouring point on S are slightly displaced laterally. Since the maxima of these two fringe systems do not coincide, the contrast in the resultant pattern is reduced. The fringes corresponding to other points on S reduce the contrast further and, unless the angle between A and S is very small, no fringes can be seen with the complete slit source.

7–5. Fresnel's Double Mirror

If light from a slit source S (Fig. 7–4) is reflected from two slightly inclined mirrors whose line of intersection is accurately parallel to S, the reflected light

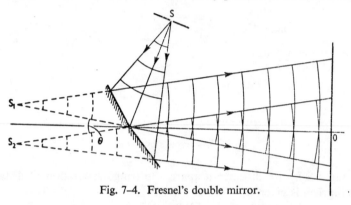

Fig. 7–4. Fresnel's double mirror.

consists of two cylindrical wave envelopes diverging from S_1 and S_2, the images of S formed by the mirrors. The conditions are obviously similar to those obtaining in the biprism experiment, and similar interference fringes are formed. The angle θ is now twice the angle between the mirrors.

7–6. Lloyd's Mirror

Similar fringes may be obtained with only one mirror used at almost grazing incidence (see Fig. 7–5). In this case, the interference is between the direct and

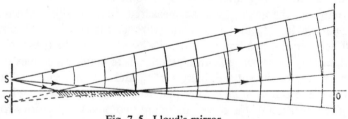

Fig. 7–5. Lloyd's mirror.

reflected disturbances. The slit must be exactly parallel to the plane of the mirror in order to render parallel the slit and its image. It will be seen from the diagram that the central fringe can be observed only by placing the receiving screen in contact with the end of the mirror. Lloyd found that this zero path difference fringe was dark instead of bright; this confirms the existence of the phase change of π undergone by light reflected at a denser medium. (See §21–4.)

7–7. Other Arrangements

Two arrangements for producing a similar fringe system are the Billet split lens (Fig. 7–6) and the biplate (Fig. 7–7). In the former, two real images S_1S_2 of the slit S are formed by separating the halves of the lens and cylindrical wave envelopes diverge from these images. In the biplate one again has two real images of the slit from which cylindrical envelopes diverge. In all these arrangements the same interference effects are observed, and in all cases the separation of the effective sources involved may be found by using a convex lens as described for the biprism.

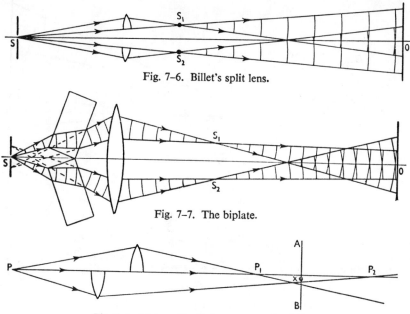

Fig. 7–6. Billet's split lens.

Fig. 7–7. The biplate.

Fig. 7–8. Halves of split lens separated axially.

Fig. 7–8 shows an interesting variation of the split lens arrangement. One now has a point source P and the halves of the lens are separated along the axis and not transversely so that the images P_1 and P_2 also are separated along the axis. It must be noted that one does not have interference between waves diverging from P_1 and P_2 and the surfaces of constant intensity are no longer hyperboloids of revolution about P_1P_2 as described in § 7–2, but are ellipsoids.

It will be seen that the disturbances associated with P_1 and P_2 are superposed only in the region between P_1 and P_2 and only on one side of the plane through P_1P_2 perpendicular to the diagram. Here interference occurs between one diverging and one converging spherical wave. The optical path difference between PP_2 and PP_1 is simply P_1P_2, so that at X the difference between the paths traversed by the superposed disturbances is $P_1X-(P_1P_2-P_2X)$ and a surface of constant intensity is generated if X moves and maintains this quantity constant. Thus (P_1X+P_2X) must remain constant, i.e. the locus of X is an ellipsoid with P_1 and P_2 as foci. The surfaces of maximum or minimum intensity consequently form a family of ellipsoids and these intersect a plane such as AB in a family of circles. Actually it can be shown that there is a phase change of π when a wave passes through a focus. This has occurred to one of the superposed waves so that a path difference of $p\lambda$ corresponds to a minimum. Since, as pointed out above, disturbances are not superposed over the whole plane AB, only approximately semi-circular fringes can be observed.

7–8. Achromatic Fringes with Lloyd's Mirror

Using a slit illuminated with white light, the fringes obtained with Lloyd's mirror (and similar arrangements) rapidly blend into a uniformly illuminated

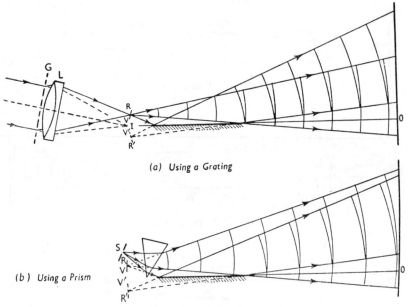

(a) Using a Grating

(b) Using a Prism

Fig. 7–9. Achromatic fringes with Lloyd's mirror.

field because the fringe separation varies with wavelength. Now the fringe separation depends also on the separation of the effective sources and if the source consists of a short spectrum with its blue end towards the mirror the effective sources have different separations for different wavelengths. If the dispersion in the spectrum is such that the distance from the mirror to any point in the spectrum is proportional to λ, the separation of the effective sources is also proportional to λ, and the fringe separations will be the same for all wavelengths. In this achromatic system the fringes are black and white. It is shown in Chapter XII that the dispersion given by a diffraction grating is almost exactly what is required in this case. The arrangement is given in Fig. 7–9(a). A collimated beam of white light is incident on the grating G which forms the usual spectra in the focal surface of the lens L. It is essential that there is no overlapping of orders and usually one of the first order spectra RV is employed. It is shown in Chapter XII that the distance from the zero order image to a first order maximum is proportional to λ so that if the zero order image I lies in the plane of the mirror the separation of the effective sources is proportional to λ and the achromatic fringes are formed.

The dispersion in the spectrum formed by a prism does not satisfy the requirements exactly, but for a prism of fairly small angle the approximation is sufficiently close for an appreciable number of white light fringes to be observable. The arrangement is shown in Fig. 7–9(b).

7–9. General Theory of Achromatic Fringes

In Young's experiment, the path difference at the centre of the fringe system is zero for all wavelengths and a white light maximum can be observed. Away from the centre no white light fringes are observed so that the pattern contains only one achromatic fringe. With the grating arrangement described in the preceding paragraph it is possible to secure a perfectly achromatic *system* of fringes.

In general one may say that at any point in a fringe system the order (not necessarily integral) usually depends on λ; if p is independent of λ one has a perfectly achromatic fringe and if this is true at every point in the pattern one has a perfectly achromatic *system* of fringes. Conditions of less complete achromatism can occur as follows:

If at any point in the pattern $\partial p/\partial \lambda = 0$ for any wavelength, then there is an appreciable range of wavelengths for which the variation of order is small and one is said to have an achromatic fringe at that point. This fringe is not perfectly achromatic and the notation is similar to that employed in geometrical optics. (A lens is said to be achromatic if $\partial f'/\partial \lambda$ is zero for one value of λ since f' is then approximately constant for a finite range of wavelengths; for a perfectly achromatic lens f' would have to be completely independent of λ.)

If at the same point in the pattern $\partial/\partial x(\partial p/\partial \lambda)$ is also zero (where x denotes position in the pattern), then $\partial p/\partial \lambda \doteqdot 0$ over a finite region of the pattern and one has a number of approximately achromatic fringes; one is said to have an achromatic system of fringes. It will be seen that the conditions for perfect achromatism may be expressed analytically as $\partial p/\partial \lambda = 0$ for all values of λ and $\partial^2 p/\partial x \partial \lambda = 0$ for all values of λ and x. Using Lloyd's mirror and a prism, one can obtain an achromatic system of fringes, but to obtain perfect achromatism a grating must be used. As stated above, a single perfectly achromatic fringe is obtained in Young's experiment.

7–10. Introduction of a Thin Transparent Plate

In Young's experiment or in one of its derivatives using monochromatic light, suppose that a transparent plate of thickness t and refractive index n is introduced in front of one of the real or virtual sources (e.g. in front of one half of the biprism). An extra optical path, $(n-1)t$, is then introduced in one of the interfering beams. This will cause the fringes to move across the field and, for example, the zero order fringe will move to the position formerly occupied by the fringe corresponding to a path difference of $-(n-1)t$, i.e. will move to the position formerly occupied by a fringe of order $p=(n-1)t/\lambda$. Since with monochromatic light it is impossible to distinguish one fringe from another, n cannot in general be found from this experiment.

If white light is used and the material of the plate is non-dispersive, the zero order fringe will remain perfectly achromatic and n may be found by observing the displacement of this fringe; it will simply move to the place which is optically equidistant from the two sources. If, however, the plate is dispersive then, for any point in the pattern, the optical distance from one source is independent of wavelength while that from the other is not. Consequently there can be no *completely* achromatic fringe. The central ("achromatic") fringe can be said to be located at the point P where $\partial p/\partial\lambda=0$ for the central wavelength in the range present. (For the wavelength to which the eye is most sensitive in the case of visual observations with white light.) This is the point of maximum agreement of order and therefore of phase difference, and is not the place where the path difference is zero for the central wavelength. Suppose the path from the source S to P is in air and of length l and let $(L-t)$ be the part of the other path which is in air, the dispersive part being of geometrical length t. Then the optical path difference at P is $(L+(n-1)t-l)$ and this divided by λ gives the order. Differentiating with respect to λ and equating to zero one has, when P is the central fringe,

$$\frac{\partial}{\partial\lambda}\cdot\frac{1}{\lambda}[L+(n-1)t-l]=0=-\frac{(L-l-t)}{\lambda^2}+\frac{t}{\lambda}\cdot\frac{\partial n}{\partial\lambda}-\frac{nt}{\lambda^2},$$

$$\therefore\quad l-L=\left(n-\lambda\frac{\partial n}{\partial\lambda}-1\right)t, \qquad . \quad . \quad . \quad . \quad (7\text{--}5)$$

and the achromatic fringe is moved to P, a point such that $(l-L)$ is given by equation (7–5). This corresponds to a shift of $(l-L)/\lambda$, i.e. $(1/\lambda)(n-\lambda\partial n/\partial\lambda-1)t$ fringes of wavelength λ. Now the zero order fringe of wavelength λ is moved a distance $(n-1)t/\lambda$ fringes so that the distance, in terms of fringes of wavelength λ, from the achromatic fringe to the zero order fringe for wavelength λ is $t\partial n/\partial\lambda$.

CHAPTER VIII

INTERFERENCE OF TWO BEAMS.
DIVISION OF AMPLITUDE

INTERFERENCE frequently results from the recombination of the two parts of a beam of light which is divided by partial reflection and refraction in a thin plate or film. The interference phenomena are of two main types: those observed with parallel-sided, moderately thin plates and those observed with very thin plates of varying thickness.

8–1. Parallel-sided Plates

Consider a ray SA from a monochromatic point source S entering a parallel-sided plate of glass of index n and thickness d (see Fig. 8–1). At each of the points A, B, C, D, E, etc., a small propor-

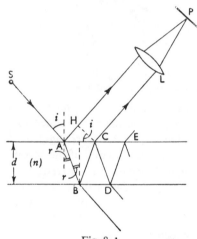

Fig. 8–1.
Reflections in a parallel-sided plate.

tion of the light is reflected and the rest is refracted. Consequently, the rays emerging at A and C have intensities which are not widely different, each having undergone one reflection. Rays from E and points farther to the right have rapidly decreasing intensities and can be ignored in a first approximation. If the disturbances travelling along the rays emerging from A and C enter the lens L they will be superposed at P in the focal plane. When these disturbances are superposed they will be out of phase by an amount determined by the optical path difference [ABC]−[AH]. The lens L may be the eye of an observer and, if the accommodation is relaxed, P will be on the retina and the resultant intensity at P will determine the brightness of the field of view in the direction concerned.

The optical path difference (o.p.d.) between the superposed disturbances is given by

$$(\text{o.p.d.}) = n(\text{AB} + \text{BC}) - \text{AH}$$

$$= \frac{2nd}{\cos r} - \text{AC} \sin i,$$

where i and r are the angles of incidence and refraction of the ray SA as in the figure. Now $\sin i = n \sin r$ and $\text{AC} = 2d \tan r$.

$$\therefore \quad (\text{o.p.d.}) = \frac{2nd}{\cos r} - \frac{2nd \sin^2 r}{\cos r} = 2nd \cos r. \quad . \quad . \quad (8\text{–}1)$$

Now it is known that when there is reflection at a denser medium there is a phase change of π (see §21–4). Here the phase change occurs at A so that the condition for the disturbances to be in phase and to give a maximum becomes

$$2nd \cos r = (p + \tfrac{1}{2})\lambda,$$

and for a minimum,

$$2nd \cos r = p\lambda,$$

. . . . . (8–2)

where p is an integer.

If the lens L has a large aperture, pairs of rays may enter after being reflected from widely different parts of the plate [Fig. 8–2(a)]. Associated with each pair of rays, there will be an optical path difference given by equation (8–1), where r is the angle of internal reflection in the plate of the ray so reflected. The constituent rays of each pair are superposed at a point in the focal plane of L so that there will be a variation of intensity across this plane. Now the usual laws of

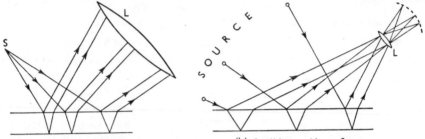

(a) Small Source and Large Lens   (b) Small Lens and Large Source

Fig. 8–2. Arrangements for observing fringes of equal inclination.

reflection must be obeyed at the plate so that, using a point source and a lens of small aperture such as the eye, the rays that can be collected are confined to a small range of directions. If one has an extended source, however, light from each point of the source enters the eye in a separate direction and, if the accommodation is relaxed, rays in each pair are brought together at a different point on the retina [Fig. 8–2(b)]. Associated with each pair of rays there is again a certain o.p.d. and this determines the brightness of the field of view in the direction concerned. To prevent obstruction of the incident light by the observer, it is obviously desirable to employ a reflector plate as indicated in Fig. 8–3. (To simplify the diagram refractions in the plane parallel plate and in the reflector have been neglected.) This figure also shows that by using a subsidiary lens S.L. it is possible to use an observing lens of small aperture (e.g. the eye) with a small source, the latter being imaged at the observing lens to give the condition known as *Maxwellian view* (see § 18–7); the diagram shows how the Maxwellian view arrangement gives the effect of an extended source.

For a particular bright or dark fringe the o.p.d. between the two superposed disturbances is constant so that one must have r=constant; the fringes are therefore known as fringes of equal inclination. Thus the fringes form a series of concentric circles in the focal plane of the observing lens, i.e. they are seen clearly in this form by an unaccommodated eye and may be said to be at infinity. Fringes of equal inclination may be seen also by transmission as indicated by Fig. 8–3 but the contrast in these fringes is poor because the ray which has undergone two reflections is much less intense than the other. The fringes by transmission are complementary to those observed by reflection but with a uniform intensity added. The fringes viewed near normal incidence are called *Haidinger fringes*.

If, in equation (8–2), p has the value p_1 for the innermost fringe, it has value $(p_1 - 1)$ for the next and $(p_1 - j + 1)$ for the jth fringe where the fringes are numbered from the centre. If r is small one may write $\cos r = 1 - (r^2/2)$ and, for the jth dark fringe by reflection, one has

$$2nd\left(1 - \frac{r_j^2}{2}\right) = (p_1 - j + 1)\lambda,$$

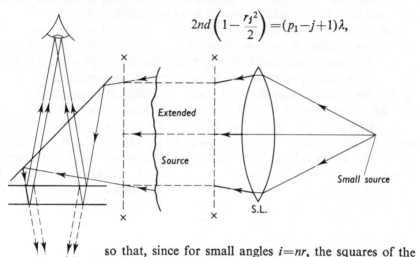

Fig. 8–3. Arrangement for observing complete fringe system.

so that, since for small angles $i = nr$, the squares of the angular diameters of the fringes vary linearly with the fringe numbers. This result applies also to the bright fringes and to the fringes viewed by transmission. The fringes viewed by reflection have an appearance similar to the fringes shown in Plate I(b) and the fringes seen by transmission resemble those shown in Plate I(c) (see §8–4). (See Miscellaneous Examples on Chapters VI–IX no. 10.)

With thick plates the fringes become very close together and it becomes necessary to observe them with the aid of a telescope of suitable magnifying power. For example, with a plate 1 cm. thick of refractive index 1·5 the angular separation of the fringes for green light is about 1′ when $r = 3°$. With very thick plates the distance CH (Fig. 8–1) may be larger than the diameter of the eye pupil, and for visual observation it is necessary to employ a telescope whose objective admits the two rays. If M is the magnifying power of the telescope, the emergent rays will be distance CH/M apart and this must be smaller than the diameter of the eye pupil. If the surfaces of the plate are semi-silvered, the telescope objective must admit rays emerging from E and points farther to the right (Fig. 8–1).

Referring again to Fig. 8–3, it will be seen that whenever the fringes are observed with a point source the rays traversing the plate at a given angle pass through a particular part of the plate, so that variations in plate thickness will distort the fringes. When the plate is very thin the effect of variation in direction becomes negligible and, even with an extended source, the intensity at a given point in the pattern characterizes the local thickness of the plate. Fringes of this type are discussed in the following sections.

8–2. Interference in Very Thin Films

If d is very small it may become impossible for cos r to vary sufficiently to make the variations in $2nd$ cos r appreciable. Certainly this will be so if only a moderately extended source is employed. If the variations of $2nd$ cos r are small compared with λ, the field of view will be of uniform intensity.

Now if d is very small the distance AC (Fig. 8–1) is very small and the total disturbance consists of two superposed wave trains along closely neighbouring paths. They may be said to interfere at what is now the single point of reflection and the resultant may be regarded as a single ray whose intensity is given by the superposition of the internally and externally reflected disturbances. (Notice here that it is convenient to combine the terminologies of ray and wave optics.)

If d varies across the plate, the o.p.d. introduced between the internally and externally reflected rays will depend upon the thickness of the film at the point of reflection and will be independent of the angle of emergence. The resulting intensity will therefore be characteristic of the local thickness of the film. This discussion is satisfactory for thin films whose thickness undergoes abrupt changes so that it consists of a number of adjacent parallel-sided films. Such films may consist of the air gap between an optical flat and, for example, the surface of a sheet of mica which exhibits abrupt changes of level.

In the case of films of continuously varying thickness, such as a wedge, the internally and externally reflected rays are not parallel and the formation of the fringes is more easily understood by considering the wavefronts reflected at the two surfaces of the film.

8–3. The Wedge Film

A common arrangement for observing the so-called *Fizeau fringes* is shown in Fig. 8–4(a). This is a particular case of the use of Maxwellian view, the wave-

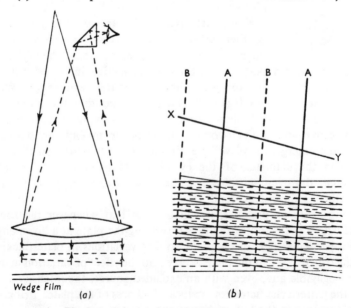

Fig. 8–4. Fringes of equal thickness in a wedge

fronts incident on the film being plane. The arrangement can be used with a
point or moderately extended source which ensures that the variation of cos r
is small. For a point source two plane waves leave the wedge, one being reflected
from each surface. Fig. 8–4(b) shows the reflected waves in the region between
the wedge and the lens; the full lines denote crests and the broken lines troughs.
Along the lines A the crests and troughs of one wave train coincide with the
crests and troughs of the other. Along the lines B the crests of one wave train
coincide with the troughs of the other and vice versa. Thus along the lines A
the disturbances associated with the two wave trains are in phase and there is
constructive interference, whereas there is destructive interference along the
lines B. It follows that in any plane XY there is a fringe system, the maxima and
minima occurring where the loci A and B, etc., intersect it. The loci of maxima
and minima intersect the film where the thickness is given by $2nd=(p+\frac{1}{2})\lambda$ and
$2nd=p\lambda$, respectively. These thicknesses cause a relative retardation of one
wavefront with respect to the other of odd and even multiples of $\lambda/2$ which
respectively give maxima and minima since there is the usual phase change of
π at the upper surface. Beyond the lens L the interfering wavefronts are, of
course, spherical.

With an extended source each point of the source gives a pair of coherent
waves which interfere in this way and the loci of the maxima and minima
emanate from · the points on the film as defined
above. Thus *it is only at the film itself that all the
fringe systems coincide*, and it is here that the aggregate
fringe system is located. The fringes are said to be
localized in this plane and may be observed conveniently with a low-power microscope as in Fig. 8–5.
If the source is not at the focus of a lens, spherical
waves are reflected by the wedge. These interfere
in the same way as plane waves to give the same
fringe system, provided the curvature of the spherical
waves is not sufficient to give an appreciable variation
in r. These *fringes of equal thickness* will be equally
spaced straight lines parallel to the edge of the wedge
and will be contours of the wedge thickness; the

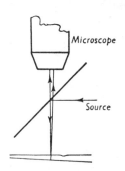

Fig. 8–5. Observation
of localized fringes of
equal thickness.

latter increases by $\lambda/2$ in going from one dark or bright fringe to the next.
Using these fringes, one can measure the angle of a very thin wedge-shaped
sheet of any transparent material or the wedge-shaped air gap between two
slightly inclined plates. These fringes are frequently viewed with a small but
finite source and, although they are localized in the plane of the wedge, the
depth of focus is large, i.e. they are not sharply localized. With a true point
source the fringes are not localized in this sense but still characterize the local
thickness of the wedge. If observations are made at oblique incidence, variations
in r may become important and the fringes may become curved and less distinct
—particularly in the thicker parts of the wedge. Plate I(f) shows the fringes
viewed by reflection with an air wedge between two optical flats. The fringes
viewed by transmission have very poor contrast (see § 8–1); it is similar to that
of the fringes shown in Plate I(c).

8-4. Newton's Rings

If an optical flat is placed in contact with a shallow convex spherical surface, a thin air film of varying thickness results (see Fig. 8–6). This film is, of course, symmetrical about the point of contact so that the localized interference fringes, which are contours of the air film thickness, become concentric circles centred on

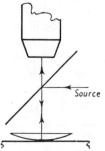

the point of contact. These fringes were first observed by Newton and are known as *Newton's rings*. One has, here, interfering spherical wavefronts and the fringes of equal thickness will again be non-localized for point sources and sharply localized for extended sources.

If R is the radius of the spherical surface and d is the thickness of the air film a distance r from the point of contact,

$$R = \frac{r^2 + d^2}{2d}.$$

Fig. 8–6. Observation of Newton's rings.

If R is large $d^2 \ll r^2$ and one has

$$d = \frac{r^2}{2R}. \quad \ldots \quad \ldots \quad (8\text{–}3)$$

Since the film is thin, any departure from normal incidence can be ignored. The film is of air so that $n=1$ and, remembering that there is a phase change of π at the lower surface, the conditions for a maximum or minimum are:

$$\text{for a minimum} \quad 2d = p\lambda,$$
$$\text{for a maximum} \quad 2d = (p+\tfrac{1}{2})\lambda.$$

Therefore the radii of the bright and dark fringes are given [from (8–3)] by

$$\left. \begin{array}{ll} \text{(dark)} & r^2 = pR\lambda, \\ \text{(bright)} & r^2 = (p+\tfrac{1}{2})R\lambda. \end{array} \right\} \quad \ldots \quad \ldots \quad (8\text{–}4)$$

Hence if λ is known the radius of curvature of the spherical surface may be determined by measuring the radii of the rings.

It will be seen that when contact at the centre is perfect p is zero at the centre and the squares of the radii of successive dark rings are proportional to the natural numbers. On plotting r^2 against the ring number for either dark or bright fringes, one obtains a straight line of slope $R\lambda$. It will be seen that the relationship between the ring numbers and ring diameters is the same as for the fringes of equal inclination described in § 8–1. Since, to a first approximation, both sets of fringes result from the interference of only two beams, the fringes are similar in appearance. They are shown in Plate I(b). As before, the fringes viewed by transmission have poor contrast; they are shown in Plate I(c). (See also the Miscellaneous Examples VI–IX no. 10.) It must be pointed out that although the fringes are similar in appearance they are somewhat different in origin. In Newton's rings the order increases as one moves away from the centre of the pattern, and in Haidinger's fringes the order decreases as one moves outwards across the fringes.

The effect of imperfect contact between the surfaces is simply to alter the intercept on the r^2 axis. This is easily shown since equation (8–3) becomes

$$d=\frac{r^2}{2R}+d_o$$

where d_o is the separation of the surfaces at the centre. The fact that with good contact the central fringe (viewed by reflection) is dark, is an important confirmation of the existence of the phase change of π associated with reflection at the denser medium. The arrangement is a special case of two spherical surfaces of different radius of curvature in contact and, in general, the radius of one surface can be found if that of the other is known. One has

$$2d=r^2\left(\frac{1}{R_1}\pm\frac{1}{R_2}\right).$$

The spherical surface in Fig. 8–6 is part of a lens, but the effect of the latter on the size of the rings can be ignored since it is in contact with the plane of the fringes and the magnification is unity.

At the point of contact in the Newton's rings arrangement (or at the edge of the wedge), the phase difference is simply π for all wavelengths so that, using white light, an achromatic dark fringe can be observed. As the thickness of the film increases, the phase difference varies with wavelength and (as shown by equation (8–4)) the rings for different wavelengths have different radii. Thus the first few bright rings are drawn out into spectra with the shorter wavelengths inside, and after a few orders the bright fringes of various colours and orders intermingle to give a white continuum as in Young's experiment (see Chapter VII). Thus for films whose thickness is not greater than a few wavelengths, polychromatic localized fringes can be observed. Such *interference colours* are frequently seen in films of oil on water or in the films of a soap bubble.

Example

A thin equiconvex lens of focal length 4 metres and refractive index 1·52 rests on and in contact with an optical flat and, using light of wavelength $5\cdot46\times10^{-4}$ mm, Newton's rings are viewed normally by reflection. What is the diameter of the fifth bright ring? What would be observed if (a) a liquid (index n) were introduced between the lens and the flat, (b) the lens were lifted slowly off the flat?

Equation (1–7) gives the radius of curvature of each face of the lens. One has $r_2=-r_1$, $f=4$ m, $n=1\cdot52$, whence $r_1=4\cdot16$ m. In equation (8–4) $p=0$ for the innermost bright or dark ring. Hence for the fifth bright ring $p=4$ and (8–4) gives (noting that the radius of the lens surface is now denoted by R and r denotes the radius of a fringe)

$$r^2=4\cdot5\times4160\times5\cdot46\times10^{-4}\text{ mm}^2$$

which gives: diameter $=2r=6\cdot39$ mm.

(a) One can say that the condition for a maximum becomes o.p.d. $=2nd$ $=(p+\frac{1}{2})\lambda_o$ where λ_o is the vacuum wavelength. (8–4) then becomes

$$r^2=(p+\tfrac{1}{2})R\lambda_o/n.$$

Alternatively, one can say that (8–4) retains its original form while λ becomes the wavelength in the liquid ($=\lambda_o/n$). Obviously the same result is obtained. Ignoring the very small difference between λ_o and λ in air, the effect of introducing the liquid is to cause the value of r^2 to be multiplied by $1/n$ i.e. the diameter of each fringe is decreased—divided by $\sqrt{n}$. An additional effect is that the fringe pattern is less bright. This is because there is a smaller change of index at each reflecting surface and the intensity of the reflected light is reduced.

(b) A particular fringe corresponds to a particular value of the film thickness d. As the lens is lifted off the flat any particular value of d occurs nearer to the centre so that the diameter of the corresponding fringe is reduced. Hence the overall effect is that the fringes move towards the centre; the relationship between r^2 and ring number remains linear, and the slope of the line remains $R\lambda$. A fringe disappears at the centre for each $\lambda/2$ movement of the lens.

8–5. Non-reflecting Films

At each refracting surface in an optical system some light is reflected. If this light does not subsequently reach the image plane, the effect is simply a reduction

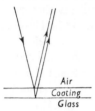

Fig. 8–7.
Non-reflecting film.

in image brightness and, for most purposes, no great harm is done. Frequently, however, this light reaches the image plane and causes an overall reduction in image contrast and/or local bright patches or *glare spots*. The amount of light reflected at any surface can be reduced considerably by coating the surface with a thin film of transparent material such that the internally and externally reflected rays interfere destructively (see Fig. 8–7). Obviously the destructive interference will be complete if the rays have equal amplitudes and are π out of phase. For a given angle of incidence, the relative amplitudes of the two rays depend on the refractive indices of the glass and thin film, and the phase difference depends on the optical thickness of the film. Now when light is incident normally on a surface separating dielectrics whose indices are n and n', the ratio of the reflected and incident amplitudes is $(n'-n)/(n'+n)$ (see §21–3). Suppose, then, that a disturbance of unit amplitude is incident on the film. Most of the light is transmitted at the interface between air and the film so that the amplitude of the light incident on the glass is approximately unity. Again, most of the light reflected at the glass surface emerges from the film. Consequently, the amplitudes of the externally and internally reflected rays are, to a first approximation, $(n_F-1)/(n_F+1)$ and $(n_G-n_F)/(n_G+n_F)$ respectively. It is easily shown that these are equal if $n_F=\sqrt{n_G}$ and this determines the necessary index for the film. The thickness of the film is determined by the fact that for a phase difference of π or an odd multiple of π

$$2n_Fd=(2p+1)\frac{\lambda}{2}.$$

Approximations have been made in the argument used here but a full analysis shows that for zero reflectance for normal incidence with a quarter-wave film the condition $n_F=\sqrt{n_G}$ is exact (see equation 21–67).

It should be noted that although the conditions for zero reflectance can be satisfied exactly for only one angle of incidence and one wavelength, the intensity of the reflected light is small for an appreciable range of angles and wavelengths. A surface treated in this way is said to be *bloomed*. It so happens that no suitable material exists at present to give a single ideal anti-reflection coating on glass—for a glass of index 1·5 one would require a film of index 1·22; a further obvious requirement is that the film must be durable. Materials that have been widely used include magnesium fluoride (1·38) and cryolite (1·36) and these are, of course, better for the higher index glasses. The films are deposited by evaporation in a vacuum. A typical bloomed glass surface reflects less that 1% of the incident white light, which is a worthwhile improvement compared with an uncoated surface. Attempts have been made to produce non-reflecting films by chemically attacking the surface of the glass, leaving a layer of lower index.

It can be shown that with available coating materials it is possible to obtain zero reflectance for one wavelength by using two layers on glass—one of high index and one of low, with the high index layer next to the glass. Unfortunately it is only in the vicinity of the "design wavelength" that the double layer is superior to the single film—typically, over a range of about $0·1\mu$ in the middle of the visible region. By using three layers the reflectance can be made to fall to zero for two wavelengths and remain less than $\frac{1}{4}\%$ over most of the visible region.

A more detailed discussion of thin films and multilayers is given in §21.11.

8–6. Further Applications of Localized Fringes

The localized fringes observed in thin films provide a simple method of testing the accuracy of optical components. The surface under test is placed in contact with a standard *test plate*. The correctness of the fit is indicated by the number of fringes of equal thickness which can be seen. On pressing the surfaces together, the air gap, if present, tends to be reduced; since a given fringe corresponds to a given air-film thickness, the fringes will move to places where the air film is relatively thicker. For example, if contact is at the centre the fringes will move outwards when the surfaces are pressed together.

The wedge fringes are used in precision engineering for comparing end gauges. An end gauge consists of an accurately known length of steel rod whose ends are polished flat and parallel. Such gauges have two important advantages when compared with the older line gauges consisting of two rulings on a metal bar. Firstly, the length of an end gauge can be measured more accurately than the distance between two rulings; and secondly, by "wringing" a number of end gauges together, a compound gauge can be made. This is done by pressing the gauges together—sometimes with a minute film of paraffin between them. If the ends of the gauges are clean and flat, they will be held together by atmospheric pressure.

In order to test an end gauge (slip gauge), the ends are first examined by means of a test plate as indicated above. If the ends are flat, one end is wrung on to the polished surface of a flat steel plate alongside a standard gauge as in Fig. 8–8. A glass optical flat is then placed across the top of the gauges as shown. If the gauges are not exactly equal in length, wedges of air occur between the gauges and the plate. If l is the separation of the appropriate edges of the gauges and

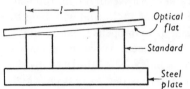

Fig. 8–8. Testing a slip gauge.

m is the number of fringes per cm. in the wedge above the standard, the difference in height between the gauges is given by

$$\delta h = \frac{m\lambda l}{2}.$$

This assumes that the ends of the standard gauge are exactly parallel. The fringes above the gauge under test will, of course, yield the same result if this gauge also has parallel ends. End gauge interferometers are discussed further in § 8–16.

In a similar way one can measure small movements of one surface towards or away from a standard surface and this has obvious applications to the measurement of small elastic deformations or thermal expansions.

8–7. The Michelson Interferometer

The essential feature of the formation of fringes in parallel-sided plates and wedges is the division of a beam of light by partial reflection at the first surface, and the subsequent superposition of the two disturbances after they have traversed unequal optical paths. In the Michelson interferometer the beam is divided (i.e. the amplitude is divided) into two beams of equal intensity by the plate A (Fig. 8–9) which is half-silvered on its rear surface. The beams are reflected at the front silvered mirrors M_1 and M_2 and recombine at plate A. The unsilvered compensating plate C is identical to A and parallel to it and is employed so that the separate beams have equal paths in glass. For the reasons given in § 7–9 this is important when white light fringes are required. Let M_2' be the virtual image of M_2 formed by reflection at the plate A. Then the conditions obtaining in the beam that has undergone reflection at M_2 are exactly the same as they would be after reflection by a mirror located at M_2'. Hence the fringes will be similar to those that would be obtained if M_1 and M_2' were reflecting surfaces enclosing the air gap which lies between them. That is to say, if A is at 45° to the incident beam and M_1 and M_2 are perpendicular and placed so that [AM$_1$] and [AM$_2$] are slightly different, the fringes are similar to those of a parallel-sided plate where the plate consists of air. Hence, with an extended monochromatic source the fringes are the usual concentric fringes of equal inclination and can be viewed with a telescope or unaccommodated eye. As

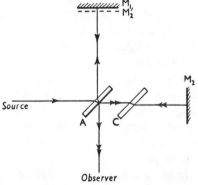

Fig. 8–9. The Michelson interferometer.

before, it is necessary to use a telescope when the "plate" is thick.

In the direction inclined to the normal to M_1 at angle θ, the path difference between the superposed disturbances is [from equation (8–1)] $2d \cos \theta$, where d is the separation $M_1 M_2'$, i.e.

$$\text{phase difference} = \delta = \frac{2\pi}{\lambda} . (2d \cos \theta).$$

Thus, if each beam is of amplitude a, the resulting intensity is [from equation (6–11)] given by

$$I = 2a^2(1 + \cos \delta)$$
$$= 4a^2 \cos^2 \left(\frac{\delta}{2}\right), \qquad \cdots \cdots \quad (8\text{–}5)$$

I reaches a maximum value when $\delta = 2p\pi$ and is zero where $\delta = (2p+1)\pi$ (p an integer). Thus maxima and minima occur at values of θ given by $\cos \theta = p\lambda/2d$ and $\cos \theta = (p + \frac{1}{2})\lambda/2d$ respectively and, as for the fringes observed with a plane parallel plate, the squares of the fringe diameters vary linearly with the fringe numbers when θ is small. As is pointed out above, these two fringe systems are formed in a similar manner but the Michelson fringes, being formed by the interference of two beams, are $\cos^2$ fringes whereas the others are only approximately $\cos^2$ fringes since disturbances undergoing additional reflections are involved. In the foregoing paragraphs the effect of the additional beams was neglected; they do not affect the positions of the maxima and minima but, as will be seen in Chapter IX, they increase the sharpness of the fringes.

Newton's rings (which are fringes of equal thickness and are formed in a rather different manner from the fringes discussed above) are also approximately $\cos^2$ fringes when viewed by reflection with unsilvered surfaces, and again the effect of silvering is an increased sharpness of the fringes. With Newton's rings it has already been seen that, if the fringes are numbered outwards from the centre, the squares of the fringe diameters vary linearly with the fringe numbers so that Newton's rings are similar in appearance (although not in origin) to the fringes of equal inclination discussed above. It has already been pointed out that with Newton's rings the order increases as one moves outwards from the centre of the pattern whereas for the fringes of equal inclination the order is a maximum at the centre. The appearance of all three fringe systems is represented qualitatively by Plate I(b) which is actually a photograph of Newton's rings by reflection.

Now, in general, if the phase difference between two superposed disturbances is $p . 2\pi$, p is called the order of interference and is not necessarily integral. With the Michelson fringes, as has just been pointed out, the order is a maximum at the centre of the pattern. Suppose, now, that the mirror M_1 is moved continuously so that d is decreased. Then the path difference is decreased over the whole field of view and a fringe of given order (i.e. corresponding to a given path difference) will move towards the centre of the pattern and disappear. A fringe will disappear each time $2d$ is decreased by λ. As d is decreased steadily, θ must decrease more and more rapidly to maintain the order. The result is that the fringes become more widely spaced (for adjacent maxima $\cos \theta_p - \cos \theta_{p-1} = \lambda/2d$). When d is zero the whole field of view is of uniform intensity and is occupied by the zero order fringe. If one continues to move the mirror so that d increases fringes appear from the centre of the pattern and spread across the field of view, becoming closer together (*a*) toward the edge of the field and (*b*) as *d* increases.

It is assumed here that the beams undergo identical phase changes at the plate A. The phase changes depend on the nature of the semi-reflecting surface, and if a phase difference is introduced between the beams, the fringes are displaced. In practice there is often a phase difference of approximately π, giving a minimum when the path difference is $p\lambda$.

8-8. Localized Fringes. White Light

When *d* is very small, suppose that one of the mirrors is tilted so that M_1 and M_2' are inclined at a small angle. The fringes will then be those of a thin wedge and, with an extended source, are straight fringes parallel to the line of intersection of M_1 and M_2' and localized in the plane of the wedge. As *d* is increased,

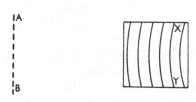

Fig. 8–10. Appearance of fringes when M_1 and M_2' form a thin wedge.

the fringes move towards the thin end of the wedge. Another effect is the introduction of a variation of path difference with θ. This causes the fringes to become curved. Since a fringe is a line of constant optical path difference, and since the path difference is smaller for larger values of θ, the fringes will curve towards the thicker part of the wedge in the outer part of the field. In Fig. 8–10

AB is the line of intersection of M_1 and M_2'; it is vertical and to the left of the field of view so that the fringes are drawn to the thicker parts of the wedge at X and Y (where θ is largest) and are thus concave towards the thicker part of the wedge. Only the zero order fringe, which lies along the line of intersection of M_1 and M_2' is exactly straight; if this lies in the field of view and the compensating plate is in position, the localized fringes can be observed with white light. This is possible because the path difference is zero for all wavelengths if it is zero for one. Thus the zero order fringe is a white light achromatic fringe. The fringes are coloured on each side of this central white light fringe, the maxima for shorter wavelengths being on the inside; in the outer part of the field of view the fringes disappear into a white continuum. The importance of the white light fringes lies in the fact that they indicate when M_1 actually intersects M_2' and therefore provide a very sensitive indicator of the position of the movable mirror M_1.

8-9. Adjustment of the Michelson Interferometer

Both mirrors can be adjusted with levelling screws and M_1 is carried on a nut threaded on an accurate screw thread so that it can be moved along the line AM_1 (Fig. 8–9). To adjust the instrument for the formation of the circular fringes of equal inclination, it is necessary to make M_2' parallel to, and a short distance from, M_1. These adjustments are carried out as follows:

1. The movable mirror M_1 is set so that $AM_1 = AM_2$ to within a millimetre or so.

2. Using a small source such as an illuminated pinhole, one looks into the instrument from the usual position of observation. The direction of vision should be roughly normal to M_1 and frequently this is ensured by looking through a suitably positioned aperture fixed to the instrument. Several images of the source might be seen but, of these, two are brighter than the others. These are formed by light which has traversed the paths indicated; the others are formed by reflections at the unsilvered surfaces of A. The levelling screws of mirror M_2 are then adjusted until these two images appear to coincide. Adjustment of M_1 may be necessary in some cases. This makes M_2' parallel to M_1 within about one minute of arc.

3. Using an extended monochromatic source, fringes should now be visible although, if the previous adjustment was not performed sufficiently accurately, further adjustment of M_2 may now be necessary before fringes are seen. Many workers find it helpful to hold, say, the tip of a pencil in front of the source and adjust M_2 until the two main images coincide; fringes should become visible as this point is reached, although they might be closely spaced initially. Further adjustment of M_2 is now required in order to secure circular fringes centred at the centre of the field. If M_2 is adjusted in the direction which broadens the fringes, this state will be reached eventually. M_2' is then parallel to M_1. When M_2' is almost exactly parallel to M_1 the fringes are almost circular and the departure from true circular shape is too small to be noticed. However, a lateral movement of the eye will move the centre of the observed pattern to a different part of M_1 and if there is a change in $M_2'M_1$ the fringes will open out or close in since the diameters of the rings depend on this distance. M_2 must then be adjusted until the fringes do not move in this way.

4. To obtain localized fringes, $M_2'M_1$ is made to approach zero. This broadens the circular fringes and, when only one or two fill the field of view, M_2 is tilted and the eye focused on M_1. The gap between M_1 and M_2' is now wedge shaped but usually M_2' does not intersect M_1 within the field of view and the fringes are curved as in Fig. 8-10. M_1 is then moved until the fringes are almost straight and, using white light, movement of M_1 is continued in the same direction until the central white light fringe is found and brought into the field. It will be obvious that the optical quality of the mirrors and the plates A and C is important since any departure from flatness or (in the case of A and C) homogeneity will introduce optical path differences and distort the fringes

8-10. Visibility of the Fringes. Structure and Width of Spectral Lines

As remarked above, the circular fringes with a monochromatic source are of good contrast. Michelson defined the *visibility* of the fringes by the relation

$$V = \frac{I_{max} - I_{min}}{I_{max} + I_{min}}. \qquad \ldots \ldots \quad (8\text{-}6)$$

where I_{max} and I_{min} are the relative intensities of a bright fringe and the neighbouring dark fringes. For a monochromatic source equation (8-5) shows that the visibility of the fringes is unity and is constant over the field of view.

Suppose now that the source gives two slightly different wavelengths (e.g. a sodium flame). For certain values of d the path difference $2d \cos \theta$ is an integral number of wavelengths for both the wavelengths emitted by the source, and for separations of M_1 and M_2' in this region the fringes for the two wavelengths coincide over an appreciable range of angles θ. The combined effect is a set of fringes of good visibility. For intermediate values of d the fringes for one wavelength will be exactly complementary to those for the other wavelength (supposing the two wavelengths are emitted with equal intensities) and the fringe visibility falls to zero. By increasing d from zero and noting the values of d corresponding to maximum and minimum visibility, the ratio of the two wavelengths may be determined. If the two wavelengths are not emitted with equal intensities, the minimum visibility is not zero; if it is measured, the relative intensities of the two wavelengths can be found. If the amplitudes of the wavelengths are a_1 and a_2, $V_{min} = (a_1{}^2 - a_2{}^2)/(a_1{}^2 + a_2{}^2)$. The corresponding method of comparing two wavelengths using the Fabry–Perot interferometer is described in §9–4.

For sources giving a single spectral line it is found that the visibility of the fringes decreases steadily as d is increased and made very large. Now it was shown in Chapter VI that no source emits completely monochromatic light and that a nominally monochromatic source actually emits light with a frequency distribution of the form shown in Fig. 8–11. Since most of the energy is concentrated into a narrow band of wavelengths, the fringe systems for the various wavelengths will be almost coincident for small values of d. As d is increased, the slightly different wavelengths produce fringe systems that are not quite coincident—the maxima for the various wavelengths gradually get out of step. When one has a continuous distribution of wavelengths, the fringes never become in step again so that a continued increase in d causes a steady decrease in the visibility of the fringes. The decrease in visibility is often explained by saying that a source of light emits wave trains of finite length and that there is no fixed phase relation between successive trains. If d is large, this would mean that light of one wave train is superposed on light of another so that there is no fixed phase relation and fringes cannot be seen. It was shown in Chapter VI that these two explanations are really the same.

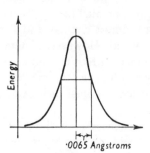

Fig. 8–11. Energy distribution in a spectral line.

·0065 Angstroms

By careful observation of the visibility of the fringes obtained with the red cadmium line, Michelson showed that the energy distribution of this line approximated to a Gaussian distribution with a *half-width* of 0·0065 angstrom. (The half-width is here defined as the distance from the mean wavelength at which the energy falls to half its maximum value; the *half-value width* is twice this.) It can be shown that the corresponding length of the finite wave trains is about 50 cm., i.e. each wave train contains about a million waves.

The effect described here was, of course, observed with a thermal source. In principle, the same effect could be observed with a laser but, since the half-width could be of the order 10^{-9}Å, a path difference of the order 1000 km could be required in order to observe a reduction in the visibility of the fringes!

For a thermal source the finite width of a spectral line is due to three main causes: natural damping, the Doppler effect, and pressure broadening. As the atomic oscillator radiates, it loses energy and therefore its amplitude decreases. This causes the amplitude of the wave to decrease as it is emitted, so that it is not a simple harmonic wave. The Doppler effect causes a variation of frequency since the radiating atoms are not stationary. Pressure broadening is associated with the fact that at high pressures other atoms disturb the radiating atom, so that the radiation is not simple harmonic. At low pressures the Doppler effect is the most important factor. This will be reduced if the speed of the atoms is reduced, so that the most highly monochromatic sources are low-pressure discharge lamps operated at low temperatures.

If the Michelson fringes are observed with a source that emits a spectral line that has a complicated structure, the fringe visibility varies in a complicated way as the path difference is varied. Rayleigh and Michelson knew that *in principle* it must be possible to determine the structure of a spectral line in this way, but they lacked the means of performing the extensive computation that is necessary. This is the problem that is solved in what is now called Fourier transform spectroscopy; furthermore, the method is now used for the analysis of complicated spectra covering a wide range of wavelengths.

8–11. Fourier Transform Spectroscopy

Michelson made his observations visually, observing the spatially displayed fringe system obtained with an extended source. In the modern form of Michelson's method—Fourier transform spectroscopy—a physical detector such as a photo-cell or thermal detector is employed to record the intensity at one

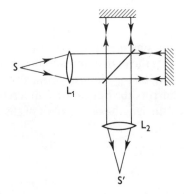

Fig. 8–12. Fourier transform spectrometer.

(usually the central) point in the field. The optical system consists of a Michelson interferometer with the addition of collimating and focusing lenses L_1 and L_2 (Fig. 8–12); this corresponds to the Twyman and Green interferometer discussed in §8–12. For the present it will be assumed that one has a polychromatic point source at S and that the intensity is recorded at the conjugate point S'.

Suppose the spectral distribution of the source and the various losses (absorption, reflection, etc.) within the interferometer are such that each of the emergent beams has intensity $B(\bar{v})d\bar{v}$ in the wave number range $\bar{v}$ to $\bar{v}+d\bar{v}$ (wave number $=1/$wavelength). Then if the interferometer introduces an optical path difference $\varDelta$ between the beams, the phase difference is $(2\pi/\lambda)\varDelta$, i.e. $2\pi\bar{v}\varDelta$, and the resultant intensity of the light in this spectral range is (from equation 8–5)

$$dI = 2B(\bar{v})d\bar{v}(1 + \cos 2\pi\bar{v}\varDelta)$$

The total intensity is simply the sum of the intensity contributions from the various spectral regions so that

$$I = 2 \int_0^\infty B(\bar{v})d\bar{v} + 2 \int_0^\infty B(\bar{v}) \cos 2\pi\bar{v}\varDelta d\bar{v}. \qquad . \quad . \quad . \quad (8\text{–}7)$$

The first term is constant, and the second term varies as the optical path difference varies; the second term is, in fact, the Fourier cosine transform of the spectral distribution in each emergent beam. (If one compares (8–7) with (6–20), taken in conjunction with the appropriate form of (6–17), it will be seen that the auto-correlation function gives the cosine Fourier transform of the spectral distribution function or power spectrum. This is known as the *Wiener-Khintchine theorem*.) If one puts

$$\phi(\varDelta) = 2 \int_0^\infty B(\bar{v}) \cos 2\pi\bar{v}\varDelta d\bar{v},$$

then $$B(\bar{v}) = 2 \int_0^\infty \phi(\varDelta) \cos 2\pi\bar{v}\varDelta d\varDelta. \qquad . \quad . \quad . \quad . \quad (8\text{–}8)$$

[See equations (6–37) and (6–38).]

It follows from (8–8) that if one measures the resultant intensity of the emergent radiation as a function of the path difference, one can calculate the spectral distribution of the source by taking the transform of $\phi(\varDelta)$. Since the light is incident normally on the mirrors the optical path difference is simply $2d$, where d is the difference between the lengths of the arms of the interferometer. If the

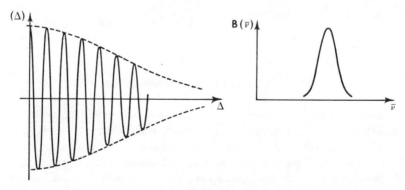

Fig. 8–13.   Single spectral line.

source emits a narrow spectral line, $\phi(\Delta)$ shows a periodic variation and the profile of the line can be obtained from the transform of $\phi(\Delta)$. Fig. 8–13 illustrates the simple case where the spectral line profile is Gaussian and $\phi(\Delta)$ has a Gaussian envelope. This corresponds to the appearance of fringes of good visibility with the classical arrangement using an extended field of view, the contrast in the fringes gradually decreasing as the path difference is increased.

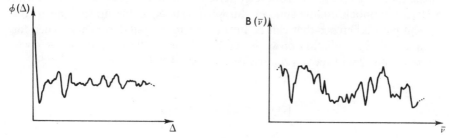

Fig. 8–14.   Complicated spectrum.

If, however, the spectral distribution, $B(\bar{\nu})$, of the light from the source extends over a wide range of values of $\bar{\nu}$ in a complicated manner, $\phi(\Delta)$ also has a complicated form over a wide range of values of Δ (see Fig. 8–14). It will be appreciated that it can be a formidable task to compute $B(\bar{\nu})$ from $\phi(\Delta)$; in fact, although both Michelson and Rayleigh appreciated the principle of Fourier transform spectroscopy before the end of the 19th century it is only recently that satisfactory digital computing techniques have been developed. For example, it was reported by J. Connes and Delouis in 1970 that a computation which could then be completed in 9 minutes had an estimated computing time of about a century 12 years earlier.

Although, as explained in § 18–12, the Fourier transform spectrometer has enormous photometric advantages compared with classical spectrometers the solution of the computational problems has been the crucial factor leading to the present widespread use of the technique.

For any type of spectroscopic instrument the chromatic resolving power is ultimately limited by the maximum optical path difference that can be introduced between the interfering disturbances that directly or indirectly produce the spectrum. (Chromatic Resolving Power is formally defined in § 11–5). Thus in the Fourier transform spectrometer there is an upper limit to the path difference that can be introduced between the two arms of the interferometer and this means that, in effect, one takes $\phi(\Delta)=0$ above a certain value of Δ. The effect of computing the transform of this truncated function is that there is a limit to the fine structure that can be shown in the calculated spectrum. The precise relationship between maximum optical path difference and chromatic resolving power is slightly different for different forms of spectrometer, some of the differences arising from the differences between the criteria for resolution that are traditionally employed—see, for example, § 9–3. However, the order of magnitude is always given by $(\lambda/\delta\lambda)\sim 2\times$ (max. o.p.d.) $\div\lambda$—see, for example, § 12–6. Interferometers have already been produced with a mirror displacement of 1 metre, i.e. with a chromatic resolving power equal to that of a grating 1 metre wide in a Littrow mounting at grazing incidence; it is claimed that instruments of considerably greater resolving power are possible.

As in all other spectrometers, the resolving power is determined by the *instrumental profile* (*instrumental* or *scanning function*), i.e. the apparent spectral distribution that the spectrometer would give for a strictly monochromatic (and hence strictly impossible!) source (in this case the spectral distribution in the spectrum computed from the interferogram that would be obtained with a strictly monochromatic source). Suppose that Δ_M is the upper limit to the optical path difference that can be introduced in the interferometer. Then for monochromatic radiation of wave number $\bar{v}_0$ the oscillating part of I is given by $\phi(\Delta) = \cos 2\pi\bar{v}_0\Delta$ for $\Delta \leqslant \Delta_M$ and $\phi(\Delta) = 0$ for $\Delta > \Delta_M$. This gives

$$B(\bar{v}) = 2 \int_0^{\Delta_M} \cos 2\pi\bar{v}_0\Delta \, \cos 2\pi\bar{v}\Delta \, d\Delta$$

which yields $\qquad B(\bar{v}) = \dfrac{\sin 2\pi(\bar{v} - \bar{v}_0)\Delta_M}{2\pi(\bar{v} - \bar{v}_0)\Delta_M} + \dfrac{\sin 2\pi(\bar{v} + \bar{v}_0)\Delta_M}{2\pi(\bar{v} + \bar{v}_0)\Delta_M}.$

The second term is identical to the first if the sign of $\bar{v}_0$ is changed, i.e. corresponds to an instrumental profile centred on a negative frequency; it can be neglected if, as is always the case, $\Delta_M \gg \lambda$ (compare with the similar discussion in §6–9). Thus the first term, of the form $(\sin \alpha/\alpha)$, gives the instrumental profile. In a prism or grating spectrum one has a profile of the form $(\sin \alpha/\alpha)^2$ and it will be seen from Fig. 11–4 that the present profile has a broader central maximum and stronger subsidiary maxima. Two of these broader maxima need to be farther apart in order to be resolved, and since in each case the first zeros occur at positions corresponding to $\bar{v} - \bar{v}_0 = \pm 1/2\Delta_M$, it follows that the spectrum obtained with the interferometer has poorer resolution than that obtained with a perfect grating, *using the same maximum optical path difference*. In fact, it is possible to modify the instrumental profile during the computation of the spectrum (a technique known as *apodization*); furthermore, the interferometer can employ path differences larger than any grating.

In practice, for all classes of spectrometer, what might be called laboratory realism, including manufacturing imperfections, limited detector sensitivity, etc., cause the resolving power to fall below the ideal limit. The overall performances of the various classes of spectrometers are discussed further in § 18–12 when it will be seen that the method of Fourier transform spectroscopy has certain photometric advantages compared with the direct recording of a spectral distribution by simply scanning the spectrum with the aid of a monochromator. These advantages are of greatest importance in the infra-red, especially the far infra-red, and although suitable materials for the lenses and beam splitter are now becoming available, these components have been the source of considerable difficulty—it is still often the beam splitter that limits the range of any particular Michelson system in the far infra-red, although there have been significant advances recently in producing infra-red beam splitters consisting of thin dielectric films. Obviously the optical system can be modified in order to employ mirrors instead of lenses; much less obviously the beam splitter can be replaced by a reflecting device also. The device, known as a *lamellar grating*, was designed and developed by Strong, McCubbin, and Vanasse (1954, 1957) and corresponds to beam division by division of wavefront instead of division of amplitude. The lamellar grating consists of narrow plane reflecting strips where the odd-numbered strips are rigidly connected to each other and fixed in

Fig. 8–15. Lamellar grating beam splitter.

space whilst the even-numbered ones are rigidly connected to each other but can be displaced as a whole in the direction perpendicular to the mirror surfaces (Fig. 8–15). The device is mounted in the position normally occupied by the grating in an Ebert type arrangement (see Fig. 12–20(a)). It must be noted that the lamellar grating is not a spatial dispersing element in the system. It is used near normal incidence so that one uses the zero order beams from the two coarse gratings that the two sets of mirrors might be said to constitute. This means that, in effect, two beams reach the detector, one from the front set of mirrors and the other from the rear set, the path difference between the beams being $2\,d$ where d is the distance between the planes of the two sets of mirrors.

The use of the lamellar grating in place of a traditional beam splitter avoids the loss of the light that in the traditional arrangement is returned to the source. (The student should satisfy himself that this loss does, in fact, occur!) Another advantage of the lamellar grating beam splitter is that the beams are unpolarized even in the very far I.R. The range of the device is limited at the short wavelength end when the optical quality of the mirrors becomes inadequate, and at the long wavelength end when the metal facets act as waveguides so that the delay depends on the cavity velocity rather than on the free space velocity. A disadvantage is that the device requires the use of a source slit that is sufficiently narrow for adjacent strips of the mirrors to be illuminated coherently. The limiting slit width depends on the wavelength, and the limiting length of the slit is equal to the diameter of the circular source that can be used with the equivalent classical Michelson system.

It must be pointed out that errors in measuring the displacement of the interferometer mirror give errors in the computed spectrum in a manner analogous to the way in which optical surface imperfections or grating ruling errors give erronious spectra in classical spectrometers—in each case one is concerned with errors of path difference. In particular, periodic errors in both the interferometer and the grating give rise to ghosts in the spectra (see § 12–16). For a high resolution interferometer the maximum displacement of the mirror must be large (Δ_M large) and the displacement must be monitored and servo-controlled in a manner analogous to that employed when translating the blank during the ruling of a high precision grating. If the displacement of the mirror is measured interferometrically using an accurately known wavelength, the absolute wavelength accuracy in the computed spectrum will be correspondingly high. For an instrument of more modest resolving power the required displacement of the mirror is smaller and one can employ less precise measurement and control of the movement. Modest displacements of the mirror have been achieved electromagnetically using a system resembling a moving coil loudspeaker. One method that has proved convenient for recording the mirror displacement employs moiré fringes as suggested in § 12–19.

If one is to compute the spectrum with a digital computer one can sample the interferogram directly at discrete values of the path difference instead of obtaining a continuous record and sampling subsequently.

Although the modern Fourier transform spectrometer was developed for use in the fairly distant infra-red, where its photometric advantages were most needed, it is now used throughout the whole of the visible and infra-red regions and, as an instrument of the very highest (virtually unlimited) resolving power giving high photometric and wavelength accuracy, it rivals all other techniques. Prominent among those who have developed the Fourier transform method to this advanced state are J. and P. Connes, Fellgett, Gebbie, Jacquinot, Mertz, Strong, and Vanasse.

A particular application of the Fourier transform spectrometer, pioneered by Chamberlain, Gibbs, and Gebbie (1963) must be mentioned although space does not permit a full discussion. It is characteristic of two beam interferometers that the white light zero order fringe becomes asymmetrical if the path of one of the beams is dispersive (see § 7–10). The extent of the asymmetry depends upon the magnitude of this dispersion and can, in fact, be used to measure it. Thus, suppose one arm of the Michelson interferometer contains a slab of dispersive material. The interferogram obtained with "white light" is no longer symmetrical about the principal maximum and the position of the latter depends upon the dispersive element— the interferogram is no longer an even function of Δ and must be recorded for both positive and negative path differences. Subsequent computation, which involves the sine as well as the cosine transforms, yields a dispersion curve for the spectral region concerned. This *asymmetric mode of operation* of the interferometer is often referred to as *amplitude spectroscopy* (as opposed to intensity or power spectroscopy) because phase information is provided.

8–12. The Twyman and Green Interferometer

The interferometer is illustrated in Fig. 8–16(a). It will be seen to resemble a Michelson interferometer in which the extended source is replaced by a monochromatic point source S at the principal focus of a well corrected objective L_1. Single plane waves are therefore reflected at M_1 and M_2 which, if the half-silvered mirror A is at 45°, are accurately perpendicular. The plane waves are superposed and brought to a focus at E by the second objective L_2. The eye is placed at this focus so that the conditions of Maxwellian view obtain. In order to adjust the instrument, images of an illuminated pinhole at S are received alongside the

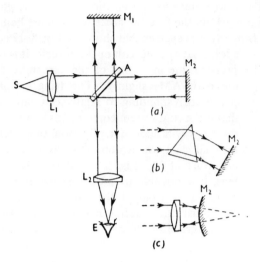

Fig. 8–16. The Twyman and Green interferometer.

pinhole itself, and also on a screen at E. The mirrors are adjusted until the two images at S coincide with the pinhole at the same time as the two images at E coincide. Frequently, one adjusts a Michelson interferometer by converting it into a Twyman and Green instrument by means of two lenses and proceeding in this way. With the mirrors adjusted as described, the superposed plane waves are exactly parallel so that, across the whole field of view, there is a constant phase difference between the superposed disturbances. The field is therefore of uniform intensity. This intensity is determined by the difference between the paths AM_1A and AM_2A and is obviously a maximum when these are equal or differ by an integral number of wavelengths. It will be seen that, with this form of Maxwellian view, different directions of the field of view do not correspond to different angles of incidence on the mirror. In consequence, the circular fringes of the Michelson interferometer are absent. By tilting M_1 or M_2, it is possible to observe the wedge fringes, but these are not localized in the sense of being in focus only in the plane of M_1. If, when M_1 and M_2 are perpendicular, the plane mirror M_2 is replaced by an optically perfect prism plus a plane mirror, or a perfect lens and spherical mirror, as suggested by Fig. 8–16(b) and (c), the field of view will remain of uniform intensity. If, however, the prism or lens is not perfect (the mirrors M_2 being perfect), the wave returning to A no longer will be plane and the phase difference between the superposed disturbances will vary across the field of view and fringes will be seen (see Fig. 8–17). These fringes, being lines of equal phase difference, are contours of the distorted wavefront, so that the imperfections of the prism or lens are found at once in terms of the wavefront aberration which they introduce (see Chapter XV). This method was originally used in the manufacture of prisms, the errors caused by inhomogeneity of the prism being corrected by local *figuring* of the surface of the prism until the field of view in the interferometer was of uniform intensity. As stated above, the fringes are not in focus in any particular plane but, if the eye is focused on a prism face as viewed through L_2, it is possible to see the fringes clearly and mark on the face the areas which must be figured. In order to see whether a fringe system such as that shown in Fig. 8–17 corresponds to an

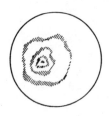

Fig. 8–17.
Appearance of a
defect.

additional optical path or the reverse, it is only necessary to press on the mounting of the mirror M_2 so as to increase the path AM_2 slightly. The fringes will then move to points corresponding to relatively smaller optical paths so that if the ring system corresponds to additional retardation the rings will open out. By mounting a lens under test on a nodal slide which automatically maintains the centre of curvature of M_2 (Fig. 8–16) at the Gaussian focus of the lens, it is possible to find the aberrations present at any point of the field of the lens. It is frequently convenient to adjust the interferometer to give the linear fringes referred to above. An imperfection in the system under test then distorts the fringes.

It is interesting to note that the use of Fizeau fringes with a test plate enables the optical quality of the surface of a component to be measured, whereas the Twyman and Green interferometer tests the optical path *through* the component.

A modified form of the Twyman and Green interferometer can be used for testing end gauges. The method is better than that described in §8–6 because it is not necessary to wring gauges on to a third surface. This is a distinct advantage since frequent wringing of one surface against another involves a risk of damage to the surfaces, especially if dust settles on them. A compensating plate is placed in the arm AM_2 so that white light fringes can be employed, and the mirror M_2 is replaced by the polished ends of two gauges which are to be compared. The mirror M_1 is tilted to give wedge fringes and adjusted until the central fringe falls across the ends of the gauges. When these ends are co-planar the sections of each fringe will be collinear. A similar procedure is carried out when viewing the other ends of the gauges which must, of course, be left undisturbed. By measuring the new positions of the central fringe on the ends of the gauges it is possible to find the difference between the lengths of the gauges. Other methods of comparing end gauges are described in §§8–16 and 9–10.

8–13. The Jamin Interferometer

A plan of the instrument is shown in Fig. 8–18(a). The glass blocks P and Q are as nearly identical as possible and are silvered on their rear surfaces. Rays from an extended source S are amplitude-divided at A and the component rays are recombined at D. The disturbances traversing the ray paths ABCD and AEFD are obviously coherent and can interfere at D. If the faces of the blocks P and Q are parallel, the optical paths ABCD and AEFD are equal and this equality will hold for the component parts of all rays from S which are divided by P. The components of all rays thus interfere constructively at D to give emergent rays of equal intensity. The field of view of the telescope T is therefore of uniform brightness.

Suppose the blocks are rotated through equal and opposite small angles about horizontal axes parallel to their respective faces. Equal optical paths are now traversed by parallel rays that lie in a horizontal plane in the space between the blocks. Now parallel rays in a given direction are focused at a point in the primary image plane of T, and the rays in different directions in a horizontal

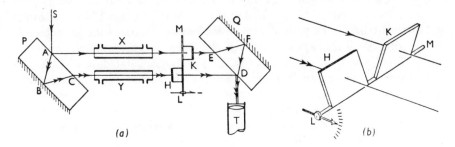

Fig. 8–18. The Jamin interferometer.

plane between the blocks are focused at points along a horizontal line in the image plane. That is, each point along this line corresponds to a zero path difference between rays internally reflected in P and those reflected in Q; the line is therefore a zero order bright fringe. Rays that are focused at points along another horizontal line in the image plane of T have optical path differences introduced between the components that entered P and the ones that entered Q. Since they are not horizontal between the blocks they make different angles with the two blocks and therefore have unequal paths in them. The rays that are focused at a non-central point on the line have optical paths that are different from those for rays that are internally reflected in the same block but are focused at the mid-point of the line, and the differences are proportional to the paths of the rays focused at the mid-point. This means that the path differences between rays internally reflected in P and those internally reflected in Q are not the same as the difference for rays focused at the mid-point. That is, the horizontal line is not a line of constant phase difference—it is not a fringe. However, the path differences between rays focused at the mid-point of the line are already small and the discrepancy between these differences and those for rays focused at other points along the line are very small. That is to say, one has a system of approximately linear horizontal fringes in the focal plane of T. A similar effect occurs with two Fabry-Perot etalons in series (see § 9–7).

Owing to the considerable thickness of the blocks, the beams are sufficiently separated to be passed through the identical gas or liquid containers X and Y. These tubes have optically flat windows at each end and vary in length from about 10 cm. upwards. Any path difference introduced between the component rays to be recombined at D will cause a movement of the fringe system. If p fringes traverse the cross-wires and l is the length of the tubes, the difference

δn between the refractive indices of the contents of the tubes is given by

$$l\delta n = p\lambda.$$

Since the o.p.d. cannot easily be introduced gradually, p cannot be found directly by observing the passage of the fringes across the field of view of the telescope. To overcome this difficulty, Jamin employed a compensator of the form shown in Fig. 8-18(b). Two glass plates H and K are fixed to a common axis LM and are inclined to one another at a small angle. The arrangement can be rotated as a whole about LM and the rotation is indicated on a circular scale. The compensator is mounted as indicated in Fig. 8-18(a) such that LM is horizontal and one plate lies in the path of each of the beams traversing the gas tubes. It will be seen that if the rays pass through the plates at different angles an o.p.d. is introduced and this is varied as the compensator is rotated as a whole. By employing monochromatic light and observing the passage of fringes across the field of view as the compensator is rotated, the scale can be calibrated to read o.p.d.'s in terms of wavelengths. By varying the angle between the compensator plates, the sensitivity of the device is changed, being greater when the inclination between the plates is reduced since a larger rotation is required to introduce a given o.p.d.

To determine the refractive index of a fluid, the tubes X and Y are evacuated and, using white light, the central fringe is brought on to the cross-wires by adjusting the compensator. The fluid is then introduced into one of the tubes and the compensator adjusted to return the central fringe to the cross-wires. The o.p.d. introduced by the fluid is then indicated on the scale of the compensator, and the required index can be found since o.p.d. $= (n-1)l$. It is, of course, necessary to employ white light since it is only in this way that the zero order fringe can be identified.

If the second plate Q is rotated about a vertical axis vertical fringes are formed, but white light fringes can be seen only by using the compensator and there will be no truly achromatic central fringe if the dispersions of the blocks and compensator plates are different (see § 7-10). A similar difficulty may occur when the dispersion of the specimen under test is very different from that of the compensator plates. It is then necessary to perform a preliminary experiment with shorter tubes or a smaller concentration of fluid in order to identify the central fringe.

One of the most important disadvantages of the Jamin interferometer has been the expense involved in obtaining identical glass blocks, accurately worked and homogeneous. This difficulty has been overcome by Kuhn and Wheatley (1945) by employing air-spaced reflecting surfaces at P and Q. This overcomes the chief disadvantage that the Jamin interferometer had compared with the Rayleigh interferometer (see § 11-14).

8-14. The Mach-Zehnder Interferometer

Fig. 8-19 shows the optical system of the Mach-Zehnder interferometer. It will be seen that the instrument is related to the Jamin and Michelson instruments. S is a moderately extended source of monochromatic light. M_1 and M_2

are front-silvered mirrors and P_1 and P_2 are semi-silvered plates. The fringes are observed from O. It will be seen that the beams are widely separated, and this has recently proved a useful feature in the study of gas flow in such apparatus as wind tunnels. One beam is made to pass through the gas to be investigated at G while the other is passed through a pair of plates C which compensate for the windows of G. It can be shown that when the paths $P_1M_1P_2$, $P_1M_2P_2$ are equal a suitable adjustment of the mirrors gives localized fringes of the type observed with the Michelson interferometer. In the present case a suitable adjustment of the mirrors causes the (virtual) fringes to be localized in a chosen plane in the vessel G in which the variations of gas pressure are to be investigated. It is this ability to localize the fringes in a chosen plane

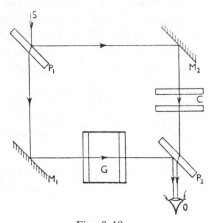

Fig. 8–19.
The Mach-Zehnder interferometer.

which is the chief advantage of the Mach-Zehnder over the Michelson interferometer. Several modified forms of the interferometer have been used to study the gas flow in the region of projectiles and aircraft.

8–15. The Wavefront Shearing Interferometer

In a Twyman and Green interferometer the wavefront under test is compared with a reference wavefront and, to produce the latter, it is necessary to employ a dividing mirror A and comparison mirror M_1 (Fig. 8–16) which are highly corrected over an aperture equal to that of the optical system under test. This may become an insuperable difficulty when it is desired to test very large aperture astronomical telescope objectives. In the *wavefront shearing interferometer* this difficulty is overcome. When one part of a wavefront is under test, a different part of the same wavefront acts as the reference wavefront. The principle of the method is as follows.

In Fig. 8–20 AB represents a wavefront which should be spherical but in fact exhibits a local error at E. A′E′B′ is an identical wavefront which is displaced circumferentially a distance δs. If the wavefronts are coherent, they will interfere and in the region of E and E′ will give patterns similar to those observed in the Twyman and Green interferometer. If the wavefronts are tilted through equal and opposite small angles about the chord AB, an eye viewing the waves under conditions of Maxwellian view will see a series of fringes parallel to AB, the

Fig. 8–20. Principle of the wavefront shearing interferometer.

fringes being distorted in the region of E and E′. As in the Twyman and Green interferometer, measurement of these fringes enables the nature of the error E to be calculated. In the wavefront shearing interferometer the interfering wavefronts are obtained by amplitude division of the wave emerging from the optical system under test. The components are recombined in such a way that one is sheared a distance δs relative to the other. Bates (1947) used the optical system of the Mach-Zehnder interferometer to effect the division and recombination of the disturbances. With this arrangement a shear can be introduced by rotating the plate P_2 (Fig. 8–19). A slit source can be used, the slit being narrow in the direction parallel to the direction of shear.

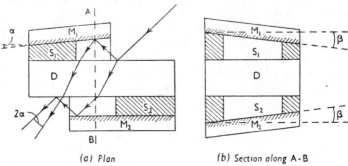

(a) Plan   (b) Section along A-B

Fig. 8–21. Interferometer with fixed shear and tilt.

Since the interferometer can be placed close to the focus of the wavefront under test, it is not necessary to have wide aperture mirrors. For example, Brown (1954) has described an instrument measuring only $2 \times 1\frac{1}{2} \times 1\frac{1}{8}$ inches. This follows a principle suggested by Drew (1951) but gives a fixed shear and tilt to the wavefront. The components can then be mounted rigidly and the instrument becomes much easier to use. The arrangement is shown in Fig. 8–21; it forms a single cemented unit. The plate D plays the part of the beam splitter and recombiner of the Mach-Zehnder system. S_1 and S_2 are glass spacers. The angle α controls the shear and β controls the tilt.

(a) Gauge in position

Wavefront shearing interferometers belong to the class known as *common-path interferometers*, in which both the test and reference beams pass through the system under examination.

8–16. Gauge Measuring Interferometers

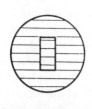

(b) Appearance of Fringes

Fig. 8–22.
Principle of the
N.P.L. gauge testing
interferometer.

Fig. 8–22(a) shows the principle of the interferometer used at the N.P.L. for measuring slip gauges. The gauge G is wrung to a base plate B and, viewing from above, one observes monochromatic fringes formed between the semi-reflecting glass plate P and the gauge, and between P and B. The plate P is tilted to obtain linear fringes across G and B as in (b). If the faces of G are parallel, the two sets of fringes are parallel and equally spaced. By observing the misfit of the fringes for four wavelengths, the thickness of the gauge can be obtained using the method of exact fractions (see §9–5).

Kösters' comparator employs a modification of the

Twyman-Green arrangement (see Fig. 8–23). The gauge G is wrung to the
mirror M_2 and the position of M_1 adjusted so that its virtual position M_1' is
across the middle of the gauge. If M_1 is then tilted, equally clear linear monochromatic fringes can be seen across G and M_2. The mis-fit of the fringes is

again found for several wavelengths and the
method of exact fractions used. Since M_1'
is approximately half-way along the gauge,
the optical path differences involved in
this method are only half those involved
in the previous method. In consequence,
the Kösters interferometer can be used for
longer end gauges than can the above
N.P.L. arrangement. In both these instruments the gauge must be wrung to a plate,
with a consequent risk of damage to the
surfaces. This difficulty is overcome in

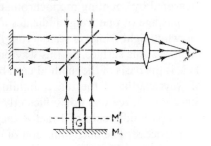

Fig. 8–23.
Principle of the Kösters interferometer.

Dowell's interferometer. As will be seen from Fig. 8–24, this is another adaptation of the Twyman-Green instrument. Fringes can be seen across the ends of
G_1 and G_2 in an obvious manner. If the gauges are symmetrically located
between the mirrors, white light fringes can be used to compare them, fractions
being estimated with monochromatic light. This instrument may be modified

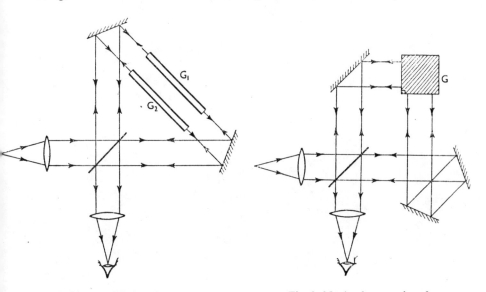

Fig. 8–24. Dowell's interferometer.   Fig. 8–25. Angle-gauge interferometer.

to give an angle-gauge interferometer—Fig. 8–25 shows the arrangement for
comparing a right-angle gauge G with a standard. The standard is put in
position at G and the mirrors adjusted until the field of view is of uniform
brightness. The standard is then replaced by the gauge under test, when the
spacing of the fringes observed indicates the error in the gauge.

The measurement of an end gauge will now be used to explain the principle of another method of using a Michelson interferometer to measure distances. The gauge G is wrung to a mirror M_2 as in Fig. 8-23. M_1 is tilted slightly and moved until the zero order white light fringe is observed across M_2. This indicates that M_1' intersects M_2. M_1 is then moved to the left and its movement measured by counting monochromatic fringes—at any point in the field of view the passage of one fringe indicates a movement of $\lambda/2$. M_1 is then moved until the white light fringe is seen across the upper surface of G, indicating that M_1' intersects this surface. In this way one can find the length of G in wavelengths. This is the basis of the method used by Michelson and Benoît to find the number of wavelengths of the red cadmium line in the length of the standard Metre. Since the source was not sufficiently monochromatic for fringes to be observed with a path difference of 1 metre, they built up to 1 metre by means of a step by step process. The determination of the Metre is discussed in §§ 9–10 and 9–11.

CHAPTER IX

MULTIPLE BEAM INTERFEROMETRY

THE present chapter is concerned with arrangements involving multiple reflections. Those involving multiple division of wavefront (e.g. the echelon) are discussed in Chapter XII. It will be shown that, in general, the effect of employing high reflectivity and therefore multiple beams is to cause a sharpening of the fringes.

Unless otherwise stated, wavelengths are measured in air.

9–1. Multiple Beam Fringes of Equal Inclination

The discussion of §8–1 was concerned with the fringes that may be observed with a parallel-sided plate or film when the surfaces have low reflectivity. In

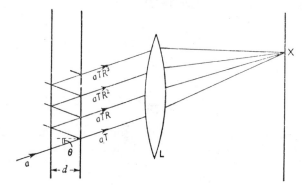

Fig. 9–1. Multiple reflections in a parallel-sided plate.

that case the interference was substantially between only two beams and the fringes observed by transmission were found to be of very poor contrast. When the boundary surfaces have higher reflectivity it is easily shown that the fringes are sharper and that those viewed by transmission have good contrast. In this case beams undergoing many internal reflections contribute to the pattern. The most important fringes are those associated with an air space between two reflecting surfaces; usually these surfaces consist of metal films deposited on glass plates. In developing the theory, the existence of the plates can be ignored because (a) their outer surfaces have low reflectivity, and (b) if the plates are plane-parallel, they will introduce no deviation in the rays.

Fig. 9–1 illustrates the successive partial reflections of a ray between two reflecting surfaces in air. If the reflections occur at a denser dielectric, there is a phase change of π at each reflection. The difference between the paths traversed by successive beams in passing to X, in the focal plane of L, is $2d \cos \theta$. Hence the phase difference at X between successive beams is

$$\delta = \frac{2\pi}{\lambda} (2d \cos \theta) + 2\pi$$

where the term 2π can, of course, be omitted. If the reflections occur at metal surfaces the phase changes are not π but the effect can be ignored in a first treatment. If T and R are the fractions of the incident energy which are transmitted and reflected at each surface, the fractional amplitudes are $T^{\frac{1}{2}}$ and $R^{\frac{1}{2}}$ and successive emergent rays have amplitudes aT, aTR, aTR^2, aTR^3, etc., where a is the amplitude of the initial incident ray. Thus if the first transmitted beam is assumed to have zero phase, the disturbances superposed at X may be represented by

$$aTe^{i\omega t}, \quad aTRe^{i(\omega t-\delta)}, \quad aTR^2e^{i(\omega t-2\delta)}, \text{ etc.} \quad . \quad . \quad \text{(see § 6–3).}$$

The complex amplitude of the resultant is therefore given by

$$\psi = aT(1+Re^{-i\delta}+R^2e^{-2i\delta}+R^3e^{-3i\delta}+ \ldots)$$

$$= \frac{aT}{1-Re^{-i\delta}}.$$

Hence the resultant intensity is (see § 6–3)

$$I = \psi\psi^* = \frac{a^2T^2}{1+R^2-2R\cos\delta},$$

i.e.

$$I = \frac{a^2T^2}{(1-R)^2} \cdot \frac{1}{1+\dfrac{4R}{(1-R)^2}\sin^2\dfrac{\delta}{2}} \cdot \quad . \quad . \quad . \quad . \quad (9\text{–}1)$$

Fig. 9–2 shows the variation of I with δ for various values of R when the maxima have the same intensity (see below). It will be seen that the sharpness

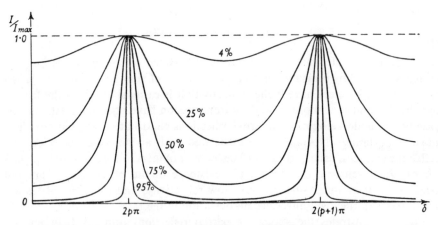

Fig. 9–2. Intensity distribution in multiple beam fringes

of the fringes increases as the reflecting power of the films is increased. For all values of R and T, the maxima and minima occur at $\cos\delta=+1$ and $\cos\delta=-1$ respectively, i.e. at $\delta=2p\pi$ and $(2p+1)\pi$ (p an integer).

Plate I(e) shows a photograph of the fringes viewed by transmission. (The multiple-beam fringes can also be observed by reflection; they are shown in

Plate I(d) beneath fringes resembling those obtained with a Michelson interferometer. It will be seen that the spacing of the maxima is the same in each case but the multiple beam fringes are very much sharper than the Michelson fringes which are formed by the interference of two beams and are the usual $\cos^2$ fringes.) (See Miscellaneous Examples VI–IX no. 10.)

From equation (9–1), on putting $\delta=2p\pi$, one has, for the intensity of the maxima,

$$I_{max} = \frac{a^2 T^2}{(1-R)^2}, \qquad \ldots \ldots \quad (9\text{--}2)$$

and, putting $\delta=(2p+1)\pi$,

$$I_{min} = \frac{a^2 T^2}{(1+R)^2}. \qquad \ldots \ldots \quad (9\text{--}3)$$

If a fraction A of the incident energy is absorbed at each reflecting surface, one has $A+T+R=1$. When there is no absorption, $T=1-R$ and

$$I_{max}=a^2,$$

$$I_{min}=a^2 \left(\frac{1-R}{1+R}\right)^2.$$

Whilst the absolute values of I_{max} and I_{min} depend on the absorption at each reflecting surface, it will be seen from equation (9–1) that the general shape of the fringes is unaffected. In particular, the ratio I_{max}/I_{min} is independent of A and is given by

$$\frac{I_{max}}{I_{min}} = \left(\frac{1+R}{1-R}\right)^2, \qquad \ldots \ldots \quad (9\text{--}4)$$

and the visibility of the fringes is given by

$$V = \frac{I_{max}-I_{min}}{I_{max}+I_{min}} = \frac{2R}{1+R^2}. \qquad \ldots \ldots \quad (9\text{--}5)$$

For unsilvered glass plates $R\dot=4\%$ and $V\dot=0\cdot08$. V reaches $0\cdot8$ when $R\dot=50\%$ and approaches unity as R approaches 100%. Although the absorption does not affect the sharpness of the fringes, it does control the absolute intensity in the pattern, and it is important to select reflecting surfaces whose absorption is as low as possible. The surfaces consist of a metal film deposited on a glass surface and are usually formed by evaporation in a vacuum. Silver and aluminium are probably the most widely used materials. The former is the better in the visible and infra-red regions but the latter is better in the ultra-violet because silver has an absorption band in the region of $\lambda 3000$ angstroms. Aluminium is frequently used in the visible region also because it is cheaper and does not tarnish easily.

The way in which an increase of the reflectivity causes a sharpening of the fringes is easily understood if one uses the graphical method of summing the disturbances. Although, in fact, a large number of disturbances is superposed, a sufficiently good indication of the state of affairs is obtained when the discussion is restricted to the interference of ten beams with two values of the reflectivity. At a maximum the vector diagrams in each case will be straight lines composed of the vectors, representing separate beams, lying end to end (see Fig. 9–3). For the higher reflect!vity the lengths of the vectors decrease less rapidly, but the total lengths of the resultants will be equal if there is no absorption since, in this case, the maxima are equally bright. At a point a short distance away from the centre of a maximum the superposed disturbances are slightly out of phase, the

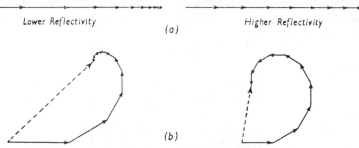

Fig. 9–3. The effect of increased reflectivity.

phase difference between successive disturbances being constant. The vector diagrams take the form indicated in Fig. 9–3(b) where the angles between successive vectors are equal and are the same for both values of the reflectivity. It will be seen that the resultant is very much smaller in the case of higher reflectivity, showing that as one moves away from the centre of a maximum the intensity falls off more rapidly.

The above discussion has related to the distribution across the fringes as illustrated in Fig. 9–2 where intensity is plotted against δ, the phase difference between successive rays. It is instructive to think, now, of the pattern actually formed in the focal plane of the lens in Fig. 9–1.

At the centre of a fringe all disturbances arrive in phase (i.e. with phase differences that are integral multiples of 2π). A fringe is sharp if the intensity falls off rapidly as one moves away from the maximum, and this will happen if

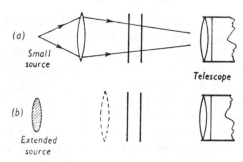

Fig. 9–4. Experimental arrangement.

strong disturbances get out of phase rapidly. The requirement is, then, that for a small displacement away from a maximum the largest possible phase difference should arise between the first transmitted ray and the last ray of appreciable amplitude. This corresponds to the largest possible plate separation and the highest possible reflectivity.

Unless the reflecting surfaces are very close together, it is necessary to use an observing telescope in order to collect sufficient multiply reflected rays (see §8–1) and one may again employ a small source under conditions resembling Maxwellian view [Fig. 9–4(a)] or an extended source with or without a collimating lens [Fig. 9–4(b)]. With no lens a very large source is required. It should be noted that the presence and position of any lenses on the source side of the film make no difference to the fringes in the region in which they can be seen, the lenses simply cause light to traverse the reflecting plates at various angles and enter the observing lens, thus increasing the field of view over which fringes can be seen. The fringes are always in focus in the focal plane(s) of whatever lenses follow the reflecting plates.

9–2. The Fabry-Perot Interferometer and Etalon

In the Fabry-Perot interferometer the separation between two semi-silvered glass plates can be varied. One plate remains stationary with respect to the frame of the instrument whilst the other is mounted on a nut threaded on an accurate screw. The nut is constrained to move exactly linearly so that, once the plates are adjusted to be parallel, they remain parallel as the separation is altered. The instrument is difficult to manufacture and must now be considered obsolescent.

The etalon consists of two semi-silvered plates rigidly held a fixed distance apart. This is obviously much more easily manufactured and serves many useful purposes. In both instruments the reflecting surfaces must be very accurate optical flats and, to keep the fringes sharp and truly circular, it is necessary for the reflecting surfaces to be flat to a small fraction of a wavelength. An accuracy of approximately $\lambda/100$ can be achieved. It is usual for the glass plates of the instruments to be slightly wedge shaped so that the outer, unsilvered, faces of the plates are not exactly parallel to the inner, silvered, faces. The wedge angle is usually between 1 and 10 minutes of arc. This is desirable in order to reduce the effect of the interference patterns formed by reflections at the unsilvered faces. The observed fringe system is not materially affected.

The adjustment of the Fabry-Perot interferometer is in many ways similar to that of the Michelson. Using a small separation, the plates are made roughly parallel by viewing a small source through the plates. The plates are adjusted until the images of the source, formed by multiple reflections, appear collinear. Next, using a monochromatic source, the mirrors are adjusted to give circular fringes centred at the centre of the field as with the Michelson interferometer. Since the plate separation is small, these can be seen with the naked eye. As before, exact parallelism of the plates is secured by adjusting the mirrors until a lateral movement of the eye does not cause the fringes to open out or close in.

One simple application of the etalon can be mentioned here: the measurement of the refractive index of air (or any other gas). The optical path difference between successive beams emerging from the etalon is $2nd \cos \theta$, where n is the refractive index of the gas between the plates. If the space between the plates is evacuated, the path difference becomes $2d \cos \theta$. That is, the optical path difference is decreased by $2(n-1)d \cos \theta$. If λ_0 is the wavelength in a vacuum, $[2(n-1)d \cos \theta]/\lambda_0$ is the change in path difference in wavelengths and is equal to the number of fringes which cross the field of view. By counting this number, n can be found if d and λ_0 are known.

9-3. Resolving Power of Fabry-Perot Instruments

As for a spectral line (§8–10) the sharpness of a fringe can be indicated by stating the half-value width, i.e. the width of the fringe at $I=\frac{1}{2}I_{max}$. In Fig. 9–5 I/I_{max} is plotted against order; referring to the fringes shown by the unbroken

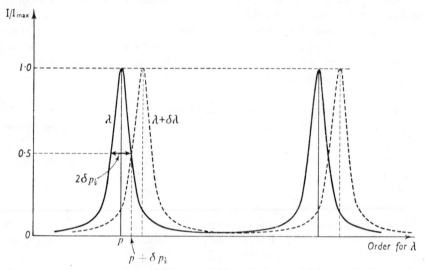

Fig. 9–5. Fringes for neighbouring wavelengths.

lines, the order p is an integer at the centre of a fringe and increases by $\delta p_{\frac{1}{2}}$ as I/I_{max} falls to $\frac{1}{2}$. The half-value width in terms of order is $2\delta p_{\frac{1}{2}}$. Equation (9–1) (known as Airy's formula) gives

$$I= \frac{I_{max}}{1+F \sin^2(\frac{\delta}{2})},$$

where
$$F=4R/(1-R)^2 \quad \text{and} \quad \delta=(2\pi/\lambda).2d \cos \theta=2\pi p$$

$$\therefore \frac{I}{I_{max}} = \frac{1}{1+F \sin^2\pi p}.$$

When $I=\frac{1}{2}I_{max}$, $p=\text{integer}+\delta p_{\frac{1}{2}}$ and $\sin \pi p=\sin \pi\delta p_{\frac{1}{2}}$.

$$\therefore \frac{1}{2}=\frac{1}{1+F \sin^2\pi\delta p_{\frac{1}{2}}},$$

or
$$F \sin^2\pi\delta p_{\frac{1}{2}}=1.$$

If the reflectivity is high, F is large, $\delta p_{\frac{1}{2}}$ is small, and the angle may be equated to its sine.

$$\therefore (\pi\delta p_{\frac{1}{2}})^2 =\frac{1}{F},$$

or
$$\delta p_{\frac{1}{2}}=\frac{1}{\pi\sqrt{F}}.$$

$$\therefore \text{ Half-value width}=2\delta p_{\frac{1}{2}}=\frac{2}{\pi\sqrt{F}}.$$

The ratio: $\dfrac{\text{separation of successive orders}}{\text{half-value width}}$ is called the *finesse*.

Now, *in terms of order* the separation of successive orders is simply unity, so that the finesse is simply the reciprocal of the half-value width:

$$\text{Finesse}=\frac{1}{2\delta p_{\frac{1}{2}}}=\frac{\pi\sqrt{F}}{2}.$$

(Fabry and Perot called F the coefficient of finesse.)

Suppose, now, that two wavelengths λ, $\lambda+\delta\lambda$, are emitted by the source with equal intensities. The bright fringes of each wavelength will be formed alongside one another as in Fig. 9–5. As $\delta\lambda$ decreases, the maxima overlap to a greater extent, and when the fringes are just resolved (i.e. just distinguishable), $\lambda/\delta\lambda$ is, as usual, referred to as the *chromatic resolving power*. The subject of resolving power is discussed more fully in §§11–5 and 11–6 and it is seen that when investigating the resolving power of prism spectroscopes Rayleigh formulated a very simple criterion for deciding when two slit images are resolved. However, the intensity distribution in the Fabry-Perot fringes for strictly monochromatic light (the so-called *instrumental profile* or *function*) is quite different from that in diffraction slit images and the simple statement of the criterion as given in §11–5 cannot be applied here. When expressed in terms of intensity in the pattern, the *Rayleigh criterion* implies that two patterns may be said to be resolved if, midway between their centres, there is a reduction of intensity to about 81% of that occurring at the centre of each separate image. Although this latter is the correct criterion to adopt here in order to be able to compare the resolution of a Fabry-Perot instrument with that of a prism or a grating, it leads to a fairly complicated calculation; the *Taylor criterion* is a useful alternative which is in the spirit of Rayleigh's, being simple in statement and giving simple results rather than being precise in the sense of being based on a careful consideration of what variations in intensity can be detected in particular circumstances. The Taylor criterion states that there is a just-detectable drop in intensity between two fringes, so that they can be said to be resolved, if the separate intensity curves intersect at $I=\frac{1}{2}I_{\max}$, i.e. if the separation of the maxima is equal to the half-value width.

It has been pointed out by Tolansky that the spectroscopist is usually concerned with the wave number $\bar{\nu}=1/\lambda$ rather than the wavelength of a line, and that when two lines are very close it is $\delta\bar{\nu}$ rather than $\delta\lambda$ that has significance. The value of $\delta\bar{\nu}$ when two fringes are just resolved Tolansky calls the *resolving limit*; it is easily obtained from the classical resolving power. One has $\bar{\nu}\lambda=1$,

$$\therefore\ \frac{\lambda}{\delta\lambda}=-\frac{\bar{\nu}}{\delta\bar{\nu}},\qquad\cdots\quad\cdots\quad\cdots\quad(9\text{–}6)$$

$$\therefore\ \delta\bar{\nu}=-\frac{\bar{\nu}}{(\lambda/\delta\lambda)}.\qquad\cdots\quad\cdots\quad\cdots\quad(9\text{–}7)$$

The minus sign has no importance and will be dropped when the result is applied below. The value $\delta\bar{\nu}$ is a direct measure of the power of the instrument to separate close components.

Suppose, now, that at the pth order maximum for $\lambda+\delta\lambda$ the order for λ is $p+\delta p$. Since this corresponds to a given point in the pattern,

$$2d\cos\theta=(p+\delta p)\lambda=p(\lambda+\delta\lambda),$$

whence

$$\frac{\lambda}{\delta\lambda}=\frac{p}{\delta p}. \qquad \qquad \ldots \ldots \ldots \quad (9\text{-}8)$$

Now, at the limit of resolution the Taylor criterion gives

$$\delta p=2\delta p_{\frac{1}{2}}$$

$$\therefore \quad \frac{\lambda}{\delta\lambda}=\frac{p}{2\delta p_{\frac{1}{2}}}=p\cdot\frac{\pi\sqrt{F}}{2}=p\times\text{Finesse.}$$

whence

$$\delta\bar{\nu}=\frac{2}{\lambda p\pi\sqrt{F}}$$

or, since $\lambda p=2d$ if θ is small,

$$\delta\bar{\nu}=\frac{1}{\pi d\sqrt{F}}, \quad \text{or} \quad \frac{\lambda}{\delta\lambda}=\frac{\pi d\sqrt{F}}{\lambda}.$$

In order to be able to compare the resolution of a Fabry-Perot instrument with that of a prism or a grating, the Rayleigh criterion will now be applied as indicated above. As before, the order at a short distance from a maximum is an integer $+\delta p$ and the intensity is given by

$$I=\frac{I_{\max}}{1+F\sin^2(\pi\delta p)}.$$

Fig. 9-6 shows the appearance of two maxima at the limit of resolution. The total intensity is indicated by the full line and, according to the criterion given above, falls at the midpoint to 0·81 of its maximum value. It will be seen that the maximum reached by the total intensity is greater than $I_{\max}$ for a single

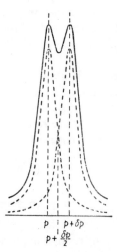

fringe because the intensity due to the adjacent fringe does not fall to zero. However, if it is assumed that at the centre of one fringe the intensity due to the other may be neglected, the total intensity midway between the maxima is simply $0·81I_{\max}$, and the intensity at this point associated with each fringe is $0·405I_{\max}$. Now, if p is the order at the centre of the maximum and δp is the smallest resolvable change in order, the order for wavelength λ at the centre of the maximum for wavelength $\lambda+\delta\lambda$ is $p+\delta p$ and that midway between the maxima is $p+\frac{1}{2}\delta p$. Thus at the mid-point one may write, for a first approximation,

Fig. 9-6. Fringes at limit of resolution.

$$I=\frac{I_{\max}}{1+F\sin^2\frac{1}{2}\pi\delta p}=0·405I_{\max},$$

$$\therefore \quad \sin^2\tfrac{1}{2}\pi\delta p=\frac{1-0·405}{0·405F}=\frac{1·469}{F}.$$

If the reflectivity of the surfaces is large, F is large and the angle may be equated to its sine. Whence

$$\delta p = \frac{1}{1\cdot301\sqrt{F}}.$$

If allowance is made for the fact that the intensity in one fringe is not zero at the centre of the other, it can be shown that one obtains, as a better approximation,

$$\delta p = \frac{1}{1\cdot49\sqrt{F}}.$$

Hence the resolving power is given by

$$\frac{\lambda}{\delta\lambda} = \frac{p}{\delta p} = p\,1\cdot49\sqrt{F}. \quad \cdots \quad (9\text{–}9)$$

The resolving limit is given by

$$\delta\bar{\nu} = \frac{\bar{\nu}}{(\lambda/\delta\lambda)} = \frac{1}{\lambda p\,1\cdot49\sqrt{F}}$$

or, since $\lambda p = 2d$ when θ is small,

$$\delta\bar{\nu} = \frac{1}{2d\,1\cdot49\sqrt{F}}. \quad \cdots \quad (9\text{–}10)$$

As will be seen in Chapter XII, the resolving power of a diffraction grating is the product of the order and the total number of apertures; for the reflection echelon the latter is the total number of plates. In comparing the properties of the echelons and the Fabry-Perot instruments it is possible, in view of equation (9–9), to regard $1\cdot49\sqrt{F}$ as the effective number of reflections in the latter instruments. It is easily shown that this increases from 8·3 when $R=70\%$ to 28·3 when $R=90\%$ and reaches 58 when $R=95\%$, so that the attainment of a higher reflectivity is an important factor in these instruments. As $R\to1$, $F\to4/(1-R)^2$ and the effective number of reflections is given approximately by $3/(1-R)$. This gives 50 for $R=94\%$ and 60 for $R=95\%$.

It is interesting to note that all spectral lines have a width which depends on the conditions obtaining in the source, so that resolution of smaller intervals is useless. With non-laser sources the maximum resolving power which can be employed usefully is about 10^7 and this is obtained with an etalon of thickness about 20 cm. when the reflectivity of the surfaces is about 80%.

Usually one is interested in examining neighbouring spectral lines over a finite range of wavelengths. The possibility then arises of the extreme wavelengths in the range being sufficiently different to cause the maximum of a certain order of one to overlap that of the next order of another. This, of course, renders examination difficult and it is important to know the *spectral range* over which an instrument can be used without this overlapping of orders. To examine a larger spectral range one would have to employ an auxiliary dispersion system to isolate manageable regions. If $\Delta\lambda$ is the difference between two wavelengths such that the pth order of one coincides with the $(p+1)$th order of the other, one has from equation (9–8) since $\delta p=1$,

$$\Delta\lambda = \frac{\lambda}{p} = \frac{\lambda^2}{2d} \quad \cdots \quad (9\text{–}11)$$

or, in terms of wave numbers,

$$\Delta\bar{\nu} = \frac{\bar{\nu}}{p} = \frac{1}{2d}.$$

It follows at once from (9–9) and (9–11) that

$$\frac{\Delta\lambda}{\delta\lambda} = \frac{1}{\delta p}.$$

Now, using the Taylor criterion, $\delta p = 2\delta p_{\frac{1}{2}}$ so that, in terms of either wave numbers or wavelengths,

$$\frac{\text{spectral range}}{\text{smallest resolvable increment}} = \frac{\Delta\bar{\nu}}{\delta\bar{\nu}} = \frac{\Delta\lambda}{\delta\lambda} = \text{Finesse}.$$

It is the relatively small free spectral range and the consequent need of a premonochromator to isolate this range that is the chief disadvantage of Fabry-Perot instruments. For this reason, in spite of the basic photometric superiority of the Fabry-Perot, high quality diffraction grating systems are often used for resolving powers up to about 10^6. The development of efficient interference filters, which are themselves Fabry-Perot systems (see §9–17), swings the balance back in favour of the Fabry-Perot since the interference filter is photometrically superior to the prism or grating monochromator. For ultra-high resolution spectroscopy the Fabry-Perot and Fourier transform systems are the only ones in common use. The relative merits of the various spectroscopic systems are discussed in §12–17 and §18–12.

9–4. Comparison of Wavelengths. Method of Coincidences. Hyperfine Structure

If the source emits two wavelengths λ_1 and λ_2, two sets of fringes will be observed with a Fabry-Perot interferometer. Suppose that, in a direction θ, the maxima for λ_1 and λ_2 coincide when the plate separation is d_1. Then

$$2d_1 \cos\theta = p\lambda_1 = q\lambda_2, \quad . \quad . \quad . \quad . \quad . \quad (9\text{–}12)$$

where p and q are integers. Thus $\lambda_1/\lambda_2 = q/p$. Proceeding outwards from the centre of the pattern, the condition for the next maxima to coincide is $\lambda_1/\lambda_2 = (q-1)/(p-1)$. Now since p and q are large ($p=40{,}000$ when $\theta=0$, $d=1$ cm., $\lambda=5000$ angstroms), the ratios q/p and $(q-1)/(p-1)$ are nearly equal. Therefore the maxima will coincide over a considerable range of angles—certainly over the field of view of the observing telescope, i.e. the complete ring systems will appear coincident.

Suppose the plate separation is increased until the two sets of fringes again coincide. The order of λ_1 has then increased by k and that of λ_2 by $k+1$ so that

$$2d_2 \cos\theta = (p+k)\lambda_1 = (q+k+1)\lambda_2.$$

Hence, from equation (9–12),

$$2\cos\theta(d_2-d_1) = k\lambda_1 = (k+1)\lambda_2, \quad . \quad . \quad . \quad (9\text{–}13)$$

$$\therefore \ (\lambda_1-\lambda_2) = \frac{\lambda_2}{k} = \frac{\lambda_1}{k+1}, \quad . \quad . \quad . \quad . \quad (9\text{–}14)$$

or, writing $\cos \theta = 1$,

$$\lambda_1 - \lambda_2 = \frac{\lambda_1 \lambda_2}{2(d_2 - d_1)}. \qquad \ldots \ldots \quad (9\text{–}15)$$

One can proceed as follows: k may be found by counting the number of λ_1 (or λ_2) fringes which pass an instrumental mark (or arise from the centre of the pattern). Equation (9–14) then gives λ_1/λ_2 or, if one of the wavelengths is known, $(\lambda_1 - \lambda_2)$ may be found. If the change in mirror separation is given by the instrument, equation (9–13) can be used to give k if λ_1 or λ_2 is known or to find λ_1 and λ_2 if k is found by counting. Alternatively, if approximate values of λ_1 and λ_2 are available, equation (9–15) gives a value for $(\lambda_1 - \lambda_2)$ which is more accurate than that known from the approximate values of the wavelengths.

For widely different wavelengths, k is relatively small and several coincidences should be recorded. The method is then very accurate but has the disadvantage of employing the costly interferometer and of requiring individual measurements for every spectral line examined. The method of exact fractions described in the next section uses the cheaper etalon and enables numerous lines to be examined relatively quickly.

One of the most common applications of Fabry-Perot instruments is to the investigation of the hyperfine structure of spectral lines. In this work one measures the small difference in wave numbers $(\Delta\bar{\nu})$ between the satellite and the main line. If θ_1 and θ_1' are the angular radii of the main line and the satellite for order p_1, and θ_2 is the angular radius for the main line of order p_2, one has, if the angles are small,

$$1 - \frac{\theta_1^2}{2} = \frac{p_1}{2\bar{\nu}d}, \quad 1 - \frac{\theta_2^2}{2} = \frac{p_2}{2\bar{\nu}d}, \quad 1 - \frac{\theta_1'^2}{2} = \frac{p_1}{2\bar{\nu}'d},$$

where $\bar{\nu}$ and $\bar{\nu}'$ are the wave numbers of the main line and the satellite respectively. If θ_1 and θ_2 correspond to adjacent maxima, $p_2 = p_1 - 1$. The above equations then give

$$\frac{\theta_2^2 - \theta_1^2}{2} = \frac{1}{2d\bar{\nu}} \quad \text{and} \quad \frac{\theta_1'^2 - \theta_1^2}{2} = \frac{p_1}{2d}\left(\frac{1}{\bar{\nu}} - \frac{1}{\bar{\nu}'}\right) = \frac{p_1}{2d} \cdot \frac{\Delta\bar{\nu}}{\bar{\nu}\bar{\nu}'},$$

$$\therefore \quad \frac{\theta_1'^2 - \theta_1^2}{\theta_2^2 - \theta_1^2} = \frac{p_1 \Delta\bar{\nu}}{\bar{\nu}'}$$

or, since $\Delta\bar{\nu}$ is small,

$$\Delta\bar{\nu} = \left(\frac{\theta_1'^2 - \theta_1^2}{\theta_2^2 - \theta_1^2}\right)\frac{\bar{\nu}}{p_1} = \left(\frac{\theta_1'^2 - \theta_1^2}{\theta_2^2 - \theta_1^2}\right)\frac{1}{2d}.$$

9–5. Method of Exact Fractions

Given an etalon whose plate separation is known approximately, and given a number of accurately known wavelengths, the method of exact fractions enables one to find the plate separation more accurately and to determine other wavelengths.

For any wavelength λ, the order of interference at the centre of the Fabry-Perot fringes is not usually an integer and may be written $(p + \varepsilon)$ where p is the order of the first bright ring and is, of course, integral. If one has, for example, three accurately known wavelengths λ_1, λ_2, λ_3, one may write, for the orders at the centre of their respective fringe systems, $(p_1+\varepsilon_1)$, $(p_2+\varepsilon_2)$, $(p_3+\varepsilon_3)$, and one has

$$2d=\lambda_1(p_1+\varepsilon_1)=\lambda_2(p_2+\varepsilon_2)=\lambda_3(p_3+\varepsilon_3). \quad . \quad . \quad . \quad (9\text{--}16)$$

The fractions ε_1, ε_2, ε_3, can be found from observation of the fringe systems (see below), but if d is unknown the values of the integers p_1, p_2, p_3, are unknown since there will be an infinite number of sets of values which will satisfy equation (9–16). However, if an approximate value of d is available, this will give a range of possible values for, say, p_1. One can then take values of p_1 within this range and, using the observed value of ε_1, calculate $(p_2+\varepsilon_2)$ and $(p_3+\varepsilon_3)$ from equation (9–16) until a value of p_1 is found which yields the observed values of ε_2 and ε_3. If the approximate value of d is very unreliable it may be found that equation (9–16) can be satisfied for more than one value of p_1 lying within the possible range. This difficulty is easily overcome since values of p_1 that satisfy equation (9–16) become less frequent as the number of standard wavelengths is increased. Thus one must measure the fractions for additional standard wavelengths until the values of p_1 satisfying equation (9–16) become sufficiently infrequent for only one value to lie within the range of possible values. In this way the value of p_1 can be found exactly. The value of ε_1 can be found to an accuracy of about ±0.01 so that if, for example, λ_1 is 6438 angstroms (cadmium red light) d can be determined to an accuracy of about 0·0000003 cm. If another wavelength is known with sufficient accuracy to determine the integral part of the order at the centre, measurement of the exact fraction as described below yields a more accurate value for the wavelength. By proceeding subsequently to thicker etalons, wavelengths can be determined with still greater accuracy (e.g. to about ±0.0001 angstrom with a 10 cm. etalon). To justify the use of the thicker etalons it is, of course, necessary to have a source of light yielding very sharp spectral lines.

Example

This discussion is intended to illustrate the principles upon which the method of exact fractions is based. There are various computational tricks that can be used to shorten the calculation in practice.

Suppose the fringes are observed for four accurately known wavelengths and for one wavelength that is known much less accurately, using an etalon whose thickness is known approximately. It is required to find both the etalon thickness and the "unknown" wavelength as accurately as possible. This example actually assumes a rather low precision for the determination of the fractions. Suppose, then, that the initial data are:

$\lambda_1 = 4358\cdot343 \pm 0\cdot001 \text{ Å}$ $\varepsilon_1 = 0\cdot95 \pm 0\cdot05$
$\lambda_2 = 5085\cdot822 \pm 0\cdot001 \text{ Å}$ $\varepsilon_2 = 0\cdot70 \pm 0\cdot05$
$\lambda_3 = 3460\cdot742 \pm 0\cdot001 \text{ Å}$ $\varepsilon_3 = 0\cdot65 \pm 0\cdot05$
$\lambda_4 = 5875\cdot6 \pm 0\cdot1 \text{ Å}$ $\varepsilon_4 = 0\cdot05 \pm 0\cdot05$
$\lambda_5 = 6438\cdot470 \pm 0\cdot001 \text{ Å}$ $\varepsilon_5 = 0\cdot60 \pm 0\cdot05$

$$d = 1\cdot234 \pm 0\cdot001 \text{ cm.}$$

It is required to find d and λ_4 as accurately as possible.

d lies between 1·233 and 1·235

Using $2d = p_1\lambda_1$, one finds that p_1 lies between 56581 and 56673.

Using $(p_1 + \varepsilon_1)\lambda_1 = (p_2 + \varepsilon_2)\lambda_2$, and inserting the known values of λ_1 and λ_2 and the measured value of ε_1, it is found that 14 values of $(p_1 + \varepsilon_1)$ between 56581·95 and 56673·95 yield calculated values of $(p_2 + \varepsilon_2)$ for which the calculated ε_2 agrees with the measured value 0·70 ± 0·05.

Using $(p_1 + \varepsilon_1)\lambda_1 = (p_3 + \varepsilon_3)\lambda_3$ (employing the known value of λ_3) it is found that of these 14 values of $(p_1 + \varepsilon_1)$ only two yield calculated values of $(p_3 + \varepsilon_3)$ for which ε_3 agrees with the measured value 0·65 ± 0·05.

In fact these are

$$(p_1 + \varepsilon_1) = 56614·95$$

(which yields $(p_2 + \varepsilon_2) = 48516·71$ and gives $(p_3 + \varepsilon_3) = 45185·69$) and

$$(p_1 + \varepsilon_1) = 56649·95$$

(which yields $(p_2 + \varepsilon_2) = 48546·71$ and gives $(p_3 + \varepsilon_3) = 45213·62$).

The first of these yields

$$(p_5 + \varepsilon_5) = 38323·91,$$

and the second yields

$$(p_5 + \varepsilon_5) = 38347·60.$$

Since 0·63 is in agreement with the observed value of 0·60 ± 0·05 for ε_5 but 0·91 is not, the correct value of $(p_1 + \varepsilon_1)$ is identified as

$$(p_1 + \varepsilon_1) = 56649·95 \pm 0·05;$$

it is the only value of $(p_1 + \varepsilon_1)$ that yields calculated values of ε_2, ε_3, and ε_5, that are in agreement with the measured values.

Using $2d = (p_1 + \varepsilon_1)\lambda_1$ one obtains

$$d = 1·234500 \pm 0·000002$$

If one now assumes $\lambda_4 = 5875·6$ one obtains $(p_4 + \varepsilon_4) = 42021·24$. Since ε_4 has been measured and is known to be 0·05 ± 0·05, one assumes that the actual value of $(p_4 + \varepsilon_4)$ is

$$(p_4 + \varepsilon_4) = 42021·05 \pm 0·05$$

which yields

$$\lambda_4 = 5875·627 \pm 0·005$$

9–6. Measurement of the Fractions

In order to measure the fractional orders at the centres of the fringe systems for several different wavelengths, it is convenient to employ a prism spectrometer

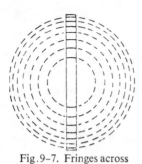

Fig. 9-7. Fringes across image of slit.

arrangement in which the etalon is placed between the telescope and collimator. For each wavelength present in the source (or successive sources) the arrangement is equivalent to that illustrated in Fig. 9–4(b) with the exception that, apart from the slit width, the source extends in one direction only. Thus in the focal plane of the telescope one gets a vertical section across a diameter of the fringe system for each wavelength, the width of the section being equal to that of the slit image. The appearance of each spectral line is therefore as shown in Fig. 9–7. If j is the ring number going out from the centre of the pattern and p is the order for the first ring, the order for the jth ring is $(p-j+1)$ and one has

$$\frac{2d \cos \theta_j}{\lambda} = p-j+1,$$

$$\therefore \quad (p+\varepsilon) \cos \theta_j = p-j+1,$$

where θ_j is the angular radius of the jth ring before subsequent magnification.

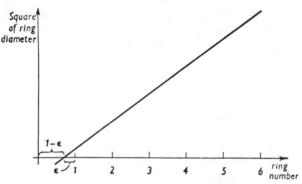

Fig. 9–8. Determination of the fraction.

The ring diameters may be measured directly using a telescope with a micrometer eyepiece. Alternatively the fringes may be photographed and the ring diameters measured on the plate or film. If, in any suitable units, D_j is the diameter of the jth ring, one may write $mD_j=2\theta$, where m is a magnification constant. This is justifiable since the angular diameters of the observed rings are small. In addition one may write

$$\cos \theta_j=1-\frac{\theta_j{}^2}{2} =1-\frac{m^2D_j{}^2}{8},$$

$$\therefore \quad (p+\varepsilon)-(p+\varepsilon)\frac{m^2D_j{}^2}{8} =p-j+1,$$

or

$$\left(\frac{p+\varepsilon}{8}\right) m^2D_j{}^2=j+\varepsilon-1,$$

so that when $D_j=0$, $j=1-\varepsilon$, and a graph of (ring diameter)^2 v. (ring number) is a straight line, the intercept on the ring number axis being $(1-\varepsilon)$ (see Fig. 9–8).

9–7. Brewster's Fringes

Fig. 9–9(a) shows a plan view of two equal etalons A and B whose normals are inclined at equal and opposite small angles, $\pm\alpha$, to the axis of the telescope T. S is an extended source of white light.

Rays directed from S parallel to the axis are amplitude-divided at the various semi-silvered surfaces. In particular, ignoring refractions in the glass plates, there will be disturbances along paths of the types (1) and (2). Such disturbances will arrive in phase in the primary image plane of T. Rays traversing the etalons in this and similar ways may be divided into pairs, one of which undergoes p reflections in A and $(p+2)$ in B (p even) and the other $(p+2)$ in A and p in B. Those of type (1) have an extra path $2d\cos\alpha$ in B, and those of type (2) an equal extra path in A. Rays not parallel to the axis but inclined at angle ϕ and lying

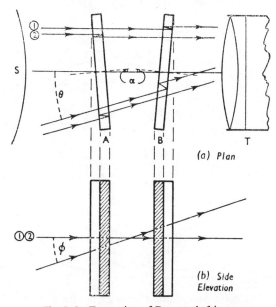

Fig. 9–9. Formation of Brewster's fringes.

in vertical planes that are parallel to the axis will pass obliquely through the etalons [see Fig. 9–9(b)]. Members of each pair of rays in a particular direction will again have equal optical paths through the etalons and will arrive in phase at a single point in the primary image plane of T. Those in different directions ϕ will arrive in phase at different points along the central vertical line in the image plane. Thus there is constructive interference for all wavelengths along the central vertical line in the field of view of T and this line will be a white light maximum.

Consider horizontal rays inclined at angle θ to the axis of the telescope. These and all parallel rays pass through A at angle $(\theta-\alpha)$ and through B at angle $(\theta+\alpha)$ (or vice versa). Thus for a ray of type (1) the extra path in B is

$2d \cos (\theta+\alpha)$, and differs from the extra path $2d \cos (\theta-\alpha)$ in A traversed by a ray of type (2). Thus, at the appropriate non-central point on the central horizontal line in the primary image plane of T, the members of each pair arrive with a phase difference of

$$\frac{2\pi}{\lambda} 2d \, [\cos (\theta-\alpha)-\cos (\theta+\alpha)] = \frac{8\pi d}{\lambda} \sin \theta \sin \alpha.$$

Consider, now, rays inclined at the same angle θ to the vertical plane containing the axis but inclined also at angle ϕ to the horizontal. The additional paths through the appropriate etalons of the members of each pair are now given by multiplying the above values by $\cos \phi$. The resulting phase difference between the members of a pair is also given by multiplying the previous value by $\cos \phi$. Rays traversing the etalons at the same angle θ but with different values of ϕ come to a focus at points along a vertical line in the primary image plane of T, and at each point the phase difference between the members of a pair of rays is that given above. Consequently, along such a vertical line the order of inter-

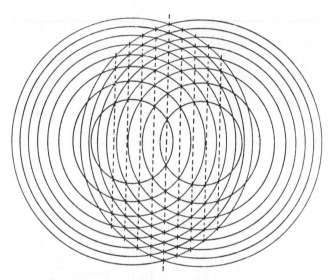

Fig. 9–10. Monochromatic fringes of superposition.

ference between the members of a pair of rays is not exactly constant. However, the variation is small since ϕ is small, and for low orders the variation is completely negligible. Thus in the field of view of T one sees a series of vertical fringes. They are known as *Brewster's fringes*.

At a maximum, the phase difference between the disturbances in each pair is $2j\pi$ (j an integer) i.e. at a maximum

$$\frac{8\pi d}{\lambda} \sin \theta \sin \alpha = 2j\pi,$$

or, since θ and α are small,

$$\theta = \frac{j\lambda}{4d\alpha}, \qquad \cdot \quad \cdot \quad \cdot \quad \cdot \quad \cdot \quad \cdot \quad (9\text{--}17)$$

so that the angular separation of the fringes is $\dfrac{\lambda}{4d\alpha}$.

With white light one sees a well defined central maximum, a number of coloured fringes as the maxima for various wavelengths intermingle, and finally a white background. With monochromatic light high order fringes of super-position can be observed; these, as shown above, are curved. With monochromatic light one can see the Fabry-Perot fringes for each etalon as in Fig. 9–10 and the difference fringes appear where the two circular fringe systems are super-posed; they are indicated by the dotted lines.

The fringes described above are a special case and, in general, fringes of super-position will be formed by interference between disturbances undergoing p reflections in A and $(p+q)$ in B and those undergoing $(p+q)$ in A and p in B (p and q even integers). The path difference between such disturbances is $qd[\cos(\theta-\alpha)-\cos(\theta+\alpha)]$ i.e. $2qd \sin\theta \sin\alpha$ and this is $j\lambda$ at a maximum. Thus, in general, the fringe separation is given by $\theta=\lambda/2qd\alpha$. In each case the central white light maximum occurs at $\theta=0$.

It is fairly obvious that if one etalon is exactly twice as thick as the other, Brewster fringes will be formed, for example, by the interference of distur-bances undergoing $(p+4)$ reflections in the thinner and p in the thicker and those undergoing p in the thinner and $(p+2)$ in the thicker. This can obviously be extended to cover any case where the ratio of the thicknesses is N_1/N_2, where N_1 and N_2 are small integers. If N_1 and/or N_2 are large, the fringes will be of poor contrast since the relevant interference will be solely between beams that have undergone many reflections and are consequently of small amplitude.

In addition, Brewster's fringes can be observed when the ratio of the thick-nesses of the etalons is *nearly* N_1/N_2. In this case the path difference between rays of types (1) and (2) can be zero when the etalons are inclined at slightly different angles to the axis of the telescope. For example, the path difference between disturbances which have undergone $2N_2$ extra reflections in A and those which have undergone $2N_1$ extra in B is

$$2N_2 d_1 \cos(\theta-\alpha_1)-2N_1 d_2 \cos(\theta+\alpha_2).$$

In particular, if one etalon has a thickness slightly less than that given by $d_1/d_2=N_1/N_2$, the fringes will be seen when this etalon is placed perpendicular to the axis of T and the other is tilted. For example, suppose that, when d_1/d_2 is slightly less than N_1/N_2, the first etalon is co-axial with the telescope and the second is tilted through angle α_2. The path difference between disturbances undergoing $2N_2$ extra reflections in the first etalon and those undergoing $2N_1$ extra in the second is

$$2N_2 d_1 \cos\theta - 2N_1 d_2 \cos(\theta+\alpha_2).$$

On putting $\theta=0$ it follows that the zero order white light maximum will be central in the field of view of the telescope if

$$2N_2 d_1 - 2N_1 d_2 \cos\alpha_2 = 0,$$

i.e. if
$$N_2 d_1 = N_1 d_2 \cos\alpha_2. \qquad \ldots \ldots \quad (9\text{–}18)$$

For a first order maximum

$$2N_2d_1 \cos \theta - 2N_1d_2 \cos (\theta + \alpha_2) = \lambda,$$

or, from (9–18),

$$2N_1d_2 \sin \theta \sin \alpha_2 = \lambda,$$

or, since θ and α_2 are small,

$$\theta = \frac{\lambda}{2N_1d_2\alpha_2}, \qquad \cdots \cdots \quad (9\text{–}19)$$

and this will be the angular separation of the Brewster fringes. Alternatively, in terms of d_1 and d_2 one has [from (9–18)]

$$\theta = \frac{\lambda}{\sqrt{8N_1d_2(N_1d_2 - N_2d_1)}}. \qquad \cdots \quad (9\text{–}20)$$

On comparing the fringes described above with those formed in the Jamin refractometer, it will be seen that the latter can be regarded as a special case of Brewster's fringes, being formed by reflection instead of transmission, the interference being between disturbances undergoing a single internal reflection in one plate.

9–8. Central Spot Scanning

Originally, Fabry-Perot instruments were used in the visible and near U.V. regions and the fringes were observed visually or photographed so that the complete fringe system in the focal plane of the lens (or a section across it) was observed or recorded simultaneously. For work in the I.R., particularly in the far I.R., other detectors must be used (e.g. photo-conductive cells or thermal detectors) and none of these give the spatial discrimination necessary for the whole fringe pattern to be recorded simultaneously. However, these detectors

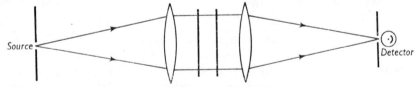

Fig. 9–11. Central spot scanning.

do give good temporal discrimination—which the photographic plate obviously does not—so that the appropriate method of examining the fringe system is by scanning across it and recording the variations of intensity with time. In fact, of course, the photographic method involves recording first and scanning (of the photograph) later, whilst what is here called the scanning method involves simultaneous scanning and recording. It is much easier to obtain reliable intensity measurements by using a detector such as a photo-cell—photography is a good method of recording positions but a relatively poor method of measuring intensities.

Now the condition for a maximum is

$$2nd \cos \theta = p\lambda_o,$$

λ_o being the vacuum wavelength. Hence for a fixed etalon the maxima for different wavelengths in a given order are found at different positions in the pattern, and this is the basis of the photographic method. In principle, it is possible to make this the basis of a scanning method in which a small scanning aperture is moved across the fringe system and the intensity variations are recorded. However, it is more convenient to use the fact that at a given point in the pattern (θ constant) different wavelengths will appear for different values of the etalon spacing d, or for different values of the refractive index n. In order to achieve maximum photometric efficiency one would need to use an annular aperture so that the detector could receive all the light corresponding to a given path difference. It is more usual to record the intensity at the centre of the pattern and vary either n or d—the former by varying the pressure of the air in the etalon, and the latter by displacing one of the plates. This is referred to as *central spot scanning* and the optical system is shown in Fig. 9–11. In principle it is not necessary to employ both pinholes, but if both are used there is a reduction in the amount of stray light reaching the detector. One is here measuring the intensity at the centre of the Fabry-Perot fringe system, but with the system shown in Fig. 9–11 no actual fringe system is formed in space; one can be said to be using the etalon as a filter or monochromator and measuring the intensity of the transmitted wavelength. The instrumental profile is the same as for the classical Fabry-Perot system. (The use of an etalon as a filter is discussed further in §9–17.) The character of the available detectors enforced the use of the scanning method in the I.R. but central spot scanning has advantages in all spectral regions.

Any method of using an etalon that involves oblique beams involves the migration of the beams across the aperture as they undergo repeated reflections. Eventually, some of these beams are lost at the edge of the aperture and this loss can be serious if the aperture is inadequate—the effect is more likely to be serious with thick etalons and has been ignored in all the preceding discussions. A loss of multiply reflected beams causes a loss of resolving power for oblique incidence; causes a reduction in intensity, which invalidates the comparison of intensities for light incident in different directions; and causes the appearance of spurious maxima analogous to the subsidiary maxima that occur with a diffraction grating which has a finite number of lines (see §§12–1 and 12–7). Thus the method of central spot scanning gives a higher resolving power, more reliable intensity measurement, and the elimination of spurious maxima. In addition, one has the advantage of obtaining a linear scale of wavelengths—see below. (The dispersion is not linear across the spatial fringe pattern.)

For normal incidence one has, using λ_o for the vacuum wavelength,

$$2nd = p\lambda_o$$

Thus $n \propto p$ if d is constant and $d \propto p$ if n is constant, so that the abscissa of Fig. 9–5 can be n or d, as appropriate. Consider, now, a source emitting a spread of wavelengths. The wavelength transmitted at maximum intensity is given by

$$\lambda_o = \left(\frac{2n}{p}\right)d = \left(\frac{2d}{p}\right)n$$

so that in a given order $\lambda_0 \propto d$ if n is constant or $\lambda_0 \propto n$ if d is constant, and the abscissa of Fig. 9–5 can be λ_0 for a fixed order. Hence if the output from the detector is fed to a recorder whose speed is adjusted so that its time-base follows the variations of n or d, the spectrum is plotted with a linear wavelength scale. If $\delta\lambda_0$ is the smallest resolvable wavelength difference, one has, for mechanical scanning or pressure scanning respectively,

$$\delta d = \frac{p}{2n}\, \delta\lambda_0 \quad \text{and} \quad \delta n = \frac{p}{2d}\, \delta\lambda_0$$

where δd and δn are the corresponding changes in d and n. The change in d or n required to scan a range of wavelengths equal to the free spectral range $\Delta\lambda_0$ is given by the change corresponding to a change of one order for wavelength λ_0. One has

$$\Delta d = \frac{\lambda_0}{2n} \quad \text{and} \quad \Delta n = \frac{\lambda_0}{2d}$$

since $\Delta p = 1$ by definition. Using these equations, one has for the chromatic resolving power,

$$\frac{\lambda_0}{\delta\lambda_0} = p \cdot \frac{\Delta d}{\delta d} \quad \text{or} \quad p \cdot \frac{\Delta n}{\delta n}$$

Using the Taylor criterion, this gives, in each case,

$$\frac{\lambda_0}{\delta\lambda_0} = p \times \text{finesse.}$$

Pressure scanning is often achieved by evacuating the space between the etalon plates and allowing the air to re-enter slowly—slowly in order to avoid heating effects. The change in index is then limited to $2 \cdot 7 \times 10^{-4}$ with the result that it may be impossible to cover the full free spectral range ($\lambda_0/2d$) if the wavelength is too long or the plate separation is too small, i.e. one may be unable to cover the larger spectral range obtained with a low resolution system. The actual range of wavelengths covered by a pressure scan of one atmosphere is $\lambda_0 - \lambda_0'$ where $2nd = p\lambda_0$ and $2d = p\lambda_0'$ i.e. $\lambda_0 - \lambda_0' = (n-1)2d/p = (n-1)\lambda$ and is seen to depend upon the spectral region in which one is working, but not upon the plate separation (i.e. not upon the resolving power being used). Pressure scanning is probably easier technically—one can adjust the gas flow into the container to give a linear wavelength scale and one avoids the difficulty of keeping the plates parallel while d is changed.

Mechanical scanning has been achieved by several methods, including the use of piezo-electric spacers in the etalon. Tolansky and Bradley (1959) have successfully used the system illustrated by Fig. 9–12 to oscillate one mirror. It will be seen that a displacement of $\lambda/2$ is sufficient to scan the free spectral range. In fact this system enabled 26 orders to be scanned before the lack of parallelism became of the same order as the lack of flatness of the plates (approximately $\lambda/80$). Rapid scanning is possible and repetitive frequencies up to 1000 cycles per second were used, corresponding to a time resolving limit of 4 microseconds

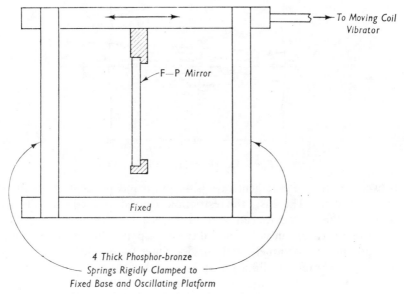

To Moving Coil
Vibrator

F—P Mirror

Fixed

4 Thick Phosphor-bronze
Springs Rigidly Clamped to
Fixed Base and Oscillating Platform

Fig. 9–12. Mechanical scanning.

for a scan of 5 orders. Bruce and Hill (1961) also used an electromagnetic vibrator for the moving mirror in a mechanical scanning system, but used a stationary coil and a moving soft iron ring assembly; they achieved a precision of 1 part in $2 \cdot 8 \times 10^9$ in setting on a fringe peak using the krypton 86 line at 6056 Å.

As indicated earlier, the small free spectral range is the chief disadvantage of Fabry-Perot systems for many purposes, but for the very highest resolving powers the Fabry-Perot and Fourier transform spectrometers are the only systems in common use (see §§ 12–17 and 18–12).

In the above discussion it was implied that one employs a point source (i.e. a pinhole of infinitesimal size). Obviously this is unrealistic since the energy reaching the detector would be infinitesimal. Alternatively, one could use an equally ruinous infinitesimal exit aperture in front of the detector. Obviously, in practice one must employ finite source and detector apertures; light enters the system in a finite range of directions close to normal incidence, with the result that disturbances reach the detector with a range of path differences. That is, a finite range of wavelengths arrive with maximum intensity so that the resolution deteriorates. This effect is reduced in the spherical Fabry-Perot described below; the photometric properties of these systems are discussed in §18–12.

9–9. The Spherical Fabry-Perot Interferometer

This instrument, introduced by P. Connes (1956), employs two spherical mirrors with a separation equal to their common radius of curvature. As indicated in Fig. 9–13, half the aperture of each mirror is totally reflecting and half semi-reflecting. It will be seen that any incident ray gives rise to an infinite number of emergent rays which are *coincident*, not simply parallel as they are with the classical Fabry-Perot. In the paraxial approximation the path difference

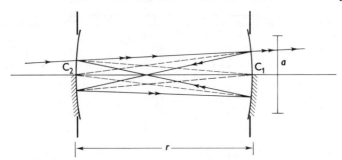

Fig. 9–13. The spherical Fabry-Perot.

between successive emergent rays is $4r$ and is independent of the direction of the incident ray. This eases the restriction referred to in §9–8 on the size of the source, so that the device can be regarded as a field-widened Fabry-Perot. The restriction is eased but not removed since the path difference is not precisely $4r$ for light incident obliquely, and one does have to impose a limit by employing circular diaphragms to limit the apertures (a) of the mirrors. The photometric properties of the device are discussed further in §18–12. Since the optical path difference between successive beams is doubled, the chromatic resolving power of a spherical Fabry-Perot is double that of a clasical etalon with the same mirror separation; obviously, for any given pair of spherical mirrors only one mirror separation is possible. That is, each device has to be manufactured specifically for the spectral range for which it is to be used. The system has also been used with both mirrors highly (but not totally!) reflecting over the whole of their apertures. In this form it has been widely used as the resonant cavity for a laser. In order to obtain maximum reflectivity with minimum absoption the mirrors usually consist of multilayer dielectric films rather than metal films (see §9–17).

In addition to the advantage of field widening, the spherical Fabry-Perot has the advantage of being easier to adjust than the classical form with plane mirrors; the adjustment of the angle between the mirrors is less critical and it is therefore easier to make the initial adjustment and to achieve the necessary stability. This becomes more important for higher resolving powers when larger mirror separations are employed and, in particular, is important in a laser. Although it is more difficult to make spherical mirrors than plane mirrors the importance of this difficulty is reduced by the fact that a spherical Fabry-Perot can utilize only small-diameter mirrors.

9–10. Comparison of Wavelengths with Mechanical Standards of Length; Method of Sears and Barrell

Prior to 1960 the legal metrical standard of length consisted of a platinum-iridium bar on which were engraved two lines whose distance apart was defined as one metre; practical measuring scales were derived from this by processes of comparison and sub-division. When interferometers such as the Michelson and Fabry-Perot instruments became available it became possible to measure distances in terms of wavelengths, and obviously it became necessary to obtain accurate determinations of the number of wavelengths in the length of the standard Metre. It was realized that the possibility would then arise of adopting the wavelength of some convenient spectral line as the ultimate standard of

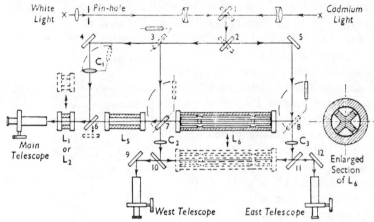

Fig. 9-14. Determination of the metre by Sears and Barrell.

length, and eventually this was done in 1960. (It was, in fact, Babinet who, as early as 1829, first suggested using a wavelength as a standard of length.)

The first attempt to relate the wavelength of a spectral line to the standard Metre was made by Michelson and Benoît, using a method based on the fact that the Michelson interferometer provides a means of measuring displacements in terms of wavelengths (see §8–16). Kösters and Lampe also used the Michelson interferometer and other workers, following Fabry, Perot, and Benoît, have used methods employing Fabry-Perot etalons. The determinations to be discussed in detail here were made by Sears and Barrell from 1933 onwards.

The method of exact fractions described in §9–5 enables one to find the thickness of an etalon in terms of wavelengths of light, so that if one could use an etalon with a plate separation of a metre one could make the required determination. Unfortunately, when this determination of the Metre was made it was impossible to observe fringes with a plate separation of a metre because there was no light source that was sufficiently monochromatic for such large path differences to be used. In consequence, the determination was carried out in three stages employing etalons whose thicknesses were approximately 1 metre, $\frac{1}{3}$ metre, and either $\frac{1}{9}$ or $\frac{1}{12}$ metre:

1. The determination of the thicknesses of the two smallest etalons in terms of the red cadmium line using the method of exact fractions.

2. The comparison of the two smallest etalons with the $\frac{1}{3}$ metre, and of the $\frac{1}{3}$ metre with the 1-metre etalon, using Brewster's fringes.

3. The comparison between the thickest etalon and the standard Metre.

The whole apparatus was made very compact and the temperature was measured to an accuracy of $0 \cdot 001°$ C. The general arrangement is shown in Fig. 9-14. The main telescope was employed to observe the circular fringes used in the measurement of the smallest etalons and also to observe the Brewster fringes when comparing etalons. The former measurement required cadmium red light with other cadmium and krypton lines as auxiliaries, and the latter required white light. The positions of the lamps and mirrors are indicated in the figure; it will be seen that the smallest etalon could be moved aside when comparing the other two etalons. In each comparison it was the $\frac{1}{3}$-metre etalon which was tilted.

In order to obtain a rigid etalon with a small expansion coefficient, a hollow invar cylinder was used as spacer in each case. The ends were made accurately parallel, and since it is difficult to obtain a clean optically polished surface on invar, they were chromium plated before being polished optically flat. Partially silvered glass or quartz plates were then wrung on to the ends of the cylinder to make an airtight joint so that the etalon could be evacuated. Quartz plates were used for the largest etalon because they were less likely to be damaged by the repeated wringing which is necessary (see below). Adjustment for parallelism and for small changes in length could be made by means of invar straining rods joining two flanges on the sides of the cylinder. Fig. 9–15 shows a scale drawing of the $\frac{1}{3}$-metre etalon.

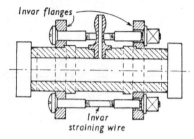

Fig. 9–15. Etalon used by Sears and Barrell.

Comparison of largest etalon with Metre

This comparison is made difficult by the fact that the standard Metre is a line gauge. The comparison was made in three stages.

Inside the largest etalon there was fitted an end gauge whose cross-section was a rectangular cross (see Fig. 9–16). This was only slightly shorter than the etalon and had optically polished ends. It was supported inside the etalon in such a way as to avoid flexure of the gauge. The gaps between the ends of this X-gauge and the etalon mirrors were measured using monochromatic fringes. These were viewed by reflection by means of the east and west telescopes, the etalon being displaced sideways to the dotted position (see Fig. 9–14). The X-gauge remained in the etalon during the earlier part of the experiment when the white light fringes were observed in the four channels alongside the X-gauge.

The X-gauge was then compared with a composite end gauge carrying rulings whose separations could be related to the overall length of the gauge. The length of this line gauge was then compared with the standard Metre using travelling microscopes. Part of the composite gauge consisted of a rod of circular cross-section and of length approximately half an inch less than 1 metre, the ends being optically polished and perpendicular to the axis of the rod. In addition there were two parallel-faced rectangular blocks each half an inch in thickness with sides equal respectively to the radius and diameter of the rod. At the middle of the half-inch face of each block a fine line was ruled parallel to the longer edge. By wringing one block on to one end of the rod, an approximately one-metre end gauge was formed, and by wringing a block on each end of the rod, a one-metre line gauge was obtained (see Fig. 9–16). The composite end gauges were compared with the X-gauge by an extension of the method used for comparing the latter with the largest etalon. The principle is as follows. If the X-gauge is placed between a pair of parallel semi-silvered mirrors, the sizes of the gaps can be measured using monochromatic fringes. If, keeping the mirrors fixed, the composite end gauge is substituted for the X-gauge and the new gaps similarly found, the difference between the gauges is obtained. Using this method, the X-gauge was compared with the composite end gauges formed by wringing each block, in turn, each way round on each end of the rod. The mean of these readings gave the difference between the length of the X-gauge and the length of the rod plus half the sum of the end blocks. (The thickness of the wringing film between the rod and a block can be neglected; it has been shown

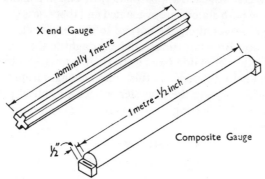

Fig. 9–16. Subsidiary gauges.

to be about 50 angstroms.) These measurements were made with the gauges occupying the position of the displaced etalon (Fig. 9–14) utilizing the east and west telescopes.

A block was then wrung on each end of the rod and the resulting line gauge compared with the standard Metre. By reversing each block in turn and then both together three more gauges were formed and the mean of the four comparisons gave the difference between the Metre and the length of the rod plus half the sum of the blocks—the length previously compared with the X-gauge. In this way the X-gauge and therefore the longest etalon was compared with the Metre. It should be noted that the lengths of the components of the composite gauge are eliminated and need not be known precisely. Comparisons between line gauges involve setting microscopes on rather broad and diffuse images and it was necessary in this investigation to take the mean of the results obtained by several observers.

Results

Sears and Barrell measured the number of wavelengths of the red cadmium line in the standard metre in a vacuum and in dry air free from CO_2 at 15° C. and 20° C. Their results were:

1. In a vacuum, 1,552,734·52 ($\therefore$ $\lambda = 6440\cdot2510 \times 10^{-10}$ metre).

2. In air at 15° C., 1,553,163·76 ($\therefore$ $\lambda = 6438\cdot4711 \times 10^{-10}$ metre).

3. In air at 20° C., 1,553,156·33 ($\therefore$ $\lambda = 6438\cdot5019 \times 10^{-10}$ metre).

These results are quoted to an accuracy justified by the optical part of the experiments. The purely metrological part reduces the accuracy so that the wavelengths are probably correct to about $\pm0\cdot001 \times 10^{-10}$ metre.

The number of wavelengths in the yard was also found:

1. In a vacuum,   1,419,818·31.
2. In air at 15°C., 1,420,210·81.
3. In air at 20°C., 1,420,204·02.

Sears and Barrell went on to measure the refractive index and dispersion of air. This was done by viewing Brewster fringes with a series of monochromatic sources using two etalons, each approximately 67 cm. in length. The principle is as follows: suppose both etalons are evacuated and that one is slightly longer than the other. If one observes along the normal to the thinner etalon the zero order (zero path difference) Brewster fringe can be brought to the centre of the field of view by tilting the thicker etalon (see §9–7). If air is now admitted to the thinner etalon, the optical path difference in this etalon changes from $2d$ to $2nd$, so that for the disturbances forming the Brewster fringes an optical path difference of $2(n-1)d$ is introduced. If p fringes move across the centre of the field as the air is introduced one has

$$2(n-1)d=p\lambda.$$

Hence, by counting fringes $(n-1)$ can be found. The method has an important advantage over the simple method described in §9–2 where one simply observes the Fabry-Perot rings in a single etalon as it is evacuated. In the present method one has, for a given d, the same *change* in the optical path difference as air is introduced, but it is not added to the already very large path difference between the disturbances forming the rings in the evacuated etalon. In consequence one can use a much longer etalon giving much greater accuracy since many more fringes are counted; each etalon can be so long that the path difference is too great for one to see the Fabry-Perot rings for the individual etalons.

9–11. Summary of Results of All Determinations. The New Metre

In 1948 Barrell gave a summary of the results obtained in the various determinations of the Metre, the wavelength of the red cadmium line being corrected where necessary to correspond to air under standard conditions, i.e. 15°C., 760 mm. pressure (g=980·665 cm. sec.$^{-2}$), and containing 0·03% by volume of CO_2. The table shows $\lambda \times 10^{10}$ in metres. It is an odd coincidence that the mean of all the above results is the same as the result announced by Benoît, Fabry, and Perot, and already accepted internationally for many purposes, although it is now known that the original result should have been given as shown in the table. It seems to be agreed that the differences between the various determinations are due partly to differences between the conditions obtaining in the various sources used, and partly to differences between various copies of the Metre; probably the least accurate part of each experiment is the setting of the microscope cross-wires on the rather diffuse images of the terminal marks of the line standards.

In 1907 the angstrom was defined by taking the result of Benoît, Fabry, and Perot, i.e. by defining the wavelength of the cadmium line to be 6438·4696 angstroms, and this has since served as the spectroscopic unit.

In 1927 the cadmium line was accepted as the provisional metrological standard and its use then became a permitted alternative to direct reference to

Date	Observers	Result	Difference from Mean
1892–3	Michelson and Benoît	6438·4691	−0·0005
1905–6	Benoît, Fabry, and Perot	6438·4703	0·0007
1927	Watanabe and Imaizumi	6438·4682	−0·0014
1933	Sears and Barrell	6438·4713	0·0017
1933	Kösters and Lampe	6438·4689	−0·0007
1934–5	Sears and Barrell	6438·4709	0·0013
1934–5	Kösters and Lampe	6438·4690	−0·0006
1937	Kösters and Lampe	6438·4700	0·0004
1940	Romanova, Wahrlich, Kartashev, & Batarchukova	6438·4687	−0·0009

Mean Value 6438·4696

the prototype Metre or its authorized copies. The search then continued for the source that would give the narrowest possible spectral line with the highest degree of reproducibility. Now a single isotope with even atomic mass and charge (zero nuclear spin) emits a line free from hyperfine structure and eventually, in 1957, it was decided to recommend the use of the orange-red line produced by a hot cathode lamp employing the krypton 86 isotope. This lamp was perfected by Engelhard and is operated at the triple point of nitrogen (63°K).

It was decided to maintain continuity with the provisional definition through the intermediary of the wavelength of the cadmium line in standard air, and the mean value of $6438\cdot4696 \times 10^{-10}$ metre was accepted. (Hart and Baird (1961) have since reconsidered the results given in the table and have concluded that some of them require slight modification to take account of what is now known about the various factors affecting the results. Fortunately, the net result is that the mean value remains almost unchanged: it becomes 6438·4695.) Since the new standard would involve a vacuum wavelength it was necessary to specify a dispersion formula for air, and that due to Edlén (1953) was adopted. Edlén derived his formula by using the absolute values of Sears and Barrell for the index and dispersion of air, in combination with the relative measurements by Traub and by Koch in the visible and U.V. For standard air Edlén's formula gives $n = 1\cdot000276381$ so that the vacuum wavelength for the cadmium line is $6440\cdot2490_7 \times 10^{-10}$ metre.

Comparisons between the cadmium and krypton lines were made in five different laboratories, using the Engelhard krypton lamp and the cadmium lamp specified by the International Committee in 1935. The results for the wavelength of the krypton line are given in the second table.

Laboratory	BIPM	PTB	NPL	CIIWM	NRC	NRC	Mean
Interferometer used	M & F–P	M	F–P	F–P	F–P	M	
Vacuum Wavelength 6057·80....	21	23	21	21	21	19	6057·8021

Wavelength in Standard air = 6056·1252.

The laboratories were:

BIPM—Bureau International des Poids et Measures, Sèvres.
PTB—Physikalisch-Technische Bundesanstalt, Braunschweig.
NPL—National Physical Laboratory, Teddington.
CIIWM—Central Inspection Institute of Weights and Measures, Tokyo.
NRC—National Research Council, Ottawa.

At the NPL the Fabry-Perot etalon, with a path difference of 125 mm., was illuminated with the cadmium and krypton lamps simultaneously and was used in series with a prism spectrograph in order to photograph the fringes for the various spectral lines. The wavelengths were compared by using the method of exact fractions to find the orders at the centres of the patterns. Settings on fringes were made in two ways: (1) using a microscope to make visual settings and (2) using a microdensitometer. As a result of the above measurements it was recommended in 1957 that the Metre should be defined as "the length equal to 1,650,763·73 wavelengths in vacuum of the radiation corresponding to the transition between the levels $2p_{10}$ and $5d_5$ of the atom of krypton 86". This was formally adopted by the General Conference on Weights and Measures in 1960. The angstrom was then defined as 10^{-10} metre and subsequently, in 1964, the yard was defined as 0·9144 metre.

In addition to the measurement of the orange-red krypton 86 line in terms of the red cadmium line, a number of other krypton and cadmium lines were measured, together with lines in the mercury 198 spectrum, and in 1962 the secondary standards shown in angstroms in the third table were recommended.

Kr^{86}	Hg^{198}	Cd^{114}
6458·0720	5792·2683	6440·2480
6422·8006	5771·1983	5087·2379
5651·1286	5462·2705	4801·2521
4503·6162	4359·5624	4679·4581

After correction for departures from the specified ideal conditions of excitation, the Engelhard lamp gives a wavelength that is reproducible to about 1 part in 10^9 but practical calibrations of existing end standards and line standards of 1 metre length in terms of the krypton line are probably not better than 10 parts in 10^9—about one order of magnitude better than can be obtained with material standards of the Metre. It will be appreciated that end standards are much easier to check interferometrically than line standards, although the accuracy of setting on line standards has been improved in recent years by employing photoelectric microscopes.

When the krypton standard was adopted it was realized that an atomic beam source using a pure isotope would be expected to give highly monochromatic radiation but it was decided to adopt the krypton standard in view of the urgent need for a new definition. A relevant development since that time is the laser; this gives very highly monochromatic light, enabling interference effects to be observed with path differences far in excess of any used previously. It is, however, proving difficult to achieve the required reproducibility in the wavelength emitted.

9–12. Two Fabry-Perot Instruments in Series

If two Fabry-Perot instruments in series are arranged to be exactly parallel, light will be transmitted only in directions which correspond to maxima for both instruments.

For a single instrument, maxima are found when $2d \cos \theta = p\lambda$, so that the angular separation of orders is given by $(d/dp)(\cos \theta) = \lambda/2d$. Thus the maxima of the thinner interferometer are more widely spaced, and maxima associated with the thicker instrument are suppressed when they do not coincide with those of the thinner. The increased range without overlapping of orders is an important advantage of the arrangement. The resolving power of the compound interferometer is better than that of the thicker component and better than each component when the components are the same. (This is fairly obvious since, if the maximum of one component coincides with a maximum of the other, the ordinates of the intensity curve of one are modified by those of the other with a consequent steepening of the intensity curve.) It can be seen also that, since the dispersion $(d/d\lambda)(\cos \theta) = p/2d = (\cos \theta)/\lambda$ is independent of d, a change in λ causes the same change in θ for each instrument so that a complete line pattern is either transmitted or cut out. (Alternatively, all maxima of a given *order* are transmitted or cut out.)

The compound interferometer has, however, some disadvantages. Between maxima, the intensity transmitted by a Fabry-Perot interferometer does not fall completely to zero so that the intermediate orders of the thicker interferometer are not completely suppressed. These *spurious* maxima may obscure weak satellites in a line or cause difficulty in identification. The latter should not be regarded as a serious difficulty because the position of the unsuppressed maxima are accurately known. Since four silvered surfaces are employed, it is essential to have reflecting films of very low absorption if weak lines are to be examined. With metal films this may necessitate a reduction in the reflectivity of each surface. The latter would cause a loss in resolving power and also a brightening of the spurious maxima since the intensity curve for the thinner interferometer would not fall to a low minimum.

9–13. Method of Edser and Butler for Calibration of a Spectroscope

The arrangement, first used by Esselbach in 1856, is illustrated in Fig. 9–17. S is a source of white light focused on to the spectroscope slit by means of the

Fig. 9–17. Position of etalon for Edser and Butler fringes.

lens L. E is an etalon which, in practice, is sufficiently thin for variations in obliquity to be ignored. Although it is usually placed in the position shown, the thin etalon may be placed anywhere between the source and the observer. Assuming normal incidence, the condition for a maximum by transmission through an etalon is $2d = p\lambda$ (p an integer), and for a minimum $2d = (p + \frac{1}{2})\lambda$. Thus in the focal plane of the spectroscope the spectrum consists of lines at those wave-lengths satisfying the first equation.

Let λ_0 be a wavelength at the red end of the spectrum at which a maximum occurs. Suppose successive maxima on the blue side of λ_0 are numbered, and let λ_j be the wavelength at which the jth bright fringe occurs. Then if p_0 is the order for λ_0, (p_0+j) is the order for λ_j and one has

$$2d = p_0\lambda_0 = (p_0+j)\lambda_j,$$

$$\therefore \quad \lambda_0 = \lambda_j + \frac{j\lambda_j}{p_0} = \lambda_j + \frac{j\lambda_j\lambda_0}{2d},$$

$$\therefore \quad \frac{1}{\lambda_j} = \frac{j}{2d} + \frac{1}{\lambda_0}.$$

Suppose a line spectrum containing a number of known lines lies alongside the white light spectrum. Then, in general, the spectral lines will not fall exactly opposite a maximum and so will correspond to non-integral values of j. If these values of j are observed (by estimating fractions of a fringe separation) for a number of known wavelengths and $1/\lambda_j$ plotted against j, a straight line will result. The slope of this line is $1/2d$ so that the thickness of the etalon may be found. Also, if the value of j is observed for any unknown line, the wavelength may be found from the graph. It will be seen that the fringes constitute a scale which can be calibrated to indicate wavelengths—obviously they give a linear scale of wave numbers or of frequency.

9–14. The Lummer-Gehrcke Interferometer

If a thick, parallel-sided glass plate (i.e. an all-glass Fabry-Perot etalon) is used near grazing emergence, it is possible to secure multiple beam fringes of equal

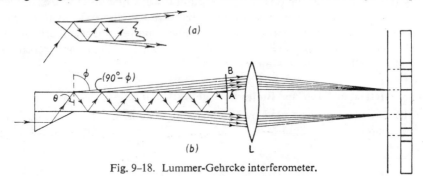

Fig. 9–18. Lummer-Gehrcke interferometer.

inclination without employing silvered surfaces. This is possible because the reflectivity of a glass to air surface increases for angles of incidence near the critical angle. However, this means that if a simple glass plate is used near grazing incidence, a rather small amount of light will enter the plate initially. In the Lummer-Gehrcke interferometer (introduced in 1903) the light is passed into the plate as in Fig. 9–18(a) or, better, by means of an auxiliary prism as in 9–18(b). The most useful plates are those made of quartz and in this case the auxiliary prism is wrung on to the plate, thereby overcoming the difficulty of finding a cement which is transparent throughout a large spectral range.

As in the etalon, the multiple beam fringes of equal inclination can be observed in the focal plane of the lens L, and if the source is a wide vertical slit, the pattern takes the form indicated. Owing to the way in which the light is passed into the plate, emerging beams on the two sides have comparable intensities and identical patterns are formed on the two sides of the plate.

As usual for an etalon, the condition for a maximum is

$$2nd \cos \theta = p\lambda, \qquad (p \text{ an integer}),$$

but the complete theory of the Lummer plate is complicated by the fact that (1) the medium between the reflecting surfaces is dispersive and (2) there is a finite number of beams to be superposed, the number depending on the length of the plate. To a first approximation the resolving power of a Lummer plate is proportional to its length. For a given resolving power a Lummer plate can give a free spectral range greater than that of an etalon, and the theoretical superiority of a Lummer plate over an etalon is greater for thinner plates and shorter wavelengths. However, the advantages of Lummer plates are outweighed by their disadvantages: the surfaces must be plane to at worst $\lambda/80$; the material must be homogeneous if ghosts are to be avoided; strains from the supports must be avoided; and the temperature should be controlled with an accuracy approaching $0.01°$C. Owing to the high cost and the inconvenience, high resolution Lummer plates are now virtually obsolete, although those fortunate enough to possess good plates might continue to use them.

9–15. Multiple Beam Fringes of Equal Thickness

It was seen earlier in the chapter that by silvering the reflecting surfaces it is possible to improve the sharpness of the fringes of equal inclination which are observed with parallel-sided plates. In addition it becomes possible to observe the fringes by transmission. By silvering the relevant surfaces of a thin wedge film the same effect is produced in the localized fringes of equal thickness. For example, multiple beam Newton's rings by transmission resemble Fabry-Perot fringes [Plate I(e)]; those seen by reflection resemble the fringes shown in (d). These multiple beam Fizeau fringes have an important application in the examination of surface topography. The surface to be examined is used as one face of the wedge and an optical flat is used for the other. The structure of the surface under examination introduces differences of thickness in various parts of the wedge and these cause a distortion of the localized fringes observed in the wedge. The movement of the fringes at various points enables the topography of the surface to be examined. The use of the sharp multiple beam fringes improves the accuracy of measurement since relatively small movements of the fringes can be measured. Plate I(g) shows the interference pattern obtained between an optical flat and a sheet of mica. If the wedge angle is vanishingly small a maximum occupies a wide region and variations in the surface under test cause a measurable change in intensity.

Fig. 9–19 shows an air wedge between two semi-silvered plane surfaces (a) and (b), the outer faces of the plate being omitted for simplicity. W is a plane wave incident normally on surface (a). The figure shows the paths of a number of internally reflected rays associated with W. As the disturbances undergo repeated internal reflections between the surfaces they move across the wedge towards the thicker side. Consequently the phases of the various disturbances that are superposed at P are characteristic of the wedge thickness *at* P only if the angle of the wedge is very small and/or the wedge is very thin. In order to study local variations in the height of, say, the second surface, it might be necessary to have the ordinary wedge fringes fairly close together. Thus an appreciable wedge angle must be used and this may be increased when necessary by tilting the reference plate or specimen. In addition, the specimen itself may show steep wedge-like structures. In consequence, it is essential in this work to employ very thin wedges. Since the disturbances move across the wedge each time they are internally reflected, the phase difference between beams undergoing successive numbers of reflections is not constant. This causes the high order beams to begin to interfere destructively with those of low order, thereby reducing the sharpness of the fringes. Since the beams will get out of phase more rapidly with thicker wedges, this effect places an upper limit on the value of d which can be used when a given number of multiply reflected beams is to be superposed. It can be shown that the path difference between the first and nth beams is approximately $[2nd-(4/3)n^3\theta^2d]$, where θ is the wedge angle and d the thickness at the point of emergence of the disturbances. The second term in this expression must not reach $\lambda/2$ if the nth beam is not to be destructive.

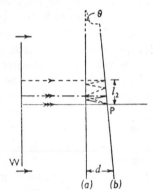

Fig. 9–19. Reflections in a wedge film.

The higher is the reflectivity of the surfaces, the less rapidly does the intensity of the beams decrease and the larger is the number of reflections before the intensity is reduced to a negligible value. In practice the reflectivity of the surfaces is such that it may be assumed that about 60 disturbances give appreciable contributions to the resultant. Thus, assuming that the 60th beam has a path difference of $\lambda/2$, the maximum permissible value of d is given by

$$\frac{4}{3}(60)^3\theta^2d_{max}=\frac{\lambda}{2}. \qquad \ldots \ldots \quad (9\text{--}21)$$

If there are X fringes per cm. across the wedge, $\theta=\lambda X/2$ so that, taking $\lambda=5\cdot5\times10^{-5}$ cm. (green light), d_{max} is given by

$$d_{max}=\frac{1}{7\cdot92X^2}. \qquad \ldots \ldots \ldots \quad (9\text{--}22)$$

Thus d_{max} is 0·0012 cm. when X is 10 and falls to 0·000012 cm. when X is 100.

As seen above, the higher order beams superposed at P have passed through a part of the wedge progressively farther from P. To a first approximation it can be shown that the linear separation on the wedge between the points of entry and emergence of the nth beam is

$$l_n=2n^2\theta d.$$

If the fringes are to give reliable information regarding the detailed topography of a small surface area it is obviously necessary for l_{60} to be as small as possible so that the intensity at any point in the fringe system is characteristic of the local height of the surface studied. If the thickness of the wedge is the maximum permitted by the phase condition discussed above, then from equation (9–21) one has, for the 60th beam,

$$l_{60} = \frac{1}{40X} = \frac{1}{40} \text{ (distance between fringes).}$$

l_{60} is reduced to 1/2000 cm. if the wedge angle is adjusted to give 50 fringes per cm. and the wedge thickness used is the maximum permitted by the phase condition (i.e. 5×10^{-5} cm. or about one wavelength). Assuming normal incidence, the angle of emergence of the 60th beam is 120θ, where θ is the wedge angle. If λ is $5 \cdot 5 \times 10^{-5}$ cm. the angle of emergence is about $10°$. If this beam is to enter the microscope used to examine the fringes, the numerical aperture of the objective must be about $0 \cdot 17$. This is roughly what can be obtained using a nominal 2-inch objective. In normal microscopy one would employ a magnifying power of 50 with such an instrument so that on the photo-micrograph the fringes would be about 1 cm. apart. It is interesting to note that the conventional limit of resolution of this microscope is $0 \cdot 0002$ cm. and is less than l_{60} ($0 \cdot 0005$ cm.).

Tolansky (1948) has summarized the conditions under which sharp multiple beam Fizeau fringes can be employed as follows:

1. The surfaces must be coated with a highly reflecting film of minimal absorption.

2. The film must contour the surface exactly and be highly uniform in thickness.

3. Monochromatic light or at most a few widely spaced monochromatic wavelengths should be used.

4. The interfering surfaces must be separated by at most a few wavelengths of light.

5. A parallel beam should be used. (Within $1°$–$3°$ tolerance.)

6. The incidence should preferably be normal.

With regard to the first condition it is important to note that, for fringes observed by transmission, absorption in the reflecting coatings causes an overall reduction in intensity in the fringe system but does not spoil the fringe sharpness (as was seen to be the case for the Fabry-Perot fringes). On the other hand, it can be shown that if one observes the fringes by reflection—as one must if one is investigating the surface of an opaque specimen—absorption in the reflecting coatings can be ruinous since it very quickly causes a severe loss of contrast in the fringes. For this reason it is common practice to use a highly reflecting dielectric multilayer (see §9–17) on the reference flat (although it is usual to retain the use of silver on the surface that is being examined).

With regard to the second condition, it has been shown that an evaporated silver film can be made uniform and the top surface can be made to follow the surface on which the film is deposited, even if the thickness of the film exceeds considerably the changes in height involved.

The importance of a very thin wedge has already been pointed out. Now the chromatic resolution of a Fabry-Perot instrument obviously becomes very poor if the surfaces are very close together. In this work the low chromatic resolution is an advantage since it means that the fringes will not be broadened by the use of rather broad spectral lines. Thus very intense sources can be used and, at least when viewed by transmission, very thick silvering of the surfaces can be tolerated; also, high power microscopes can be employed.

The need for good collimation of the light arises simply from the fact that the whole point of the multiple-beam method is that very small changes of optical path can be detected, and lack of collimation would itself introduce variations in path. Thus for a film of thickness d the path difference in an air film changes by $2d(1 - \cos \theta)$ when one changes from normal to oblique incidence. It will be seen that the tolerance on θ is reduced if d is increased—which re-emphasises the need to use a very thin air film.

Fig. 9–20 represents the appearance of the fringes when one of the surfaces has a sharp step along AB from one flat region to another. It is possible to obtain fringes that are so fine that the fringe width is only 1/50th of an order and the

Fig. 9–20. Fringes at a step.

measurable displacement is 1/5th of a fringe width, i.e. 1/250th of an order. Since the film thickness changes by $\lambda/2$ between orders, this means that if one uses visible light of wavelength 5×10^{-5} cm one can measure down to $(1/250)$ $(2 \cdot 5 \times 10^{-5})$ cm $= 10^{-7}$ cm, or 10 Å. If the wedge angle is made to approach zero, a single fringe expands across a wide area and it is possible to measure the change in intensity that results from a change in surface level instead of measuring a fringe displacement. In this way one can measure topographical features down to 3 Å. For measurements of this kind the multiple-beam fringes of equal thickness provide a far more sensitive method than the phase-contrast microscope. However, it must be remembered that, whilst one obtains extremely good resolution *in depth*, the lateral resolution is not always good (as is indicated by the value of l_{60}).

Usually, multiple beam interferometry is preferable to phase contrast microscopy if the surface under investigation has coarse lateral structure since phase contrast works well only if the specimen gives a wide diffraction spread (see § 14–14). A further advantage of using multiple beam interferometry rather than phase contrast microscopy is that it is almost always easier to make quantitative as opposed to simply qualitative observations.

9–16. Fringes of Equal Chromatic Order

In Fig. 9–21 S is a source of white light illuminating a small circular aperture A. Light from A passes through L_1 and falls near normal incidence on a thin air wedge W between semi-silvered optical flats. A magnified image of the wedge film is projected on to the slit of the spectroscope by means of the achromatic lens L_2, the line of intersection of the flats being parallel to the slit. Now along each line parallel to the edge of the wedge the latter passes only those wave-lengths that satisfy the condition

$$2d=p\lambda \qquad (p \text{ integral}). \qquad . \quad . \quad . \ (9\text{–}23)$$

Hence the light entering the slit consists of the wavelengths satisfying (9–23), d being the thickness of the strip of the wedge that is imaged on to the slit. Consequently the pattern observed through the spectroscope consists of a series

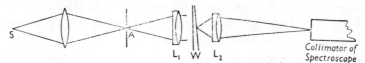

Fig. 9–21. Observation of fringes of equal chromatic order.

of spectral lines resembling the Edser-Butler bands. If the wedge film is rotated about the axis of the collimator the strip imaged on to the slit has a thickness that increases along its length with the result that as one moves along the length of the slit there is a gradual change in the series of wavelengths received, i.e. for each value of the integer p there is a gradual change in λ as one moves along the slit. Each value of p corresponds to a fringe and, since the wavelength varies in the horizontal direction in the field of view, the fringes observed through the spectroscope are inclined to the vertical. The same effect is observed with the arrangement shown in Fig. 9–17 when E is a wedge film and is close to the slit.

Suppose now that one of the boundary surfaces of the air film is not a plane. Again the film thickness varies along the length of the strip that is imaged at the slit and for each value of p there is a variation in the wavelength entering the slit.

In general d does not vary linearly along the strip of the film and hence the wavelength does not vary linearly with distance along the slit. In general, there-fore, each fringe observed through the spectroscope will be curved, the wave-length varying along it. Since p is constant along any fringe, d/λ is constant. The fringes, which were first described by Tolansky (1945) are known as *fringes of equal chromatic order* (and hence sometimes as FECO fringes) since although, as always, the order is constant along a fringe, a fringe is, in general, chromatic. If one boundary of the film is an accurate optical flat, the fringes can be used to investigate the surface microtopography of the other boundary—if the surfaces are nominally parallel, each fringe is effectively a profile of the strip of the surface that is imaged on to the slit of the spectroscope. To examine an extended surface area it is necessary to displace the specimen so that, in effect, its surface is scanned.

When using multiple beam fringes of equal thickness one needs at least two fringes in the field of view (usually more are desired) and the higher the lateral magnification employed the more closely spaced must be the fringes on the wedge itself. This requires the use of a larger wedge angle which tends to increase the transverse migration of the light and the phase defect referred to in §9–15. The result is that there is a loss of fringe sharpness and, in addition, the use of the high magnification is itself invalidated. The situation is very different with

FECO fringes: the fringe spacing is determined solely by the film thickness and does not depend on the wedge angle. (One must remember that the observer's field of view is extended horizontally simply by the dispersion of the spectroscope—one is actually examining just a narrow strip of the specimen, not an extended area.) Since one does not need a large number of fringes in the field to extract the information that is available, one can use a small film thickness and can reduce the wedge angle nominally to zero (leaving only the wedge structure of the surface being examined). In consequence, the migration and phase defect become extremely small, so that the FECO fringes are inherently even sharper than the fringes of equal thickness. The accuracy of measurement with the former is similar to the best that can be obtained with the latter. One advantage of FECO fringes can be pointed out: in Fig. 9–20 one cannot tell at a glance whether the lower set of fringes is displaced a small fraction of an order to the right or a large fraction to the left, i.e. one cannot tell whether the air film is thinner above or below AB; if AB lies across the strip being examined the FECO fringes give this information at once since the fringe spacing is larger where d is smaller.

An interesting application of FECO fringes is to the measurement of small surface angles giving rise to very small wedge angles. Thus an increase of $\lambda/2$ in film thickness in distance y along the strip (wedge angle $\lambda/2y$) gives a horizontal movement of one fringe width (x) in a vertical distance my in the field of view, m being a magnification factor determined by lens L_2 and the spectroscope. The inclination of the fringe to the vertical is x/my, so that the angle of the wedge is effectively magnified by $(x/my) \div (\lambda/2y) = 2x/m\lambda$, and this can be made very large—for the middle of the visible region 10^4 is typical.

9–17. Use of Highly Reflecting Dielectric Films. Interference Filters

It was pointed out in §8–5 that a single quarter-wave film of index $\sqrt{n_G}$ gives zero reflectivity. If one uses a quarter-wave film of high index, the internally and externally reflected waves are in phase and the reflectivity is increased because the phase change of π on reflection occurs only at the outer surface. For example, it follows from the more complete discussion given in §21–11—using equation (21–66)—that if one has a quarter-wave film of index 2·5 on glass of index 1·5, the reflectivity is increased to 0·38 for the wavelength concerned. It is shown in §21–11 that by employing what is known as a *quarter-wave multilayer stack* in which the refractive indices of the films alternate between high and low values, one can achieve very high reflectivities. These high reflectivities occur because the disturbances reflected from successive boundaries are in phase when they return to the front surface of the first film; those which are retarded by an odd multiple of π by the extra optical path length (i.e. extra films) traversed also have a phase change of π on reflection. The quantitative discussion given in §21–11 is incomplete because the existence of the substrate is ignored although in practice the multilayer must be deposited on a substrate—usually glass. However, it is found that multilayers used in this way can give very high reflectivities with very low absorptions and that it is possible to maintain the high reflectivity over a fairly wide spectral range (although this range is usually smaller than the range that can be achieved with a metal coating). A complicating fact is that a multilayer does not introduce a constant phase change.

It is beyond the scope of this section to discuss in detail the design of multi-layer dielectric films for Fabry-Perot mirrors; typically, 9 layers might be used, giving a reflectivity of more than 98%, with an absorption of less than 0·1%. The choice of materials is an important factor and it will be obvious that this must depend upon the spectral region in which the interferometer is to be used. Usually it is more difficult to produce a multilayer than a metal film with the required precision—the flatness of the mirrors is often the limiting factor in the performance of a Fabry-Perot interferometer. However, when high reflectivity and low absorption are of over-riding importance, multilayers are used. For example, they are often used as the reflectors in etalons that form laser cavities. Here one aims at extreme reflectivity, and Perry (1965) has measured a re-flectivity of 99·8% with a 25-layer stack using zinc sulphide (index 2.3) and cryolite (index 1.35); such reflectors could be used in, say, a Helium-Neon gas laser.

If a single collimated beam of light is incident normally on a Fabry-Perot etalon, only those wavelengths are transmitted which satisfy the relation $2d=p\lambda$. As d is reduced, the transmitted wavelengths become widely spaced. When d is between 2×10^{-5} and 6×10^{-5} cm. it is easily verified that the film transmits only one maximum in the visible region. For example, if $d=5000$ angstroms the fringes are at $\lambda\lambda$ 10,000, 5000, 3333 of which only $\lambda 5000$,is in the visible region. With a reflectivity of 0·94 for the surfaces of the etalon, it follows from the discussions of §9–3 that the fringe half-value width is about 1/50th of an order. Hence the half-value width of the transmitted maximum in the visible region is about 50 angstroms.

If the source of light contains only a few lines, it is frequently possible to separate very close components by employing a rather larger value of d in order to reduce the fringe width, care being taken to ensure that the other transmitted wavelengths do not coincide with lines emitted by the source. For example, it is possible to separate the two mercury yellow lines which are 20 angstroms apart, transmitting one and cutting out the other together with the green line. It is worthy of note that it is not necessary to adjust the etalon thickness precisely since, if it is made slightly larger, the effective thickness $d\cos\theta$ can be reduced to the desired value by tilting. (N.B.—it cannot be *increased*.) Colour filters of this type may be made by employing a layer of cryolite between reflecting surfaces instead of an air-spaced etalon. A silver film is first evaporated on to a glass plate followed by a layer of cryolite and then a second silver film. It is not necessary for the glass plate to be an optical flat because the evaporated layers follow the contours of the glass, the layers remaining of uniform thickness. By using two filters in series, the narrow transmission band of the thicker can be coupled with the small number of bands associated with the thinner. By evaporating first a dielectric layer and then a semi-reflecting metal film on to a perfectly reflecting mirror, a reflection filter can be produced and this can have a zero reflectivity for chosen wavelengths.

In using etalons as filters, it is obvious that in order to obtain a filter that will transmit very narrow bands of wavelengths, it is necessary to employ surfaces of very high reflectivity. This necessitates a rather thick metal film and the resulting absorption causes the overall transmission to be rather small. Again the solution is to replace the metal film by a highly reflecting dielectric multilayer. The problem of maintaining the correct etalon spacing can be solved by using an *all-dielectric filter*. For example, one can employ two quarter-wave stacks as reflectors with, for example, a half-wave dielectric spacer layer between them. Fig. 9–22 represents a 15-layer filter of this type where layers 1–7 and 9–15 form the reflectors and layer 8 is the spacer. The odd layers are of zinc sulphide and the even layers of cryolite. As suggested above, the thickness of the spacer

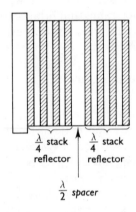

Fig. 9–22. All-dielectric multilayer interference filter.

can be increased to a multiple of $\lambda/2$ in order to reduce the transmitted band width. If extremely narrow band widths are required it may be impossible to produce the necessary spacers by evaporation, and disks of fused silica have been used as spacers, although these, too, are difficult to produce. For example, for a band width of $\frac{1}{2}$Å the thickness of the spacer is about $100\,\mu$ and this has been achieved using a silica disk—one is in the region where, bearing in mind the precision required, the thickness is too great to be produced easily as an evaporated layer and too small to be produced easily as a separate disk! More complicated all-dielectric filters have been produced which consist of a single series of layers equivalent to two filters in series, e.g. reflector stack—spacer — reflector stack—spacer—reflector stack.

An alternative method of securing high reflectivity without absorption is the use of frustrated total reflection. The electromagnetic theory of light indicates that when light is incident at an angle greater than the critical angle from the

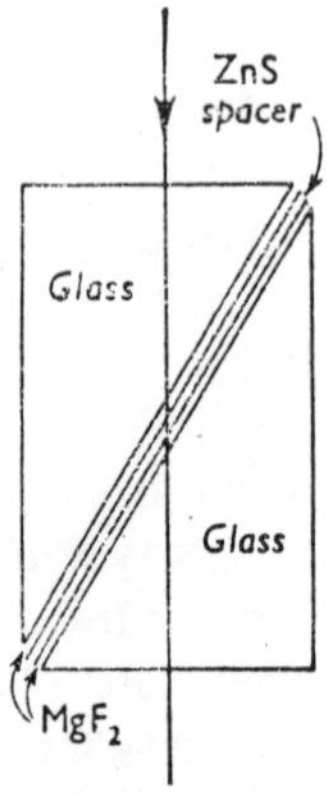

Fig. 9–23. Frustrated total reflection filter.

denser side of a boundary between two media, energy penetrates a short distance into the second medium (see §21–6). It is found that if a second high-index medium is brought close to the first, light passes into the last medium. The thinner is the layer of low index, the larger is the transmitted fraction of the incident energy. One is said to have *frustrated total reflection*. If the second high-index layer is followed by another low-index layer and another high-index medium, these give a second partial reflection. Therefore the high-index layer is effectively between two partially reflecting films, the reflectivities of which depend on the thicknesses of the low-index layers. Such an arrangement can be used as a filter in the same way as an etalon. The band width is determined by the reflectivity of the surfaces and in this case is obviously controlled by the thicknesses of the low-index films. The transmitted wavelengths are controlled by the thickness of the high-index layer which plays the part of the air space of the etalon. A frustrated total reflection filter is illustrated in Fig. 9–23.

9–18. Summary of Types of Fringes

In this section there is given a brief summary of the conditions giving rise to the various types of fringes in systems employing division of amplitude. One has, in effect, interference between disturbances reflected from the two boundary surfaces of a medium, and the basic equation is

$$2nd \cos \theta = p\lambda,$$

the symbols having their usual meanings. Usually, this equation corresponds to a maximum or a minimum, depending on the phase changes that occur at the various reflecting surfaces in the system, but it is possible that a further effective path difference has to be added. In any system a fringe is the locus of points corresponding to a constant phase difference. In each case the use of highly reflecting surfaces results in a multiple beam system, giving sharper fringes. The following possibilities arise:

1. λ, n, and d are constant. θ varies

One has monochromatic light incident on a film of constant thickness. The path difference depends on θ and one has fringes of equal inclination, such as the circular fringes in the Michelson or Fabry-Perot interferometers. The fringes observed with the Lummer plate are also of this type. When d is large slightly different wavelengths correspond to significantly different values of θ, and this is why these fringes are useful in high resolution spectroscopy.

2. d and θ are constant. λ (and hence n) varies

One has collimated polychromatic light incident on a film of constant thickness. The phase difference depends on λ and one has the basis of an interference filter. Edser-Butler fringes demonstrate this. If d is large one has maxima for closely spaced wavelengths, and if d is small one has maxima for widely separated wavelengths.

3. λ, n, and θ are constant. d varies

One has collimated monochromatic light incident on a film whose thickness varies from point to point. The path difference depends on d and one has fringes of equal thickness such as those obtainable with a wedge film, e.g. with a Michelson interferometer with inclined plates; Newton's rings are of this type. If d is very small moderate variations in θ have little effect. The fringes are contour lines of the film thickness and are the ones used in studying surface topography.

4. θ is constant. d and λ (and hence n) vary

One has collimated polychromatic light incident on a film whose thickness varies from point to point. Along any fringe p is constant, and hence nd/λ is constant. One refers to fringes of equal chromatic order.

CHAPTER X

HUYGENS' PRINCIPLE AND THE
SCALAR THEORY OF DIFFRACTION

10-1. Early Geometrical Picture

IN order to explain the propagation of a disturbance as a wave motion, Huygens assumed that each point on a wavefront becomes the source of so-called secondary waves, and that the effect of these waves is mainly confined to their forward envelope. As shown in §§1–5, 1–6, 1–7, this simple picture is sufficient to account for the passage of light through very small holes, the approximately rectilinear propagation of light through large apertures, and the classical laws of reflection and refraction. It is not sufficient, however, to explain in detail the departures from exactly rectilinear propagation of light which are encountered in cases of diffraction. It is not, of course, made clear in the simple theory why the secondary sources produce no effect in the reverse direction (e.g. on the backward envelope), but it must be remembered that at the time of Huygens light was thought of as a disturbance in an aether consisting of closely packed perfectly elastic spheres; the impact of one sphere on another would then result in the total forward transfer of momentum.

10-2. Fresnel's Extension

As originally propounded (c. 1678), Huygens' principle was meant to apply to isolated waves or pulses. In order to extend the principle to cover the propagation of an extended train of simple harmonic waves, Fresnel (1815–18) assumed that the secondary waves interfere with one another according to the principle of superposition. In the case of the passage of light through an aperture in a screen (see Fig. 10-1), Fresnel simply assumed that each element of that part of the wave which is not obstructed acts as a secondary source. The secondary disturbances from these sources are then superposed at P to give the

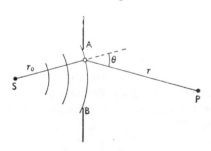

Fig. 10-1. Passage of a wave through an aperture.

resultant disturbance. If the secondary disturbances are simple harmonic waves, and the sources are in phase on the primary wavefront, the phase of each disturbance when it arrives at P will depend on the optical path from P to the point on the wavefront concerned. It follows that there may be a maximum or a minimum at P, according to the degree of constructive or destructive interference which occurs. In this way it is possible to account for the so-called diffraction fringes which are observed.

Fresnel assumed that the amplitude of the disturbance at P due to the sources in any elementary area of the primary wavefront is:

1. Proportional to the amplitude of the disturbance at the element.
2. Proportional to the area of the element.
3. Inversely proportional to the distance from the element to P.

In addition, it was found necessary to assume that the contribution of the sources in any elementary area depends on the angle θ between the normal to the primary wavefront and the direction of P. This *obliquity* or *inclination factor* is essential to the theory as will be shown below. The phase of the disturbance at P due to any element of the wavefront was assumed to be that of a disturbance which has travelled the total distance from the source to P via the element of wavefront concerned.

The next six sections can be omitted by the student whose mathematics is weak. A simplified discussion of Fresnel's method is given in Chapter XIII.

10–3. Propagation of a Spherical Wave

The disturbance from a point source in an isotropic medium consists of expanding spherical waves, and the most obvious test to apply to the Fresnel

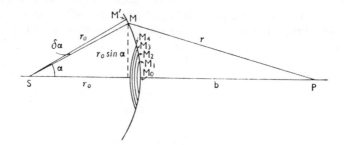

Fig. 10–2. Half-period zones.

theory is to see whether the secondary sources located on a spherical wavefront at one instant produce a similar wave of larger radius at a subsequent time.

As was seen in § 1–3, a spherical wave expanding from the monochromatic point source S (Fig. 10–2) may be represented by

$$\phi = \frac{a}{r_o} \cos k(ct - r_o). \qquad \ldots \ldots \quad (10\text{--}1)$$

Each point on the wavefront (radius r_o at time t) is, according to Fresnel, to be considered as a source of disturbances which interfere according to the principle of superposition to yield the subsequent disturbance. To calculate the disturbance at P, Fresnel divided the wavefront into annular zones centred on the axis SP as illustrated in the figure. The central zone extends to M_1, where $r_1 = M_1P = M_0P + \lambda/2 = b + \lambda/2$. The next zone extends from M_1 to M_2, where $r_2 = M_2P = r_1 + \lambda/2 = b + \lambda$, and so on, such that the boundaries of successive zones are $\lambda/2$ farther from P. These zones are called *Fresnel zones* or *half-period zones*. The disturbance at P is calculated in two stages. Firstly, that due to the sources in each zone is found, and secondly, the effects of the various zones are summed.

Any area over which the factors enumerated in the previous section are constant can be taken as an elementary area, and to find the effect of the nth Fresnel zone, one may take, as an element of that zone, an elementary annulus which subtends an angle $\delta\alpha$ at S. The boundaries of this element are represented by M and M', where $MP=r$, and $M'P=r+\delta r$. It is assumed that the obliquity factor is a slowly varying factor and that it may be assumed to be approximately constant over a single half-period zone. The contribution of the sources within this elementary area to the disturbance at P is, therefore,

$$d\phi = \frac{a}{r_o} \cdot d\sigma \cdot \frac{1}{r} \cdot f_n(\theta) \cdot \cos k[ct-(r_o+r)], \qquad . \quad . \quad (10\text{–}2)$$

where $f_n(\theta)$ is the obliquity factor for the nth zone, and $d\sigma$ is the area of the elementary annulus; the term (r_o+r) occurs because the phase must be appropriate to the total distance SMP traversed.

Now, $$d\sigma = 2\pi(r_o \sin \alpha)r_o d\alpha.$$

But, from $\triangle$SMP,

$$r^2 = r_o^2 + (r_o+b)^2 - 2r_o(r_o+b) \cos \alpha,$$

$$\therefore \quad 2rdr = 2r_o(r_o+b) \sin \alpha d\alpha,$$

$$\therefore \quad d\sigma = \frac{2\pi r_o r}{r_o+b} dr,$$

$$\therefore \quad d\phi = 2\pi f_n(\theta) \frac{a}{r_o+b} \cos k[ct-(r_o+r)]dr.$$

Hence the disturbance due to the nth zone is given by

$$\phi_n = \int_{n\text{th zone}} d\phi = \frac{2\pi f_n(\theta)a}{(r_o+b)} \int_{r_{n-1}}^{r_n} \cos k[ct-(r_o+r)]dr$$

$$= \frac{f_n(\theta)a\lambda}{(r_o+b)} \left\{ \sin k[ct-(r_o+r_{n-1})] - \sin k[ct-(r_o+r_n)] \right\},$$

since $2\pi/\lambda = k$. By expanding the sines separately and putting

$$r_{n-1} = b + (n-1)\frac{\lambda}{2},$$

$$r_n = b + \frac{n\lambda}{2}.$$

it is easily shown that this reduces to

$$\phi_n = (-1)^{n+1} \cdot \frac{2f_n(\theta)a\lambda}{(r_o+b)} \sin k[ct-(r_o+b)]. \qquad . \quad . \quad (10\text{–}3)$$

That is, ϕ_n is positive or negative when n is odd or even respectively. The importance of the obliquity factor now becomes evident because, in its absence, successive zones would, according to (10–3), give alternately positive and negative contributions of equal magnitude and the total disturbance would oscillate between 0 and ϕ_n according to the number of zones summed; it would never agree with the known disturbance at P as given by the appropriate form of equation (10–1).

Fresnel assumed that $f_n(\theta)$ gradually decreases as θ (and therefore n) increases until it is zero when $\theta = 90°$. Thus one must construct Fresnel zones in the wavefront (Fig. 10–2) until PM is tangential to the wavefront. The disturbance at P is then given by the sum of the finite series $\Sigma\phi_n$ where, although it may not be known how many terms are taken, it is known that $|\phi_n|$ steadily decreases to zero.

The series to be summed is

$$\phi_p = |\phi_1| - |\phi_2| + |\phi_3| - |\phi_4| + \ldots , \qquad \ldots \quad (10\text{--}4)$$

and in order to perform this summation it is necessary to rearrange the terms of the series. For example, if n is odd, the series is written in the forms

$$\left.\begin{aligned}
\phi_p &= \frac{|\phi_1|}{2} + \left(\frac{|\phi_1|}{2} - |\phi_2| + \frac{|\phi_3|}{2}\right) + \left(\frac{|\phi_3|}{2} - |\phi_4| + \frac{|\phi_5|}{2}\right) \\
&\quad + \ldots + \left(\frac{|\phi_{n-2}|}{2} - |\phi_{n-1}| + \frac{|\phi_n|}{2}\right) + \frac{|\phi_n|}{2}, \quad (a) \\[2mm]
\text{or} \quad \phi_p &= |\phi_1| - \frac{|\phi_2|}{2} - \left(\frac{|\phi_2|}{2} - |\phi_3| + \frac{|\phi_4|}{2}\right) - \left(\frac{|\phi_4|}{2} - |\phi_5| + \frac{|\phi_6|}{2}\right) \\
&\quad - \ldots - \left(\frac{|\phi_{n-3}|}{2} - |\phi_{n-2}| + \frac{|\phi_{n-1}|}{2}\right) - \frac{|\phi_{n-1}|}{2} + |\phi_n|. \quad (b)
\end{aligned}\right\} \quad (10\text{--}5)$$

Now it is not clear exactly how $f_n(\theta)$ must vary with n and two possibilities arise: either $|\phi_j|$ is greater than, or it is less than, the arithmetic mean of $|\phi_{j-1}|$ and $|\phi_{j+1}|$, and each case must be considered.

(a) If $|\phi_j| > \frac{1}{2}(|\phi_{j-1}| + |\phi_{j+1}|)$, each bracket in (10–5(a) and (b)) is negative, i.e.

$$\phi_p < \frac{|\phi_1|}{2} + \frac{|\phi_n|}{2}$$

and

$$\phi_p > |\phi_1| - \frac{|\phi_2|}{2} - \frac{|\phi_{n-1}|}{2} + |\phi_n|.$$

If it is assumed that $f_n(\theta)$ varies slowly, $|\phi_j|$ and $|\phi_{j-1}|$ are never widely different. These inequalities indicate that ϕ_p is only slightly smaller than $\frac{1}{2}(|\phi_1| + |\phi_n|)$. Consequently, to a first approximation one may write

$$\phi_p = \frac{|\phi_1|}{2} + \frac{|\phi_n|}{2}. \qquad \ldots \ldots \quad (10\text{--}6)$$

(b) If $|\phi_j| < \frac{1}{2}(|\phi_{j-1}| + |\phi_{j+1}|)$ the above inequalities are reversed and again yield

(10–6). If n is even, a suitable rearrangement of terms in the series gives, in a similar way,

$$\phi_p = \frac{|\phi_1|}{2} - \frac{|\phi_n|}{2}. \qquad \cdots \cdots \quad (10\text{–}7)$$

If it is assumed that Fresnel zones are constructed until $\theta = 90°$ (when $f_n(\theta) = 0$), one has $\phi_n = 0$, and $\phi_p = \frac{1}{2}\phi_1$. Thus the disturbance at P is half that due to the first Fresnel zone. In this sense it is possible to look upon the energy as travelling along SP, i.e. along the ray. It will be possible to look at the problem in this way provided one can draw the necessary number of Fresnel zones so that $\phi_n = 0$. This will be possible only if P is not in the vicinity of a focus. If P is at or near the focus of a converging wave (see Fig. 10–3), all points on the wavefront are approximately equidistant from P, the obliquity factor is almost constant over the whole wave, and it is impossible to construct the necessary zones.

Fig. 10–3.
Converging wave.

10–4. Defects of Fresnel's Theory

Returning to an expanding spherical wave, it will be seen that the present theory gives, for the disturbance at P (Fig. 10–2),

$$\phi_p = \frac{f_1(\theta)a\lambda}{(r_o + b)} \sin k[ct - (r_o + b)]. \qquad \cdots \quad (10\text{–}8)$$

But, on replacing r_o by $(r_o + b)$, equation (10–1) must give

$$\phi_p = \frac{a}{(r_o + b)} \cos k[ct - (r_o + b)]. \qquad \cdots \quad (10\text{–}9)$$

On comparing equations (10–8) and (10–9), the following difficulties are encountered:

1. In order to get the correct amplitude one must put $f_1(\theta) = 1/\lambda$, but it is difficult to justify an obliquity factor which is not unity when the obliquity is zero. The alternative seems to be to retain $f_1(\theta) = 1$ and assume that the secondary sources radiate with an amplitude $1/\lambda$ times the amplitude of the primary disturbance which excited them.

2. The theory yields the wrong phase for the disturbance at P since the sine occurs in place of the cosine. This could be put right by assuming that the secondary sources vibrate a quarter of a period ahead of the primary disturbance.

3. To these difficulties regarding the nature of the secondary sources there is added the difficulty of explaining the existence of the obliquity factor and, in particular, why the sources do not radiate backwards.

In spite of these difficulties the Fresnel theory does yield results which are in very good agreement with experimental observations in many diffraction problems. The reasons why the above defects do not show up are that one is not normally interested in absolute intensities in a pattern nor in the absolute phases of the disturbances which are superposed, but only in relative intensities and phases. In addition, one can easily "forget" the backwave.

10–5. Circular Aperture. Circular Obstacle. Irregular Obstacle

For a circular aperture or obstacle, the results given by Fresnel's theory are contrary to the idea of rectilinear propagation. At first it was claimed that this disposed of the theory, but experiment confirmed the predictions. Thus, if a circular obstacle is placed centrally on SP, construction of the half-period zones may be commenced at the edge of the obstacle and the disturbance at P will be half that due to the new first zone. Hence there will be a finite intensity at all points on SP beyond the obstacle. This result was deduced by Poisson (1818) and seemed to be absurd. However, the phenomenon was soon observed by Arago—see Plate III(a). Suppose, now, that an opaque screen is interposed between S and P and that it contains a circular aperture, equal in size to the central zone, and centred on SP. From Fresnel's theory one would expect the whole zone to be effective so that the presence of the screen causes the amplitude at P to be doubled and the intensity to be increased fourfold. An increase is, in fact, observed.

For an irregular obstacle, the principle of rectilinear propagation is less

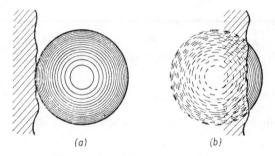

(a)   (b)

Fig. 10–4. Irregular obstacle.

noticeably violated. Suppose that an irregular obstacle is interposed between S and P. Figure 10–4 shows the appearance of the wave, divided into Fresnel zones and viewed from P. In (a) P is inside and in (b) it is outside the geometrically illuminated area. In (a) the central zones are unaffected and the outer zones are partially obstructed so that their contribution to the disturbance is reduced. This causes higher terms in series 10–4 to decrease in magnitude and the intensity at P is not appreciably affected. In (b), however, the central zones are obstructed and the next zones are partially obstructed by decreasing amounts so that the terms in the series now decrease in magnitude at both ends and the sum is nearly zero. Thus when P is in the geometrical shadow the intensity at P is small, and when it is in the geometrically illuminated area the intensity is almost unaffected by the presence of the obstacle. This is roughly in accordance with the principle of rectilinear propagation.

The Fresnel theory was very successful in explaining these and other simple diffraction phenomena but there remain fundamental objections to the theory as described above. Kirchhoff gave an analytical formulation of Huygens' principle which overcomes most of these difficulties. The theory is based on the general equation of wave motion which must be satisfied by the function used to represent the light disturbance.

10–6. Kirchhoff's Formulation of Huygens' Principle

It is assumed that the disturbance can be represented by a single scalar function ϕ, which satisfies the general equation of wave motion:

$$\nabla^2\phi = \frac{1}{c^2}\frac{\partial^2\phi}{\partial t^2}, \quad \ldots \ldots \ldots \quad (10\text{–}10)$$

where ∇^2 is the operator $\left(\dfrac{\partial^2}{\partial x^2} + \dfrac{\partial^2}{\partial y^2} + \dfrac{\partial^2}{\partial z^2}\right)$ and c is the speed of propagation of the disturbance. If the disturbance is not monochromatic it may be resolved into its components, each of which satisfies (10–10). Consider, then, the component of frequency v, wavelength $\lambda = 2\pi/k$. As shown in §6–3, one may split the disturbance into its space and time parts and write

$$\phi = \psi(x, y, z)e^{ikct},$$

where ψ is the complex amplitude, and the physical disturbance is now the real part of ϕ.

$$\therefore \quad \nabla^2\phi = e^{ikct}\nabla^2\psi.$$

Also

$$\frac{\partial\phi}{\partial t} = \psi ikce^{ikct},$$

$$\therefore \quad \frac{1}{c^2}\frac{\partial^2\phi}{\partial t^2} = -\psi k^2 e^{ikct}.$$

Thus, from (10–10),

$$\nabla^2\psi = -k^2\psi. \quad \ldots \ldots \ldots \quad (10\text{–}11)$$

If ψ_1 and ψ_2 are any two continuous single-valued functions whose derivatives are continuous, Green's theorem states

$$\iiint (\psi_2\nabla^2\psi_1 - \psi_1\nabla^2\psi_2)\,dv = \iint \left(\psi_2\frac{\partial\psi_1}{\partial n} - \psi_1\frac{\partial\psi_2}{\partial n}\right)dS,$$

where the volume integral extends over the volume enclosed by the closed surface S, and $\partial/\partial n$ denotes differentiation along the outward normal.

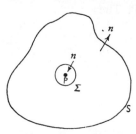

Fig. 10–5. Exclusion of P from volume of integration.

Suppose one requires the disturbance at the point P. One can take the closed surface S to surround P and put $\psi_1 = \psi$, the complex amplitude. ψ_2 may be any well-behaved function; it proves convenient to put $\psi_2 = e^{-ikr}/r$, where r is measured from P as origin. This at once causes some difficulty since ψ_2 becomes infinite at P and, in order to apply Green's theorem, P must be excluded from the volume of integration. To do this, take a small sphere Σ centre P. Green's theorem may now be applied to ψ_1 and ψ_2, where the volume of integration is the volume enclosed between S and Σ which, together, constitute the surface of integration (see Fig. 10–5).

It is easily shown that $\nabla^2\psi_2 = -k^2\psi_2$. To verify this, consider first $\partial^2\psi_2/\partial x^2$

Now
$$\frac{\partial}{\partial x} = \frac{\partial r}{\partial x} \cdot \frac{\partial}{\partial r} = \frac{x}{r}\frac{\partial}{\partial r} \quad \text{since } r^2 = x^2 + y^2 + z^2.$$

$$\therefore \quad \frac{\partial \psi_2}{\partial x} = -\frac{ikxe^{-ikr}}{r^2} - \frac{xe^{-ikr}}{r^3},$$

whence

$$\frac{\partial^2 \psi_2}{\partial x^2} = -\frac{k^2 x^2 e^{-ikr}}{r^3} + \frac{2x^2 ike^{-ikr}}{r^4} - \frac{ike^{-ikr}}{r^2} + \frac{x^2 ike^{-ikr}}{r^4} + \frac{3x^2 e^{-ikr}}{r^5} - \frac{e^{-ikr}}{r^3}$$

with similar expressions for $\partial^2 \psi_2 / \partial y^2$ and $\partial^2 \psi_2 / \partial z^2$. On collecting terms and simplifying, it follows that

$$\nabla^2 \psi_2 = -k^2 \psi_2. \quad \cdots \cdots \quad (10\text{--}12)$$

This result follows much more easily if one writes ∇^2 in spherical polar co-ordinates. For a function such as ψ_2, where there is symmetry about the origin, one has $\nabla^2 \psi_2 = \partial^2 \psi_2 / \partial r^2 + (2/r)\partial \psi_2 / \partial r$.

On substituting the values of $\nabla^2 \psi_1$ and $\nabla^2 \psi_2$ from (10–11) and (10–12), the volume integral in Green's theorem becomes zero. Hence the surface integral is zero.

Consider the surface integral over Σ. Over this surface the external normal from the volume of integration is, of course, towards P so that $\partial/\partial n$ becomes $-\partial/\partial r$.

$$\therefore \quad \iint_\Sigma \left(\psi_2 \frac{\partial \psi_1}{\partial n} - \psi_1 \frac{\partial \psi_2}{\partial n} \right) dS = \iint_\Sigma \left\{ \frac{e^{-ikr}}{r} \left(-\frac{\partial \psi}{\partial r} \right) - \psi \left[-\frac{\partial}{\partial r}\left(\frac{e^{-ikr}}{r} \right) \right] \right\} dS$$

$$= \iint_\Sigma \left\{ \frac{e^{-ikr}}{r}\left(-\frac{\partial \psi}{\partial r} \right) - \psi \left(\frac{e^{-ikr}}{r^2} + \frac{ike^{-ikr}}{r} \right) \right\} dS.$$

Suppose the surface element dS on Σ subtends, at P, a solid angle $d\omega$. Then $dS = r^2 d\omega$,

$$\therefore \quad \iint_\Sigma = \int_{\text{round P}} \left\{ re^{-ikr}\left(-\frac{\partial \psi}{\partial r} \right) - \psi e^{-ikr} - r\psi ike^{-ikr} \right\} d\omega.$$

Assuming ψ is always finite, only the second term in the integrand does not approach zero as $r \to 0$. If Σ collapses round P, the value of ψ on and inside Σ may be assumed to be constant and equal to ψ_P, its value at P. Then

$$\iint_\Sigma \to -\psi_P \int e^{-ikr} d\omega$$

$$= -\psi_P \int d\omega \text{ since } e^{-ikr} \to 1 \text{ as } r \to 0$$

$$= -4\pi\psi_P.$$

Now since $\iint_{S+\Sigma}$ is zero it follows that $\iint_S = +4\pi\psi_P$, i.e.

$$4\pi\psi_P = \iint_S \left\{ \frac{e^{-ikr}}{r} \cdot \frac{\partial \psi}{\partial n} - \psi \frac{\partial}{\partial n}\left(\frac{e^{-ikr}}{r} \right) \right\} dS$$

$$= \iint_S \left\{ \frac{e^{-ikr}}{r} \cdot \frac{\partial \psi}{\partial n} - \psi e^{-ikr} \frac{\partial}{\partial n}\left(\frac{1}{r} \right) + \frac{ik\psi e^{-ikr}}{r} \cdot \frac{\partial r}{\partial n} \right\} dS. \quad (10\text{--}13)$$

This equation is frequently referred to as Helmholtz's equation (c. 1859).

Since $\phi = \psi e^{ikct}$,

$$\phi_P = \frac{1}{4\pi}\iint_S \left\{ \frac{e^{ik(ct-r)}}{r}\cdot\frac{\partial\psi}{\partial n} - \psi e^{ik(ct-r)}\frac{\partial}{\partial n}\left(\frac{1}{r}\right) + \frac{ik\psi e^{ik(ct-r)}}{r}\cdot\frac{\partial r}{\partial n} \right\} dS,$$
$$\cdots \quad (10\text{--}14)$$

and this expression gives the disturbance at P at time t. In the integrand the factor $\psi e^{ik(ct-r)}$ can be written $\psi e^{ikc(t-r/c)}$. Obviously, it is the value of ϕ at the element dS at time $(t-r/c)$; it is referred to as the *retarded value* of ϕ and is written $[\phi]_{t-r/c}$. In the same way the first and third terms may be written in terms of the retarded values of $\partial\phi/\partial n$ and $\partial\phi/\partial t$. One has

$$\phi_P = \frac{1}{4\pi}\iint_S \left\{ \frac{1}{r}\left[\frac{\partial\phi}{\partial n}\right]_{t-r/c} - \frac{\partial}{\partial n}\left(\frac{1}{r}\right)[\phi]_{t-r/c} + \frac{1}{cr}\frac{\partial r}{\partial n}\left[\frac{\partial\phi}{\partial t}\right]_{t-r/c} \right\} dS,$$
$$\cdots \quad (10\text{--}15)$$

which is usually known as Kirchhoff's formula.

Thus the disturbance at P can be calculated from a knowledge of the conditions obtaining over the closed surface S. Physically, one may say that since the conditions at any element dS can affect ϕ_P only by means of a disturbance which travels from dS to P, the value of ϕ_P at any time t depends on the disturbance at dS at the previous time $(t-r/c)$, r/c being the time taken for a disturbance to travel the distance r from dS to P. Thus the appearance of the retarded values is explained.

10-7. Application to Spherical Waves

Equation (10–15) was derived for a surface S which enclosed the point P but not the source (since it was assumed in applying Green's theorem that ψ remains finite throughout the volume of integration). It can be shown that the same result follows if S encloses the source but not the point P. Thus in the case of a single point source the closed surface S may be taken as a spherical wavefront.

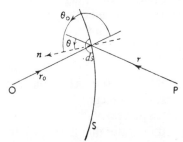

Fig. 10-6. Expanding spherical wave.

Consider first the case of a point source at O (Fig. 10–6), where S is any surface enclosing O. (n is the outward normal from the space containing P.) One may put

$$\phi = \frac{a}{r_o} e^{ik(ct-r_o)},$$

where r_o is the radius vector from O as origin.

Thus
$$\psi = \frac{a}{r_o} e^{-ikr_o}.$$

One has $\dfrac{\partial\psi}{\partial n} = \dfrac{\partial\psi}{\partial r_o}\cdot\dfrac{\partial r_o}{\partial n}$, where $\dfrac{\partial r_o}{\partial n} = \cos\theta_o$. Also, $\dfrac{\partial r}{\partial n} = \cos\theta$.

In equation (10–14), one has

$$\frac{e^{ik(ct-r)}}{r} \cdot \frac{\partial \psi}{\partial n} = -\cos\theta_0 \cdot \frac{a}{rr_0} \cdot ike^{ik[ct-(r_0+r)]} - \cos\theta_0 \cdot \frac{a}{rr_0{}^2} \cdot e^{ik[ct-(r_0+r)]},$$

$$-\psi e^{ik(ct-r)} \cdot \frac{\partial}{\partial n}\left(\frac{1}{r}\right) = +\cos\theta \cdot \frac{a}{r^2 r_0} e^{ik[ct-(r_0+r)]},$$

$$ik\psi \frac{e^{ik(ct-r)}}{r} \cdot \frac{\partial r}{\partial n} = +\cos\theta \cdot \frac{ika}{rr_0} e^{ik[ct-(r_0+r)]}.$$

Whence $\phi_P = \dfrac{1}{4\pi}\displaystyle\int\int_S e^{ik[ct-(r_0+r)]} \cdot \dfrac{a}{r_0 r}\left\{\cos\theta\left(ik+\dfrac{1}{r}\right) -\cos\theta_0\left(ik+\dfrac{1}{r_0}\right)\right\}dS.$

If $\lambda \ll r_0$ or r, one may neglect $1/r_0$ and $1/r$ in comparison with k so that

$$\phi_P = \frac{1}{4\pi}\int\int_S \frac{ika}{r_0 r}(\cos\theta - \cos\theta_0)\, e^{ik[ct-(r_0+r)]}dS. \qquad . \quad (10\text{–}16)$$

Now $k=2\pi/\lambda$ and $e^{i\pi/2}=i$, and *if* S *is taken to be a wavefront*, $\cos\theta_0 = -1$.

$$\therefore \quad \phi_P = \int\int_S \frac{a}{r_0} \cdot \frac{1}{r} \cdot \frac{1}{\lambda}\left(\frac{1+\cos\theta}{2}\right)e^{ik[ct-(r_0+r)]+i\frac{\pi}{2}}dS. \qquad . \quad (10\text{–}17)$$

It will be seen that this integral has the form of a summation of disturbances at P due to sources distributed over the surface S. Now it will be recalled that in the simple theory of Fresnel, difficulties were encountered concerning the amplitude and phase of the disturbance at P, and that an obliquity factor was introduced in a rather arbitrary manner (see § 10–4). These difficulties could be resolved only by making certain assumptions regarding the nature of the secondary sources. Examination of equation (10–17) reveals that these difficulties no longer exist.

1. The integrand includes an obliquity factor $\frac{1}{2}(1+\cos\theta)$ which is unity in the forward direction ($\theta=0$) and zero in the reverse direction ($\theta=\pi$). The amplitude factor $1/\lambda$ is also present as was seen to be necessary.

2. The presence of the factor $i\pi/2$ in the exponent indicates that the secondary sources vibrate a quarter of a period ahead of the primary disturbance, as was seen to be necessary to account for the phase of the disturbance at P.

3. The absence of the direct backwave is taken care of by the obliquity factor, although it is interesting to notice that the disturbance becomes zero only when $\theta=\pi$ (not when $\theta=\pi/2$ as assumed earlier).

If the above expression for ϕ_P is evaluated for an expanding spherical wave by means of Fresnel's zones, it is obvious that the result will be similar to that obtained in § 10–4 but with the correct phase and amplitude.

10–8. Kirchhoff's Diffraction Formula. St. Venant's Principle

One is frequently interested in the passage of light through an aperture or a number of apertures in an opaque non-reflecting screen. Usually the screen can be assumed to extend to infinity and then to enclose P, in which case the surface S in (10–14, 15, and 16) can be taken to consist of the screen plus open surfaces which bridge the diffracting apertures. According to St. Venant, one may assume that ψ and $\partial\psi/\partial n$ are zero over the part of S covered by the screens, and that over the surfaces which bridge the gaps they have the values which would obtain in the absence of the screens. Although this is a natural assumption to make, it is fundamentally unsound since it implies that ψ and $\partial\psi/\partial n$ are discontinuous at the edges of the apertures. This is physically impossible and contrary to the assumptions made when applying Green's theorem. Nevertheless, it leads to many results that are in agreement with observation. Experimental diffracting screens are not totally non-reflecting but the dark side of an opaque screen will give a completely negligible contribution and the remote surface enclosing P can be the non-reflecting walls of the laboratory.

For diffraction of light from a point source one then obtains

$$\phi_P = \int\int_\Sigma \frac{a}{2\lambda r_o r}(\cos\theta - \cos\theta_o)e^{ik[ct-(r_0+r)]+i\frac{\pi}{2}}\,dS, \quad . \quad (10\text{–}18)$$

where the integral is taken over any surfaces Σ which bridge the gaps in the diffracting screen. This is the Kirchhoff diffraction integral (1883).

10-9. Classification of Diffraction Patterns

In terms of the simple Fresnel theory, the disturbance at any point in the presence of a diffracting aperture is given by the summation of the effects of the secondary sources over the unobscured part of the wavefront or over any convenient surface which bridges the aperture. The contribution of any element of the wavefront has an amplitude proportional to the area of the element dS, proportional to the disturbance at the element (a/r_0), inversely proportional to the distance r of the element from the point in the pattern, and dependent on an inclination factor. The phase of the contribution is appropriate to the total distance (r_0+r) from the source to the field point in the pattern via the wavefront element. The Kirchhoff theory shows the nature of the obliquity factor and confirms the fact that this simple Fresnel theory will give the correct relative intensities in the pattern.

The most important diffraction patterns result from diffraction by an opaque plane screen containing one or more apertures. It will be assumed that the screen lies in the xy plane. Frequently, the surface Σ over which the diffraction integral is taken will simply be that part of the xy plane not covered by the screen; sometimes it is the unobstructed part of a wavefront. Usually the variations of r and r_0 are very small compared with r and r_0, and the amplitude term $1/r_0 r$ may be taken to be constant. This means that, apart from the effect of the obliquity factor, equal areas dS give contributions with equal amplitudes. Small variations in (r_0+r) can make a significant difference to the phase factor since (r_0+r) is multiplied by the large quantity $k(=2\pi/\lambda)$—it is obvious that a change in (r_0+r) of only a fraction of a wavelength will significantly affect the phase of the contribution from any element.

In general there may be a lens system between the source and the diffracting screen, and between this and the receiving screen. In this case (r_0+r) in the phase term becomes the optical path from the source to dS plus that from dS to P, the field point in the pattern. This is denoted by $[r_0+r]$ and will depend upon the co-ordinates (x, y) of the element dS. The quantity $[r_0+r]$ may be expressed as a power series in these variables. If $[r_0+r]$ is a linear function of x and y the pattern is referred to as a *Fraunhofer pattern*; otherwise it is a *Fresnel pattern*. It will be shown in the next section that patterns formed in planes conjugate to the source are Fraunhofer patterns; they were observed by Fraunhofer in 1821. As a special case, Fraunhofer patterns are formed when, in the absence of lenses, the source and receiving screen are infinitely distant from the diffracting screen. The Fresnel patterns usually discussed are those observed when there are no lens systems involved and the source and/or receiving screen are at finite distances from the diffracting screen. It is very often stated that in Fraunhofer diffraction the wavefronts incident on the diffracting screen must be plane so that the source and receiving screen are at infinity or in the focal planes of two

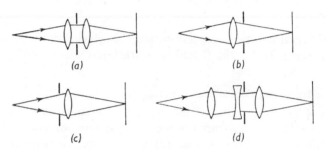

(a)   (b)

(c)   (d)

Fig. 10–7. Formation of Fraunhofer patterns.

lenses between which lies the diffracting screen [see Fig. 10–7(a)]. In the paraxial approximation this restriction is not necessary, however, and the arrangements in (b), (c), and (d), would give similar patterns in the plane conjugate to the source (assuming the lenses are perfect). It will be seen that images formed by optical instruments are, in fact, Fraunhofer diffraction patterns. One has, in the image space, a converging spherical wave diffracted by the exit pupil.

Further investigation quickly shows that for any particular point in a Fraunhofer pattern, the inclination factor is constant over the whole surface Σ if this is taken to be a wavefront, whereas this is never (strictly) true in Fresnel patterns.

The above theory is found to yield results which are in good agreement with observation for moderate angles of diffraction and the Fraunhofer patterns predicted by the theory agree with the observed patterns for systems of moderate aperture. But it has been assumed that the light disturbance at any point can be represented adequately by a single scalar quantity ϕ, whereas it follows from the electromagnetic theory of light that in order to specify the disturbance one must specify the six components of the electric and magnetic field vectors. A scalar theory cannot, for example, take any account of the phenomenon of polarization. However, within the limitations mentioned above, the scalar theory of diffraction does yield a useful first approximation.

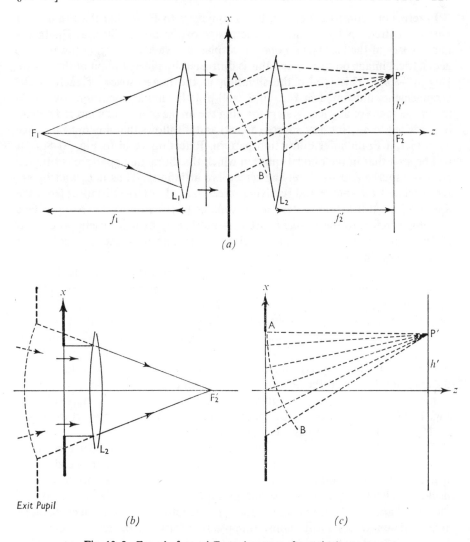

Fig. 10–8. Fraunhofer and Fresnel patterns for a single aperture.

10–10. Fresnel and Fraunhofer Patterns for a Single Slit

In Fig. 10–8(a) there is a point source at the first principal focus F_1 of lens L_1 so that a plane wave is incident on a rectangular aperture extending from $x=-a/2$ to $x=+a/2$ and from $y=-b/2$ to $y=+b/2$. Now, within the limits of paraxial theory, a converging lens causes *any* incident plane wave to converge towards a point focus. Hence, within these limits, for *any* point P′ in the focal plane of L_2 there will be a plane AB such that all points on AB are optically equidistant from P′. It will be seen at once that the optical distance from P′ to a point in the aperture in the xy plane is linearly dependent on the co-ordinate x of the latter point (and independent of y) so that the pattern in the second focal plane of L_2 is a Fraunhofer pattern. Furthermore, the figure shows that the obliquity factor is the same for all points in the aperture.

In terms of geometrical optics, F_2' is conjugate to F_1 so that the Fraunhofer pattern centred on F_2' constitutes the image of the point object at F_1. In the image space of the lens system one has a spherical wave converging towards the geometrical image point F_2', and this is diffracted by the exit pupil of the system [Fig. 10–8(b)]. This is also the situation in the image space of each of the arrangements illustrated in Fig. 10–7, and exactly similar conditions obtain in the image space of any lens system forming the image of a point source. This can be extended to any form of source (object) and it follows that any optical image is, in fact, a Fraunhofer diffraction pattern. Referring again to Fig. 10–8(a), it will be seen that in the optical space in which the diffracting aperture is situated both the source and the receiving screen are at infinity. Thus if L_1 and L_2 are removed and the source and receiving screen placed at large distances from the aperture, the diffraction pattern approximates to the Fraunhofer pattern. One sometimes refers to the *far-field* pattern—and it may be more useful to express the intensity in the pattern as a function of direction from the aperture rather than position in a receiving screen.

Suppose now that lens L_2 is removed. Fig. 10–8(c) shows that the distance from P′ to a point on the wave passing through the aperture is no longer linearly dependent on x. Points on the arc AB are equidistant from P′ and it is easily shown that if h' is small compared with the distance from the aperture to the receiving screen, the distance from P′ to a point in the aperture is dependent on $(h'-x)^2$, i.e. x^2 if the origin is moved to A. It follows that one has a Fresnel pattern, and it will be seen from the figure that the obliquity is not constant across the aperture. However, the obliquity factor varies only slightly and the variation is neglected in the discussion given below. With both lenses removed from Fig. 10–8(a) one has a diverging spherical wave incident on the aperture and it is obvious that one again has a Fresnel pattern.

The optical distance from P′ to any elementary area in the aperture affects both the amplitude and the phase of the disturbance at P′ due to the element. It has already been pointed out that small variations in this distance produce a negligible effect on the amplitude but an important effect on the phase. To find the disturbance at P′ one may divide the aperture into any convenient elementary areas and sum their contributions. Suppose that for the arrangements shown in Fig. 10–8(a) and (c) the aperture is divided into very narrow equal strips parallel to the y axis. If the variations of the obliquity factor can be ignored, it may be assumed that each strip produces a contribution of the same amplitude at P′, and that this is the same for all positions of P′ near the z axis.

In the case of the Fraunhofer pattern, the phase of the disturbance due to any strip is linearly dependent on x so that the phase difference between successive strips is constant. (If $l=ax+b$, $dl/dx=a$.) In the Fresnel case, with an appropriate choice of origin, the phase corresponding to each strip is proportional to x^2 so that the increment in phase in going from one strip to the next is proportional to x. [$(d/dx)(kx^2)=2kx$.] It follows that if one uses the graphical method to sum the contributions of the various strips, one gets an equilateral equiangular vector polygon in the Fraunhofer case [Fig. 10–9(a)]. In the Fresnel case the polygon is equilateral but not equiangular, the angle between successive vectors being proportional to x, or to the strip number if the strips are numbered from the origin of x [Fig. 10–9(b)]. If the strips in each case are made infinitesimally small, the vector polygons become arcs of a circle and a spiral respectively.

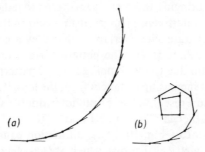

Fig. 10–9. Vector polygons for Fraunhofer and Fresnel patterns.

The more complicated geometrical figure that results in the Fresnel case makes this pattern more difficult to analyse.

10–11. Fresnel and Fraunhofer Patterns for a Double Slit

Further aspects of the relationship between Fresnel and Fraunhofer patterns may be seen by brief consideration of the patterns formed with two slits, using a slit source.

Figs 10–10(a) and (b) illustrate respectively the formation of the Fresnel and Fraunhofer patterns with a so-called double slit S_1S_2. For one narrow slit the

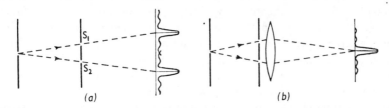

Fig. 10–10. Fresnel and Fraunhofer patterns for a double slit.

Fresnel pattern can be shown to consist of a central maximum on either side of which there are subsidiary maxima of rapidly decreasing intensity. (The intensity does not fall completely to zero between the maxima.) Using a double slit, the pattern associated with each slit is centred on the line through the source and diffracting slit as indicated in the figure. The outer maxima of the two patterns will obviously overlap and in this region the light in the two patterns can interfere. Consequently, superposed on the overlapping diffraction patterns, there will be the usual double slit interference fringes [see Plate III(i)]. The separate patterns actually extend to infinity so that interference does occur in the central maxima. This may pass unobserved because the intensity associated with one pattern may be very small at the centre of the other.

When the lens is used as in Fig. 10–10(b), the Fraunhofer pattern for each slit is centred on the geometrical image of the source slit. In general, the Fraunhofer pattern for a single slit is quite different from the Fresnel pattern but for very narrow slits the two patterns are similar [see Chapters XI and XIII, and Plates

II(*b*) and III(*d*)]. The main difference then is that the intensity falls to zero at the minima of a Fraunhofer pattern. Now it is obvious from the figure that one may look upon the lens as bringing the separate patterns together so that the central maxima coincide. The interference fringes again occur in the region in which the patterns due to the separate slits overlap so that they are now observed in the central maximum—see Plate II(*d*). The pattern formed on a screen held fairly close to the lens is again due to overlapping Fresnel patterns, and by examining the pattern in planes at increasing distances from the lens, it is possible to observe the transition from the Fresnel to the Fraunhofer double slit pattern. The patterns are discussed further in Chapters XI and XIII.

With the arrangement of Fig. 10–10(*a*), diffraction causes fringes to appear at the edges of the geometrical shadows which are cast by the boundaries of the apertures. This effect is particularly noticeable when the apertures are fairly large—see § 13–8. In the arrangement of Fig. 10–10(*b*) one has fringes around the geometrical image of the source. These effects—fringed shadows and fringed images—are characteristic of Fresnel and Fraunhofer diffraction respectively.

10–12. Babinet's Principle

Two screens, S_1 and S_2, are said to be *complementary* if S_2 is the screen which is formed when the transparent parts of S_1 are made opaque and the opaque parts made transparent. If P is a point at which, in the absence of any diffracting screens, the disturbance is zero, then Babinet's theorem states that the intensity at P caused by the introduction of S_1 is the same as that caused by the introduction of S_2. Since it is the intensity which is normally observed in a diffraction pattern, a reasonable alternative statement of the theorem is: complementary diffracting screens produce identical diffraction patterns in regions that are not illuminated when there are no screens.

The theorem is easily proved as follows: Let ϕ_1 and ϕ_2 be the disturbances at P in the presence of S_1 and S_2 respectively. The disturbances are due to the light that passes through the apertures of these screens, so that if the apertures of both screens are present simultaneously, the disturbance at P would be $\phi_1 + \phi_2$. But if both sets of apertures are present simultaneously, there are no opaque regions, i.e. there is, in fact, no diffracting screen and, by assumption, the disturbance at P is zero.

i.e.
$$\phi_1 + \phi_2 = 0 \quad \cdots \cdots \quad (10\text{–}19)$$

or
$$\phi_1 = -\phi_2 \quad \cdots \cdots \quad (10\text{–}20)$$

∴
$$|\phi_1|^2 = |\phi_2|^2 \quad \cdots \cdots \quad (10\text{–}21)$$

Equation (10–21) shows that S_1 and S_2 produce equal intensities at P. In view of equation (10–20) it is sometimes said that complementary screens produce *complementary* patterns.

It would seem that there is no *exact* application for Babinet's principle in this form because if no diffracting obstacles of any kind are present anywhere, a single source will illuminate all space. However, when one has an optical system of large aperture, the intensity is very small at points outside the geometrical image of the source and, apart from the region near this image, approximately equal intensity distributions do result from the successive interposition of complementary screens. Thus the theorem can be applied for Fraunhofer patterns (see §11–9). A simple variation of the original form of Babinet's principle enables one to deal with these problems more exactly. Thus, suppose one has a diffracting screen in which there is a large aperture A, and that ϕ_A is the disturbance at P due to this aperture. Suppose, now, that S_1 and S_2 are screens that are *complementary over the area of* A. If ϕ_1 and ϕ_2 are the disturbances produced at P by S_1 and S_2, the argument given above yields, in place of (10–19),

$$\phi_1 + \phi_2 = \phi_A . \quad . \quad . \quad . \quad . \quad . \quad . \quad (10\text{--}22)$$

In what is probably the most common application of Babinet's principle, A is the aperture of a lens and S_1 and S_2 are screens which are opaque beyond the edge of A but which contain complementary groups of apertures within the area A.

10–13. Young's Edge-Wave Interpretation of Diffraction

Before Fresnel advanced his very successful formulation of Huygens' principle, Young had suggested that the incident light undergoes some kind of reflection at the edge of a diffracting aperture. He assumed that the diffraction pattern was then the result of the interaction between the unobstructed light and an *edge wave* or *boundary wave* which can be regarded as originating from the rim of the aperture.

As a result of experimental investigations, Fresnel concluded that the diffraction pattern did not depend upon the degree of polish or sharpness of the edge and the success of Fresnel's theory caused Young's ideas to be forgotten for a time. Many years later work by Sommerfeld and by Rubinowicz put Young's ideas on to a firm basis.

For any diffracting aperture, the surface Σ in the Kirchhoff integral [equation (10–18)] can be any surface which bridges the aperture. That is to say, the only requirement is that the rim L of the open surface Σ must be the boundary of the aperture. Consequently, the integral depends only upon the curve L and it must be possible to transform the surface integral over Σ into a line integral round L. This transformation has been effected by Rubinowicz and the result indicates that within the geometrical shadow the disturbance can be regarded as being due to radiation from the boundary of the aperture. In the geometrically illuminated region the disturbance is due to the superposition of the edge wave and the disturbance which would be present in the absence of the diffracting screen. The effect of the edge wave is discussed further in §§13–7, 8, 9, and 14–13.

CHAPTER XI

FRAUNHOFER DIFFRACTION PATTERNS

11–1. Rectangular Aperture and Point Source. Location of the Minima

WHEN investigating the patterns formed by diffraction at a rectangular aperture, it is usually convenient to employ a slit source parallel to one of the edges of the aperture. It must be pointed out that each point of the slit source emits spherical waves and, although these have a cylindrical envelope, the resultant is

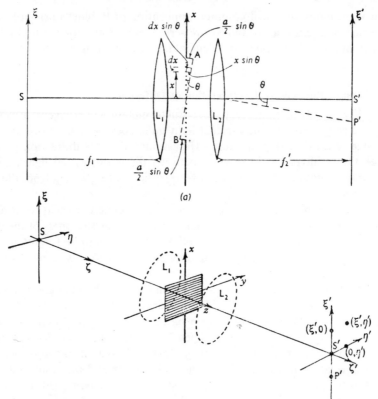

(a)

Fig. 11–1. Diffraction by a rectangular aperture. Path difference for extreme strips

$$2\frac{a}{2}\sin\theta = a\sin\theta.$$

not a cylindrical wave. A true cylindrical wave is produced by placing a point source at the focus of a cylindrical lens; it is seldom encountered in practice. For this reason it is necessary to commence the analysis with an investigation of the pattern formed when a point source is used.

Suppose that the monochromatic point source S (Fig. 11–1) is at the principal focus of a perfect lens L_1 and that the receiving screen is in the focal plane of a second lens L_2, the diffracting aperture being located between the lenses and centred on their common optical axis. As was shown in §10–10, this arrangement of the lenses is not essential and the same pattern results whenever the screen is conjugate to the plane of the source.

With the present arrangement a plane wave is incident on the rectangular aperture which will be assumed to have limits $x = \pm a/2$, $y = \pm b/2$. The elements of the wave which lie within these limits act as secondary sources, and it is necessary to find their combined effect at the point P' on the receiving screen. Initially it is assumed that P' lies in the plane of diagram (a). As shown in § 10–10, all points on the plane AB [Fig. 11–1(a)] are optically equidistant from P', and there occur only small variations in the optical paths from P' to points on the wavefront. Consequently, it may be assumed that equal elementary areas of the wavefront give disturbances of equal amplitude at P'.

Suppose the unobstructed part of the wavefront is divided into narrow strips parallel to the y-axis. Points on a single strip are optically equidistant from P' and are sources of coherent secondary waves. Consequently, each strip gives a disturbance at P' whose amplitude is proportional to the area of the strip. It follows that if the strips are of equal width they contribute disturbances of equal amplitude at P', the phase of the contribution of any strip depending on its optical distance from P'. The positions of the minima are easily found as follows. When θ is such that the difference between the paths from P' to the strips at the edges of the aperture is λ, the path difference for the first strip and the first past the middle is $\lambda/2$. At P' these strips then give contributions of equal amplitude but opposite phase; they therefore interfere destructively. Similarly, the disturbance from the second strip and the second past the middle cancel, and so on. Thus the resultant at P' is zero. Now, from Fig. 11–1, it is seen that the path difference for extreme strips is $a \sin \theta$, so that a minimum occurs where $a \sin \theta = \lambda$. By a similar argument it can be shown that another minimum occurs when $a \sin \theta = 2\lambda$. In this case the aperture can be divided into two halves. *For each half* the optical distances from P' to the extreme strips differ by λ so that the disturbance from the first strip interferes destructively with that from the first strip past the middle, and so on. By an obvious extension it can be shown that minima occur when $a \sin \theta = p\lambda$, where p is an integer.

In Fig. 11–1 the receiving screen is conjugate to the plane containing the source and, as has been pointed out before, the Fraunhofer diffraction pattern constitutes the image of the source. In the image space one has a converging spherical wave which is diffracted by the rectangular exit pupil. Obviously these conditions will obtain whenever one has an optical system with a rectangular aperture stop. Fig. 11–2 illustrates the diffraction of the wave in the image space for any system of this type. When P' is at the first minimum in the pattern the distances of the extreme edges of the wave from P' differ by λ. If all points on AB are the same distance from P', and θ and U' are small, one may write

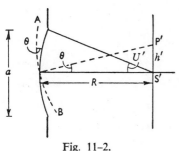

Fig. 11–2.
Diffraction of wave in image space.

$$h' = R\theta = R\frac{\lambda}{a},$$

where R is the radius of the converging wave. As usual, U' is the angular semi-aperture in the image space; one has

$$\frac{a}{2} = R \sin U'.$$

Whence
$$h' = \frac{\lambda}{2 \sin U'}.$$

In this equation λ is the wavelength in the image space. If n' is the refractive index of this space and λ is the wavelength in a vacuum, one has

$$h' = \frac{\lambda}{2n' \sin U'}. \qquad \cdots \cdots \quad (11\text{--}1)$$

11–2. Intensity Distribution. Use of the Graphical Method

For a detailed examination of the complete pattern it is convenient to take the central strip as giving a contribution of zero phase at P' (Fig. 11–1). This is justified because only relative phases are important. Since points on AB are optically equidistant from P' it is obvious that the strip at x is $x \sin \theta$ farther from P' than is the central strip, so that the former gives, at P', a contribution whose phase is $(2\pi/\lambda)x \sin \theta$. If each strip is of width dx, the difference between the phases of the contributions of adjacent strips is $(2\pi/\lambda)dx \sin \theta$, and is constant. If the contributions of the strips are added by means of the graphical method, the vector polygon is an equilateral figure since each strip gives a vector of the same length; the angle between successive vectors is constant and equal to $(2\pi/\lambda)dx \sin \theta$. The total number of sides in the polygon is equal to the total number of strips; i.e. it is determined by a, the total width of the aperture. In the limiting case where the strips are made infinitesimal in width, the polygon becomes the arc of a circle, the length of the arc being proportional to a (Fig. 11–3). The resultant amplitude is given, as usual, by the length of the chord, and the intensity is the square of that length. Since the number of unobstructed

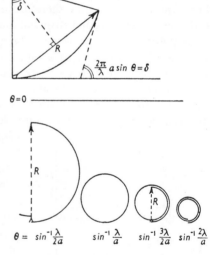

Fig. 11–3. Vibration curves.

strips is constant, the arc length is constant for all field points P'. For different positions of P' the curvature of the arc varies, being determined by the increment in phase between successive strips. Thus when $\theta = 0$ P' is on the axis and the arc commences as a straight line since, in this position, all points on the wavefront are optically equidistant from P'. As θ increases, the curvature of the arc increases. When $(2\pi/\lambda)a \sin \theta = 2\pi$ (i.e. $a \sin \theta = \lambda$) the first and last strips in the aperture give contributions which are 2π out of phase, and the arc is bent into a complete circle. As θ increases further, the arc executes more than one complete revolution; it again becomes a closed curve when $a \sin \theta = 2\lambda$ and, in general, when $a \sin \theta$ is an integral number of wavelengths (see Fig. 11–3).

∴   at a minimum $\sin \theta = \dfrac{p\lambda}{a}$ (p integral).   . . . . .   (11–2)

For certain intermediate values of θ the disturbance reaches maximum values. Fig. 11–3 shows the vibration curves midway between the minima, but the maxima do not lie *exactly* midway between the minima. It can be seen at once that successive maxima are weaker as P′ recedes from the axis because the vibration curve winds up. When $\theta = 0$ the vibration curve is simply a straight line, so that if one takes the central maximum to have unit amplitude, the length of the arc is fixed as unity. As the arc becomes curved, the angle subtended at the centre is the phase difference between contributions from extreme strips of the aperture and is $\delta = (2\pi/\lambda)a \sin \theta$. The radius of the arc is given by

$$r = (\text{arc length}) \div (\text{angle at centre})$$

$$= \frac{1}{(2\pi/\lambda)a \sin \theta}.$$

The resultant is defined by the chord and (see Fig. 11–3) is given by

$$R = 2r \sin \frac{\delta}{2}$$

$$= 2 \cdot \frac{1}{\dfrac{2\pi a \sin \theta}{\lambda}} \cdot \sin \frac{\pi a \sin \theta}{\lambda}$$

$$= \frac{\sin \left(\dfrac{\pi a \sin \theta}{\lambda} \right)}{\left(\dfrac{\pi a \sin \theta}{\lambda} \right)}.$$

∴   Intensity $= \left[\dfrac{\sin \left(\dfrac{\pi a \sin \theta}{\lambda} \right)}{\left(\dfrac{\pi a \sin \theta}{\lambda} \right)} \right]^2 = (\text{say}) \left(\dfrac{\sin \alpha}{\alpha} \right)^2.$ (11–3)

This gives the intensity at any point along a line through S′ parallel to the x-axis (i.e. along the ξ'-axis). An exactly similar expression (in which a is replaced by b and θ by ϕ, the corresponding angle in the yz plane) gives the intensity along the η'-axis.

The minima occur when the numerator in (11–3) is zero, i.e. when α is an integral multiple of π. That is, minima occur at angles given by

$$\theta = \sin^{-1} \frac{p\lambda}{a} \ (p \text{ integral}),$$

which is the result given above [equation (11–2)].

The positions of the maxima may be found by equating $dI/d\alpha$ to zero. One has

$$\frac{dI}{d\alpha} = 2\left(\frac{\sin \alpha}{\alpha}\right)\left[-\frac{\sin \alpha}{\alpha^2} + \frac{\cos \alpha}{\alpha}\right] = 0.$$

Hence at a maximum

$$\tan \alpha = \alpha.$$

This equation may be solved graphically by plotting $y = \alpha$ and $y = \tan \alpha$ and finding the points of intersection of the two curves (see Fig. 11–4). It is found

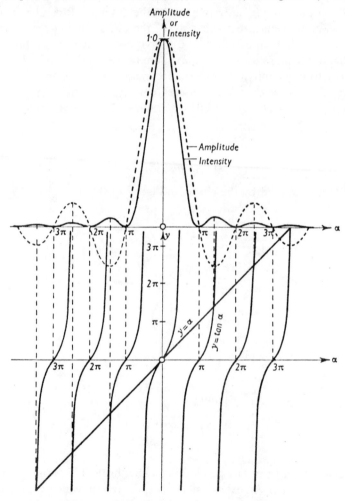

Fig. 11–4. Graphical solution for single slit pattern.

that the maxima occur at $\alpha = 0$, $1\cdot4303\pi$, $2\cdot4590\pi$, $3\cdot4707\pi$. . . . The amplitude and intensity distributions are shown in the figure. It will be seen that after the first few the maxima are approximately midway between the minima.

It should be noted that the obliquity factor has been neglected so that the above results are not accurate for large values of θ. The intensity should be multiplied by $\frac{1}{4}(1 + \cos \theta)^2$.

Suppose now that each of the strips of the unobstructed part of the wavefront is divided into elements by means of lines parallel to the x-axis. The difference between the disturbances at the points $(\xi',0)$ and $(0,0)$ is due to the phase differences introduced between elements whose y co-ordinates are the same. When P' moves from $(0,\eta')$ to (ξ',η') the same phase differences are introduced. Thus the variations of relative intensity are the same along all lines parallel to the ξ'-axis. By a similar argument it follows that the variations of intensity along any line parallel to the η'-axis is given by an equation of the same form as (11–3), a and θ being replaced by b and ϕ. Thus the general expression for the intensity in the $\xi'\eta'$ plane is

$$
I = \left[\frac{\sin\left(\dfrac{\pi a \sin\theta}{\lambda}\right)}{\left(\dfrac{\pi a \sin\theta}{\lambda}\right)} \right]^2 \left[\frac{\sin\left(\dfrac{\pi b \sin\phi}{\lambda}\right)}{\left(\dfrac{\pi b \sin\phi}{\lambda}\right)} \right]^2 . \qquad (11\text{–}4)
$$

θ and ϕ are, of course, measured in the $\xi'\zeta'$ and $\eta'\zeta'$ planes respectively. A photograph of the pattern is shown in Plate II(a). The lines of zero intensity are at $\theta = \sin^{-1} p\lambda/a$ and $\phi = \sin^{-1} p\lambda/b$ where p is an integer. For small angles one has $\theta = p\lambda/a$, $\phi = p\lambda/b$, i.e. in the $\xi'\eta'$ plane the lines of zero intensity are given by

$$
\xi' = p \cdot \frac{f_2'\lambda}{a},
$$

$$
\eta' = p \cdot \frac{f_2'\lambda}{b},
$$

where f_2' is the focal length of the lens L_2 (Fig. 11–1).

The pattern is centred on the origin in the $\xi'\eta'$ plane and, in terms of geometrical optics, this is conjugate to the source. That is to say, the above Fraunhofer pattern is the physical image of a point source as formed by an optical system having a rectangular aperture. It will be seen that this image becomes smaller as the size of the aperture is increased—the pattern is more compressed in the direction in which the aperture is wider. The larger is the aperture the more closely does the pattern approximate to the ideal geometrical point image.

11–3. Use of the Kirchhoff Integral

Equation (11–4) can also be deduced directly from Kirchhoff's diffraction integral. Equation (10–18) gives

$$
\phi_{\mathrm{P'}} = \int\!\!\int_{\Sigma} \frac{a}{2\lambda} (\cos\theta - \cos\theta_0) \frac{e^{ik[ct-(r_0+r)]+i\frac{\pi}{2}}}{rr_0} \, dS,
$$

where Σ can be any surface which bridges the aperture and, in the present case, is taken to be simply the unobstructed part of the wavefront. In the exponential r_0 and r become, respectively, the optical paths from the source, and from the field point P' to the current point on Σ, r_0 becoming a constant. It may be assumed, as above, that the obliquity factor is constant and that the amplitude term,

$1/r_0 r$, also becomes a constant. In addition, the phase factor $i(\pi/2)$ may be omitted since only relative phases are important. The complex amplitude is then given by

$$\psi_{P'} = (\text{constant}) \int\int_\Sigma e^{-ikr} dS.$$

Using the complex representation of disturbances, this equation also follows from the simple Fresnel theory since it corresponds to a summation of the effects of the secondary sources distributed over the exposed part of the wavefront.

If the optical path from the centre of the aperture to P' is denoted by R,

$$r = R + x \sin \theta + y \sin \phi \qquad (\text{see Fig. 11–1}).$$

If the edges of the aperture are $x = \pm a/2$, $y = \pm b/2$, and the constant term is omitted, one has

$$\psi_{P'} = \int_{-a/2}^{+a/2} e^{-ikx \sin \theta}\, dx \int_{-b/2}^{+b/2} e^{-iky \sin \phi}\, dy$$

$$= \left(\frac{e^{\frac{i\pi a \sin \theta}{\lambda}} - e^{\frac{-i\pi a \sin \theta}{\lambda}}}{2 \frac{i\pi \sin \theta}{\lambda}} \right) \left(\frac{e^{\frac{i\pi b \sin \phi}{\lambda}} - e^{\frac{-i\pi b \sin \phi}{\lambda}}}{2 \frac{i\pi \sin \phi}{\lambda}} \right)$$

$$= \left(\frac{i \sin \left(\frac{\pi a \sin \theta}{\lambda} \right)}{\frac{i\pi \sin \theta}{\lambda}} \right) \left(\frac{i \sin \left(\frac{\pi b \sin \phi}{\lambda} \right)}{\frac{i\pi \sin \phi}{\lambda}} \right),$$

so that if the expression is *normalized* to give unit intensity at the centre of the pattern, one has

$$\text{Intensity} = \left[\frac{\sin \left(\frac{\pi a \sin \theta}{\lambda} \right)}{\left(\frac{\pi a \sin \theta}{\lambda} \right)} \right]^2 \left[\frac{\sin \left(\frac{\pi b \sin \phi}{\lambda} \right)}{\left(\frac{\pi b \sin \phi}{\lambda} \right)} \right]^2,$$

which is the result obtained in § 11–2.

11–4. Slit Aperture and Slit Source

If, in the above example, there are other point sources along the η-axis, each will give a pattern centred on its geometrical image point on the η'-axis. For a large number of closely spaced point sources along the η-axis, the patterns will overlap and there will be no variation of intensity along lines parallel to the η'-axis. Along lines parallel to the ξ'-axis, however, the variation in intensity will still be given by an equation of the form of (11–3). For example, the lines of zero intensity parallel to the η'-axis are the same for each point source. Thus a slit or line source along the η-axis gives linear fringes parallel to the η'-axis, the intensity distribution across the fringes being given by (11–3). Moreover, the relative intensities along a line across the fringes are independent of the length of the aperture in the y-direction. This result justifies the common procedure of restricting the discussion for slit source and aperture to one plane. A photograph of the fringes obtained with a narrow slit source parallel to a single diffracting slit or rectangular aperture is shown in Plate II(*b*).

If the width of the source slit is increased, each line element produces the above pattern so that the combined effect is the superposition of a number of fringe systems at slightly different positions along the ξ'-axis. It follows that if the slit source is too wide the overlapping fringes produce a continuous band of light and no fringes are distinguishable.

11–5. Chromatic Resolving Power of a Prism Spectroscope. The Rayleigh Criterion

Fig. 11–5 shows the optical system of a simple prism spectroscope. If the narrow slit S is illuminated by a source which emits a number of discrete wavelengths, a line spectrum is formed in the focal plane of L_2. If the rectangular

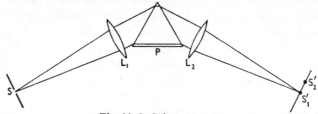

Fig. 11–5. Prism spectrum.

cross-section of the prism constitutes the limiting aperture of the system, each line in the spectrum is a Fraunhofer pattern for a slit source and rectangular aperture. (For any spectroscopic instrument the intensity distribution across a monochromatic spectral line is referred to as the *instrumental profile* or *function*.) Suppose the source emits wavelengths λ and $\lambda + \delta\lambda$ with equal intensities, and that S_1' and S_2' are the geometrical line images for these wavelengths. The two wavelengths must have been emitted by different sources (on the atomic scale) so that the disturbances are incoherent. Consequently, the total intensity at any point in the focal plane of L_2 is simply the sum of the intensities for each wavelength. (It will be appreciated that it is impossible to distinguish any colour difference between λ and $\lambda + \delta\lambda$.)

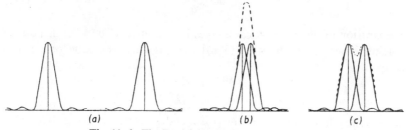

Fig. 11–6. The Rayleigh criterion for resolution.

If $\delta\lambda$ is fairly large, $S_1'S_2'$ is large and the central maxima of the diffraction patterns are clearly distinguishable [see Fig. 11–6(*a*)]. They are said to be resolved. If, however, $\delta\lambda$ is very small, the separate patterns may overlap to such an extent that the total intensity shows no drop between S_1' and S_2' [Fig. 11–6(*b*)]. It is then impossible, by casual observation, to see whether the pattern corresponds to a single line or to two neighbouring lines—the lines are not resolved. Obviously it is important to know how close the lines can be before they cease to be resolved. That is, one needs to know the minimum separation of S_1' and S_2' for which there is a just perceptible decrease in intensity between them. According to Rayleigh, it is just possible to see, by casual observation, that the resultant pattern consists of two overlapping patterns if the centre of one pattern falls on the first minimum of the other [Fig. 11–6(c)]. It will be appreciated that by careful analysis of the intensity distribution, it is possible to detect the existence of two overlapping patterns when their centres are closer than this, but the Rayleigh criterion provides a useful simplification of the problem.

When the lines are just resolved, the intensity I midway between S_1' and S_2' may be found from equation (11–3). It is twice the intensity in each central maximum midway between its centre and its edge, and if the intensity at the centre of each pattern is unity, $I=0.81$. When the intensities of the two spectral lines are different, the Rayleigh criterion as stated above is not directly applicable. That is, when the first minimum of one pattern falls on the centre of the other there might not be an easily detectable drop in intensity between the patterns. In such cases it is usually assumed that the patterns are resolved if, at any point between the maxima, the intensity falls to about 0.81 of that of the weaker maximum. This latter criterion is also used when discussing overlapping interference fringes (see §9–3). If the spectral lines for wavelengths λ and $\lambda+\delta\lambda$ are just resolved, ($\lambda/\delta\lambda$) is the *chromatic resolving power*. Its reciprocal is the fractional difference in wavelength between the two lines when they are just resolved.

When the lines are just resolved, the separation of their centres (the geometrical line images) is equal to the distance from the centre to the first minimum of one of the lines. It was shown in §11–1 that the angular separation of the first minimum and the centre of the pattern is $\theta=\sin^{-1}(\lambda/a)$, or λ/a if the angle is small. Consequently, the spectral lines are resolved if the angular separation of S_1' and S_2' is given by

$$\delta\theta=\frac{\lambda}{a}. \qquad \ldots \ldots \ldots (11\text{–}5)$$

The separation of the geometrical images is, of course, due to the dispersion of the prism. Now [from equation (5–6)] the angular dispersion of a prism is given by

$$\frac{\delta\theta}{\delta\lambda}=\frac{t}{a}\cdot\frac{dn}{d\lambda},$$

where n is the refractive index of the prism and t is the length of its base. Hence, from (11–5),

$$\frac{\lambda}{\delta\lambda}=t\cdot\frac{dn}{d\lambda}. \qquad \ldots \ldots \ldots (11\text{–}6)$$

It will be seen that in order to resolve very close spectral lines a large prism is necessary.

In the above discussion it has been assumed that the width of the slit is infinitesimal. If this is not so, the widths of the spectral lines are increased and a larger difference in wavelength is necessary before neighbouring lines can be resolved. If $\Delta\lambda$ is the wavelength difference when two lines can be resolved with a certain slit width, $\lambda/\Delta\lambda$ is sometimes called the *purity*. As the slit width is decreased, $\Delta\lambda$ decreases and the purity increases. It reaches its maximum value (the chromatic resolving power) when the slit width is zero. For wide slits, the slit width is obviously the dominant factor governing the purity. With a continuous spectrum it is the purity which determines the range of wavelengths found at a given point in the spectrum. (See also §18–12.)

11–6. Limit of Resolution with a Slit Aperture

A problem similar to the one discussed above arises when neighbouring object points are imaged by an optical system. In the image plane one has a Fraunhofer pattern centred at each geometrical image point, and it is necessary to know the separation of the object points when the images are just resolved. This is referred to as the *limit of resolution* of the instrument. Suppose that an optical system having a rectangular aperture images two independent (i.e. incoherent) point sources and that the line joining the sources is perpendicular to the edges of the aperture. Along the line joining the geometrical image points the variation in intensity is similar to that across a pair of spectral lines. Consequently, one can adopt the same criterion for resolution. That is, the objects are resolved if the separation of their geometrical images is equal to the distance from the centre of one pattern to the first minimum. Thus, from equation (11–1), the objects are resolved if the separation of the images is given by

$$h' = \frac{0 \cdot 5\lambda}{n' \sin U'}, \qquad \cdots \cdots \cdots \quad (11–7)$$

where U' is the angular semi-aperture of the instrument in the image space and n' is the refractive index of that space.

Using this result one can find the value of the limit of resolution h, but the discussion would have very limited significance since optical instruments do not, in general, have rectangular apertures. The Fraunhofer pattern for a point source and a circular aperture is investigated in the next two sections and the limit of resolution of optical instruments is discussed in Chapter XIV.

11–7. Circular Aperture and Point Source. Qualitative Discussion

In Fig. 11–7, a circular aperture is placed between lenses L_1 and L_2. The point source S in the focal plane of L_1 causes a plane wave to be incident on the aperture, and the Fraunhofer pattern is formed in the focal plane of L_2 and is centred on S', the geometrical image of S. In order to find the intensity at a point P', divide the unobstructed part of the wavefront into strips of equal width perpendicular to S'P'. If the graphical method is used to sum the contributions of the strips, the vibration polygon will be equiangular since the phase difference between contributions from successive strips is constant (as for a rectangular aperture). However, the polygon is not equilateral because, as will be seen from the figure, the strips are not of equal length. When P' is at S', the polygon is, as before, a straight line [see Fig. 11–7(*a*)]. As P′ moves away from S′, the polygon takes the form shown in (*b*). When the difference between the optical paths from P′ to the extreme edges of the aperture is $\lambda/2$, the extreme vectors are parallel but in opposite directions as in (*c*). When the path difference

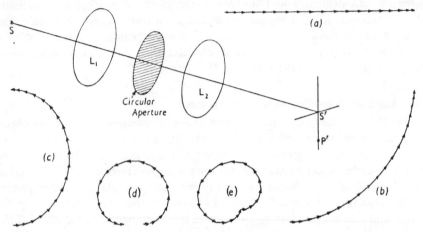

Fig. 11–7. Strip division for a circular aperture.

is λ they are parallel as in (*d*) but the polygon is not a closed figure as it was for a rectangular aperture. It can be shown that the polygon does not become a closed figure until the path difference is 1.22λ as in (*e*). Thus, for a circular aperture, the first minimum occurs at P′ where the optical paths from P′ to the extreme edges of the aperture differ by 1.22λ. The direction of the minimum is therefore given by

$$a \sin \theta = 1.22 \lambda$$

or

$$\theta = 1.22 \frac{\lambda}{a} \text{ since } \theta \text{ is small.}$$

From the symmetry of the arrangement about the axis, it is obvious that the fringe system is circular so that, since $\theta = S'P'/f_2'$, there is a dark ring centred at S′ of radius

$$S'P' = f_2' \cdot \frac{1.22\lambda}{a}.$$

In order to deduce the exact nature of the pattern it is convenient to use an annular division of the wavefront as in the next section.

11–8. Derivation of the Airy Pattern

It has already been pointed out that when an optical system forms an image of a point source, one has, in the image space, diffraction of a converging spherical wave by the exit pupil.

In Fig. 11–8 ABCD is an annular zone, radius ϱ, of a spherical wave emerging from the circular exit pupil and converging to the geometrical image point G′ on the optical axis. *ds* is an element of the annular zone in the azimuth ϕ referred to the plane containing the field point P′ and the optical axis. AB and

CD are, respectively, perpendicular to and parallel to this plane. P′K is perpendicular to the line joining ds to G′. (ds)H is perpendicular to AB.

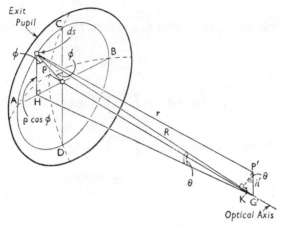

Fig. 11–8. Wave emerging from exit pupil.

If $(ds)\text{P}'=r$, the complex amplitude at P′ due to the emergent wave is, from Kirchhoff's formula, given by

$$\psi_{\text{P}'}=(\text{constant})\int_{\text{wave}} e^{-ikr}ds.$$

This may be written

$$\psi_{\text{P}'}=(\text{constant})\int e^{-ik(r-R)}ds,$$

where R is the radius of curvature of the wave $[=\text{G}'(ds)]$. This is justified since R is a constant and one is interested in relative phases only.

From the figure,

$$(R-r)=\text{KG}'=h'\sin\theta=h'\frac{\varrho\cos\phi}{R},$$

$$\therefore\ \psi_{\text{P}'}=(\text{constant})\int_{\phi=0}^{\phi=2\pi}\int_{\varrho=0}^{\varrho=\varrho_{\max}} e^{i\left(\frac{kh'\varrho}{R}\right)\cos\phi}\varrho\,d\varrho\,d\phi,$$

which may be written

$$\psi_{\text{P}'}=(\text{constant})\int_{\varrho=0}^{\varrho=\varrho_{\max}}\varrho\,d\varrho\cdot\frac{1}{2\pi}\int_{\phi=0}^{\phi=2\pi} e^{i\left(\frac{kh'\varrho}{R}\right)\cos\phi}\,d\phi,$$

$$=(\text{constant})\int_0^{\varrho_{\max}}\varrho\cdot J_0\left(\frac{kh'\varrho}{R}\right)d\varrho.\text{*}$$

* $J_0(x)$ and $J_1(x)$ denote Bessel functions of order 0 and 1 respectively. Their properties are given in standard texts and their values have been tabulated for various values of x. Two standard results which are used here are: $\dfrac{1}{2\pi}\displaystyle\int_0^{2\pi} e^{ix\cos\phi}\,d\phi=J_0(x)$ and $\displaystyle\int_0^z xJ_0(x)dx=zJ_1(z)$.

It is convenient to put

$$x = \frac{kh'\varrho}{R} \quad \text{and} \quad z = \frac{kh'\varrho_{max}}{R}.$$

Since $\dfrac{k\varrho_{max}}{R}$ is constant, z can be used as a measure of the distance P'G'. Then

$$\varrho = \frac{R}{kh'}x, \quad \text{and} \quad d\varrho = \frac{R}{kh'}dx.$$

$$\therefore \quad \psi_{P'} = \text{(constant)} \left(\frac{R}{kh'}\right)^2 \int_{x=0}^{x=z} x J_0(x) dx$$

$$= \text{(constant)} \left(\frac{R}{kh'}\right)^2 z \cdot J_1(z)$$

$$= \text{(constant)} \, \varrho_{max}^2 \cdot \frac{J_1(z)}{z},$$

and the intensity is $|\psi_{P'}|^2$.

Now ϱ_{max} is constant, and $\dfrac{J_1(z)}{z} \to \tfrac{1}{2}$ as $z \to 0$. Hence, on normalizing the expression for the intensity distribution in the pattern to give unit intensity at the centre, one has

$$I = \left[\frac{2J_1(z)}{z}\right]^2. \qquad \cdots \cdots \quad (11\text{--}8)$$

This result, first derived by Airy in 1834, has an infinite number of zeros, and the variation of I along a line through G' is of the form shown in Fig. 11–9.

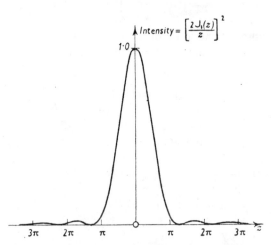

Fig. 11–9. Intensity distribution across Airy pattern.

The complete pattern is formed by rotating this about the central ordinate and obviously consists of a bright central disc (the so-called Airy disc) surrounded by rings of rapidly decreasing intensity. The first zero occurs at $z = 1\cdot22\pi$. Now $z = (kh'\varrho_{max})/R = (2\pi/\lambda) h' \sin U'_{max}$ where U'_{max} is, as usual, the semi-angle of the emergent cone of light. Thus the first zero occurs at $h' = 0\cdot61\lambda/(\sin U'_{max})$. In this expression λ is, of course, the wavelength in the medium of the image space. If one takes λ to be the wavelength in a vacuum, one has

$$\text{Radius of Airy disc} = \frac{0.61\lambda}{n' \sin U'_{max}} . \quad . \quad . \quad . \quad (11\text{–}9)$$

Further minima and maxima occur for the following values of the numerical factor in the numerator: minima for 1·12, 1·62, 2·12 . . . and maxima for 0·82, 1·33, 1·85. . . . If the central maximum has unit intensity, the intensities of the next three maxima are 0·017, 0·004 and 0·0016, i.e. the intensity falls off rapidly in the outer parts of the pattern; it can be shown that about 84% of the energy falls within the first dark ring. The appearance of the pattern is shown in Plate IV(c).

11–9. Young's Eriometer

In Fig. 11–10 S is a monochromatic point source, L a lens, and AB a screen containing a large number of equal small circular apertures randomly distributed. Each aperture acting alone would give a broad Airy pattern centred at

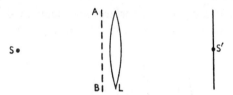

Fig. 11–10. Screen with small holes.

S′, the geometrical image of S. Now, since the apertures are the same size, the bright and dark rings have the same radii in every pattern. Consequently, the resultant pattern shows the same rings. Suppose the screen at AB is replaced by its complementary screen. Then, according to Babinet's principle, the two screens give the same pattern apart from the region of the geometrical image S′. By

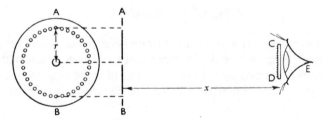

Fig. 11–11. Young's eriometer.

measuring the ring diameters it is therefore possible to measure the diameters of small particles if a number of similar particles are scattered over a glass plate and held at AB. A convenient experimental arrangement is shown in Fig. 11–11. AB is a screen containing a small aperture 1–2 mm. in diameter at the centre of a ring of small pinholes. The actual radius r of this ring depends on the size of the particles to be measured. The screen is illuminated from the left by a monochromatic source. CD is a glass plate on which are scattered the particles whose mean diameter is required (e.g. lycopodium dust). This is held in front of the observer's eye E. The observer views the central hole in the screen AB and the Fraunhofer pattern as described above is formed on the retina since this is conjugate to the screen AB. Thus a number of halos are seen. The angular diameter of the rings can easily be found by adjusting the distance x until one of the halos appears to fall on the ring of pinholes. (The pinholes do not transmit sufficient light for their own halos to be seen.) The angular radius of the first bright ring is $1 \cdot 64\lambda/a$ where a is the diameter of each particle, and when this appears to coincide with the ring of pinholes one has

$$\frac{r}{x} = \frac{1 \cdot 64\lambda}{a}$$

or

$$a = \frac{1 \cdot 64\lambda x}{r}.$$

If the second bright ring is brought into coincidence with the ring of pinholes the numerical factor in the numerator becomes $2 \cdot 66$; for the third bright ring it is $3 \cdot 7$.

11–10. The Double Slit

In Fig. 11–12 the two equal slit apertures are illuminated by monochromatic light from a narrow slit source S located in the focal plane of L_1. Each aperture

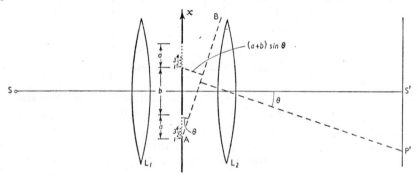

Fig. 11–12. The double slit.

acting alone would give rise to the usual Fraunhofer fringe system centred on S′, the geometrical image of S, in the focal plane of L_2. The complete pattern in this plane is therefore formed by the superposition of these two Fraunhofer single slit patterns which can interfere because the disturbances originate from the same source.

From the discussions of the earlier sections it will be seen that the variation in intensity along a line perpendicular to the direction of the slits may be found by discussing the case of a point source which causes a single plane wave to be incident on the slit apertures. Suppose that each aperture is of width a and that the opaque screen between them is of width b. At any point P′ the disturbance due to each aperture alone is given by equation (11–3). The effect of the superposition of the disturbances due to the two apertures may now be found if the phase difference between these disturbances can be found. It is convenient to divide the apertures into equal elementary strips and to number the strips from the lower edge of each aperture. It will be seen that since points on AB are optically equidistant from P′, the optical path from P′ to the first strip in the upper aperture is $(a+b) \sin \theta$ greater than the optical path from P′ to the first strip in the lower aperture. Hence the phase difference at P′ between the disturbances from these two strips is $(2\pi/\lambda)(a+b) \sin \theta$. This is also the phase difference between the disturbances from the second strips in the two apertures and so on, for all pairs of strips. Consequently, it must also be the phase difference at P′ between the total disturbances due to the two separate slit apertures. It follows at once that, whatever the slit width, minima occur when the path difference between disturbances passing through the middles of the slits is $(p+\frac{1}{2})\lambda$. In general, the disturbances to be added are each of amplitude $(\sin \alpha)/\alpha$ and the phase difference between them is $(2\pi/\lambda)(a+b) \sin \theta$.

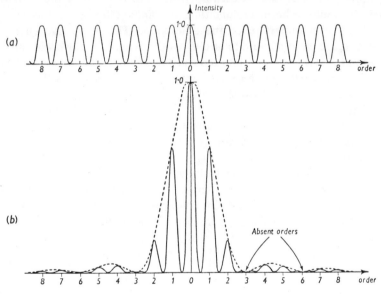

Fig. 11–13. Double slit pattern.

From equation (6–11) the resulting intensity is given by

$$I = \frac{4 \sin^2 \alpha}{\alpha^2} \cdot \cos^2 \left(\frac{\pi(a+b) \sin \theta}{\lambda} \right).$$

If this expression is normalized to give unit intensity at the centre of the pattern,

$$I = \frac{\sin^2 \alpha}{\alpha^2} \cdot \cos^2 \left(\frac{\pi(a+b) \sin \theta}{\lambda} \right). \qquad \cdot \; \cdot \; \cdot \quad (11\text{–}10)$$

Fig. 11–13(b) shows the intensity variation across the pattern for the case $b=2a$. The first term in equation (11–10) gives the intensity distribution in the Fraunhofer pattern for a single aperture and in the present problem gives the slowly varying envelope (shown dotted) within which lie the fringes given by the second term. The $\cos^2$ fringes are characteristic of the interference between two point or line sources $(a+b)$ apart, and the fringe separation does not depend on the width of the apertures. It is easily shown that the $\cos^2$ fringes do, in fact, correspond to those observed in Young's experiment. [From § 10–9 it will be seen that in Young's experiment when D is large the interference pattern is, by definition, the Fraunhofer pattern for the narrow slits.] In Fig. 7–2 it is necessary to write, when θ is not small, $S_2A = d \sin \theta$, whence

$$I = 4a^2 \cos^2 \left(\frac{\pi d \sin \theta}{\lambda} \right), \qquad \ldots \ldots \quad (11\text{–}11)$$

which, when normalized, is identical to equation (11–10) when the apertures are very narrow and the envelope term becomes unity.

The second term in (11–10) represents the effect of the *interference* between disturbances from the two apertures and the first term specifies the amount of light thrown, by *diffraction* at each aperture, into the part of the pattern considered. It should be borne in mind that, as pointed out in § 7–1, there is no real distinction between interference and diffraction. That is to say, when, in the present case, the pattern is said to be due to diffraction by the two wide apertures, this *means* that it is due to the interference between the secondary wavelets from the unobstructed parts of the wavefront. Fig. 11–13(a) shows the cos² interference fringes for two narrow slits $3a$ apart (i.e. $b = 2a$); the dotted curve in (b) is the Fraunhofer pattern for an aperture of width a and the full curve is the combined effect for two apertures of width a whose centres are $3a$ apart. Plate II(d) shows a photograph of a double slit pattern for which $b = 3a$, i.e. for slits whose centres are $4a$ apart.

In the double slit pattern the intensity will fall to zero if either term in (11–10) is zero. One may say that the resultant is zero either if there is no light directed from each aperture to the field point concerned or if there is destructive interference between the disturbances from the two apertures. These two causes lead, respectively, to zeros at

$$\sin \theta = \frac{p\lambda}{a}, \qquad \ldots \ldots \ldots \quad (11\text{–}12)$$

and

$$\sin \theta = (p + \tfrac{1}{2}) \frac{\lambda}{a+b}. \qquad \ldots \ldots \quad (11\text{–}13)$$

The positions of the maxima are more difficult to calculate exactly, but at places where the variation due to the first term (i.e. the envelope) is small, the maxima are approximately at the interference maxima given by $(a+b) \sin \theta = p\lambda$. Following the usual notation, p may be called the order of interference. If the relation between a and b is such that an interference maximum occurs at a diffraction minimum, i.e. where the envelope to the fringes falls to zero, there will be no well defined maximum. In this way it is possible for interference maxima of certain orders to be missing from the pattern. (Actually each might

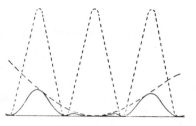

Fig. 11–14.
Appearance of "missing" orders.

be observed as a pair of weak maxima as indicated by Fig. 11–14.) Now, as stated above, interference maxima occur at $\sin \theta = p\lambda/(a+b)$ and diffraction minima at $\sin \theta = p\lambda/a$. It is possible for these conditions to be satisfied simultaneously when b is an integral multiple of a. Certain orders will then be missing. Thus in Fig. 11–13, for which $b=2a$, the missing orders are 3, 6, 9, etc., and in Plate II(d), for which $b=3a$, the missing orders are 4, 8, 12, etc.

If white light is used with a single or double slit, the central fringe is achromatic since this is the position of zero path difference for all wavelengths. In addition, a few coloured fringes can usually be observed where the path differences are small.

11–11. Derivation of the Double Slit Pattern from Kirchhoff's Integral

The analysis is identical to that of § 11–3 except that the integral must be taken over two apertures. If the intensity variation is required along a line perpendicular to the slits, the integration involving y may be omitted (i.e. included in the constant). It is again assumed that a disturbance from $x=y=0$ would have zero phase at P'. One has (see Fig. 11–12),

$$\psi_{P'} = (\text{constant}) \left[\int_{x=\frac{b}{2}}^{x=a+\frac{b}{2}} e^{-ikx\sin\theta}\, dx + \int_{x=-(a+\frac{b}{2})}^{x=-\frac{b}{2}} e^{-ikx\sin\theta}\, dx \right]$$

$$= \frac{(\text{constant})}{ik\sin\theta} \left[e^{-ik\frac{b}{2}\sin\theta} - e^{-ik(a+\frac{b}{2})\sin\theta} + e^{ik(a+\frac{b}{2})\sin\theta} - e^{ik\frac{b}{2}\sin\theta} \right]$$

$$= \frac{2\,(\text{constant})}{k\sin\theta} \left[\sin\left\{ k\left(a+\frac{b}{2}\right)\sin\theta \right\} - \sin\left\{ k\frac{b}{2}\sin\theta \right\} \right]$$

$$= \frac{2\,(\text{constant})}{k\sin\theta} \cdot 2\sin\left(\frac{ka\sin\theta}{2}\right)\cos\left(\frac{k(a+b)\sin\theta}{2}\right).$$

That is, $$\text{Intensity} = \left[\frac{\sin\left(\dfrac{\pi a\sin\theta}{\lambda}\right)}{\left(\dfrac{\pi a\sin\theta}{\lambda}\right)} \right]^2 \cos^2\left(\frac{\pi(a+b)\sin\theta}{\lambda}\right),$$

where, as before, the expression is normalized to make $I_{\text{max}}=1$.

11–12. Double Source and Source of Finite Width

In Fig. 11–15 S_1 and S_2 are two independent point or line sources imaged geometrically at S_1' and S_2' by the lens L, and A_1 and A_2 are diffracting slits. Each source produces a Fraunhofer double slit pattern centred on its geometrical image and the total intensity at any point is simply the sum of the intensities due to the separate sources. If α, the angular separation of the sources, is very small, the two patterns are nearly coincident and the brightness of the fringes is increased [Fig. 11–15(a)]. As α increases the resultant fringes become less distinct [Fig. 11–15(b)] until, when the first minimum of one pattern coincides with the central maximum of the other, the fringes disappear leaving a uniform intensity across the image plane [Fig. 11–15(c)]. A further increase in α causes the fringes to reappear. If A_1 and A_2 are narrow apertures, the superposed fringes are approximately cos² fringes and they will again disappear when the central maximum of one pattern coincides with the second and any subsequent minimum of the other. If A_1 and A_2 are wide apertures, these subsequent disappearances of the fringes are incomplete, since the envelope to the cos² fringes reduces the intensity of the outer maxima relative to the central maximum.

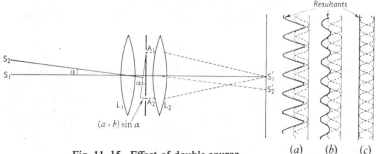

Fig. 11–15. Effect of double source. (a) (b) (c)

With very narrow slits each source gives cos² fringes, and when the central maximum of one set of fringes falls on the first minimum of the other set the total intensity is of the form

$$I=A^2 \cos^2 (\delta/2)+A^2 \cos^2 (\delta/2+\pi/2)$$
$$=A^2 \cos^2 (\delta/2)+A^2 \sin^2 (\delta/2)$$
$$=A^2$$

That is to say, one has uniform illumination across the field. In general, the two sets of fringes are exactly out of step when the maximum of one set falls on any minimum of the other. The intensity then has the form

$$A^2 \cos^2 (\delta/2)+A^2 \cos^2 [\delta/2+(p+\tfrac{1}{2})\pi]=A^2, \text{ as before.}$$

Now at the first minimum of a pattern the disturbances that have travelled via A_1 and A_2 arrive π out of phase. If S_1 is on the axis, A_1 and A_2 are equidistant from S_1' so that if the disturbances from S_2 are π out of phase at S_1' they must be π out of phase when they reach A_1 and A_2. That is, they traverse optical paths that differ by $\lambda/2$. It follows at once from Fig. 11–15 that the path difference is $(a+b)$ sin α so that the fringes disappear when $(a+b)$ sin $\alpha=\lambda/2$ or, for small angles, when $\alpha=\lambda/2(a+b)$. At the $(p+1)$th minimum the disturbances have a phase difference of $(2p+1)\pi$ corresponding to a path difference of $(p+\tfrac{1}{2})\lambda$. Thus in general the fringes disappear when

$$\alpha=(p+\tfrac{1}{2}) \frac{\lambda}{(a+b)}, \quad \cdot \ \cdot \ \cdot \ \cdot \ \cdot \ \cdot \ \cdot \quad (11\text{–}14)$$

Consider now the cases where the source is a line S_1S_2 or a slit of finite width S_1S_2. In these two cases, respectively, each point or line element of the source gives the usual Fraunhofer pattern and the resultant intensity at any point is the sum of the separate intensities. The result will be a uniform intensity along the line $S_1'S_2'$ if, for every fringe system along that line, there is a complementary system. The point or line element at one edge of the source, and the first one past the middle, may be regarded as a pair of independent sources whose angular separation is half the angular width of the source. Equation (11–14) shows that if the angular separation of these source elements is $\lambda/2(a+b)$ they give complementary fringes. That is, they give complementary fringes when the angular width of the complete source (α) has twice this value. When this occurs, the second point or line element and the second one past the middle of the source also give complementary fringes, and so on, and the fringes disappear. The narrowest source for which this occurs is that whose angular width is $\alpha = \lambda/(a+b)$. When $\alpha = 2\lambda/(a+b)$ the source can be divided into two halves and, *in each half*, the first element and the first past the middle give complementary fringes. By dividing the source into three parts it will be seen that the fringes again disappear when $\alpha = 3\lambda/(a+b)$. Thus, in general, the fringes disappear for a single wide source when its angular width satisfies the equation

$$\alpha = \frac{p\lambda}{(a+b)}. \qquad \cdots \cdots \quad (11\text{–}15)$$

i.e., when the paths from S_2 to A_1 & A_2 differ by $p\lambda$. If S_1S_2 is a circular disk the first disappearance occurs when the path difference $(a+b)\alpha$ is $1\cdot22\lambda$. If A_1A_2 and L_2 are replaced by the system of Fig. 11–17 one has $D\alpha = 1\cdot22\lambda$.

The above discussion is concerned with a simple form of extended source. In general, it follows from the discussion of partial coherence in §6–5 that when one has an extended source the disturbances emerging from A_1 and A_2 are partially coherent, and the fringes disappear as described above when the degree of coherence is zero. If A_1 and A_2 are equal small apertures, they give equal amplitudes in the image plane. If each gives amplitude a, it follows from equation (6–24) that the maximum and minimum intensities in the fringe pattern are given by

$$I_{max} = 2a^2(1 + V_{12})$$
$$\text{and} \quad I_{min} = 2a^2(1 - V_{12}).$$

Thus the fringe visibility is given by

$$\text{Visibility} = \frac{I_{max} - I_{min}}{I_{maz} + I_{min}} = V_{12}. \qquad \cdots \cdots \quad (11\text{–}16)$$

It is now necessary to calculate the complex degree of coherence (γ_{12}) for the general case of light received from an extended source. In Fig. 11–16 the points P_1 and P_2 receive light from the self-luminous quasi-monochromatic source Σ which lies in a plane parallel to P_1P_2. The origin of (x,y) is at P_2 and the origin of (ξ,η) is at the foot of the perpendicular from P_2. Let $l(\xi,\eta)$ be the luminance of Σ (see §18–1). The luminous intensity of the source element $d\sigma$ is $l(\xi,\eta)d\sigma$ and the intensities at P_1 and P_2 are $l(\xi,\eta)d\sigma/r_1{}^2$ and $l(\xi,\eta)d\sigma/r_2{}^2$ respectively. Hence one may write, for the complex amplitudes at P_1 and P_2 due to $d\sigma$,

$$u_1\sqrt{d\sigma} = \frac{\sqrt{l(\xi,\eta)d\sigma}}{r_1}e^{-ikr_1},$$

$$u_2\sqrt{d\sigma} = \frac{\sqrt{l(\xi,\eta)d\sigma}}{r_2}e^{-ikr_3}.$$

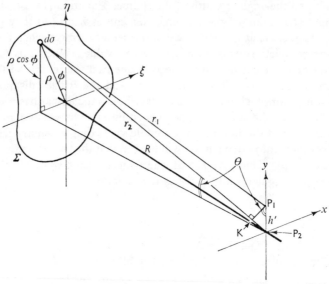

Fig. 11–16. Finite quasi-monochromatic source.

Hence the phase coherence factor γ_{12} for the disturbances at P_1 and P_2 is given by equation (6–22):

$$\gamma_{12} = \frac{1}{\sqrt{I_1 I_2}} \int\int_{\Sigma} u_1 \, u_2^* d\sigma$$

$$= \frac{1}{\sqrt{I_1 I_2}} \int\int_{\Sigma} \frac{I(\xi,\eta)}{r_1 r_2} e^{ik(r_2-r_1)} \, d\sigma,$$

where I_1 and I_2 are the total intensities at P_1 and P_2. It follows from Fig. 11–16 that

$$r_2-r_1 = KP_2 = h' \sin\theta = \frac{h'\rho \cos\phi}{r_2}.$$

If r_1, $r_2 \gg \rho$, h', the fractional variation in r_2 is small. Hence if one puts $r_2 = R$, the perpendicular separation of the (x,y) and (ξ,η) planes, one introduces only a small fractional error into (r_2-r_1). One then writes

$$\gamma_{12} = \frac{1}{R^2 \sqrt{I_1 I_2}} \int\int_{\Sigma} I(\xi,\eta) e^{i\left(\frac{kh'\rho}{R}\right) \cos\phi} \, d\sigma. \quad . \quad . \quad (11\text{–}17)$$

On referring to §11–8, it will be seen that if one had a diffracting aperture whose boundaries were the same as those of the present source area Σ, and if one had a spherical wave emerging through the aperture and converging on to P_2, and if the *amplitude* across the wave were $I(\xi, \eta)$, then the complex amplitude at P_1 would be given by

$$\psi_{P_1} = (\text{constant}) \int\int_{\Sigma} I(\xi,\eta) e^{i\left(\frac{kh'\rho}{R}\right) \cos\phi} \, d\sigma.$$

This has the same form as (11–17) and can be normalized to make $|\psi_{P_1}| \to 1$ as $P_1 \to P_2$; in equation (11–17) $|\gamma_{12}| \to 1$ as $P_1 \to P_2$, as it obviously should. That is to say, *the phase coherence factor for two points on a plane illuminated by a parallel plane quasi-monochromatic self-luminous surface is equal to the normalized complex amplitude at one of the points in the Fraunhofer diffraction pattern associated with a spherical wave centred on the other point when the wave passes through an aperture whose boundary is the same as that of the source, the variation of amplitude across the wave being the same as the variation of luminance across the source.* This is referred to as the *van Cittert-Zernike theorem.* If the extended source is a uniform disc centred at $\xi=\eta=0$ the related diffraction pattern is the Airy pattern and one has

$$\gamma_{12}=\frac{2J_1(z)}{z}, \quad \cdots \cdots \cdots \quad (11\text{--}18)$$

where $z=\dfrac{2\pi}{\lambda} \cdot P_1P_2 \sin \alpha$, 2α being the angular diameter of the source seen from P_2. $|\gamma_{12}|$ falls to 0·88 when $z=1$, i.e. when

$$P_1P_2=\frac{0\cdot 16\lambda}{\sin \alpha}. \quad \cdots \cdots \cdots \quad (11\text{--}19)$$

One can take 0·88 to be a convenient "tolerance" so that (11–19) gives the diameter of the *coherence patch* over which disturbances are coherent for many practical purposes. Frequently (11–19) is used to set a limit to the size of the source that can be used in an interference or diffraction experiment where a given area must be coherently illuminated.

11–13. Michelson's Stellar Interferometer

Suppose a double star is examined with a telescope and a double slit of variable slit separation is placed in front of the objective so that the slits are perpendicular to the line joining the stars. Then, in general, fringes will be observed in the focal plane of the telescope. If the separation of the slits is now increased from a very small initial value, the fringes will be seen to become narrower but less distinct until, when equation (11–14) is satisfied, i.e. (putting $p=0$) when $(a+b)=\lambda/2\alpha$, the fringes disappear. The angular separation of the two stars can then be calculated from a measurement of $(a+b)$. This experiment can be performed provided the aperture of the telescope objective is sufficiently large. It will be seen that, for an objective of diameter d, the fringes disappear for a double star of angular separation $\lambda/2d$, but a decrease in the visibility of the fringes is detectable for somewhat closer stars. Thus the double nature of a star can be *inferred* from these observations even though it is not resolved by the telescope used in the normal way—when the limit of resolution is $1\cdot22\lambda/d$ (see §14–6). It must be remembered that the double star is not resolved by the present method although its double nature can be detected. In practice this method is not often used since more sensitive methods are available (see §6–15).

The use of the double slit arrangement is of more importance in determining the diameters of single stars. If the star radiates uniformly so that, viewed from the telescope, it can be regarded as equivalent to a uniform disc, then it follows from the discussion of §11–12 that the disturbances arriving at the two slits are partially coherent, the degree of coherence decreasing as the slit separation D is increased. The degree of coherence, and hence the fringe visibility, falls to zero when $z = 1 \cdot 22\pi$. One has

$$1 \cdot 22\pi = \frac{2\pi}{\lambda} D \sin \alpha.$$

In practice α is small and when the fringes disappear the angular diameter of the star is given by

$$2\alpha = \frac{1 \cdot 22\lambda}{D}. \qquad \cdots \cdots \cdots \quad (11\text{--}20)$$

It is worth noting that if D is increased beyond this value the fringes reappear and (11–18) gives a small negative value for γ_{12}. This corresponds to the reappearance of fringes of poor visibility and with reversed contrast, i.e. with maxima and minima interchanged. If one continues to increase D, the fringes repeatedly reappear and disappear, corresponding to the oscillating value of γ_{12} given by (11–18).

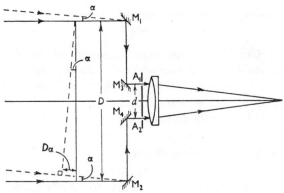

Fig. 11–17. Michelson's stellar interferometer.

In practice, stellar diameters are so small that the values of D corresponding to the first disappearance are more than 20 ft., i.e. larger than the diameter of the largest telescope objective. In addition, one would have very closely spaced fringes that would be difficult to observe. (Equation (11–13) shows that for a slit separation D the angular separation of the fringes is λ/D.) These difficulties are overcome in the Michelson stellar interferometer (Fig. 11–17). Light from the star falls on the outer mirrors M_1 and M_2 and is then directed via M_3 and M_4 to the slits A_1 and A_2. The degree of coherence between the interfering disturbances is determined by the distance D between M_1 and M_2, and the fringe spacing is determined by the slit separation A_1A_2. The separation of the outer mirrors is increased until the fringes disappear, when the angular diameter of the star is given by (11–20). Referring to Fig. 11–17, it will be seen that the fringes disappear when the path difference $D\alpha$ is $1 \cdot 22\,\lambda$ where α, not 2α, is the angular diameter of the star. (See p. 251.)

The use of the relatively small mirror and slit apertures helps to reduce the effect of atmospheric inhomogeneity which, in telescopes of large aperture, causes unsteadiness of the image.

11–14. The Rayleigh Refractometer

Suppose that, in the double slit arrangement, a path difference is introduced between the disturbances that pass through one slit and those that pass through the other. There will be no change in the pattern produced by each aperture acting alone since this depends upon the relative phases of disturbances from various parts of the same aperture. Thus no change is produced in the slowly varying envelope in the fringe system. The $\cos^2$ interference fringes will, however, be affected since these are controlled by the phase differences between disturbances from corresponding elements of the two apertures. Very small

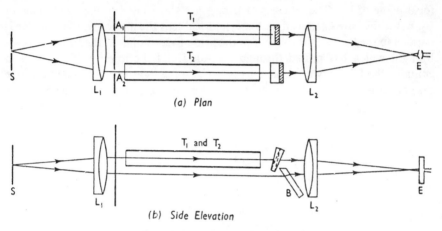

(a) Plan

(b) Side Elevation

Fig. 11–18. Rayleigh refractometer.

changes in optical path can be measured in this way. For example, if a path difference of only one wavelength is introduced between the disturbances passing through one slit and those passing through the other, the $\cos^2$ fringes move a distance equal to the fringe spacing. This fact is utilized in the Rayleigh refractometer which is used to measure the refractive indices of liquids and gases.

A plan of the optical system is shown in Fig. 11–18(a). Light from a slit source S in the focal plane of L_1 passes through the slit apertures A_1 and A_2 and along the tubes T_1 and T_2 which have glass ends. Since A_1 and A_2 are fairly wide, most of the light passes almost straight along the tubes. Hence the usual Fraunhofer double slit pattern is formed in the focal plane of L_2 if the optical paths through T_1 and T_2 are identical. Since the separation of A_1 and A_2 is fairly large, the fringes are very close together and a high magnification is required. The magnification is, however, only required in the direction perpendicular to the fringes so that a cylindrical lens can be used. In this way a considerable economy in light is effected, resulting in brighter fringes. However, it is essential for another reason to use a cylindrical lens (see below). In practice a glass rod is used as eyepiece at E.

If each tube is of length t and the liquids or gases contained in them differ in refractive index by δn, the optical path difference introduced is $t\delta n$. As in the Jamin refractometer, it is impracticable to observe the movement of the fringes, but compensating plates of the same type can be used to return the fringes to their original positions. As before, monochromatic fringes cannot be distinguished from one another and white light fringes are employed. There is, however, one important difference between the two instruments in the method of setting. The fringes in the Rayleigh instrument are so fine that it is impossible to produce a sufficiently fine cross-wire. For this reason a separate stationary fringe system is used as a fiducial mark. It is also worth noting that a cylindrical eyepiece could not be used with cross-wires. The arrangement is shown in Fig. 11–18(*b*) which is a side elevation of the system. It will be seen that only the top half of the aperture is occupied by the tubes so that light that passes through the lower half of the aperture gives a fixed double slit pattern in the focal plane of L_2. The two sets of fringes are superposed in the focal plane of L_2, but if they are examined with a cylindrical lens the axis of whose cylinder is parallel to the fringes, the top half of the field of view contains the fringes formed by light which has passed through the tubes and the lower half of the field contains the fiduciary system. This is because the light forming the two fringe systems comes from different halves of the aperture of L_2 and passes on as

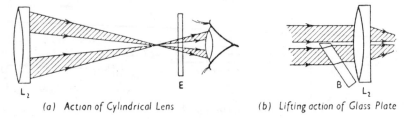

(a) Action of Cylindrical Lens   (b) Lifting action of Glass Plate

Fig. 11–19. Components of the Rayleigh refractometer.

indicated in Fig. 11–19(*a*). It will be seen that the cylindrical lens does not deviate the rays in the vertical plane. The plate B [Figs. 11–18(*b*) and 11–19(*b*)] raises the light forming the fiduciary pattern until the two pencils of light touch. In the field of view the upper edge of the fiduciary pattern then touches the lower edge of the other fringes and the setting is made easier.

After calibrating the compensators, the procedure is to introduce, into one of the tubes, the gas or liquid whose refractive index is required (the other being evacuated or containing air at known pressure). The compensator is then adjusted until the central white light fringe is returned to lie exactly above that of the fixed fringe system. It is found that many observers can make this setting more accurately than the setting of a fringe on a cross-wire. Displacements of about one-fortieth of a fringe width can be detected.

With a dispersive material it is obviously desirable to measure the refractive index for one wavelength rather than the mean value for white light. The procedure is to make an initial setting using white light in order to identify the central fringe. On changing to monochromatic light, the necessary fine adjustment can be made to bring the central fringes into exact coincidence. When a large tilt of the compensator is used, and the dispersive power of the material under test is very different from that of the material of the compensators, it may become difficult to identify the central fringe with white light (see § 7–10). It is then necessary to make a preliminary measurement with a smaller concentration or with shorter tubes. This involves a smaller rotation of the compensator and enables the central fringe to be identified.

Rotation of the Jamin compensator causes the fringe systems to be slightly inclined to one another, and this is important in the present instrument which employs a cylindrical eye lens to view the fringes (which must be parallel to its axis). The defect is reduced to an unimportant magnitude by employing one fixed and one movable compensating plate which consequently has to be moved through a very small angle by means of a micrometer screw arrangement.

To obtain maximum sensitivity it is obviously desirable to employ long tubes since a given change in refractive index will produce a relatively large change in total optical path. However, the length of the tube that can be used is limited by the uniformity that can be achieved in the fluid contained therein. It is necessary to use relatively short tubes for liquids. The instrument has been used for checking the purity of a gas (e.g. of air in a coal mine) and is used in chemical analysis.

In the Rayleigh refractometer use is made of the Fraunhofer double slit diffraction pattern but the important fringes in that pattern are those associated with the interference between disturbances from the two slits. As pointed out in § 10–11, the purpose of the lenses is the concentration of the light into a single area so that the interference fringes are bright.

11–15. Further Discussion of Fraunhofer Patterns

Consider the arrangement shown in Fig. 11–20. If the lenses have very wide apertures and there is no diffracting screen at Σ, the plane wave W gives an approximate point focus on the axis. If a diffracting screen containing a slit aperture is interposed at Σ so as to limit the extent of the plane wave, a Fraunhofer pattern is formed on the receiving screen S. Now each point along the ξ' axis is the focus of a plane wave entering L_2. That is to say, when the incident plane wave W is limited by the slit Σ, it becomes equivalent to a continuous distribution of plane waves travelling in all directions (θ) through L_2, the relative amplitudes being given by the relative amplitudes on the receiving screen. The

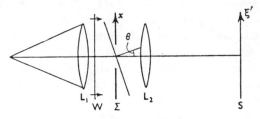

Fig. 11–20. Analysis of Fraunhofer pattern.

complex amplitude at a point on the ξ' axis defined by the angle θ is given (from § 11–3) by

$$\psi(\theta) = \int_{\Sigma} a e^{ikx \sin \theta} \, dx.$$

In discussing the phenomenon of diffraction it has usually been assumed, as above, that the phenomenon arises as a result of the passage of light through an aperture, i.e. as the result of the limitation of the width of a wavefront. The general theoretical situation is that diffraction occurs whenever one has a wavefront over which the amplitude and/or the phase is not uniform. However, in all practical situations the widths of wavefronts must, ultimately, be limited; the simplest example, which is the case discussed in detail already, is the sharp limitation of the width of a *uniform* wavefront, and the generalisation which commonly occurs is the sharp limitation of the width of a *non*-uniform wavefront. To include this last more general (but still practical) case, suppose that the disturbance passing through the slit aperture is a non-uniform wave. Then in the above formula a is replaced by $f(x)$. Since $f(x)$ is zero outside the aperture of the slit, one can extend the limits of integration to $\pm\infty$. Now, since θ defines position in the diffraction pattern, one can put $(\sin\theta)/\lambda = u'$ and use u' in place of ξ' as "co-ordinate" in the image plane.

One can then write

$$\psi(u') = \int_{-\infty}^{+\infty} f(x)e^{2\pi i u' x}dx. \quad . \quad . \quad . \quad . \quad . \quad (11\text{–}21a)$$

From §6–8 this implies

$$f(x) = \int_{-\infty}^{+\infty} \psi(u')e^{-2\pi i u' x}du'. \quad . \quad . \quad . \quad . \quad (11\text{–}21b)$$

That is to say, for a Fraunhofer pattern the *pattern function* is the Fourier transform of the *aperture function* and vice versa. This means that if one thinks of $f(x)$ as being analysed into (synthesized from) simple harmonic Fourier components, the amplitude at u' in the image plane gives the amplitude of the component of $f(x)$ of spatial frequency u'. Alternatively, if $\psi(u')$, the distribution of light across the image plane, is analysed into spatial frequencies, $f(x)$ is the amplitude of the component of frequency x. Thus there is a one-to-one correspondence between position in one plane and spatial frequency in the other.

The results expressed by (11–21(a)) and (11–21(b)) have emerged as a result of the application of Huygens' principle. In fact, there is no need to use this principle or to make any other detailed assumptions about the mechanism by means of which waves are propagated. Once one has decided to ignore the actual physical processes occurring at the edge of the diffracting screen, and made the approximation that the screen simply imposes a sharp limitation on the width of the wavefront without affecting the distribution of light across the unobstructed area, the pattern function follows directly from the principle of superposition and general Fourier theory.

An unlimited uniform plane wave whose normal is in the xz plane and is inclined to the z-axis at angle θ gives, along the x-axis, a disturbance which can be represented by

$$a(\theta)e^{i(\omega t - kx\sin\theta)}$$

where $a(\theta)$ is the amplitude of the wave. This would give zero phase at the origin. If one has similar waves in various directions θ, with different phases at the origin, an additional phase term $e^{-i\delta(\theta)}$ should be included. This can be done by replacing the real amplitude $a(\theta)$ by a complex amplitude $\psi(\theta)$ so that the wave in direction θ is represented by

$$\psi(\theta)e^{i(\omega t - kx \sin \theta)}$$

which, on omitting the simple harmonic time factor, can be written as

$$\psi(u')e^{-2\pi i u' x}.$$

Hence, by the principle of superposition, the total disturbance along the x-axis is given by

$$f(x)= \int\limits_{-\infty}^{+\infty} \psi(u')e^{-2\pi i u' x}du',$$

and Fourier's theorem gives

$$\psi(u')= \int\limits_{-\infty}^{+\infty} f(x)e^{2\pi i u' x}dx.$$

Thus $(11-21(a))$ and $(11-21(b))$ are obtained at once.

It is left as an exercise for the reader to verify by direct evaluation of the integral that if $f(x)$ is the "top hat" function defined by $f(x)=1$ for $-\frac{1}{2}a\leqslant x\leqslant\frac{1}{2}a$ and $=0$ for $|x|>\frac{1}{2}a$ the transform is $\psi(u')=a$ sinc $(\pi u'a)$. (The calculation is, in fact, simply a different approach to the one-dimensional version of the calculation given in §11–3.)

For a wave limited in two dimensions the results are

$$\psi(u',v')= \int\limits_{-\infty}^{+\infty}\!\!\int f(x,y)e^{2\pi i(u'x+v'y)}dxdy, \quad . \quad . \quad . \quad (11-22a)$$

$$f(x,y)= \int\limits_{-\infty}^{+\infty}\!\!\int \psi(u',v')e^{-2\pi i(u'x+v'y)}du'dv'. \quad . \quad . \quad (11-22b)$$

It was seen in Chapter VI that if one has a wave train of finite *length* it can be analysed into a continuous distribution of simple harmonic waves with all *frequencies* between $\pm\infty$. The above analysis shows that if a wave is limited in *width* it can be analysed into a continuous distribution of waves in all *directions*. In the former case the frequency distribution is narrower if the wave train is longer, and in the latter case the angular spread becomes smaller if the wave is wider.

In all the arrangements discussed in detail so far there has been a uniform disturbance across the diffracting aperture, i.e. $f(x, y)$ has been a constant. $f(x, y)$ can be made to vary across the aperture by inserting a screen whose absorption and optical thickness varies from point to point. It is obvious that if this is done the diffraction pattern is altered. In particular, it is possible to decrease the intensities of, or to remove, the subsidiary maxima in the slit images formed by a spectroscope. The process is known as *apodization*—

"removal of the feet". (See also pages, 164, 291, 333, 359.) If the subsidiary maxima are suppressed in this way the central maximum is broadened slightly. An important development of these ideas is discussed in §14–16.

In the discussion given above the object was thought of as being in the first focal plane of L_1 with the image in the second focal plane of L_2. To a first approximation the Fourier transform relationship exists between the image or pattern function and the aperture function regardless of the actual position of the aperture, which is the aperture stop of an image-forming system (see also §10–9). It can be shown that this transform relationship exists to a greater accuracy if the aperture is in the first focal plane of L_2. One can now think of using the same optical system in an entirely different way. Suppose an object structure is placed in the plane of what was previously the aperture stop, the object structure being coherently illuminated by means of a point source in the first focal plane of L_1. The pattern in the second focal plane of L_2 then gives the Fourier transform of this object structure, and the system can be regarded as an analogue computer yielding the spatial frequency spectrum of the structure.

The existence of the Fourier transform relationship between the Fraunhofer pattern function and the aperture function provides a simple demonstration of a well-known theorem in the theory of Fourier transforms: The Rayleigh-Parseval theorem states that if $F(u, v)$ is the Fourier transform of $f(x, y)$,

$$\int\int_{-\infty}^{+\infty} |f(x, y)|^2 dx dy = \int\int_{-\infty}^{+\infty} |F(u, v)|^2 du dv.$$

Now, for a Fraunhofer diffraction pattern, if $f(x, y)$ is the aperture function and $F(u, v)$ is the pattern function, the left hand side of the equation gives the total energy passing through the aperture and the right hand side gives the total energy in the diffraction pattern; the theorem then follows at once from the principle of the conservation of energy.

CHAPTER XII

DIFFRACTION GRATINGS

ANY periodic arrangement of diffracting bodies may be referred to as a diffraction grating. One of the simplest types of grating consists of a number of identical equidistant slit apertures in an opaque screen. This is seldom reproduced in practice although the early gratings used by Fraunhofer were of this type. These consisted of a grid formed by winding fine wires on two parallel screw threads. Present-day gratings usually consist of equidistant diamond rulings on a plate or mirror or a replica of such a ruled grating. When the elements of the grating are narrow slits, they act as sources of disturbances and, since they are narrow, they radiate uniformly. When the grating elements are not narrow slits the elements can still be regarded as sources but in this case they do not radiate uniformly. The result is that maxima of almost exactly the same shape are formed in the same parts of the pattern as before but with different relative intensities. Thus many of the important properties of diffraction gratings (e.g. dispersion, resolving power, etc.) are associated with the interference effects between disturbances from corresponding parts of the separate elements. That is, they are concerned with the periodicity of the diffracting elements rather than the shape of the individual elements. Most of these results can be deduced without extensive mathematical analysis. For example, it is not necessary to find the complete form of the intensity distribution in the pattern. It is therefore useful to begin by discussing, in simple terms, a grating which consists of a number of equidistant very narrow slits.

12–1. Gratings with Narrow Apertures

Light from the monochromatic slit source S [Fig. 12–1(a)] is collimated by the lens L_1 and is incident normally on the grating G which consists of N very

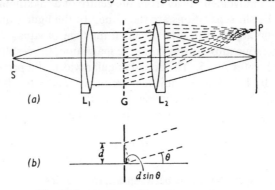

Fig. 12–1. Grating with narrow apertures.

narrow slits parallel to S and distance d apart. The Fraunhofer diffraction pattern is formed in the focal plane of the lens L_2. It will be seen from (b) that the phase difference at P between disturbances from adjacent slits is

$$\delta = \frac{2\pi}{\lambda} (d \sin \theta),$$

and when this is a multiple of 2π, the disturbances from all slits interfere constructively at P. There is therefore a maximum at P when

$$\frac{2\pi}{\lambda} (d \sin \theta) = p2\pi \ (p \text{ an integer}),$$

i.e. for a maximum

$$d \sin \theta = p\lambda. \ . \ . \ . \ . \ . \ . \ . \ . \ (12\text{–}1)$$

For these and only these values of θ, the disturbances from all slits are exactly in
phase at P. Hence there cannot be other maxima as strong as these; they are
therefore called the *principal maxima*. Since the slits are
narrow, the amplitude from each slit is independent of θ
so that all the principal maxima have the same intensity.
Now (12–1) does not contain N so that, *for a given slit
separation*, the positions of the principal maxima are
independent of the number of slits in the grating.

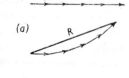

(a)

If the contributions · of the slits are compounded
by employing the usual graphical method, it is
easy to show that the maxima become narrower as the
number of slits is increased. At the centre of a principal
maximum the vibration polygon is simply a straight
line whereas a short distance away from the centre
each vector is inclined to the previous one, the polygon
being equiangular. In Fig. 12–2 (a) and (b) show
the vibration polygons at a principal maximum and

(b)

Fig. 12–2. Vibration
polygons near maxima.

at a short distance away for $N=5$ and $N=10$ respectively. The angle between
successive vectors is δ, the phase difference between disturbances from adjacent
slits. For a given movement away from the centre of a maximum, δ is the same
in each case if the slit separations are the same. In the figure, the lengths of the
vectors have been adjusted to give principal maxima of equal intensity. It will
be seen that when one moves away from the maximum, the resultant R is smaller
for $N=10$ than it is for $N=5$. The actual angular width of a principal maximum
is easily found by locating the adjacent minima. The vibration polygon is a
straight line when $\delta = p2\pi$. This corresponds to the principal maximum of order
p and one has, at the centre of this principal maximum,

$$\frac{2\pi}{\lambda} (d \sin \theta) = p2\pi.$$

Now as one increases the angle between adjacent sides, a polygon of N sides
first becomes a closed figure when the angle reaches $2\pi/N$. Therefore, at a
minimum adjacent to a principal maximum

$$\delta = \frac{2\pi}{\lambda} (d \sin \theta') = p \cdot 2\pi + \frac{2\pi}{N}.$$

Thus the increment in $\sin \theta$ in going from the centre of a principal maximum to the first minimum is given by

$$\sin \theta' - \sin \theta = \frac{\lambda}{Nd},$$

and the corresponding value of $d\theta$ is half the angular width of the maximum. Since $d(\sin \theta) = \cos \theta d\theta$ one has

$$d\theta = \frac{\lambda}{Nd \cos \theta}. \quad \cdot \quad \cdot \quad \cdot \quad \cdot \quad \cdot \quad \cdot \quad (12\text{–}2)$$

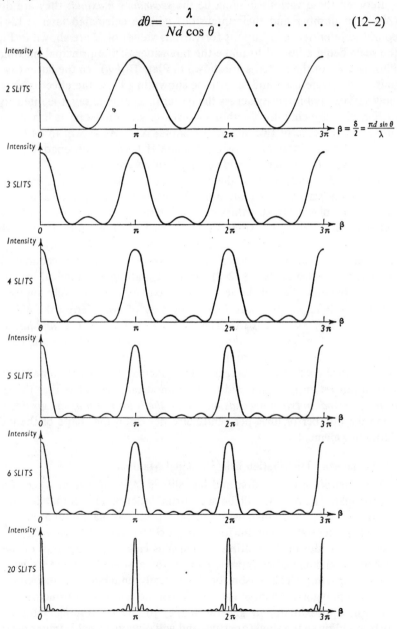

Fig. 12–3. Intensity distributions with narrow slits.

Minima also occur when the angle between successive vectors is $q.(2\pi/N)$ (q an integer) since, in this case, the vibration polygon executes q complete turns and is a closed figure. δ has the value $p.2\pi + q(2\pi/N)$, and when $q = N$ one has the principal maximum of the next order—given by $\delta = (p+1)2\pi$. $q = 0$ represents the original maximum so that $(N-1)$ values of q represent minima between adjacent principal maxima.

Between these various minima lie the *secondary maxima*; they are obviously $(N-2)$ in number and their intensities will be calculated later (§ 12–7). The complete intensity distributions for various values of N are shown in Fig. 12–3, the scale being adjusted to make the intensities of the principal maxima equal. Photographs of the patterns are shown in Plate II(*d–h*). In the latter case slits of finite width were used and, as will be shown in § 12–8, the effect is to impose a slow variation of intensity across the pattern. That is to say, the intensity curves of Fig. 12–3 are contained within a slowly varying envelope as has already been proved in the case of two slits (see Chapter XI). As shown in § 12–8, the 4th, 8th, 12th, etc., orders are missing in Plate II because the opaque parts of the gratings were three times as wide as the transparent parts.

In Fig. 12–3 and Plate II, the separation of adjacent slits is the same in all cases and it is seen at once that, as pointed out above, the principal maxima occur in the same places for all values of N but become more narrow as N increases. Other results derived above and easily seen from the figure are that there are $(N-1)$ minima and $(N-2)$ secondary maxima between adjacent principal maxima. It can also be seen that as N increases most of the secondary maxima become much weaker (see § 12–7). In most gratings of practical importance N is very large and most of the secondary maxima are completely unobservable.

When N is large, the more interesting way of considering the effect on the patterns of increasing N is to make *the separation of the first and last slits constant*. This corresponds to taking a grating of fixed overall size and considering the effect of increasing the number of slits per centimetre. Equations (12–1) and (12–2) then show that as N increases the principal maxima become more widely spaced but retain the same width. It is important to realize that the *sharpness of the principal maxima depends only upon the total width of the grating* and not upon the number of lines per centimetre nor upon the shape of the individual diffracting elements.

12–2. Intensity Distribution near Principal Maxima

The separation of the first and last slits is $(N-1)d$ which, when N is large, approximates to Nd. It will be seen from equation (12–2) that for low orders when $\cos\theta \fallingdotseq 1$, the angular width of each principal maximum approximates to that of the central maximum for a single slit aperture whose width is equal to the overall width of the grating. When N is large, the intensities of the nearby secondary maxima also approach those for a single aperture as wide as the complete grating. The reason for this is easily understood if one considers the vibration polygon for a large value of N. In the region of a principal maximum, this will approximate, in general shape, to the continuous vibration curve corresponding to the single aperture and will obviously yield similar maxima and minima. For example, at the centre of a principal maximum for a grating, and at the centre of the pattern for a single aperture, the vibration curve is a straight line, and in each case the polygon curls up as one moves away from the centre of the maximum. As one recedes from a principal maximum of a grating, one would therefore expect to get maxima and minima resembling those of the single aperture, but there is nothing in the latter case which corresponds to the subsequent increase in the intensities of the secondary maxima and the eventual appearance of another principal maximum. This is because, for a continuous vibration curve, there is nothing corresponding to the way in which discrete vectors become parallel when their phase difference is $p2\pi$. Consequently, the continuous vibration curve is never again a straight line. For large values of N, the various slits give contributions in a manner similar to the infinitesimal strips of the single aperture. For example, it was shown in §11–1 that with a single aperture the first minimum occurs where the contributions from strips at the extreme edges of the aperture are 2π out of phase, i.e. when the optical paths from the edges of the aperture to the field point in the pattern differ by λ. Now when the first and last slits of the grating yield a path difference of λ, one has $(N-1)d \sin \theta=\lambda$. The discussion in §12–1 shows that the minimum adjacent to the zero order principal maximum actually occurs when $Nd \sin \theta=\lambda$. The disagreement between these results is small when N is large, and the error in the present result is due to the fact that when N is odd the contribution of one slit goes unbalanced, whilst when N is even there is no slit exactly at the centre of the grating. When N is small, the analogy given above obviously breaks down, and when N is moderately large, the limited resemblance between the patterns referred to is due to the overlapping of the secondary maxima associated with neighbouring principal maxima.

12–3. Oblique Incidence. Minimum Deviation. Reflection Gratings

In §12–1 it was assumed that the light was incident normally on the grating.

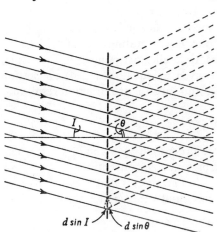

Fig. 12-4 shows a collimated beam of light of wavelength λ incident at angle I on a grating with spacing d. It will be seen from the figure that the path difference introduced between disturbances passing through neighbouring slits is $d(\sin I+\sin \theta)$. But using, for θ and I, the same sign convention as that employed in geometrical optics for angles of incidence and refraction, I is negative in the figure. Therefore the angle of diffraction θ_p for the pth order principal maximum is given by

$$d(\sin \theta_p-\sin I)=p\lambda. \quad (12\text{-}3)$$

Fig. 12-4. Oblique incidence.

Equation (12–3) can be generalized to include the possibility of the refractive indices on the two sides of the grating being different. If these are n and n', the principal maxima are located by the equation

$$d(n' \sin \theta_p - n \sin I) = p\lambda. \quad \ldots \ldots \quad (12\text{--}4)$$

For the zero order maximum one has $p=0$ and $n' \sin \theta_0 = n \sin I$, which is Snell's law and defines the path of the direct or undiffracted light.

The light in the pth order maximum is deviated through an angle D where

$$D = \theta_p - I.$$

This reaches a maximum or minimum when $d(D) = 0$ or

$$d\theta_p - dI = 0.$$

From (12–4) one has

$$n' \cos \theta_p d\theta_p - n \cos I\, dI = 0,$$

so that at maximum or minimum deviation, when $n = n'$,

$$\cos I = \cos \theta_p$$

or, since both I and θ_p are acute,

$$\theta_p = \pm I.$$

The solution $\theta_p = I$ is trivial since it applies only to the zero order light where there is no diffraction. It is easily proved that for a given order, the result $\theta_p = -I$ does, in fact, correspond to a *minimum* value of D. Thus for minimum deviation the angles of incidence and diffraction must be numerically equal and

$$D_{\mathrm{min}} = 2\theta_p, \quad \ldots \ldots \ldots \quad (12\text{--}5)$$

and, from (12–3),

$$2d \sin \left(\frac{D_{\mathrm{min}}}{2} \right) = p\lambda. \quad \ldots \ldots \quad (12\text{--}6)$$

Hitherto it has been assumed that the gratings are of the transmission type, i.e. light is transmitted through equally spaced slits and absorbed at all other points. The theory applies equally well to reflection gratings which reflect light from equally spaced line elements and absorb at all other points. For example, the positions of the principal maxima are given by equation (12–4) on putting $n' = -n$, the usual method of converting a transmission or refraction result into one for reflection. One has,

$$d(\sin I + \sin \theta_p) = p\lambda.$$

In addition, it will be seen that for the zero order light $p=0$ and $\theta_p = -I$, which is the algebraic law of reflection with the present sign convention.

One important advantage of using reflection gratings is that they can be used for wavelengths that are absorbed by glass and other materials with which transmission gratings can be constructed.

12–4. Grating Spectra

Suppose that, in Fig. 12–5, the slit source S emits light of several discrete wavelengths. The collimated beam of light from the achromatic lens L_1 is incident on the grating G, and for each wavelength present in the source, the usual pattern is formed in the focal plane of the achromatic lens L_2. The central (zero order) maximum lies on the axis for every wavelength, but the first order maxima for the various wavelengths do not coincide. These are formed at angles defined by equation (12–3) if it is assumed that $n'=n$. In most cases of practical importance the total number of slits in the grating is sufficiently large to render the secondary maxima negligibly weak. Thus, on one side of the axis, each wavelength gives a single first order maximum, and the maxima of the several wavelengths fall alongside each other and constitute a line spectrum. It has been

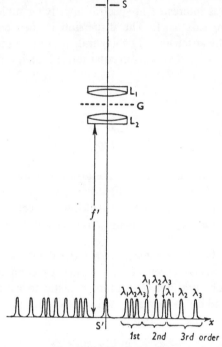

Fig. 12–5. Grating spectra

shown that each principal maximum for a diffraction grating resembles the central maximum in the Fraunhofer pattern of a single slit aperture. Therefore the lines in a diffraction grating spectrum resemble those obtained with a prism spectroscope having a rectangular aperture of width equal to the overall width of the grating. It will be seen later that prism and grating spectra differ in one important respect: they have significantly different dispersions. Each wavelength will also give a maximum in each of the higher orders so that several line spectra are formed. One refers to the first, second, third order spectra, etc.

12–5. Dispersion. Overlapping of Orders

Equation (12–3) defines the angles of diffraction for each wavelength, and on differentiating with respect to λ, the angular dispersion is at once found to be given by

$$\frac{d\theta}{d\lambda} = \frac{p}{d \cos \theta_p}. \qquad \cdots \cdots \quad (12–7)$$

This gives the angular separation $\delta\theta$ of the spectral lines whose wavelengths are λ and $\lambda+\delta\lambda$.

For normal incidence higher orders correspond to larger angles of diffraction and for a given spectral region (12–7) shows that the dispersion increases with increasing order as indicated in Fig. 12–5. In a spectrum of given order θ_p increases as λ increases and (12–7) shows that the dispersion is greater in the red (longer wavelength) end of the spectrum. This contrasts with the prism spectrum where the dispersion is always greater at the blue (shorter wavelength) end (see § 5–3).

In some cases one is interested in spectral lines lying sufficiently close to the axis to justify writing $\cos \theta = 1$. The dispersion is then constant throughout the relevant part of the spectrum. The linear dispersion in the focal plane of the lens L_2 is $dx/d\lambda$ where $x=f'\theta$ and, from equation (12–7), it follows that when θ is small

$$\frac{dx}{d\lambda} = \frac{pf'}{d},$$

or
$$x = \frac{pf'}{d}\lambda. \quad \ldots \ldots \ldots \quad (12\text{–}8)$$

Thus in a spectrum of any given order, the distance from the zero order image to the spectral line for wavelength λ is directly proportional to λ. A spectrum of this type, where there is a linear scale of wavelengths, is called a *normal spectrum*. A prism spectrum can be regarded as normal over only a very limited range of wavelengths.

Equation (12–7) indicates that *for a given order* the dispersion depends upon the grating interval d. However, if one wishes to compare the dispersions obtained with different gratings it is more realistic to assume that the same angles of incidence and diffraction are used and not that the same order is used. Using (12–3), equation (12–7) becomes

$$\frac{d\theta}{d\lambda} = \frac{\sin \theta_p \pm \sin I}{\lambda \cos \theta_p}, \quad \ldots \ldots \quad (12\text{–}7a)$$

the positive sign corresponding to a reflection grating. For the special case of a grating used in an arrangement of the Littrow type (see §§ 5–18 and 12–12), $I=\theta_p$ and one has

$$\frac{d\theta}{d\lambda} = \frac{2 \tan I}{\lambda}. \quad \ldots \ldots \ldots \quad (12\text{–}7b)$$

If one wishes to use particular angles of incidence and diffraction with any particular wavelength one is restricted to certain values of d (i.e. those satisfying the equation $d(\sin \theta_p \pm \sin I)=p\lambda$), but it will be seen that the dispersion does not depend upon which of these gratings is chosen.

Within the range of wavelengths represented, the spectra of various orders are completely separated in Fig. 12–5. However, each spectrum does in fact commence on the axis for $\lambda=0$ so that, in general, the spectra of various orders overlap. For example, it will be seen at once from (12–8) that the p_1th order line for λ_1 will coincide with the p_2th order line for λ_2 when $p_1\lambda_1 = p_2\lambda_2$. The visible spectrum extends from about 4000 angstroms to about 8000 angstroms so that, as far as visible light is concerned, there is no overlapping of the first and second order grating spectra.

However, in general, one's interest is not confined to the visible region and one may need to record the spectrum photographically or by using some other physical detector such as a photo-cell or thermopile. There is then no colour discrimination and if two or more spectra overlap it may be difficult to identify the various spectral lines. In a situation of this kind it is important to know the free spectral range $\Delta\lambda$ lying between successive orders of any given spectral line. If the source radiates over a wider spectral range one can employ auxiliary apparatus (such as a filter, a prism, or a second grating) to prevent overlapping by arranging that only the range of wavelengths $\Delta\lambda$ is incident on the grating. The spectral range is such that the $(p+1)$th order of λ coincides with the pth order of $\lambda+\Delta\lambda$ so that one has

$$d(\sin \theta_p \pm \sin I) = (p+1)\lambda = p(\lambda+\Delta\lambda).$$

Hence

$$\Delta\lambda = \frac{\lambda}{p},$$

as for the Fabry-Perot instruments. In terms of wave numbers ($\bar{\nu} = 1/\lambda$) one has

$$\Delta\bar{\nu} = \frac{1}{\lambda} - \frac{1}{\lambda+\Delta\lambda} = \frac{\Delta\lambda}{\lambda(\lambda+\Delta\lambda)}.$$

(One cannot proceed as in §9–3 because the spectral range of a grating is not small.) One has

$$\Delta\bar{\nu} = \frac{1}{p(\lambda+\Delta\lambda)} = \frac{1}{d(\sin \theta_p \pm \sin I)},$$

or, with a Littrow system,

$$\Delta\bar{\nu} = \frac{1}{2d \sin I}.$$

It will be shown in the next section that the resolving power is determined by the total width of the grating and not by the interval d, so that an increase in the resolving power (decrease in resolving limit) does not cause a decrease in the free spectral range.

12–6. Resolving Power

The chromatic resolving power of an instrument at wavelength λ is defined as $\lambda/\delta\lambda$ where $\delta\lambda$ is the wavelength difference between fringes or spectral lines when they are just resolved. Since, when N is large, the lines in a grating spectrum closely resemble those of a prism spectrum (i.e. the instrumental profiles of prism and grating spectrometers have the same form), it is possible to use again the Rayleigh criterion for resolution. This states that two neighbouring lines will be just resolved when the centre of one maximum falls on the

first minimum of the adjacent line, i.e. when the centres of the maxima are
separated by half the width of the lines. From equation (12–2) it follows that
at the limit of resolution the angular separation of the neighbouring lines is

$$\delta\theta = \frac{\lambda}{Nd \cos \theta}$$

and this is not restricted to the case of normal incidence. From (12–7) the
corresponding increment in wavelength is given by

$$\delta\lambda = \frac{d \cos \theta}{p} \cdot \delta\theta.$$

Therefore the chromatic resolving power is given by

$$\frac{\lambda}{\delta\lambda} = pN. \qquad \cdots \cdots \cdots \cdots \quad (12\text{–}9)$$

Hence in a given order the resolving power is proportional to the total number
of slits or, for gratings of the same overall width, proportional to the number of
slits per centimetre. The resolving limit is given by

$$\delta\bar{\nu} = \bar{\nu} \cdot \frac{\delta\lambda}{\lambda} = \frac{1}{\lambda pN}.$$

Equation (12–3) shows that for a given wavelength and a given angle of
incidence, the order corresponding to a particular angle of diffraction is given by

$$p = \frac{d (\sin \theta_p - \sin I)}{\lambda}$$

and (12–9) shows that the resolving power and resolving limit are

$$\frac{\lambda}{\delta\lambda} = \frac{Nd (\sin \theta_p - \sin I)}{\lambda}, \quad \text{and} \quad \delta\bar{\nu} = \frac{1}{Nd (\sin \theta_p - \sin I)}.$$

Thus for a given wavelength the resolving power *for a given angle of incidence
and diffraction* depends on Nd, the total overall width of the grating, but not upon
the number of slits contained in that width. For coarser gratings a given angle
of diffraction corresponds to a higher order and the overlapping of orders
associated with the higher order spectra make such gratings less convenient for
many purposes. Taking finesse $= \Delta\lambda/\delta\lambda$ (see page 184) gives finesse $= N$.

On putting $-I = \theta = 90°$, it will be seen that the resolving power of a grating
cannot exceed $2Nd/\lambda$, i.e. twice the width of the grating measured in wavelengths.
In practice the intensity would be vanishingly small under such conditions.

For a reflection grating used in a Littrow mounting one has

$$\frac{\lambda}{\delta\lambda} = \frac{2Nd \sin I}{\lambda},$$

and if the grating is in a medium of index n this becomes

$$\frac{\lambda}{\delta\lambda} = \frac{2Ndn \sin I}{\lambda}.$$

That is, the resolving power is proportional to the index of the medium in which
it is mounted.

12–7. Intensity Distribution in the Pattern for Narrow Slits

The positions and widths of the principal maxima have already been discussed in §§ 12–1 and 12–2. An expression will now be derived for the intensity distribution throughout the Fraunhofer pattern for a grating with narrow slit apertures. It will then be shown how it is possible to modify these results to allow for the finite width of the apertures. The theory will then be extended to cover gratings which do not consist simply of apertures in opaque screens.

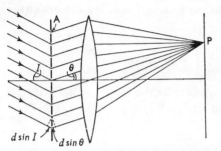

Fig. 12–6. Paths of superposed disturbances.

In Fig. 12–6 a collimated beam of light is incident at angle I on a grating consisting of very narrow slits which may be assumed to diffract equally in all directions. When the angle of diffraction is θ, the phase difference at P between disturbances from successive slits is (since I is negative in the figure)

$$\delta = \frac{2\pi}{\lambda} d(\sin \theta - \sin I).$$

If one takes the first slit, A, to give a disturbance of unit amplitude and zero phase at P, the complex amplitudes of the disturbances from the adjacent slits are $e^{-i\delta}$, $e^{-2i\delta}$, ... If there are N slits, the complex amplitude of the resultant is given by

$$\psi = 1 + e^{-i\delta} + e^{-2i\delta} + \ldots + e^{-(N-1)i\delta}.$$

(See § 6–3.)

i.e.

$$\psi = \frac{1 - e^{-iN\delta}}{1 - e^{-i\delta}}$$

and the resultant intensity at P is given by

$$
\begin{aligned}
\text{Intensity} &= \psi\psi^* \\
&= \left(\frac{1 - e^{-iN\delta}}{1 - e^{-i\delta}}\right)\left(\frac{1 - e^{+iN\delta}}{1 - e^{+i\delta}}\right) \\
&= \frac{1 - e^{iN\delta} - e^{-iN\delta} + 1}{1 - e^{i\delta} - e^{-i\delta} + 1} \\
&= \frac{1 - \cos N\delta}{1 - \cos \delta} \\
&= \frac{\sin^2\left(\dfrac{N\delta}{2}\right)}{\sin^2\left(\dfrac{\delta}{2}\right)} \\
&= \frac{\sin^2\left(\dfrac{\pi Nd(\sin \theta - \sin I)}{\lambda}\right)}{\sin^2\left(\dfrac{\pi d(\sin \theta - \sin I)}{\lambda}\right)} = \frac{\sin^2 N\beta}{\sin^2 \beta}, \text{ (say),} \quad (12\text{--}10)
\end{aligned}
$$

where $\beta = \pi d (\sin \theta - \sin I)/\lambda$, or $(\pi d \sin \theta)/\lambda$ for normal incidence.

It has already been shown that for normal incidence the so-called principal maxima occur at values of θ given by $\sin \theta = p\lambda/d$. For these values of θ both numerator and denominator of (12–10) are zero so that the fraction is indeterminate. However, at points very close to the principal maxima the quantities $(\pi d \sin \theta)/\lambda$ and $(\pi N d \sin \theta)/\lambda$ are small.

$$\therefore \quad \text{Intensity} \rightarrow \left[\frac{(\pi N d \sin \theta)/\lambda}{(\pi d \sin \theta)/\lambda} \right]^2 = N^2 \text{ as } \sin \theta \rightarrow \frac{p\lambda}{d}.$$

Thus to normalize (12–10) to give unit intensity at the centre of the principal maxima, it is necessary to multiply by $1/N^2$.

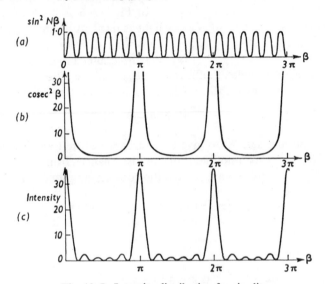

Fig. 12–7. Intensity distribution for six slits.

The numerator of (12–10) falls to zero more frequently than the denominator, giving minima between the principal maxima. Between these minima lie the secondary maxima. Fig. 12–7 gives an indication of the relative intensities of the various maxima for a grating having six slits. Fig. 12–7(a) shows the function $\sin^2 N\beta$ and (b) shows $1/(\sin^2 \beta)$ (i.e. $\operatorname{cosec}^2 \beta$). The intensity at any point is then given by the product of the corresponding ordinates. This resultant is shown in (c). (It is again seen that at the principal maxima the product is "zero × infinity" and is indeterminate.) Fig. 12–3 showed the intensity distribution in the pattern for a number of values of N, the intensities being adjusted to make those of the principal maxima equal.

A point worth noting is that for a grating, which has a finite number of lines, the series that gives the complex amplitude in the pattern has a finite number of terms, whereas the corresponding series that gives the complex amplitude in the fringe pattern of a Fabry-Perot interferometer has an infinite number of terms (see §9–1); the former yields a pattern containing subsidiary maxima—Fig. 12–7, but the latter gives no subsidiary maxima—Fig. 9–2. (See also §9–8.) If one has the idealised case of a grating of infinite width, one has an infinite series; the principal maxima have zero widths and there are no subsidiary maxima.

Now the pattern formed with the grating is a Fraunhofer diffraction pattern and it has been shown (§11–15) that the pattern function is the Fourier transform of the aperture function, the latter being the function that describes the structure of the grating. For a real grating, which has a finite number of slits, the aperture function is not, in the strict sense, periodic (since it does not extend to infinity), and the Fourier transform indicates a continuous distribution of spatial frequencies, giving a continuous distribution of light in the diffraction pattern: principal maxima with finite widths and subsidiary maxima between them. The infinite grating is represented by a Fourier *series*, which has discrete terms, indicating the presence of discrete spatial frequencies giving discrete maxima of zero width with no subsidiary maxima between them. This question arises again in §14–11 when considering the formation of the image of a grating.

12–8. Wide Slits and Other Diffracting Elements

In Fig. 12–8 G is a grating consisting of a number of apertures of finite width a in an opaque screen. Parallel light is again incident at angle I. To calculate the variation in intensity in the image plane along a line in the plane of the diagram, it is convenient to divide each aperture into a number of very narrow strips perpendicular to the plane of the diagram. For a given point P in the image plane, the phase difference between disturbances from *corresponding strips of adjacent apertures* is again $(2\pi/\lambda)d(\sin \theta - \sin I)$, where $d = a + b$ and b is the width of the opaque portion separating adjacent apertures. Consequently, this is also the phase difference between the resultant disturbances from adjacent apertures. Now it follows from the discussion of §11–2 that at any point P in the image plane the resultant amplitude due to a single slit aperture of width a is $(\sin\alpha)/\alpha$, where $\alpha = \pi a(\sin \theta - \sin I)/\lambda$. On summing the contributions of N such apertures, the resultant amplitude is $(\sin \alpha)/\alpha$ times the resultant for N very narrow slits (the resultant again being normalized to give unit intensity at the centre of the pattern). The resultant intensity for narrow slits must therefore be multiplied by $(\sin\alpha/\alpha)^2$ to give the intensity for slits of finite width. That is, for a grating having N slits of width a separated by opaque elements of width b

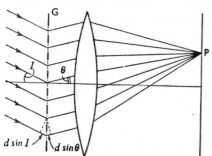

Fig. 12–8. Wide slits.

$$\text{Normalized Intensity} = \frac{1}{N^2}\left(\frac{\sin \alpha}{\alpha}\right)^2 \left(\frac{\sin N\beta}{\sin \beta}\right)^2 . \quad . \quad (12\text{–}11)$$

where

$$\alpha = \frac{\pi a(\sin \theta - \sin I)}{\lambda}, \quad \beta = \frac{\pi(a+b)(\sin \theta - \sin I)}{\lambda}.$$

Fig. 12–9 shows the intensity distribution for $N=6$. It will be seen that, as with the double slit, the effect of the first term is the appearance of the slowly varying envelope beneath which the intensity curve is contained. As before, it may be said that the first term gives the amount of light from each aperture reaching any point in the pattern and that the second term gives the result of the interference between disturbances from N apertures.

If a principal maximum, as given by the second term, coincides with a minimum of the envelope or first term then, as for the double slit, the principal maximum of that order is absent. If it is absent for one wavelength it will be absent for all wavelengths. For example, for normal incidence the principal maxima occur at angles θ defined by

$$\frac{\sin \theta}{\lambda} = \frac{p}{(a+b)}, \qquad \cdots \cdots \quad (12\text{--}12)$$

and the minima of the envelope term occur at angles θ defined by

$$\frac{\sin \theta}{\lambda} = \frac{q}{a}. \qquad \cdots \cdots \cdots \quad (12\text{--}13)$$

Therefore when

$$\frac{p}{a+b} = \frac{q}{a} \ (p \text{ and } q \text{ integers})$$

the principal maximum of order p is absent. Thus when $a=b$ the missing orders are given by $p=2q$. That is, all the even order spectra are absent from the pattern for a grating whose transparent and opaque elements have equal width.

Similarly, if $a=\frac{1}{2}b$, $p=3q$ defines the missing orders and every third order is missing; in Fig. 12–9 and Plate II(d–h) $a=\frac{1}{3}b$ and every fourth order is seen to be missing. It will be seen that an order is missing if the disturbances from corresponding parts of adjacent apertures are in phase when disturbances from the extreme edges of a single aperture are in phase. (See also § 11–10.)

The wire gratings used by Fraunhofer were transmission gratings, and consisted of apertures separated by opaque elements. Most present-day gratings are reflection gratings, and are aluminized plastic replicas of master gratings ruled on aluminium coatings on glass plates. For many of the older ruled gratings and replicas the ruled grooves are simply diffuse scattering elements and the intervals between them correspond respectively to the perfectly transparent or reflecting elements of a transmission or reflection grating. With a grating of this type a large proportion of the energy falls in the zero order maximum where there is no dispersion. It would obviously be an advantage if a larger proportion of the light was diffracted into orders other than zero since the spectra would become brighter and more easily examined. Now, as remarked above, the first term in (12–11) is characteristic of each separate element of the grating, and the second is associated with the interference between disturbances from different elements and is characteristic of the periodicity of the rulings. Whatever the nature of the separate elements, provided they are all the same, the phase difference between disturbances from corresponding parts of adjacent elements remains equal to $2\pi d(\sin \theta - \sin I)/\lambda$. As before, this will also be the phase difference between the *resultant* disturbances from adjacent elements. The third term in (12–11) gives the resultant of N such disturbances and will remain unchanged so that the positions of the principal maxima will be as before. The relative intensities of

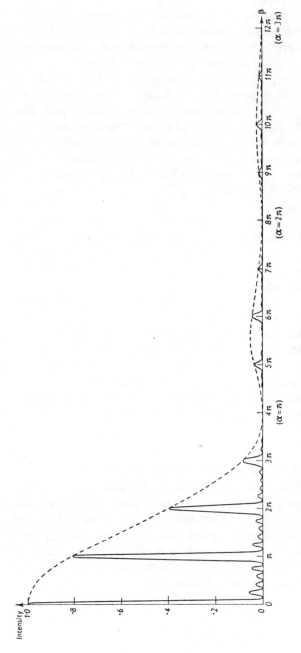

Fig. 12–9. Intensity distribution for grating with six wide slits ($a = \frac{1}{3}b$).

these maxima will be determined by the amount of light they receive from each element of the grating. Thus, if the elements diffract the incident light differently from a single slit aperture, the second (envelope) term of (12–11) will be changed since it gives the diffraction pattern for one element; the relative intensities of the spectra of different orders will then be altered. In Fig. 12–9 this corresponds to a displacement of the envelope so that its maximum occurs at an order other than zero. It would be an advantage if the central maximum of the envelope were made fairly sharp so that only one or two orders fell within it; this would maximize the brightness of the favoured orders. Gratings of this kind are referred to as *blazed* gratings and the possibilities of such gratings were first pointed out by Rayleigh. It should be noted that the dispersion and resolving power will not be affected by the structure of the grating elements since these quantities are determined by the third (interference) term. (The envelope is not sufficiently steep to make an appreciable difference to the shape of the individual principal maxima.)

It is important to realize that the periodic interruptions or distortions imposed across a wavefront by a grating are not present on the component wavefronts into which the transmitted light is resolved. The discussion of §11–15 shows that the modulated wavefront emerging from the grating is equivalent to an angular distribution of uniform wavefronts, each corresponding to a particular point in the pattern formed by the grating.

If, in Fig. 12–8, the grating is translated in its own plane the pattern in the focal plane of the lens remains stationary. The regular variations of refractive index in a medium through which an ultrasonic wave is passing constitute a moving grating of this kind, and a stationary pattern can be observed. (A stationary grating could be produced by setting up a standing wave pattern.) The acoustic grating is not confined to a plane such as G in Fig. 12–8 since the ultrasonic beam has finite width and, although it can be seen that the beam will act as a grating, an exact analysis of the intensity distribution in the pattern is difficult.

12–9. Blazed Gratings, Echelons, Echelettes, Echelles

The fundamental need for blazed gratings has been explained above. What is required is that each groove in the grating should direct as much light as possible into one of the grating spectra, and the space between the rulings should be as small as possible since this plane transmitting or reflecting surface contributes maximum intensity to the zero order image. It is now possible to rule a grating having a pre-determined groove form that concentrates most of the diffracted light within a narrow range of directions.

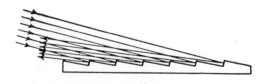

Fig. 12–10. Grating with wedge elements.

A grating having the form shown in Fig. 12–10 will be blazed with the maximum possible efficiency, maximum intensity occurring in the direction corresponding to specular reflection from the *face of each wedge*. (It is specular reflection from the plane of the unruled blank that gives a maximum intensity in the zero order.) One could, of course, utilize reflection from the shallow face of each wedge, corresponding to relatively small angles of incidence and diffraction and hence to a lower order, lower dispersion, and lower resolving power, but greater free spectral range. The angle of inclination of the face of the groove that is used is called the *blaze angle*; it is equal to the angle of incidence and diffraction when the grating is used in a Littrow mounting.

In the early years of grating spectroscopy it was impossible to produce a ruled grating with accurately spaced, identical, smooth, wedge-shaped grooves of this form, but in 1898 Michelson suggested that a coarse grating of this type could be produced by arranging a number of glass plates in step formation as shown in Fig. 12–11, the steps being silvered. It is convenient to use a *reflection echelon* grating of this type in a Littrow arrangement where the light is incident on each step nearly normally as shown. When the suggestion was originally made by Michelson in 1898, the manufacture of such a grating was held up by the difficulty of placing the plates in contact without unequal films of air between them. Such gaps would introduce unequal phase differences between the various adjacent disturbances. By using a *transmission echelon*, this difficulty was overcome although the order of interference was reduced. The arrangement is shown in Fig. 12–12. Collimated light passes normally through the echelon and each step acts as an aperture of width s and produces a single aperture pattern centred on the direction of normal transmission. In the focal plane of the lens L, this pattern provides the envelope beneath which there is formed the pattern resulting from the interference between disturbances passing through different steps. In

Fig. 12-11.   Reflection echelon.

the direction of normal transmission, the path difference between the disturbances passing through adjacent steps is $(n_{glass} - n_{air})t$ whereas, for the reflection echelon, it would be $2n_{air}t$, or between 3 and 4 times greater. It will be seen that if, in the transmission echelon, there is an extra large air film between one pair of plates, an equal path is added to each disturbance so that the phase *differences* are unaffected. The width of the grating which the reflection echelon constitutes is the length of the dotted line in Fig. 12–11, and this will be somewhat longer

Fig. 12–12.   Transmission echelon.

than Nt where N is the number of plates and t the thickness of each. (s is usually about 1 mm., and t about 1 cm.) Thus a grating of 40 plates has width over 40 cm.—a width that has been achieved only very recently with a ruled grating (Harrison and Thompson, 1970). For normal incidence and diffraction from each step, the order is $2d/\lambda$ or roughly 40,000 for green light; with 40 plates the resolving power is $1{\cdot}6 \times 10^6$. Williams subsequently (1926) produced a satisfactory reflection echelon by wringing together quartz plates. With such high orders the overlapping of orders must be overcome by employing an auxiliary dispersion system to isolate a small range of wavelengths. The performance and use of the echelon will be discussed further below.

The blaze of an echelon is the most efficient possible and it is very difficult to produce ruled gratings with an efficiency approaching this. The first successful attempt to rule a blazed grating was by R. W. Wood in 1910. Initially, he produced gratings with between 2000 and 3600 lines per inch with a blaze angle of about 20° (see §12–15). With visible light these gave very little energy in the zero order but distributed most of the light in a group of orders—typically from the 12th to the 16th. These gratings were more suitable for the infra-red and, for a selected wavelength, could be blazed to give most of the energy in the first order. Wood called these gratings *echelettes* because their properties were intermediate between those of the classical unblazed ruled gratings and the echelons. Wood subsequently ruled gratings with up to 15,000 lines per inch blazed for the first order in the visible region.

It should be noted that a grating is blazed for particular angles of incidence and diffraction—for a single angle of incidence if it is assumed that a Littrow mounting is used. When, as in the echelon, there is no "space" between adjacent grooves, one has the sharpest possible envelope (maximum blaze efficiency). It is shown below that the central maximum of the envelope then contains either one or two orders for any given wavelength so that the blaze of the grating embraces the whole free spectral range, and hence all wavelengths, since successive ranges are superposed. (The blaze will, of course, be a maximum for those wavelengths falling at the centre of the envelope.)

In recent years there has been a very great improvement in the quality of ruled gratings—both in constancy of spacing and in groove form (see §12–15). In particular, it has proved possible to rule gratings that approach closely the ideal form of Fig. 12–10 where the steep side of the groove is utilized. These gratings, known as *echelles*, have the properties of an echelon with a large number of thin plates. Echelles have been ruled with from about one hundred to several thousand lines per inch. It is, perhaps, worth emphasizing that in principle echelles, echelettes, and echelons, are simply efficiently blazed gratings, echelons being the extreme form. The distinction between modern echelles and Wood's echelettes is that the echelles utilize the steep side of the wedge whereas the echelettes used the shallow side.

The echelon gratings are very difficult to manufacture and the transmission echelon must now be considered obsolete since the Fabry-Perot etalon can give better resolution at a much smaller cost. Alternatively, the interferometer can do the work of several different transmission echelons—again with less expense. On the other hand, the reflection echelon does give high resolution in all spectral regions—it is not limited by the transmission of the material of the plates. In both echelons, the plates must be figured very accurately although for the reflection echelon the homogeneity of the material is unimportant. For the latter instrument, fused silica is usually employed for the plates. This has a comparatively high thermal conductivity and a low thermal expansion and can therefore be given a very good polish. In addition, these thermal properties facilitate the attainment of thermal equilibrium when the echelon is in use. As mentioned in the discussion above, the plates of a reflection echelon must be wrung together; the contact is improved by heating. A considerable saving in material is effected by employing identical plates which are aligned as indicated by Fig. 12–13. After mounting, the front surfaces of the plates are aluminized. In

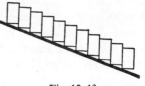

Fig. 12- 13.
Equal plates in echelon.

principle, the reflection echelon is similar to any other grating and the dispersion and resolving power are given by the usual formulae. In Fig. 12–14 the usual grating interval is d and the width and depth of each step is s and t respectively. For a collimated beam incident at angle I (defined in the usual way), the path

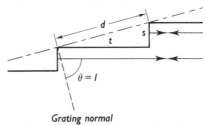

Grating normal

Fig. 12–14. Reflection echelon.

difference between disturbances from corresponding points in adjacent steps for an angle of diffraction θ is seen from the figure to be given by the usual value

$$d(\sin \theta + \sin I).$$

Thus for a principal maximum

$$n_{air} \, d(\sin \theta_p + \sin I) = p\lambda, \quad \ldots \ldots \quad (12\text{–}14)$$

where p is an integer and λ is the vacuum wavelength. The order p is very large (typically 40,000) so that the separation of the orders, $\Delta\theta$, is small and one may write, since $\Delta p = 1$,

$$n_{air} \, d \cos \theta \, \Delta\theta = \lambda \quad \ldots \ldots \quad (12\text{–}15)$$

since s is fairly large (e.g. 1 mm), the light is concentrated in the region of specular reflection from each step. Thus for a Littrow mounting one can put $d \cos \theta = s$

i.e.
$$\Delta\theta = \frac{\lambda}{n_{air} \, s}. \quad \ldots \ldots \ldots \quad (12\text{–}16)$$

As usual, the principal maxima are contained beneath the envelope corresponding to the diffraction pattern for one element of the grating. Here one has, by reflection, the pattern for a single aperture of width s, and for normal incidence the pattern will be centred on the normal to the step face. The first minima will be in directions $\pm \lambda/s$ relative to the normal or, in terms of the vacuum wavelength, one has for the (small) angular width centred on the normal

$$\delta\theta = \pm \frac{\lambda}{n_{\text{air}} s}. \qquad \ldots \ldots \ldots \quad (12\text{--}17)$$

Now the angular separation of the principal maxima is $(\lambda/n_{\text{air}}s)$ so that there is room for only two principal maxima in the central maximum of the envelope. In general, these will fall in positions such as those indicated by Fig. 12–15(a). When $2tn_{\text{air}}$ is an integral number (p) of wavelengths, the line of order p is central and is the only line observed since the adjacent lines coincide with minima of the envelope; one speaks of the *single-order position* [Fig. 12–15(b)]. When $2tn_{\text{air}}=(p+\frac{1}{2})\lambda$, one has the *symmetrical position*; two lines of equal intensity are observed at $\delta\theta = \pm \lambda/2n_{\text{air}}s$ relative to the normal [Fig. 12–15(c)].

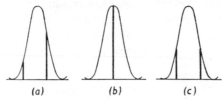

| (a) | (b) | (c) |

Fig. 12–15. Principal maxima with echelon.

One can mount the echelon in an air-tight chamber and effect the transformation by changing the air pressure; this changes the optical path $2n_{\text{air}}t$ until it is $p\lambda$ or $(p+\frac{1}{2})\lambda$ as required.

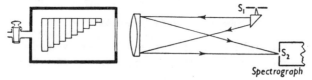

Fig. 12–16. Mounting of reflection echelon.

The plan of a convenient arrangement for a reflection echelon is shown in Fig. 12–16. As shown, the high quality slit S_1 is vertical and the dispersion of the echelon either assists or opposes that of the auxiliary spectrograph whose slit S_2 is also vertical. The pressure is adjusted to give the symmetrical position of the lines and S_2 must be sufficiently wide to admit two orders. By changing the direction of the dispersion of the echelon, one can decide whether a satellite line is on the long or short wavelength side of the main line. For a spectrum rich in lines, it is necessary to make S_1 horizontal and to rotate the echelon so that its plates are horizontal and its dispersion is along the slit S_2 which can be made narrow. By adjusting the pressure one can then switch to the single-line position. If the echelon chamber is airtight, the density of the air is unaffected by temperature and, since the expansion of the fused silica echelon plates is negligible, the pattern will remain steady although the laboratory temperature may change. In this way long exposures may be given.

The use of an echelon with an auxiliary dispersion system is very similar to the use of a Fabry-Perot etalon. In each case the order is large and the free spectral range $\Delta\lambda(=\lambda/p)$ is small. The auxiliary system isolates the very small spectral range to be examined in detail by the high resolution device. Owing to the large spectral range of an echelle, the combination of a coarse echelle with an auxiliary system has some special features and will be dealt with in §12–12.

12–10. The Concave Grating

By measuring the angle of incidence and diffraction for a given order, a grating ruled on a plane surface may be used to determine wavelengths if the grating spacing is known or measured with a travelling microscope. Small gratings can be used in this way if mounted on the prism table of a conventional prism spectrometer. If the wavelength of one line is known accurately, others may be determined by locating the position where the line of one order for the known wavelength falls on or near the line of another order for the unknown (e.g. the fourth order lines for wavelengths in the region of 4400 angstroms will fall near the third order sodium line, 5890 angstroms). It is possible to allow for the small separation of the lines compared in this way. In performing such an experiment, an important difficulty is encountered: the lens of the observing telescope (or camera) cannot have exactly the same focal length for all wavelengths, so that the lines to be examined cannot be exactly in focus simultaneously. This difficulty is overcome by using a concave grating. In 1882 Rowland discovered that gratings ruled on a concave spherical reflecting surface could form focused spectra without the use of lenses. There is then no chromatic aberration, and the lines of different wavelengths in overlapping orders are exactly in focus. In addition, it becomes possible to examine lines in the U.V. and I.R. regions beyond the limits of transmission of lenses. It is possible to secure a normal spectrum in which the linear separation of two lines is almost exactly proportional to the wavelength difference. Rowland's gratings were ruled on polished speculum metal but more recent reflection gratings have usually been ruled on aluminized glass plates. It is found very convenient to use concave gratings in the following manner. In Fig. 12–17 AB represents the concave spherical reflecting surface on which a grating is ruled with lines perpendicular to AB, *equidistant along the direction of the tangent at the mid-point* O. The circle whose centre is C touches the grating at O and has radius half that of the surface of the grating, i.e. the diameter of the circle is the radius of the grating. This is known as the *Rowland circle* and it is found that if the slit source is located at any point on this circle, the spectral lines of all wavelengths in all orders are in focus on the same circle. This fact is made use of in almost all methods of employing concave gratings. A brief account of the theory of the concave grating is given in the next section.

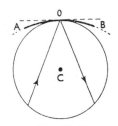

Fig. 12–17.
The Rowland circle.

In order to obtain adequate linear dispersion on the plate the radius of the grating must be large, and if it is pointed out that 21 feet has been a widely used value for the radius of curvature of the grating, it will be realized that the method of mounting the grating and auxiliary equipment is of considerable importance. Mountings for gratings are described in §12–12.

12–11. The Theory of the Concave Grating

A very extensive account of the properties of the concave grating has been given by Beutler (1945). In the simplified treatment given here the analysis is restricted to two dimensions and the aberrations will not be discussed in detail. In Fig. 12–18 O is the mid-point of the grating and is taken as the origin of co-ordinates. C is the centre of curvature of the grating and R is its radius. The point P(x',y') receives light from the source S(x,y) after diffraction at the grating. For light to and from O the angles of incidence and diffraction are I and θ respectively and, with the usual convention, I is negative in the figure.

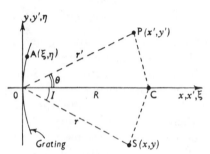

Fig. 12–18. The concave grating.

Consider a disturbance which is diffracted to P from the point A(ξ,η) on the grating. Now it is assumed that the rulings of the grating are equidistant along the tangent at O (i.e. along the η axis) and that there is a diffracting element at O. Consequently, if there is a diffracting element at A, η must be an integral multiple of d, the grating interval. In other words, η/d is an integer and is the number of rulings from O to A. If there is a principal maximum at P, disturbances diffracted from adjacent elements of the grating traverse paths which differ by pλ where p is an integer—the order. Thus disturbances via O and A have paths which differ by (η/d)pλ. Therefore the condition that a principal maximum is formed at P is

$$(SO+OP)-(SA+AP)=\left(\frac{\eta}{d}\right)p\lambda, \quad \cdot \ \cdot \ \cdot \quad (12\text{–}18)$$

and it is now necessary to evaluate the left-hand side of this equation.

One has

$$\left. \begin{array}{l} SA^2=(x-\xi)^2+(y-\eta)^2, \\ AP^2=(x'-\xi)^2+(y'-\eta)^2. \end{array} \right\} \quad \cdot \ \cdot \ \cdot \ \cdot \quad (12\text{–}19)$$

If OS=r and OP=r' one has

$$\left. \begin{array}{ll} x=r\cos I, & x'=r'\cos\theta, \\ y=r\sin I, & y'=r'\sin\theta. \end{array} \right\} \quad \cdot \ \cdot \ \cdot \ \cdot \quad (12\text{–}20)$$

Since A lies on the circle, centre (R,0), radius R, one has

$$(\xi-R)^2+\eta^2=R^2.$$

$$\therefore \quad \xi^2-2R\xi+\eta^2=0, \quad \cdot \ \cdot \ \cdot \ \cdot \ \cdot \quad (12\text{–}21)$$

or

$$\xi=R\pm[R^2-\eta^2]^{\frac{1}{2}}$$

where only the negative sign is significant because the positive sign corresponds to points on the diametrically opposite side of the circle. On expanding the square root, one obtains

$$\xi = \frac{\eta^2}{2R} + \frac{\eta^4}{8R^3} . \qquad \cdots \cdots \quad (12\text{--}22)$$

Now (12–19) gives

$$SA^2 = x^2 + y^2 + \xi^2 + \eta^2 - 2x\xi - 2y\eta$$

or, using (12–20) and (12–21),

$$SA^2 = r^2 + 2R\xi - 2x\xi - 2y\eta.$$

Substituting for x, y, and ξ, from (12–20) and (12–22), one has

$$SA^2 = r^2 - 2r\eta \sin I + \left(1 - \frac{r \cos I}{R}\right)\left(\eta^2 + \frac{\eta^4}{4R^2} + \cdots\right)$$

$$= r^2 \left\{1 + \left[-\frac{2\eta \sin I}{r} + \frac{\eta^2}{r^2} - \frac{\eta^2 \cos I}{rR}\right]\right\} + \cdots,$$

$$\therefore \quad SA = r - \eta \sin I + \frac{\eta^2}{2r} - \frac{\eta^2 \cos I}{2R} - \frac{\eta^2 \sin^2 I}{2r}$$

$$+ \frac{\eta^3 \sin I}{2r^2} - \frac{\eta^3 \sin I \cos I}{2rR} - \frac{\eta^3 \sin^3 I}{2r^2} + \cdots,$$

where terms involving powers of η higher than the third are omitted.

$$\therefore \quad SA = r - \eta \sin I + \frac{\eta^2}{2}\left(\frac{\cos^2 I}{r} - \frac{\cos I}{R}\right) + \frac{\eta^3 \sin I}{2r}\left(\frac{\cos^2 I}{r} - \frac{\cos I}{R}\right) + \cdots.$$

Obviously AP is given by a similar expression involving r' and θ. The condition that disturbances via O and A arrive in phase at P (i.e. that there is a principal maximum at P) is obtained by substituting these values of SA and AP into equation (12–18). Remembering that $SO = r$ and $OP = r'$, one has

$$\eta(\sin I + \sin \theta) - \frac{\eta^2}{2}\left[\left(\frac{\cos^2 I}{r} - \frac{\cos I}{R}\right) + \left(\frac{\cos^2 \theta}{r'} - \frac{\cos \theta}{R}\right)\right]$$

$$- \frac{\eta^3}{2}\left[\frac{\sin I}{r}\left(\frac{\cos^2 I}{r} - \frac{\cos I}{R}\right) + \frac{\sin \theta}{r'}\left(\frac{\cos^2 \theta}{r'} - \frac{\cos \theta}{R}\right)\right] + \cdots = \left(\frac{\eta}{d}\right)p\lambda$$

$$\cdots \quad (12\text{--}23)$$

This equation is satisfied if the first term on the left-hand side is equal to $(\eta/d)p\lambda$ and the other terms are zero. Since r, r', and R, are large compared with η, the terms on the left-hand side are in order of decreasing magnitude and, to a first order approximation, there is a maximum at P if $\eta(\sin I + \sin \theta) = (\eta/d)p\lambda$, i.e. if

$$d(\sin I + \sin \theta) = p\lambda. \qquad \ldots \ldots \quad (12\text{–}24)$$

This is the usual grating equation which has already been seen to give the directions of the principal maxima for a plane reflection grating (see § 12–3). On putting $R = \infty$ in (12–23), it will be seen that the other terms become zero if $r = r' = \infty$. Thus for collimated light incident on a plane grating, the disturbances will arrive exactly in phase to give principal maxima at infinity (or in the focal plane of a lens) if (12–24) is satisfied.

For a concave grating, it will be seen that both the second and third terms of (12-23) become zero if

and
$$\left.\begin{array}{l} r = R \cos I \\ r' = R \cos \theta. \end{array}\right\} \qquad \ldots \ldots \quad (12\text{–}25)$$

Reference to Fig. 12–18 will show that if these equations are satisfied, angles OSC and OPC are right angles. Consequently S and P lie on the circle whose diameter is OC. As mentioned in the previous section, this is known as the Rowland circle, and for any position of a slit source S on this circle, the spectral lines of all orders are in focus on the circle. More detailed analysis shows that some of the higher order terms in (12–23) are also zero when the properties of the Rowland circle are utilized but that a large amount of astigmatism is present, the maxima being astigmatic focal lines. The presence of this astigmatism causes the light to be spread out in the direction of the lengths of the spectral lines with a consequent decrease in intensity (see § 16–7). The higher order terms in (12–23) cause a deterioration in the focus as the width of the grating is increased, and the optimum width of the grating can be shown to be larger for coarser gratings and smaller for higher order spectra. In the next section an account is given of a number of methods of utilizing the properties of the Rowland circle.

It can easily be verified that the second term in (12–23) becomes zero if

$$r = \infty$$

and
$$r' = \frac{R \cos^2 \theta}{\cos \theta + \cos I}. \qquad \ldots \ldots \quad (12\text{–}26)$$

Thus if $r = \infty$, (12–26) defines the surface on which the maxima are most sharply focused for a given value of I. The residual path differences are given by the remaining terms in (12–23). If, in addition, one has $\theta = 0$, the third term of (12–23) is zero and it can be shown that this also makes some of the higher terms, including the astigmatism, zero. That is to say, anastigmatic maxima of good definition are formed on the grating normal when the source is at infinity. However, it is obvious that for a given angle of incidence only one wavelength in each order can fall exactly on the normal. In the *Wadsworth* mounting, the incident light is collimated and the spectral lines are observed in the region of the grating normal, but it must be realized that as one moves away from the normal the definition deteriorates. The optimum width of the grating is again determined by the higher terms in (12–23) and decreases as θ increases. The wavelength range for which a satisfactory grating width can be used, and a satisfactory resolving power obtained, grows with increasing angle of incidence or, for θ constant, with increasing values of λ and of p. For this reason the Wadsworth mounting has been most widely used for the infra-red region. The arrangement is shown in the next section.

12–12. Mountings for Gratings

For visual observations small plane transmission and reflection gratings are conveniently mounted on the prism table of an ordinary prism spectrometer (Fig. 5–5). For spectral regions in which a lens can be used plane reflection

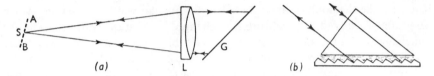

Fig. 12–19. Littrow mountings for plane gratings.

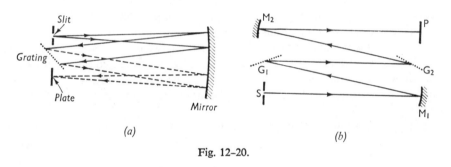

Fig. 12–20.

gratings can be mounted in spectrographs of the Littrow type shown in Fig. 12–19(a), the slits being mounted slightly above or below the plate AB. The grating G can be rotated to bring the required spectral region on to the plate.

As shown in §12–6, the resolving power of a grating is increased by increasing the index of the medium in which it is mounted. Hulthén and Neuhaus (1954) have achieved this improvement by placing a cover glass in immersion contact with the grating. If the grating is to be used at a large angle of incidence the cover glass is replaced by a prism as in Fig. 12–19(b). Fig. 12–20(a) shows a plan of the Ebert mounting first proposed in 1889. In 1930 Czerny and Turner used a modification of the Ebert mounting in which the collimating and focusing were performed by two separate mirrors instead of by the two halves of the large mirror, and Fastie (1952) replaced the plate by an exit slit to form a monochromator. It is worth noting that in the Ebert mounting the two mirrors (i.e. the two portions of the single mirror) are used off-axis by equal and opposite amounts. To a first approximation this symmetry eliminates the coma so that sharp spectral lines are formed. However, this correction applies only to the line that is the same distance off-axis as the entrance slit, so that only a small spectral range is sharp for any orientation of the grating. In consequence, the Ebert mounting is unsatisfactory for a spectrograph in which it is required to photograph a wide spectral range but is satisfactory for a monochromator if the exit slit is positioned correctly. The mirror does introduce some astigmatism and Fastie minimized its effect, and also compensated for spectral line curvature for all wavelengths, by using curved slits that formed arcs of the same circle. Fastie also used an "over-and-under" variation of the Ebert mounting in which the plate was mounted above, and the slit was mounted below, the level of the grating.

If two gratings are placed in series their dispersions add. Hence, if one wishes to double the resolving power of a grating system, one can double the width of the grating, or one can employ two identical gratings in series (or the same grating twice). When the grating width is doubled, the angular dispersion is unchanged while the doubled aperture halves the angular width of each line; when two gratings are used, the aperture is unchanged, so that the width of each line is unchanged, but the angular dispersion is doubled. A doubling of the aperture and a narrowing of each spectral line obviously leads to brighter lines. An improvement in the photometric efficiency also results from the use of two gratings in series since for a given *linear* dispersion the focal length of the focusing system can be halved. (See § 18–5.) The use of two gratings leads to the more compact system and, in any case, it may be impossible to produce a grating with twice the width of the original. (The required doubling of the aperture of the mirrors is not a trivial matter.) If there were no losses the use of two gratings in series, with a halved projection distance, would give a four-fold improvement in the photometric efficiency of the system. The best modern gratings are so good that, even after allowing for the losses of light at the second grating, the factor is between 2 and 3.

It is worth noting that if one uses the same grating twice, an extra mirror reflection is required to return the light to the gratings and the photometric efficiency of the system is reduced. Fig. 12–20(*b*) shows a double grating system due to Stroke (1958).

In connection with photometric efficiency it is important to note that while the resolving power of a grating in a Littrow mounting is proportional to sin I the angular dispersion is proportional to tan I. Thus, although there is no great improvement in resolution, an increase in obliquity from, say, $I = 64°$ to $I = 76°$ approximately doubles the angular dispersion, and for a given linear dispersion the projection distance is halved. The aperture of the system ($\propto \cos I$) is almost halved, but if the efficiency of blaze is retained the photometric efficiency of the system is approximately doubled.

Mountings for Concave Gratings

In the past most of the mountings used for concave gratings made use of the properties of the Rowland circle, although the original arrangement used by Rowland is now obsolete. For photographic work the plate is curved to fit this circle of focus.

1. *Paschen Mounting*. In this widely used arrangement the grating and slit are fixed and the plate is curved to fit any convenient portion of the Rowland circle. A typical arrangement is shown in Fig. 12–21. The approximate positions of the

visible spectra are shown for a typical
grating. Usually the slit and grating are
mounted on separate piers while the
plate mounting is extremely rigid. The
spectra will be nearly normal where the
angle of diffraction is small. It is frequently useful to provide facilities for
mounting the slit in more than one position on the Rowland circle. The Paschen
mounting can cover a very wide spectral
range and some extra large installations
(e.g. 10 metres) have been employed; the
system is bulky and requires great
mechanical precision.

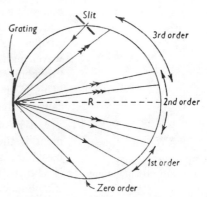

Fig. 12–21. Paschen mounting.

2. *Eagle Mounting.* A very compact and much used arrangement is that
introduced in 1910 by Eagle (Fig. 12–22) and this may be regarded as a Littrow
type of mounting. In principle the slit S is mounted alongside or just above or
below the plate, and the properties of the Rowland circle are utilized as follows.
Light from the slit S is reflected towards the grating G by the total reflection
prism. S′, the virtual image of the slit, is at the centre of the plate P which is
curved to fit the Rowland circle. The grating can be moved to and from the
plate and, as it is moved, both G and P are rotated so that P remains on the

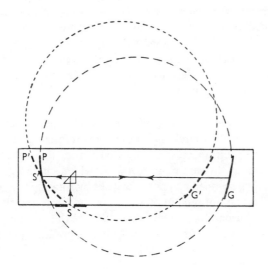

Fig. 12–22. Eagle mounting

Rowland circle. For example, when G is translated and rotated to G′, P is
rotated to P′. With this arrangement $\theta = I$ and, for a maximum,

$$2d \sin I = p\lambda \quad (p \text{ integral})$$

so that wavelengths up to $2d/p$ can be examined. The working range is therefore

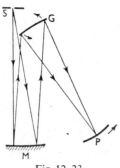

Fig. 12–23.
Wadsworth mounting.

double that of the Rowland mounting. The relative compactness of the Eagle mounting causes it to be widely used in vacuum spectrographs since a much smaller volume has to be evacuated. (Air absorbs wavelengths below about 2000 angstroms.) Temperature control is also easier with this smaller apparatus. (This is important since a variation of temperature causes a significant variation of the grating interval.)

3. *Wadsworth mounting.* The Wadsworth arrangement is the one important mounting which is not based on the Rowland circle; the optical system is shown in Fig. 12–23. Light from the slit S is collimated by the mirror M (originally Wadsworth used a lens) and falls on the concave grating G. The plate P is placed on the grating normal, and to investigate different spectral regions the angle of incidence must be varied to bring the middle of the range on to the normal. As was shown in § 12–11, the surface on which the spectral lines are in focus has a shape which depends upon the angle of incidence employed [equation (12–26)]. This means that the plate holder must be such that the curvature of the plate can be varied. The Wadsworth arrangement is confined within little more than half the area occupied by the Eagle mounting. (A similar arrangement has been used with a plane grating placed in light rendered convergent by M.) Another important advantage of the Wadsworth arrangement is that the astigmatism is very much reduced (see § 12–11). In addition, the grating is used with a relatively large angular aperture and this, plus the absence of astigmatism, renders the lines more intense. For spectrochemical analysis the Wadsworth mounting is more convenient than the mountings due to Rowland and Paschen which were widely used in the earlier days when gratings were employed for the accurate measurement of wavelength. The Eagle mounting has been widely used in all branches of spectroscopy.

4. *The Seya and Johnson-Onaka Mountings.* The mechanical linkages used with the various Rowland circle mountings are impracticable for monochromators where the spectrum must be scanned across the exit slit. In the Seya mounting (1952) both slits are close to the Rowland circle for the zero order. The spectrum is then scanned across the exit slit by rotating the grating about its mid-point. The slits do not remain on the Rowland circle and there is some loss of focus. In the Johnson-Onaka mounting (1957–8) the grating is rotated about an axis some distance away from the centre of the grating in order to minimize the loss of focus.

Mountings for Echelles

The spectral range of an echelle in a Littrow-type mounting is $1/(2d \sin I)$ wave numbers. If this is less than the total range of the spectrum to be examined, the orders must be separated by means of an auxiliary dispersion system. Harrison and Bausch (1950) have described a system in which an echelle in an arrangement of the Ebert type is used to illuminate the slit of a large quartz prism Littrow spectrograph, the dispersion of the echelle being at right angles to that of the prism. If the dispersion of the echelle is in a vertical plane, that of the prism is in a horizontal plane, and the entrance slit for the echelle is horizontal (i.e. parallel to the rulings) while the prism spectrograph slit is vertical. To understand how the system works, first take the prism spectrograph alone and open the slit until its image on the plate extends over a range equal to the spectral range of the echelle. Now insert the echelle system in front of the prism slit. Successive echelle spectral ranges are superimposed on one another, each extending along the prism spectrograph slit, but on the plate these are spread out horizontally. The appearance of the pattern on the plate is represented by Fig. 12–24. AB represents a strip across part of a spectrum as formed by the prism spectrograph alone, using a narrow slit. The pattern below represents that photographed with the echelle in position. The echelle expands each spectral range vertically. In this way a very large spectral range can be photographed with high resolution on a single plate. The spectrograph is comparable with the best that can be produced with a very much more cumbersome concave grating installation using many more exposures. It will be appreciated that the particular arrangement described can be used only if the spectrograph slit can be made long enough to admit the dispersed spectral range of the echelle.

A simpler arrangement described by Harrison and Bausch for crossing an echelle with a prism substitutes the echelle for the mirror behind the prism in a Littrow system.

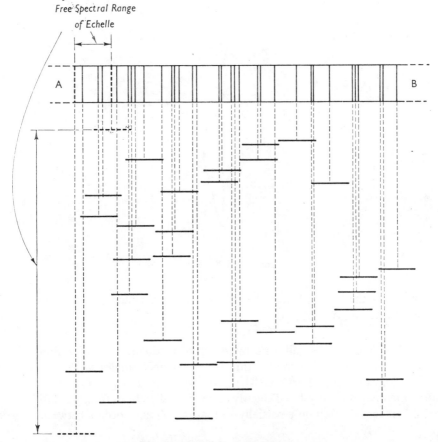

Fig. 12–24. Echelle spectrogram.

The auxiliary dispersion for an echelle can also be provided by a grating, and echelles, like any other gratings, can be used in series. For example, Harrison and Stroke (1960) have described a system using two echelles in series and in series with a concave grating which supplies the auxiliary dispersion. The layout resembles that shown in Fig. 12–20(*b*) which now represents a vertical section. The mirror M_2 is replaced by the concave grating with its lines vertical. The spectrogram again has the form illustrated by Fig. 12–24.

SISAM

This arrangement, due to Connes (1957), derives its name from: *s*pectromètre *i*nterferéntiel à *s*eléction par l'*a*mplitude de *m*odulation. It consists of a Twyman-Green interferometer (§8–12) in which the mirrors are replaced by dispersing elements such as diffraction gratings (see Fig. 12–25). Suppose the point source S emits light of wavelength λ and that the gratings are rotated into the Littrow position, i.e. $2d \sin I = p\lambda$. Then plane waves of wavelength λ enter lens L_2 to give coincident rectangular aperture diffraction patterns centred on S'. If the phase difference δ between the wavefronts is varied by displacing one of the

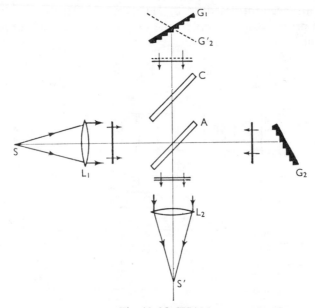

Fig. 12-25. SISAM.

gratings along the axis, or by rotating the compensator plate C, the flux received by the detector is given by

$$E = E_0(1 + \cos \delta)$$

i.e. the flux varies periodically and the depth of modulation is unity. Suppose, now, that the gratings are rotated through equal small angles so that G_1 and G_2' rotate in opposite directions. The plane waves enter L_2 obliquely and the diffraction images are displaced slightly in opposite directions from S'. It follows that the images overlap only partially so that interference occurs between only part of the light in the two patterns. In consequence, the variation of δ causes a variation of flux with a reduced depth of modulation M:

$$E=E_0(1+M \cos \delta). \quad . \quad . \quad . \quad . \quad (12\text{–}27)$$

M falls to zero as the patterns for wavelength λ move away from S', and the next spectral line (λ') is detected by the increase in M as the gratings rotate to bring the corresponding images together at S'. It is important to note that it is the modulation depth in the total flux E and not the intensity at S' that is recorded. As the grating is rotated to scan a monochromatic spectral line in an ordinary single-beam Littrow system, the instrumental profile, i.e. the variation in intensity at a stationary scanning point is of the form $(\sin \alpha/\alpha)^2$ where α is proportional to the rotation of the grating. It can be shown that in the present method the instrumental profile, which is now the variation of the modulation depth as the two gratings are rotated, is of the form $(\sin \alpha/\alpha)$. It follows that the subsidiary peaks in M are proportionately much stronger than the subsidiary maxima in the intensity distribution in a single image (20% of the maximum instead of 4%—see Fig. 11–4). It can be shown that these subsidiary peaks can be suppressed if the aperture of the system is made diamond-shaped, the diamond being inscribed within the rectangular apertures of the dispersers, i.e. the system can be apodized (see §11–15). When apodized, SISAM gives a resolving power corresponding to the gratings used in an ordinary Littrow mounting with a slit of infinitesimal width. In practice one must use a finite source with SISAM, just as one must use finite slits in conventional systems. The advantage of SISAM is that for a given resolving power its photometric efficiency is superior to that of a conventional grating spectrometer; it is intended for use with weak sources in the infra-red region. The relative merits of the various classes of spectrometers are discussed in §§ 12–17 and 18–12.

12–13. Multislit and Grille Spectrometers

In the accounts of traditional grating spectrometers given so far it has been assumed that a grating monochromator employs entrance and exit slits that have infinitesimal widths. Obviously this is an unrealistic assumption since the detector would receive infinitesimal power. In practice the slits must have finite widths, with the result that the resolving power deteriorates. The photometric properties of the various classes of spectrometer are discussed further in §18–12 but it is obvious that if one is to obtain maximum resolving power, the slit widths must be kept to a minimum; it is equally obvious that if a monochromator is to transmit maximum radiant power the entrance and exit apertures must have the largest possible areas. In 1949 Golay pointed out that it would be possible to replace the slits in a traditional form of monochromator by a row of slits if one were able to utilize only the radiation that passed through a "conjugate" pair of slits in the entrance and exit rows, i.e. discard radiation that passed through the "wrong" exit slit since it would be of the wrong wavelength. He extended the idea to arrays of entrance and exit apertures and showed that the principle was workable, but the mode of operation of the system was complicated.

More recently, Girard (1963) has replaced the traditional entrance and exit slits by two-dimensional grilles. Although, like the Golay systems, a grille spectrometer transmits more power because it uses extended entrance and exit apertures, in operation it is more closely related to the SISAM since one measures a flux modulation rather than the flux itself. The form of the grilles which would replace vertical slits is illustrated by Fig. 12–26. Alternate opaque and transparent regions are bounded by rectangular hyperbolas as shown although in practice the number of zones is very much larger than is indicated by the figure—30 mm square grilles have been made with 600 hyperbolas.

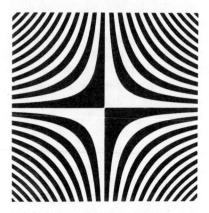

Fig. 12–26. Girard grille.

The grille spectrometer can be operated in several ways. In one method of operation the opaque portions of the exit grille are made reflecting and by means of a suitable optical and electrical system the detector outputs due to reflected and transmitted radiation are combined and the difference measured. Suppose, then, that monochromatic light enters the system. If the image of the entrance grille misses the exit grille completely there is no reflected light and no transmitted light and the difference signal is zero. If the monochromator is adjusted so that the image of the entrance grille begins to cross the exit grille the reflected and transmitted powers remain almost equal, giving a near-zero difference signal until the entrance grille is imaged exactly on the exit grille. The transmitted light then rises to a sharp maximum, the reflected light falls to a minimum, and the difference signal rises to a sharp maximum.

In an alternative method of operation the image of the entrance grille is oscillated in a direction parallel to the length of the slits of a classical system. This causes the transmitted flux to be modulated for the wavelength for which the vertical asymptotes of the exit grille and the entrance grille image coincide.

For a given resolving power, the radiation transmitted by a grille monochromator exceeds that transmitted by a traditional instrument employing single entrance and exit slits by a factor approaching 100.

12–14. Mock Interferometry

The Mock interferometer, introduced by Mertz (1960), is so-named because it simulates the output from an interferometer although it is not an interfero-

meter itself. The theory and use of the Mock interferometer have been discussed in detail by Selby and Ring (1966). To understand how the device works
consider first two identical grids, each consisting of alternate opaque and transparent parts of equal widths, one being placed on top of the other. In Fig.
12–27(a) one grid is drawn with broken lines and the opaque parts of the two
grids are shaded in different directions. It will be obvious that if one grid is
translated, without rotation, relative to the other, the radiation transmitted by
the combination varies between a maximum, when the transparent strips of
one grid fall exactly on those of the other, and zero when the opaque parts of
one fall on the transparent parts of the other. The modulation profile is triangular, and the frequency is proportional to the component of the translation
velocity perpendicular to the lines, and to the number of lines per mm.

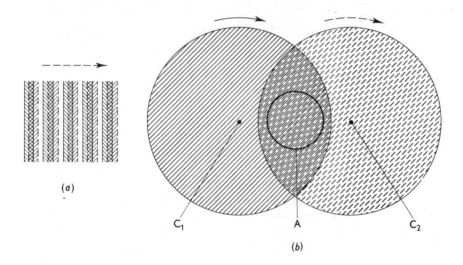

(a)

C_1　　　　　　A　　　　　　C_2

(b)

Fig. 12–27. Grids for the Mock interferometer.

Suppose, now, that one has two identical grids that rotate at the same speed
and in the same direction about the points C_1 and C_2 (Fig. 12–27(b)) so that the
grid lines are always parallel. In effect, one grid is again translated across the
other, and the radiation transmitted by an aperture at A is modulated with a
frequency that will remain constant if the grids are rotated with an angular
velocity proportional to $\sin \theta$ where θ is the angle between the grid lines and the
line joining their centres of rotation; the radiation is again modulated by a
triangular function. The important point to note is that the modulation
frequency is proportional to the separation of the centres of rotation of the grids.
   In practice one employs a single grid and the effect of two grids is obtained
by imaging one region of the single grid on to a neighbouring region. Thus
suppose one has a spectrometer of the Littrow type with the grid lying in the
focal plane. Fig. 12–28 represents the focal plane, the dispersion of the Littrow
system being in the horizontal direction. Suppose light of wavelength λ enters
through the area A and emerges through A′. The region of the grid at A is
imaged at A′ and if C is the centre of rotation of the real grid, C′ will be the

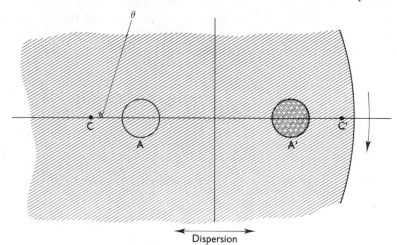

Fig. 12–28. Mock interferometer in Littrow mount.

centre of rotation of the grid image for wavelength λ and as the grid is rotated the image grid will be translated across the actual grid at A' and the emerging radiation will be modulated. The modulation frequency will be constant if $\sin \theta$ varies linearly with time. For other wavelengths the centre of rotation of the image grid will be displaced horizontally from C' so that each wavelength is modulated at a different frequency. Apart from the fact that the modulation profile is triangular, one obtains an output resembling that from a Michelson interferometer. The spectral range of the system is limited by the extent of the dispersed grid images that can be accommodated. With a simple Littrow system two important sources of error are the usual line curvature, and variation across the field of the lateral magnification. However, if the field is restricted good resolution can be achieved and one obtains a *multiplex gain* (see § 18–12). The triangular modulation profile can be taken into account when the spectrum is computed.

12–15. The Production of Gratings

The resolving power of a grating is given by

$$\frac{\lambda}{\delta\lambda} = pN = \frac{Nd(\sin \theta_p \pm \sin I)}{\lambda}.$$

Hence the requirement for maximum resolving power is that the total overall width of the grating should be as large as possible. If one rules few lines in this width a high order will be used and there will be serious overlapping of orders— the free spectral range will be small (see § 12–5). To obtain maximum spectral range with maximum resolving power one must rule lines as closely as possible over the maximum width. The problem, then, is to rule thousands of parallel and equally-spaced grooves per inch over a width of several inches—using a Littrow mounting close to grazing incidence a perfect 10-inch grating is needed to achieve a resolving power of 10^6 at a wavelength of about 5000 angstroms.

Most gratings have been produced by ruling a glass or metal surface with a diamond tool. Speculum metal blanks were widely used in the past for the production of reflection gratings but most modern gratings are plastic replicas of master gratings ruled on aluminized glass blanks, reflection gratings being replicas with an evaporated metal film.

Rowland (1882) was the first to produce good-quality gratings of appreciable size. His machine moved the diamond across the blank for each groove and moved the blank between rulings. Until relatively recently this has been the most usual method, although Strong displaced the diamond between rulings and moved the blank past the stationary diamond for each groove. Rowland's machine ruled 14,438 lines per inch. This gives $d = 1 \cdot 6932 \times 10^{-4}$ cm., and equation (12–1) shows that with normal incidence the angle of diffraction becomes 90° when λ is 16,932, 8466, and 5644 angstroms in the first, second, and third orders respectively, so that the complete visible spectrum can be observed in only the first two orders. An idea of the dispersion can be obtained from the fact that in the first order the angles of diffraction are about $13 \cdot 7°$ and $28 \cdot 2°$ for wavelengths 4000 angstroms and 8000 angstroms, so that if the spectrum is formed in the focal plane of a lens of focal length 100 cm., the length of the visible spectrum is about 25 cm. If the grating were perfect the resolving power in the first order would be $14,438 \times$ (width of grating in inches), i.e. about 86,600 for a six-inch grating. Enormous numbers of copies have been made of gratings ruled on Rowland's machine. Rowland ruled concave gratings in addition to plane gratings.

Wood ruled his early blazed echelettes on gold-plated copper plates, using the natural edges of selected carborundum crystals. Collodion copies were mounted on glass and sputtered to give reflection gratings which were better than the originals because the glass mounts were flatter than the copper plates. Later, Wood replaced the carborundum tool by a diamond specially ground and polished to produce grooves of the desired form. Rulings were made on copper plates which were chromium plated after ruling. Subsequently, Wood ruled better echelettes on aluminized glass plates.

The groove depth, determined by the loading of the diamond, as well as the groove form, determined by its shape, is relevant to the efficiency of the grating. If the ruling is too light the grooves will be shallow, and therefore narrow, and there will be strips of the plane surface of the blank between grooves; these strips give light in the useless zero order. If the ruling is too heavy each groove spoils the shape of the previous one.

The resolution of a diffraction grating is reduced if the rulings are not exactly straight, parallel, and equally spaced, and if the blank is not accurately plane (or spherical). In consequence, a great deal of effort has been put into the design of ruling machines. Very great improvements have been made in recent years by checking the operation of the ruling machine interferometrically. For example, it is claimed that the machine at M.I.T. developed by Harrison and Stroke is producing gratings whose spectroscopic quality is almost 100 times that of Rowland's gratings, and efficiently blazed gratings are now available with resolving powers in the 1 to 2 million range in the visible and near ultra-violet. It was found that it was not satisfactory to apply interferometric control of a stop-start machine in which the blank was stationary during the ruling of each groove and was moved between rulings. The new procedure is to advance the blank uniformly and continuously so that it continues to advance slowly as the diamond moves uniformly across it. The motion of the blank carriage is watched continuously by employing a Twyman-Green interferometer system with one of the mirrors attached to the carriage. The output from the interferometer operates a servo-mechanism to advance or slow the carriage. In this way the movement of the grating blank is synchronized with the movement of the diamond to within about $\pm 0.003\mu$. In addition to controlling the translation of the blank carriage, an interferometric servo-mechanism is used to control the rotation of the blank about a vertical axis. In this way parallelism of the grooves is maintained to $\pm 1/200$ second of arc.

Since the diamond may be called upon to rule several hundred thousand grooves it is important to mount it with its axis orientated so that wear is reduced to a minimum. (Aluminized blanks are used because the metal is soft.) As remarked earlier, the loading of the diamond is also of great importance. Some idea of the magnitude of the problem of ruling a good grating can be obtained from the fact that it often takes more than a day to load and orientate the diamond; that the engine runs for about 18 hours to reach equilibrium before ruling starts; that the engine makes only about 10 strokes per minute; and that the temperature of the laboratory is controlled to within about $\pm 0.02°C$.

The interferometric control of the ruling machine is now so good that it is more difficult to obtain a sufficiently flat aluminium coating on the glass blank before ruling starts than to maintain a corresponding regularity of spacing. For the highest resolving powers the coated blank should be flat to less than $\lambda/10$, and it is particularly difficult to obtain the thick aluminium coatings (up to 30μ) necessary to accommodate the deep grooves of coarse gratings. This, together with the heavy loading required on the diamond, makes it impracticable to rule efficiently blazed high resolution gratings on aluminized glass blanks with less than about 20 grooves per mm. Coarser gratings for use in the infra-red have been ruled on magnesium metal blanks.

The reflecting gratings supplied to users are not the masters actually produced by the ruling machine, but aluminized plastic replicas of these gratings mounted on flat (or concave) glass blanks. Transmission gratings simply consist of non-aluminized replicas. Bausch and Lomb have produced a transmission grating for use down to 2000 angstroms by employing a special resin for the replica and mounting it on a fused quartz blank. A step of major importance in the production of a replica grating is the clean separation of the replica from the master grating. It is claimed that a replica grating is actually more efficient than the master from which it is derived because in the replication process it is possible to eliminate the effect of the ridge thrown up by the diamond on the edge of each groove. Considerable effort is being made to increase the size of the gratings that can be made and Harrison and Thompson (1970) reported a ruling machine at M.I.T. with a capacity of 450×635 mm.

A novel method of producing plane gratings was devised by Merton (1950). There are three main operations:

1. Upon one half of a polished metal cylinder is cut a helix having a pitch

equal to that of the required grating. The diamond tool is mounted on a suitably geared screw-cutting lathe. The helix as a whole must be free from progressive errors in pitch but periodic errors in the helix due to the gearing and lead-screw are corrected in the next stage.

2. A corrected helix is ruled by a diamond which is guided from the original helix by means of a nut lined with strips of soft elastic material such as cork. This *Merton Nut* effectively averages out the errors of the first helix.

3. The corrected helix is coated with a solution of a suitable plastic which, after hardening and annealing, forms a film which can be removed and opened out with rulings down on a moist gelatine-coated glass plate. When the gelatine is dry the original film can be removed, leaving the gelatine moulded into a plane grating which can be hardened and aluminized.

Up to the present the Merton process has not produced high resolution gratings equal to the best now available.

A completely different method of producing gratings is by, in effect, photographing straight-line interference fringes. The fringes are produced by the interference of two collimated beams of monochromatic light. The $\cos^2$ fringes are converted into grating grooves on an optical flat by recording the fringes in a layer of photoresist. When aluminized, the surface becomes a reflection grating with an approximately sinusoidal groove form. It is only recently that by using high power laser light sources and high resolution photoresists that it has been possible to produce gratings with the precision required for spectroscopic use. These so-called *holographic gratings* have no periodic errors and hence no ghosts. Another advantage is that it is likely that this method can produce finer gratings than any ruling process—which is an advantage for ultra-violet and X-ray spectroscopy. Gratings have been produced with aberrations of less than $\lambda/4$ in the visible region, and, by improving the optical quality of the interferometer, it should be possible to produce gratings of the even higher quality necessary for shorter wavelengths.

12–16. Irregularities in Gratings. Ghosts

In the foregoing theory it has always been assumed that all the rulings in a grating are identical and that the separation of adjacent rulings is constant. In practice this will never be true exactly, and errors in the grating spacing can cause troublesome effects. These errors are of three types:

1. *Random errors.* An error which is random in magnitude and direction gives a background illumination across the spectra and is relatively unimportant.

2. *Progressive errors.* An error in spacing which increases progressively across the grating can be shown to cause the spectral lines to be sharper in planes other than the focal plane of the optical system. Errors of this kind can occur over certain portions of the grating and they become very important in high resolution gratings.

3. *Periodic errors.* In the past these have been the most troublesome errors; they cause the appearance of additional *ghost* lines in the grating spectra which, for a perfect grating, would indicate the presence of other wavelengths.

The ghosts are of two main types: *Rowland* ghosts, which are close to, and symmetrically disposed about, the parent line and give the impression of fine structure, and *Lyman* ghosts which are far removed from the parent line and can be mistaken for weak lines in a different spectral region. Neither of these constitute a serious problem in the best gratings currently available.

Until recently there was little attempt to use gratings for high resolution spectroscopy requiring resolving powers in the region of those corresponding to perfect gratings working at high angles of incidence. In consequence, more attention was paid to the existence of ghosts that were separated from the parent line than to the existence of spurious lines—sometimes referred to as "satellites", sometimes as "grass"—affecting the sharpness or structure of the true spectral lines. Stroke (1960) has made a full investigation of the effect of grating imperfections on the spectral quality of gratings and has pointed out that since the ideal grating would give a plane diffracted wave, the grating defects are best thought of as sources of wavefront aberration causing loss of sharpness in the image (see §15–1). Stroke has pointed out that departures from flatness on the grating blank, as well as ruling errors, cause loss of resolution. A ruling defect of particular significance is a cumulative ruling error over a portion of the grating, the line interval increasing linearly across the defective region. This gives a wavefront error similar to that arising from a portion of the grating ruled perfectly on a region of the blank inclined to the main area. The wavefront from a grating can be examined in the usual way—with a Twyman-Green interferometer.

It is worth noting that, while the resolving power of a perfect grating in a Littrow mounting is proportional to sin I, the quality of the grating needs to increase with sin$^2 I$ in order to achieve the theoretical improvement resulting from an increased angle of incidence. The best modern gratings used in Littrow mounts at 64° or 76° have no satellites stronger than 1/100th of the intensity of the main line in the visible, and the Rowland ghosts are weaker than 1/2000th at those angles; resolving powers approaching 2×10^6 have been achieved.

12–17. Preliminary Review of the Various Classes of Spectrometers

In a number of sections various spectroscopic systems have been described more or less in isolation. This has occurred because optical systems of many different kinds can be used in spectroscopy. It is desirable now to look at the field from the spectroscopic point of view. It is not appropriate at this stage to look at every aspect and attempt to give a full account of the relative merits of the various classes of spectrometers. When choosing a spectrometer for a particular purpose the photometric (radiometric) properties of the system are an important, and sometimes a decisive, factor influencing the choice. The various classes of spectrometer are looked at from this point of view in §18–12.

Although the various systems—prisms, two beam and multiple beam interferometers, and diffraction gratings—might appear to have little in common, there are, in fact, certain features common to them all. As has been pointed out by Gebbie (1969), in effect all spectroscopic systems take the radiation to be analysed, divide it into a number of beams, introduce systematic phase differences between the beams, and then recombine the beams. The result depends upon the spectral distribution, and the latter can therefore be obtained by analysing the former. In every class of instrument the resolving power is determined by the maximum phase difference that is introduced. Looked at from this point of view a prism is seen as transmitting an infinite number of beams—it transmits an undivided wavefront which is regarded as an infinity of adjacent elements. For a Fabry-Perot the effective number of beams is taken as equal to the finesse —for a grating the resolving power is ($N \times$ order), and for a Fabry-Perot it is (finesse × order).

With a complicated spectrum only a prism (∞ beams) gives an unambiguous presentation of the spectrum directly; a grating and a Fabry-Perot (N beams) require an additional component to separate the orders; and a Michelson (2 beams) requires the whole additional step of Fourier transformation to get the spectrum. The above sequence of instruments is in order of decreasing number of beams and of increasing complexity of presentation of the spectral information; it will be seen in §18–12 that it is also in order of increasing photometric efficiency.

At least in cases where the spectrum is displayed directly the need for a large maximum path difference can be appreciated immediately by noting that the requirement is that a displacement in the image should give rise to the maximum possible phase difference between extreme disturbances in order that the intensity should decrease rapidly as one moves away from the centre of a spectral line or fringe. Thus the requirement is for a thick etalon with highly reflecting plates; for a long Lummer plate; for a wide grating or echelon; and for a large prism.

A decade or so ago improvements in the production of ruled diffraction gratings took them into the 10^6 resolving power class in the visible and ultra-violet regions. The ability of a grating spectrograph to record a broad spectral band with a single exposure is an obvious attraction, but a grating mono-chromator is photometrically inefficient. Modern instruments employing gratings and achieving better photometric properties with little loss of resolu-tion include multislit or grille spectrometers, which retain the basic layout of a classical grating system, and SISAM, which involves rather more complex instrumentation.

One of the attractions of a grating or an echelle is the fact that one has a large free spectral range which is not reduced by using a wider grating to increase the resolving power. For example, a resolving power of 10^6 at λ 5000 Å (resolving limit 0·02 cm^{-1}) can be obtained with a 1 cm etalon having a finesse of 25, and this has a free spectral range of $\Delta\lambda = 0·125$ Å ($\Delta\bar{\nu} = 0·5$ cm^{-1}), which is insuffi-cient to study the hyperfine structure of, for example, the λ 5461 mercury green line without a pre-monochromator. Approximately the same resolving power can be obtained with a 10-inch grating or echelle in a Littrow-mounting at 76°. A grating with 300 grooves/mm would have a spectral range of about 400Å (~ 1500 cm^{-1}), and an echelle with 30 grooves/mm would have a spectral range of about 40 Å (~ 150 cm^{-1}). A further disadvantage of the etalon is that one often needs to adjust the parallelism of the plates. On the other hand the etalon has superior photometric properties.

The concave grating has the advantage of not requiring any additional optical components, but the importance of this fact has decreased with the development of aluminized mirrors. The need for auxiliary dispersion is eliminated by having a large number of rulings and using a low order. The concave grating is still widely used in the vacuum ultra-violet and in fixed-position spectrographs, but in most applications calling for a grating a plane grating is preferable. Low resolution concave gratings have been used to provide the auxiliary dispersion for high resolution components. The advantage of an echelle crossed with a low dispersion spectrograph over a classical large concave grating installation has already been discussed in § 12–12. Attention has already been drawn to the advantage of using a small grating interval, enabling a low order to be used. However, it must be pointed out that if the groove aspect as "seen" by the light is appreciably less than a wavelength, the grating will act as a plane mirror, reflecting the light into the zero order. Thus one cannot use very low orders at high angles of incidence.

A prism cannot be classified as a high resolution device—a dense flint prism can give a resolving power in the visible of the order $1000 \times$ the length of the prism base in cm. Prism instruments are, of course, widely used to provide auxiliary dispersion for high resolution devices and also for many instruments of moderate resolution (see §§ 5–16 to 5–18). However, owing to the recent improvements in the quality and availability of gratings it seems likely that gratings will replace prisms in many applications—it must not be forgotten that in much routine spectroscopy one is not attempting the ultimate in resolving power. In the past one important photometric advantage of a prism was that it gave only one spectrum; gratings are now blazed so efficiently that this advantage is small. The relative simplicity of the design and operation of a grating spectrometer leads one to the conclusion that for the foreseeable future gratings will be used in a very wide range of spectrometers of moderate resolution.

For most scanning instruments of moderate resolution—spectrometers, monochromators, and spectrophotometers—a plane grating is preferable to a concave grating since no focus adjustment is required if the grating is rotated to scan the spectrum. In addition, a plane grating is cheaper than a concave grating.

It has already been pointed out that if they give adequate resolving power, grating instruments have certain advantages but that they have poor photometric properties. Even for quite low resolving powers this is often a decisive factor causing the classical system to be replaced with a modern system such as a grille spectrometer, a SISAM, or a straightforward interferometer (see § 18–12). One must not forget that there are occasions when one requires a beam of near-monochromatic light as such, and a grating monochromator, which is the core of a grating spectrometer (as opposed to a spectrograph), is likely to be indicated; a possible alternative for some purposes is a narrow pass-band interference filter.

If one wishes to achieve resolving powers much in excess of 10^6 an interfero-meter must be used and in practice this almost always means either a Fabry-Perot etalon or a Fourier transform spectrometer. The Lummer plate, which was widely used in early high resolution spectroscopy, is very difficult to manu-facture with the required accuracy and, together with the transmission echelon (which offers little in return for its very high cost) must now be regarded as obsolete. The reflection echelon, which is really a highly blazed grating, is extremely expensive and only a few large echelons have ever been made; it works in a very high order and hence has a small spectral range, and the overall optical system is photometrically inefficient. However, it does give very high resolution in every spectral region (about 0.01 cm^{-1} in the visible) and can be used in the far ultra-violet. With the possible exception of its use in the far ultra-violet it is difficult to see any future for the reflection echelon.

For many years after it was realized that in principle a Fourier transform spectrometer could provide a system of very high chromatic resolving power, the magnitude of the computation involved made it impracticable. Thus the Fabry-Perot etalon became the recognized maximum resolution device. The spatially displayed fringe system was well suited to visual observation or photographic recording and was an attractive feature in the era of high precision wavelength measurements in visible and ultra-violet emission spectroscopy. The development of the central spot scanning technique and multilayer dielectric mirrors gave increased resolving power and an instrument suitable for intensity measurements. Even now, when the computational problems associated with the Fourier transform method can reasonably be said to be solved, there are many who prefer the more direct display of the spectrum that one can obtain with the Fabry-Perot in its various forms. On the other hand, the Fourier transform method can be used with very high resolution over a wider spectral range and is probably the best instrument for the examination of broad complicated spectra; there is, in fact, no serious competitor for the examination of a broad spectrum at high resolution throughout most of the infra-red. In the *very* far infra-red and sub-mm region, and in the visible and ultra-violet, the photometric advantages are smaller and might even disappear. Another reason why the Fourier transform method is less advantageous in the short wavelength regions is that to achieve high resolution very great precision is required in moving the interferometer mirror. The corresponding precision has to be produced only once with a grating: when it is ruled. If the accuracy of the mechanism rotating the grating in a monochromator lacks precision, the accuracy of the absolute wavelength scale is lowered, but there is no corresponding loss in resolving power. On the other hand, the Fourier transform method is capable of virtually unlimited resolving powers—certainly of resolving powers far beyond the reach of any grating and, unlike the Fabry-Perot, no pre-monochromator is required. It seems likely that in the realm of maximum resolution spectroscopy the Fourier transform method will predominate in the future. Its advantages are greatest for the investigation of complicated broad spectra and for the analysis of weak sources; in the infra-red its advantages for such work are overwhelming—even if very high resolving powers are not required.

For absorption spectroscopy a radical change might occur in the future arising out of the development of tunable lasers (see §20–5). These already exist as infra-red sources, and if the wavelength could be held constant at a known value throughout a wide spectral range, classical monochromators would become obsolete. Rapid advances are also being made in the development of very sensitive infra-red detectors such as very low temperature ($<1°$K) and super-conducting bolometers. It has been suggested by Richards (1971) that the Fourier transform method may not be the permanent solution for infra-red spectroscopy (see also § 18–12). The development of new laboratory sources is, of course, relevent only to absorption spectroscopy.

12–18. Two- and Three-dimensional Gratings

Suppose a collimated beam of light is incident on a series of equidistant diffracting point elements along the x-axis (Fig. 12–29). In the notation adopted hitherto, the direction of a maximum is given by

$$d_x (\sin \theta - \sin I) = p_1 \lambda, \quad \ldots \ldots \quad (12\text{–}28)$$

where d_x is the distance between elements. The figure indicates that the diffracted

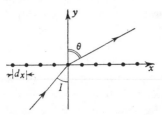
Fig. 12–29. Diffracting elements along the x direction.

light lies in the xy plane but this is a special case and disturbances from these diffracting elements will be in phase in all directions which are inclined to the x-axis at the same angle $(\pi/2 - \theta)$. Again, maxima are formed in these directions for incident light in any direction that is inclined to the x-axis at the same angle $(\pi/2 - I)$. Consequently, it is better to write (12–28) as a relation between the angles which the incident and diffracted rays make with the x-axis. Writing α_o and α for the appropriate direction cosines of the incident and diffracted light, (12–28) becomes

$$d_x(\alpha - \alpha_o) = p_1 \lambda. \quad \ldots \ldots \quad (12\text{–}29)$$

It will be seen that for a given value of α_o the directions in which a maximum of given order is found form the generators of a cone whose axis is the x-axis and whose semi-vertical angle is $(\pi/2 - \theta)$. Maxima are found in these directions for equally spaced diffracting elements along any line parallel to the x-axis.

For diffracting elements along the y-axis with spacing d_y, the directions of the maxima are given by

$$d_y(\beta - \beta_o) = p_2 \lambda, \quad \ldots \ldots \ldots \quad (12\text{–}30)$$

so that the directions in which a maximum of given order is found form the generators of a cone whose axis is the y-axis. (As before, maxima are found in parallel directions for similarly spaced elements along any parallel line.) Suppose, now, that both sets of diffracting elements are present together. One has a two-dimensional grating and maxima will be formed in directions which are specified by the values of the direction cosines α and β which satisfy (12–29) and (12–30) (remembering that $\alpha^2 + \beta^2 \not> 1$). For given values of the integers p_1 and p_2, each of the direction cosines, α and β, determines a cone about the relevant axis so that for the two-dimensional grating the maxima are formed in directions given by generators which are common to those two cones. Obviously, a pair of values for p_1 and p_2 will not correspond to a maximum if the corresponding values of α and β define cones which have no common generators.

If one has, in addition, a set of equidistant diffracting elements along the z

direction, one has a three-dimensional grating. Maxima for a periodicity in the z-direction are defined by the relation

$$d_z(\gamma-\gamma_o)=p_3\lambda, \qquad \cdots \cdots \cdots \quad (12–31)$$

and if (12–29), (12–30), and (12–31), can be satisfied simultaneously, the values of α, β, and γ, determine the direction of a maximum for the three-dimensional grating. Now since one must always have

$$\alpha^2+\beta^2+\gamma^2=1, \qquad \cdots \cdots \cdots \quad (12–32)$$

it follows that α and β completely determine a direction in space and, in general, the directions of the maxima for the two-dimensional grating will not be directions corresponding to maxima for the grating interval in the z-direction. That is to say, when values of α and β which satisfy (12–29) and (12–30) are substituted into (12–32), the resulting values of γ cannot, in general, satisfy (12–31) (i.e. do not give an integral value of p_3). It is only for particular values of α_o and β_o that it is possible for all the conditions to be fulfilled. That is, it is only for particular directions of the incident light that one gets any maxima with a three-dimensional grating. The necessary directions of incidence obviously depend upon the wavelength.

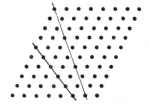

Fig. 12–30. Reflection planes.

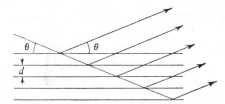

Fig. 12–31. Reflection from parallel planes.

Three-dimensional gratings are very important since a crystal lattice constitutes a structure of this type. Since the grating intervals are very small, a crystal provides a grating suitable for use with waves of very short wavelength, i.e. X-rays. This opens up the vast subject of X-ray crystallography about which little more can be said here. One point of importance which should be mentioned is the interpretation, by Bragg, of the directions of the incident light which will give diffraction maxima.

According to Bragg, any plane in the crystal which contains a relatively large number of lattice points can be regarded as a reflecting plane. Fig. 12–30 indicates two such planes in a monoclinic crystal. Thus if a parallel beam of X-rays is incident on such a plane in a direction near to grazing incidence, a small fraction of the beam is reflected. In the lattice there will be a large number of equidistant parallel planes and reflections occur at these also (Fig. 12–31). The path difference between beams reflected from adjacent layers is $2d \sin \theta$ and there will be a "reflection maximum" if the reflected beams are exactly in phase. That is to say, there will be a maximum if the glancing angle of incidence θ satisfies the *Bragg equation*

$$2d \sin \theta=p\lambda. \qquad \cdots \cdots \cdots \quad (12–33)$$

For a slightly different angle of incidence there will be a small phase difference between disturbances reflected from adjacent planes, but since there are a very large number of such planes, the resultant intensity is very small. That is, maxima can be observed only if the disturbances are incident at an angle which satisfies (12–33) exactly. Alternatively, for a given angle of incidence, maxima will be observed for certain wavelengths only. In the Laue method of observing X-ray diffraction spectra, the incident beam contains a wide range of wavelengths; in the Bragg method the angle of incidence is varied until the reflection condition is satisfied; and in the Debye-Scherrer-Hull method one employs a powder in which small crystals are orientated at random, thus giving all possible angles of incidence.

X-ray diffraction patterns are Fraunhofer patterns, and in the *optical diffracto-meter* they are simulated optically. In principle the arrangement is the same as that shown in Fig. 12–1, although in practice long focus (e.g. 150 cm.) lenses are used. One proceeds, by trial and error, to find an optical diffracting screen that gives a pattern resembling the X-ray diffraction pattern produced by the crystal whose structure is required. The atoms in the crystal are represented by small circular holes in an opaque screen, the array of holes representing a projection of the crystal on to a specific plane. When the optical pattern resembles the X-ray pattern the relative positions of the holes in the screen are similar to the relative positions of the atoms in the crystal.

12–19. Crossed Gratings—Moire Fringes

Suppose one has two identical transmission gratings, each of which consists of alternate opaque and transparent elements of equal width. Then, if these

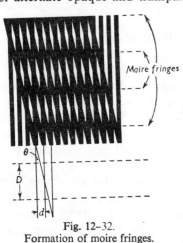

Fig. 12–32.
Formation of moire fringes.

gratings are placed face to face with their rulings inclined at a small angle, and viewed against a bright background, no light will be transmitted where the opaque parts of one grating fall on the transparent parts of the other. The appearance is of a set of dark fringes crossing the gratings as indicated by Fig. 12–32. These are known as *moire* fringes, and can be observed with various textiles by viewing through two layers. It will be seen that if D is the separation of the moire fringes and d is the grating interval of each grating, $D = d/\theta$, where θ is the angle between the rulings of the two gratings and is assumed to be small.

Suppose, now, that one of the gratings is kept stationary and the other is moved in the direction perpendicular to its rulings. Reference to the figure will show that the moire fringes move in the direction perpendicular to the fringes themselves. They move a distance D as the grating moves a distance d. That is, one moire fringe crosses a point in the field of view each time the grating advances a distance d. Consequently, the movement of the moire fringes can be used to measure the movement of the grating and the magnification, $1/\theta$, can be made very large. This method of measuring linear displacements has found an application in the realm of automation. In operating an automatic lathe, for example, one grating is attached to the work and moves across a stationary grating. Light that has passed through the gratings is received by a photo-cell and varies in intensity as the movement of the work causes a moire fringe to cross the field. The output from the cell then controls the next displacement of the work in a predetermined manner.

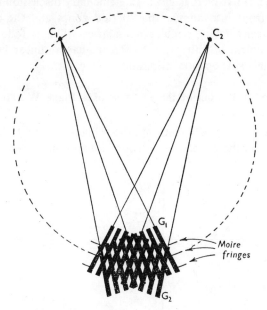

Fig. 12-33. Moire fringes with radial gratings.

By employing radial gratings, it is possible to employ moire fringes to measure rotations also. In Fig. 12–33, the rulings of the gratings G_1 and G_2 radiate from C_1 and C_2 respectively. Each moire fringe is now the arc of a circle which passes through C_1 and C_2, and if one grating is rotated, the fringe moves along the radius of the circle.

It is not essential to employ gratings of the opaque-transparent type, but not all gratings give moire fringes of good definition. It is beyond the scope of this book to investigate the conditions under which clear fringes are formed with ruled gratings.

CHAPTER XIII

FRESNEL DIFFRACTION PATTERNS

As stated in Chapter X, the Fresnel patterns most commonly investigated are those in which no lens system is involved. Such patterns are the subject of the present discussion. First there is given an elementary discussion of the division of a plane wavefront into annular half-period zones and the application to diffraction by screens having circular boundaries. This is followed by a discussion of the problem of diffraction by screens having linear boundaries: the straight edge, single slit, opaque strip, and double slit.

13-1. Elementary Discussion of the Division of a Plane Wavefront into Half-period Annular Zones

In Fig. 13-1 AB is a plane wavefront advancing to the right. In order to find the disturbance at P, the wavefront is divided into annular zones as follows: the

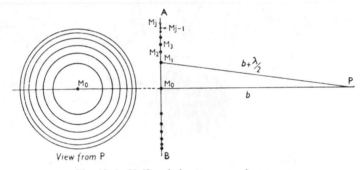

Fig. 13-1. Half-period zones on a plane wave.

central zone extends as far as M_1 where $M_1P=M_0P+\lambda/2$ and the second extends to M_2 where $M_2P=M_1P+\lambda/2=M_0P+\lambda$. Thus if $M_0P=b$ the jth zone extends from M_{j-1} to M_j where

$$M_{j-1}P=b+(j-1)\lambda/2$$

and

$$M_jP=b+j\lambda/2.$$

The area of the jth zone is obviously given by

$$\text{Area} =\pi M_jM_0^2-\pi M_{j-1}M_0^2$$
$$=\pi[(b+j\lambda/2)^2-b^2]-\pi[\{b+(j-1)\lambda/2\}^2-b^2]$$
$$=\pi[\lambda b+(\lambda^2/4)(2j-1)] \qquad \cdots \cdots \cdots \quad (13\text{-}1)$$
$$\doteqdot\pi\lambda b$$

if the λ^2 term is neglected. That is, the areas of all the zones are approximately equal. Now, according to Fresnel, the contribution that any element of the wavefront makes to the disturbance at P has an amplitude which is proportional to the area of the element and to the amplitude at the element, and is inversely proportional to the distance from the element to P. In addition, the amplitude depends upon an obliquity factor. It is assumed that the obliquity factor varies slowly and that it may be regarded as constant over a single half-period zone. Now the amplitude of the primary disturbance is constant across the wavefront, and the distance from P varies only slightly across a single zone. Therefore, in a first approximation, a half-period zone can be taken as an element which gives, at P, an amplitude proportional to its area. Furthermore, the distances of successive zones from P are not widely different and, to a first approximation, the zones have equal areas. Consequently, it may be assumed that, apart from the obliquity factor, each zone gives a contribution at P of the same amplitude. Equation (13–1) shows that the areas of the zones actually increase slightly. In addition, the successive zones do get slightly farther away from P. Further analysis shows that these two factors exactly balance one another so that, apart from the obliquity factor, each zone does give a contribution at P of the same amplitude (see § 10–3). Now since a given zone is ($\lambda/2$) farther from P than the

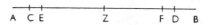

Fig. 13–2. Contributions of half-period zones.

previous one, it gives a disturbance of opposite phase. The effect of the obliquity factor is simply to cause the amplitudes of successive zones to decrease gradually. Thus if a_1, a_2, a_3 . . . etc., are the amplitudes at P due to the half-period zones, $|a_1|>|a_2|>|a_3|$ . . ., and the resultant is given by

$$R=|a_1|-|a_2|+|a_3|-|a_4|+. . . .$$

If one constructs zones over the complete wave and assumes that $a_n \to 0$, this series sums to (see § 10–3)

$$R=\frac{a_1}{2}. \qquad . \quad . \quad . \quad . \quad . \quad . \quad (13–2)$$

The superposition of the contributions of the zones can be represented graphically in the usual way. The contribution of the central zone is represented by AB (Fig. 13–2). The contribution of the next zone has opposite phase and slightly smaller amplitude; it can be represented by BC. The next zone gives CD, and so on. The sequence of points ABCDE . . . converges on Z, the mid-point of AB, and the resultant $=AZ=\frac{1}{2}AB$.

13–2. Circular Obstacle and Circular Aperture

Suppose a spherical wave from a point source S (Fig. 13–3) is diffracted by a circular obstacle centred on the line SP joining S to the field point P. When S is at a large distance from the obstacle one has diffraction of a plane wave. In each case one may divide the exposed part of the wavefront into half-period zones commencing at the edge of the obstacle. The contributions of successive zones to the disturbance at P will again alternate in sign and decrease in amplitude, so that the effect at P is again half that of the first zone and is only slightly less than that found in the absence of the obstacle. As one moves away from P the intensity decreases rapidly. For larger obstacles the first zones become increasingly narrow and any irregularities in the surface of the obstacle will cause a serious reduction in the effect produced by the zones they partially obstruct. In consequence, the bright spot at P can be observed only if the obstacle is smooth and exactly circular.

For an accurately circular obstacle, point sources near S will give bright spots near P, so that if a brightly illuminated object is placed at S, an image is formed at P. The effect at P for a point source is shown in Plate III(*a*). The obstacle was a small steel ball.

If the obstacle subtends an appreciable angle at P the maximum at P is very narrow. For a source of finite area (even as small as most laboratory "point" sources), neighbouring points of the source give neighbouring maxima at P so

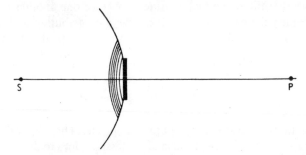

Fig. 13–3. Circular obstacle.

that, although the area of the maximum is increased, its intensity is not. Outside the area of the geometrical shadow the intensity increases as the area of the source increases and, consequently, the intensity of the central maximum within the shadow becomes a smaller fraction of that outside the shadow; it is therefore more difficult to observe.

It has been pointed out in § 10–5 that if one places on SP a circular aperture which obstructs all but the central zone, the whole of that zone is effective; the amplitude is double that found when the whole wave is unobstructed [see (13–2)] and the intensity at P is increased fourfold. If the aperture admits the first two zones, the intensity at P is very small since these zones give disturbances of opposite phase and almost equal amplitude. Inclusion of the third zone will almost cancel the effect of the second zone and the intensity at P is again approximately that of the central zone, and so on. Now the dimensions of the zones depend on the distance of P beyond the screen. Suppose P moves from a great distance towards the circular aperture. The intensity at P gradually increases until the central zone is sufficiently small to be admitted in its entirety. As P moves nearer to the aperture, the zones become smaller and the intensity at P falls to a minimum when the first two zones are admitted. It rises to a maximum again when three zones are unobstructed and, as P continues to approach the aperture, the intensity varies between a minimum and four times that due to the unobstructed wave. To find the intensity at intermediate positions along the axis SP, it is necessary to calculate the effects of the zone partially admitted. It, may be shown that the off-axis pattern consists of a series of concentric rings [see Plate III(*b*) and (*c*)].

13–3. The Zone Plate

As shown in §13–1, the jth zone on a plane wave approaching a point P extends from M_{j-1} to M_j (Fig. 13–1), where

$$M_0M_j = \sqrt{(b+j\lambda/2)^2-b^2}$$
$$= \sqrt{jb\lambda+j^2\lambda^2/4}.$$

If j is not large, the second term may be neglected, and it will be seen that for a particular position of P in front of the wavefront, the radii of successive zones are proportional to $\sqrt{j}$.

If a number of concentric circles are drawn whose radii increase in this way, and alternate spaces are made black and white, it is possible, by photographing the drawing, to construct a diffracting screen which obstructs alternate Fresnel zones for particular values of b and λ. Such a screen is known as a *zone plate* (Fig. 13–4). The intensity at P will be very large if such a screen is interposed between S and P at the appropriate distance b, since all the transmitted zones give contributions at P which are of the same sign. (A still greater intensity results at P from a so-called *phase reversal* zone plate in which alternate zones are phase-retarded by π instead of being obstructed.) Point sources near S give additional very bright spots near P so that the zone plate acts as a lens in concentrating light to form an image.

Fig. 13–4. Zone plate.

When P is distance $\frac{1}{3}b$ from the zone plate, each transparent part of the plate transmits three zones, of which two give approximately equal and opposite contributions at P. It will be seen that the effect at P is the sum of the contributions of every sixth zone. These are of the same sign so that there is a maximum at P. Additional maxima occur at distances $b/5$, $b/7$, etc., from the zone plate, while at distances $b/2$, $b/4$, the disturbance is very small since each transparent area of the plate transmits an even number of zones.

Thus a zone plate has a primary focal length b, and a number of subsidiary focal lengths at which weaker maxima occur.

In the case of a lens, disturbances through every point of the aperture arrive at the focus in phase, but in the case of a zone plate, each point of a single zone is not equidistant from P. Thus, although the zones give appreciable disturbances at P, the zone plate is considerably less efficient than a lens in forming images.

In the discussion given above a zone plate was thought of as a means of transmitting selected half-period zones of a plane wavefront and thereby producing foci from such a wavefront; its behaviour in this respect was likened to that of a lens. In fact the similarity between the properties of a zone plate and those of a lens are even closer: a zone plate, like a lens, gives "image points" for point sources at finite distances and, furthermore, the relationship between the object and image distances is the same as that for a lens. Thus in Fig. 13–5 B is a point source and a zone plate, constructed as indicated above, is such that for the axial point B' $(r_o + r')$ exceeds $(l + l')$ by $\lambda/2$, r_o and r' being the distances from B and B' to the edge of the central zone of the plate, and l and l' the axial distances from B and B' to the plate.

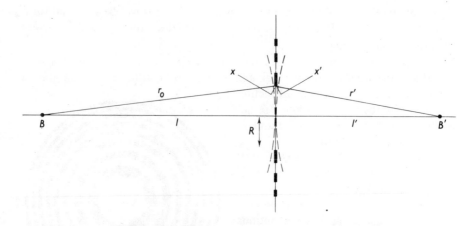

Fig. 13–5. Image formation by a zone plate.

It will be seen that disturbances from B passing through the transparent zones of the zone plate will be in phase and give a high intensity at B', so that B' can be regarded as the image of B. Let R be the radius of the central zone of the plate. The dotted curves are circles centred on B and B' so that one has

$$x' + x = \lambda/2.$$

Since x' and x are small, one has, to a good approximation (see eq. 4–2),

$$\frac{R^2}{2l'} + \frac{R^2}{2l} = \frac{\lambda}{2}$$

or

$$\frac{1}{l'} + \frac{1}{l} = \frac{\lambda}{R^2}.$$

The right hand side is a constant for monochromatic light so that, if one adopts the sign convention that was used in geometrical optics, one has

$$\frac{1}{l'} - \frac{1}{l} = \frac{1}{f'} = -\frac{1}{f} = \frac{\lambda}{R^2} \qquad \cdot \quad \cdot \quad \cdot \quad \cdot \quad (13\text{–}3)$$

where $l = f$ when $l' = \infty$ and $l' = f'$ when $l = \infty$. This equation has the same form as the equation relating the object and image distances for a thin lens in air, i.e. the zone plate behaves as a lens with focal length f' given by

$$f' = \frac{R^2}{\lambda}.$$

It will be seen from the discussion given earlier that the zone plate will also give subsidiary image points corresponding to the subsidiary focal lengths of the plate—see Miscellaneous Examples X–XIII, 13 and 14.

13–4. Rectangular Aperture. Cornu's Spiral and Fresnel's Integrals

It is very often convenient, when investigating diffraction by a rectangular aperture, to employ a slit source parallel to one of the linear boundaries of the aperture. However, it is necessary, in a proper analysis, to commence with a discussion of the pattern formed with a point source. (See also Chapter XI.)

Fig. 13–6 illustrates the diffraction of a spherical wave from the point source S by a plane screen. To find the disturbance at P due to the secondary sources

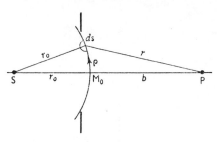

Fig. 13–6.
Lunar division of a spherical wave.

in the element ds of the wavefront, it is necessary to calculate the distance r from ds to P. In the present theory it is assumed that the apertures are not far removed from the axis SP, so that the obliquity factor is constant over the unobstructed parts of the wavefront and the variation in r is not large. As explained in § 10–10, this means that the variations in r do not affect the amplitude factor, so that equal elements of area give disturbances at P which have equal amplitudes. However, the phase of the disturbance due to each element ds depends on r, and this distance must be found.

From the figure $r^2 = (r_o + b)^2 + r_o^2 - 2r_o(r_o + b) \cos(\varrho/r_o)$, where r_o, r, b, and ϱ, are as in the figure, ϱ being the length of the arc from M_0 to ds.

If ϱ is not large, powers of (ϱ/r_o) higher than the second may be neglected so that one may write

$$\cos\left(\frac{\varrho}{r_o}\right) = 1 - \frac{\varrho^2}{2r_o^2} .$$

Hence

$$r^2 = b^2 + \frac{r_o + b}{r_o} \varrho^2,$$

$$\therefore \quad r = b\left(1 + \frac{r_o + b}{r_o b^2} \varrho^2\right)^{\frac{1}{2}}$$

$$= b + \frac{r_o + b}{2b r_o} \varrho^2, \quad \cdot \quad \cdot \quad \cdot \quad \cdot \quad \cdot \quad \cdot \quad (13\text{–}4)$$

if powers of ϱ higher than the second are neglected.

It is now necessary to sum the effects of the unobstructed elements of the wavefront. When discussing the diffraction of spherical waves by circular apertures and obstacles, it is convenient to divide the wavefront into annular zones because an elementary zone is either passed or obstructed in its entirety.

If, however, the diffracting aperture has linear boundaries, it will pass fractions of elementary zones and the effects of these fractions are not easily computed.

Fig. 13-7(*a*) illustrates the diffraction of a spherical wave by a slit or rectangular aperture whose edges are parallel to the axes of *x* and *y*. It is now convenient to divide the wavefront into lunes by means of great circles whose planes are perpendicular to the *xz* plane. Viewed from the direction of P, the wavefront appears as in Fig. 13-7(*b*). There is now no merit in dividing the wavefront into elements analogous to *half-period* zones. Suppose, then, that the wavefront is divided into a large number of *equal* lunes. In the present theory it is assumed that only the central part of the wave is unobstructed so that one is concerned

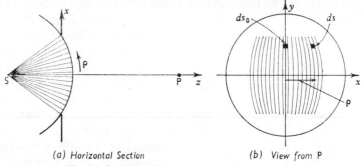

(*a*) Horizontal Section   (*b*)   View from P

Fig. 13-7. Diffraction at a rectangular aperture.

with only relatively short sections of the lunes lying close to the axis SP. Suppose the arc length ϱ is measured along the wave in the *xz* plane. Consider equal lunes at the origin and at ϱ. The elements ds_0, ds [Fig. 13-7(*b*)] of these lunes, lying between planes y, $y+\delta y$, are equal and produce equal amplitudes at P. Equation (13-4) shows that the element ds in the lune at ϱ is $[(r_0+b)/2r_0b]\varrho^2$ farther from P than is the element ds_0 in the central lune. (The former is $[(r_0+b)/2r_0b](\varrho^2+y^2)$ and the latter $[(r_0+b)/2r_0b]y^2$ farther from P than is the pole of the wave.) It will be seen that each element in the lune at ϱ is $[(r_0+b)/2r_0b]\varrho^2$ farther from P than is the corresponding element in the central lune, so that if one takes the *total* contribution of the central lune to have zero phase at P, the phase of the total contribution of the lune at ϱ is $(2\pi/\lambda)[(r_0+b)/2r_0b]\varrho^2$, the two lunes giving equal amplitudes. For any lune, each element of area is proportional to the width of the lune so that the amplitude at P due to that lune is also proportional to the width. Consequently, if one divides the wavefront into lunes of equal width $d\varrho$, one may say that the lune at ϱ produces a disturbance at P whose amplitude is proportional to $d\varrho$ and whose phase is

$$\frac{\pi(r_0+b)}{\lambda r_0 b}\varrho^2.$$

If one plots a vibration curve to represent the effects at P of these elementary lunes, the element of arc representing the amplitude of the lune at ϱ is of length ds (not to be confused with the area ds referred to above); it is proportional to $d\varrho$, and is inclined to the zero-phase line at angle $\psi=[\pi(r_0+b)/(\lambda r_0 b)]\varrho^2$. Suppose the vibration curve is plotted in the XY plane with OX as the zero-phase line (Fig. 13-8) on such a scale that $ds=[\sqrt{2(r_0+b)}/(\lambda r_0 b)]d\varrho$. If the vibration curve commences at the origin with the contribution of the central elementary lune, the element ds at distance $s=[\sqrt{2(r_o+b)/(\lambda r_o b)}]\varrho$ along the arc represents the contribution of the lune at ϱ and is inclined to the X-axis at angle $\psi=(\pi/2)s^2$. From the usual relations of differential geometry, i.e. $dX/ds=\cos\psi$, $dY/ds=\sin\psi$, one has, since $X=Y=0$ when $s=0$,

$$X=\int_0^s \cos\frac{\pi s^2}{2}\,ds\,, \qquad \ldots\ldots \quad (13\text{--}5(a))$$

$$Y=\int_0^s \sin\frac{\pi s^2}{2}\,ds. \qquad \ldots\ldots \quad (13\text{--}5(b))$$

These integrals are known as *Fresnel's integrals* and define a curve known as *Cornu's spiral*. The integrals have been evaluated in series form and have been tabulated for various values of the upper limit. The spiral is shown in Fig. 13–9; it has two branches and has asymptotic points J_1, J_2, at $X=Y=\pm\frac{1}{2}$. The values of s are indicated along the curve. For any point P, the upper and lower branches of the spiral are the vibration curves for the disturbances at P due to the lunes lying above and below the axis SP [Fig. 13–7(a)]. It was assumed initially that

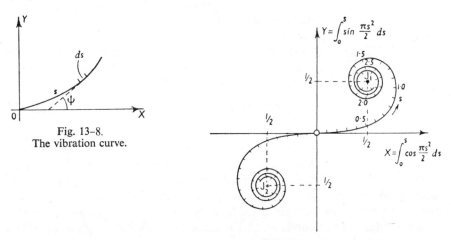

Fig. 13–8.
The vibration curve.

Fig. 13–9.   Cornu's spiral.

only that part of the wave near the axis SP is unobstructed, so that Cornu's spiral is not accurately the vibration curve in the region of the asymptotic points. Within these limitations, the part of the spiral which lies between $s=s_1$ and $s=s_2$ is the vibration curve for lunes which lie between $\varrho=\varrho_1$ and $\varrho=\varrho_2$, where the relation between s and ϱ is as given above.

Consider the diffraction of a spherical wave from the point source S by a rectangular aperture in an opaque screen lying in the xy plane, the edges of the aperture being parallel to the axes of x and y. One is usually interested in the variation of intensity along the line AB parallel to the x-axis (see Fig. 13–10). The present theory can be applied only if P is not far removed from the z-axis.

In this case, to the present accuracy, the vibration curve for the complete wave is the same complete Cornu spiral for all positions of P and a certain section of the spiral will be the vibration curve for the unobstructed lunes. Now one measures ϱ from the axis SP, so that for any position of P, the unobstructed lunes are those between $\varrho=\varrho_1$ and $\varrho=\varrho_2$, and the vibration curve is the corresponding arc of the Cornu spiral. As P moves along AB the values of ϱ_1 and ϱ_2 change, and different parts of the spiral become the vibration curve for the disturbance at P. Now the aperture always admits the same lunes although they are, effectively, different lunes for different positions of P. Hence the effective arc of the spiral will have the same length for all positions of P. Alternatively, one may simply say that $(\varrho_2-\varrho_1)$ is the same for all positions of P so that (s_2-s_1) is also constant. If one imagines a string of fixed length sliding along the spiral it will become the vibration curve for successive positions of P. This will be discussed in more detail later.

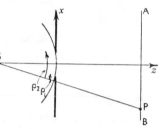

Fig. 13–10. Determination of effective part of spiral

If, for a particular position of P, the values of ϱ at the edges of the aperture are ϱ_1 and ϱ_2, the corresponding values of s define the appropriate section of the spiral and the square of the distance between the ends of the arc gives the intensity at P. An approximate value for the intensity at P can be obtained by direct measurement if an accurate plot of the spiral is available. Alternatively, the tabulated values of Fresnel's integrals can be employed. If X_1, Y_1, and X_2, Y_2, are the values of the Fresnel integrals corresponding to s_1 and s_2 [from equation (13–5)], the intensity at P is given by

$$I=(X_2-X_1)^2+(Y_2-Y_1)^2. \quad . \quad . \quad . \quad . \quad (13–6)$$

The advantage of dividing the wavefront into lunes is that, in the case of diffraction by rectangular apertures, an elementary lune is either obstructed or passed in its entirety. Whereas in the case of annular division of the wavefront it was convenient to employ half-period zones, there is no particular advantage in using half-period lunes. If, however, this method of lunar division is employed, Cornu's spiral emerges as the limiting case of a vibration polygon which is equiangular but not equilateral. The advantage of employing the variable s is that the spiral is plotted on the same scale for all problems.

13–5. Derivation of Fresnel's Integrals from Kirchhoff's Formula

The above results follow at once from Kirchhoff's diffraction formula. As has already been pointed out, it may be assumed that the obliquity factor and the amplitude factor are constant so that, from equation (10–18),

$$\phi_p=(\text{constant})\int\int_{\Sigma} e^{ik[ct-(r_o+r)]+i\frac{\pi}{2}}\,ds.$$

Equation (13–4) gives the value of r for any element ds distance ϱ from the axis PS. If SP is always taken as the z-axis (Fig. 13–11), one may write $\varrho^2=x^2+y^2$

and, since one is concerned with relative phases and relative intensities only,

$$\psi_p = \int\!\!\int_\Sigma e^{-ik\frac{r_0+b}{2r_0b}(x^2+y^2)}\,dx\,dy.$$

This integral may be factorized into its x- and y-dependent parts. Now the integral with respect to y is the same for all points P along a line parallel to the x-axis. Hence, if one is interested only in the variations in intensity along such a line, the integral with respect to y is constant and one may write

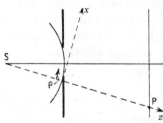

Fig. 13–11. Use of rotating axes.

$$\psi_p = \int_{x_1}^{x_2} e^{-ik\frac{r_0+b}{2r_0b}x^2}\,dx,$$

where x_1, x_2, are the values of x at the edges of the rectangular aperture. Or putting $s=[\sqrt{2(r_0+b)/(\lambda r_0 b)}]x$ and again omitting the constant term,

$$\psi_p = \int_{s_1}^{s_2} e^{-i\frac{\pi}{2}s^2}\,ds = X+iY \text{ (say)}.$$

The intensity at P is $\psi_p\psi_p^* = X^2 + Y^2$, where

$$X = \int_{s_1}^{s_2} \cos\frac{\pi}{2}s^2\,ds,$$

$$Y = \int_{s_1}^{s_2} \sin\frac{\pi}{2}s^2\,ds.$$

Since $\int_{s_1}^{s_2} = \int_0^{s_2} - \int_0^{s_1}$, X and Y may be written in terms of the Fresnel integrals as defined above.

Ignoring the inaccuracy of the spiral in the region of the asymptotic points, the unobstructed wave gives an amplitude represented by the distance $J_1J_2=\sqrt{2}$ since these points are at $X=Y=\pm\frac{1}{2}$. Thus if the intensity in the absence of diffracting screens is taken as unity, intensities calculated from Cornu's spiral must be halved.

13–6. The Use of a Slit Source

The above calculations give the variation in intensity along a line parallel to the x-axis. It will be seen that as far as relative intensities along a particular line are concerned, the result is independent of the y co-ordinate of the line for a given value of b provided all points of the diffracting aperture are fairly close to the axis SP. Thus the pattern consists of linear fringes parallel to the y-axis although there may be variations of intensity along the fringes themselves. If one employs a slit source through S parallel to the y-axis, an infinite number of these patterns are formed alongside one another in the y-direction and the intensity variations parallel to the y-axis disappear. Thus the above theory gives the relative intensity across the fringes in the case of a slit source parallel to one pair of edges of a rectangular diffracting aperture. Such an arrangement is common in practice since it greatly increases the amount of light in the pattern. It is obviously essential, in order to obtain clear fringes in such cases, for the slit to be accurately parallel to the edges of the aperture. If this is not so, different elements of the source slit will give fringes which are displaced laterally.

13–7. Diffraction at a Straight Edge

Suppose an opaque plane screen covers the region x negative of the xy plane [Fig. 13–12(a)]. S may be a point source or a slit source parallel to the y-axis.

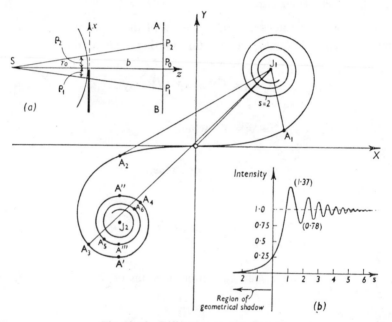

Fig. 13–12. Diffraction at a straight edge.

For any point P on the screen AB, one branch of the Cornu spiral corresponds to elements of the wave above SP, and the other to those below SP. (The inaccuracy of the spiral in the region of the asymptotic points will be neglected.)

The disturbance at P_0, on the z-axis, is due to exactly half the wave so that the amplitude at P_0 is half, or the intensity one quarter, of that due to the unobstructed wave (taken as unity). One complete branch of the spiral is effective and the intensity is represented by $(OJ_1)^2 = \frac{1}{4}(J_1J_2)^2 = \frac{1}{4}$ if the scale is adjusted to make $(J_1J_2)^2 = 1$.

When P moves below the z-axis to P_1, the part of the wave below SP_1 is completely obstructed so that the lower branch of the spiral is again ineffective. In addition, the elements of wavefront immediately above SP_1 are obstructed so that the portion near O of the upper branch of the spiral is ineffective also. The effective part of the spiral commences at A_1, a distance s_1 along the spiral from O where s_1 is determined by ϱ_1: $s_1=[\sqrt{2(r_o+b)/(\lambda r_o b)}]\varrho_1$. Thus the effective part of the spiral is that between A_1 and J_1, and the intensity at P_1 is proportional to $(A_1J_1)^2$ [i.e. $\frac{1}{2}(A_1J_1)^2$]. As P_1 moves away from P_0, ϱ_1 (and therefore s_1) increases, and A_1 moves round the spiral so that $(A_1J_1)^2$ gradually decreases. That is to say, the intensity at P_1 gradually decreases as P_1 recedes from P_0. This is shown in Fig. 13–12(b). If P is above the z-axis at a point such as P_2, the whole of the upper part of the wave (that above SP_2) is unobstructed so that the whole of the upper branch of the spiral is effective. In addition, part of the wave below SP_2 is unobstructed. The corresponding part of the spiral is that between O and A_2 where A_2 is distance s_2 along the spiral, s_2 being determined, in the usual way, by ϱ_2. The total effective part of the spiral is, therefore, that between A_2 and J_1 so that the intensity at P_2 is $\frac{1}{2}(A_2J_1)^2$. As P_2 recedes from P_0, ϱ_2 (and therefore s_2) increases and A_2 moves round the lower branch of the spiral. It will be seen that the intensity at P_2 increases and reaches a maximum when A_2 reaches A_3. The intensity is then $\frac{1}{2}(A_3J_1)^2$ and is greater than that due to the unobstructed wave; it is approximately 1·37. As P_2 moves farther above the z-axis, A moves farther round the spiral. The intensity passes through a minimum (approximately 0·78) when P_2 reaches the point for which A reaches A_4. It passes through successive maxima and minima $(A_5J_1)^2$, $(A_6J_1)^2$, etc., as P_2 recedes from P_0. As P moves farther above the z-axis the terminal point of the effective part of the spiral approaches J_2 and the intensity approaches that due to the unobstructed wave, i.e. $\frac{1}{2}(J_1J_2)^2$. Fig. 13–12(b) shows how the intensity passes through a series of maxima and minima and approaches unity, and Plate III(g) shows a photograph of the fringes using a slit source. In Fig. 13–12(b) the intensity is plotted against s, the distance moved round the spiral by the point A. Now if $r_o=b=100$ cm. and $\lambda=5000$ angstroms, $\varrho=[\sqrt{(\lambda r_o b)/2(r_o+b)}]s=0\cdot0354s$. For the point P_0 in the pattern, A is at O, and as P moves away from the z-axis a distance x, A moves round the spiral a distance s. From the figure one has $x=\varrho(r_o+b)/r_o$ or $x=0\cdot0708s$. This relates the abscissa in Fig. 13–12(b) to distances moved from the edge of the geometrical shadow. In particular, when A_1 moves round the upper branch of the spiral to the point $s=2$, the intensity falls to about 0·025. P has then moved a distance $P_0P_1=0\cdot1416$ cm. That is to say, the intensity falls to about 0·025 when one moves approximately 1·4 mm. inside the geometrical shadow.

A common procedure in which a cylindrical wave is divided into half-period strips (that are erroniously assumed to have approximately equal areas) gives maxima when the edge of the wave gives a disturbance at P_2 which is π out of phase with that from the pole of the wave. This corresponds to the occurrence of maxima when A is at A', A''' ... and minima when A is at A'' ... etc.

It was pointed out in § 10–13 that a diffraction pattern can be looked upon as the result of the superposition of (1) light propagated according to the laws of geometrical optics, and (2) light which can be considered to have come from the edges of the diffracting screen. When one is not very close to the diffracting edge, the pattern formed with a slit source and a single straight edge can be regarded as the result of the superposition of part of the original cylindrical wave envelope and a cylindrical wave envelope spreading out from the edge (Fig. 13–13). It must not be supposed that an exact cylindrical wave envelope actually spreads out from the edge—that would give infinite amplitude at the line source along the axis of the cylinder, i.e. at the diffracting edge. However, when one is far removed from the edge, the *edge wave* behaves like a cylindrical envelope. The above discussion has shown that the intensity decreases as one passes into the geometrical shadow. In this region only the edge wave is effective and its

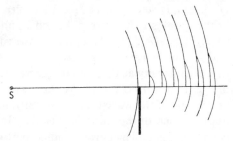

Fig. 13–13. The edge wave.

intensity must therefore decrease with increasing obliquity. In addition, the occurrence of the maxima and minima in the geometrically illuminated region is caused by the interference between the original wave and the edge wave. The amplitude of the latter must decrease as one passes into the geometrically bright region in order to explain the decreasing amplitude of the variations in intensity [Fig. 13–12(b)]. The existence of an edge wave of general cylindrical form can be demonstrated by viewing the diffracting edge with a low-power microscope placed just inside the region of geometrical shadow. A line of light will be observed along the edge of the screen and this is the effective source of the edge wave.

13–8. Single Slit

Fig. 13–14(a) represents diffraction by a slit or rectangular aperture whose length is considerably greater than its breadth. S is a point source or, more

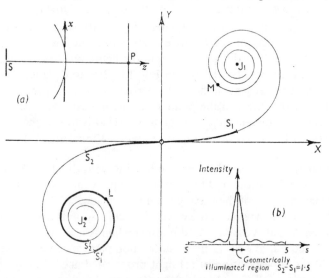

Fig. 13–14. Single slit.

usually, a slit source parallel to the diffracting slit. For any point P the effective section of Cornu's spiral is that which lies between $s=s_1$ and $s=s_2$, and is determined by ϱ_1 and ϱ_2, the values of ϱ at the edges of the slit. As remarked earlier, $(\varrho_2-\varrho_1)$ is constant for all positions of P so that (s_2-s_1) is constant, and the effective arc of the spiral is always of the same length. When P is on the z-axis, S_1 and S_2 are equal distances from O along the spiral. As P moves above or below the z-axis, S_1 and S_2 move along the lower or upper half of the spiral respectively, the arc length S_1S_2 remaining constant. For each position of P, $(S_1S_2)^2$ gives the intensity. In the figure the arc length S_1S_2 is small and refers to a narrow slit. The greatest intensity obviously occurs when P is on the z-axis and the complete pattern is of the form shown in (b). The minima are not exactly zero but occur when S_1 and S_2 are opposite one another on successive rotations of the spiral, e.g. $S_1'S_2'$. The pattern for a narrow slit is shown in Plate III(d).

For certain slit widths, a minimum may occur on the z-axis. For example, if in that position S_1 and S_2 are at L and M, the intensity is at a minimum. This follows at once since, if one moves away from the z-axis so that M moves round towards J_1, L will move round towards O and LM will increase. The complete intensity pattern would then be of the form shown in Fig. 13–15. The relation between s and distance moved across the receiving screen is as given in § 13–7. That is, if $r_0=b=100$ cm., $x=0\cdot0708s$. It will be seen that a minimum occurs in the middle of the geometrically illuminated region. Here the edge waves from the edges of the slit are out of phase with the "direct" light. A pattern of this type is shown in Plate III(e). For wider slits a number of maxima and minima occur in the geometrically illuminated region, and when the slit is very wide the pattern resembles two straight-edge patterns which meet in the middle of the bright region [see Plate III(f)].

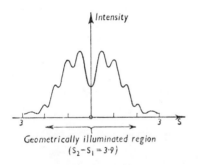

Fig. 13–15. Single slit pattern.

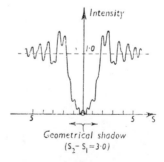

Fig. 13–16. Opaque strip pattern.

13–9. Opaque Strip

Suppose the screen and aperture of the preceding example are interchanged. Then the whole of the Cornu spiral is effective at any point P except the arc S_1S_2 (Fig. 13–14) where, as P moves away from the z-axis, S_1 and S_2 slide along the arc exactly as before. The resultant amplitude at any point is then the vector sum of J_1S_1 and S_2J_2. Typical intensity patterns are shown in Fig. 13–16 and Plate III(h). One fairly easily observed feature is the set of fringes within the geometrical shadow. These approximately equally spaced fringes can be regarded as being caused by the interference of the two edge waves. In effect, these spread into the geometrical shadow from the two edges of the opaque strip to form fringes as in Young's double slit interference experiment.

It is worth noting that the patterns for the single slit and the opaque strip are not the same—the conditions for the first form of Babinet's theorem do not obtain.

13–10. Double Slit

The effective portion of the Cornu spiral now consists of an arc corresponding to each slit, separated by a fixed arc length corresponding to the opaque screen between the slits. For any point P, if S_1S_2 and T_1T_2 represent the effective arcs of the spiral, the amplitude is the vector sum of the chords S_1S_2 and T_1T_2. As P moves across the receiving screen, S_1S_2 and T_1T_2 slide along the spiral, the arc lengths S_1S_2, S_2T_1, and T_1T_2, remaining constant. A double slit pattern for very narrow slits is shown in Plate III(i).

CHAPTER XIV

THE PHYSICAL THEORY OF IMAGE FORMATION IN OPTICAL INSTRUMENTS

14–1. Imagery by Perfect Optical Systems

IT has been remarked (p. 17) that in terms of geometrical optics a perfect optical system gives a point image of a point object. In terms of rays, the emergent rays in the image space pass exactly through the image point; in terms of waves, the emergent wave is exactly spherical. Thus the optical paths from object point to image point are independent of the zone of the aperture traversed and all the disturbances arrive in phase at the image point.

In terms of physical optics, it has been shown that one does not have an infinite concentration of energy at the image point, and that in the plane perpendicular to the axis the image formed by a perfect optical system is a Fraunhofer diffraction pattern. In particular, for an optical system having a circular aperture, the pattern corresponding to the image of a point source was found (§ 11–8). This pattern is centred on the image point indicated by geometrical optics and, if the angular aperture in the image space is not large, was seen to consist of a central maximum (the so-called Airy disc) surrounded by an infinite series of concentric bright rings. Although one must obviously not depend upon the rays to give the intensity distribution in the image, they are of interest because they are the wave normals and the condition for the formation of the Airy pattern is the condition for perfect geometrical imagery—concurrency of the rays. Moreover, away from a focus it is legitimate to think of the energy as travelling along the ray paths.

14–2. Optical Paths Along Neighbouring Rays

As stated in Chapter I, one may regard Fermat's theorem as the fundamental law of ray optics. It was seen that this theorem states that the ray path along which a disturbance travels is such that the optical path has a stationary value. Therefore, if one takes a fictitious ray path which everywhere lies close to an actual ray path, then the optical paths along the two paths are equal. In particular, if two actual ray paths lie close together everywhere, the optical paths along the neighbouring rays are equal. (The term "neighbouring rays" will be used to describe two rays which are everywhere close together so that at no point are they inclined to one another at a large angle.) It was shown in Chapter I that Snell's law of refraction follows from Fermat's principle, refractive indices being defined in terms of the speed of light. Adopting the alternative procedure of defining the refractive indices by means of Snell's law (a common practice), one can quickly prove the equality of optical paths along neighbouring rays in the case of refraction through a single surface.

In Fig. 14–1, $A_1B_1C_1$ and $A_2B_2C_2$ are neighbouring rays so that A_1A_2 and C_1C_2 may be regarded as perpendicular to either or both rays. B_2P and B_1Q are perpendiculars as shown so that $A_2Q = A_1B_1$ and $B_2C_2 = PC_1$. Thus, using square brackets to denote optical paths, one may write

$$[A_2B_2C_2]-[A_1B_1C_1] = [QB_2]-[B_1P]$$
$$= nQB_2 - n'B_1P.$$

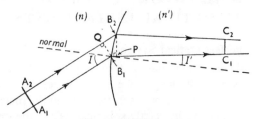

Fig. 14–1. Neighbouring rays.

Since the rays are neighbouring, B_1B_2 is small so that $\angle QB_1B_2$ and $\angle B_1B_2P$ may be equated to I and I', the angles of incidence and refraction respectively.

$$\therefore \quad QB_2 = B_1B_2 \sin I \quad \text{and} \quad B_1P = B_1B_2 \sin I',$$

$$\therefore \quad [A_2B_2C_2]-[A_1B_1C_1] = B_1B_2 \,(n \sin I - n' \sin I')$$
$$= 0.$$

Thus, along the neighbouring rays, the optical paths between A_1A_2 and C_1C_2 are equal.

14–3. The Sine Relation and the Sine Condition

The earlier discussions of optical systems commenced with the assumption that geometrically perfect imagery is possible and it was shown that, within the paraxial region, an optical system of spherical refracting surfaces satisfies the necessary conditions. That is, a system of infinitesimal aperture and field can give geometrically perfect imagery. The present section is concerned with the conditions under which small objects lying close to the optical axis, in a plane perpendicular to the axis, are imaged accurately by an optical system of *large* aperture. The discussion commences with the assumption that this is possible.

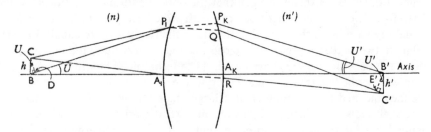

Fig. 14–2. Imagery of small object perpendicular to the axis.

In Fig. 14–2, A_1P_1 and A_KP_K represent the first and last surfaces of an optical system. BC is a small object of height h perpendicular to the optical axis BA_1 ... A_KB'. $B'C'$ is its image, of height h'.

The ray BP_1 emerges along P_KB', the ray CA_1 along RC', and the ray CP_1 along QC'.

Since B′ is the image of B, the optical path from B to B′ is independent of the zone of the aperture traversed, i.e.

$$[\text{BA}_1 \dots \text{A}_K\text{B}'] = [\text{BP}_1 \dots \text{P}_K\text{B}'],$$

or $\qquad n\text{BA}_1 + [\text{A}_1 \dots \text{A}_K] + n'\text{A}_K\text{B}' = n\text{BP}_1 + [\text{P}_1 \dots \text{P}_K] + n'\text{P}_K\text{B}',$ (14–1)

and since C′ is the image of C,

$$[\text{CA}_1 \dots \text{RC}'] = [\text{CP}_1 \dots \text{QC}'],$$

i.e. $\qquad n\text{CA}_1 + [\text{A}_1 \dots \text{R}] + n'\text{RC}' = n\text{CP}_1 + [\text{P}_1 \dots \text{Q}] + n'\text{QC}'.$ (14–2)

Now, since h is small, CA_1 and BA_1 are neighbouring rays. Therefore R, the point of emergence of ray CA_1, must be near A_K, and the optical path from A_1 to A_K for a disturbance from B must be equal to the optical path from A_1 to R for a disturbance from C.

Similarly, rays BP_1 and CP_1 are neighbouring rays and emerge from neighbouring points P_K and Q respectively, so that $[\text{P}_1 \dots \text{P}_K]$ for a disturbance from B is equal to $[\text{P}_1 \dots \text{Q}]$ for a disturbance from C.

Thus, on subtracting (14–2) from (14–1), the terms in the square brackets cancel and one has

$$n(\text{BA}_1 - \text{CA}_1) + n'(\text{A}_K\text{B}' - \text{RC}') = n(\text{BP}_1 - \text{CP}_1) + n'(\text{P}_K\text{B}' - \text{QC}').$$ (14–3)

In addition, since h and h' are small,

$$\text{BA}_1 = \text{CA}_1 \quad \text{and} \quad \text{A}_K\text{B}' = \text{RC}'$$

and, dropping perpendiculars CD and B′E′ as in the figure,

$$\text{CP}_1 = \text{DP}_1 = \text{BP}_1 - \text{BD} \quad \text{and} \quad \text{P}_K\text{B}' = \text{QE}' = \text{QC}' - \text{E}'\text{C}'.$$

Thus, from (14–3),

$$0 = n(\text{BD}) + n'(-\text{E}'\text{C}').$$

Also, from the figure,

$$\text{BD} = h \sin U \quad \text{and} \quad \text{E}'\text{C}' = h' \sin U',$$

$$\therefore \quad nh \sin U = n'h' \sin U'. \qquad \dots \dots \quad (14\text{–}4)$$

This is known as the *Sine Relation*.

Now a geometrically perfect optical system is defined as one which gives geometrically perfect imagery for a wide aperture and a wide field for *all object positions*. It was pointed out in Chapter II that if one has perfect imagery for *two* pairs of conjugates, the system is perfect; and it was shown that for such a system the Helmholtz equation will be satisfied, i.e.

$$nh \tan U = n'h' \tan U'. \qquad \dots \dots \quad (14\text{–}5)$$

It will be seen that, except within the paraxial region where the sines and tangents may be equated, the Sine Relation and the Helmholtz equation are contradictory. As has just been proved, the Sine Relation must be satisfied in order to achieve imagery of a single small object with a wide aperture system. Therefore the Helmholtz equation cannot be satisfied and an optical system cannot be perfect in the sense defined above. That is to say, the most that can be hoped for in an optical system is geometrically perfect imagery for one pair of conjugates. In practice, it is seldom possible to achieve even this exactly; on the other hand, it can often be approached very closely. It must also be borne in mind that, although the imagery cannot be *perfect* for *two* pairs of conjugates, it might prove to be possible to make the imperfections small in each case. A further discussion of this problem is given in Chapters XV and XVI.

In deducing equation (14–4), the discussion was restricted to rays passing through a single zone of the system defined by the angles U and U'. Usually an optical system has a circular aperture centred on the axis and admits a solid cone of rays from each object point. Thus rays are admitted at all values of U up to the value for the so-called marginal ray which passes through the extreme zone of the aperture.

Now, for a given object, the size of the image formed by rays passing through any given zone of the aperture is given by

$$h' = h \cdot \frac{n \sin U}{n' \sin U'},$$

and, for a good image, this must be the same for each zone. In particular, an image of the same size must be formed by rays lying within the paraxial region. If u and u' refer to paraxial rays, the paraxial image height is

$$h' = h \cdot \frac{nu}{n'u'},$$

so that one may equate the values of h'.

$$\therefore \quad \frac{n \sin U}{n' \sin U'} = \frac{nu}{n'u'},$$

or
$$\frac{\sin U'}{u'} = \frac{\sin U}{u}. \qquad \ldots \ldots (14\text{–}6)$$

This is known as the *Sine Condition* and must be satisfied for all zones of the aperture if the image is to be of the same size for rays traversing each of these zones. The significance of the Sine Condition is discussed further in Chapter XV.

14–4. Herschel's Condition

Suppose an optical system of large aperture forms an image of an axial point object B. Herschel's condition is the condition that must be satisfied if the system is to form, simultaneously, a good image of a neighbouring axial object point C.

In Fig. 14–3 A_1P_1, A_KP_K are the first and last surfaces of an optical system which is assumed to form a good image of both B and C—disturbances from B and C arrive in phase at B' and C' respectively.

Proceeding as in the previous paragraph, one has

$$nBA_1 + [A_1 \ldots A_K] + n'A_KB' = nBP_1 + [P_1 \ldots P_K] + n'P_KB' \quad (14\text{–}7)$$

and
$$nCA_1 + [A_1 \ldots A_K] + n'A_KC' = nCP_1 + [P_1 \ldots Q] + n'QC'. \quad (14\text{–}8)$$

Again $[P_K \ldots P_K] = [P_1 \ldots Q]$ since these are optical paths along neighbouring rays. Hence, subtracting (14–8) from (14–7),

$$n(BA_1 - CA_1) + n'(A_KB' - A_KC') = n(BP_1 - CP_1) + n'(P_K'B' - QC').$$

Dropping perpendiculars CD and B'E' as in the figure, one has, as before,

$$DP_1 = CP_1, \quad \text{and} \quad QE' = P_K B'$$

$$\therefore \quad n(BC) + n'(-B'C') = n(BD) + n'(-E'C'),$$

i.e. $\qquad\qquad n(BC-BD) = n'(B'C'-E'C').$

But $\qquad\qquad E'C' = B'C' \cos U' \quad \text{and} \quad BD = BC \cos U$

$$\therefore \quad nBC(1-\cos U) = n'B'C'(1-\cos U'). \quad . \quad . \quad . \quad (14\text{–}9)$$

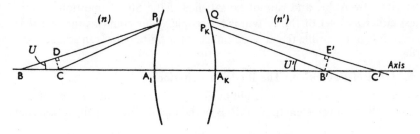

Fig. 14–3. Imagery of small axial object.

Now, since B' and C' are good images of B and C respectively, they must be the images as formed by paraxial rays in addition to rays traversing the zone defined by U, U'. If m is the paraxial transverse magnification for planes at B and B', the paraxial form of Helmholtz's equation (which has been shown to apply to a system of spherical refracting surfaces) gives [from equation (2–8)]

$$m = \frac{nu}{n'u'}, \quad . \quad . \quad . \quad . \quad . \quad . \quad (14\text{–}10)$$

where u and u' refer to paraxial rays through B and B'. Also, for imagery by paraxial rays, it has been found that, for conjugate increments dx and dx' along the axis [from equation (2–11)],

$$\frac{dx'}{dx} = m^2 \cdot \frac{n'}{n}.$$

In the present problem dx'/dx becomes B'C'/BC. Hence, from (14–10),

$$\frac{B'C'}{BC} = \frac{nu^2}{n'u'^2},$$

$$\therefore \quad nBCu^2 = n'B'C'u'^2, \quad . \quad . \quad . \quad . \quad (14\text{–}11)$$

and, dividing (14–9) by (14–11), one has

$$\frac{1-\cos U}{u^2} = \frac{1-\cos U'}{u'^2},$$

or $\qquad\qquad \dfrac{\sin^2 \dfrac{U}{2}}{u^2} = \dfrac{\sin^2 \dfrac{U'}{2}}{u'^2},$

which is usually written in the form

$$\frac{\sin \dfrac{U}{2}}{\dfrac{u}{2}} = \frac{\sin \dfrac{U'}{2}}{\dfrac{u'}{2}}, \qquad \ldots \ldots (14\text{--}12(a))$$

or
$$n \sin \frac{U}{2} = n'm \sin \frac{U'}{2}, \qquad \ldots \ldots (14\text{--}12(b))$$

where m is the transverse linear magnification. This is *Herschel's condition* in the form given by Abbe, and cannot be satisfied if the Sine Condition is satisfied unless either $m = \pm 1$ or the system is telescopic. This confirms the fact that, in general, it is not possible to get good imagery for finite objects in adjacent object planes.

14–5. Limit of Resolution. The Rayleigh Criterion

In Fig. 14–4 P_1 and P_2 are neighbouring incoherent point sources of monochromatic light of wavelength λ. P_1P_2 is perpendicular to the axis. It is assumed

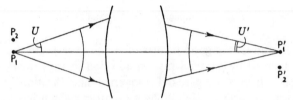

Fig. 14–4. Imagery of neighbouring points.

that disturbances from P_1 and P_2 arrive in phase at the geometrical image points P_1' and P_2' respectively, and that $P_1'P_2'$ is also perpendicular to the axis. On the image plane each point source will give rise to an Airy pattern centred on its geometrical image point. Since the sources are incoherent, the total intensity at any point on the image plane is simply the sum of the intensities associated with the separate patterns. If the distance $P_1'P_2'$ is large compared with the radius of the Airy disc, the separate patterns can be distinguished easily [Fig. 14–5(a)]. The images are said to be resolved. On the other hand, if $P_1'P_2'$ is very small the central maxima will overlap to such an extent that the total intensity shows no decrease at the mid-point and the resulting pattern will have roughly the same appearance as that for a single point source, i.e. the images are not resolved

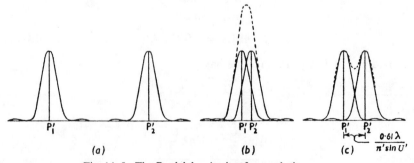

Fig. 14–5. The Rayleigh criterion for resolution.

[Fig. 14–5(*b*)]. Obviously it is important to find the smallest separation of the images for which the resultant pattern can be seen to consist of two neighbouring patterns which are superposed.

The same problem arose in §11–5 when discussing the chromatic resolving power of a prism spectroscope. In that case it was stated that, according to Rayleigh, two patterns with equal intensities may be said to be resolved when the central maximum of one pattern falls over the first minimum of the other. It is convenient to adopt the same criterion for the resolution of Airy patterns [Fig. 14–5(*c*)]. Thus the images are said to be just resolved if their centres (i.e. the geometrical images points) are separated by a distance equal to the radius of the first dark ring, i.e. $0.61\lambda/(n' \sin U')$, where λ is the wavelength of the light, U' is the semi-angle of the emergent cone of rays, and n' is the refractive index of the image space (see §11–8). As pointed out in Chapter XI, the criterion is adopted largely because of its convenience and there is no doubt that detailed analysis of the intensity distribution would make it possible to decide whether the image was that of a single or double source for considerably smaller separations of the images. In using an optical instrument one is interested in the separation of the object points when the images are just resolved. This is referred to as the *limit of resolution* of the instrument and will be seen to be inversely proportional to the aperture of the system. Thus, to resolve finer detail in an object, an instrument must have a larger aperture and/or employ light of a shorter wavelength.

14–6. Limit of Resolution of a Telescope Objective

In the case of a telescope objective, the objects are usually at large distances so that the actual linear separation of the objects is very large when their images are just resolved. One consequently refers to α, their angular separation.

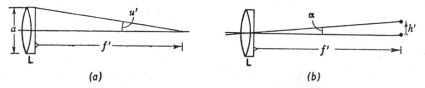

(*a*)   (*b*)

Fig. 14–6. Limit of resolution of a telescope objective.

In Fig. 14–6(*a*) and (*b*) L represents a telescope objective imaging two distant point sources whose angular separation, α, is small. (Since $n'=n=1$, the nodal points are at the principal points.) h' is the separation of the geometrical image points and therefore of the centres of their Airy patterns. Usually, u' is small so that one can say that the sources will be resolved when

$$h' = \frac{0.61\lambda}{u'}.$$

From Fig. 14–6(*b*) $\alpha = h'/f'$, so that the angular separation of the sources when the images are just resolved is

$$\alpha = \frac{0.61\lambda}{f'u'}.$$

But, from Fig. 14–6(a), $f'u'=a/2$

$$\therefore \quad \alpha = \frac{1 \cdot 22 \lambda}{a} \quad \cdots \cdots \cdots \quad (14\text{–}13)$$

Hence, for a given wavelength, the angular limit of resolution of a telescope objective depends only upon the linear diameter of the objective.

14–7. Limit of Resolution of a Microscope Objective. (Incoherent Object Points)

In the case of a microscope, one needs to know the separation of the object points when the images are just resolved. In the present theory it is assumed that one is viewing two neighbouring incoherent object points. These conditions would obtain if one had a self-luminous object but may be approached, at least for low powers, if a diffusing screen is placed behind the object and is illuminated from the rear. Again the angle u' is small, although U is usually large.

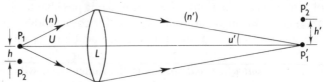

Fig. 14–7. Limit of resolution of a microscope objective.

In Fig. 14–7 P_1' and P_2' are the geometrical images of the point sources P_1 and P_2. If the objective L gives good images of both P_1 and P_2, the Sine Relation must be satisfied for all zones of the aperture. In particular, for the extreme marginal zone, one has

$$nh \sin U = n'h' \sin U'.$$

As before, the images are just resolved when $h'=(0\cdot 61\lambda)/(n' \sin U')$ so that, if h is the separation of the objects when the images are just resolved,

$$h = \frac{n'h' \sin U'}{n \sin U} = \frac{0\cdot 61\lambda}{n \sin U}.$$

The quantity $(n \sin U)$ is referred to as the *numerical aperture* of the objective and is written (N.A.), i.e.

$$\text{Limit of resolution} = \frac{0\cdot 61\lambda}{(\text{N.A.})}. \quad \cdots \cdots \quad (14\text{–}14)$$

It will be seen that the limit of resolution depends upon the refractive index of the object space, and it is largely for this reason that *immersion objectives* are employed (see §16–9). Neither the size of the image patterns nor the separation of the geometrical images at the limit of resolution are affected explicitly by the index of the object space, but the magnification is affected and the just-resolved images correspond to the closer object points.

It follows from (14–13) and (14–14) that *for a given object distance* the limit of resolution of a lens (e.g. a camera lens) depends on its *linear* aperture. (At a given distance, the angle subtended by an object is proportional to its linear size, and the (N.A) of the lens is determined by its linear aperture.)

Equation (14–14) also shows that the limit of resolution is proportional to the wavelength of the image-forming radiation. This provides motivation for the development of a U.V. microscope using the shortest possible wavelength. The electron microscope is the extreme example of the use of very short wavelengths and although the apertures of electron microscopes are relatively small the gain in resolution is enormous; limits of resolution of a few angstroms have been achieved. Considerable difficulties are encountered in designing U.V. microscopes and there is the further complication of the possibility of the specimen being damaged by the image-forming radiation (see §14–15). The design of an electron microscope is, of course, quite different from that of an optical instrument, the electron beam being focused electromagnetically. The risk of damage to the specimen is very much greater, and with the very low limit of resolution that can be achieved it is often difficult to interpret the image that is observed.

14–8. Necessary Magnifying Power of a Microscope

For the standard observer, the near point is about 25 cm. from the eye, and for many purposes one can assume that at this distance the limit of resolution is about 0·01 cm. That is to say, objects can be resolved by the eye when they subtend an angle of about one minute. Thus for two optical images to be resolved by the observer, they must have an angular separation not less than this. If an optical instrument resolves two neighbouring object points in the sense of the previous section, the observer will be quite unable to appreciate the fact unless the magnifying power of the instrument is such that the physically resolved images subtend an angle of at least one minute at the observer's eye. Thus, for any visual instrument, there is a minimum magnifying power below which the eye will be unable to resolve the objects even though the images as formed by the instrument itself may be clearly resolved. This minimum magnifying power is usually called the *necessary magnifying power*. A magnifying power excessively greater than this is of no advantage since the eye will never be able to resolve images that the instrument does not resolve. Psychologically, this additional magnification is a disadvantage because it renders visible the diffuse nature of the separate Airy discs so that one gets an impression of poor image quality. Such additional magnification, greatly in excess of the necessary value, is called *empty magnification*.

The necessary magnifying power of a microscope is easily calculated from the results given above. The smallest resolvable separation of point objects, h, is given by (14–14) and the magnifying power must be such that the images subtend an angle of $0·01/D_v$.

$$M = \frac{\text{angle which images must subtend}}{\text{angle subtended at } D_v \text{ by just resolvable objects}}$$

$$= \frac{0·01/D_v}{h/D_v}$$

$$= \frac{0·01}{h}$$

$$= \frac{0·01 \times (\text{N.A.})}{0·61\lambda}. \qquad \qquad \qquad \text{(14–15)}$$

Now it has already been seen [equation (3–7)] that the magnifying power of a compound microscope is given by

$$M = \left(\frac{g}{f_0'}\right)\left(\frac{D_v}{f_E'}\right). \qquad \cdots \cdots (14\text{–}16)$$

Using a given objective, the numerical aperture is fixed. Hence the necessary magnifying power may be calculated from (14–15). In addition, g and f_0' are given, so that, from (14–16), it is possible to calculate D_v/f_E'; this is the power of the eyepiece which must be employed with the given objective in order to give the necessary overall magnification.

It must be remembered that the value quoted above for the limit of resolution of the human eye is an average value. It is determined by a number of factors including the structure of the retina, the refractive condition of the eye, etc. Many authorities consider that for comfortable vision, especially during prolonged observation with a microscope, it is advisable to assume an angular limit of resolution of roughly $2'$ ($\frac{1}{5}$th mm. at the least distance of distinct vision).

Using equation (14–15), and assuming that the observations are carried out with sodium light ($\lambda = 0.000059$ cm.), one has, for the necessary magnifying power,

$$M = \frac{0.01 \times (\text{N.A.})}{0.61 \times 0.000059} \doteqdot 300 \,(\text{N.A.}), \qquad \cdots (14\text{–}17)$$

or 600 (N.A.) if one uses the more conservative estimate for the resolution of the eye.

The above discussion is an over-simplification since it has been assumed that the limit of resolution of the eye is constant at about one minute whereas it does, in fact, depend on the size of the eye pupil, and deteriorates considerably for pupil diameters below about 2 mm. (See §17–7.) If the exit pupil of the microscope is smaller than the actual eye pupil, the limit of resolution of the eye is determined by the area of the eye pupil that is filled with light, i.e. corresponds to an eye pupil equal to the exit pupil. If the final image is at infinity the Sine Relation becomes

$$nh \sin U = -n'y'u'_{pr},$$

where $2y'$ is the exit pupil diameter and u'_{pr} is the angle (assumed to be small) subtended by the image of h (see §2–10). Ignoring the sign and taking $n' = 1$, this becomes

$$h(\text{N.A.}) = y'u'_{pr},$$

and one has

$$M = u'_{pr} \div h/D_v = \frac{D_v.(\text{N.A.})}{y'}.$$

If $M = 300$ (N.A.), $y' \doteqdot 1$ mm., and the diameter of the exit pupil is approximately 2 mm. This suggests that there would be no advantage in increasing M since the diameter of the exit pupil would decrease and the resolution of the eye would deteriorate. One is led to the conclusion that for normal observers magnifying powers below about 300 (N.A.) are usually too small, whilst values above about 600 (N.A.) seldom reveal extra detail; within these limits the choice of magnifying power is largely a matter of personal taste. The points of substance are:

1. It is necessary to employ an objective whose numerical aperture is sufficiently large to give images which are physically resolved.

2. It is necessary to employ an eyepiece which will enable the physically resolved primary images to be resolved by the observer's eye.

In practice, it is found that large values of the numerical aperture can be obtained only with objectives of short focal length, so that the primary magnification increases rapidly as the numerical aperture increases. As a result, it is found that the necessary eyepiece power decreases for higher power objectives.

The necessary magnifying power for a telescope is discussed in the Miscellaneous Examples on Chapters XIV–XVI nos. 4 and 6.

14–9. Depth of Focus. The Rayleigh Quarter-wave Limit

If the image formed by an optical system is received on a screen placed some distance away from the actual image plane, the image received on the screen will

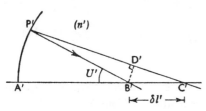

Fig. 14–8. Depth of focus.

be of poor quality. It is said to be out of focus, and the maximum movement away from the ideal image plane which may be made without causing a serious deterioration of the image is called the *depth of focus*.

In Fig. 14–8 A'P' is a spherical wave converging to the geometrical image point B'. An Airy pattern will be formed on the plane perpendicular to the axis at B', and a slightly different pattern will be formed on a screen at C'. For a screen at C', points on A'P' are no longer equidistant from the central point in the pattern so that the disturbances do not arrive in phase, and the intensity of the central maximum is therefore reduced.

According to Rayleigh, there will be no appreciable deterioration of the image, i.e. no marked change from the Airy pattern, provided the maximum phase difference between disturbances arriving at the centre of the pattern does not exceed $\pi/2$. It will be seen later that a very similar criterion exists concerning the presence of aberrations. Aberrations in a system also introduce phase differences between disturbances arriving at the Gaussian image point from various zones of the aperture, and again it is found that, provided the phase differences do not exceed $\pi/2$, the definition is not seriously impaired. Theory indicates that these tolerances correspond to a reduction in intensity at the centre of the pattern from unity to about 0·8. (In the Rayleigh criterion for resolution, a reduction of 26·4% occurs midway between the images with a circular aperture.)

If $\delta l'$ (Fig. 14–8) is the depth of focus, disturbances from A' and P' will arrive at C' with a phase difference of $\pi/2$. Since A'P' is a wavefront, the disturbances at A' and P' are in phase and the phase difference at C' is caused by the inequality of the optical paths A'C' and P'C'. The optical path difference is obviously $\lambda/4$.

It is assumed that $\delta l' \ll$ A'B' so that each of the angles at B' and C' may be written as U'.

In the figure, let B′D′ be perpendicular to P′C′. Then, to the present order of accuracy,

$$P'D'=P'B'=A'B'.$$

The optical path difference $[A'C']-[P'C']$ is given by

$$n'(A'C'-P'C')=n'(B'C'-D'C')$$

$$=n'\delta l'(1-\cos U')$$

$$=2n'\delta l' \sin^2 \frac{U'}{2}$$

$$=\frac{\lambda}{4} \text{ when } \delta l' \text{ is the depth of focus.}$$

Hence the depth of focus is given by

$$\delta l'=\pm \frac{\lambda}{8n' \sin^2 \dfrac{U'}{2}}, \qquad \cdots \cdots \quad (14\text{--}18)$$

where "$\pm$" has been inserted because a movement towards A′ introduces the same optical path difference as an equal movement away from A′.

The deterioration of the image resulting from a movement of the receiving screen through the distance given by (14–18) will be noticeable only if the image is of good quality initially and if the image is examined very critically. In many instruments the state of correction is not good and a considerably greater deterioration can be tolerated. For large movements away from the best focus it is sufficiently accurate to rely on geometrical or ray optics. Thus, in Fig. 14–9, the image of a point source is assumed to be formed in the region of B′. A movement of the receiving screen to C′ then causes the "image" to become an approximately circular patch of light and if one assumes that the image will be satisfactory for the purpose in question provided its radius does not exceed a certain value r, one obviously gets, for the depth of focus,

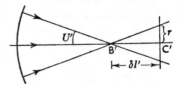

Fig. 14–9. Depth of focus.

$$\delta l'=\pm \frac{r}{\tan U'}. \qquad \cdots \cdots \quad (14\text{--}19)$$

The depth of focus is the permissible movement of the receiving screen for a stationary object plane. If the receiving screen remains stationary and the object plane is moved, the conjugate image plane moves away from the receiving screen and the image on the screen is out of focus. The permissible movement (δl) of the object plane is called the *depth of field*. It is given by an equation of the same form as (14–18) but involving undashed quantities. The depth of field is obviously the movement of the object plane which causes the image plane to move a distance equal to the depth of focus. That is, the depth of focus and the depth of field are conjugate axial distances. For small movements along the axis one has $\delta l = (\delta l'/m^2)(n/n')$ where m is the transverse magnification [see equation (2–11)]. Frequently the object distance is large, m is small, and the depth of field is large; the above relation does not hold but δl remains conjugate to $\delta l'$.

It was pointed out in § 11–15 that it is possible to modify the image formed by an optical system by the introduction of a screen across the aperture. In a similar manner it is possible to change the way in which the pattern varies as the receiving screen is moved along the axis. For example, it is possible to arrange that as the receiving screen is moved away from the ideal position the image deteriorates more rapidly than it does with an ordinary lens system. That is, the depth of focus can be made much smaller by apodization. This can be a useful method of improving the accuracy of longitudinal setting of a microscope.

14–10. Imagery by a Microscope of Non Self-luminous Objects

The foregoing theory assumes that the object is self-luminous and that neighbouring points are independent sources emitting non-coherent disturbances which cannot interfere. This is justifiable for telescopes but not for microscopes. With a low-power microscope, one may place an opal glass screen immediately behind the object (assuming it to be a transparency) and, in this case, one does get a close approximation to self-luminous objects emitting non-coherent light. Usually, however, a cone of light emerges from the substage condenser, and neighbouring points of the object are not incoherently illuminated. In the past it has been argued that if one images a diffuse source in the object plane by means of a perfect substage condenser whose numerical aperture is considerably larger than that of the objective, the Airy discs in the object plane (corresponding to points in the source) are small and below the limit of resolution of the objective. It was then suggested that, in effect, neighbouring object points are incoherently illuminated. This argument is unsound because the Airy patterns do, in fact, spread to infinity, and a portion of the light incident at one point of the object plane will always be coherent with a portion of the light falling on a neighbouring point. Because of the above interpretation of the imagery by a microscope, it became common practice to image a diffuse source in the plane of the object and, since this focusing of the source was thought to be critical, the arrangement was known as *critical illumination*. More recent analysis has shown that there is nothing particularly critical about it and that, within limits, any form of illumination will suffice provided the condenser supplies a wide cone of light. That is, the numerical aperture of the condenser is important, but the focusing and freedom from aberration are relatively unimportant. However, the arrangement for critical illumination is a convenient one to use; it is shown in Fig. 14–10. Usually the numerical aperture of the substage condenser can be varied by means of the iris diaphragm I_1. In addition, since the source is imaged in the plane of the object, a second iris I_2 immediately in front of the source can be used to control the area of the object which is illuminated. It is always advisable to illuminate only that part of the object which is to be examined since in this way one reduces the amount of stray light entering the microscope.

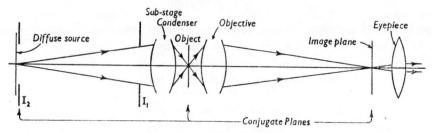

Fig. 14–10. Critical illumination.

Another convenient form of illumination is that usually termed *Köhler illumination*. The arrangement is shown in Fig. 14–11. The source S is imaged in the back focal plane of the substage condenser by means of the auxiliary lens L. Thus there emerges from the condenser an array of collimated pencils at various

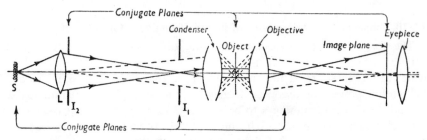

Fig. 14–11. Köhler illumination.

angles and there is no doubt that neighbouring object points are partly coherently illuminated. It is frequently arranged so that the iris I_2 is imaged in the plane of the object for the reason given above; the substage iris I_1 again controls the obliquity of the illumination.

Whatever the nature of the illumination, there is obviously a two-fold diffraction of the light:

1. Incident light is diffracted by the object.
2. Light is further diffracted by the aperture of the objective.

Consequently, every point of the object, as well as every point of the aperture of the objective, contributes to the disturbance at any point in the image plane.

14–11. Abbe Theory of the Microscope

One of the few cases that can be discussed in detail is the imagery of a regular grating of opaque and transparent elements. Initially, it is assumed that the illumination is restricted Köhler illumination where, in effect, a point source S is located in the back focal plane of the condenser (Fig. 14–12). This gives rise to a single collimated beam of light incident parallel to the axis; the light strikes the grating G and is diffracted in the usual way. If the objective has a sufficiently large aperture, the diffracted beams are collected as indicated in the figure. The back focal surface of the objective is conjugate to the source so that the diffraction maxima are in focus in this surface and constitute a Fraunhofer pattern. The light passes on to interfere and give the image in the plane conjugate to the grating.

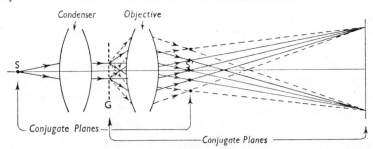

Fig. 14–12. Imagery of a grating.

Abbe was the first to realize that to obtain the best possible image the objective must have an aperture wide enough to admit all the light diffracted by the object, i.e. the complete Fraunhofer pattern must be formed behind the objective and light from the whole of this pattern must contribute to the image. This remains true with any kind of object structure. It follows that if light from only part of the Fraunhofer pattern contributes to the image the resulting image will correspond to an object whose complete diffraction pattern is identical to the utilized part of the actual object pattern. In ordinary microscopy the part of the Fraunhofer pattern most likely to be lost is the outer part, corresponding to large angles of diffraction, and it will be lost if the numerical aperture of the objective is too small. This results in a loss of sharpness in the image, i.e. to a deterioration of the resolution and a loss of the finer detail since it is the sharper boundaries and finer detail which diffract light at the larger angles.

Consider now how the image of a grating is built up. It follows from the discussion of §11–15 that an interrupted, distorted, or limited wavefront produced by any diffracting screen is equivalent to an angular distribution of uniform wavefronts. For an extended grating at G (Fig. 14–12) one has a uniform wavefront for each of the diffracted orders. It is important to realize that the interruptions imposed across the incident wavefront by the grating elements are not present on the component wavefronts into which the interrupted wavefront is resolved. Thus, in particular, the zero order beam is a uniform plane wave that proceeds to a focus S' at the second principal focus of the objective and then diverges to give uniform illumination across the image plane. There will be two first order beams diffracted at equal angles above and below the axis and producing a pair of maxima behind the objective. These constitute two coherent sources from which disturbances travel to the image plane where they interfere to give the usual two-beam sinusoidal amplitude distribution (i.e. $\cos^2$ fringes). Diffraction maxima of higher orders give effective sources of proportionally greater separation behind the objective and sinusoidal amplitude distributions with proportionally more cycles per unit distance across the image plane. The uniform amplitude produced by the zero order beam, and the sinusoidal amplitude distributions produced by the various pairs of diffracted beams, are superposed in the image plane to give the image. The synthesis of the image from these sinusoidal amplitude distributions occurs in exactly the same way as the synthesis of a square wave form from its Fourier components as described in §6–7. The student may be more familiar with the idea of analysing a complicated periodic motion into simple harmonic motions. In the present situation the Fourier components are simple harmonic variations of amplitude with distance across the image plane and not variations with time; in consequence, one speaks of *spatial frequencies*. Thus, the amplitude distribution across the image plane can be analysed into spatial frequencies in the same way as a complicated periodic motion or wave form can be analysed into simple harmonic oscillations or waves.

For example, in §6–7 the square wave has equally wide crests and troughs and, if a constant term a is added, this corresponds to the amplitude transmitted by a grating having equally wide opaque and transparent elements. In §6–7 the even Fourier components are zero and in §12–8 it is shown that for the corresponding grating the even orders are missing. It also follows from §12–8 that the relative amplitudes of the various odd orders diffracted by the grating are the same as the relative amplitudes of the odd Fourier components of the square wave form illustrated in §6–7. That is to say, the amplitude distribution transmitted by the grating object can be resolved into spatial frequencies, each of which gives a pair of maxima behind the objective and is reproduced in the image. (There will, of course, be a magnification factor between object and image.) In fact, this follows from the discussion of §11–15 where it is shown that, in general, for a Fraunhofer diffraction pattern the pattern function is the Fourier transform of the aperture function. In the present problem one has two stages: (1) the object diffracts to give a Fraunhofer pattern behind the objective, the pattern function being the transform of the aperture (i.e. object) function, and (2) the lens aperture can be said to diffract the light from the object (looked upon as the source) to give a Fraunhofer pattern in the image plane. In the second stage the pattern function describes the image and is the transform of the aperture function describing the distribution of light behind the objective. (Here it is implied that the lens pupil is in the back focal plane, but this is not a serious restriction.) The aperture function of stage (2) is the pattern function of stage (1) and it will be seen that if no light is lost the image is the transform of the transform (the synthesis of the analysis) of the object, and is identical to it. A grating object is periodic (strictly speaking, only if the grating is infinitely wide —see §6–9 and §12–7) and the analysis gives a Fourier series of discrete terms corresponding to integral orders in the diffraction pattern. A non-periodic or irregular object structure gives a Fourier transform corresponding to a diffraction pattern without discrete orders. A grating of finite width gives weak secondary maxima between principal maxima, i.e. contains the intermediate frequencies, but with small amplitudes.

It is now possible to understand the effect on the grating image of the limitations imposed by the aperture of the objective. The grating structure analyses into a discrete spectrum of spatial frequencies where each frequency corresponds to a diffracted beam of given order, and higher orders correspond to larger angles of diffraction. Diffracted beams above a certain order will fail to enter the lens, and the corresponding spatial frequencies in the object will not be reproduced in the image. The distribution of light in the image will be represented by the sum of the first p terms of the Fourier series, p being the number of diffracted orders entering the lens (see §6–7).

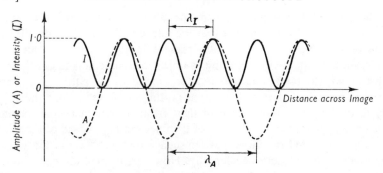

Fig. 14–13. Fourier component.

An irregular object structure analyses into a continuous distribution of spatial frequencies (referred to as a continuous Fourier spectrum) and, as before, the light diffracted at the larger angles corresponds to the higher frequencies. Again, if this light is not transmitted by the lens the corresponding frequencies will be missing from the image and this finer detail will not be observable.

One is now led to a new definition of limit of resolution: it is simply the highest spatial frequency in the object that is reproduced in the image. It is, of course, the reciprocal of this—the wavelength—that should be compared with the earlier definition in terms of resolved point objects. Attention must be drawn to a common source of confusion: in the above discussions the illumination across the object is coherent and in a theoretical treatment it is natural to discuss amplitudes in object and image. However, the eye detects intensities, and this can introduce a factor of 2 when discussing spatial frequencies in the image. This is illustrated in Fig. 14–13. The broken curve shows a simple harmonic variation of amplitude and the wavelength is λ_A; the square of the amplitude gives the intensity which is shown by the unbroken curve and the wavelength is λ_I. In what follows the limit of resolution is the smallest intensity wavelength that is imaged.

Return, now, to the simple grating object consisting of alternate opaque and transparent elements. At first sight, in view of what has been said in the earlier discussion, it might be thought that in order to be able to observe the periodicity of the grating the two first order beams must contribute to the image to reproduce the fundamental or lowest spatial frequency component in the amplitude distribution. However, one is, in fact, more interested in intensities, and it is easy to show that if the zero order and *one* of the first order beams enter the lens, giving two effective sources in the focal plane, the intensity distribution across the image plane exhibits maxima corresponding to the grating spaces, and the rulings of the grating are resolved. This need to transmit the zero order plus one of the first order beams is sometimes referred to as the *Abbe principle* and the limit of resolution is the grating interval of the finest grating that can be resolved in this way. Finer gratings give larger angles between the zero and first orders and so require lenses of larger numerical aperture in order to be resolved.

By interposing stops in the back focal plane of a microscope objective it is easy to vary the effective aperture of the lens, and Plate IV(*a*) and (*b*) shows the effect on the resolution of reducing the numerical aperture of the objective.

In addition to the possibility of some details in the object being absent from the image, there is also the danger that the image will show false detail, i.e. will indicate detail which is not, in fact, present in the object. As an extreme example, suppose that, with a grating as object, alternate maxima in the pattern are obstructed; the remaining maxima correspond to the pattern for a grating whose spacing is half that of the actual grating employed. Since the image is simply formed by the interference of light from the unobstructed maxima, it will have the appearance of an image of the finer grating instead of the real one. An obstruction of the Fraunhofer pattern such as that just described is, of course, extremely unlikely to occur accidentally. However, it must always be borne in mind that under certain conditions false detail may appear in the image. This is most likely to occur when extra stops and diaphragms, etc., are employed as in dark-ground and phase-contrast microscopy (see § 14–14).

The above discussion of the mechanism of image formation with non self-luminous objects follows that given by Abbe (c. 1873). The same problem is approached rather differently in the following method due to Rayleigh. Light from the substage condenser falls on the object and, from each point of the latter, a certain resultant disturbance emerges. The light from each point in the object then gives an Airy pattern in the conjugate image plane. The final image is formed by the superposition of these Airy patterns, and the effect of the superposition can be calculated if the phase relations and degrees of coherence between the various disturbances are known. These are determined by the structure of the object and the form of the illumination. In this treatment of the problem, the Airy patterns result from diffraction by the objective and the structure of the object is taken into account by inserting the phase relations between the Airy patterns. In the Abbe theory, diffraction by the object gives the Fraunhofer pattern behind the objective and diffraction by the objective itself is taken account of by the way in which its aperture limits the extent of the Fraunhofer pattern. Thus in both methods of approach one takes account of diffraction by the object and by the objective.

It is natural that, associated with these two methods of approaching the problem, one has the two different definitions of the limit of resolution: the separation of two just resolvable points, and the spacing of a just resolvable grating structure (or a spatial frequency). The Abbe theory has been greatly extended more recently and it is interesting to note that the results are of very great practical importance in considering the performance of almost all lenses except microscope objectives! This is because, as will be seen later, the results are relevant to the assessment of imperfect lenses, whereas for most purposes microscope objectives can be considered to be aberration-free on axis (§ 14–17).

14–12. Derivation of Limit of Resolution from Abbe Principle

According to the Abbe principle given above, the minimum requirement for any degree of resolution of a grating to be apparent is that the objective must collect the direct (zero order) beam and at least one of the first order beams.

Consider first the simple form of Köhler illumination where a single collimated beam is incident along the axis [Fig. 14–14(a)]. The angle of diffraction for the first order maximum is defined by

$$n'd \sin \theta_1 = \lambda,$$

where d is the grating interval and n' is the refractive index of the medium between the grating and the objective. Now, according to the Abbe principle, the limit of resolution of the objective is the value of d for which the first order

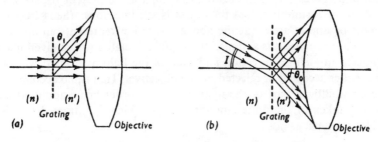

Fig. 14–14. Limit of resolution from the Abbe principle.

beam can just enter the objective. It is the value of d when θ_1 is the semi-angle of the largest cone of rays which the objective can admit, i.e. when $n' \sin \theta_1$ is the numerical aperture of the objective. Hence the limit of resolution is given by

$$d = \frac{\lambda}{(\text{N.A.})}. \qquad \ldots \ldots \quad (14\text{--}20)$$

Suppose now that the condenser is capable of illuminating the object by means of oblique beams. For the beam incident at angle I [Fig. 14–14(b)], the direction of the pth order maximum is given by the usual grating formula

$$d(n' \sin \theta_p - n \sin I) = p\lambda. \quad \ldots \ldots \quad (14\text{--}21)$$

For the zero order beam $d(n' \sin \theta_0 - n \sin I) = 0$, which simply gives Snell's law of refraction. For the first order beam,

$$d(n' \sin \theta_1 - n \sin I) = \lambda,$$

or
$$d(n' \sin \theta_1 - n' \sin \theta_0) = \lambda. \quad \ldots \ldots \quad (14\text{--}22)$$

Now the Abbe principle is that the zero order beam and one first order beam must enter the objective. For a given objective this can occur for the smallest possible value of d (i.e. for the largest angle between the beams) when the incident light is so directed that the zero order and one of the first order beams enter diametrically opposite extremities of the objective aperture, i.e. when $\theta_0 = -\theta_1$. Putting $\theta_0 = -\theta_1$ in (14–22), one obtains

$$2dn' \sin \theta_1 = \lambda,$$

i.e. the limit of resolution is given by

$$d = \frac{0 \cdot 5\lambda}{(\text{N.A.})}. \qquad \ldots \ldots \quad (14\text{--}23)$$

This result confirms that the limit of resolution depends upon the refractive index of the object space; it indicates also that the numerical aperture of the condenser is relevant to the limit of resolution of the objective since the ability of the condenser to illuminate the object with an oblique beam depends on its numerical aperture.

More detailed analysis of this and similar problems confirms that the aperture of the condenser is important, but indicates that, in general, its influence is rather less than that represented by the factor of two indicated by equations (14–20) and (14–23). It is interesting to compare the present results with that derived on the assumption that the object is self-luminous. (See § 14–7.)

It must be appreciated that the above discussion represents an approximate theory. For example, it is assumed that light associated with a given maximum is diffracted in a single direction and that all the light associated with a given order is either accepted or rejected by the objective. That is, no account is taken of the possibility of part of the light in a given order being admitted, and no account is taken of the finite amount of light which is diffracted between the principal maxima.

14–13. Dark-Ground, Phase-Contrast, and Interference Microscopy

As the numerical aperture of the microscope objective is increased, more diffracted light contributes to the image and the distribution of amplitude and phase across the primary image plane of the microscope approaches that immediately behind the object. In the present discussion it will be assumed that the imagery is perfect in the sense that the distribution is exactly the same in these two planes. Although this condition can never be attained exactly, it provides a useful starting point in explaining the principles of the various methods of microscopy.

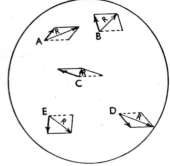

Fig. 14–15. Disturbances in image of phase object.

Suppose, now, that the object is completely transparent but has an optical thickness which varies from point to point. Such an object is known as a *phase object*—it introduces phase differences between disturbances which pass through different parts of it. Consequently, the disturbances immediately behind the object, and in the conjugate image plane, have the same amplitude at all points but will show variations in phase from point to point. Now the human eye is sensitive to intensity only and cannot detect changes of phase so that the field of view appears uniformly bright. In Fig. 14–15 the lengths and directions of the vectors shown as thick, full lines represent the amplitudes and phases of the disturbances at the various points A,B,C, . . . in the image plane; they will have equal lengths but different directions. Suppose that, in some way, a disturbance of constant amplitude and phase (represented by the dotted vectors) is added to every point in the primary image plane, and that this disturbance is coherent with the existing disturbances. At each point the two disturbances will interfere and the resulting intensity (R^2) at any point will depend on the phase difference between them as indicated in the figure. The variations in optical thickness in the object cause variations in intensity in the image so that the phase object is rendered "visible". This method of converting differences in phase into differences in intensity is employed in some interference microscopes and also forms the basis of the dark-ground and phase-contrast techniques. Thus, in one

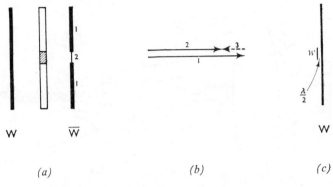

Fig. 14–16. Amplitude object.

class of interference microscope a wavefront that has been modulated by having been passed through (or reflected by) the object structure interferes with an unmodulated reference wavefront which constitutes the added constant "background" disturbance. In the dark-ground method, a disturbance is *subtracted* from every point in the field of view, and in the phase-contrast method introduced by Zernike (*c.* 1935), part of the light at each point in the field is phase-shifted. Since phase-shifting the light is equivalent to removing it and then adding light of different phase, this process also is equivalent to adding a constant background disturbance across the field. In a second class of interference microscope the system resembles a wavefront shearing interferometer so that light that has passed through one part of the object interferes with light that has passed through another part. The principle of the phase-contrast microscope can be understood by comparing the imagery of a small absorbing particle with the imagery of a small phase object. Consider first a purely amplitude particle, i.e. one that does not introduce a phase change compared with its surroundings. In Fig. 14–16(*a*) the particle consists of a small, slightly absorbing region surrounded by non-absorbing material of the same refractive index. Suppose the incident light from the substage condenser consists of a single plane wave W. The wave $\overline{W}$ that emerges from the object has a region of reduced amplitude, represented in the figure by the thinner line. The disturbance at 1 which has not passed through the particle can be represented by the vector 1 in Fig. 14–16(*b*). The disturbance at 2 which has passed through the particle has slightly reduced amplitude but has the same phase as 1; it is represented by vector 2. It will be seen that in (*b*) $2 = 1 + 3$, which means that the disturbance in the region 2 is equivalent to the disturbance in region 1 plus a weak disturbance 3 having a phase lag of π. This out of phase disturbance corresponds to a wave travelling a distance $\lambda/2$ behind the wave 1. That is to say, the wave $\overline{W}$, which has the region of reduced intensity, is equivalent to a uniform wave W plus a small wave w following on $\lambda/2$ behind—Fig. 14–16 (*c*).

Consider these waves entering the microscope objective (Fig. 14–17). W passes through a region of focus at the principal focus $F_0{}'$ and then spreads across the image plane. w, a wavefront of limited width (known as the diffracted wave), is equivalent to a broad angular distribution of disturbances that enters over the

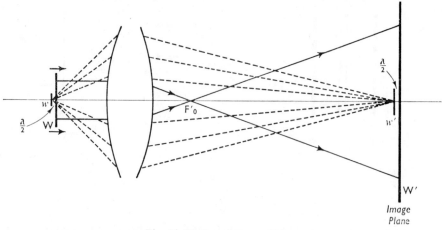

Fig. 14–17. Imaging a particle.

whole aperture of the lens. This is readily appreciated once it is realized that it is a wave such as w which would be transmitted by a very narrow slit, earlier discussion (§§ 11–1 and 11–2) having shown that light from such a slit is diffracted over a wide angle. (See also § 11–15.) The transmission of this light through the lens is equivalent to the formation of the image of such a slit, and if the lens is well corrected and the aperture is wide enough to transmit most of the light involved, the image formation is almost perfect (§ 14–11). That is to say, a small wave w' will be found $\lambda/2$ behind the "background wave" W'. An argument that is the reverse of that given above referring to Fig. 14–16 then shows that across the image plane one has a wave of the form $\overline{W}$ [Fig. 14–16(a)]. On viewing this image plane the observer will detect the less bright region which is the image of the amplitude particle; the particle is seen.

Consider now the imagery of a phase object consisting of a small phase-retarding but non-absorbing particle [Fig. 14–18(a)]. If the particle introduces a small phase lag the transmitted wave $\overline{W}$ is of uniform amplitude but has a slightly retarded region (2). The disturbance (1) that has missed the particle is represented by vector 1 in (b) and the disturbance (2) that has passed through the particle is represented by vector 2, which is the same length as 1 but makes a small angle δ with it, δ being the phase retardation introduced by the particle. Obviously 2=1+3. If δ is small, 3 is a weak disturbance approximately $\pi/2$ out of phase with 1. That is to say, $\overline{W}$ in (a) is equivalent to a uniform wave W plus a small wave w travelling $\lambda/4$ behind it—Fig. 14–18 (c). It will be seen at once that if these enter the microscope objective there will be found at the image plane a wave W' plus a small wave w' as in Fig. 14–17 except that one now has w' $\lambda/4$ behind W'. On inverting the argument referring to Fig. 14–18, one sees that across the image field one will have uniform intensity, and the particle will not be detectable. It will be seen that the conditions obtaining in the image plane with the phase particle closely resemble those with the amplitude object, except that with the phase object the diffracted wave lags in phase by $\pi/2$ whereas with the amplitude object it lags by π. Obviously, the appearance of the image of the phase object

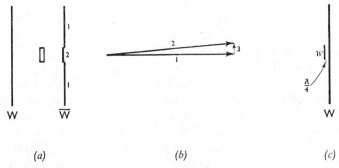

(a)   (b)   (c)

Fig. 14–18. Phase particle.

can be made to resemble the image of the amplitude object if an additional $\pi/2$ phase difference can be introduced between the diffracted wave and the background wave. This is what is done in the phase-contrast microscope. It can be done by inserting into the back focal plane of the objective a *phase plate* of some transparent material that has a reduced thickness over a small region surrounding the principal focus. In this way the diffracted light, which is distributed widely over the aperture of the lens, is retarded by $\pi/2$ compared with the background light which is concentrated into the region of the focus (see Fig. 14–17). The total phase difference between the background light and the diffracted wave becomes π and the particle is "seen".

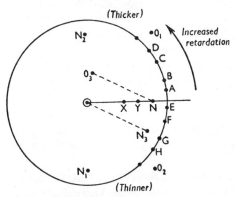

Fig. 14–19. Phase contrast.

Suppose now that the whole object plane is occupied by an extended phase object, the optical thickness varying from point to point. Following Zernike, divide the object plane into a large number of equal small areas and suppose, as before, that the incident light is a plane wave. The disturbances emerging from the various elementary areas A, B, C, . . . are represented in Fig. 14–19 by the vectors OA, OB, OC, . . . which, for convenience, are drawn from the common origin O so that for the pure phase object the points A, B, C, . . . lie on the circle centred at O. Since the phase-contrast method is chiefly of use when the variations of phase introduced by the object are small (less than about $\pi/4$), it is assumed that the points A, B, C, . . . lie on one side of O. In the figure, an anticlockwise rotation of a vector about O corresponds to increased retardation of the disturbances. That is, the vector OD corresponds to the thickest, and OH to the thinnest part of the object. Each of the disturbances OA, OB, OC, . . . can be looked upon as the disturbance ON plus the disturbance NA, NB, NC, ... Hence, emerging from the object plane, one has, in effect, a uniform plane wave represented by ON plus diffracted waves represented by NA, NB, NC, . . . from the various object elements. As before, the broad plane wave is brought to a focus at F_0' and then spreads over the image plane. In addition, each of the diffracted waves NA, NB, NC, . . . gives a small contribution in the region of F_0', but if N is the centroid of the points A, B, C, . . . the resultant of all these small contributions is zero. Thus, since only *relative* amplitudes and *relative* phases are important, the total zero order contribution from F_0' to each point in the image plane can be represented by ON. As before, for a wide aperture well corrected objective, the vectors representing the disturbances at various parts of the object plane also represent those across the image, and the non-zero order light which does not pass through F_0' gives contributions represented by NA, NB, NC, . . . at the various image points, the resultant disturbances in the image being represented by OA, OB, OC, ... Absorbing elements in the object plane (amplitude objects) of average optical thickness would give vectors such as OX, OY, along the mean phase line.

In *central dark-ground* illumination the zero order light is removed without affecting the diffracted light. (With the arrangement shown in Fig. 14–17 this could be done by placing an obstruction at F_0'.) This corresponds to subtracting, from each point in the field of view, a disturbance represented by ON. In the figure, this corresponds to the addition of a vector NO to each vector OA, OB, etc., so that the resultants are OA', OB', OC', etc., where A', B', C', etc., lie a distance NO to the left of A,B,C, etc. The points A', B', C', etc., are not shown in the figure because the state of affairs can be represented equally well by moving the origin a distance ON to the right, i.e. to N. That is to say, with central dark-ground illumination the disturbances at the various points in the field of view are represented by NA, NB, NC, etc. Since these vectors are of different lengths, there will be variations in intensity across the field of view. However, a vector such as NC, which corresponds to a relatively thick part of the object, will have the same length as, say, NG, which corresponds to a relatively thin part, and since the intensity depends only upon the length of the vector, the two parts of the field appear equally bright. Furthermore, the more absorbing part of the object which corresponds to the point X, will now appear brighter than the more transparent part corresponding to Y; obviously this is a disadvantage.

In the phase-contrast method, the phase of the direct or zero order light is changed with respect to the diffracted light. As an example, suppose the phase of the direct light is advanced through $\pi/2$ with respect to the diffracted light. Obviously this is equivalent to removing the direct light (which transfers the origin from O to N) and then adding a disturbance which is $\pi/2$ ahead of it. In the figure, ON_1 is a disturbance of the same amplitude as ON but $\pi/2$ ahead. When the origin is at N, the effect of adding ON_1 is to move the points A,B,C, etc., a distance ON_1 downwards or to move the origin from N to O_1. The disturbances in the image plane are then represented by O_1D, O_1C, O_1B, etc., and it will be seen that the thicker parts of the object give smaller intensities than the thinner parts. This is known as *positive phase-contrast*. By retarding the phase of the zero order light with respect to the diffracted maxima, one obtains *negative phase-contrast* in which the thicker parts of the object appear brighter. If the retardation is $\pi/2$ the origin is at O_2.

By securing a progressive decrease (or increase) in intensity as the thickness of the object increases one overcomes the first disadvantage of central dark-ground illumination mentioned above. On the other hand, one still has $O_1Y<O_1X$ so that the more absorbing part of the object (corresponding to X) appears brighter. This difficulty can be overcome for a fairly wide range of absorptions, while retaining the gradation of intensity with thickness, by advancing (or retarding) the phase of the direct light by something less than $\pi/2$. For example, ON_3 is less than $\pi/2$ ahead of ON, and if this is added when the origin is at N, the points A,B,C, etc., are moved "to the south-east" or the origin is moved to O_3. The disturbances in the image plane are then represented by O_3A, O_3B, etc., and the intensity decreases as the thickness of the object increases. In addition $O_3X<O_3Y$ so that the more absorbing part of the object appears less bright. It is, however, not possible to decide whether the reduction in intensity observed in a part of the field of view is due to absorption in the object or simply to increased retardation. This can be decided if one can examine the object under ordinary conditions of illumination since the only remaining reduction in intensity will be that caused by absorption. Plate IV(d) and (e) shows some flakes of mica immersed in glue with normal illumination and with phase-contrast. By absorbing some of the zero order light in addition to changing its phase, one can move the origin to any point on the lines NO_1, NO_2, NO_3, etc., thereby changing the ratio of, for example, O_1C/O_1B, and altering the contrast in the image. This is useful when the object shows only very small variations in thickness, but one must be careful to avoid placing the origin among the points A,B,C, . . . since this would reintroduce the unwanted reversal of contrast observed with central dark ground.

In the form of dark-ground illumination referred to above, the zero order light is removed but all the diffracted light contributes to the image. In addition, the discussion really concerns the intensity over an elementary area of the image. For example, if the transparent object consists of a number of flakes of mica, the above discussion gives the intensity over the image of each flake. In the form of dark-ground illumination usually employed in practice (see next section), the illumination is oblique so that the zero order and all the diffracted orders on one side of it are ineffective. In addition, the most interesting effect is the appearance of the image of the actual boundary between adjacent elements of the object. In § 10–13 it was shown that after diffraction by an object the light can be analysed into two components: the direct light and the edge waves. In oblique dark-ground illumination one collects only part of that portion of the edge wave which spreads to one side of the direct light. That is, the light entering the objective has the form of disturbances that have originated at the discontinuities in the object. These edge waves come to a focus in the image plane so that discontinuities in the object show up as bright lines in the image (see Plate IV(g)).

14–14. Experimental Arrangements

(a) Bright-field, Dark-ground, and Phase-contrast

In the previous section, dark-ground illumination is introduced as being the result of obstructing the zero order light. Normally it is only oblique dark-ground illumination that is used in practice and the simplest way of excluding the zero order light is to arrange that it does not enter the objective. This can be done if the object is illuminated by a hollow cone of light as indicated by Fig. 14–20. For low power objectives one can employ an ordinary condenser with an annular aperture in its back focal plane as indicated in the figure. For higher

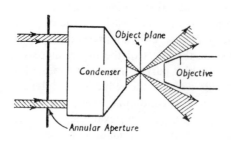

Fig. 14–20.
Low-power dark-ground illuminator.

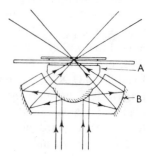

Fig. 14–21.
Dark-ground illuminator.

powers (higher apertures) it is necessary to employ a reflecting condenser such as the Beck focusing dark-ground illuminator shown in Fig. 14–21. The upper component A is oiled to the slide and the lower component B can be moved to enable the illuminator to be used with slides of various thicknesses. In many fixed focus condensers a similar system is employed except that, in effect, A and B form a single block of glass.

An extreme example of oblique dark-ground illumination is the arrangement known as the *ultra-microscope*. Here an intense beam is incident at right angles to the optical axis of the microscope. Discontinuities in the object appear bright as usual and it is possible to detect the presence of particles that are considerably below the limit of resolution of the objective. The arrangement has been used to observe particles in colloidal suspensions. A small volume of the suspension is illuminated so that only particles in that region are observed.

The diagrams given in §14–10 illustrate critical and Köhler illumination of transmitting objects. Exactly similar arrangements are possible for the illumination of opaque specimens, the objective performing the function of the substage condenser as well as forming the image. This is achieved by inserting a semi-reflecting plate behind the objective as in Fig. 14–22. It will be seen that a plane mirror at the object plane, which is equivalent to a blank slide in transmission, reflects the light back through the objective so that the system gives ordinary bright-field illumination.

For opaque objects absorbing and phase-retarding structure correspond respectively to variations in reflectivity and departures from surface flatness. Oblique dark-ground illumination for opaque objects is achieved in a manner exactly analogous to that used for transmitting objects, the zero order light reflected from a plane mirror object falling outside the objective aperture.

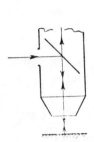

Fig. 14–22. Illumination
of opaque specimens.

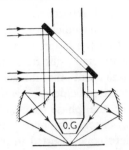

Fig. 14–23.
Dark-ground illuminator
for opaque specimens.

Fig. 14–23 shows a suitable arrangement; in some instruments the two mirrors are the outer surfaces of a solid block of glass similar to component B in Fig. 14–21 except, of course, that there is a central aperture through which the objective passes.

As explained in §14–13, for phase-contrast it is necessary to introduce a phase difference between the zero order and the diffracted light. The discussion in §14–13 referred to a simple form of Köhler illumination with a point source whereas in practice one wishes to use a more extended source in order to obtain a brighter image. It can be shown that with an extended source each plane wave from the condenser gives the same effect so that the net result is that the extended

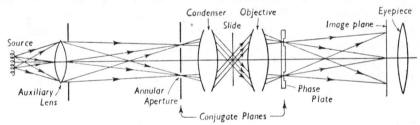

Fig. 14–24. Arrangement for phase contrast.

source simply raises the intensity level in the image without changing the relative intensities in different parts of it. This, of course, assumes that for each source point one can introduce the required phase difference between the zero order and diffracted light. Fig. 14-24 shows the arrangement usually employed for transmitting objects. An annular aperture is placed in the back focal plane of the substage condenser and the *phase plate* is conjugate to this, in the back focal plane of the objective. For positive phase-contrast, the phase plate has an annular depression as indicated, and this is exactly conjugate to the annular aperture. The figure shows the path of the zero order light and it will be seen that it passes through the thinner part of the phase plate. Consequently, its phase is advanced with respect to the diffracted light which passes through the thicker parts of the plate. There is no doubt that some of the diffracted light will also pass through the thinner annulus in the plate but this is unavoidable; it causes a slight reduction in image contrast. If t is the depth of the depression and n is the refractive index of the phase plate, the zero order light is advanced in phase by

$(2\pi/\lambda)(n-1)t$ with respect to most of the diffracted light. It will be appreciated that, for maximum efficiency, all the direct light must pass through the thin annulus but that the latter must be as small as possible in order to reduce the amount of diffracted light transmitted. It is there-fore important that the image of the annular aperture should coincide exactly with the annular depression in the phase plate. In practice this adjustment is checked by observation with a small auxiliary microscope inserted in the eyepiece tube. It is not essential to employ an annular aperture, and an aperture in the form of a cross has also been used. The relative merits of these arrange-ments cannot be discussed here. Since the back focal plane of a microscope objective frequently falls within the lens system, the phase plate must be "built in" as indicated by Fig. 14–25.

Fig. 14–25. Location of phase plate in a 4 mm. objective.

Before leaving the subject of phase contrast, mention must be made of a very simple method of rendering visible phase objects which show relatively large variations in thickness. When the direct and all the diffracted light is effective under ordinary illumination, the distribution of amplitude and phase in the image plane is an exact copy of that immediately behind the object. In the phase-contrast method this condition is upset, and the object rendered visible, by introducing a phase difference between the direct and diffracted beams. Now the direct and diffracted beams traverse different zones of the objective aperture and it was shown in § 14–9 that disturbances traversing different zones of the aperture of an optical system are out of phase in planes slightly removed from the ideal image plane. That is to say, a phase difference between the direct and diffracted beams can be introduced simply by slightly defocusing the micro-scope. This is sufficient to render visible variations in optical thickness which are greater than about $\lambda/8$.

(b) Interference Microscopes

In an interference microscope one examines the form of the wavefront leaving a phase object. This is done by superposing a reference wavefront, coherent with the first, and observing the interference pattern. As has already been mentioned in § 14–13, the second wavefront can be a wavefront that has not been modu-lated by the object structure, or it can be produced by a wavefront shearing interferometer. An example of the first class of instrument is the arrangement used by Linnik (1926) for opaque objects. The system resembles a Michelson or Twyman-Green interferometer. Identical microscope objectives are placed in the two arms, one being focused on a plane mirror at M_1 (Fig. 8–16) and the other on the specimen which replaces M_2. Equivalent systems for transmitting objects employ arrangements similar to the Jamin or to the Mach-Zehnder interferometer. For example, in Fig. 8–19, G and C can be replaced by (object and objective) and (blank slide and objective) respectively. This principle has been used in the Leitz microscope. A disadvantage of the systems referred to above is that each requires a pair of objectives that are optically identical. Alternative arrangements, such as that employed in the Watson interference

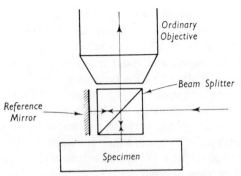

Fig. 14–26. Watson interference objective.

objective for opaque objects, avoid this difficulty by using a Michelson system with the beam splitter and reference mirror mounted between the object and the objective (see Fig. 14–26). One obvious difficulty in this system is the need for an objective with a long working distance in order to accommodate the beam splitter; the difficulty increases as the power and aperture are increased.

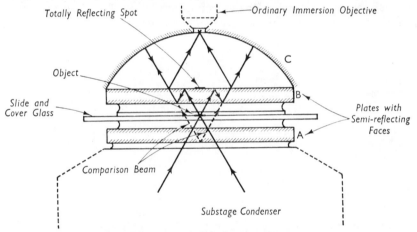

Fig. 14–27. Dyson's interference microscope.

Another system that produces an independent reference beam is that due to Dyson (Fig. 14–27). This is a unit magnification interferometer system added below an ordinary microscope objective. Light from the condenser is amplitude-divided at the upper surface of plate A so that the comparison beam is displaced sideways and misses the object. The other beam passes straight on, passes through the object, and is reflected in plate B to recombine with the comparison beam. The combined disturbances are then reflected at the spherical surface C and again at the upper surface of plate B before entering the objective. C is, in fact, the upper surface of a block of glass that is cemented to the plate B; contact between the other components is made with an immersion fluid as indicated. The totally reflecting spot on the upper surface of B obstructs the axial rays of both beams; this is necessary because both would have passed through the object. The ray paths shown in the figure indicate that, in effect, the objective is focused on both the object and the lower surface of plate A. That is to say, the interference occurs between light that has *passed through* the object and light that has been *reflected by* the lower surface of A acting as a reference surface—one is observing phase differences between disturbances in a beam that has passed through the object and those in an identical beam that has been reflected by a plane mirror occupying the position of the object. The plates A and B are slightly wedge-shaped (wedge angle 5 minutes) and one can vary the path difference between the two beams by displacing plate A laterally. If the path difference is made zero fine phase particles will appear dark on a bright background, and if the path difference is $\lambda/2$ they will appear bright on a dark background. It should be noted that if the lateral extent of the object is considerably larger than the field of view, the reference beam will traverse the outer region of the object structure and the system cannot be classified as one that produces an unmodulated reference beam.

It was stated earlier that there is a second class of interference microscope, which uses the principle of the wavefront shearing interferometer (see §8–15). There are two sub-classes: those that introduce a shear that is larger than the lateral extent of the object structure being examined, and those that introduce a very small shear comparable with the limit of resolution. Thus Françon refers to *image duplication* methods and *differential* methods.

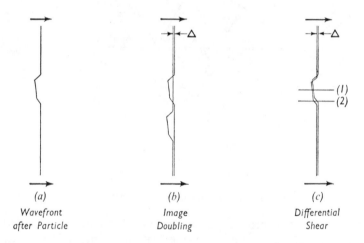

(a)	(b)	(c)
Wavefront	Image	Differential
after Particle	Doubling	Shear

Fig. 14–28. Wavefront shearing.

Fig. 14–28(a) illustrates the form of a wavefront emerging from a phase particle. A shearing interferometer system giving image duplication will produce the effect shown in (b) and a differential system will give the effect shown in (c). $\triangle$ is the instrumental path difference.

It will be seen at once that if one is examining a single small particle with the duplication method the effect is the same as with the first class of interferometer where a completely independent reference wavefront is produced (except of course, that two "images" of the particle will be seen). Obviously, the pattern is difficult to interpret when the object field contains other particles since the various images become superimposed. Referring now to the differential method, it will be seen from (c) that the path differences at positions (1) and (2) depend upon the slope of the wavefront and upon the extent of the shear and not upon the actual retardation introduced by the particle. That is to say, the differential method reveals optical path gradients in the object structure, and if the shear is smaller than the limit of resolution no double contours will be seen in the image. Differential systems are not restricted to the examination of isolated particles.

A large number of wavefront shearing interference microscopes have been produced employing birefringent double image prisms to introduce the shear. For example the image duplication system due to F. H. Smith and the differential system due to Nomarski employ double image prisms of the Wollaston type (see §22–7). In practice, it is usual to duplicate the wavefront and introduce a shear before passing through the object. The shear is then removed to give a net result equivalent to introducing a single shear after the object (see Fig. 14–29).

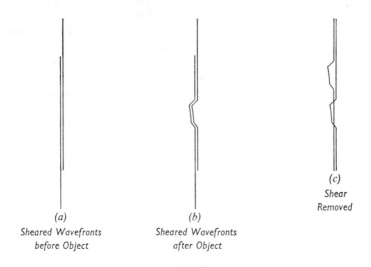

(a)
Sheared Wavefronts
before Object

(b)
Sheared Wavefronts
after Object

(c)
Shear
Removed

Fig. 14–29. Wavefront shearing.

The reason for this is as follows: a Wollaston prism introduces an optical path difference between the two emergent rays and this depends on the part of the aperture traversed, being zero at the mid-point where the thicknesses of the two component prisms are equal. Hence if there is only one shearing prism in the system the aperture must be made small. However, if one has a prism behind the condenser and a second one behind the objective, the optical path difference in the first can be balanced in the second, and a wide aperture can be used. If the limitation of aperture is accepted, one can simply insert a wavefront shearing interferometer system (an interference eyepiece) behind the objective of an ordinary microscope.

Another form of interference microscope that should be mentioned is the axial image duplication system due to Smith. A doubly refracting plate is added to the objective so that it has two planes of focus—the actual object plane and

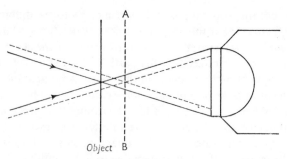

Fig. 14–30. Smith's axial duplication system.

the plane AB (Fig. 14–30). Thus these two planes are imaged into the same image plane and the reference "wavefront" is the distribution of light disturbances across the plane AB. This is equivalent to saying that interference occurs between an in-focus image and an out-of-focus or axially displaced image, instead of a laterally displaced image as before. As in the shearing systems a second image-doubling plate is added to the substage condenser, but again this does not affect the principle of the method.

In conclusion it should be pointed out that the phase-contrast microscope is not satisfactory if the lateral structure of the object is coarse. This is because the object structure will diffract only a small amount of light at an angle large enough for it to be well separated from the zero order light on reaching the phase plate. It will then be impossible for the phase plate to introduce a phase difference between the zero order and most of the diffracted light, and the contrast will be poor. Another disadvantage of the phase-contrast microscope is that images are often surrounded by halos caused by diffraction at the phase plate itself. Interference microscopes do not have these two disadvantages and, although they are often more complicated and difficult to adjust, they often have a variable path difference, giving the benefits of a variable phase plate and (since the path difference can be made zero) the possibility of using white light and colour contrast.

14–15. The Flying-spot Microscope

When one forms an image in an ordinary microscope one records simultaneously the way in which each element of the object modulates the incident light. In a *flying-spot microscope* the object is scanned by a very small spot of light, and one records sequentially the extent to which the light is modulated by successive object elements. The scanning spot is produced initially on the screen of a cathode ray tube which is imaged on to the microscopic object by using an ordinary microscope system in reverse. In this way the very small image spot is made to scan the object and, since different object elements correspond to different instants of time, no spatial resolution is required in the detector; the detector is a single photo-cell or photo-multiplier that occupies the position of the substage condenser of an ordinary microscope, and collects light from all parts of the object. The output from the detector is then amplified and fed to a second C.R.T. which is the display tube—and normally much larger than the scanning source tube. The image is synthesised on the display tube in the same way as a television picture is synthesised, and the contrast in the image can be controlled electronically, which in this respect gives the system a distinct advantage over an ordinary microscope. Although one can obtain an improved contrast, the ultimate limit of resolution cannot be better than that of a microscope that has the same numerical aperture and is used in the normal way; it is determined by the size of the scanning spot and this, in turn, is determined by the aperture of the microscope—assuming that the source tube itself does not impose a higher limit.

One useful application of the flying-spot microscope is in particle size analysis. The size of a particle in the field of view can be determined by recording the length of time for which the scanning spot is obstructed. The method can be used to count the number of particles in the field of view and give the distribution of sizes. The electronic timing mechanism can be arranged to count only those particles that are greater than a given size. This is done by recording only those signal interruptions that are longer than the appropriate value—determined by the time base. One can avoid counting particles twice by arranging not to repeat the recording of a signal interruption from the same part of the field until a positive signal has been received, indicating that the scanning line has cleared the particle. Difficulties do arise with particles of some complicated shapes, and also with overlapping particles, but a statistical allowance can be made for these effects.

One can employ ultra-violet illumination although there are obvious instrumental difficulties of the kind that are encountered in conventional ultra-violet microscopy (§ 16–13). The flying-spot technique has the advantage over conventional ultra-violet microscopy that the whole specimen is not illuminated throughout the effective period of observation, so that the risk of radiation damage is very much reduced.

An analogous technique is used in the *scanning electron microscope* and has been extended in what is sometimes called the *scanning X-ray microscope* or *electron-probe micro-analyser*. This complicated technique involves analysing the spectral distribution of the X-rays that are generated by the scanning electron beam. In this way one obtains information about the chemical composition of the region of the specimen in which the X-rays are generated.

14–16. Image Formation with Coherent Illumination. Spatial Filtering

In the simplified arrangement for image-formation by a microscope represented by Fig. 14–12 the object structure is coherently illuminated and the light that is responsible for reproducing a particular spatial frequency in the image passes through a particular point in the back focal plane of the objective. It was seen in § 14–11 that the limit of resolution of the objective is set by its aperture since this determines the highest spatial frequency that can be transmitted. In other words, an aberration-free microscope objective gives perfect transmission of all spatial frequencies up to a certain maximum. The purpose of the present section is simply to point out that in such a system one can deliberately introduce into the back focal plane a form of transparency which will modulate the spatial frequencies in a predetermined way. Such a procedure is referred to as *spatial filtering*; in this language, an aberration-free lens with coherent illumination is simply a low-pass spatial frequency filter. (It will be seen that a phase-contrast microscope can be said to be a spatial filtering system, the phase plate being the filter.)

It often happens that either the overall quality or some features of particular interest in a photograph can be improved by suppressing certain frequencies (and, by implication, emphasising others). The subject of spatial filtering has developed considerably during the past two decades and although a detailed discussion cannot be included here, a brief description will be given of a possible optical system for introducing the filter.

The general procedure of spatial filtering can be carried out with an optical system of the type shown in Fig. 14–12, but in this system the second lens is used to form the spatial frequency spectrum and also to form the final (filtered) image, i.e. it is used at two different pairs of conjugates so that the spectrum cannot be formed on a plane surface. It is better to use the system shown in Fig. 14–31, which does not suffer from this disadvantage. The spatial frequency spectrum of the object is formed in the plane of the spatial filter and the image is synthesised from the modified spectrum. The positions of the various components are as indicated because, as pointed out in §11–15, the relevant Fourier

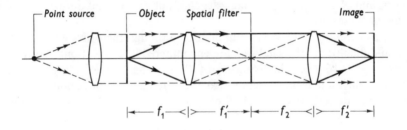

Fig. 14–31. Spatial filtering.

transform relationships are more accurate if the spectrum is formed in the focal plane of the appropriate lens. In the figure the coherent illumination is represented as being derived from a point source at the focus of a collimating lens; obviously one can use a laser with whatever lens system (often telescopic) is required to give a beam of the required width.

14–17. Image Formation with Incoherent Illumination. The Optical Transfer Function

In the present section most of the discussion refers to the formation of the image of an incoherently illuminated object, and is of special importance when the image-forming system is imperfect. One then needs to be able to specify the image-forming properties of a lens so that it is possible to compare the performances of different lenses. From the designer's point of view, one needs to be able to specify the performance required from a particular lens and then to specify tolerances for its aberrations.

Suppose, then, that a lens with a circular aperture forms an image of an object plane, where it may be assumed that the disturbances from various parts of the object are completely incoherent. These conditions are approached very closely in most object fields presented to camera lenses. If the lens is aberration-free, each point in the object gives an Airy diffraction pattern and the image results from the superposition of these Airy patterns, the total intensity at any point being the sum of the separate intensities. The expression giving the intensity distribution in an Airy diffraction pattern (equation 11–8) is such that in general it is extremely difficult to find the intensity distribution in the image of an

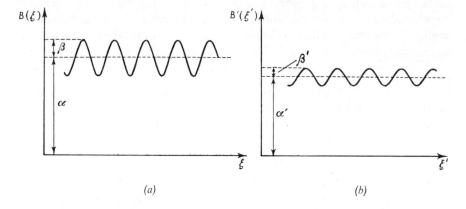

Fig. 14–32. Imagery of sinusoidal component.

extended object. For a lens with aberration the diffraction image of a point is much more complicated than the Airy pattern and it becomes virtually impossible to calculate the form of an extended image. If one cannot find the effect of the aberration on the image it is impossible to specify an aberration tolerance appropriate to a given image quality. In fact, it is difficult even to specify the image quality or lens performance required if the problem is approached in the above manner. One thing that could be specified is the limit of resolution but it will be seen below that in general this is quite inadequate. Furthermore, it is usually not even possible to give an aberration tolerance corresponding to a stated limit of resolution, and the best that can be done is to state the magnitude of the aberration that causes the standard of image formation to become just noticeably inferior to that of an aberration-free system. That is to say, this method of approach to the problem is satisfactory only for very highly corrected systems such as microscope objectives. The reason for the failure of this approach for most camera lenses is that the object is thought of as being composed of a collection of points, each giving a complicated pattern. An alternative approach is suggested by the Abbe theory of image formation in a microscope. This leads one to think of the intensity distribution across the object as the superposition (the synthesis) of a number of sinusoidal distributions. The enormous advantage of this is that provided certain conditions are satisfied—and usually they are— the image of a sinusoidal intensity distribution also has a sinusoidal form that is independent of the presence of other sinusoidal components. That is to say, a lens is a linear filter of spatial frequencies, and this remains true, with certain limitations, when the lens has aberrations. The situation is as follows: if one has an object consisting of a line structure across which the intensity distribution is of the form

$$B(\xi)=a+\beta \cos 2\pi R\xi \qquad \ldots \ldots \quad (14\text{--}24)$$

[see Fig. 14–32(a)], then the intensity distribution across the image is given by

$$B'(\xi')=a'+\beta' \cos (2\pi R'\xi'+\theta) \qquad \ldots \ldots \quad (14\text{--}25)$$

[see Fig. 14–32(b)]. The contrast in the image is always lower than that in the object and, unless there is no asymmetrical (comatic) aberration, there is a spatial phase shift or a displacement of the image structure from the position corresponding to an aberration-free system. The ratio of the modulation or contrast in the image to that in the object ($=\beta'/a'\div\beta/a$) is called the *modulation* (or *contrast*) *transfer factor* and is denoted by T; one then puts

$$D=Te^{i\theta}, \qquad \ldots \ldots \ldots \quad (14\text{--}26)$$

where D is called the *optical transfer function* (or *frequency response*). D depends on the spatial frequency, R. It is convenient in this work to define "optical co-ordinates" (u, v), (u', v') in the object and image planes by the relations

$$u=\frac{n \sin U_{\max}}{\lambda} \xi, \quad v=\frac{n \sin U_{\max}}{\lambda}\eta, \quad \ldots \ldots \quad (14\text{--}27)$$

with similar expressions for u' and v', where (ξ, η) and (ξ', η') are the ordinary cartesian co-ordinates and the other symbols have their usual meanings. Corresponding to these new co-ordinates one has new spatial frequency variables in the object and image planes defined by

$$s=\frac{\lambda}{n \sin U_{\max}} R \text{ and } s' =\frac{\lambda}{n' \sin U'_{\max}} R'. \quad \ldots \quad (14\text{--}28)$$

It follows at once from the sine relation (equation 14–4) that the point at (u, v) is imaged at $u'=u$, $v'=v$, i.e. in these new co-ordinates all lenses have unit magnification and $s'=s$, i.e. the spatial frequency variable is the same for the object and the image. The performance of the lens is then specified by giving the

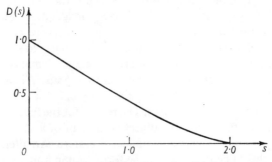

Fig. 14–33. $D(s)$ for aberration-free lens.

$T(s)$ and $\theta(s)$ curves. The limit of resolution is given by the value of s for which $T(s)$ falls to zero. For any aberration-free lens the limit of resolution corresponds to $s=2$, $\theta(s)=0$ and $D(s)$ [$=T(s)$] has the form shown in Fig. 14–33. It should be noted that $s=2$ corresponds to different actual frequencies for different lenses

—for $\lambda = 5 \times 10^{-4}$ mm. it corresponds to 182 lines per mm. in the image plane for an $f/11$ lens and 1000 lines per mm. for an $f/2$ lens. (It can be seen from the discussion of §14–11 that for an aberration-free lens with coherent illumination $D(s) = 1$ up to the limit of resolution when it falls suddenly to zero. This is because the diffraction maxima are either transmitted by the objective to reproduce the corresponding spatial frequency without loss, or they fall outside the objective aperture.) Always the effect of aberration is to cause a reduction in the value of the modulation transfer function below that for an aberration-free lens. Even for the lower spatial frequencies the presence of a small amount of aberration causes a significant reduction in the value of $T(s)$. This means that a lens may need to be highly corrected even if it is to be used solely to image relatively coarse detail—the fact that a lens can actually resolve 200 lines per mm. does not automatically mean that it is satisfactory for all purposes for a

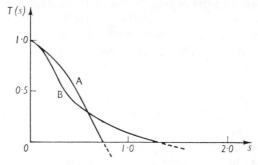

Fig. 14–34. Two imperfect lenses.

frequency of 100 lines per mm.! E.g. many old 405 line TV cameras were concerned only with frequencies up to 10 lines per mm. in the image plane but since one requires an extremely high performance over this range one must use a well corrected wide aperture lens over frequencies that are very low compared with the limit of resolution—with an $f/2$ lens $s \leqslant 0.02$.

The lens designer must be told the range of frequencies for which the lens is to be used and the value of $D(s)$ required. It is inevitable that a lens will have some aberrations; it is the task of the designer to reduce them as far as possible and to achieve the best balance between the unavoidable residuals (see Chapters XV and XVI). The aberrational balance that gives the best imagery of the finer detail and the best ultimate limit of resolution may be different from the compromise that gives the best performance for the lower frequencies. Consider, for example, two lenses A and B whose $T(s)$ curves have the forms shown in Fig. 14-34. Lens A will be better for imaging lower frequencies whilst lens B is better for imaging fine detail and has the better ultimate limit of resolution. By giving the $T(s)$ and $\theta(s)$ curves one gives a complete statement of the image-forming properties of the lens; stating the limit of resolution merely gives the value of s for which $T(s)$ falls to zero. Fig. 14-34 indicates that beyond the limit of resolution $T(s)$ becomes negative. This corresponds to a reversal of contrast—maxima occurring where there should be minima. The $T(s)$ curve can oscillate beyond the limit of resolution before finally falling to zero at $s = 2$. This explains the "spurious resolution" of detail below the limit of resolution of the lens.

It was stated above that provided certain conditions are satisfied a lens system is a linear filter of spatial frequencies. It will now be shown that the conditions are that over the relevant region of the field of view of the lens the aberrations must not vary rapidly, so that the movement of a point source within this region causes a corresponding movement of the diffraction image without significant change of structure. It is essential that the aberrations remain approximately constant over the area within which the diffraction image of a point source has appreciable intensity. A region in the field over which the aberration may be taken as constant is called an *isoplanatism patch*—the lens is *isoplanatic* over that region.

Now any two-dimensional distribution of intensity in the object plane can be represented as the superposition of line structures having various spatial frequencies and orientations. There is therefore no loss of generality in restricting the discussion to the imagery of line structures. In what follows the co-ordinates u, u' are those defined above. Let $G(u')\,du$ be the intensity distribution in the image plane when there is unit intensity along a strip of width du at $u=0$ in the object plane. $G(u')$ is often called the (line) *spread function*. If one remains in an isoplanatism patch a displacement of the line source merely gives a bodily displacement of the image so that unit intensity along a strip of width du at u in the object plane gives an intensity distribution $G(u'-u)du$ in the image. Suppose now that in the object plane there is, in fact, a one-dimensional variation of intensity (line structure) given by $B(u)$. Then the strip of width du at u produces (at u') in the image plane an intensity contribution $dB(u')= B(u)G(u'-u)du$. To obtain the total intensity at u' one adds the intensities due to neighbouring object strips. It is true that, strictly speaking, object strips far removed from u do give contributions, but only those close to the strip conjugate to u' (and hence within the isoplanatism patch) have spread functions with values at u' significantly different from zero. Hence one may write for the total intensity at u'

$$B'(u')=\int_{-\infty}^{+\infty} B(u)G(u'-u)du, \qquad \ldots \ldots \quad (14\text{–}29)$$

the regions beyond the isoplanatism patch giving a negligible contribution. Now the object and image are looked upon as being synthesized from Fourier components, the amplitudes of the components of frequency s being given by

$$b(s)=\int_{-\infty}^{+\infty} B(u)e^{-2\pi ius}du \qquad \ldots \ldots \ldots \quad (14\text{–}30)$$

$$\text{and } b'(s)=\int_{-\infty}^{+\infty} B'(u')e^{-2\pi iu's}du' \qquad \ldots \ldots \ldots \quad (14\text{–}31)$$

(see §6–8).

Similarly, the transform of the spread function is given by

$$g(s)=\int_{-\infty}^{+\infty} G(u')e^{-2\pi iu's}du'. \qquad \ldots \ldots \ldots \quad (14\text{–}32)$$

From equation 14–29 one has

$$b'(s) = \int\limits_{-\infty}^{+\infty} \left[\int\limits_{-\infty}^{+\infty} B(u)G(u'-u)du \right] e^{-2\pi i u's} du'$$

$$= \int\limits_{-\infty}^{+\infty} B(u)\left[\int\limits_{-\infty}^{+\infty} G(u'-u)e^{-2\pi i(u'-u)s}e^{2\pi i us}du' \right] du$$

$$= \int\limits_{-\infty}^{+\infty} B(u)e^{2\pi i us}\left[\int\limits_{-\infty}^{+\infty} G(u'-u)e^{-2\pi i(u'-u)s}d(u'-u) \right] du$$

$$= \int\limits_{-\infty}^{+\infty} B(u)e^{2\pi i us}g(s)du$$

$$= g(s)b(s). \qquad \ldots \ldots \quad (14\text{–}33)$$

This shows that each spatial frequency in the object appears in the image modulated by $g(s)$, the transform of the line spread function. That is, provided the isoplanatism condition is satisfied a lens is a linear filter of spatial frequencies in the object, and the optical transfer function (frequency response) is given by the Fourier transform of the spread function. Since uniform illumination in the object gives uniform illumination in the image $D(s)$ is normalized to make $D(0)=1$, i.e. $D(s)=g(s)/g(0)$. Thus the optical transfer function can be found by measuring the intensity distribution in the image of a slit source and calculating the Fourier transform of the measured spread function. Alternatively, one can scan the image of a grating structure to obtain the transfer function directly. A square wave grating can be used and the result for a sine-wave structure obtained by selecting only the fundamental from the detector output.

It was pointed out in §11–15 that by means of the process of apodization one can change the aperture function and, therefore, the spread function. In other words, one can apodize an optical system in order to change the relative amplitudes and phases of the Fourier components from which an image is synthesized, i.e. one can change the optical transfer function and introduce spatial filtering.

If one is concerned with coherently illuminated objects, one usually discusses amplitudes rather than intensities, and it has already been seen (§14–16) that an aberration-free lens gives perfect transmission of all spatial frequencies up to a certain limit. In the language of the present section, then, an aberration-free lens used with coherent illumination gives an optical transfer function which is unity up to the limit of resolution, where there is a sharp cut-off; the optical transfer function, which is the Fourier transform of the image pattern function for a point object, is, when normalized, simply equal to the aperture function (often referred to in this context as the pupil function).

14–18. Wavefront Reconstruction (Holography). Gabor's Original Method

If one wishes to obtain a permanent record of the appearance of a field of view, one traditional procedure is to take an ordinary photograph. Although this provides a permanent record of the appearance of the field, it is severely limited in a very important respect: in taking the photograph the camera lens has to be focused on a particular plane in the field of view, and it is only the appearance of this plane that is recorded. All other planes are out of focus to a greater or lesser extent, and there is nothing that one can do subsequently to enable one to obtain from the photograph a clear view of any other plane in the field. The fundamental reason for this is that the photograph simply records an intensity distribution. If one could record both the amplitude and the phase distribution in any plane between the object and the observer, it would be possible in principle to obtain from this complete record the amplitude and phase distribution (and hence the intensity distribution—which is what an observer actually sees) across any other surface in the object field, so that one could make a complete investigation of the field of view at any convenient subsequent time. This is what is done in holography.

One could, of course, take a series of photographs with an ordinary camera focused at different distances. However, this is a clumsy procedure and, furthermore, if there are moving objects in the field neither a series of stills nor a cine film can record the appearance in different planes in the field at *the same instant*. No classical procedure gives a single, sharp, three-dimensional record of the scene as is obtained by holography.

The holographic method of image recording and retrieval was introduced in 1948 by Gabor. The principle of Gabor's method is as follows: suppose light from a point focus S (Fig. 14–35) is incident on the small object O located a short distance away. A small proportion of the light is diffracted by O, and on the plane AB the secondary disturbances from O are superposed on the strong coherent background provided by the primary wave from S. Since the primary

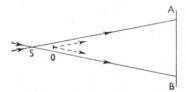

Fig. 14–35. Production of a hologram.

wave is uniform and much stronger than the secondary wave, the variations in intensity across AB are not determined simply by the variations in intensity across the secondary wave, but depend upon the variations of phase across it. That is to say, the presence of the strong coherent background enables one to record information about the phase of the diffracted light—a general technique introduced by Zernike. Gabor called the pattern on AB the *hologram*; if the hologram is recorded photographically, and the plate developed by reversal, or printed, the transmission at any point depends upon the phase of the secondary wave. It can be shown that if the object is removed and the hologram is inserted at AB and illuminated by S alone, the disturbance transmitted by the hologram contains as a major component a reconstruction of the original wave. If one then looks through AB towards S, the object will appear to be in position at O. The reconstructed wave is only a component of the light transmitted by the hologram but Gabor has shown that the remainder gives an effect which can be removed almost completely. One then sees an accurate reconstruction of the object. Actually the effective intensity in the hologram corresponds to a given phase *difference* between secondary disturbance and coherent background, and is unaffected by a change in the sign of that difference. That is, an identical object structure symmetrically placed behind S would give the same hologram as O. In consequence, this second object also appears to be in position on looking through the hologram. If the object O was originally three-dimensional, one plane after another can be viewed in the reconstruction simply by adjusting the focus of the observing lens.

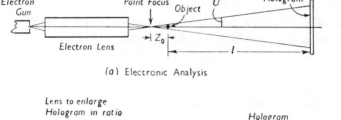

(a) Electronic Analysis

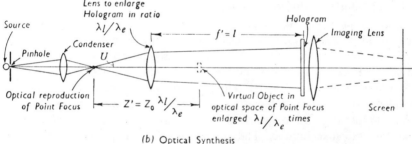

(b) Optical Synthesis

Fig. 14-36. Arrangements for analysis and synthesis.

The application originally proposed for this technique depends on the fact that light of one wavelength can be used in the *analyser* to obtain the hologram and light of a longer wavelength used in the *synthesizer* to obtain the reconstruction. It is simply necessary for the dimensions of the synthesizer and analyser to be in the same ratio as the wavelengths employed. This, of course, involves a similar enlargement of the hologram. Thus the analysis can be carried out with electrons and the synthesis with light. This involves an enormous magnification factor (10^5, say) and the necessary magnification of the hologram is obtained by the introduction of an additional lens into the synthesizer. This has a focal length equal to the distance from the object to the photographic plate in the analyser and is placed between the point focus and the hologram in the synthesizer so that the effective distance of the hologram is increased by a factor of 10^5 when viewed from the optical space of the point focus. The arrangements for electronic analysis and optical synthesis are shown in Fig. 14-36. It should be noted that the synthesizer must provide an exact imitation of the original (electron) wavefront in the plane of the hologram. Hence it is necessary for the spherical aberration of the electron lens to be copied in the optical system. The limit of resolution is determined by the angle U, being of the order $\lambda/\sin U$ where λ is the wavelength of the electrons.

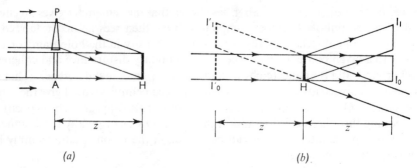

Fig. 14–37. Holographic Imagery of a Transparency.

In an ordinary electron microscope the difficulty of correcting the spherical aberration is an important factor limiting the resolution obtainable. The method of wavefront reconstruction described above enables one to circumvent this difficulty.

14–19. The Basis of Modern Holography

Although the method of wavefront reconstruction was originally proposed by Gabor with its application to electron microscopy in mind, it has never been applied seriously in this field—possibly because conventional electron microscopes were greatly improved before the problem of obtaining high quality holographic images was solved. However, during the past decade very great improvements have been achieved using new *optical* methods for the production of both holograms and images, and holography is now firmly established as a powerful optical technique. The subject is still undergoing extremely rapid development and all that can be done here is to give a short account of some of the basic concepts and methods and a very brief indication of the fields in which the new techniques are being applied.

It was essential in Gabor's original arrangement to have a strong coherent background upon which the Fresnel diffraction pattern of the object was superposed to produce the hologram. The arrangement had shortcomings in principle: firstly, the method of producing the hologram, including the strong coherent background, restricted the application of the technique to object structures of small lateral extent, and, secondly, the phase of the secondary wave was not recorded unambiguously—causing the twin reconstruction which always appeared as an out of focus background when the reconstructed object was viewed, and whose effect could not be removed completely. There were also serious practical limitations imposed by the nature of the radiation sources available at the time. The need for an illuminating beam of good lateral coherence required a pinhole source (leading to a shortage of light) and the lack of longitudinal coherence in the sources available restricted the method to object structures of fairly small longitudinal extent and also ruled out strongly diffusing objects.

The technique used by Leith and Upatnieks (1962) succeeded in producing a hologram in which the amplitude and phase were separately and unambiguously recorded so that the quality of the reconstructed object was greatly improved.

The original experiments employed a classical thermal source but later experiments used a laser, which gave a beam of light in which excellent lateral coherence was combined with high intensity. A laser also gives a long coherence length, and this made possible the extension of the technique to three-dimensional and diffusely reflecting objects. It will be assumed throughout the remainder of this discussion that unless otherwise stated the light sources are lasers.

Fig. 14–37 illustrates the principle of the method of Leith and Upatnieks as applied to an object consisting of an ordinary transparency. For the production of the hologram (Fig. 14–37(a)), the incident light consists of monochromatic plane waves. The hologram is produced at H as a result of the superposition of light that has passed through the object transparency at A and a uniform reference beam that is supplied via the prism P. (This serves to illustrate the principle of the method; in practice P is replaced by a mirror system.) With the object removed, the interference pattern at H consists of equidistant linear fringes of uniform contrast; with the object in position, the amplitude variations in the pattern due to the object cause variations in the contrast in the fringes, and the phase variations in the object pattern cause variations in the phase of the fringes, i.e. departures from linearity. The regular pattern obtained with the object absent is analogous to the carrier wave of a communication system, and the modulation of the fringe system caused by the presence of the object corresponds to the amplitude and phase modulation of the carrier wave by a signal. It will be seen that, although it is recorded photographically so that only an intensity distribution is recorded, the hologram records unambiguously both the amplitude and the phase of the light that has passed through the object, i.e. it contains the information that is required to specify the object structure uniquely.

The reconstruction system is shown in Fig. 14–37(b). The hologram H is illuminated by collimated monochromatic light and, due to its fine, basically regular, structure, acts as a diffraction grating. It can be shown that in the zero order there are poor quality real and virtual images formed at I_0 and I_0' respectively due to the weak modulation of a strong background. However, in the first order one obtains a reconstructed real image of high quality on one side of the zero (I_1) and a high quality virtual image on the other side of zero (I_1').

The principle of the method can be demonstrated adequately by a discussion that is restricted to two dimensions, i.e. one dimension in the plane of the hologram. Referring to Fig. 14–37(a), if the uniform plane reference wavefront from P has an angle of incidence θ at H the phase at any point at position x on H can be written as $(2\pi/\lambda)x \sin \theta = \alpha x$ (say). The total complex amplitude at any point x in the hologram is then given by

$$\psi_x = A_0 e^{-i\alpha x} + A_x e^{-i\delta_x}$$

where A_0 is the (uniform) amplitude of the reference wave and A_x and δ_x are the amplitude and phase of the disturbance at x due to the object structure. The intensity in the hologram is then given by

$$I_x = |\psi_x|^2 = |A_0 e^{-i\alpha x} + A_x e^{-i\delta_x}|^2$$
$$= A_0^2 + A_x^2 + 2A_0 A_x \cos(\alpha x - \delta_x) . \quad . \quad . \quad (14\text{–}34)$$

The third term in (14–34) represents the fringe system in the hologram and it will be seen that the contrast in the fringes, determined by $A_o A_x$, is in fact determined by A_x since A_o is constant, and the fringe spacing is determined by δ_x, that is to say, the hologram contains a record of both the amplitude and the phase of the light received from the object. If the photographic processing is suitably controlled the *amplitude* transmission factor in the hologram inserted into the reconstruction system can also be represented by (14–34). (One is here assuming an appropriate linear characteristic curve for the developed photograph and omitting a constant factor multiplying I_x and also omitting a factor which would appear as a constant subtracted from I_x—since no part of an emulsion can have 100% transmission.) Hence in the reconstruction the first two terms in (14–34) correspond to the zero order beam. Any reconstructed object would have poor contrast if $A_x^2 \ll A_o^2$, and the phase is not determined. The third term is the grating structure that gives rise to the two diffracted beams; it may be written as

$$A_o A_x e^{i(\alpha x - \delta_x)} + A_o A_x e^{-i(\alpha x - \delta_x)} \quad \ldots \ldots \quad (14\text{–}35)$$

of which the second term is

$$A_o e^{-i\alpha x}. A_x e^{i\delta_x} \quad \ldots \ldots \ldots \quad (14\text{–}36)$$

and the first is

$$A_o e^{+i\alpha x}. A_x e^{-i\delta_x}. \quad \ldots \ldots \ldots \quad (14\text{–}37)$$

In (14–37) $A_x e^{-i\delta_x}$ is the complex amplitude distribution produced on the original hologram by the original object structure. If this were present alone, then the view obtained by looking through the hologram in the reconstruction arrangement would be the same as the view obtained by looking directly at the object in the arrangement used to produce the hologram. In fact the whole process is simply equivalent to introducing a time delay between the light leaving the object and reaching the observer. Thus one would see a high quality image of the original object, and since the original object was to the left of the plane of the hologram, the image would be to the left of the hologram in the reconstruction, and therefore would be virtual. The term $A_o e^{+i\alpha x}$ superimposes a phase shift corresponding to a collimated beam inclined at angle θ. It simply indicates that the light has been deviated through angle θ and indicates that it is necessary to look through the hologram from the direction θ in order to see the virtual image at I_1'. In (14–36) $A_x e^{i\delta_x}$ has the phase corresponding to the original object but with the sign reversed—a phase advance in place of a phase lag—and indicates that the corresponding image is an equal distance to the right of the hologram in the reconstruction, i.e. is a real image. The term $A_o e^{-i\alpha x}$ indicates that this image is to be found in the direction $-\theta$, i.e. at I_1.

It will be seen that the reconstructions are spatially separated so that each can be observed without any interference due to the presence of the other. If one places a screen at I_1 one obtains an image entirely without the use of lenses— one could, for example, obtain a high quality photograph of the original object simply by placing a film at I_1.

The virtual image I_1' can be recorded by means of a lens suitably positioned to the right of the hologram. Alternatively, one can view I_1' simply by looking through the hologram as though it were a window, but one sees the image as though one were viewing a transparency with a point source behind it, i.e. for any given eye position one sees only a very small part of the image. In the above discussion normal illumination was used. One can, instead, use oblique illumination so that the first order beam emerges normally from the hologram and one looks normally through the hologram in order to see the image. An important improvement was introduced by Leith and Upatnieks and by Stroke (1964): suppose a diffusing screen is placed immediately before the object in Fig. 14–37(a). The reconstructed image then corresponds to a transparency backed by a diffusing screen, and all parts of the image can be seen from a single point of observation. The use of the diffuser has other very important advantages. With the diffuser in position there is virtually no correspondence between position in the hologram and position in the image, with the result that a dust particle or similar defect on the hologram does not give a localized defect on the reconstructed image as it does when no diffuser is used. This is specially important if the object contains areas of roughly uniform luminance since the defects referred to are particularly noticeable in such regions. Another advantage follows from the lack of correspondence between points on the object and points on the hologram: the light from an extra bright element of the object does not produce an extra bright spot on the hologram but is distributed over a wide area; this reduces the range of intensities that the photographic process is required to record. Finally, since all parts of the image can be seen by viewing through any portion of the hologram if a diffuser is used, the remainder of the hologram can be removed. Similarly all parts of the real image can be obtained from a fragment of the hologram! Although no specific region of the image is lost by using only part of the hologram there is an overall loss of resolution, as the utilized area is reduced. It should, perhaps, be noted here that to obtain high resolution in holographic image formation one must record the hologram on a high resolution emulsion.

Throughout the above discussion it has been assumed that the object consisted of a simple transparency, i.e. was two dimensional. The longitudinal coherence of a laser source makes it possible to use an exactly similar procedure for a three-dimensional object provided the depth of the object does not exceed

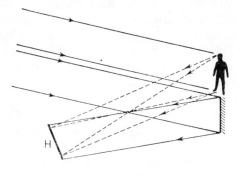

Fig. 14–38. Three-dimensional object.

the coherence length of the laser. Thus a hologram can be produced from a diffusely reflecting three-dimensional object as indicated by Fig. 14–38. One then obtains the advantages of employing a diffuser as indicated above. The reconstructed image is also three-dimensional and exhibits perspective and parallax, depending on the viewing position. It is worth noting that whereas normal parallax is observed with the virtual image, which occupies a position equivalent to that of the original object, interesting effects are obtained if the aerial real image is viewed directly from a point beyond it. Although the observer views the reconstructed object field from behind, each object in the field is seen as illuminated from the front, i.e. the front of each object is seen. On the other hand, if the observer moves, the parallax between different reconstructed objects in the field is appropriate to the actual relative distances of the various images from the observer, i.e. corresponds to the viewpoint behind the images. That is to say, the parallax is the opposite of what one would have had if the original object field had been viewed from the front; one refers to a *pseudoscopic* image. Another aspect of the same phenomenon is the fact that if the observer moves to a point from which two reconstructed opaque objects appear to be in line, it is what the parallax indicates as the nearer that is obscured, whilst the farther seems to be seen through a hole created by the disappearance of the nearer!

Another interesting point is that, whereas for Gabor's original reconstructions (corresponding to the zero order images) it was necessary for the hologram to be a photographic positive, in the present method both positive and negative holograms produce the same first order images if the photographic processing is correct in each case. This is because the change from a positive to a negative hologram interchanges the bright and dark fringes in the grating-like structure but does not cause any changes in the contrast and the shape of the fringes, which are the two things that determine the structure of the image.

The holographic images discussed so far have been monochromatic. If the object were illuminated with lasers emitting the three primary colours and a reference beam were provided for each colour, the hologram would contain three incoherently superimposed holograms—one for each colour. If the same three lasers were used in the reconstruction process each would give three images since each would send light onto the two "wrong" holograms as well as the "correct" one. Hence there would be nine virtual and nine real images but the "correct" three of each would be coincident and give the required coloured images. (The "correct" images are those formed by the correct combination of laser and hologram, e.g. the red image formed from the hologram that was produced with red light. The "wrong" images—such as the red image formed by the hologram that was produced with green light—would be formed in different directions. The holograms are, of course, recorded with an ordinary black and white emulsion.)

The spurious images can be suppressed, and white light can be used in the reconstruction process, by forming holograms in thick emulsions. For each of the three primary colours the reference beam and the "signal" beam are inclined to each other at a large angle. Apart from the fine structure imparted by the signal beam, each hologram then consists of a parallel oblique *planes* of silver within the emulsion (instead of parallel lines as in the normal hologram). Such a *volume hologram* is a three-dimensional grating which, for a given angle of incidence, will diffract particular wavelengths in particular directions (see §12–18). It is then possible to illuminate a suitable produced hologram with white light in such a way that the required wavelengths are diffracted to reconstruct a coloured image. Monochromatic beams are not required in the reconstruction process because the hologram itself selects the wavelengths.

The experimental arrangement discussed above the the production of holographic images involved the use of the near-field or Fresnel diffraction pattern to form the hologram with coherent illumination, and served to illustrate the basic principles involved in holography. Holograms can be produced with other geometrical arrangements and can even be produced with incoherently illuminated objects.

14–20. Applications of Holographic Image Formation

In recent years holography has been applied in a wide variety of different ways; only a few can be mentioned here. Several applications have been made in the field that has come to be known as *holographic interferometry*; the applications are often grouped into three classes: *real-time, time-lapse,* and *time-averaged* interferometry. The principle of real-time holographic interferometry is as follows. Suppose a hologram of an object is made in the usual manner and the object is then changed or distorted in some way. Suppose then, that the object is inserted in the position of the virtual image in the reconstruction arrangement. On looking through the hologram one will see the original object apparently superimposed on the modified object and, since the two are viewed with coherent light, difference fringes will appear and indicate the changes that have occurred in the object since the hologram was made. In fact the object can be distorted or moved in some way whilst it is actually in position and the changes can be observed and measured as they occur—in "real time". The method is an extremely powerful one since it is applicable for objects that have rough and uneven surfaces which would make them completely unsuitable for traditional interferometric study. This is because the holographic method measures the *differences* between the two images, no matter how complicated are the individual images. Moreover, the method can be applied to objects that are, for example, in an enclosure in which they are viewed through a window of poor optical quality or through a non-homogenious intervening medium; the poor quality of, for example, the window will be present for both images and will not show any *difference* fringes—the method automatically ignores irrelevent detail. In the time-lapse technique a hologram formed after the object has been modified is superimposed on a hologram formed before the event. The double exposure hologram reconstructs both images simultaneously and again one can see difference fringes. In the time-averaged method a hologram is obtained by using a long exposure whilst, for example, a specimen is vibrating. This enables one to observe nodes, and the amplitudes of vibration at various places over the surface of the specimen.

Holography finds a useful application in the microscopic examination of certain kinds of specimen. If high magnification is required, the lateral field of view of a microscope is small, but the depth of field is *very* small. There are the two specially difficult problems with which traditional microscopy cannot cope satisfactorily: (1) If one wishes to make a prolonged examination of, say, a small specimen that is suspended in a medium, one would have to re-focus the microscope continuously as the specimen drifts about; at best, this is inconvenient. (An ordinary single short-exposure photomicrograph will not solve the problem if, as is usually the case, the specimen itself has a thickness that is greater than the depth of field of the microscope.) (2) One may wish to be able to examine the appearance of the whole of a continuously changing three-dimensional region simultaneously, and this is literally impossible in traditional microscopy. (In principle, both problems can arise in fields other than microscopy but in microscopy they are especially severe.) Both problems can be solved by forming a holographic image. A single hologram produced with a very short exposure can give a complete three-dimensional image of the whole field, each part of which can be examined subsequently. Obviously, the image of a single particle can be the subject of prolonged microscopic examination and, although one must refocus the observing microscope to examine different planes in the volume, one is able to observe the appearance of the whole volume at the instant when the hologram was made. If one is dealing with moving objects as indicated above, the exposure used to produce the hologram must be very short so that the movement is much less than a wavelength during the exposure; this is possible with a pulsed laser.

Holography also provides a high-capacity system for image storage and retrieval. One can record several holograms on one plate using carrier fringes in various orientations and with various spatial frequencies. The individual holograms are produced, and the individual images retrieved from the composite hologram, by using reference and reconstruction beams in the appropriate directions: in various orientations for correspondingly oriented fringes, and inclined to the signal beam at various angles for various fringe spacings.

Leith and Upatnieks (1965) have described another experiment which at first seems rather startling and which has some interesting possible applications. They placed a translucent screen between the object and the hologram during the construction process. If one uses the reconstruction system in the usual way, one simply sees an image of the diffusing screen— Leith and Upatnieks then placed the diffusing screen exactly in the position of its *real* reconstructed image and found that a sharp clear image of the original object was found beyond this position. This happens because if the phase term introduced by the diffuser itself is $e^{-i\delta x}$, the phase term in its real image is $e^{+i\delta x}$ [see equations (14–36) and (14–37)]. Hence, by placing the diffuser in the position of its real image, the phase effects are cancelled and the effect of the diffuser is removed from the reconstruction so that the reconstructed object is no longer obscured by it. Obviously the cancellation is not complete unless the reconstruction is effected with the unchanged diffuser in exactly the correct position. This provides a method of detecting a change in the structure of a diffusing screen, which is a difficult task by most techniques; alternatively, it provides a method of storing the information contained on, say, a confidential document—the document can be read only by somebody who has both the hologram and the diffusing screen and knows the correct geometrical set-up.

14–21. The Foucault Test. Schlieren Systems

As mentioned in §4–7, the Foucault knife-edge test provides a method of examining the form of a mirror surface. In terms of geometrical (ray) optics, the reflected rays are not concurrent if the mirror surface is not a perfect sphere. It will be seen from Fig. 14–39 that when the knife edge is in the region of focus, rays reflected by some parts of the mirror are obstructed while those from other parts are not. Thus parts of the mirror such as A appear dark and parts such as B appear bright. The shadows on the mirror surface resemble those that would be obtained if a relief map of the mirror were illuminated at almost grazing incidence from the direction of the point X. This geometrical interpretation of the test is fairly satisfactory when the mirror exhibits gross errors of form but is unsatisfactory when the errors are very small.

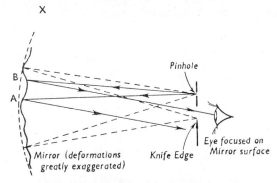

Fig. 14–39. The Foucault knife-edge test.

In terms of wave optics, the incident wave is spherical but the reflected wave is a deformed sphere. It resembles the wave one obtains in the image space of the lens L_2 (Fig. 11–20, p. 257) if there is a phase structure across the diffracting aperture; that is, it resembles the wave emerging from the exit pupil of a microscope objective used. to examine a phase object. It will be seen that the disturbances reflected from the imperfect mirror can be analysed into zero and higher order maxima in the plane of the knife edge—the disturbances in this plane (which is conjugate to the source) do, in fact, constitute a Fraunhofer pattern and the knife edge itself obstructs part of this pattern. If the knife edge obstructs the image of the source as formed by the ideal spherical mirror, the arrangement resembles that used in dark-ground microscopy with oblique illumination—the direct light and all orders on one side are obstructed. In this analogy the eye lens focused on the mirror surface plays the part of the microscope objective focused on the phase object. To increase the brightness of the field of view in the knife-edge test the pinhole can be replaced by a short slit source parallel to the knife edge. In the *phase-contrast test* the knife edge is removed and a small phase plate is placed in the position of the zero order maximum. This test bears the same relation to the phase-contrast method in microscopy as does the knife-edge test to the oblique dark-ground method.

The knife-edge method was applied by Toepler (1864) to the examination of variations in the refractive index of a medium. The principle of his method—

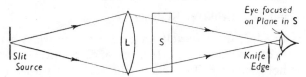

Fig. 14-40. Principle of the Toepler schlieren method.

known as the *schlieren method*—is as illustrated by Fig. 14-40. The slit source is imaged along the knife edge by the lens L. The medium whose variations in refractive index are to be investigated is located at S and the variations in optical path through S cause the emergent wavefronts to be distorted. These converging wavefronts then resemble those reflected from a mirror being examined by the knife-edge test and the shadows reveal the variations in refractive index in S. Since a shadow reveals an angular displacement of the rays, it indicates the rate of change of index across the region concerned.

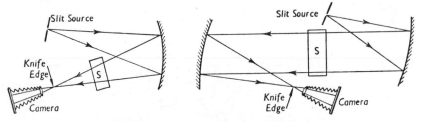

Fig. 14-41. Schlieren systems employing mirrors.

To obtain a large field of view a large aperture lens L would be required. It is then more convenient to replace the lens by an appropriate mirror system. One then avoids chromatic errors and effects caused by striations in the lens itself. The mirrors must be used with oblique illumination in order to avoid double passage through the medium under investigation. Typical schlieren systems are shown in Fig. 14-41.

14–22. Fibre Optics

A fibre, or a bundle of fibres, provides a useful method of conveying light from one place to another. Each fibre consists of a long thin cylinder of a transparent material such as glass. The fact that light can be trapped within a light pipe of this kind has been known for very many years but the modern era of fibre optics can be said to have started in 1951 with the work of H. H. Hopkins and Kapany and of van Heel. If the diameter of the cylinder is large compared with the wavelength of the light, the passage of the light down the fibre can be discussed in terms of geometrical optics; if the diameter is comparable with the wavelength, the fibre acts as a waveguide and a geometrical treatment of the problem is inadequate. The present discussion will be restricted to the former case and will deal only with meridional rays within the fibre; the full geometrical theory includes a discussion of the paths of skew rays and is too long and complicated to be included here. Although most of the light travels along skew ray paths, a discussion that is restricted to meridional rays does provide an adequate introduction to the principles of fibre optics. Fig. 14-42 indicates the path of a meridional ray along a straight fibre of refractive index n_F situated in air. The ray is incident on the flat end face at angle I, is refracted into the fibre, and strikes the wall at angle ϕ. The ray will be totally internally reflected if ϕ is greater than

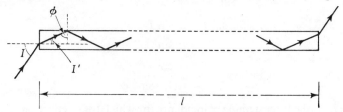

Fig. 14–42. Ray path along a fibre.

the critical angle θ between glass and air. In practice $\theta(\mathrm{cosec}^{-1} n_F)$ is always less than $45°(\mathrm{cosec}\ 45° = 1\cdot414)$. Hence even for grazing incidence $(I = 90°)$, $I' < 45°$ so that $\phi > 45°$ and one has total reflection. If $I < 90°$, ϕ is increased. Hence *all* the light that enters the fibre is "captured", i.e. suffers successive total internal reflections and passes to the far end of the fibre. One simple general result is worth noting at once: the path length of a ray that is multiply reflected down a straight fibre is independent of the diameter of the fibre—it is $l \sec I'$ where l is the length of the fibre; for light incident at 40° in air on a fibre of index 1·5, it is only about 10% greater than the length of the fibre. If the fibre were straight and homogeneous and isolated in air, and the walls were smooth and clean, the only significant loss of light would be due to absorption by the glass itself, and would be well below 1% per inch. However, if the fibre were closely surrounded by neighbouring fibres in a bundle, one would lose light through frustrated total reflection (see §21–6); in addition, one would fail to get *total* internal reflection if the walls were not clean. The losses arising from these causes can be large—for a 25μ diameter fibre of index 1·5 light entering at an angle of incidence in air of 10° suffers over 100 reflections per inch. To avoid these losses it is usual to employ a fibre that consists of a core of high-index glass with a cladding of low-index glass to protect the fire-polished surface of the core and to isolate it from neighbouring fibres. (Van Heel originally employed a plastic coating.)

It was seen above that for a glass fibre in air all the light entering the fibre is "captured"; this is not necessarily the case for a clad fibre. Consider, then, the general case illustrated by Fig. 14–43 where n_F is the index of the fibre itself, n_C the

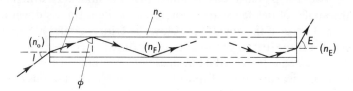

Fig. 14–43. A clad fibre.

index of the protective coating and n_0 the index of the medium from which the light is incident. The angle of incidence at the wall is given by

$$\phi = \pi/2 - I'.$$

If $\phi \geqslant \phi_C$, the critical angle between the fibre and the coating, one has total internal reflection. ϕ_C is given by

$$\sin \phi_C = n_C/n_F$$

One has
$$n_o \sin I = n_F \sin I'$$
$$= n_F \sin (\pi/2 - \phi)$$
$$= n_F \cos \phi$$
$$= n_F(1 - \sin^2 \phi)^{1/2}$$

When $I = I_{max}$, the largest angle of incidence for which total reflection is obtained, $\phi = \phi_C$. Hence one has

$$n_o \sin I_{max} = n_F[1 - (n_C/n_F)^2]^{1/2}$$
$$= (n_F{}^2 - n_C{}^2)^{1/2}.$$

$n_o \sin I_{max}$ is referred to as the numerical aperture (N.A.) of the fibre and, although the analysis has been confined to meridional rays, it gives an indication of the acceptance angle of the fibre for efficient transmission. If $n_F = 1.79$ and $n_C = 1.48$ (which are realistic values) the N.A. just exceeds unity, i.e. light incident in any direction in air would be captured. If one does not demand such a large entrance cone, the core can be of glass of somewhat lower index and usually this will mean of somewhat better transparency. As a ray travels down a fibre its angle of inclination to the axis remains constant but changes sign with each reflection. Since the end faces of a straight fibre are parallel, $n_o \sin I_{max} = n_E \sin E$ where E is the angle of emergence, i.e. the N.A. of the exit cone of rays is equal to that of the incident cone. It can be noted here that if one has a finite pencil incident at angle I the skew rays will cause the light to disperse in azimuth gradually so that for a long fibre the emergent light will fill an annulus of a cone of semi-angle E.

To manufacture a clad fibre one inserts a rod of higher index glass into a tube of the lower index glass and the clad fibre is drawn in a furnace. It is, in fact, possible to draw a number of such fibres simultaneously in such a way that the high index cores are embedded in a continuous low index matrix (Kapany 1959). Typically, a multiple fibre of this type can have an overall cross-section diameter of 45μ and can contain 150 cores, each being about $2\frac{1}{2}\mu$ in diameter, all the cores being aligned with great precision.

In the above discussion it was assumed that the fibre was straight. If it is curved, the angles of incidence are changed as the light travels down the fibre, and losses will occur if the angles fall below the critical angle; it can be shown that skew rays are the first to be lost as the curvature is increased. The curvature that can be tolerated is increased if one does not attempt to utilize the maximum theoretical numerical aperture. Typically, one can tolerate a radius of curvature down to about twenty times the fibre diameter.

So far, the discussion has referred to cylindrical fibres, but brief mention must now be made of conical fibres. It is obvious that if all the light that enters the thicker end of a conical fibre reaches the thinner end, there is a gain in flux per unit area. Fig. 14–44 illustrates two important points. Firstly, the angle of incidence at the wall of a conical fibre gradually decreases as the ray proceeds, so that there is the possibility that one will cease to get total internal reflection; this difficulty could be at least partially overcome by employing a metal coating. However, there is the possibility that eventually the ray will strike the wall at, or

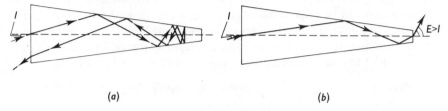

(a)   (b)

Fig. 14–44. Conical fibres.

"beyond", normal incidence and return towards the thicker end of the fibre—Fig. 14–44(a). Secondly, but arising from the same fundamental geometry, the solid angle of the cone of rays that does emerge from the thinner end is always greater than that of the cone formed by the same rays when they entered at the thicker end (Fig. 14–44(b)).

Since light can be conveyed efficiently from one end of a fibre to the other, a bundle of fibres can be used to convey a distribution of light from one surface to another. For example, images can be conveyed provided the fibres in the bundle are aligned accurately. Fibre bundles are of two types: rigid bundles of relatively short fibres, and flexible bundles of long fibres (Fig. 14–45).

A short rigid bundle of large cross-section can have one face curved and the other flat. By arranging the curved face to fit the curved image surface of an optical system, one can produce a field flattener (Kapany and R. E. Hopkins

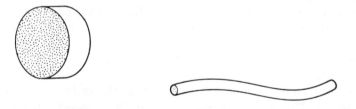

Fig. 14–45. Fibre bundles.

1957) (see §15–9). The *Focon* (Kapany 1961) is a field flattener that also corrects distortion (§15–10) by changing the relative positions of the fibres in the bundle between one face and the other; by using a conical bundle one can, at the same time, increase the illumination of the image. Another, not dissimilar, application of a short broad rigid fibre bundle is to convey the image from a cathode ray tube to a photographic film. This can also correct any distortion in the system, but the more important feature is the improvement in the photometric efficiency. The luminescent screen of a C.R.T. radiates in all directions and an ordinary camera lens some distance in front of the screen can collect only a small fraction of this light whereas a fibre bundle in contact with the screen can collect virtually all the light since the acceptance angle of the bundle is virtually 180°. One can fuse the front face of an air-tight bundle against the face of the C.R.T. or one can deposit the luminescent layer directly onto the front face of the bundle; the rear face of the bundle is placed against the photographic plate or film.

Thin flexible fibre bundles have been used in medicine as endoscopes (i.e. cystoscopes, bronchoscopes, etc.) where it is required to make observations in an enclosure that can be approached only along a curved path. In some flexible fibrescopes of this kind the outer fibres in a composite bundle are used to convey light to illuminate the object, a lens is used to form an image, and the inner fibre bundle is used to convey this image to the remote end (see Fig. 14–46).

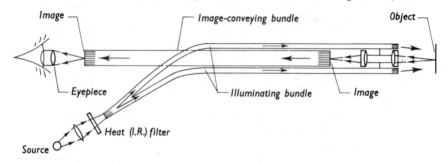

Fig. 14–46. Flexible fibrescope.

Fibre bundles of both types have been used as light funnels to collect light from an aperture of one shape and convey it through an aperture of a different shape, e.g. to funnel light from an astronomical telescope or from a source of awkward shape and size into the slit of a spectrometer. Fibre bundles can be used also in coding devices—for example, by intertwining the fibres in a bundle an image can be "scrambled".

When an image is formed on one end of a fibre bundle and conveyed to the other end, the limit of resolution is determined in an obvious way by the diameters of the fibres. One refers to *static scanning*. It is found that *dynamic scanning*, in which the two ends of the bundle are given exactly synchronized rapid random transverse displacements equal to a few fibre diameters, effectively smooths out the fibre bundle structure from the final image and improves the resolution. The scanning period must be small compared with the response time of the detector. With dynamic scanning light from a given element of the object is not conveyed solely by a particular fibre, so that the effect of a broken fibre is less obvious. A similar advantage is obtained in a system employed by Koester (1968). A dispersed image of the object is projected on to one end of the fibre bundle using a system resembling a direct vision spectroscope, and a similar system removes the dispersion and reconstitutes the polychromatic image at the far end of the bundle. Again several fibres are involved in conveying light from each object element—and each fibre is concerned with many object elements.

The above examples of the application of fibre bundles simply serve to illustrate the wide variety of uses to which these devices have been put; they have been used in very many analogous situations. An application that is somewhat different in kind is in refractometry. For a given angular aperture of the light entering an unclad rod or fibre there will be a reduction in the flux emerging from the far end as the index of the surround approaches that of the rod, and this has been used to detect small changes in the index of the surround. To apply the method to a wide range of indices it is necessary to use a series of rods of different indices.

All the fibres and fibre bundles that have been referred to above have been *passive*, that is to say, they simply convey light. *Active fibres* have also been produced in which light is actually generated and then conveyed to the end of the fibre. For example, scintillating fibres have been used in tracking high energy particles. Fibre lasers have also been developed (Snitzer, 1961), which make available a flexible high intensity laser probe.

CHAPTER XV

THE MONOCHROMATIC ABERRATIONS

15-1. Introduction. Ray and Wavefront Aberration

FOR an optical system of finite aperture, the Gaussian image is defined as that which would be formed if, in terms of geometrical optics, the system gave perfect imagery and had the cardinal planes of its paraxial region. When defined in this way, the Gaussian image can be found for any object although both object and image may extend beyond the paraxial region. If, geometrically, the system does give perfect imagery, the rays from each point object intersect at the Gaussian image point and a spherical wave converges to the latter so that disturbances which have passed through different zones of the aperture arrive exactly in phase. However, it was proved in §§ 14–3 and 14–4 that, for a given optical system, such imagery is not possible for more than one position of the object plane. That is to say, outside the paraxial region a completely perfect optical system is impossible. In practice, for large aperture and field, it is usually impossible to secure *perfect* imagery for even one position of the object, although a close approach to perfection can be attained in many instances. Thus, in general, an image will show some defects caused by the aberrations of the optical system.

The defect of the image may be said to be caused by the failure of the rays from a point source to unite at the Gaussian image point, or by the failure of the emergent wavefront to be a sphere converging on this point and the consequent failure of the disturbances to arrive exactly in phase. The aberrations are frequently discussed in terms of their effect on the ray paths, but it has been pointed out that, near a focus, the rays do not give a good indication of the distribution of light in the image. To obtain information concerning the latter, it is necessary to consider diffraction by the aperture stop of the instrument, and it was found convenient to discuss this in the image space of the system. For a perfect instrument and a point source, one then has the diffraction of a converging spherical wave by the exit pupil of the system. In order to calculate the intensity at any point in the image plane, it was necessary to find the relative phases of the disturbances from each element of the wave. In the presence of aberration the emergent wave is no longer a sphere and these relative phases are changed. The new phase differences can be calculated if the shape of the wavefront is known, and it is convenient to specify this by means of the amount by which the wave differs from the ideal spherical shape. The deviation of the wave from the ideal spherical form, measured as an optical path length along the wave normal, is called the *wavefront aberration*; it is positive when the wave lies in front of the ideal position.

For regions which are not close to a focus it is possible to assume that the rays give a good indication of where the energy goes, and since there is no exact focus when the aberrations are large, it is possible in this case to get a fairly good idea of the distribution of light from the ray paths. In such cases it may be convenient to describe the aberrations by their effect on the rays, but since most optical systems are of little use unless the aberrations are small, it is usually better to discuss the shapes of the wavefronts.

The wavefront aberration in wavelengths gives the phase difference introduced in multiples of 2π, so that the importance of an aberration can be seen at once. On the other hand, the ray aberrations frequently give a very misleading idea of how significant the aberration is. In Fig. 15–1 B' is the Gaussian image point at a large distance from the exit pupil E'. PQ is a wave with wavefront aberration W' at the edge of the aperture and, even when W' is small, the so-called marginal ray aberration $\delta l'=$B'M' may be very large. This would give the impression that the image is very poor, whereas the small value of W' indicates that on the Gaussian plane at B' the image is almost perfect. If the Gaussian image is at

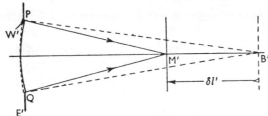

Fig. 15–1. Ray and wavefront aberration.

infinity, the ray aberration must always be infinite which, in addition to the impression of seriousness, gives certain computational difficulties. Thus, except in special cases, the wavefront aberration gives a more reliable and useful measure of the imperfection of an optical instrument. Since the wavefront aberration is a measure of the retardation or advancement of the wave with respect to the ideal wave, the aberrations arising in different parts of a system are, for small errors, simply additive. On the other hand, the ray aberration after passage through one part of a system is not simply added to that which arises later.

15–2. Choice of Reference Surface

As defined above, the wavefront aberration is measured from a reference sphere centred on the Gaussian image point. Frequently it is convenient to choose an alternative spherical reference surface, and it is now necessary to find how the wavefront aberration is affected by a change in the reference surface. A change in the radius alone simply adds a constant to the aberration at all points on the wavefront, and is of little interest. A movement of the centre of the reference sphere is more significant. As originally defined, the wavefront aberration indicates the extent to which the intensity distribution in the region of the Gaussian image differs from that for a perfect instrument. If measured from a reference sphere that is not centred at the Gaussian image point, the aberration gives a measure of the phase differences between disturbances arriving at the centre of *that* sphere, i.e. it measures the perfection of the image in that region. In other words: a movement of the centre of the reference sphere corresponds to a change of mind concerning the region in which the wavefront is considered to be coming to a focus. For example, with one reference sphere the aberration may be large, whereas with another it is small. Instead of saying that the image in the region of the centre of the first reference sphere is poor, one may say that the image is at the centre of the second sphere and is good—the disturbances are more nearly in phase at the second point, which therefore comes closer to being a focus. Two special cases are important in practice: a movement of the reference sphere centre (a) along, and (b) perpendicular to, the principal ray of a pencil proceeding to a region of focus. (It must be noted that although the rays do not give a reliable indication of the distribution of energy in the region of focus, they are the wave normals, and their paths remain significant.)

15–3. Longitudinal Shift of Focus

A movement of the reference sphere centre along the principal ray corresponds to a longitudinal movement of the point which is regarded as the focus. In

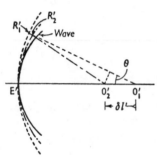

Fig. 15–2.
Longitudinal shift of focus.

Fig. 15–2 $E'O_1'$ represents the principal ray of an image-forming pencil—it may, as a special case, be the optical axis of the system. $E'R_1'$ is a reference sphere centred at O_1', and W' is the wavefront aberration measured from this sphere. The curve shown in full represents the actual wave; the rays are not shown. Suppose the centre of the reference sphere is moved a distance $\delta l'$ to O_2' and that its radius is changed by the same amount so that the sphere is $E'R_2'$. The wavefront aberration is now $W' + \delta W_L'$, where $\delta W_L'$ is negative and equal to $-n'(R_1'R_2')$ where n' is the refractive index of the image space. It is assumed that $R_1'R_2'$ is small, so that $\angle O_2'R_2'O_1'$ is small and $\angle E'O_2'R_2' \doteqdot \angle E'O_1'R_1' = \theta$ where θ is also small.

From the figure $$R_1'O_2' = R_1O_1' - \delta l' \cos \theta$$

and $$R_2'O_2' = E'O_2' = E'O_1' - \delta l'.$$

Now $R_1O_1' = E'O_1'$ since $E'R_1'$ is a sphere centre O_1'.

$$\therefore \quad R_1'R_2' = (R_1'O_2' - R_2'O_2') = \delta l'(1 - \cos \theta)$$
$$= \tfrac{1}{2}\theta^2 \delta l' \text{ since } \theta \text{ is small,}$$
$$\therefore \quad \delta W_L' = -\tfrac{1}{2}n'\theta^2\delta l'.$$

If the aberration is not large, θ may be replaced by the corresponding angle made by the ray. In terms of the linear aperture, ϱ, $\delta W_L' \propto \varrho^2$.

15–4. Transverse Shift of Focus

In Fig. 15–3 $E'O_1'$ again represents the principal ray. Suppose the centre of the reference sphere (radius $E'O_1' = R$) is moved from O_1' a distance $\delta h'$ to O_3'. At any point on the wave, the resulting change $\delta W_T'$ in the wavefront aberration will depend on the azimuth ϕ of that point with respect to the plane $E'O_1'O_3'$.

The figure shows the section of the wave and the old and new reference spheres in the plane of azimuth ϕ. $\delta W_T'$ is negative and is given by $-n'(R_1'R_3')$. One has

$$R_1'R_3'=R_1'O_3'-R_3'O_3'$$
$$=\frac{(R_1'O_3')^2-(R_3'O_3')^2}{R_1'O_3'+R_3'O_3'}.$$

Now R_1' and O_3' are the points (x,y,z) and $(R,0,-\delta h')$ respectively, with axes as shown.

$$\therefore \quad (R_1'O_3')^2=(x-R)^2+y^2+(z+\delta h')^2.$$

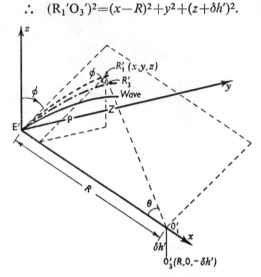

Fig. 15–3. Transverse shift of focus.

Since R_1' lies on the sphere, centre O_1' $(R,0,0)$, radius R, one has

$$(x-R)^2+y^2+z^2=R^2,$$
$$\therefore \quad (R_1'O_3')^2=R^2+(\delta h')^2+2z\delta h'.$$

In addition, since $E'R_3'$ is the sphere whose centre is O_3',

$$(R_3'O_3')^2=(E'O_3')^2=R^2+(\delta h')^2.$$

Hence $$(R_1'O_3')^2-(R_3'O_3')^2=2z\delta h'.$$

It is sufficiently accurate to write $(R_1'O_3'+R_3'O_3')=2R$.

$$\therefore \quad R_1'R_3'=\frac{z}{R}\delta h'.$$

Now $$z=\varrho\cos\phi,$$
$$\therefore \quad \delta W_T'=-\frac{n'}{R}\varrho\delta h'\cos\phi.$$

15–5. The Aberration Function

It will be assumed that each refracting surface in the optical system is a surface of revolution about a common axis of symmetry—the optical axis. It is not necessary to assume that the surfaces are spherical. When the object point lies on the optical axis, the emergent wavefront will also be a surface of rotation about the optical axis, and the wavefront aberration at any point P on the wave will depend only upon the aperture ϱ—i.e. its distance from the axis. For an off-axis object point, the aberration depends also upon the azimuth of P since the wave will not, in general, be a surface of rotation about the principal ray. It is the purpose of this section to deduce the form of the dependence of W' on the aperture and azimuth, and also the dependence of W' on the position of the object point. It is

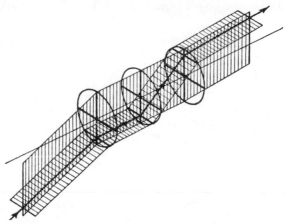

Fig. 15–4. Sagittal and tangential planes.

convenient to define two planes as follows: (1) The *meridian* or *tangential plane*, from which the azimuth ϕ is measured, is the plane containing the optical axis and the off-axis object point (and Gaussian image point); throughout the system the principal ray lies in the meridian plane. (2) In the object, image, or any particular intermediate space, the *sagittal plane* is the plane which contains the principal ray and is perpendicular to the tangential plane. It will be seen that, whereas there is only one meridian plane for a given object point, the sagittal plane is different in different spaces through the system. Fig. 15–4 illustrates the passage of a principal ray through a system of three refracting surfaces; the tangential plane is shaded vertically and the sagittal planes are shaded "horizontally".

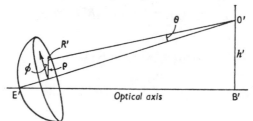

Fig. 15–5. Wave in image space.

Fig. 15–5 shows the reference sphere in the image space. E' is the centre of the exit pupil and O' is the Gaussian image point at height h' from the optical axis. The wavefront aberration of the element of wavefront associated with the ray through R' will depend on the aperture ϱ, and on the azimuth ϕ measured from the tangential plane E'B'O'. It is convenient to replace the variables ϱ and h' by their fractional values $r=\varrho/\varrho_{max}$, and $\sigma=h'/h'_{max}$; ϱ_{max} is the radius of the exit pupil and is therefore the maximum value of ϱ which can occur, and h'_{max} is the off-axis image height at the edge of the field of view of the instrument. For a particular ray in a pencil traversing the system, r remains constant whereas ϱ varies as the diameter of the pencil expands and contracts. In a similar way, σ is constant through the system for a particular extra-axial object point whereas, in intermediate image spaces within the system, h' has different values. Thus W' depends on σ, r, and ϕ, and $W'(\sigma,r,\phi)$ is referred to as the *aberration function*. Whilst it is very difficult to find $W'(\sigma,r,\phi)$ for any but the simplest systems, it is possible, from considerations of symmetry, to deduce the general form which the function can take. In those cases where the function does not contain ϕ, symmetry about the principal ray and about the optical axis require W' to be an even function of σ and r. That is, in the general form of the function the terms which do not contain ϕ must be functions of σ^2 and/or r^2. In addition, there is symmetry about the meridian plane, so that W' must remain unchanged if ϕ is replaced by $-\phi$. This condition is satisfied if terms involving ϕ are functions of $\cos \phi$. Again, when O' is an equal distance below the axis, the value of W' for the point corresponding to R' must be the same as for R' itself, i.e. W' is unchanged if r is constant and ϕ is changed to $\pi \pm \phi$ while the sign of σ is reversed. Alternatively, W' is unchanged if the sign of r and σ is changed, keeping ϕ constant. This means that terms which contain ϕ must also contain σ and r; the conditions are satisfied if these terms are functions of $(\sigma r \cos \phi)$. Thus the aberration function contains the variables in the forms σ^2, r^2, $\sigma r \cos \phi$, and can be written as the most general power series in these variables. One has

$$W'(\sigma,r,\phi) = {}_0C_{00} + ({}_0C_{20}r^2 + {}_1C_{11}\sigma r \cos \phi + {}_2C_{00}\sigma^2)$$
$$+ ({}_0C_{40}r^4 + {}_1C_{31}\sigma r^3 \cos \phi + {}_2C_{22}\sigma^2r^2 \cos^2\phi$$
$$+ {}_2C_{20}\sigma^2r^2 + {}_3C_{11}\sigma^3r \cos \phi + {}_4C_{00}\sigma^4) + \ldots .$$

It is possible to show that some of these terms may be omitted.

It has already been pointed out that a change in the radius of the reference sphere adds a constant wavefront aberration to each part of the wave. If the radius of the reference sphere is always adjusted so that $W'=0$ along the principal ray, one has ${}_0C_{00}={}_2C_{00}={}_4C_{00}= \ldots =0$, since these are the only terms which remain on putting $r=0$. For an axial object, $h'=0$ and one has

$$W' = {}_0C_{20}r^2 + {}_0C_{40}r^4 + \ldots .$$

Now it has been shown that a longitudinal shift of focus introduces a wavefront aberration proportional to the square of the aperture, so that such a shift can remove the first term. The movement required is the same for all points in the field of view, so that this term simply moves the image plane without changing the image quality; it corresponds to an error in the assumed position along the axis of the Gaussian image plane. In a similar way, the term ${}_1C_{11}\sigma r \cos \phi$ can be removed by a transverse shift of focus, since such a movement of the reference sphere centre introduces a term proportional to $(r \cos \phi)$. This term will be removed for all image heights simultaneously if one moves the centre of the reference sphere transversely by an amount proportional to σ; i.e. proportional to the existing height of the centre. Such a movement of the centre of the reference sphere for all object points therefore changes the size of the image but in no way affects the image quality. It corresponds to a change in the value of the Gaussian transverse magnification. Thus the terms $_0C_{20}\,r^2$ and $_1C_{11}\sigma r \cos \phi$ do not represent aberrations or defects of the image, but can be removed by changing the assumed values of the Gaussian constants of the system. There are other terms which correspond to longitudinal and transverse shifts of focus, but these do affect image quality (see below).

The remaining terms in the aberration function are as follows:

$$W'(\sigma,r,\phi)=_0C_{40}r^4+_1C_{31}\sigma r^3 \cos \phi+_2C_{22}\sigma^2 r^2 \cos^2 \phi$$
$$+_2C_{20}\sigma^2 r^2+_3C_{11}\sigma^3 r \cos \phi+ \ldots \quad . \quad (15\text{--}1)$$

The terms given here are those in which the sum of the powers of σ and r does not exceed 4. They represent the so-called *primary aberrations*. When expressed as transverse ray aberrations, the aperture and field dependence leads to their being referred to as the *third order* aberrations. They are also called the *Seidel* aberrations since much early work on the evaluation of these errors was done by von Seidel.

The terms of (15–1) are, respectively, *primary spherical aberration, coma, astigmatism, field curvature,* and *distortion*; they will now be discussed in turn. When discussing each aberration it will be assumed that the others are absent.

15–6. Spherical Aberration

For an axial object point $\sigma=0$, and the aberration function reduces to

$$W'=_0C_{40}r^4+_0C_{60}r^6+_0C_{80}r^8+ \ldots .$$

These terms represent the primary, secondary, tertiary, ... spherical aberration. The present discussion will be restricted to systems of moderate aperture so that only the primary aberration is important. Fig. 15–6 illustrates the shape of wavefront exhibiting positive primary spherical aberration. W' is proportional

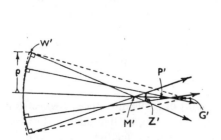

Fig. 15–6. Spherical aberration.

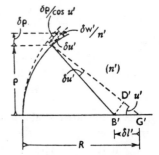

Fig. 15–7. Ray and wavefront aberration.

to the fourth power of the aperture ϱ, and since the rays are simply the wave normals, the effect on the ray paths can also be seen. The rays in the paraxial region, where ϱ is vanishingly small, come to a focus at the Gaussian image point G′, and those from the marginal zone of the aperture intersect the axis at the so-called marginal focus M′. Rays from intermediate zones come to a focus between M′ and G′.

In Fig. 15–7 G′ is the Gaussian focus, and B′ is the point of intersection of the ray at aperture ϱ. $\delta l'$ is the longitudinal ray aberration, and $\delta u'$ the angular ray aberration. R is the radius of the reference sphere, and the wavefront aberrations at apertures ϱ and $\varrho + \delta \varrho$ are W' and $W' + \delta W'$. From the figure, remembering that W' is an optical path length, one has

$$\delta u' = \left(\frac{\delta W'}{n'} \right) \Big/ \frac{\delta \varrho}{\cos u'}$$

$$= \frac{\cos u'}{n'} \cdot \frac{\delta W'}{\delta \varrho}$$

In addition, $\qquad\qquad \varrho = R \sin u'.$

$$\therefore \quad \delta u' = \frac{1}{n'R} \cdot \frac{\partial W'}{\partial u'}.$$

Now for primary spherical aberration one may write

$$W' = Cu'^4,$$

$$\therefore \quad \delta u' = \frac{4Cu'^3}{n'R}.$$

If the aperture is not large, one may write

$$B'D' = R\delta u' = u'\delta l',$$

$$\therefore \quad \delta l' = \left(\frac{4C}{n'} \right) u'^2.$$

That is, the longitudinal ray aberration is proportional to the square of the aperture.

It will be seen that in the presence of spherical aberration there is no real focus, i.e. no point where all the disturbances are in phase and all the rays intersect. In terms of rays, there will be a circular patch of light on any screen perpendicular to the axis. The size of this patch reaches a minimum when the screen is at Z' which divides M'G' in the ratio 1:3 (Fig. 15–6). This minimum patch is sometimes called the disc of least confusion and, in terms of geometrical optics, it is here that one has the closest approach to a focus. However, the disturbances from various zones of the aperture are most nearly in phase at the point P', midway between M' and G'. If the aberration is small, the best focus will certainly be at P' since this is the point where there is the least possible destructive interference. When the aberration is large, so that phase differences exceeding 2π are involved, there can be no simple discussion of the effect of the superposed disturbances, and it is likely that as the aberration increases the best focus moves towards Z'. This is in accordance with the general result that in the absence of a focus the energy travels approximately along the ray paths.

For the image of a point source, the effect of a small amount of spherical aberration is to cause the outer maxima to become more intense. In the absence of aberration, the patterns slightly inside and outside the focus consist of a series of rings, but with a small amount of positive spherical aberration, the pattern outside the focus quickly changes to a more diffuse distribution of light. This provides a useful test for the presence of aberration and a wavefront aberration of a quarter of a wavelength can be detected easily.

15-7. Coma

The term $_1C_{31}\sigma r^3 \cos \phi$ represents the aberration known as primary coma. Since it varies linearly with σ, it is the first aberration to appear as one goes off-axis. It is convenient to consider the shape of the wavefront in the sagittal and tangential planes. In the former, $\phi=\pi/2$ and $W'=0$, so that the wavefront coincides with the reference sphere centred at the Gaussian image point. In the tangential plane, $\phi=0$ and, for a given extra-axial object point, W' is proportional to the cube of the aperture.

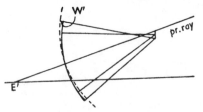

Fig. 15-8. Coma.

Fig. 15-8 shows, for positive coma, the shape of the wavefront in the tangential section and shows also the ray paths. It will be seen that the rays at a given aperture come to a focus below the principal ray, the error being a linear measure of the coma in the tangential section; it can be shown to be proportional to r^2. Although, in the sagittal section, $W'=0$, the wavefront is tilted so that the rays do not intersect on the principal ray. This is illustrated by Fig. 15-9(a). For a given aperture, rays in the tangential section intersect at T' and those in the sagittal section at S'. Rays from diametrically opposite elements of the wave in other azimuths come to a focus at other points on the circumference of the circle indicated. A similar coma circle gives the points of focus for rays from diametrically opposite elements of each annulus of the wave. The pattern of ray intersection points on the Gaussian plane is shown in Fig. 15-9(b), the

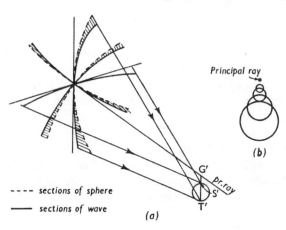

Fig. 15-9. Coma.

radius of each coma circle being half the distance of its centre from the principal ray—G'T'$=$3G'S'; one may say that the linear tangential coma is three times the linear sagittal coma. Obviously the effect of coma is more easily discussed in terms of wavefront aberration! Again the pattern indicated by geometrical optics is observed only with very large aberrations—its comet-like appearance being the origin of the term coma. The appearance of the image in the presence of coma is shown in Plate IV; (f) is for about $\lambda/2$ and (h) is for about 5 wavelengths of aberration.

15–8. Astigmatism

As one proceeds farther off-axis, two more aberrations become significant. These are astigmatism and field curvature and the former, represented by the term $_2C_{22}\sigma^2r^2 \cos^2 \phi$, will be discussed first. As with coma, it is convenient to consider the sagittal and tangential sections of the wavefront, and again $W'=0$ in the sagittal plane where $\phi=\pi/2$. In the tangential plane, $\phi=0$ and, for a given object point, W' is proportional to the square of the aperture. Thus in this plane astigmatism corresponds to a shift of focus and, as indicated by Fig. 15–10, the

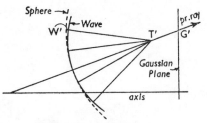

Fig. 15–10. Astigmatism.

rays intersect at a point T′ somewhat removed along the principal ray from the Gaussian focus G′. In the sagittal plane the rays are focused exactly at G′. Usually field curvature accompanies astigmatism and the rays in the sagittal section are then focused at a point S′ which is also removed from G′. That is, in the presence of both field curvature and astigmatism, the wavefront is a circle in both sagittal and tangential sections. The circles differ from each other and differ from the circular section of the reference sphere. The shape of the wavefront in the presence of astigmatism and field curvature is shown in Fig. 15–11. In terms of geometrical optics, one obtains line foci at T′ and S′, the lines being perpendicular to the tangential and sagittal planes respectively and centred on

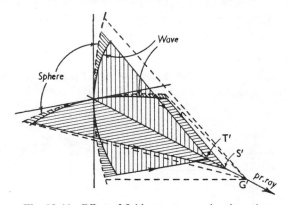

Fig. 15–11. Effect of field curvature and astigmatism.

the principal ray. To visualize the paths of the rays, it is convenient to divide the cone of light entering the optical system into fans of rays as follows. In Fig. 15–12 G is the object point. E and E′ are the entrance and exit pupils. The cone of light from G which enters E can be divided into fans of rays which intersect E along lines parallel to the tangential plane; these are known as tangential fans of rays. A typical tangential fan of rays is shaded vertically. It emerges from E′ and proceeds to a focus as shown. In the same way the cone of light from G could be divided into sagittal fans which enter E along lines parallel to the sagittal plane. A typical sagittal fan is shaded "horizontally". It is found that each tangential fan comes to a focus at a point on the so-called *tangential focal line* and each sagittal fan is focused at a point on the *sagittal*

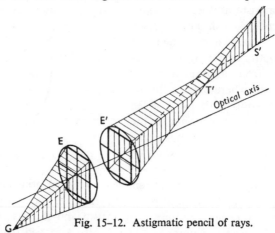

Fig. 15–12. Astigmatic pencil of rays.

focal line. The general appearance of an astigmatic pencil of rays is therefore as illustrated. The maximum cross-section of the pencil is least midway between the focal lines; here the cross section is a circle. Since the point on the principal ray midway between the focal lines is also the point where the disturbances from all elements of the emergent wave are most nearly in phase, there is little doubt that this is the position of the best focus. The appearance of the focal line in the presence of about λ and $2\frac{1}{2}\lambda$ of astigmatism is shown in Plate IV (*i*) and (*j*). (*k*) and (*l*) show the appearance of the image midway between the focal lines for $2\frac{1}{2}$ and $6\frac{1}{2}$ wavelengths of astigmatism.

15–9. Field Curvature

Since it is proportional to the square of the aperture, the term $_2C_{20}\sigma^2r^2$ corresponds to a longitudinal shift of focus. The wavefront is therefore spherical,

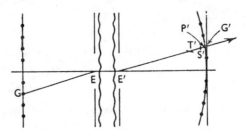

Fig. 15–13. Effect of field curvature and astigmatism.

but focused on a point removed along the principal ray from the Gaussian image plane. Since the term is also proportional to σ^2, the shift of focus is proportional to the square of the image height. Therefore, to a first approximation, if one has a series of point objects such as G in a plane perpendicular to the axis (Fig. 15–13), the emergent wavefronts are focused on points such as

P' lying on a spherical surface which touches the Gaussian image plane; i.e. the image field is curved. In practice, the various constant coefficients in the aberration function are not completely independent; e.g. $_2C_{20}$ and $_2C_{22}$ are related. The curvature of the image in the absence of astigmatism is called the Petzval field curvature, and is positive when as shown in the diagram. As far as the image quality on the Gaussian plane is concerned, one has a simple out of focus effect corresponding to the distance P'G'. If astigmatism and Petzval field curvature are present together, the coefficient $_2C_{20}$ is changed, and in the sagittal section the wavefront is not focused on P'. For positive astigmatism one has focal lines at S' and T'. It can be shown that P'T'=3P'S', where P' is defined by the value of $_2C_{20}$ when $_2C_{22}=0$. Thus, as the astigmatism decreases, S' and T' approach P', and for negative astigmatism T' and S' lie to the right of P', the three to one relationship being maintained.

15-10. Distortion

The first three aberrations affect the quality of the image of a point source, and the fourth affects its position. However, since the latter introduces an out of

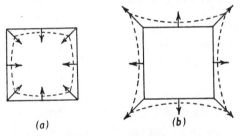

(a)   (b)

Fig. 15-14. Distortion.

focus effect, all four may be said to affect the quality of the image on the Gaussian plane. The fifth primary aberration is defined by the term $_3C_{11}\ \sigma^3 r \cos \phi$ and, for a point source, causes a displacement of the image in the Gaussian plane without loss of image quality. This can be seen at once, since any term proportional to $r \cos \phi$ corresponds to a transverse shift of focus. The displacement is proportional to σ^3 and causes a distortion of the shape of the image without loss of definition. The aberration is known simply as distortion and is the one aberration which is best thought of in terms of geometrical optics. (For a point object there is little sense in saying that the image at G' is poor if a perfect image is, in fact, formed a short distance away.) The nature of distortion is best illustrated by its effect on the image of a square in a plane perpendicular to the axis. For positive distortion, each point in the image is moved towards the optical axis by an amount which is proportional to the cube of its distance as indicated by Gaussian optics. For negative distortion, the displacement of each image point is away from the axis. The effects are shown in Fig. 15-14(a) and (b) for positive and negative distortion respectively; the figures shown dotted are the distorted images of the squares whose Gaussian images are as shown. For obvious reasons, positive and negative distortions are referred to as barrel and pin-cushion respectively.

15–11. Coma and the Sine Condition

It was shown in Chapter XIV that in the absence of spherical aberration an optical system of large aperture will give good imagery of points near the axis for one pair of conjugates if the Sine Condition is satisfied for all zones of the aperture. Now for points near the axis one can neglect the aberrations which depend on powers of σ higher than the first, and, for the inner zones of the aperture, primary coma is the only significant aberration if spherical aberration is absent. That is to say, by satisfying the Sine Condition for the inner zones, one ensures elimination of primary coma. For a somewhat larger aperture there will be the possibility of an aberration $_1C_{51}\sigma r^5 \cos \phi$, and for a still larger aperture $_1C_{71}\sigma r^7 \cos \phi$ may become important. If the Sine Condition is satisfied for all zones of a wide aperture system, there can be no significant aberration for small values of σ, and all these coma type aberrations must be absent. Thus the satisfaction of the Sine Condition provides a more highly corrected system than the elimination of primary coma.

In deducing the Sine Condition, it was assumed that the magnification was the same for all zones of the system. Reference to Fig. 15–8 will show that this must eliminate the coma, since this aberration causes a variation with aperture of the height above the axis of the point of intersection of the rays. It is possible to formulate a Sine Condition of much wider significance than that given above. In particular, it is useful to investigate the significance of the condition in the presence of spherical aberration. Such analysis is beyond the scope of the present discussion.

15–12. Tangent Condition for the Elimination of Distortion

In Fig. 15–15 it is assumed that the plane at B′ is the Gaussian image of the plane at B. The entrance and exit pupils are at E and E′ and it is assumed that

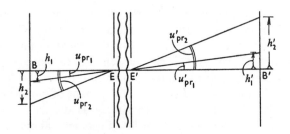

Fig. 15–15. The tangent condition.

there is no spherical aberration of the principal rays from object points in the plane at B.

Referring to the figure, one has

$$\frac{h_1}{h_2} = \frac{\tan u_{pr1}}{\tan u_{pr2}}$$

and

$$\frac{h_1'}{h_2'} = \frac{\tan u'_{pr1}}{\tan u'_{pr2}}.$$

If
$$\frac{h_1}{h_2} = \frac{h_1'}{h_2'}$$

one has
$$\frac{\tan u_{\mathrm{pr1}}}{\tan u'_{\mathrm{pr1}}} = \frac{\tan u_{\mathrm{pr2}}}{\tan u'_{\mathrm{pr2}}},$$

i.e. there is no distortion if

$$\frac{\tan u_{\mathrm{pr}}}{\tan u'_{\mathrm{pr}}} = \text{a constant.}$$

This result is frequently quoted but is of limited application since it is true only if there is no aberration of the pupil points E and E′. In practice, the spherical aberration is corrected for the object and image planes and not for the pupil planes.

15–13. Astigmatic Correction in the Presence of Petzval Field Curvature

Most simple optical systems of positive power have positive Petzval field curvature. It follows that in the absence of astigmatism the image of a point

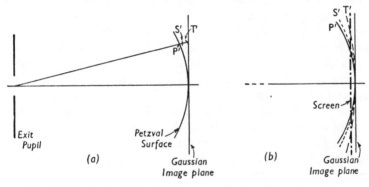

Fig. 15–16. Positive Petzval field curvature and negative astigmatism.

object is at a point such as P′ [see Fig. 15–16(a)], and the image is out of focus on the Gaussian plane. If negative astigmatism is introduced into the system, the focal lines S′ and T′ will lie on the principal ray in positions such as those indicated. As more negative astigmatism is introduced, the best focus, which lies midway between the focal lines, moves towards the Gaussian plane. As it does so, however, the best focus itself deteriorates since its quality depends on the magnitude of the astigmatism alone. Thus, whilst the out of focus effect associated with the Petzval field curvature is reduced, the astigmatic effect is increased. It can be shown that the disturbances are most nearly in phase on the Gaussian plane when the tangential focus lies on that surface. Since both field curvature and astigmatism are proportional to σ^2, it follows that the whole tangential field is then flat and, for a plane object, the image on the Gaussian plane is as good as possible. This would be the correction for maximum resolving power; it is not always the state of correction actually required (see §14–17).

In the absence of spherical aberration, the axial image will be perfect and the image quality on the Gaussian plane will deteriorate as one moves towards the edge of the field. It may be desirable to sacrifice the image quality in the centre of the field in order to improve it farther out and it is then desirable to introduce rather less negative astigmatism and to move the receiving plane to the position indicated in Fig. 15–16(*b*).

15–14. Permissible Aberration

It has been seen that for each point in the object the effect of aberration is to distort the emergent wavefront from its ideal spherical shape. This means that there is no longer an image point at which all the disturbances arrive in phase and, in consequence, there is a deterioration of the image quality. The appearance of the image of a point source in the presence of aberration was discussed in §15–6 to §15–8. If the aberration is small the phase differences between disburbances arriving at the centre of the point source image are small, and the image retains the general form of a strong central maximum surrounded by weak secondary maxima. The intensity at the centre of the pattern is slightly lower than that at the centre of the ideal Airy pattern and in general there is a slight broadening of the central maximum and a slight increase in the intensities of some of the outer maxima. A similar effect results from the introduction of phase differences by a small shift of focus, and in discussing depth of focus for an aberration-free system it was stated that, according to Rayleigh, there is no appreciable deterioration of the image if the phase differences do not exceed $\pi/2$. It seems reasonable to take the same criterion in the present problem. Thus one may say that the image quality will not be seriously impaired provided the wavefront aberration does not exceed $\lambda/4$, it being assumed that the maximum aberration of the wave is reduced as far as possible by appropriate choice of the reference sphere—it is at the centre of this sphere that the image may be said to be located. An alternative approach is to assume that provided the general form of the point source image remains unchanged the intensity at the centre of the pattern—known as the Strehl intensity—provides a measure of the state of correction of the lens. It is then assumed that the image quality will not be noticeably inferior to that for an aberration-free lens provided the Strehl intensity does not fall below $0.8 \times$ the intensity at the centre of an ideal Airy pattern.

The Rayleigh quarter-wave limit and the limit of 0.8 on the Strehl intensity ratio give very similar aberration tolerances. These tolerances correspond to very highly corrected systems and, provided they are not exceeded, the image quality for any kind of object will not be appreciably inferior to that for an aberration-free system. It is found that if the aberrations exceed these tolerances there is usually a rapid deterioration of the image quality. Whereas it is necessary for a microscope objective to be highly corrected and to give near-ideal imagery for all spatial frequencies up to the limit of resolution, there are other systems which are required to be near-perfect over a more limited frequency range. One could, for example, require that over the frequency range concerned the modulation transfer factor should not fall below $0.8 \times$ the value for an aberration-free system. It is worth noting that for frequencies in the range $s=0.2$ to $s=1.4$ (see §14–17) this criterion gives tolerances similar to those given by the Strehl criterion, and for frequencies outside this range the aberration tolerances are larger. For lower standards of image formation one can set more liberal tolerances on the frequency response over the range of interest.

For systems of relatively low quality it may be satisfactory to discuss the imagery in terms of geometrical optics and to define the permissible aberration by placing a limit on the size of the light patch formed by the intersection of the pencil of rays with the image plane.

CHAPTER XVI

THE OCCURRENCE AND CONTROL OF
ABERRATIONS IN OPTICAL SYSTEMS

16–1. Factors Influencing the Monochromatic Aberrations

ABERRATIONS appear immediately one goes outside the paraxial region of an optical system, i.e. whenever one has a finite aperture, or finite field, or both. In terms of geometrical optics, the appearance of the defects is associated with the occurrence of large angles of incidence, and large angles of inclination between rays and the optical axis. When one has a finite angle of incidence at a refracting

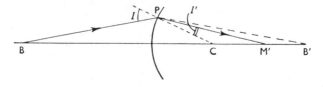

Fig. 16–1. Refraction at a single surface.

surface, the path of the refracted ray is determined by Snell's law ($n \sin I = n' \sin I'$), and the deviation of the ray given by this equation is greater than that which would be given by the paraxial form ($ni = n'i'$). Thus for a single refracting surface as in Fig. 16–1, the ray BP, incident at the large angle I, is refracted along PM′ instead of along PB′ as it would be if the paraxial form of Snell's law was applicable. In this way spherical aberration is introduced, and one may say that there is a general tendency for large angles of incidence to cause large aberration. There are exceptions to this [see, for example, § 16–4(d)]. For any system of refracting surfaces, the angles of incidence obviously depend upon the orientation of the initial incident ray, so that the aberrations of a system depend upon the position of the object.

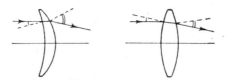

Fig. 16–2. Refraction through a lens.

The purpose of a lens is to deviate the rays and, as indicated in Chapter II, the deviation produced is proportional to the power of the lens. Now there are an infinite number of possible shapes for a lens of given power and refractive index, and, in giving the desired deviation to the ray, the angles of incidence involved depend upon the lens shape as indicated in Fig. 16–2. Thus, in general, the aberration introduced by a lens system depends upon the shapes of the components. Moreover, for a given deviation of a ray at a refracting surface, the necessary angle of incidence depends on the change of refractive index. One may generalize again and say that the aberrations of a system will depend upon the refractive indices of the components. The angles of incidence at the various surfaces of a system depend also upon the separations of the surfaces, so that this is another factor which influences the aberrations. One of the most important factors which influences the aberrations of extra-axial images is the position of the aperture stop. As indicated in Fig. 16–3, this controls the part of a lens through which the oblique pencils pass. It does not, of course, affect the spherical aberration of an axial pencil of rays. Although an exact analysis of the effects

Fig. 16–3. Influence of stop position.

of a shift of stop cannot be given here, it is possible to give a simple summary of the results. One tabulates the primary aberrations as follows:

> Spherical Aberration
> Coma
> Astigmatism
> Petzval Field Curvature
> Distortion.

(It should be noted that this is the order in which they appear as one increases the field.) It can be shown that, in general, a particular aberration is affected by a movement of the stop if one or more of those present above it in the list is present in the system. The exception is field curvature, which is never affected by a stop shift; it is included in the list because distortion will be affected if field curvature is present. In many simple and inexpensive systems (e.g. cheap camera lenses of low aperture), it is customary to leave some spherical aberration in the system in order to be able to remove the coma by a suitable positioning of the stop. A system which is free from spherical aberration and coma is said to be *aplanatic*; it will be seen that the astigmatism of an aplanat cannot be controlled by moving the stop. Again, if a system has a flat field free from astigmatism it is said to be *anastigmatic*; if it is also aplanatic the distortion is independent of the stop position. The variation of coma and distortion is illustrated in the next two sections.

For a given position of the object, the factors which influence the aberrations of a system may be summarized as follows:

> The distribution of the power between the components
> Shapes, thicknesses, and separations, of the components
> Stop position
> Glass types employed.

16–2. Variation of Coma with Stop Position

It was stated in the last section that if a system has spherical aberration the coma depends upon the position of the aperture stop. This fact is demonstrated by Fig. 16–4(a). With the stop in the first position, E_1' is the centre of the exit pupil and $E_1'P'$ is the principal ray. The other rays belong to the pencil of rays from an off-axis object point, it being assumed that there is spherical aberration

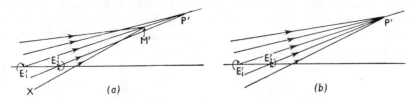

Fig. 16–4. Variation of coma with stop position.

but no coma. If the stop is moved so that E_2' is the centre of the exit pupil, $E_2'P'$ is the new principal ray, and the rays adjacent to it (i.e. those from E_1' and X) are now the rays at about half aperture. Their point of intersection, M', does not lie on the principal ray $E_2'P'$, and negative coma is now present. Fig. 16–4(b) shows that if there is no spherical aberration, coma is not introduced by such a movement of the stop.

16–3. Variation of Distortion with Stop Position

In Fig. 16–5 it is assumed that the point B is imaged without spherical aber-ration or coma by the wide aperture thin lens L which has astigmatism and field curvature. The tangential focus is B′. If one introduces the aperture stop (1) in contact with the lens, ray (1) is the principal ray and, since it passes undeviated through the thin lens, P_1' is the Gaussian image point and there is no distortion. When the stop is moved to position (2), the ray marked (2) is the principal ray and obviously there is barrel distortion since P_2' lies closer to the axis than the Gaussian image P_1'. Similarly, there is pincushion distortion when the stop is

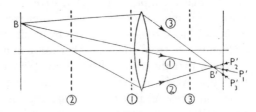

Fig. 16-5. Variation of distortion with stop position.

at (3). This diagram represents a special case but shows the manner in which the distortion is affected by the stop position. It will be seen that the effect is dependent upon the position of the tangential focus B′. For example, if B′ is to the right of the Gaussian plane, stop (2) gives pincushion, and stop (3) barrel distortion. One particularly important result emerges: the distortion is not affected by a stop shift if the system is an aplanat with a flat tangential field.

16–4. The Aberrations of a Single Spherical Refracting Surface. The Aplanatic Points

For a surface of given power separating two given media, the angles of incidence (and therefore the aberrations) can depend only upon the positions of the object and the aperture stop. There are fewer variables than there are with a lens since, with the latter, one is free to choose the refractive index of its material, and its shape and thickness. There are, however, a number of cases of special interest.

(a) Stop at the Centre of Curvature

If the stop is at the centre of curvature, every principal ray is a radius of the surface (see Fig. 16–6) and all pencils are effectively axial pencils so that coma,

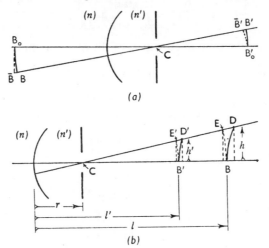

Fig. 16–6. Field curvature.

astigmatism, and distortion, are absent. Spherical aberration is still present, since this does not depend upon the stop position, and there is also Petzval field curvature. The fact that the field is curved is shown by Fig. 16–6(a). B_o' is the image of the axial object B_o, formed by a surface of radius r separating media n and n'. (The spherical aberration is removed by using a small aperture at C.) If B_o is moved round a circle (centre C) to B, the image will move round a concentric circle to B' since this corresponds to an unchanged object and image distance measured from the refracting surface. If B is moved back to $\bar{B}$ where $B_o\bar{B}$ is perpendicular to the axis, the increased object distance must result in a decreased image distance, and the image will move from B' to a position such as $\bar{B}'$. Thus, corresponding to the plane object $B_o\bar{B}$, one has the curved image $B_o'\bar{B}'$.

In general, the refracting surface will be part of a system of surfaces and the object field will be curved. In order to deduce a relationship between the object and image field curvatures, it is convenient to change to an all-positive diagram as in Fig. 16–6(b). It is assumed that the stop at C is small and that B'D' is the image of BD; BE and B'E' are arcs centred on C. Thus E and E' are also conjugate points, and the elements ED and E'D' are conjugate distances along the auxiliary axis CD. They are therefore related by equation (2–11) so that one has

$$\frac{E'D'}{ED} = \left(\frac{h'}{h}\right)^2 \frac{n'}{n},$$

or

$$\frac{h^2 E'D'}{n'} = \frac{h'^2 ED}{n}. \qquad \ldots \ldots (16\text{–}1)$$

Now if h and h' are small, one may write ED and E'D' in terms of the sags in the circular arcs BD, BE, B'D', and B'E'. One has

$$ED = \frac{h^2}{2(l-r)} + \frac{h^2}{2R}$$

and

$$E'D' = \frac{h'^2}{2(l'-r)} + \frac{h'^2}{2R'},$$

where R and R' are the radii of curvature of the object and image fields. Therefore, from (16–1),

$$\frac{1}{n'(l'-r)} + \frac{1}{n'R'} = \frac{1}{n(l-r)} + \frac{1}{nR}. \qquad \ldots \ldots (16\text{–}2)$$

Now l, l', and r, are related by the usual paraxial equation

$$\frac{n'}{l'} - \frac{n}{l} = \frac{n'-n}{r}.$$

This gives

$$l' = \frac{n'lr}{nr + (n'-n)l},$$

i.e.

$$l' - r = \frac{nr(l-r)}{nr + (n'-n)l}.$$

On substituting this into equation (16–2), one obtains

$$\frac{1}{n'R'} - \frac{1}{nR} = \frac{1}{n(l-r)} - \frac{nr + (n'-n)l}{n'nr(l-r)},$$

whence

$$\frac{1}{n'R'} - \frac{1}{nR} = -\left(\frac{n'-n}{nn'r}\right).$$

To conform to the sign convention for aberrations used in the previous chapter, it is necessary to regard R and R' in this example as negative. One then has

$$\frac{1}{n'R'} - \frac{1}{nR} = \frac{n'-n}{nn'r}. \qquad \ldots \ldots (16\text{–}3)$$

Thus for a given object curvature, the image curvature is independent of the object and image distances. It can be shown that it is also independent of the position of the aperture stop. When one has a system of refracting surfaces, equation (16–3) can be applied to each in turn. Consequently, for the complete system one obtains

$$\frac{1}{n'R'} - \frac{1}{nR} = \sum \left(\frac{n'-n}{nn'r}\right),$$

where this so-called *Petzval sum* extends over every surface in the system.

(b) Object at the Centre of Curvature

If the axial point of the object is at the centre of curvature, all the rays of the axial pencil are normal to the surface, and the axial object and image points coincide—there is obviously no spherical aberration. If the stop coincides with the surface, it can be seen from Fig. 16–7 that for a small object h, the angle i_{pr} is small, so that $ni_{pr}=n'i'_{pr}$ and $h'/h=i'_{pr}/i_{pr}=n/n'$. Since $u'=u$, the Sine Condition is satisfied, so that there is no coma for this position of the stop, and since there is no spherical aberration, there will be no coma for other stop positions. There will, however, be astigmatism, field curvature, and distortion, the astigmatism and field curvature being independent of the stop position.

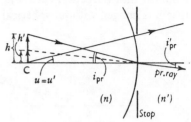

Fig. 16–7.
Object at centre of curvature.

(c) Object Coincident with Surface

When the axial point of the object coincides with the pole of the refracting surface, the Gaussian image coincides with the object. Each object point lies close to the refracting surface so that, approximately, the wavefronts diverge from points on that surface, and its presence cannot alter the shapes of the wavefronts. Consequently, there is no spherical aberration, coma, or astigmatism. In effect, the fact that these conditions do not hold exactly gives rise to field curvature. (As stated above, this is the same for all object positions.) Associated with the deviation imposed on the principal rays, there is distortion. An approach to these conditions occurs in field lenses.

(d) The Aplanatic Points

In addition to the above, there is one position of the object for which there is no spherical aberration, and for which the Sine Condition is satisfied for a large aperture. This, and the corresponding image point, are called the aplanatic points, and one is said to have aplanatic refraction—there is no spherical aberration or coma. One of the aplanatic points is virtual, and the points are defined by

$$\left. \begin{array}{l} l = \left(\dfrac{n'+n}{n}\right)r, \\[2mm] l' = \left(\dfrac{n'+n}{n'}\right)r. \end{array} \right\} \qquad \ldots \ldots \quad (16\text{–}4)$$

Fig. 16–8(a) shows the aplanatic points B and B', assuming B is a real object and B' a virtual image; a reversal of the arrows gives a virtual object B' and real image B. The positions of B and B' can be more easily described and remembered for the case when $n'=1$ and one has a sphere of material of index n. (The more general case follows at once on putting n/n' for n since this is the index of the first medium with respect to the second.)

If one has a sphere of glass, centre C, radius r, of index n, as in Fig. 16–8(b), the positions of the aplanatic points B and B′ can be specified by drawing concentric circles centre C with radii r/n and rn. B and B′ are then as shown, it being assumed that refraction occurs at the right-hand side of the sphere. To prove that B is imaged without spherical aberration at B′ and that the Sine Condition

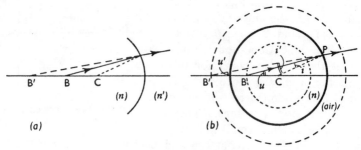

Fig. 16–8. The aplanatic points.

is satisfied, one proceeds as follows. Take any point P on the refracting surface and join CP, BP, B′P. The angles i, i', u, and u', are as shown. If it can be proved that $n \sin i = \sin i'$, then a ray BP is refracted along B′P produced; if this is true for any point P, B′ is the image of B, and if ($\sin u/\sin u'$) is constant, the Sine Condition is satisfied.

Now in $\triangle$'s BPC, B′PC one has

$$\frac{BC}{PC} = \frac{PC}{B'C} \quad \text{(by construction)}$$

and $\angle$BCP is common.

Therefore the triangles are similar.

$$\left. \begin{array}{r} \therefore \quad u = i', \\ u' = i. \end{array} \right\} \quad \ldots \ldots \ldots \quad (16\text{–}5)$$

and

Now from $\triangle$ BPC,

$$\frac{BC}{\sin i} = \frac{PC}{\sin u},$$

or, from (16–5),

$$\frac{\sin i'}{\sin i} = \frac{PC}{BC} = n.$$

Since P was chosen at random, it follows that B is imaged at B′ without aberration. Further, from (16–5) one has

$$\frac{\sin u}{\sin u'} = n.$$

That is, the Sine Condition is satisfied. Therefore, for object and image planes at B and B′, spherical aberration and coma are absent. This will be true for all stop positions. When the stop is at C, there can be no astigmatism, and since the system is aplanatic, it is anastigmatic for all positions of the stop. There will, as always, be field curvature, and there will also be distortion. Since no restriction was placed on the size of u, it follows that all orders of spherical aberration are absent, and from the Sine Condition, all the terms linearly dependent on σ are absent.

One frequently finds that one or more surfaces in a system can be positioned so that one or more of the conditions (a)–(d) can be approached, if not reached exactly. For example, the properties of the aplanatic points described in (d) are frequently utilized for at least one surface in a microscope objective; condition (c) is approached in field lenses; and condition (a) in certain wide-angle lenses. These points will be discussed further in §§ 16–8 to 16–12.

16–5. The Aberrations of a Single Thin Lens

(a) Spherical Aberration

For a thin lens of given power, one is free to choose the refractive index and the shape of the lens, and for a given position of the object, both these factors will affect the spherical aberration since they affect the angles of incidence involved. It is necessary to introduce a parameter to specify the shape of the lens; it is convenient to introduce a second parameter to specify the relative positions of object and image. The shape and object position parameters are defined respectively by the relations

$$X = \frac{r_2 + r_1}{r_2 - r_1},$$

$$Y = \frac{l + l'}{l - l'},$$

r_1 and r_2 being the radii of curvature of the lens, and l and l' the conjugate distances. It will be seen that these parameters are dimensionless—their values are independent of the units in which lengths are measured. (There are, in fact, certain advantages to be gained by defining X and Y in terms of the angles α_1, α_2, u_1, u_2', where these have their usual meanings. For example, $X = (\alpha_1 + \alpha_2)/(\alpha_1 - \alpha_2)$ and $Y = (u_2' + u_1)/(u_2' - u_1)$ are more convenient if a conjugate is infinite or a surface is plane.) Figs. 16–9 and 16–10 show the lens shapes and object positions corresponding to various values of X and Y respectively. In Fig. 16–10 a reversal of the arrows does not change Y.

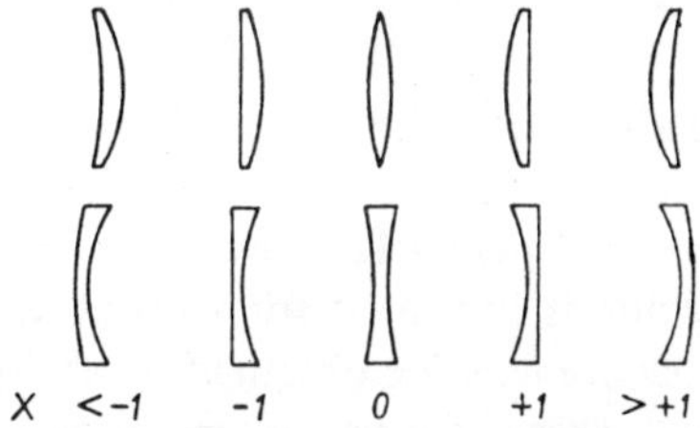

X   <-1   -1   0   +1   >+1

Fig. 16-9. The shape parameter.

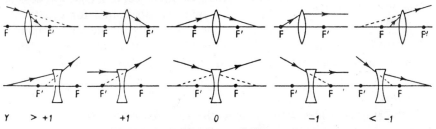

Y   > +1   +1   0   -1   < -1

Fig. 16-10. The conjugate distance parameter.

For any given object position, and for a lens of given aperture and refractive index, the variation of the spherical aberration with X follows a parabolic curve. The curves in Fig. 16–11 show the spherical aberration parabolae for $Y=0$ for thin convex lenses of refractive index 1·5 and 1·7 (in air), the lenses having unit power. The figure shows that the aberration is reduced to a minimum by distributing the bending of the rays equally between the surfaces of the lens, and

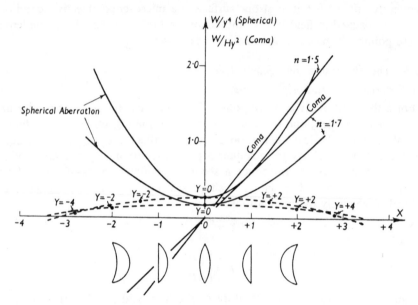

Fig. 16–11. Spherical aberration and central coma for thin lens.

is also reduced by increasing the refractive index. For other values of Y, each parabola retains the same shape but moves bodily, its vertex following the inverted parabola shown as a broken curve. In each case, the position of the vertex is indicated for various values of Y. The corresponding curves for a concave lens are obtained by changing the sign of the ordinates. It will be seen that, except for heavily meniscus shapes and one real and one virtual conjugate

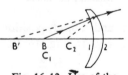

Fig. 16–12. Use of the aplanatic points.

so that $|Y|$ is fairly large, the spherical aberration has the same sign as the power of the lens. That is, it is positive for converging lenses and negative for diverging. It is easy to show from the results of §16–4 that, for a virtual object or image, it is possible to have zero spherical aberration for a particular meniscus shape lens. Thus if the object point **B** (Fig. 16–12) is at the centre of curvature of surface (1) and at the aplanatic point of surface (2), there will be no spherical aberration or coma.

(b) Coma

If the stop coincides with the lens, the so-called *central coma* for a given aperture and field is a linear function of both X and Y. The coma lines for a lens of unit power are shown in Fig. 16–11 for refractive indices 1·5 and 1·7 for $Y=0$. It will be seen that the coma is zero for the same value of $X(X=0)$ as the spherical aberration is a minimum. This is exactly true only when $Y=0$, but for small values of $|Y|$, the central coma is small when the lens shape is adjusted to give minimum spherical aberration.

It has been stated that when spherical aberration is present, the coma depends upon the position of the aperture stop; this fact is frequently used to correct the coma of a single lens (see §16–11).

(c) Astigmatism

Astigmatism is caused by the fact that the lens has different powers in the sagittal and tangential sections, and it is easily shown that when the stop coincides with the lens, this difference in power depends only on the angle of field; that is, it is independent of the lens shape and refractive index for a lens of given power.

Suppose a collimated beam of light is incident centrally at angle U_{pr} on a thin lens of focal length f'. Figs. 16–13(a) and (b) show the sections in the tangential

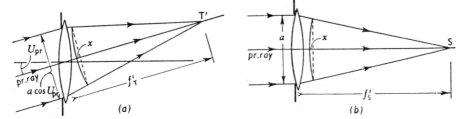

Fig. 16–13. Astigmatism of a thin lens at the stop.

and sagittal planes respectively. The incident wave is plane, and if there is no spherical aberration or coma, the emergent wave has a circular section in each of these planes; the radius in each case is the focal length in that section. The curvature of the emergent wave is caused by the fact that the optical path through the centre is greater than that through the edge of the aperture; this introduces a retardation of the centre of the wave and causes the sags (x) in the circular sections of the wavefront. The sags in the sagittal and tangential sections are obviously equal, but they correspond to chords of different lengths. Thus, if a is the linear aperture of the lens, this may be taken as the length of the chord in the sagittal section, but in the tangential plane the chord is of length $a \cos U_{pr}$. Now if x is the sag of a circular arc of radius r, for a chord of length $2c$, one may write $r=c^2/2x$ if x is small. Thus for circular arcs having equal sags, r is proportional to the square of the length of the chord. If f_S' and f_T' are the focal lengths in the sagittal and tangential sections, they are the radii of the circular sections of the wavefront, and one has

$$\frac{f_S'}{f_T'} = \frac{a^2}{a^2 \cos^2 U_{pr}}, \quad . \quad . \quad . \quad . \quad . \quad (16\text{–}6)$$

$$\therefore \quad f_T' = f_S'(1-\sin^2 U_{pr}).$$

Hence one may write

$$\delta f' = f' \sin^2 U_{pr},$$

where $\delta f' = f_S' - f_T'$, and is the separation of the focal lines.

In terms of powers, one has, from (16–6),

$$\frac{F_T}{F_S} = \frac{1}{\cos^2 U_{pr}},$$

whence
$$F_T - F_S = F \sin^2 U_{pr}, \quad \cdot \quad \cdot \quad \cdot \quad \cdot \quad (16\text{--}7)$$

so that the astigmatic difference of power depends only upon the paraxial power F and the angle of field. That is, as pointed out above, it does not depend on the lens shape or refractive index.

(d) Field Curvature

On applying equation (16–3) to each surface of the lens in turn, one has, for a lens of index n in air,

$$\frac{1}{R'} - \frac{1}{R} = \frac{n-1}{nr_1} + \frac{1-n}{nr_2}$$

$$= \frac{(n-1)}{n} \left(\frac{1}{r_1} - \frac{1}{r_2} \right)$$

$$= \frac{1}{nf'} = \frac{F}{n}, \quad \cdot \quad \cdot \quad \cdot \quad \cdot \quad \cdot \quad \cdot \quad (16\text{--}8)$$

so that the field curvature is dependent only upon the power of the lens and the refractive index of its material; that is to say, it does not depend on the object position, stop position, or lens shape. It will be seen that the Petzval field curvature of a thin lens is reduced if the refractive index is increased. Now, in order to reduce the chromatic aberration, it is desirable to use a glass having a

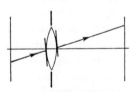

small dispersion. These two factors indicate the desirability of having glasses with high indices and low dispersions, and considerable advances have been made in recent years in the production of such glass types.

Fig. 16–14. Central passage of a principal ray.

(e) Distortion

To a first approximation the principal rays emerge parallel to their original directions when a thin lens is coincident with the stop. The lens acts as a plane parallel plate and, in this case, there is no distortion (see Fig. 16–14). For other positions of the stop there will, in general, be distortion present.

16–6. The Achromatic Doublet

(a) Residual Chromatic Aberration

It was shown in Chapter III (§ 3–9) that a combination of two thin lenses in contact has the same power for two different colours R and B if the powers of the component lenses for an intermediate colour Y are given by

$$\left.\begin{array}{l} F_{1Y}=F_Y\left(\dfrac{V_1}{V_1-V_2}\right), \\[2mm] F_{2Y}=-F_Y\left(\dfrac{V_2}{V_1-V_2}\right), \end{array}\right\} \quad \cdots \cdots \quad (16\text{–}9)$$

where V_1 and V_2 are the values of $V=(n_Y-1)/(n_B-n_R)$ for the materials of the lenses.

If R_1 and R_2 are the total curvatures of each lens $[R=(1/r_a-1/r_b)]$, $F=(n-1)R$ and the above equations give

$$\left.\begin{array}{l} R_1=\dfrac{1}{f_Y'(V_1-V_2)(n_{1B}-n_{1R})}, \\[3mm] R_2=\dfrac{-1}{f_Y'(V_1-V_2)(n_{2B}-n_{2R})}. \end{array}\right\} \quad \cdots \cdots \quad (16\text{–}10)$$

It was pointed out in Chapter III that a combination of two lenses which satisfies these equations is not completely free from chromatic aberration, and

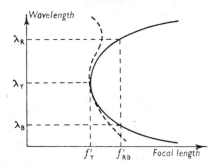

Fig. 16–15. Achromatic and apochromatic correction.

for a doublet employing hard crown and dense flint glass, the variation of focal length with wavelength is as indicated by the full curve in Fig. 16–15. The vertex of the curve lies in the region of the wavelength Y and a good indication of the magnitude of the residual aberration (*secondary spectrum*) is obtained if the difference between f_Y' and f_{RB}' is found. One has

$$\frac{1}{f_Y'}=(n_{1Y}-1)R_1+(n_{2Y}-1)R_2,$$

and

$$\frac{1}{f_{RB}'}=(n_{1B}-1)R_1+(n_{2B}-1)R_2,$$

$$\therefore \quad \frac{1}{f_{RB}'}-\frac{1}{f_Y'}=\frac{f_Y'-f_{RB}'}{f_Y'f_{RB}'}$$

$$=(n_{1B}-n_{1Y})R_1+(n_{2B}-n_{2Y})R_2,$$

$$\therefore \quad \frac{f_Y'-f_{RB}'}{f_{RB}'}=(n_{1B}-n_{1Y})R_1f_Y'+(n_{2B}-n_{Y2})R_2f_Y'.$$

Substituting for R_1 and R_2 from (16–10), one has

$$\frac{f_Y' - f_{RB}'}{f_{RB}'} = \frac{\left(\dfrac{n_{1B} - n_{1Y}}{n_{1B} - n_{1R}}\right)}{V_1 - V_2} - \frac{\left(\dfrac{n_{2B} - n_{2Y}}{n_{2B} - n_{2R}}\right)}{V_1 - V_2}$$

$$= \frac{\beta_1 - \beta_2}{V_1 - V_2} \text{ (say).} \qquad \cdots \cdots \quad (16\text{–}11)$$

This equation gives the fractional difference in focal length for wavelength Y and the wavelengths R and B for which achromatism was effected. It will be seen that in the region of the wavelength Y, approximately midway between B and R, the variation of focal length with wavelength is small. For visual instruments this should occur in the middle of the visible spectrum where the eye is most sensitive. In consequence, it is convenient to use the C (λ 6563) and F (λ 4861) lines of hydrogen for R and B respectively, and the helium d (λ 5876) line for Y. For visual observations, f_d' is then the effective focal length of the lens and equation (16–11) gives, approximately, the maximum focusing error in the visible range. $(n_F - n_C)$ is usually called the *mean dispersion* and $(n_F - n_d)$ is a *partial dispersion*. (For any other wavelength λ, $(n_F - n_\lambda)$ is the corresponding partial dispersion.) The quantities β_1 and β_2 are called the *relative partial dispersions* and, in glass catalogues, are tabulated for certain wavelengths. Considerable effort has been made to produce crown and flint glasses whose relative partial dispersions are not widely different. It has proved possible to reduce the secondary spectrum considerably and the broken curve in Fig. 16–15 indicates the state of correction attainable. In such cases, where the focal length has the same value for three wavelengths, the lens is said to be *apochromatic*. An apochromatic system is frequently achieved by the use of a triplet with three different glasses.

In the early days of photography, the plates were sensitive over a relatively small spectral range at the blue end of the spectrum. In consequence, the lenses were achromatized for the sodium D (5893) and hydrogen G' (4341) lines so that the visual and photographic foci were the same. With present day emulsions that are sensitive over a wide range, the achromatic correction must be improved; an apochromatic correction is even more desirable here than with visual instruments.

(b) Monochromatic Aberrations

The most obvious advantage of a doublet over a single thin lens is the fact that the former can be achromatic; an equally important advantage is that it is possible to exercise a far greater control over the monochromatic aberrations. For any position of the object, it is possible to make a doublet achromatic and aplanatic. Achromatism is achieved by adjustment of the relative powers of the components, and aplanatism by adjustment of the lens shapes. The spherical aberration and central coma of each component vary with shape in the manner described in the previous section, and the aberrations of one component can be balanced by those of the other. For example, it was seen that for most object positions a converging lens gives positive, and a diverging lens negative, spherical aberration.

If the lens is to be a cemented doublet, one has fewer degrees of freedom since the shapes of the components cannot be adjusted independently. However, by a suitable choice of glasses, it is possible for a cemented doublet to be achromatic and aplanatic. When suitable glasses are used, the variation of spherical aberration and coma with lens shape is of the form indicated by Fig. 16–16. The curves relate to a ten-inch telescope objective covering a 6° field at $f/5$; they show the spherical aberration at the edge of the aperture and the coma in the tangential section at the edge of the aperture and field, and indicate that an aplanat is possible. Such an achromatic aplanat may be possible with several pairs of

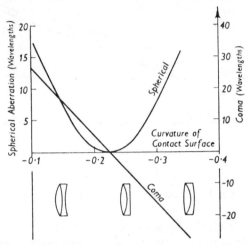

Fig. 16–16. Spherical aberration and coma for a cemented doublet.

glasses, and the final choice will be influenced by consideration of the secondary spectrum and the steepness of the lens surfaces. For example, a doublet which contains steeply curved surfaces should be rejected because there will be a tendency for higher order aberrations to be introduced at those surfaces. If the components are not cemented together, one has a much greater choice of glasses since the lens shapes can be varied independently. With uncemented doublets, there are more air-glass surfaces and therefore greater losses of light and, in addition, the lens is less easily mounted and handled. For systems of moderate aperture, cemented combinations are usually employed.

If the doublet is thin, the central astigmatism will be given by equation (16–7) and will again be independent of the choice of glasses and lens shapes. It will also be independent of the stop position if the doublet is an aplanat. The Petzval sum of the doublet will be given [from equation (16–8)] by

$$P = \frac{F_1}{n_1} + \frac{F_2}{n_2}.$$

∴   from equation (16–9)

$$P = \frac{F}{(V_1 - V_2)}\left[\frac{V_1}{n_1} - \frac{V_2}{n_2}\right]. \qquad \cdot \quad \cdot \quad \cdot \quad (16\text{–}12)$$

In order to reduce the magnitude of the term in square brackets, it is necessary that the glass with the higher V-number should have the higher refractive index.

The original crown and flint glasses were very unsatisfactory in this respect, since the more dispersive flints also had the higher indices. The position has been improved by the production of additional glasses although, in many cases, the resulting doublet has rather steep curves. A considerable advance in the production of glass types was made by Abbe and Schott in Jena (*c.* 1886). They pointed the way to the production of glasses with a high index and low dispersion (e.g. Dense Barium Crowns), and with a low index and high dispersion

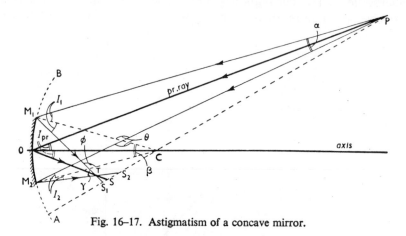

Fig. 16–17. Astigmatism of a concave mirror.

(Light Flints). However, the Petzval sum of a doublet is still usually positive so that it is desirable to introduce some negative astigmatism. This cannot be done if the doublet is a thin aplanat since it will always have the astigmatism given by (16–7). If spherical aberration is not corrected, it is possible to choose a lens shape and stop position to give zero coma and negative astigmatism, and this will be a more desirable state of correction if the lens is to cover a wide field.

The uses of an achromatic doublet are discussed further in §§16–8, 16–9, and 16–11.

16–7. The Astigmatism of a Concave Mirror

In Fig. 16–17 M_1M_2 is a concave mirror of small aperture, forming part of the spherical surface AB, radius r, centre C. P is an off-axis object point. As far as the larger surface AB is concerned, P may be taken as a point on the axis AC and, from this point of view, the rays PM_1S_1, PM_2S_2 may be taken as rays which exhibit the small difference in longitudinal spherical aberration S_1S_2. If the diagram is rotated about ACP through small angles in both directions, the rays PM_1 and PM_2 trace out sagittal fans of rays incident on the mirror M_1M_2, and since S_1 and S_2 remain stationary during this rotation, S_1S_2 is the sagittal focal line. PM_1M_2 is obviously a tangential fan of rays, incident on M_1M_2, and T is the focus which, during the rotation mentioned, traces out the tangential focal line.

Let s' and t' be the distances measured along the principal ray from O to the sagittal and tangential focal lines. Since the object is free from astigmatism, s

and t are the same and are conveniently denoted by a (not by l since this is always measured along the axis—in this case along OC).

From the figure,

$$\triangle POS = \triangle POC + \triangle COS,$$

$$\therefore \quad \tfrac{1}{2}as' \sin 2I_{pr} = \tfrac{1}{2}ar \sin I_{pr} + \tfrac{1}{2}rs' \sin I_{pr},$$

where I_{pr} is, with the usual notation, the angle of incidence of the principal ray,

i.e. $$2as' \sin I_{pr} \cos I_{pr} = ar \sin I_{pr} + rs' \sin I_{pr},$$

whence, on dividing throughout by $ars' \sin I_{pr}$, one obtains,

$$\frac{1}{a} + \frac{1}{s'} = \frac{2 \cos I_{pr}}{r}. \qquad \ldots \ldots \quad (16–13)$$

If I_1 and I_2 are the angles of incidence at M_1 and M_2, and angles $\alpha, \beta, \gamma, \theta,$ and ϕ, are as shown in the figure, one has

$$(\pi - \theta) = \alpha + I_1 = \beta + I_2, \quad \text{i.e. } I_1 - I_2 = \beta - \alpha,$$

$$(\pi - \phi) = \gamma + I_2 = \beta + I_1, \quad \text{i.e. } I_1 - I_2 = \gamma - \beta,$$

$$\therefore \quad \alpha + \gamma = 2\beta. \quad \ldots \ldots \ldots \quad (16–14)$$

Now $$\beta = \frac{M_1 M_2}{r}$$

and, since $M_1 M_2$ is small, one may write

$$\alpha = \frac{M_1 M_2 \cos I_{pr}}{a}$$

and $$\gamma = \frac{M_1 M_2 \cos I_{pr}}{t'}.$$

Hence, from (16–14), one obtains

$$\frac{1}{a} + \frac{1}{t'} = \frac{2}{r \cos I_{pr}}. \qquad \ldots \ldots \quad (16–15)$$

With a concave grating, one has rulings on $M_1 M_2$ and a slit at P perpendicular to the plane of the diagram; it will be seen that each point of the slit gives a focal line at T. If the slit is twisted, the focal lines for each point are not collinear, and the width of the spectral line is increased.

16–8. The Design of Small Objectives

The lenses considered in this section are the small aperture objectives such as those used in most prismatic binocular telescopes, in low-power microscopes, and in the telescopes and collimators of spectrometers, etc. In systems of this kind, the aperture rarely exceeds $f/5$ and the field of view is very small. Consequently, the most important aberrations are primary spherical, coma, and the chromatic errors. The lens is usually a cemented achromatic aplanatic doublet and the lens holder itself is usually the aperture stop. There is little point in providing a separate diaphragm since its position will not affect the astigmatism if the system is aplanatic. The astigmatism will have the large positive value given in § 16–5 and the Petzval sum will also be positive, so that the image quality rapidly deteriorates as one moves away from the axis; little can be done to improve it. One can place either the crown or the flint component on the long conjugate side. The former is known as the Fraunhofer, and the latter as the Steinheil type (see Fig. 16–18). The Fraunhofer type usually has shallower surfaces and is therefore easier to produce, and, for a telescope objective, it has the more stable crown glass on the exposed side. The Steinheil type can sometimes give better correction. A third form of doublet is shown in Fig. 16–18; it is known as the Gauss type. With these uncemented meniscus lenses, one can achieve better correction for spherical aberration at the expense of coma. (The spherical aberration can be corrected for two different wavelengths.)

It should be noted that in a system such as a visual telescope containing an objective and an eyepiece, the lens system should be designed as a whole. For

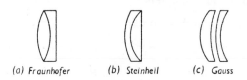

(a) Fraunhofer   (b) Steinheil   (c) Gauss

Fig. 16–18. Achromatic doublets.

example, the objective and eyepiece need not be corrected separately for, say, spherical aberration, provided that the aberration of the eyepiece is corrected by the objective. This fact is often overlooked. For measuring instruments which include a graticule in an intermediate image plane, it is obviously necessary to correct separately those parts of the system which precede and follow the graticule. The apertures of these simple systems are ultimately limited by the appearance of higher order aberrations.

16–9. Microscope Objectives

Microscope objectives have a large aperture and a small field, so that spherical aberration, coma, and the chromatic aberrations, are again the most important. A doublet is satisfactory for many of the low power ($f'=2$ inches or 1 inch) smaller aperture objectives found in measuring microscopes. If a somewhat better state of correction is required, a cemented triplet lens may be used. This has less residual chromatic aberration, and since it has less steeply curved surfaces, a larger aperture can be used before higher order aberrations begin to appear. It was pointed out in Chapter XIV that the limit of resolution of a microscope objective depends on the numerical aperture, $n \sin U$. That is to say, the objective must admit the largest possible cone of light. In order to collect a wide cone, it is necessary to employ objectives of short focal length—high power. Lenses commonly used in microscopy have nominal focal lengths of (a) $\frac{2}{3}$ inch or 16 mm., (b) $\frac{1}{6}$ inch or 4 mm., (c) $\frac{1}{12}$ inch or 2 mm. Many 16 mm. objectives consist of two cemented doublets as shown in Fig. 16–19(a). It was discovered in 1830 by Lister that a doublet having a plano-convex outer form can have two pairs of conjugates for which there is no spherical aberration. For one pair, one conjugate is virtual, and for the other, both are real. It will be seen that the objective shown in Fig. 16–19(*a*) employs a doublet working under each of these conditions. It was found that the coma was of opposite sign in the two cases and it has proved possible to employ a system of this type to give an achromatic aplanatic objective having a numerical aperture of about 0·28. In practice, some deviation from the plano-convex outer form of each component may be an advantage. Since the aperture is now fairly large, secondary spherical aberration is likely to appear, and this must be examined by finding the exact paths of the rays in the axial pencil. In addition, it is desirable to ensure that the Sine Condition is satisfied for all zones of the aperture rather than examine the primary coma alone (see § 15–11).

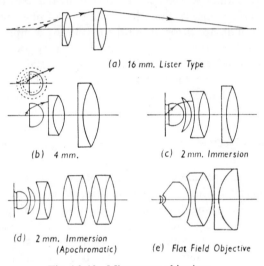

(a) 16 mm. Lister Type

(b) 4 mm.

(c) 2 mm. Immersion

(d) 2 mm. Immersion (Apochromatic)

(e) Flat Field Objective

Fig. 16–19. Microscope objectives.

Fig. 16–19(*b*) illustrates the optical system of many 4 mm. objectives. The use of the hyperhemispherical front component followed the work of Amici (*c.* 1850) and, at the rear surface, one has approximately aplanatic refraction [see § 16–4(*d*)]. The front component has large power and gives considerable chromatic aberration which must be removed in the lenses that follow. One of the most troublesome surfaces in this system is the plane front surface, and it is largely the appearance of higher order aberration at this surface which limits the numerical aperture to about 0·8. (The small size of the front component makes it very undesirable to deviate from the plano form.)

The chromatic aberration can be reduced, and the general state of correction improved, by employing more complex systems. By using fluorite as a crown glass in these more complicated systems *apochromats* can be produced—the image distance is the same for three, or even four, wavelengths and therefore shows little variation over a wide range. With the simpler achromatic objectives the balance between the primary and higher order spherical aberrations is such that the total aberration is very small at the edge of the aperture. For an intermediate zone, however, there is some residual aberration. In the apochromatic objectives this zonal aberration can be reduced considerably. Furthermore, with a system such as a microscope objective, in which it is important to secure a very high degree of correction, it becomes desirable to consider the chromatic variation of the monochromatic aberrations. For example, the spherical aberration may be different for different wavelengths, and usually it is only corrected for one. In the apochromatic objectives the spherical aberration is corrected for two wavelengths (and is consequently small over a wide range) in addition to the improved longitudinal chromatic correction. It is found that many apochromatic objectives (especially those of higher power) give a chromatic difference of magnification; this can be corrected by using a *compensating eyepiece* (see §16–12). By employing fluorite as the crown component of a doublet, the secondary spectrum and the zonal aberration can be reduced. It is then possible to improve the performance of the achromatic type of objective; the correction is inferior to that of an apochromat but the lens is less complicated.

A further factor must be considered when seeking very good correction: many specimens to be examined with a microscope are contained beneath a cover glass. Now it can be shown that a parallel-sided glass plate introduces negative spherical aberration. This can be balanced by the positive aberration introduced if the objective is used at a slightly shorter tube length than that for which it is designed to be corrected. (Usually objectives are designed for a tube length of 160 mm.) Many objectives are specially designed for a standard cover glass thickness of 0·18 mm.

Now the numerical aperture is $n \sin U$ so that by filling the space between the object and the objective with a material medium, the numerical aperture may be increased without increasing the obliquity of the rays entering the objective. It will be seen below that by interposing a suitable material medium, it is also possible for the objective to collect a wider cone of rays without increasing the aberration, so that the numerical aperture can be increased in two ways.

Although the use of an immersion fluid was first suggested much earlier by Brewster, high power immersion objectives were not produced until about 1877, when prominent workers in the field were French, Abbe, and Tolles. The system shown in Fig. 16–19(c) represents an achromat of 2 mm. focal length and numerical aperture about 1·25. The space between the cover glass of the slide and the front lens is filled with a liquid of refractive index approximately equal to that of the cover glass; usually cedar oil, index 1·517, is employed. The refractive index of the front component is not very different from this so that the rays pass almost undeviated from the object to the rear surface of the first lens where there is, as before, approximately aplanatic refraction. The rays are then incident normally on the first surface of the meniscus lens and therefore pass undeviated to the rear surface where there is again aplanatic refraction. Thus the first two components impart a considerable deviation to the rays with the introduction of little spherical aberration or coma. Since these two components are non-achromatic and of large power, considerable chromatic aberration is introduced, and this must be removed in the rear lenses. In order to obtain a large amount of negative chromatic aberration from the rear components, it is necessary to employ rather steeply curved surfaces and this renders difficult the avoidance of high-order aberration. As stated above, there is approximately aplanatic refraction at the second and fourth surfaces, and normal incidence at the third. However, it must always be remembered that, in practice, there is some slight deviation from exact aplanatism and exactly normal incidence.

As with the 4 mm. objective, it is possible to increase the numerical aperture and obtain an apochromatic lens of better all-round correction by employing more complicated systems. An example is shown in Fig. 16–19(*d*) which illus-trates an objective of about 2 mm. focal length, the numerical aperture being about 1·35. It is not possible to discuss these objectives further in simple terms.

Frequently it is desirable to examine objects under water and, for this purpose, water immersion objectives are designed in addition to the more usual oil immersion systems.

As stated at the beginning of this section, it is the spherical aberration and coma which are the most important monochromatic aberrations in microscope objectives. Usually these systems have a large Petzval sum and positive astig-matism. For microscopes that are to be used for plotting the tracks of atomic particles in photographic emulsions, and similar work, it is desirable to have an objective with a flat field. These are very much more difficult to design and manufacture. An example is shown in Fig. 16–19(*e*); it has a numerical aperture of 0·85 and gives a primary magnification of 30.

16–10. Substage Condensers

It was seen in § 14–12 that it is necessary for the substage condenser to supply a wide cone of light. However, it is not necessary for the condenser to be as highly corrected as an objective, and a popular system is the Abbe condenser shown in Fig. 16–20. If oiled to the slide, a numerical aper-ture of 1·2 is attainable. As indicated, this system has very large spherical aberration (especially when used "dry") and is unachro-matic. Frequently it is desirable to employ a more highly corrected system; the condenser then resembles an achromatic objective used in reverse.

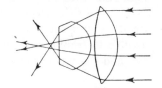

Fig. 16–20.
The Abbe condenser.

If an annular aperture is placed behind the condenser, a hollow cone of light emerges and this is suitable for dark-ground illumination with objectives of moderate power (see § 14–14). For higher powers, reflecting condensers are available (see § 14–14).

16–11. Photographic Lenses

(a) Simple Meniscus

If it is intended to use a single thin lens as a photographic objective, it is obviously impossible to correct its spherical aberration, chromatic aberration, or Petzval field curvature. Since the lens must cover a fairly wide field, it is necessary to concentrate on reducing the effect of the off-axis aberrations. The presence of spherical aberration enables the coma to be varied by moving the aperture stop, and it is usual to choose the stop position for zero coma. It has been shown (§ 15–13) that the effect of positive Petzval field curvature is reduced by the introduction of negative astigmatism. For a system with positive spherical aberration, it can be shown that the astigmatism reaches a minimum value for the stop position that removes the coma. The exact value of this minimum astigmatism depends upon the shape of the lens. Fig. 16–21 shows the form of the astigmatic fields for three meniscus lens shapes employing the stop position for zero coma. In order to show the effect more clearly, the deviations from the Gaussian plane are increased by a factor of about 2·5. The astigmatism is

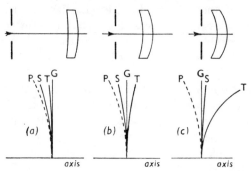

Fig. 16–21. Astigmatic fields for coma-free meniscus lenses.

negative and therefore reaches a maximum numerical value. It will be seen that as the lens is made more heavily meniscus, more negative astigmatism is introduced, and one chooses the shape which gives an approximately flat tangential field with zero coma [i.e. shape between (a) and (b)]. It will be seen that no attempt is made to choose a lens shape for minimum spherical aberration, and the effect of this defect must be reduced by employing a rather small aperture. Meniscus lenses of this type usually have an aperture of about $f/12$ or less. It is possible to employ a meniscus convex to the object with the stop behind the lens, and this has less spherical aberration.

(b) The Achromatic Doublet

The most obvious development from the simple meniscus is the use of an achromatic doublet. As pointed out in § 16–6, it is possible to control the astigmatism and coma if the spherical aberration is not corrected. Now, as with the simple meniscus, the Petzval field curvature is positive, and again it is usual

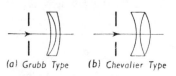

Fig. 16–22. Landscape lenses.

to choose the lens shape and stop position to give zero coma and a flat tangential field. For obvious reasons this system is frequently referred to as a landscape lens; it can take either of the forms shown in Fig. 16–22. In the Grubb type, considerable negative secondary spherical aberration is introduced at the contact surface, and this to some extent reduces the effect of the residual positive primary aberration. Originally these lenses used hard crown and dense flint glasses, but when further glass types were evolved, these were employed also. The chief advantage of these newer glasses is that the Petzval sum is reduced (see § 16–6) so that less negative astigmatism is required. The correction in the outer parts of the field is therefore improved, although the spherical aberration is often worse than with the older glasses. The general appearance of the lens is similar to that of the Chevalier type. Before describing the modern photographic lens types, it must be pointed out that it is much easier to design a satisfactory lens of short focal length than it is to produce a long focus system. A design for a short focal length will not produce a long focus lens simply by being scaled up, because its aberrations will be scaled up also. Thus a wavefront aberration which was a small fraction of a wavelength (and therefore unimportant) becomes a larger fraction and more harmful.

(c) Symmetrical Lenses

If a lens system is completely symmetrical about the aperture stop, there can be no coma, distortion, or chromatic difference of magnification. This, however, implies that the object and image distances are equal. In a photographic lens

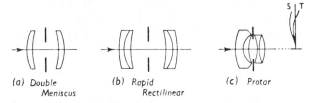

(a) Double   (b) Rapid   (c) Protar
    Meniscus   Rectilinear

Fig. 16–23. Developments from the double meniscus.

the object distance is usually much larger than the image distance; nevertheless, it is found that the control of the above aberrations is facilitated by employing a system of approximately symmetrical construction. It must be remembered, however, that small deviations from exact symmetry may be advantageous.

The simplest system of this type employs two non-achromatic meniscus lenses as in Fig. 16–23(a). The design of this lens proceeds along lines similar to those for a single meniscus. Frequently the power of the double meniscus can be varied for focusing purposes by changing the separation of the components. The obvious development from the double meniscus is the use of two achromatic doublets. (As was proved in § 3–10, a system of two separated thin lenses can, in general, be free from both chromatic aberrations only if the lenses are separately achromatic.) Fig. 16–23(b) shows an early lens of this type, employing hard crown and dense flint components; it is known as the *Rapid Rectilinear* lens, and was produced by Dallmeyer in 1866. The symmetry of the system automatically reduces the coma, distortion, and transverse chromatic aberration. The spherical aberration for each component is reduced but each has coma which is used to control the astigmatism by varying the distance of the component from the stop. Since each component has a large positive Petzval sum, it is necessary to introduce negative astigmatism. The lens covered a 25° semi-field with an aperture of about $f/8$ and marked a great advance in the performance of photographic objectives.

When the newer glasses were produced at Jena, they were at once substituted into lenses of the Rapid Rectilinear type, the order of the convex and concave components being reversed. The reduced Petzval sum enabled the correction to be improved in the outer part of the field, but the spherical aberration was increased and the overall performance was not markedly improved.

The next major advance was made by Rudolph in 1890 with the lens shown in Fig. 16–23(c). The front component was an old glass achromat of small or even negative power, and the rear component employed the newer glasses. This enabled the Petzval sum to be reduced considerably. (As shown by equation (16–12), the Petzval sum of an old glass achromatic doublet can be small only if the power is small, and negative only if the power is negative.) The contact surface in the front component gave an important contribution of negative spherical aberration, and the rear cemented surface was important in the control of the astigmatism. A number of versions of this lens have been produced and it has been made to cover a semi-field of 30° at $f/8$. It must be pointed out that it was now becoming possible to exercise some control over more than just the

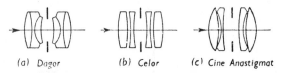

(a) Dagor   (b) Celor   (c) Cine Anastigmat

Fig. 16–24. Developments from the symmetrical type.

primary aberrations. For example, the secondary spherical aberration was made to balance the primary in the outer zones of the aperture, and the higher order astigmatism caused the astigmatic fields to bend back towards the focal plane after deviating from it in the usual way [see Fig. 16–23(c)]. This lens was, in fact, the first *anastigmat* and was known by that name originally. However, when the term came to be used to describe any lens which has a substantially flat field free from astigmatism, the name *Protar* was adopted for the original lens.

It will be appreciated that in designing a lens the desirability of exercising greater control over the several aberrations calls for more degrees of freedom. Thus the doublets of the Protar lens become triplets, as in the *Dagor* designed by von Höegh [Fig. 16–24(a)]. For a focal length of about 10 cm., this lens covered a field of 50°–60° at $f/6.8$. There have been a number of similar types— some having four cemented components on each side of the stop, and some having each half of the system corrected separately. With the latter type, one can use one half alone to give a lens of half power.

Another type of lens of approximately symmetrical construction is the *Celor* designed by von Höegh in about 1897 [Fig. 16–24(b)]. For a focal length of 10 cm. this had an aperture of about $f/5$ and covered a field of 50°–60°. A few years later a modified Celor lens was produced; this was not exactly symmetrical and had rather larger separations between the positive and negative components. It had an aperture of $f/3.5$. A large number of similar lenses have been designed: some for small field and large aperture, and some for large field and small aperture. The $f/1.9$ *Cine Anastigmat* (8 or 16 mm. cameras) can be regarded as an extreme example of the former kind, although it is probably better regarded as a development of the Petzval type (see below). It is interesting to compare the general appearance of these lenses [Fig. 16–24(b) and (c)] with the wide-angle type shown in Fig. 16–28(b).

A further important advance in symmetrical lenses was made by Lee in 1920 with the *Cooke Opic* lens shown in Fig. 16–25. This covered a field of about 48° at the very large aperture of $f/2$. The Opic resembles the Celor in so far as it consists of two diverging components (now doublet menisci) between two converging components. This construction is also responsible for the Opic and its many derivatives being referred to as *double-Gauss* type lenses; they can be said to resemble two Gauss-type objectives [Fig. 16–18(*c*)] with the stop between them. The resemblance is not always obvious at first sight because the components are sometimes complex. Wynne (1944) has described a double-Gauss lens of relative aperture $f/1\cdot0$ in which the rear diverging meniscus is reduced to a single thick component. Most modern wide-aperture lenses covering a reason-ably large field are of the double-Gauss form; there are also double-Gauss or Celor type lenses that cover a very much wider field of view with somewhat smaller apertures.

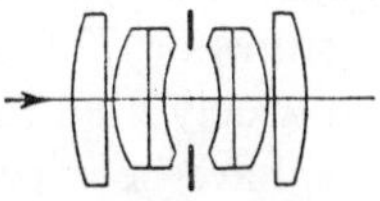

Fig. 16–25.
The Opic lens.

(d) The Cooke Triplet and its Derivatives

It was pointed out in § 3–10 that a system of three separated, thin, un-achromatic lenses can be free from longitudinal chromatic aberration and chromatic difference of magnification. The *Cooke Triplet* designed by H. D. Taylor (*c.* 1895) is corrected in this way [Fig. 16–26(*a*)]. In addition to the relative simplicity of the chromatic correction, the use of a diverging lens between two converging lenses has another important aspect. When a ray is incident at height y on a thin lens of power F, the ray is deviated through an angle Fy. It will be seen that the height of incidence of a ray in an axial pencil is relatively small at the second component, so that the lens can have a relatively high negative power. This gives a relatively large negative contribution to the Petzval field curvature, with a corresponding improvement in the off-axis correction. Triplet lenses are still widely used. They can have apertures of about $f/4\cdot5$ for a field of about 50° for a 10 cm. lens. For longer focal lengths or wider fields, the aperture is smaller, and for smaller fields the aperture can be increased to $f/3\cdot5$. For shorter focal lengths, the aperture can be increased still further

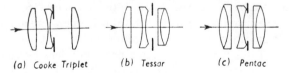

(a) Cooke Triplet   (b) Tessar   (c) Pentac

Fig. 16–26. Triplet lenses.

and, as an extreme example, a triplet has recently been used in an 8 mm. cine camera with an aperture of $f/1\cdot9$ and a 26° field. The triplet is one of the few lenses which was originally designed analytically rather than empirically. It will be seen that the Celor lens can be regarded as a derivative of the triplet.

The *Tessar* [Fig. 16–26(*b*)], introduced by Rudolph in 1902, may be considered as a development of the triplet. It can also be regarded as a development of the

Protar, the front component becoming uncemented. The state of correction of a Tessar is rather better than that of a triplet; it has a flatter field and the spherical aberration is smaller. Almost every manufacturer produces Tessars and, for apertures of $f/2\cdot8$ or $f/3\cdot5$, they may be regarded as the standard lenses for the better grade cameras.

Another lens which may be regarded as a derivative of the triplet or of the symmetrical types is the *Pentac*, designed by Booth in 1919 [Fig. 16–26(c)]. Apertures of $f/2\cdot9$ have been achieved with fields of about 50° and the lens has been widely used for aerial cameras. Pentacs with focal lengths of 12 inches or more, have been produced.

Further developments of the symmetrical and Triplet types are very numerous and very complicated; a discussion of these lenses is beyond the scope of the present book.

(e) Lenses of the Petzval Type

There are several applications for which one can sacrifice the size of the field in order to obtain a lens of wide aperture. In such cases the spherical aberration,

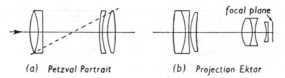

(a) Petzval Portrait   (b) Projection Ektar

Fig. 16–27. Portrait and projection lenses.

coma, and chromatic aberrations, are the most important aberrations to consider. Lenses for portrait photography and certain projection systems and cine cameras are of this type. In the original *Petzval Portrait* lens [Fig. 16–27(a)], designed by Petzval in 1840, one cemented and one uncemented doublet were used. This had large aperture ($f/3\cdot4$ with a 20° field) in order to reduce the exposure required with the rather slow plates available. In many cases, no special aperture stop was provided and the field of view was limited by the lens holder as indicated. More recent lenses have employed two uncemented components, and the lens shown in Fig. 16–24(c) can be regarded as one of this type.

It was pointed out by Piazzi-Smyth (1874) that a concave component close to the focal plane will introduce negative Petzval field curvature and help to flatten the field without upsetting the other corrections; it is a field lens [see § 16–4(c)]. A field flattener of this type was used in a projection lens by von Rohr (1911). Fairly recently, a 16 mm. projection lens (the Projection Ektar) employing a field flattener has been described by Schade. This is shown in Fig. 16–27(b); it has a focal length of $2\frac{1}{4}$ inches and, with an aperture of $f/1\cdot5$, gives extremely good definition over a 14° field—it will resolve 90 lines per mm. over the entire flat 16 mm. gate.

(f) Wide-angle Lenses

The main consideration in the design of wide-angle lenses is the reduction of the angles of incidence in the oblique pencils. Consequently, there is a tendency for the components to be heavily meniscus around the stop. A simple example is the *Hypergon* designed by von Höegh [Fig. 16–28(*a*)]. For a focal length of 10 cm. this lens covered a field of 140° at $f/22$ or less. It has a very flat field free from distortion.

Many lenses of the Celor type have been designed to cover a wide field. As they are made more meniscus, they cover a larger field at a smaller aperture. An extreme example is the *Topogon* shown in Fig. 16–28(*b*). For a 15 cm. lens, this

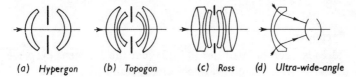

(a) Hypergon   (b) Topogon   (c) Ross   (d) Ultra-wide-angle

Fig. 16–28. Wide-angle lenses.

covers a 90° field at $f/8$–$f/11$. It has been widely used as an aerial survey lens. Another wide-angle lens is shown in Fig. 16–28(*c*). This is made by Ross and covers a field of 70° at about $f/4.5$.

Several modern ultra-wide-angle wide aperture lenses employ a diverging meniscus element a considerable distance in front of the other components as shown schematically in Fig. 16–28(*d*); it will be obvious from the figure that this element reduces the effective field angle for the other components. The large incidence height at the front element enables one to obtain a considerable deviation of the oblique pencils with an element of moderate power (§2–21). A system of the type represented by Fig. 16–28(*d*) is sometimes referred to as an *inverted telephoto lens*. Several wide angle lenses have employed a similar diverging component on the image side, i.e. have retained a symmetrical form. Systems of this type, pioneered by Roosinov, have enabled considerable coma to be introduced deliberately for the pupil planes (without upsetting the correction for the actual image plane) in order to make the illumination in the image plane more uniform (see §18–5).

(g) Telephoto Lenses

The principle of the telephoto lens was described in §3–4. If the system consists of two separated, fairly thin components it is necessary, as usual, to achromatize each component separately. One would expect the simplest telephoto lenses to consist of two separated achromatic doublets, the first of positive, and the second of negative power. The *Dallon* lens designed by Booth is of this type [Fig. 16–29(*a*)]. It has an aperture of $f/5.6$. In order to secure a large aperture, the telephoto magnification is now usually restricted to 2–3.

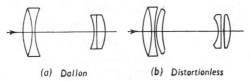

(a) Dallon   (b) Distortionless

Fig. 16–29. Telephoto lenses.

A simple positive lens placed before the aperture stop gives pincushion distortion; so, also, does a negative lens placed after the stop, and, for many years, designers were unable to produce distortionless telephoto lenses. The problem was first solved by Lee, who produced the system shown in Fig. 16–29(*b*). The broken contact surface in the rear component enables one to introduce barrel distortion and positive astigmatism which are balanced in the front component. This lens had an aperture of *f*/5 and had a telephoto magnification of 2·3.

16–12. Eyepieces

It has already been pointed out (§ 3–12) that the Huygens eyepiece gives less chromatic difference of magnification than the Ramsden, but that the latter must be used if a graticule is employed. Since each system consists of two single converging lenses, each has considerable positive Petzval field curvature. It is therefore desirable to introduce negative astigmatism. (As far as the complete instrument—objective and eyepiece—is concerned, the negative astigmatism is also desirable since the objective gives astigmatism of positive sign.) It is found that the Ramsden eyepiece has better astigmatic correction and is also better corrected for spherical and the other monochromatic aberrations. Little advantage is to be gained by deviating from the simple plano-convex form of the Huygens or Ramsden eyepieces. The chromatic correction of the Ramsden eyepiece can be improved considerably, and the general usefulness of the system

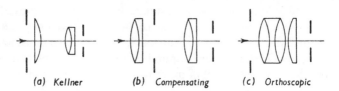

(a) Kellner   (b) Compensating   (c) Orthoscopic

Fig. 16–30. Eyepieces.

greatly extended, by employing a doublet eye lens. This *achromatized Ramsden* or *Kellner* eyepiece is shown in Fig. 16–30(*a*), the field stop and eye point being as indicated. This system covers a rather wider field of view than the simple eyepieces. The Huygens eyepiece is widely used in microscopy where the relative aperture and field of view are small. It has already been remarked that some microscope objectives give chromatic difference of magnification. This can be removed by a *compensating* eyepiece in which each lens of the Huygens system is a doublet [Fig. 16–30(*b*)]. With this eyepiece, it is possible to introduce negative transverse chromatic aberration.

Fig. 16–30(*c*) shows the system frequently referred to as an *orthoscopic* eyepiece. This system, due to Abbe, is widely used in systems whose aperture and field are beyond the capabilities of a Kellner system. It can now cover a visual field of up to 70°, and gives a very good eye clearance. There are many other

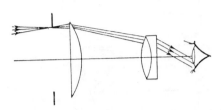

Fig. 16–31. Effect of spherical aberration of principal rays.

forms of eyepiece but these cannot be described in simple terms and will not be discussed here.

A defect whose importance is not always appreciated is spherical aberration of the principal rays. This is often severe in eyepieces of the Ramsden and Kellner types and arises chiefly at the curved surface of the field lens. The beams emerging at large angles are displaced, and in some cases enter the eye in part only (see Fig. 16–31). This results in a decrease in the apparent image brightness and, for certain eye positions, can cause isolated shadows to appear in the field of view.

16–13. Reflecting Systems

Apart from the complete absence of chromatic aberration, a mirror has certain advantages over a lens. It has been pointed out several times that a formula describing refraction at a single surface can be converted into the corresponding formula for reflection by putting $n'=-n$. A reflecting surface therefore corresponds to a change in refractive index of 2 compared with 0·5–0·7 for most refracting surfaces. Consequently, one would expect some of the aberrations of a reflecting surface to be less than those of a refracting surface under similar conditions. Thus it is found that a concave spherical mirror has considerably less spherical aberration than a thin converging lens of the same focal length. Now if the aperture stop is located at the centre of curvature of the mirror (Fig. 16–32), all principal rays are incident normally and, in each pencil of rays, there is symmetry about the principal ray in the same way as there is about the axis for an axial pencil. Consequently, there can be no coma or astigmatism. There is field curvature as illustrated, but the image surface is convex to the oncoming light, as opposed to concave as it is for a convex lens. This is obviously an advantage if a concave mirror and a convex lens are combined, since the combination will have a reduced Petzval sum. The obvious

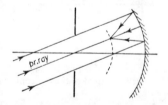

Fig. 16–32. Concave mirror with stop at centre of curvature.

difficulty associated with reflecting systems is the obstruction of the incident light by the image surface or by subsequent components. The image-forming pencils consequently consist of hollow cones of rays and this can upset the image quality although it seldom does so to a serious extent. The loss of light is usually of the same order of magnitude as the loss by absorption in a purely refracting system of comparable aperture.

For a concave mirror with the stop at the centre of curvature, spherical aberration is the only outstanding defect apart from field curvature. Now with positive spherical aberration the wavefront lies in front of the reference sphere in the outer zones of the aperture. The wave can be rendered spherical if the outer parts can be retarded, and this was done very effectively in the *Schmidt*

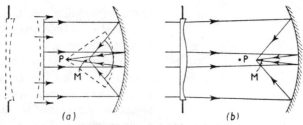

Fig. 16-33. Schmidt systems.

system by placing a corrector plate at the aperture stop as in Fig. 16–33(*a*). Disturbances in the outer zones of the aperture are retarded by an amount proportional to ϱ^4 so that the spherical aberration of the mirror is corrected. In terms of rays, the latter are deviated by the plate so that, after reflection, they pass through the paraxial focus P. It will be seen that the corrector has the general form of a diverging lens (although the second surface is not spherical) so that chromatic aberration is introduced. This dispersive effect is reduced to a minimum if the maximum ray deviation introduced by the plate is as small as possible. In the system shown in Fig. 16–33(*b*), the rays at the centre and marginal zones of the aperture are made to pass through a point between the marginal and paraxial foci of the mirror. This point is chosen to be the place where the maximum transverse ray aberration is as small as possible and is at the disc of least confusion—one quarter of the way from the marginal to the paraxial focus. It can be shown that the oblique aberrations introduced by the corrector plate are smaller for the second system. By introducing a suitable corrector plate, it is obviously possible to correct the secondary as well as the primary spherical aberration.

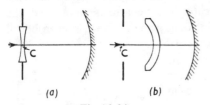

Fig. 16-34.
Development of the concentric system.

Now a diverging lens with spherical surfaces introduces negative primary spherical aberration, so that this aberration can be removed from the system shown in Fig. 16–34(*a*). The amount of spherical aberration introduced by the lens depends on its shape, and it is found that by employing a meniscus lens its power (and therefore its chromatic aberration) can be reduced. Proceeding in this way, one finds that it is possible to employ a lens for which the centre of curvature of each surface coincides with that of the mirror. This *concentric system* is shown in Fig. 16–34(*b*) and has an important advantage over the Schmidt system: since each surface is centred at C each principal ray is incident normally at all surfaces if the stop is also at C. Thus there is again no coma or

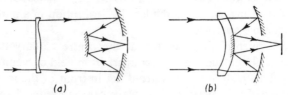

Fig. 16-35. Cameras of the Cassegrain type.

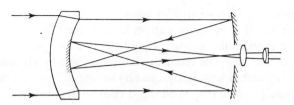

Fig. 16–36. Gregorian telescope.

astigmatism. For small systems such as cameras and terrestrial telescopes, the meniscus system is very useful, but for large astronomical telescopes the meniscus is more difficult to make than the Schmidt plate.

The optical systems of the classical reflecting telescopes were shown in Fig. 3–18. The Herschellian type has the disadvantage of employing the mirror off-axis. For terrestrial instruments, it is undesirable for the observer to look at right angles to the direction observed as in the Newtonian instrument. In recent years, several systems resembling the Gregorian and Cassegrainian instruments have appeared, some using one or more non-spherical mirrors, and some incorporating correctors of the Schmidt or meniscus type. Figs 16–35(a) and (b) show cameras suggested by Baker and by Bouwers and Maksutov; Fig. 16–36 shows a Gregorian telescope described by Bouwers.

When an attempt is made to design a purely refracting microscope objective for use in the ultra-violet, it is found that it is impossible to secure a satisfactory

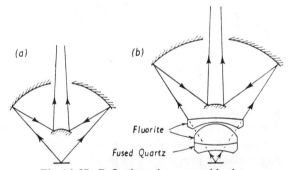

Fig. 16–37. Reflecting microscope objectives.

correction of the chromatic aberration with the transparent media available at present. This difficulty can be overcome by employing a reflecting or a cata-dioptric objective. Fig. 16–37(a) shows a simple reflecting objective, and (b) shows a catadioptric objective described by Grey; in the latter the second mirror obstructs a relatively small proportion of the aperture.

16–14. Zoom Lenses

A class of lens system that has been developed very extensively during the past 20 years or so is that in which it is possible to vary the magnification continuously over a fairly wide range without changing the positions of the object and image planes and without introducing a noticeable variation in image quality. Lenses of this type are known as *zoom lenses*; they have been developed for microscopes and terrestrial telescopes and also for cameras, especially television cameras. It is by using a zoom lens on the T.V. camera that one is able to achieve the familiar gradual change from a distant to an apparently close-up view of a scene. The required change in the magnification (i.e. in the focal length for a camera lens) is achieved by having at least one component in the system whose position can be changed relative to the others. Some zoom camera lenses have used a more or less traditional camera lens in front of which there is a telescopic system of variable power. An early successful T.V. zoom lens was that due to H. H. Hopkins (1946). The system employed two fixed converging components between which there were two diverging components whose separation was varied as they moved between the converging components (Fig. 16–38). In this way the magnification was varied by a factor of 5. In recent years many zoom lenses have been designed, some of which are extremely complex, and focal ratios of 10 or more have been achieved.

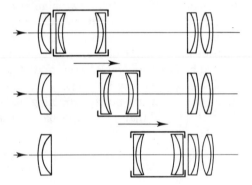

Fig. 16–38. Zoom lens.

16–15. Computerised Lens Design

The advent of the modern fast high capacity computer has revolutionized the practice of lens design and made possible many automatic lens design procedures. It is now possible to program a computer so that it changes systematically and automatically the various parameters of a lens system, such as radii of curvature, thicknesses and separations, refractive indices, and dispersions (checking that impossible values are not introduced) until, according to some pre-determined criteria, the aberrations are "corrected" or the best compromise is achieved. Major decisions to be taken when writing the computer program include how to judge the quality of the lens at any stage and how to vary the numerous parameters in order to converge as rapidly as possible on the optimum design. In effect one is usually searching for the deepest of several valleys on a surface in a many-dimensional space. Sometimes the designer is required to develop a lens system of an entirely new type and in such cases it may be necessary to start with paraxial theory and, in the light of experience select a promising class of system. In what is known as an *optimization* program a lens of known general type is taken and a search for the best version is made. Thus double-Gauss camera lenses have been investigated very thoroughly in recent years by, for example, Wynne and Kidger. If the program permits the computer to make major changes such as changing the number of surfaces in the system it is possible that the end product is barely recognizable to the inexpert eye as being a derivative of the initial design.

As was remarked above, a major problem is selecting the criteria or *merit function* by which the lens is to be judged. One can, for example, aim to correct or reduce below pre-determined tolerances individual primary and higher order aberrations, or one can minimise, say, the sum of the squares of the residuals, perhaps with some built-in weighting factors. Another procedure is to study the pattern of ray-intersection points in the image plane (the *spot diagram*) for various points in the object plane; this gives the total transverse ray aberration for rays passing through various parts of the aperture. Alternatively, one can calculate the O.T.F. (§14–17) or (for highly corrected systems) the Strehl intensity (§15–14) at various points in the field. In many cases one has the difficulty of judging the relative importance of different parts of the field of view so that one can take a form of weighted mean to construct a merit function.

It will be obvious that one is dealing with a highly complex problem; one can do little more here than indicate that this is the case.

CHAPTER XVII

VISUAL OPTICS

17–1. General Structure of the Eye

FIG. 17–1 represents an approximately full size horizontal section through a human right eye. The greater part of the eyeball is bounded by a tough, opaque, white membrane known as the *sclera*. At the front, this merges into a transparent, rather horny layer known as the *cornea*. The cornea has a thickness of about 0·5 mm. and the radius of curvature of its front surface is about 7·8 mm. The general spherical shape of the eyeball is maintained by the internal excess pressure which is some two or three cm. of mercury. Behind the cornea is the

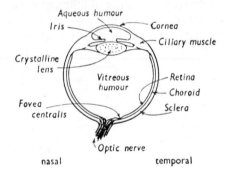

Fig. 17–1. Horizontal section of a right eye.

so-called *anterior chamber* which has an axial thickness of about 1·3 mm. and is filled with a watery, slightly saline fluid of index 1·336, known as the *aqueous humour*. At the rear of the anterior chamber is the *crystalline lens* whose aperture, usually referred to as the *pupil*, is controlled by the *iris* which is strongly coloured and almost opaque. The pupil size is varied by the iris from about 2 mm. diameter in very bright light, to about 8 mm. diameter in darkness. The crystalline lens has a bi-convex outer form, the front and back surfaces having normal radii of about 10 mm. and 6 mm. respectively, and the axial thickness being about 3·6 mm. The lens is not homogeneous but is built up of successive fibrous layers whose refractive indices vary from about 1·37 in the outer parts to about 1·42 in the middle. The *posterior chamber* behind the crystalline lens contains a transparent jelly-like substance known as the *vitreous humour* (index about 1·336). The inner surface of the rear part of the eyeball contains the receiving screen known as the *retina*. This consists effectively of a mosaic of light-sensitive elements upon which the image is formed when the eye is in use. The retina lies on the *choroid*, a layer which is dark brown in colour and richly supplied with blood vessels. Nerve impulses from the retinal receptors are communicated to the brain through the *optic nerve*.

There are no receptors where the optic nerve leaves the eyeball and that region is the so-called *blind spot*. This renders invisible a field measuring about 6° horizontally and 8° vertically. This fact often passes unnoticed—partly because the dark spot of one eye lies in the field of view of the other, and partly because the eyes are constantly rotating to explore the field of view. The existence of the blind spot can be observed as follows. Using only the right eye, look at A (Fig. 17–1). On varying the distance from eye to figure, a position will be found at which B disappears from view. Its image is now falling on the blind spot.

The resolving power of the retina reaches a sharp maximum in the region of a small depression known as the *fovea centralis*. When it "looks at" an object the eye is rotated until the image falls on the fovea. When a "normal" eye is at rest the optical system brings parallel rays to a focus on the retina, i.e. distant objects are clearly seen. In order to view nearer objects the power of the optical system of the eye is increased—a process known as *accommodation*. Although most of the power of the optical system of the eye is derived from refraction at the cornea, the *changes* in power during accommodation are effected by means of a change in the shape of the crystalline lens. The greatest observable change is in the curvature of the front surface of the lens and is controlled by the *ciliary muscle*. Some observers state that the shape of the eyeball changes also, the curvature of the cornea increasing.

17–2. The Optical System of a Normal Unaccommodated Eye

The optical system of the eye is complicated by the non-homogeneous nature of the crystalline lens. Gullstrand suggested a "schematic" eye which employed a lens of only two parts—an outer layer of index 1·386 and a core of index 1·406. This would give properties very similar to those of a real eye. He suggested, also, a simpler schematic eye in which the cornea was replaced by a simple refracting surface separating air and the aqueous humour. In addition, the lens was replaced by a simple one (see Fig. 17–2). The specification is:

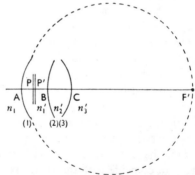

Fig. 17–2. Optical system of the eye.

$$(n_1=1\cdot000\text{—air})$$

(cornea)　$r_1=+0\cdot78$ cm.

　　　　　　　　$d_1'=0\cdot36$ cm.　　$(n_1'=1\cdot336\text{—aqueous humour})$

　　　　　⎧$r_2=+1\cdot00$ cm.

(lens)　⎨　　　　　　　$d_2'=0\cdot36$ cm.　　$(n_2'=1\cdot413\text{—crystalline lens})$

　　　　　⎩$r_3=-0\cdot60$ cm.

$$(n_3'=1\cdot336\text{—vitreous humour})$$

The positions of the cardinal points are best found using the deviation method (§ 2–22) or the matrix method (§ 2–25). One obtains:

$$f = -1\cdot678 \text{ cm.}$$
$$f' = +2\cdot242 \text{ cm.}$$
$$F = 59\cdot59 \text{ dioptres.}$$
$$AP = +0\cdot147 \text{ cm.}$$
$$CP' = -0\cdot545 \text{ cm.}$$
$$AN = +0\cdot711 \text{ cm.}$$
$$CN' = +0\cdot019 \text{ cm.}$$

The separation of the principal points (and therefore of the nodal points) is only 0·028 cm. so that, for many purposes, one can assume that the principal points coincide and that the nodal points also coincide.

The above results are for a normal eye with relaxed accommodation but, as pointed out in the previous section, the power of the eye is changed in order to view objects at finite distances. The maximum change in power which the eye can effect is called the *amplitude of accommodation* and, for the "average person", this varies from about 14 dioptres at the age of 10 years to about 4 dioptres at the age of 40, falling almost to zero at age 75. The object point for which the image is in focus when the optical system has minimum power is called the *far point*, and the *near point* is focused when the eye has maximum power. For the average observer, the far point is at infinity until the age of about 50. Thus, at age 40 the near point is 25 cm. in front of the eye and this has come to be taken as the standard *least distance of distinct vision* and is written "D_v". Beyond the age of about 50 the power of the optical system of the eye tends to decrease so that, when the accommodation is relaxed, light entering the eye must be con-vergent in order to be focused on the retina; the far point is then virtual. The associated loss of the ability to accommodate causes the near point to recede from the observer until it, too, becomes virtual.

As pointed out above, the eye rotates to bring the image on to the fovea for critical examination, and the line joining the fovea, the nodal point, and the so-called fixation point, may be called the *visual axis*. The optical system of the eye does not constitute an exactly centred system but for most purposes one may assume that it does. The axis approximating most closely to a true optical axis does not coincide with the visual axis but is inclined to it at an angle, α, of 5°–6° in the approximate direction of the figure 4 when the fixation point is the centre of the clock face. The centre of rotation of the eyeball is about 14 mm. from the front surface of the cornea.

The iris obviously constitutes the aperture stop, but there is no field stop. Although the retina can give good resolution near the fovea only, the complete field of view of the eye is very large. It extends laterally beyond even 90° on the temporal side and over 60° on the nasal side; it extends vertically about 50° upwards and 70° downwards. The retina cannot give good resolution in the outer parts of the field and the oblique aberrations are unimportant. The eye has the usual axial aberrations—chromatic and spherical—but the effect of the former is reduced by the fact that the eye is mainly sensitive to a relatively small range of colours. For many purposes, the aberrations of the "average" eye can be neglected. (See §§ 17–6 and 17–7.)

17-3. Defects in the Optical System. Spectacles

If, with a normal amplitude of accommodation, the power of the optical system is too great, the far point is at a finite distance and the near point is abnormally close to the eye. This condition is referred to as *myopia* or *short sight* and can be corrected by placing a concave lens in front of the eye in order to balance its excess positive power. If the power of the optical system is too small—a condition known as *hypermetropia* or *long sight*—the correction is effected by the use of a convex lens. With simple myopia and hypermetropia the effect of the spectacle lens is simply to shift the range of powers covered by the accommodation of the eye so that it covers the usual region. A reduction of the power of accommodation below about 4 dioptres is referred to as *presbyopia*.

If the far point remains at infinity or at a fairly large distance, the effect of presbyopia is a rather remote near point and this causes a difficulty with reading and other close work; it can be overcome by the use of a convex lens. As pointed out above, the lack of accommodation over the age of about 50 years is usually associated with an all-round reduction in the power of the optical system and it is then necessary to use a pair of spectacles with convex lenses to see distant objects, and a second pair with stronger convex lenses to see near objects. Frequently a spectacle lens of conventional appearance

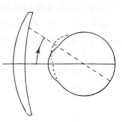

Fig. 17–3.
Vision through a
spectacle lens.

has the upper part of its aperture of one power and the lower half of a higher power since it is chiefly when reading that the eyes are directed downwards.

During use the eyes are rotated in their sockets and, as indicated by Fig. 17–3, the light entering the eye will pass obliquely through the spectacle lens. In order to reduce the effect of this oblique passage it is obviously desirable to employ a meniscus lens as indicated. This need is less marked when using spectacles for reading since the eye rotates through relatively small angles only.

In addition to the above defects many eyes exhibit axial astigmatism. The power of the optical system is then different in different azimuths, even when the object is on the axis. The variation of power with azimuthal angle is not always similar to the $\cos^2 \phi$ variation associated with primary oblique astigmatism in a system which has axial symmetry, but the defect can be reduced considerably by using a cylindrical lens which does introduce astigmatism of the conventional type. For example, if the astigmatism is such that the eye has the correct power in one azimuth and maximum incorrect power in an azimuth roughly at right angles, the defect is reduced considerably by the use of a cylindrical lens whose maximum power lies in the direction of the maximum error of the eye. Simultaneous myopia or hypermetropia can be corrected by using a lens of which one surface is spherical and one cylindrical. If a meniscus lens is desired, it is necessary for one surface to have a variation of curvature with azimuthal angle which may be described as a superposition of sphere and cylinder —a *toric* or *toroidal* surface.

In recent years *contact lenses* have been used to correct the refractive errors of the eye. These are small lenses which are placed in direct contact with the eyeball.

17–4. The Retina

The structure of the retina is extremely complex and it is sufficient for the present purpose to regard it simply as a mosaic of light-sensitive elements. The receptors are of two main types: *rods* and *cones*. These terms came into use because microscopic examination of the retina revealed the presence of bundles of rod-like and narrow conical elements, the light being incident end-on. The cones are responsible for so-called *photopic* vision, under ordinary bright conditions such as daylight, and the rods are responsible for vision in dim light such as moonlight (*scotopic* vision). There are no rods in the centre of the fovea where one has the maximum resolving power under conditions of photopic vision. The fact that there are no scotopic receptors in the fovea is easily demonstrated by attempting to look fixedly at a luminous figure on the dial of a watch in a dark room. It is found that as one tries to fixate on the figure it disappears from view. The sensation of colour is entirely dependent upon the cones so that one cannot distinguish colours in very dim light. The greatly increased resolving power of the fovea is associated with the fact that in that region the cones are very tightly packed. As one moves away from the fovea, the number of cones decreases as they intermingle with an increasing number of rods. Moreover, in the extra-foveal regions it is found that several receptors appear to be connected to a common nerve fibre; presumably this reduces the resolving power.

The first effect when light falls on a receptor is probably a photo-chemical reaction. Thus the rods are found to secrete a purple fluid known as *rhodopsin* or *visual purple* which is bleached by light and is assumed to play the leading part in the scotopic visual process. One cause of night-blindness (inability to see in dim light) is known to be vitamin A deficiency and this is also known to affect the secretion of visual purple. For cones the subject is more complex (see § 17–10).

17–5. Adaptation

A given stimulus does not always produce the same visual response. For example, the brightness level in a moderately well lit room appears very high if one enters from a dark room, but appears relatively dim if one enters from a considerably brighter room. After a time the impression of excessive brightness or excessive dullness disappears—one's eyes have become *adapted* to the level of illumination. The effect is particularly noticeable (although rather slow in action) when one leaves a well lit house on a dark night. At first one can see almost nothing, but after a few minutes a few outlines can dimly be seen, and one's vision continues to improve over a period of more than half an hour. One speaks, broadly, of light and dark adaptation when referring to the two extremes, although there are an infinite number of states of adaptation. The changes in the sensitivity of the eye are to a small extent due to changes in the size of the pupil which is increased involuntarily by the iris in dim light. By far the greater part of the process of adaptation is associated with a variation in the sensitivity of the retinal receptors. The process of dark adaptation can be explained readily if it is assumed that the sensitivity of the rods depends on the magnitude of the photo-chemical effect which occurs when light falls on the visual purple. In bright light the visual purple is quickly decomposed and, if the regeneration process is slow, the concentration is kept at such a low level that the photo-chemical effect is negligible and the rods are inactive. On entering a dark place, the relatively insensitive cones are not stimulated and the initial low concentration of visual purple causes the rods also to be insensitive. After a time, more visual purple is generated and the increased concentration gives a greater photo-chemical effect with the feeble stimulus; the visual response therefore increases. This increase in sensitivity continues until a balance is achieved between the processes of regeneration and photo-chemical decomposition. There are therefore an infinite number of degrees of dark adaptation, each corresponding to different magnitudes of the stimulus. Similarly, there are many states of adaptation corresponding to different levels of bright light when the cones alone are active. The eye becomes adapted to a decrease in stimulus in the region of photopic vision far more quickly than in scotopic vision. Assuming there are similar photochemical reactions in the cones, measurements of adaptation times in relatively bright light indicate that the regeneration process is complete in a few minutes. As the stimulus is reduced from a very large value the cones become dark adapted, i.e. more sensitive. When the stimulus is reduced below the so-called *threshold* for the cones they can no longer respond and the rods alone are active. Further reduction in the stimulus allows the rods to become dark adapted and more sensitive until, when the stimulus is reduced below the threshold of visibility, even the rods cannot be sufficiently sensitive and nothing can be seen. Between the true photopic and scotopic regions both rods and cones are active and one may say that the cones are dark adapted and the rods light adapted. Light adaptation is obviously caused by a reduction in concentration of the photo-chemical substances resulting from rapid decomposition and causing reduced sensitivity. It occurs much more quickly than dark adaptation. When a steady stimulus is suddenly applied to the dark adapted eye, the sensation rises to a peak and then falls to a fairly steady value within less than a fifth of a second. In addition, when the stimulus is removed suddenly, the sensation takes a finite time to disappear. This latter *persistence of vision* is responsible for the effectiveness of the stroboscope and cine film; it makes possible the removal of the flicker which would otherwise result from the repeated cessation of the stimulus. (See also the flicker photometer, § 18–10.)

The so-called *after-image* which is "seen" when a stimulus is suddenly removed takes a form which varies considerably with the conditions of adaptation, and the strength and duration of the stimulus, etc. If, for example, a strong white light is observed for about a second by a moderately dark adapted eye, the after-image often changes colour through red, green, and blue, becoming less definite in form. Again, if one looks at a strong source of one colour and then suddenly looks at a weakly illuminated white screen, an after-image is seen whose colour is complementary to the original stimulus. (For the meaning of the term "complementary colour", see § 17–9.) This corresponds to a light adaptation (i.e. a reduced sensitivity) for the colour originally observed. Again, if the eye is adapted to red light and then receives light which under normal conditions appears yellow, it appears green.

17–6. Spectral Sensitivity of the Eye

For the average observer, a visual sensation is produced by electromagnetic waves whose wavelengths lie between about 0.4μ and 0.75μ. Different wavelengths produce different sensations of colour, and this subject is discussed later. In addition, the sensitivity of the eye varies with wavelength and equal amounts of energy of different wavelengths will, in general, produce different sensations of brightness. Thus, if a spectrum is viewed and the energy received per second is the same for all wavelengths, the middle of the visible region appears brighter than the red and blue ends. When the intensity is high (i.e. for photopic vision) the maximum brightness occurs at about $\lambda 0.555\mu$.

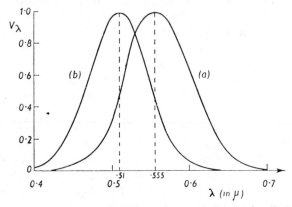

Fig. 17–4. Relative luminous efficiency curves. (a) Photopic. (b) Scotopic.

Under ordinary conditions of observation, the brightness of a source of a given wavelength (due to light emitted, reflected, refracted, or scattered) depends upon the energy per second received per unit area of the retina, and upon the state of adaptation of the eye. It also depends upon the size of the illuminated field of view and upon the psychological state of the observer. Thus in order to measure the spectral sensitivity of the eye it is necessary to consider the energy flux received per unit area of the retina, to maintain a constant state of adaptation, and to employ a constant field size.

Suppose one observes two luminous elements of equal area of which one (known as the *standard patch*) gives light of a standard wavelength λ_0, and the other (the *test patch*) gives light of any other wavelength λ. Suppose the standard patch is maintained unchanged while, for a number of values of λ, one finds the energy per second which the eye must receive in order to make the brightness of the test patch equal to that of the standard patch. If this energy flux is e_λ, $1/e_\lambda$ is a measure of the sensitivity of the eye to the wavelength λ. That is, $1/e_\lambda$ measures the effectiveness of light of wavelength λ in producing a visual response. This is usually plotted on a scale such that e_λ is unity at the wavelength of maximum sensitivity. $1/e_\lambda$ is then known as the *relative luminous efficiency* of light of wavelength λ and is denoted by V_λ. This was formerly called the *visibility* or *relative luminosity* (see below). The variation of V_λ throughout the visible spectrum is shown for photopic vision in Fig. 17–4(a). It should be noted that when the wavelengths of the standard and test patches are widely different, the colour difference makes it difficult to judge equality of brightness, and it may be desirable to proceed through the spectrum step by step comparing λ_1 with λ_2, λ_2 with λ_3, etc. The curve (a) applies to photopic vision, and for observations under rather lower levels of illumination it is found that the maximum of the relative luminous efficiency curve moves towards the region of shorter wavelength. For observation under true scotopic conditions one obtains curve (b), having a maximum at about $0\cdot51\mu$. This shift of the maximum is known as the *Purkinje effect* (1825). Obviously curves (a) and (b) can be associated with the cone and rod mechanisms respectively and are associated with the foveal and peripheral regions of the retina. The Purkinje effect is responsible for the relatively dark appearance of red objects under low levels of illumination since it causes a decrease in the value of V_λ in the long wavelength region. It will be appreciated that different observers will have different V_λ curves so that it is necessary to specify agreed curves for the "standard observer".

Referring to Fig. 17–4(a) it will be seen that $V_\lambda = 0\cdot5$ when λ is about $0\cdot51\mu$. This does not mean that a test patch giving light of wavelength $0\cdot555\mu$ appears twice as bright as a patch giving the same amount of energy of light of wavelength $0\cdot51\mu$. (Sensations such as brightness cannot easily be measured in this way.) What it does mean is that it is necessary to have twice as much energy of wavelength $0\cdot51\mu$ in order to make the patches appear equally bright. Thus if one has energy e_λ of light of wavelength λ entering the eye per second, it is the weighted energy flux $e_\lambda V_\lambda$ which determines the magnitude of the visual response in the sense that equal values of $e_\lambda V_\lambda$ give equal responses. When an amount of energy is weighted in this way one speaks of the quantity of light. The weighted flux $e_\lambda V_\lambda$ is the light flux or *luminous flux* and is sometimes referred to as the *luminosity* of the energy flux e_λ. (It will be seen that if one has a spectrum giving the same energy flux for all wavelengths, V_λ measures the relative luminosity of wavelength λ.) If one has light of wavelengths λ_1, λ_2, λ_3, etc., entering the eye, and the energy fluxes are e_{λ_1}, e_{λ_2}, e_{λ_3}, etc., respectively, one would expect that the brightness experienced would depend upon $\Sigma e_\lambda V_\lambda$ in the sense that equal values of the sum give equal sensations (and *not* that different values of Σ give sensations proportional to Σ). Usually it has been assumed that luminous fluxes are, in fact, additive in this way although a given value of $\Sigma e_\lambda V_\lambda$ corresponds to different colours if the relative values of e_{λ_1}, e_{λ_2}, etc. are changed. However, there now appears to be some doubt about this additivity of luminous fluxes. Thus, for stimuli having equal values of $\Sigma e_\lambda V_\lambda$ calculated from the V_λ curve the sensations of brightness are somewhat different if the colours of the mixtures are very different. It may be that the reasons are largely psychological, but the result does mean that the basis of photometry is undermined to some extent. It is usual to ignore this uncomfortable fact because the photometric comparisons of greatest importance seldom involve widely different colours. Fortunately, it seems certain that luminous fluxes *are* additive when they sum to stimuli that correspond to identical or similar colours. Hence the basis of colorimetry (§ 17–10) is *not* undermined.

In photometry one is concerned with the measurement of the total light flux (see Chapter XVIII). In this work one usually assumes that the conditions of photopic vision obtain.

17-7. Visual Acuity

Assuming that the eye is stationary, the following conditions must be satisfied if two neighbouring independent point sources are to be resolved by the observer: firstly, the optical system of the eye must give physically resolved images, and secondly, the mosaic of retinal receptors must have a structure sufficiently fine to record the resolved images. With regard to the first requirement, taking an average value of 4 mm. for the pupil diameter (assumed circular), one finds that the angular limit of resolution of the optical system is about half a minute. With regard to the second requirement, it is reasonable to assume that if the eye is to resolve two neighbouring retinal images, there must be an unstimulated receptor between them. In the fovea it is estimated that a cone subtends an angle of about half a minute at the nodal point, i.e. has just the order of magnitude necessary to take advantage of the resolving power of the optical system. These estimates of the limit of resolution which an eye might be expected to have are

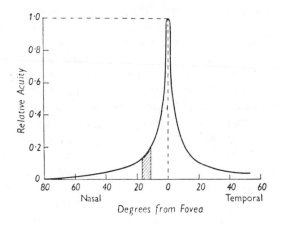

Fig. 17-5. Variation of visual acuity across the retina.

in good agreement with the observed value for an eye pupil of about 4 mm. diameter, and confirm the long accepted rough estimate of one minute of arc (corresponding roughly to a linear distance of about a tenth of a millimetre at the least distance of distinct vision—D_v=25 cm.). It seems that for pupil diameters of about 4 mm. the full theoretical resolution is attained so that the effect of the aberrations of the eye is practically negligible. The deterioration in acuity to more than one minute for pupils below about 2 mm. is explained by the increased size of the Airy pattern on the retina, and the poorer resolution for larger pupils is probably caused by the increased aberrations.

As one moves away from the fovea, the visual acuity for photopic vision decreases rapidly as indicated by Fig. 17-5. This is caused by the decreased resolving ability of the retina and also by the oblique aberrations of the optical system. For scotopic vision the fovea is inactive and the region just beyond it gives the maximum sensitivity and also maximum acuity.

The simple discussion given above is certainly not adequate to explain all the phenomena associated with the resolving power of the eye. One is not justified

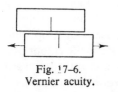

Fig. 17–6.
Vernier acuity.

in assuming that the eye is stationary since, in practice, it is constantly executing small movements. This ceaseless scanning of the object may be an important factor contributing to the very good *vernier* or *contour* acuity which is observed. This is concerned with the extraordinary ability of the eye to line up two straight lines as indicated by Fig. 17–6. It is found that the lines may be adjusted to be collinear to an accuracy corresponding to less than 10 seconds of arc. It is interesting to note that this implies that one can confidently use a simple vernier scale to measure linear movement as small as 1/50th mm. using the naked eye but that it is necessary to employ an observing lens for smaller distances. It is also found that the eye is able to detect very small movements of an object, and this ability does not decrease as rapidly as the simple limit of resolution as one moves away from the fovea. The detection of movement is associated with the sensitivity of the retinal receptors to small changes in stimulus, and it is found that the sensitivity to an intensity flicker is greater in the peripheral regions.

17–8. The Stiles-Crawford Effect

It was found by Stiles and Crawford that light entering the eye near the centre of the pupil is usually more effective in producing a visual response than that entering near the periphery. In scotopic vision the effect is found only with wavelengths greater than $0·58\mu$, but under conditions of photopic vision the effect is found throughout the visible region and is the same in the foveal and extrafoveal regions; it has been attributed to a directional property of the retinal receptors. It is obvious, from Fig. 17–7(*a*), that rays passing through the marginal zones of the eye pupil are

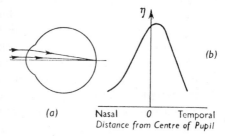

Fig. 17–7. The Stiles-Crawford effect.

incident on the retina more obliquely than rays lying close to the axis. Fig. 17–7(*b*) shows how the relative luminous efficiency, η, varies across the aperture of the eye pupil.

Suppose the eye receives a beam of light and that the energy is evenly distributed across the eye pupil. As far as the effective flux reaching the retina is concerned, an eye pupil of area a in the presence of the Stiles-Crawford effect is equivalent to an area Sa with the effect absent. For an eye pupil of diameter 2 mm. the *Stiles factor* S is 0·95, but it falls to 0·82 when the diameter of the eye pupil is 4 mm., and 0·66 when the diameter is 6 mm.

17–9. Colour Vision. Additive Mixtures

As pointed out in § 17–6, radiations of different wavelength produce different sensations of colour. For a normal eye the hues corresponding to the various regions of the visible spectrum are approximately as follows: $0·40\mu$ to $0·45\mu$ violet, $0·45\mu$ to $0·48\mu$ blue, $0·48\mu$ to $0·51\mu$ blue-green, $0·51\mu$ to $0·55\mu$ green, $0·55\mu$ to $0·57\mu$ yellow-green, $0·57\mu$ to $0·59\mu$ yellow, $0·59\mu$ to $0·63\mu$ orange, above

0.63μ red. A stimulus which does not contain a marked excess of any one group of wavelengths produces a sensation of white light. A fairly wide range of sources of this kind are referred to loosely as white (e.g. a tungsten filament lamp, sunlight, etc.). For many purposes it is convenient to define white as the colour of a source emitting an equal energy spectrum. If one wavelength is slightly dominant the sensation is of a pale or *desaturated* colour. As the proportion of the dominant wavelength is increased the colour is said to become more saturated until, for monochromatic light, complete saturation is reached. Thus, by adding white light to a saturated colour, a desaturated or pale colour is produced.

No attempt will be made to give an account of the various theories concerning the mechanism of colour vision, but it should be pointed out that colour cannot be distinguished under conditions of scotopic vision; it is therefore concluded that the cones are responsible for colour vision. Since the distribution of rods and cones varies across the retina, it became usual, in work on colour vision, to employ a field of view of not more than 2° so that only the foveal region is employed. This is also made desirable by the fact that in the region surrounding the fovea the retina is covered by a yellow pigment which probably influences the perception of colour. However, with a small field there is poor sensitivity to colour differences, and recent workers have used a 10° field.

It is found that a wide range of colours can be produced by an additive mixture of red, green, and blue light. It must be emphasized that this mixture is the superposition of red, green, and blue *light,* and must not be confused with the mixing of pigments which is discussed later. Convenient stimuli are monochromatic radiation with wavelengths 0.70μ, 0.5461μ, 0.4358μ. Using these it is possible to secure a perfect match between the additive mixture and many colours which are not highly saturated. It is found that a mixture of red and green light gives a fairly saturated orange, yellow, or yellow-green, according to the proportion of red and green employed; a mixture of blue and green gives various blue greens; and red plus blue gives various purples. A certain mixture of red, green, and blue, matches the colour of an equal energy source. If the proportion of one of the three components is increased, one obtains a desaturated red, green, or blue. If the proportion of any two of the three is greater than that which gives white, the result is a desaturated yellow, blue-green, or purple.

It is found that an additive mixture of red, green, and blue, as described above, cannot match very saturated colours. This is most noticeable in the blue-green region. In order to secure a match involving a saturated (e.g. monochromatic) blue-green, it is found that it is necessary to add some red light to it and to match this combination with a mixture of blue and green. One can look upon the saturated blue-green as being matched by a mixture of blue, green, and a negative quantity of red.

It is not essential to use the red, green, and blue radiations as given above although it is found that these enable a wide range of colours to be matched by a positive mixture. It is found that if negative quantities are allowed a colour C can be matched by a mixture of *any* three colours, which need not be monochromatic but which must be independent in the sense that it must be impossible to match one by mixing the other two. Thus, apart from the pure spectral colours, a given colour does not correspond to a unique spectral distribution (distribution of energy through the spectrum) of light entering the eye. A given spectral distribution can, however, give only one colour. The latter is simply the result of the additive mixture of the spectral colours concerned, although it could have been produced by other spectral distributions.

It may happen that an additive mixture of light of two colours produces white. The colours are then said to be *complementary*. From what has already been said, it will be seen that the fact that two colours are complementary does not necessarily mean that the superposition of their constituent radiations gives an equal energy spectrum. For example, it has been pointed out that a certain mixture of monochromatic red, green, and blue, can match the colour of an equal energy white. If the blue is omitted from this mixture, the remaining combination of red and green gives a yellow. Now this latter, when mixed with the blue, gives white; hence the blue and the yellow are complementary.

Fig. 17–8 illustrates a simple arrangement for demonstrating complementary colours. S is a slit illuminated by a filament lamp and, by means of the lens L_1 and the prism P, a spectrum is produced at A. Except for the edges, the patch of light B produced by lens L_2 contains a mixture of the spectral colours and appears white. If a region of the spectrum is obstructed, the patch at B becomes

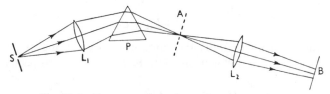

Fig. 17–8. Demonstration of complementary colours.

coloured, and its colour is complementary to that of the radiation obstructed. In this way it is easy to observe the colour which is complementary to an approximately monochromatic colour.

17–10. The Trichromatic System of Colour Measurement

The trichromatic system of colour measurement is based on two well-established experimental results:

1. As described above, with the restrictions mentioned, any colour can be matched by a suitable mixture of any three stimuli.

2. If two colours are matched in turn by mixtures of three stimuli, then the sum of the two colours will be matched by the sum of the two mixtures.

It follows that if three standard stimuli are chosen it is possible to specify a colour by stating the amount of each standard required to produce a match. It is convenient to choose, as standards, the red, green, and blue stimuli described above, but it is now necessary to define the units in which amounts of red, green, and blue light are to be measured. The units are chosen so that equal amounts of red, green, and blue, are required to match white, where white can be defined as the colour of, for example, a black body source operating at a given temperature, or of a theoretical idealized equal energy source.

Suppose a light flux (i.e. weighted energy flux or luminosity—see §17–6) L_W' of white light is matched by the superposition of light fluxes of red, of green, and of blue. Then one may write:

$$L_W'(W) \equiv L_R(R) + L_G(G) + L_B(B).$$

where brackets distinguish a symbol describing a *kind* of stimulus as opposed to an amount, and the symbol $\equiv$ denotes a match. Thus this equation is actually read as "Flux L_W' of white is matched (in brightness and hue) by the mixture of flux L_R of the red light R plus flux L_G of the green light G plus flux L_B of the blue light B". Now when there is a colour match light fluxes are additive so that

$$L_W' = L_R + L_G + L_B.$$

This is an ordinary equation relating quantities. In order to measure (W) in the new so-called *trichromatic units* it is obviously necessary to multiply the measure of red light by $1/L_R$, that of green light by $1/L_G$, and that of blue by $1/L_B$. Thus, in all similar measurements, the quantities of (R), (G), and (B), must be modified in this way. Suppose that a light flux L_C' of colour (C) is matched by L_R' of (R), L_G' of (G), and L_B' of (B). Then the equation in terms of light fluxes is

$$L_C'(C) \equiv L_R'(R) + L_G'(G) + L_B'(B),$$

where $\qquad\qquad\qquad L_C' = L_R' + L_G' + L_B'.$

In terms of the trichromatic units, the amounts of (R), (G), and (B), necessary to match a light flux L_C' of (C), will therefore be r', g', and b', respectively, where

$$r' = \frac{L'_R}{L_R}, \qquad g' = \frac{L'_G}{L_G}, \qquad b' = \frac{L'_B}{L_B}.$$

The colour, as opposed to the intensity, of (C) determines the relative amounts of (R), (G), and (B), the absolute values of r', g', and b', being determined by the *amount* of (C) concerned. One now defines a unit amount of (C) as the amount matched in intensity and colour by the additive mixture of (R), (G), and (B), when their quantities in trichromatic units have been scaled down to make their sum unity.

Thus,

$$1 \cdot 0(C) \equiv r(R) + g(G) + b(B) \qquad \ldots \ldots \quad (17\text{–}1)$$

where one must have $r + g + b = 1$, i.e.

$$r = \frac{r'}{r' + g' + b'}, \qquad g = \frac{g'}{r' + g' + b'}, \qquad b = \frac{b'}{r' + g' + b'}.$$

Equation (17–1) is known as the *unit trichromatic equation* and r, g, and b, are the *trichromatic coefficients* or co-ordinates. The amount of the colour (C) represented by equation (17–1) is known as one *trichromatic or T. unit*.

If, in a unit trichromatic equation, two of the coefficients are known, the third can be calculated since $r + g + b = 1$. The expression for (C) is therefore a function

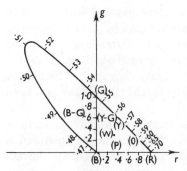

Fig. 17–9. The spectral locus on the RGB chromaticity diagram.

of only two variables and can be represented on a plane diagram such as Fig. 17–9. This is known as the *Chromaticity diagram*. The red radiation (R) is represented by the point (1,0), (G) by (0,1), and (B) by (0,0). The white colour of an equal energy source is the point (W) (1/3, 1/3). (B)(G), (B)(R), and (G)(R), are the loci of colours which contain no (R), (G), and (B), respectively. Thus, along (B)(G) one has blue-green (B–G); along (B)(R) purples (P); and along (G)(R) yellow-green (Y–G), yellow (Y), and orange (O). It is found that spectral colours cannot be matched exactly by positive mixtures of the (R), (G), and (B), standard stimuli, but, if negative values of r, g, and b, have the significance explained in the previous section, one can plot the locus of the spectral colours on Fig. 17–9, as shown. An indication of the wavelength of the monochromatic radiation represented is shown at various points along the locus. It will be seen that all the saturated blue-greens correspond to negative values of r.

Suppose p T. units of colour (C_1) and q T. units of colour (C_2) are mixed additively to give colour (C_3). For p units of (C_1), one has $p(C_1) \equiv pr_1(R) + pg_1(G) + pb_1(B)$ and, for q units of (C_2), $q(C_2) = qr_2(R) + qg_2(G) + qb_2(B)$ where r_1, g_1, b_1, and r_2, g_2, b_2, are the unit trichromatic coefficients, i.e.

$$r_1 + g_1 + b_1 = 1 = r_2 + g_2 + b_2.$$

On adding these two colours the sum of the coefficients is $(p+q)$ so that the mixture gives $(p+q)$T. units of (C_3). The unit equation for (C_3) is therefore given by

$$1 \cdot 0(C_3) \equiv \frac{pr_1 + qr_2}{p+q}(R) + \frac{pg_1 + qg_2}{p+q}(G) + \frac{pb_1 + qb_2}{p+q}(B).$$

From the geometry of the chromaticity diagram, the point representing (C_3) must lie on the line $(C_1)(C_2)$ such that it divides $(C_1)(C_2)$ in the ratio $q:p$.

Thus, if one imagines masses p and q at the points representing (C_1) and (C_2), the point representing (C_3) is at the centre of gravity. This can be extended to the mixture of any number of colours. It will be seen that (W) is at the centre of gravity of equal weights at (R), (G), and (B). For any colour it is also easy to find the complementary colour. For example, various blues and yellows form complementary pairs.

It will be seen from Fig. 17–9 that the spectral locus is always concave inwards. Now all colours must be mixtures of the spectral colours, so that only points which lie in the area bounded by the spectral locus and the line joining its extremities correspond to colours with physical significance. Points outside this area do not correspond to colours of actual physical stimuli. As far as the geometry of the figure is concerned, there is no reason why the position of the point corresponding to a given colour should not be referred to three reference points lying outside this area. In the system introduced by the Commission Internationale de l'Éclairage (the so-called C.I.E. system), three such reference points are chosen. These do not correspond to three physical stimuli, but in terms of these "theoretical" stimuli all colours are matched by positive mixtures.

The colour matches which have been referred to throughout the above discussions are those accepted as corresponding to a normal observer. Thus, the location of the spectral locus in the RGB chromaticity diagram will be slightly different for different observers and internationally agreed results have now been adopted.

An instrument that is designed for the measurement of colour is referred to as a colorimeter. In a *visual tristimulus colorimeter* the colour to be measured is simply matched visually by a mixture of three convenient stimuli such as the red, green, and blue, already mentioned. In a *photo-electric tristimulus colorimeter* three photo-cells are used which, in combination with appropriate filters, have three different spectral response curves. The outputs from these three photo-cells take the place of the amounts of the three coloured stimuli required for a visual match. A third possibility is to use a photoelectric *spectrophotometer* (*spectroradiometer*) to measure the complete spectral distribution of the radiation whose colour is required. Space does not permit a detailed account to be given here of the precise manner in which the trichromatic specification of the colour is obtained from these measurements using physical detectors but it should be obvious that it is possible in principle because each method gives the relative intensities of the various colours present.

Apart from the small variations in the matching of colours by different observers with more or less "normal" colour vision, it is found that some so-called *colour-blind* persons have distinctly defective colour vision. In some cases the observer is able to match all the colours he can see by mixing only two standard stimuli; he may, for example, be quite unable to distinguish between red and green.

The fact that a normal eye can match almost any colour by an additive mixture of saturated red, green, and blue, led to the suggestion that the average observer has three primary mechanisms for the perception of colour. This theory, associated with Young, Helmholtz, and Maxwell, has received support in recent years from the microscopic examination of the retina—microspectro-photometry indicates that there are probably three kinds of cones.

17–11. The Colour of Non Self-luminous Objects

If a source emitting an approximately equal energy spectrum illuminates an object which reflects equally at all wavelengths, the object will appear white or grey according to its reflectivity. If the object absorbs some wavelengths and reflects others it will appear coloured. For example, a pigment which reflects only wavelengths above about 0.60μ will appear red. If it is highly reflecting for all wavelengths above about 0.5μ while for shorter wavelengths its reflectivity is low, it will appear yellow. Blue-green pigments reflect well in the short wave-length region but their reflectivity falls off above about 0.5μ. Thus a mixture of these yellow and blue-green pigments does not reflect a mixture of yellow and blue light, but only those wavelengths which are reflected by both constituents. In practice, the region 0.5μ to 0.55μ is partially reflected by both, so that a mixture of blue-green and yellow pigments produces a green pigment. This will be true if the illuminating source emits all wavelengths; for other cases it must be remembered that the colour which an object appears to have is determined by the wavelengths which it reflects to the eye. The following is an example: a surface whose spectral reflectivity curve shows a broad peak centred at the middle of the visible spectrum would appear as a desaturated green when illuminated by an equal energy white source, but could appear red, yellow, or blue, if illuminated by monochromatic sources of these colours respectively. Again, by referring to Fig. 17–9 one can see that a surface whose reflectivity curve shows a broad peak stretching from about 0.51μ to about 0.70μ could appear yellow-green when illuminated by an approximately equal energy source such as daylight. If this surface is illuminated by a source which emits narrow bands of wavelengths in the regions of 0.49μ and 0.65μ, it will appear red although the source itself may appear white. It will be seen that one must be careful to distinguish between the colour of a source and its colour rendering properties. The source just referred to is white but has poor colour rendering properties in the sense that an object illuminated by it may have a colour which is different from its colour in daylight. The colour rendering properties of a source can be completely satisfactory only if it emits fairly uniformly throughout the visible spectrum.

17–12. Subtractive Colour Mixture

It was pointed out above that a mixture of yellow and blue-green pigments gives a green pigment. It is interesting to consider the possibility of producing a wide range of colours from, say, three "primary" pigments. In order to produce a wide range of colours one needs to be able to control the amounts of red, green, and blue light which enter the eye, and in additive mixing one has direct independent control of these three stimuli. Now a pigment of a certain colour reflects some wavelengths and absorbs the rest. By varying the proportion of this pigment in a mixture, one can vary the intensity of the wavelengths it absorbs but not those which it reflects since these will remain at maximum intensity. Consequently, one obtains control of the red, green, and blue light which enters the eye by varying the proportions of those pigments which absorb the spectral region corresponding to these three colours. Thus in additive mixing one starts with no light and *adds* chosen amounts of red, green, and blue; in the case of pigments one starts with light of all wavelengths and *subtracts* chosen amounts of red, green, and blue. One speaks of a *subtractive mixture* and the primary pigments may be referred to as minus red, minus green, and minus blue. The colours of these pigments are blue-green, magenta, and yellow, respectively. (Loosely referred to as blue, red, and yellow.) The control which one can exercise in a subtractive mixture is less complete than in an additive mixture since the wavelengths nominally reflected by a pigment are actually partially absorbed and the absorption bands of the various pigments overlap to some extent.

17–13. Stereoscopic Effect

When viewing a scene with one eye there are a number of factors which help one to decide the relative distances of various objects in the field of view. Firstly, one finds that different degrees of accommodation are necessary to view the various objects clearly; secondly, one draws on one's experience concerning the relative sizes of familiar objects; thirdly, one can interpret the variations of light and shade. However, these factors alone do not enable one to get an impression of three dimensional "solidity" when viewing an object. For this *stereoscopic* effect, binocular vision is essential.

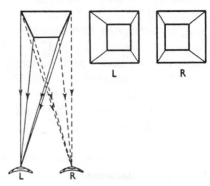

When using both eyes to view a scene, two slightly different perspective retinal images are formed, the perspective centre of each eye being its first nodal point. When the brain "fuses" the two images an impression of depth or relief is obtained. For example, Fig. 17–10 shows how an observer's two eyes obtain slightly different views of a truncated pyramid.

Fig. 17–10.

Perspective views for left and right eyes

When one "looks at" a given point with both eyes they are rotated in their sockets until, in each eye, the image point falls in focus on the fovea. It is therefore necessary to adjust the accommodation of the eyes and the convergence of their visual axes. These adjustments are, of course, made unconsciously.

The fact that it is necessary to have two different perspective retinal images to produce a stereoscopic effect explains why this effect must necessarily be absent from a simple photograph. However, as was originally pointed out by Wheatstone, if one has two photographs taken by identical cameras from positions representing those of an observer's eyes, a stereoscopic effect is produced if each eye views the appropriate photograph. The effect will be precisely the same as that obtained by binocular observation of the original scene if the camera lenses were separated by the correct interocular distance (approximately 65 mm.) and the photographs are viewed at the correct viewing distance (see §3–4). The method of viewing the photographs is of some importance. For example, it is possible to place them side by side and to view one with each eye directly, a screen being positioned so that each eye can see only the correct photograph. The necessary accommodation and convergence are obviously determined by the positions of the two photographs. However, many observers find it difficult to employ combinations of accommodation and convergence which are very different from those employed in normal vision, and they would have considerable difficulty in fusing the images. This difficulty can be reduced by employing viewing lenses as shown in Fig. 17–11. An arrangement such as this is called a *stereoscope*. Suppose the photographs are in the focal planes of the lenses so that the accommodation is relaxed. For a given separation of the photographs there will be, in the stereoscopic view obtained by the observer, a certain apparent object distance for which the convergence of the eyes is also relaxed. When the observer fixates on points which appear to be at other distances there will be some convergence without accommodation. However, it is found that the combinations of accommodation and convergence which occur with this arrangement can be made less objectionable than when the lenses are omitted.

Another way of producing a stereoscopic picture is as follows: one of the photographs, printed in red, is superimposed (with very small displacement) on the other, printed in blue. If the combination is viewed through a red filter, the dark parts of the red photograph still reflect red light and are indistinguishable from the bright parts. The dark parts of the photograph printed in blue do not

Fig. 17–11.
Simple stereoscope.

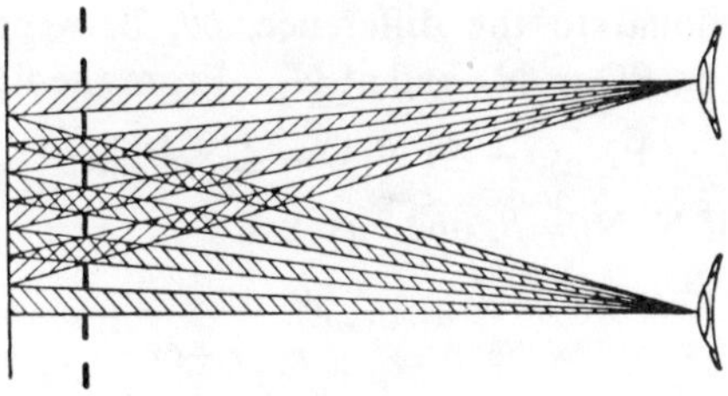

Fig. 17–12.
The grid system for stereoscopic pictures.

reflect red, and they appear dark; the contrast in the photograph is seen. Similarly, when viewing through a blue filter, only the red photograph can be seen. Thus, by placing red filter in front of one eye, and a blue filter in front of the other, one can ensure that each eye sees only the correct perspective view. On fusing the two images, a stereoscopic picture is seen.

A further method is to print parts of the two photographs on alternate narrow strips of the screen and to view this through a grid in such a way that each eye sees alternate strips and, therefore, only the correct image (see Fig. 17–12). A similar effect can be obtained by placing, immediately in front of similarly prepared photographs, a lenticular grid whose linear elements are actually long and very narrow cylindrical lenses. Refraction by these elements causes each eye to see only the correct perspective image.

A stereoscopic effect is produced in cinemas by projecting two images on to the screen. Each image is formed by plane polarized light (see Chapter VI), the planes of vibration being mutually perpendicular and inclined at 45° to the vertical. Each member of the audience wears spectacles which place a polaroid analyser in front of each eye so that only the correct image is seen.

It must be pointed out that projections on an extra-wide screen do not give a true stereoscopic picture. Their chief advantage is that the eye is presented with a wide field of view as in normal vision.

It will be seen that binocular vision involves a co-ordination of convergence and accommodation. Normally the visual axes of the eye are parallel when the muscles are at rest. If they are not parallel when the muscles are relaxed (a condition known as *heterophoria*), it may be difficult to fuse the images properly. Frequently this defect can be overcome by decentring the spectacle lenses to obtain a slight prismatic effect. If the muscle imbalance is very marked, it may be impossible to make the visual axes converge on the point to be viewed. This is known as *strabismus* or *squint*. The images cannot be fused but, usually, one of them is suppressed so that, in effect, one has monocular vision.

17-14. Stereoscopic Range. Binocular Telescopes

Suppose an observer views points P and Q which are at distances r and $r+\delta r$ and lie close to the perpendicular bisector of the line joining N_L and N_R, the first nodal points of the left and right eyes (see Fig. 17-13). It is assumed that the ability to detect the difference δr depends upon the difference in size between the retinal images of PQ. This is proportional to the difference, $\delta\theta$, between the angles subtended by PQ at N_L and at N_R. From the figure one has

$$\angle PN_RQ - \angle PN_LQ = \angle N_LPN_R - \angle N_LQN_R.$$

Thus, if $N_LN_R = b$, one has

$$\delta\theta = \frac{b}{r} - \frac{b}{r+\delta r} \qquad (17\text{-}2)$$

$$= \frac{b\delta r}{r^2} \text{ if } \delta r \text{ is small.}$$

Alternatively, $\qquad \delta r = \dfrac{r^2\delta\theta}{b}\,.$

Fig. 17-13.
Binocular
parallax.

Hence the minimum detectable difference δr is determined by the minimum detectable difference $\delta\theta$. This latter is found to be about $\frac{1}{2}$ minute.

If r is large equation (17-2) shows that δr is large and, when $r = b/\delta\theta$, δr becomes infinite. This means that for objects situated at distances greater than $r_{max} \doteq b/\delta\theta$ (i.e. about 450 metres) it is impossible to decide, from the stereoscopic effect alone, which of the points—P or Q—is nearer. A knowledge of the actual, and observation of the apparent, size does, in practice, play a more important part in such cases.

The distance r_{max} is called the *stereoscopic range*, and if one places mirrors in

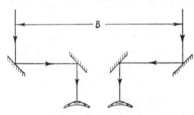

Fig. 17-14. Mirror system for enhanced stereoscopic effect.

front of the eyes as in Fig. 17-14, the difference between the perspective images is increased and the range is increased to $B/\delta\theta$. In many binocular instruments (e.g. prismatic binocular telescopes) this effect is accompanied by a magnification of the retinal images. If M is the magnifying power of the telescopes the difference $(\angle PN_RQ - \angle PN_LQ)$ is magnified M-fold. Thus if $\delta\theta$ is again the minimum value detectable by the observer, the range is extended to

$$r_{max} = \frac{MB}{\delta\theta},$$

where B is the separation of the centres of the telescope objectives. Similarly, for nearer objects the minimum detectable distance δr is decreased to

$$\delta r = \frac{r^2}{MB}\delta\theta\,.$$

17–15. Binocular Microscopes

It must be pointed out at once that not all binocular microscopes give stereo-scopic vision. Non-stereoscopic binocular microscopes are widely used and have a distinct advantage over monocular instruments when it is necessary to employ a microscope for long periods. It is found that a binocular instrument is more comfortable and less tiring.

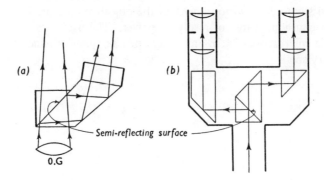

Fig. 17–15. Beam-splitting systems for binocular microscopes.

For low powers it is possible to combine two complete monocular instruments. Now a normal microscope gives an inverted image, and if two instruments are combined, one has an inverse stereoscopic effect. This difficulty can be over-come if each instrument incorporates an erecting system; in the Greenough microscope, prism erectors are used. The erect image and stereoscopic effect are very useful if the microscope is used in dissecting, etc.

For high powers it is impossible to position two objectives alongside one another and sufficiently close to the object. Consequently a single objective must be used in conjunction with a beam splitting device and two eyepieces. Typical systems, for non-stereoscopic instruments, are shown in Fig. 17–15(*a*) and (*b*). In the first, due to Beck, the beam splitter is placed behind the objective. It is necessary to employ a convergence of the eyes and it is therefore convenient to bring the image to the near point. One eyepiece is co-axial with the objective and, by removing the prism system, the instrument is easily converted into a monocular microscope. In the second system the optical axes of the eyepieces are parallel and the images are viewed with relaxed accommodation.

To obtain a stereoscopic effect, the two eyes must have slightly different views of the object, and this must be achieved by employing one half of the objective aperture for each eye. No erecting system is necessary if both eyes look through a single inverting objective-eyepiece system. As will be seen from Fig. 17–16, the effective positions of the eyes is reversed and this, plus the inversion of the image, gives a normal stereoscopic effect. When one employs a beam splitter and two eyepieces, one has, in effect, two simple objective-eyepiece systems; the left eye must use the left half of one exit pupil aperture, and the right eye must use the right-hand side of the other exit pupil. The right eye then uses the left side of the entrance pupil, and vice versa, as in Fig. 17–16. This can be ensured

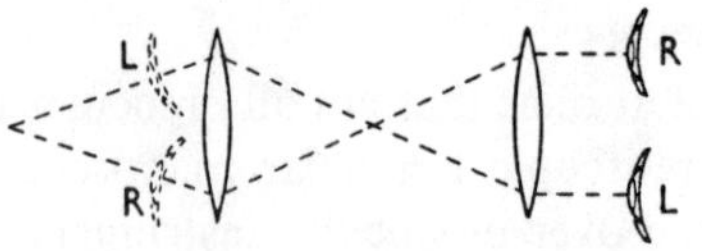

Fig. 17–16. Binocular observation with a single inverting system.

if suitable screens are positioned in the exit pupils. It will be seen that with high powers and stereoscopic vision only half the objective aperture is used by each eye, with a consequent loss of resolving power. The usefulness of high power stereoscopic microscopes is, in any case, open to question since the depth of focus is very small.

RADIOMETRY, PHOTOMETRY, AND THE RADIOMETRIC AND PHOTOMETRIC PROPERTIES OF IMAGE-FORMING AND SPECTROSCOPIC INSTRUMENTS

STRICTLY, radiometry is concerned with the measurement of radiant energy, and photometry with the measurement of radiant energy weighted according to the visual response it produces. Similarly, one should distinguish between spectrophotometry and spectroradiometry but many writers and instrument manufacturers fail to do so, and take spectrophotometry to embrace spectroradiometry.

In what follows various fundamental radiometric and photometric quantities will be defined and the relationships between them discussed. Since the relationships that exist between the various photometric quantities are identical to those that exist between the corresponding quantities in radiometry generally, there is no need to duplicate the discussion. Although radiometry is more general than photometry in the sense that it is concerned with radiation in all spectral regions, the main discussion will, in fact, be in photometric terms. This is because a discussion in photometric terms can embrace all the problems encountered in radiometry whereas the reverse would not be true. That is to say, parts of the discussion have no equivalents in radiometry; for example, when discussing luminance the question of subjective brightness arises. When a photometric term is introduced, the corresponding term in radiometry will be mentioned. Thus the discussions and relationships that follow will be in photometric terms but can be converted into their equivalents in radiometry by replacing every photometric quantity involved by its general radiometric equivalent. There is one departure from the approved symbols: e_λ is used for spectral radiant energy flux. The approved symbol for radiant energy flux is Φ_e but this would have led to triple subscripts in §17–6 and double subscripts here; Φ_e is, however, used to denote total radiant energy flux.

18–1. Fundamental Photometric and Radiometric Quantities

(a) *Luminous flux* Φ (corresponding term in radiometry generally: *radiant flux* Φ_e). Luminous flux was defined in §17–6 as the energy flux weighted according to its efficiency in producing a visual response. It was seen that if one has radiant energy flux e_λ of wavelength λ, the luminous flux may be defined as $e_\lambda V_\lambda$, where V_λ is the relative luminous efficiency of light of wavelength λ, and that if one has light of several wavelengths, the total luminous flux is $\Phi = \Sigma e_\lambda V_\lambda$. Since one never has light of a number of discrete wavelengths one should say that, for a spectrum giving energy flux $e_\lambda \delta\lambda$ in the range $\lambda - \lambda + \delta\lambda$, the luminous flux is $\int V_\lambda e_\lambda d\lambda$. The idea of luminous flux is convenient because, under identical conditions, equal luminous fluxes produce equal sensations of brightness, but it must be remembered that it is not implied that different fluxes produce sensations of brightness in direct proportion. It must be remembered also that a flux is a rate of flow so that if one says the energy flux entering the eye is e_λ, this represents the energy entering per second; it can be measured in ergs per second or joules per second (watts). Employing any convenient units of power such as these, $\int V_\lambda e_\lambda d\lambda$ can, without modification, be used as the measure of luminous flux. Although in theory it is not essential to do so, an independent system of units is employed in Photometry. This is described in §18–8.

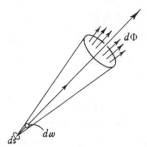

Fig. 18–1. Radiation from an elementary source.

(b) *Luminous Intensity I* (in radiometry: *radiant intensity I_e*). In Fig. 18–1 *ds* is a small source, or element of a source, which does not necessarily radiate equally in all directions. The luminous flux emitted per unit solid angle in any given direction is called the *luminous intensity* in that direction. That is, if in any direction the source emits flux $d\Phi$ into the elementary solid angle $d\omega$, the luminous intensity in that direction is given by

$$I = \frac{d\Phi}{d\omega}. \qquad \cdots \cdots \cdots \quad (18\text{–}1)$$

In the past the luminous intensity has frequently been called the *candle power*. This is unfortunate since the term candle power was also used for the unit itself. Many sources can, for practical purposes, be regarded as point sources. One speaks of a uniform point source if the luminous intensity is the same in all directions. The total flux emitted is then given by

$$\Phi = \int I d\omega = I \int d\omega = 4\pi I,$$

i.e.

$$I = \frac{\Phi}{4\pi}. \qquad \cdots \cdots \cdots \quad (18\text{–}2)$$

For a non-uniform point source which emits a total flux Φ, $\Phi/4\pi$ is called the *mean spherical candle power*; it is obviously the average value of the luminous intensity in all directions.

(c) *Illumination E* (or *illuminance*; in radiometry: *irradiance E_e*). If a luminous flux is incident upon a surface, it is said to be illuminated and the flux received per unit area at any point on the surface is called the *illumination* (or intensity of illumination or *illuminance*. Illuminance seems to be the "recommended" term; it maximises the chances of confusion with *luminance* (see below) but has what is deemed to be the over-riding advantage of ending —ance. Irradiance invites similar confusion with *radiance*). That is, if an element of area *ds* at the point concerned receives a flux $d\Phi$, the illumination E is given by

$$E = \frac{d\Phi}{ds}. \qquad \cdots \cdots \cdots \quad (18\text{–}3)$$

In Fig. 18–2 the elementary area *ds* receives light from the point source P. The distance from P to *ds* is *r* and the normal to *ds* is inclined at angle θ to the direction

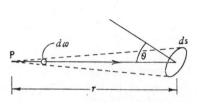

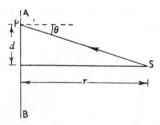

Fig. 18–2. Illumination of a surface element.   Fig. 18–3. Illumination of a plane.

ds—P. The solid angle subtended at P by the element ds is $d\omega=(ds \cos \theta)/r^2$ so that, if the luminous intensity of P in the direction of ds is I, the flux radiated into the solid angle $d\omega$ is

$$d\Phi=Id\omega=\frac{Ids \cos \theta}{r^2},$$

and the illumination at ds is given by

$$E=\frac{d\Phi}{ds}=\frac{I \cos \theta}{r^2}. \qquad \ldots \ldots \quad (18\text{--}4)$$

This equation may be said to embody the two fundamental laws of photometry: the *inverse-square law* and the *cosine law of illumination*. Frequently the luminous intensities of two sources are compared by assuming the inverse-square law and comparing the distances at which the sources produce equal illuminations on a screen (see § 18–9). In Fig. 18–3 S is a uniform point source distance r from a plane AB. The illumination at the point P distance d from the foot of the perpendicular from S to AB is given by

$$E=\frac{I \cos \theta}{PS^2}=\frac{I \cos \theta}{r^2+d^2},$$

but

$$\cos \theta=\frac{r}{\sqrt{(r^2+d^2)}},$$

$$\therefore \quad E=\frac{I \cos^3 \theta}{r^2}. \qquad \ldots \ldots \quad (18\text{--}5)$$

If S is not a uniform point source but has luminous intensity I_θ in direction θ,

$$E=\frac{I_\theta \cos^3 \theta}{r^2}.$$

(*d*) *Luminance L* (in radiometry: *radiance* L_e). In general, a surface will reflect some of the light that is incident upon it and, unless the surface is highly polished, the reflected light will be distributed over a wide range of directions. Thus a reflecting surface of this type resembles a self-luminous surface, and each element of such a surface will have a certain luminous intensity in each direction. At any point on a surface the luminous intensity per unit projected area in a given direction is called the *luminance* at that point in the direction concerned. That is, if an elementary area ds of a surface has a luminous intensity dI in a direction inclined to the normal at angle θ, the projected area in direction θ is $ds \cos \theta$ and the luminance is given by

$$L= \frac{dI}{ds \cos \theta} \qquad \cdots \cdots \cdots \quad (18\text{--}6)$$

(see Fig. 18–4).

It is important to distinguish clearly between illumination and luminance. The

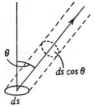

former measures the flux received by the surface, and the latter measures the flux emitted in a given direction.

When one looks at a surface which reflects or emits light it is the luminance which, for a given state of adaptation, determines the subjective brightness of the surface. Thus, under identical conditions, surfaces of equal luminance appear equally bright. However, if the luminance of one surface is double that of another, this does not mean that it appears twice as bright. Consequently, it is necessary to distinguish between luminance and *subjective brightness*

Fig. 18–4. Luminance of an elementary source.

although, until recently, the term brightness was used for both. The luminance is a measure of the luminous flux emitted, the latter being defined, as above, in terms of the standardized values of the relative luminous efficiency and the physical energy radiated. The subjective brightness describes the actual sensation experienced and cannot be measured quantitatively. For a surface of given luminance the subjective brightness will depend upon the observer's state of adaptation, etc. Although subjective brightness certainly cannot be measured quantitatively with accuracy, it is usually assumed that for a constant state of adaptation it is roughly proportional to the logarithm of the physical stimulus. This logarithmic law seems to apply to other sensations, e.g. the loudness of a sound.

18–2. Uniformly Diffusing Surfaces

Observation of self-luminous and diffusely reflecting surfaces indicates that in many cases the brightness of the surface is almost independent of its angle of inclination to the direction of observation. It will be shown in the next section that this will occur if the luminance of the surface is the same in all directions. Equation (18–6) shows that this requires dI, the luminous intensity of any element, to be proportional to $\cos \theta$. This is known as Lambert's *cosine law of emission.* A surface which obeys this law is called a *uniformly diffusing surface* (or *uniform diffuser*) or, sometimes, a *Lambert radiator.* The concept of a Lambert radiator is also used in radiometry generally although, of course, no question of brightness arises. Although no surfaces are completely uniform diffusers a large number of so-called "matt" surfaces ("good" diffusers) approach this condition.

A reflecting surface which obeys the cosine law of emission for all directions of the incident light is called a *uniformly diffuse reflector,* and if such a surface reflects all the incident light it is said to be a *perfect diffuser.* A surface which is coated with a thick layer of freshly prepared magnesium oxide is an almost perfect diffuser. Some translucent surfaces such as opal glass provide almost uniformly diffuse transmission, i.e. the transmitted light is radiated approximately according to the cosine law.

In Fig. 18–5 ds_1 is an element of a uniformly diffusing surface whose luminance in all directions is L. ds_2 is an elementary area illuminated by ds_1. Now the luminous intensity of ds_1 in the direction of ds_2 is $Lds_1 \cos \theta_1$, and this is the flux emitted per unit solid angle in direction θ_1. Now ds_2 subtends a solid angle $(ds_2 \cos \theta_2)/r^2$ at ds_1, so that the total flux received by ds_2 from ds_1 is

$$d\Phi = \frac{Lds_1 ds_2 \cos \theta_1 \cos \theta_2}{r^2} \qquad \ldots \ldots \quad (18\text{–}7)$$

and the illumination of ds_2 is

$$dE = \frac{d\Phi}{ds_2} = \frac{Lds_1 \cos \theta_1 \cos \theta_2}{r^2}.$$

(See Miscellaneous Examples XVII and XVIII no. 2.)

Fig. 18–5. Radiation from one
surface element to another.

Fig. 18–6. Observation of a
surface element.

18–3. Brightness of a Diffuse Radiator

Under given conditions of observation the brightness is determined by the luminous flux received by each retinal receptor, i.e. by the retinal illumination. In Fig. 18–6 ds_1 is an element of a uniformly diffusing surface of luminance L, and ds_2 is the pupil of an eye viewing ds_1. From equation (18–7) it will be seen that the flux from ds_1 entering the eye is given by

$$d\Phi = \frac{Lds_1 ds_2 \cos \theta_1}{r^2},$$

since $\theta_2 = 0$. The flux reaching the retina is $\tau d\Phi$ where τ is the transmission factor of the eye, and this flux is spread over the area ds_1' which is the retinal image of ds_1. The retinal illumination is therefore $\tau d\Phi/ds_1'$. Now the projected area of ds_1, normal to the visual axis, is $ds_1 \cos \theta_1$. Consequently, if l' is the distance of the retina from the second nodal point of the eye, the area of the retinal image is given by

$$ds_1' = (ds_1 \cos \theta_1) \frac{l'^2}{r^2},$$

$$\therefore \quad \text{Retinal Illumination} = \frac{\tau Lds_2}{l'^2}. \qquad \ldots \ldots \quad (18\text{–}8)$$

It will be seen that the retinal illumination, and therefore the subjective brightness of the element, ds_1, is independent of θ_1 and of r. Thus a uniformly diffusing surface appears equally bright for all distances and directions of observation. The Stiles-Crawford effect need not be considered here because the conditions of observation, including the diameter of the eye pupil, remain constant.

18–4. Total Flux Radiated by a Uniformly Diffusing Surface Element

In Fig. 18–7 ds is an element of a uniformly diffusing surface of luminance L.

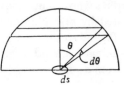

Fig. 18–7. Total radiation
from a surface element.

The luminous intensity of the element in direction
θ is therefore $Lds \cos \theta$. The flux radiated from ds
between the cones of semi-angles θ and $\theta+d\theta$ is
given by the product of the luminous intensity and
the solid angle involved. This is the solid angle
subtended at ds by the annular strip of area
$2\pi r \sin \theta.rd\theta$ on the hemisphere (radius r) described
above ds. Hence the solid angle is $2\pi \sin \theta d\theta$ and
the flux radiated into it is given by

$$d\Phi = 2\pi Lds \sin \theta \cos \theta d\theta. \qquad \ldots \ldots \quad (18\text{–}9)$$

The flux radiated into the cone of semi-angle θ is then given by

$$\Phi = \int_0^\theta 2\pi Lds \sin \theta \cos \theta d\theta,$$

$$\therefore \quad \Phi = \pi Lds \sin^2 \theta \quad \ldots \ldots \ldots \quad (18\text{–}10)$$

and the total flux radiated from ds is found by putting $\theta = \pi/2$,

i.e.   Total Flux $= \pi Lds$. $\ldots \ldots \ldots$ (18–11)

The total flux emitted per unit area has, in the past, been referred to as the
luminous emittance (*radiant emittance*). A committee has recommended *luminous*
(or *radiant*) *exitance*, denoted by M. The next committee recommendation is
awaited eagerly: *come-ance* and *go-ance*, perhaps.

**18–5. The Luminance and Illumination of the Image Formed by an Optical
System**

In Fig. 18–8 EP and EP′ are the entrance and exit pupils of an optical system
forming an image ds' of a uniformly diffusing surface element ds of luminance

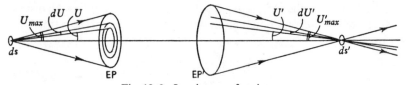

Fig. 18–8. Luminance of an image.

L held perpendicular to the optical axis. From equation (18–9), the flux entering
the system in the annular zone of the aperture between U and $U+dU$ is given by

$$d\Phi = 2\pi Lds \sin U \cos U dU \quad \ldots \ldots \quad (18\text{–}12)$$

and flux $\tau d\Phi$ emerges between U' and $U'+dU'$, where τ is the transmission
factor of the instrument. Now, for good imagery of linear object elements
perpendicular to the axis, the Sine Relation must be satisfied for all zones of the
aperture.

$$\therefore \quad nh \sin U = n'h' \sin U',$$

where n and n' are the refractive indices of the object and image spaces. It will be seen that for conjugate elements of area ds and ds' one has

$$n^2 ds \sin^2 U = n'^2 ds' \sin^2 U' \quad . \quad . \quad . \quad . \quad (18\text{–}13)$$

and, differentiating, one obtains

$$n^2 ds . 2 \sin U \cos U \, dU = n'^2 ds' . 2 \sin U' \cos U' dU'.$$

Therefore, from equation (18–12), it follows that the flux $\tau d\Phi$ emerging from the system between U' and $U' + dU'$ is given by

$$d\Phi' = \tau d\Phi = \tau L \left(\frac{n'}{n}\right)^2 ds' 2\pi \sin U' \cos U' dU'. \quad . \quad (18\text{–}14)$$

This flux passes beyond ds' between U' and $U' + dU'$ and is exactly the flux which ds' would give if it radiated according to the cosine law and had luminance $L' = \tau L (n'/n)^2$. Obviously no light emerges from the system at an angle greater than U'_{max} but, within this cone, the image ds' behaves as a uniformly diffusing surface element. Thus, for an eye placed beyond ds' and entirely within the cone of light emerging from the system, the brightness of the image will correspond to that of a uniform diffuser of luminance $\tau L (n'/n)^2$. Obviously, if the eye is outside this cone the image will not be seen, and if the eye pupil is partly within the cone it will see an image whose brightness depends on the fraction of the eye pupil which receives light.

It should be noted that in the above discussion the value of τ was assumed to be the same for all zones of the aperture of the optical system. In practice, τ is not exactly constant so that the image does not obey the cosine law of emission exactly. Now τ must be less than unity and, except for cases where $(n'/n)^2$ is greater than $1/\tau$, the luminance of the image must be less than that of the object.

As indicated by the above discussion, it is the luminance of the image which is important in visual instruments. If, however, the image is received on a screen as with a photographic lens, it is the illumination of the image rather than its luminance which is important.

From equation (18–10), the total flux entering the instrument is

$$\Phi = \pi L ds \sin^2 U_{max}$$

and the flux leaving the instrument is $\Phi' = \tau \Phi$. Using equation (18–13) one has

$$\Phi' = \tau \pi L \left(\frac{n'}{n}\right)^2 ds' \sin^2 U'_{max} \quad . \quad . \quad . \quad (18\text{–}15)$$

and the illumination of the image is given by

$$E = \Phi'/ds',$$

i.e.
$$E \propto \sin^2 U'_{max}.$$

Hence it is important for a system such as a camera lens to have a large angular aperture. The linear aperture of a camera lens is usually given as a fraction of its focal length, e.g. f/N. The f-number, N, obviously gives U'_{max}. These results show that the illumination (irradiance) of the image, and hence the exposure time required, depends upon the luminance (radiance) of the object and the f-number of the lens but not upon the object distance.

Suppose a photographic lens images an evenly illuminated and perfectly diffusing object plane perpendicular to the axis. The flux $d\Phi$ received by the entrance pupil from an extra-axial surface element in a direction inclined to the axis at angle θ is given by equation (18–7) (p. 449), where ds_2 is now the entrance pupil, $\theta_1 = \theta_2 = \theta$, and $(1/r^2) \propto \cos^2 \theta$. It will be seen that $d\Phi \propto \cos^4 \theta$. If τ is the transmission factor of the lens, flux $\tau d\Phi$ reaches the image element. Now for a perfect lens equal elements in different parts of the object plane give equal image elements. It follows that the illumination across the image plane is proportional to $\cos^4 \theta$. This factor causes an appreciable reduction in illumination at the edge of the field of a wide-angle lens.

In 1941 Slussareff pointed out that the $\cos^4 \theta$ law applies only to a perfect lens, and that if the aperture stop is imaged to the entrance pupil with a large amount of coma, the effective area of the entrance pupil can be increased for oblique pencils. This can be done without upsetting the image quality. In this way it is possible to reduce the falling-off of the illumination away from the centre of the field. The illumination across the image plane can also be varied by introducing distortion of the image since this will cause the magnification to vary across the field of view.

18–6. Observation with Visual Instruments

(a) *Extended objects.* It was pointed out in the previous section that, when an observer views an object with the aid of an optical instrument the eye pupil may be only partially filled with light. In addition, the Stiles-Crawford effect will reduce the effectiveness of the light which enters through the outer zones of the eye pupil. Consequently, the subjective brightness of any part of the field of view does not depend solely upon the luminance of the image formed by the instrument. The *apparent luminance* of any object or image is now defined as the luminance of a uniformly diffusing surface which, when the Stiles-Crawford effect is absent and the eye pupil is unobstructed, would give the same brightness as the object or image under identical conditions of adaptation, etc. The apparent luminance will be proportional to the retinal illumination produced by the object or image viewed, and to the appropriate Stiles factor (see §17–8), assuming that the beam of light which enters the eye is symmetrical about the axis. From equation (18–8) it will be seen that the apparent luminance is proportional to the luminance of the object or image viewed, the area of the eye pupil which receives light, and the Stiles factor.

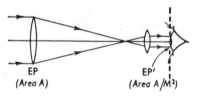

EP
(Area A)

EP′
(Area A/M²)

Fig. 18–9. Vision with a telescope.

Consider now the observation of an extended object of luminance L with a telescope whose magnifying power is M, whose transmission factor is τ, and whose objective has area A. The area of the exit pupil is A/M^2, and if this is larger than ds, the area of the eye pupil, the latter is filled with light (see Fig. 18–9). From above, the apparent luminance of the object using the telescope is then proportional to τ, L, ds, and S_0, where S_0 is the Stiles factor for an eye pupil (assumed circular) of area ds. The apparent luminance of the object viewed without the telescope is proportional to L, ds.

and S_0, assuming that the size of the eye pupil and the state of adaptation are unchanged.

$$\therefore \quad \frac{\text{Apparent luminance of object using telescope}}{\text{Apparent luminance of object viewed directly}} = \tau. \quad (18\text{--}16)$$

If the exit pupil of the telescope is smaller than the eye pupil, an area A/M^2 of the latter is effective and the apparent luminance of the object using the telescope is now proportional to τ, L, A/M^2, and S_1, where S_1 corresponds to a circular eye pupil of area A/M^2.

$$\therefore \quad \frac{\text{Apparent luminance of object using telescope}}{\text{Apparent luminance of object viewed directly}} = \frac{\tau A S_1}{ds M^2 S_0}.$$
$$\cdot \quad \cdot \quad \cdot \quad (18\text{--}17)$$

When an object is observed with a microscope it is easily shown that the apparent luminance is proportional to $[(\text{N.A.})/M]^2$. The magnification which makes the apparent luminance of the object the same with and without the instrument is sometimes called the *normal magnification*. This has little significance since, in microscopy, the illumination can be adjusted to any desired value, and the desirable magnification is determined by the limit of resolution.

(b) *Point sources.* An object such as a star is effectively a point source and the retinal image is the Airy diffraction pattern. No finite magnification can increase the size of the geometrical point image so that the same Airy pattern is formed on the retina when a telescope is used. Consequently, when viewing the star the subjective brightness depends upon the total flux which enters the eye. For any telescope, all the light which enters the objective will enter the eye if the exit pupil of the telescope is not larger than the eye pupil. It will be seen that it is desirable to increase the magnifying power of the telescope until the exit pupil is reduced to the size of the eye pupil. When this stage is reached,

$$\frac{\text{Apparent luminous intensity of star viewed through telescope}}{\text{Apparent luminous intensity of star viewed with naked eye}}$$

$$= \frac{\text{Flux entering eye with telescope}}{\text{Flux entering eye without telescope}}$$

$$= \tau \times \frac{\text{area of telescope objective}}{\text{area of eye pupil}}, \quad \cdot \quad \cdot \quad \cdot \quad (18\text{--}18)$$

where τ is the transmission factor of the telescope, and the *apparent luminous intensity* for a point source is defined in the same way as apparent luminance for extended objects. The Stiles-Crawford effect can be neglected here since the size of the eye pupil is assumed to be constant. If the Stiles-Crawford effect is present and the exit pupil is made smaller than the eye pupil, the light enters the latter closer to the axis and produces a larger visual response. That is, the apparent luminous intensity is increased.

By using a telescope objective of large linear aperture the brightness of the image of a star is increased and, for example, stars which are invisible to the naked eye can be seen with the aid of a telescope using a sufficiently large objective. The way in which the range of a telescope depends upon the size of its objective is easily seen. Suppose that in order to see a star a certain flux Φ

must enter the eye. For an eye pupil of radius p, the solid angle subtended at a star at distance d_1 is $\pi p^2/d_1^2$ and if the luminous intensity of the star is I the flux entering the eye is $I\pi p^2/d_1^2$. The star can be seen when $I\pi p^2/d_1^2 = \Phi$. Suppose a star of the same intensity is viewed at distance d_2 by means of a telescope whose objective has diameter $2a$. Then the solid angle subtended at the star by the telescope objective is $\pi a^2/d_2^2$ and, if the transmission factor of the telescope is τ and the magnifying power is such that the exit pupil is not larger than the eye pupil, the flux which enters the eye is $\tau I\pi a^2/d_2^2$, and the star will be just observable when

$$\frac{\tau I\pi a^2}{d_2^2} = \Phi.$$

The distances d_1 and d_2 at which the star can be seen without and with the telescope are therefore related by the expression

$$d_2 = \frac{ad_1\sqrt{\tau}}{p}. \qquad \ldots \ldots \quad (18\text{–}19)$$

That is to say, the range of an astronomical telescope is directly proportional to the diameter of its objective. It is interesting to note that a telescope enables many stars to be observed during the day. As shown by the above discussions, the telescope increases the brightness of the star but not that of the continuous background.

18–7. The Maxwellian View

If an optical system forms an image of a small source at the pupil of an observer's eye, the whole aperture of the system appears to have a uniform

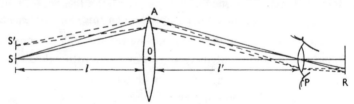

Fig. 18–10. Maxwellian view.

luminance proportional to the luminous intensity of the source in the direction of view. This arrangement is frequently useful in photometry and elsewhere. It is a convenient way of obtaining an extended field of view of uniform luminance, and is useful in the photometry of weak sources because the luminance of the field is much greater than that of the diffuse reflector which the optical system replaces.

A simple arrangement for *Maxwellian view* is shown in Fig. 18–10. S is a source of luminous intensity I at distance l from the lens O. The eye pupil P is placed at the conjugate distance l' and the eye is looking at O so that the image of O falls on the retina R. It is assumed that S and the aperture of O are small compared with l, and that the image of S is smaller than the eye pupil. The flux reaching an element (area a) of the lens aperture at A is Ia/l^2, and if τ is the transmission factor of the lens the flux reaching P from S via A is $\tau Ia/l^2$. If O were replaced by a uniformly diffusing surface of luminance L, the flux reaching P from A would be paL/l'^2 where p is the area of the eye pupil. Equating this to the flux actually reaching P from A one obtains

$$L = \frac{\tau I l'^2}{p l^2}.$$

This is the same for all elements of the lens aperture if l is large, so that the aperture of the lens appears as a uniform diffuser having this luminance.

The figure shows that the neighbouring source S' illuminates the same element of R via the same element of aperture at A but via a different part of the eye pupil P. It will be seen that if S is very small any one part of P is transversed by light from one point in the field of view. Consequently the uniformity of the field of view is marred by any imperfections in P. For use in photometry it is important to ensure that the eye pupil cannot act as a stop of variable aperture. One can either make the image of S very small so that the eye pupil never obstructs light or one can employ a diffuse source whose image is considerably larger than P, thereby ensuring that the eye pupil is always filled. The existence of the Stiles-Crawford effect makes the latter preferable.

18–8. Radiometric and Photometric Units. Absolute Measurements

In radiometry radiant energy is now measured in joules, and radiant power in watts. The other radiometric quantities are simply expressed in terms of these, together with metres and steradians for lengths and solid angles as required. Many of the corresponding photometric quantities, which could be measured simply in terms of the unit of flux together with units of distance and solid angle, have units that are given separate names.

It was pointed out in §18-1 that one may define luminous flux as $\int e_\lambda V_\lambda d\lambda$, where e_λ is in joules per second (watts) and the values of V_λ are those for a standard observer. However, it is convenient to define a separate system of units based not upon a fundamental unit of flux but upon the luminous intensity of a standard source. The two systems of units can then be related. This is a reasonable procedure, but there does seem to be an undesirable proliferation of named units for the derived quantities such as illumination and luminance (see below). For many years the standard source was a candle of standard composition. The luminous intensity or candle power of a source was then given by comparison with that of the standard candle. A number of attempts were made to replace this rather unsatisfactory standard by more easily controlled flame standards. Foremost among them were the Vernon-Harcourt pentane lamp which burnt a mixture of pentane vapour and air from a wickless burner under controlled conditions, and the Heffner lamp which burnt amyl acetate. The horizontal luminous intensity of the pentane lamp was about ten times that of a standard candle. The *international candle* was a standard of luminous intensity corresponding roughly to that of a standard candle but was maintained by carbon filament lamps of special construction. None of these standards gave satisfactory precision and the standard source is now a black-body radiator working at the freezing point of platinum. This has the advantage that its radiation, which is independent of small variations in construction, closely follows definite and well established laws concerning the total energy emitted and the spectral distribution. The major difficulty is the need for very precise temperature control. (At 2000° K a change of 10° produces a change of 5–6 % in the luminous intensity of a black body.) The construction of a suitable black-body radiator is shown in Fig. 18–11. The source is a small hollow cylinder of pure fused thoria about 45 mm. long with internal diameter about 2·5 mm. and wall thickness 2–3 mm. The bottom of this cylinder is packed with powdered fused thoria to a depth of 10–15 mm. and it is supported vertically in a fused-thoria crucible of about 20 mm. internal diameter nearly filled with pure platinum. This crucible has a lid with a small hole in the centre about 1·5 mm. in diameter, which acts as the source of light. The crucible is embedded in powdered fused thoria and the

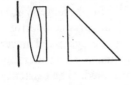

platinum is heated by a high-frequency induction furnace. The power necessary to melt the platinum is about 7 kilowatts at about 1 megacycle per second. The temperature can be controlled so closely that it takes over 20 minutes for the platinum to solidify. When the standard is in use a reflecting prism and cemented doublet lens are positioned as indicated in the figure.

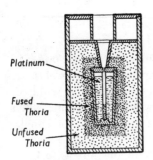

Platinum

Fused Thoria

Unfused Thoria

Fig. 18–11.
Black-body primary source.

When a satisfactory black body was produced a new unit of luminous intensity was introduced. This was called the *candela* and the luminous intensity per square cm. of a black body at the temperature of solidification of platinum was defined as 60 candelas. This unit was introduced on 1st January 1948, and the numerical factor 60 was chosen because it is the nearest convenient whole number which will make the candela and the old international candle approximately equal. (For precise equality the numerical factor would have been about 58·9.) Units for the other photometric quantities can now be defined in terms of the above primary unit.

The unit of luminous flux is the *lumen* and is defined as the flux emitted into unit solid angle by a uniform point source whose luminous intensity in all directions is one candela. It will be seen that the total flux emitted by such a uniform point source is 4π lumens. It was pointed out above that the luminous flux can be defined as $\int V_\lambda e_\lambda d\lambda$. It is found that if the luminous flux is calculated in this way (measuring e_λ in watts) it corresponds to a unit of flux equal to about 682 lumens. In particular, if one has monochromatic radiation at the wavelength of maximum visibility, $V_\lambda = 1$ and it will be seen that an energy flux of 1 watt gives a luminous flux of 682 lumens. At other wavelengths the luminous efficiency of the radiation is less and the same energy flux gives a smaller luminous flux.

There are several units of illumination. The usual metric unit is the *lux* and the illumination is 1 lux when the surface receives 1 lumen per square metre. The lux has been called a *metre-candle* and defined as the illumination on the surface of a sphere of radius 1 metre due to a uniform point source of 1 candle placed at its centre. The equivalence of the two definitions is easily seen since the total area of the sphere is 4π and the total flux from the source is 4π lumens so that each square metre receives 1 lumen. An alternative metric unit is the *phot* or lumen per square cm. but this unit is inconveniently large. On the British system the illumination is expressed in lumens per square foot or *foot-candles*. These two are synonymous since the latter is defined in the same way as the metre-candle.

Luminance is luminous intensity per unit projected area and is expressed in candelas per square cm. (known as *stilbs*) or candelas per square metre (known as *nits*). It will be seen that the primary standard is, in reality, a standard of luminance. On the British system luminance is expressed in candelas per square foot or per square inch and these units have (mercifully perhaps) no other names.

The retinal illumination is said to be one *troland* when a surface of luminance one candela per square metre is viewed with an eye whose entrance pupil (natural or artificial), after correction for the Stiles-Crawford effect, has an effective area of 1 mm^2.

In illumination engineering the luminance of any surface in any direction is measured by comparing it with the luminance of a uniform diffuser which emits a total flux of 1 lumen per unit area. Thus a uniform diffuser which emits a total flux of 1 lumen per square cm. is said to have a luminance of 1 *lambert* (1000 millilamberts). One which emits a total of 1 lumen per square foot has a luminance of 1 *foot-lambert*, and a total emission of 1 lumen per square metre corresponds to a luminance of 1 *apostilb*.

Now equation (18–11) shows that a uniform diffuser of luminance L emits a total flux πL per unit area. Hence if it emits a total flux of 1 lumen per unit area, its luminance in all directions is $1/\pi$. Thus a lambert is equivalent to $1/\pi$ candelas per square cm. (i.e. $1/\pi$ stilb), a foot-lambert is equivalent to $1/\pi$ candelas per square foot, and an apostilb corresponds to $1/\pi$ candelas per square metre.

In the sections that follow there are discussed methods of comparing various radiometric and photometric quantities such as fluxes and source intensities by what might be termed matching techniques, that is, techniques that involve adjusting for equality of either visual or physical responses. It must be noted that when comparing radiometric quantities, which are based on units of energy, visual methods are complicated by the fact that the eye is not equally sensitive to all wavelengths. Furthermore, visual radiometry is obviously seriously limited by the fact that the eye is sensitive to a relatively narrow region of the spectrum.

The fact that physical detectors do not, in general, have the same spectral response as the eye causes similar difficulties if physical detectors are to be employed for photometric comparisons. For accurate work the same difficulty also applies to visual photometry because in general the spectral response of a particular observer's eye differs from that of the "standard eye" (and owing to adaptation processes, depends upon the visual tasks undertaken previously).

Further major difficulties arise if one wishes to make absolute measurements rather than comparisons. There are two main lines of approach: to use standard or calibrated sources, or to use absolute or calibrated detectors.

Absolute radiometry requires measurement in terms of radiant energy or power in joules or watts. One can use a standard black body source since its radiance is known for all wavelengths. However, it is not always convenient to employ the primary standard source, and specially made and calibrated incandescent lamps are widely used as secondary standard sources. A primary absolute detector consists of a non-selective (i.e. thermal) detector such as a thermopile or bolometer. The procedure is to measure the electrical power that has to be dissipated in a heating element to produce the same heating effect as the radiation to be measured; again secondary standards can be calibrated.

Similarly, absolute photometry can be based on the black body primary standard source as indicated above, or on calibrated secondary standard sources. Physical detectors can be combined with filters so that the combination has the spectral response of a "standard eye" and the combination can be used in matching experiments or calibrated to measure, say, luminous flux.

18–9. The Visual Photometer Head and Bench

A convenient method of comparing the luminous intensities of two sources is to compare the distances at which they produce equal illuminations. From the inverse square law it then follows that the ratio of the luminous intensities is equal to the inverse ratio of the squares of the distances. The usual method involves the use of an optical bench at opposite ends of which the sources are mounted. Between the sources is a so-called *photometer head* which includes two identical screens each of which is illuminated by one of the sources. The distances of the sources are varied (by moving the sources and/or the photometer head) until the screens are equally illuminated. If the screens are identical and are viewed under identical conditions, this will be when they appear equally bright to the observer. One can avoid the need for a symmetrical photometer head by mounting the sources (A and B) to be compared, in turn, on one side of the head (H) and obtaining successive brightness matches with a third source C on the other side of H. One can fix A, B, and H, and adjust the distance of C for each match; fix H and C and adjust the distances of A and B; or fix A and B, and move H and C, keeping their separation constant. The intensity of C need not be known. Various photometer heads have been designed in order to facilitate the comparison of the illuminated surfaces. In order to obtain accurate results, the fields of view whose brightnesses are compared must have a sharp demarcation line, care being taken to ensure that there is neither a gap between them nor a region of overlap. To avoid errors caused by deviations from the cosine law of emission, the illuminated surfaces should be viewed at the same angle. The luminance of these surfaces depends on their illumination and if the light is not incident on each at exactly the same angle, an error is introduced since the illumination is proportional to $\cos \theta$. Now $\cos \theta$ has a stationary value at $\theta = 0$, so that the possibility of this error is reduced if, nominally, the light is incident normally on each surface. It is also found desirable for the field of view to be not smaller than about $10°$.

The most satisfactory equality-of-brightness photometer head is that usually ascribed to Lummer and Brodhun. This is shown in Fig. 18–12 and will be seen to satisfy the requirements enunciated above. Light from the sources S_1 and S_2

is incident normally on the two sides of A which are good diffusers. (Frequently A is a thick disc of plaster of Paris.) The diffusely reflected light from each side of A is reflected by the mirrors M_1 and M_2 to the prism system PQ. In this system, originally due to Swan, Q is an ordinary 45°–45°–90° prism but the hypotenuse face of P is replaced by a ground spherical surface having a polished flat central region which makes good contact with Q. The observer views the contact surface by means of the eyepiece system L and it will be seen that the outer part of the field of view is illuminated by light from S_2 while the central part is illuminated by light from S_1. When the sources are adjusted to give equal

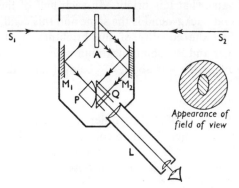

Fig. 18–12. The Lummer-Brodhun head.

illumination at A the inner and outer parts of the field of view are equally bright and the demarcation line practically disappears. With this arrangement one has normal incidence at A, the two surfaces at A are viewed under similar conditions, and a sharp demarcation line can be obtained.

The Lummer-Brodhun photometer can be converted into a *contrast head* by a suitable modification of the prism system PQ (Fig. 18–12). The new prism system is shown in Fig. 18–13. Contact between the hypotenuse faces is broken, at the places indicated, by etching prism P. Light incident from the right is then totally reflected at those parts of Q where contact is broken, and light entering P from the left is transmitted through Q at points where contact between the prisms is good. The shape of the etched portions is such that, if the beams are

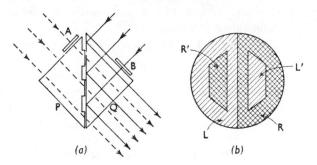

Fig. 18–13. The Lummer-Brodhun contrast head.

not equally intense, the appearance of the field of view is as in (*b*). Patch R′

and background R are illuminated by light entering the prisms from the right and appear equally bright. Similarly L and L' have the same brightness. The field of view would be of uniform brightness if the beams entering from left and right were equally intense.

Two glass plates A and B are now added in the positions shown in (a). Owing to the additional air-glass surface, A causes the luminance of the patch L' to be about 8 % below that of the background L, and B causes an equal reduction in the luminance of the patch R' below that of the background R. At the position of balance the backgrounds L and R still appear equally bright but the patches L' and R' appear somewhat less bright. In each half of the field the contrast between the patch and background is the same and this condition can be judged more accurately than the equality of the brightnesses of the backgrounds. According to Lummer and Brodhun, maximum sensitivity would be obtained if the luminance of each patch was about 4 % below that of its background.

18–10. Flicker Photometers

The intensities of sources of slightly different colour can be compared fairly accurately by using a Lummer-Brodhun contrast head. However, neither of the photometers described above is really suitable for comparing sources of widely different colours since it is difficult for an observer to judge brightness when there are colour differences.

Now if two fields of view of different colour and slightly different luminance are presented alternately to an observer's eye, and the frequency of alternation gradually increased, it is found that a stage is reached where the colour difference is not observed. By adjusting the luminances of the two fields to be equal, the flicker can then be made to disappear completely. In this way one can adjust the illuminations of the two screens of a suitable photometer head to be equal, and the luminous intensities of two sources can be compared in the usual way. The flicker caused by a difference in brightness between the two fields of view can be removed also by increasing the frequency of alternation, and the smaller the brightness difference the lower is the required frequency. In order to make the adjustment of luminance as accurate as possible, it is necessary to be able to detect the smallest possible brightness difference. Consequently, it is necessary to employ the lowest possible frequency of alternation and one increases the frequency from a low value until it is just possible to remove the flicker completely by an adjustment of luminance.

It has been found that in a flicker photometer the following conditions should be fulfilled:

1. The illumination of the comparison surfaces should be at least 25 lumen m^{-2}.

2. The photometric field should be about 2° in angular diameter and surrounded by a field of about 25° diameter maintained at approximately the same luminance as the photometric field, there being no dark demarcation line separating them.

3. When the moving parts of the flicker head are rotated slowly no shadows should be seen in the field.

4. The two halves of the field should be visible for equal lengths of time during each rotation, i.e. uniform speed of rotation is essential.

It is found that observers cannot use a flicker photometer for periods exceeding about an hour without undue fatigue and loss of accuracy.

Several flicker photometers have been designed which fulfil the conditions described above. That due to Guild is illustrated in Fig. 18–14. Light from the source S_1 passes through the prism P and is incident normally on the diffusing surface A. The sector disc B which rotates about the axis C has 90° sectors which have diffusing surfaces. When a sector is suitably positioned, light from the source S_2 is incident normally upon it. The illuminated surfaces of A and B are viewed alternately from E through the aperture D which gives a 2° field. The inner surface of the tube F is a diffusing surface which is illuminated by the source G. It will be seen that the photometric comparison surfaces A and B are both viewed at 45°. These surfaces and the inside of F (which provides the necessary bright surround) are coated with magnesium oxide. The aperture D has a sharp edge to avoid a discontinuity in the field of view. The necessity of finding the transmission of the prism P can be avoided by using the substitution method, i.e. by keeping S_1 unchanged and placing the sources to be compared at S_2 in turn.

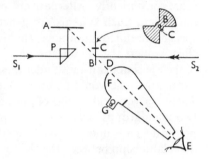

Fig. 18–14. Guild's flicker photometer.

It is obvious that the results obtained in the visual comparison of sources of different colour depend upon the spectral sensitivity of the observer's eye, and some check must be made to ensure that this does not deviate markedly from that of a "standard observer". This difficulty is overcome by using a physical receptor such as a photo-cell in place of the eye.

18–11. Physical Photometry, Radiometry, Spectrophotometry and Spectroradiometry

A physical detector has the enormous advantage of being capable of calibration so that different radiant inputs can be compared or measured, i.e. one is not restricted, as one is with visual methods, to judging when two signals are equal. In principle, any physical detector whether it be a selective detector such as a photo-cell, or a non-selective detector such as a thermopile, can be calibrated, in combination with some form of filter and meter or recorder, to measure directly in arbitrary or absolute radiometric or photometric units. The calibration can be based on the use of either a standard source or a standard detector. Many detectors have the advantage of having a calibration curve that is linear to a very good approximation. Good non-selective detectors need no calibration for non-absolute radiometry or spectroradiometry. For photometric measurements any detector can, in principle, be combined with a filter to obtain an overall spectral response that is the same as that of the standard eye. For photometric measurements involving fairly large fluxes a photo-voltaic (barrier layer) cell is very convenient since an approximately correct spectral response curve is easily obtained by means of a filter; its output can be measured directly with a micro-ammeter, and no other equipment is required. Until fairly recently most photographic exposure meters employed photo-voltaic cells. However, such cells have rather low sensitivities and often show fatigue after prolonged exposure. For applications requiring greater sensitivity a photo-emissive cell (which requires an external e.m.f.) is better since the output can be amplified more conveniently. Alternatively, one can obtain high sensitivity with a photo-multiplier. Such systems are generally more reliable. A physical detector is specially useful for photometry at low levels of illumination because the Purkinje effect renders the eye unsatisfactory as a detector; since the spectral sensitivity of the eye is not constant under these conditions, brightness matches with poly-chromatic light would be unreliable—even if there were no colour differences. There is a very wide range of physical detectors and these can be classified into two main groups: selective, for which the electrical output for a given radiant power depends upon the wavelength, and non-selective (or thermal), for which the relationship between the electrical output and radiant power received is the same for all wavelengths. Selective detectors include photo-voltaic cells, photo-emissive cells, photo-multipliers, and photo-conductive cells; non-selective detectors include thermocouples and thermopiles, bolometers, and pneumatic detectors such as the Golay cell. A reasonable generalization is that most of the selective detectors are useful from the near infra-red through to the ultra-violet whilst thermal detectors are most commonly used from the far infra-red through to the visible.

It is beyond the scope of this book to describe in detail the construction of these various classes of detectors or to discuss the physical principles underlying their operation and the factors that determine their sensitivities. However, attention must be drawn to some general points relating to the detection of radiant energy. When a nominally steady radiant flux is received by a detector, the output from the detector exhibits very small random fluctuations which are referred to as *noise*. A detailed discussion of the origin of the noise in the detector output cannot be given here but it is necessary to refer briefly to two main types of noise encountered in radiometry. These are known as *photon noise* and *detector noise*. Photon noise arises from the fact that a nominally steady radiant flux itself exhibits random fluctuations; detector noise, on the other hand, originates within the detector itself and arises from phenomena such as thermal motion within, for example, a thermocouple element. If the noise in the detector output is predominantly photon noise, the noise will increase if the incident radiation level is increased (it is proportional to the square root of the radiation level). On the other hand, if the noise is predominantly detector noise it is virtually independent of the incident flux. It will be appreciated that although amplifiers can amplify the output from a radiation detector, they will amplify the noise together with the signal, and no fundamental improvement is achieved; the important quantity is not the signal but the so-called *signal-to-noise ratio*. It can be shown that the signal-to-noise ratio can be improved by increasing the exposure time t since the signal integrates in direct proportion to t whereas the the noise integrates in proportion to $\sqrt{t}$.

Another phenomenon sometimes classified as noise consists of the (usually random and often slow) variations that occur in the overall level of the incident flux. It can arise from variations in the mean output from the source itself or, for stellar sources, it can be caused by atmospheric scintillation; it is sometimes called *modulation* or *scintillation* noise. The effect of these variations can be eliminated by monitoring the total input radiation level—a technique sometimes known as *ratio-recording*. Double beam spectroradiometers employ this technique.

Photometric and radiometric comparison of sources using physical detectors can be made by a simple substitution technique if the detector is calibrated, or by a substitution/matching technique, where the source distances are adjustable to give equal detector outputs, if the detector is uncalibrated. Alternatively, one can employ the flicker technique where the fluxes to be compared are received alternately and the source distances adjusted for zero a.c. output from the detector. Care is required in the design of the optical system in order to avoid spurious signals as one interchanges the beams that are directed to the detector (c.f. precautions 3 and 4 for visual flicker photometry). Although the fluxes being compared might be large, it is the difference that is being detected in these methods so that the problem of noise is often important. Instead of varying the source distances and using the inverse square law (which is directly applicable only if the dimensions of the source are small compared with the source distance) one can control the light reaching a detector from a source by means of a variable aperture (assuming the spatial distribution of the radiation across the aperture is uniform or, at worst, known). Alternatively, one can employ a calibrated neutral filter of variable transmission (see Fig. 18–15). §18–12 gives a discussion of the radiometric properties of spectroscopic instruments in which the emphasis is on the attainment of the very highest performance in terms of resolution and

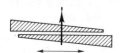

Fig. 18–15. Neutral wedges.

sensitivity. For a great deal of routine analytical work one does not require high performance in this sense—but one does demand a high degree of stability and reliability. There is a very wide selection of commercial spectroradiometers covering a wide range of spectral regions and a wide range of resolutions; those intended for large scale routine work employ various automatic recording techniques. In emission spectroscopy the optical systems employed include prism and grating monochromators identical to, or equivalent to, those described in §§5–17, 5–18, 12–12. The detector is located beyond the exit slit and the spectrum is scanned across the slit by adjusting the monochromator. In absorption spectroradiometry one can use a source that gives a continuous spectrum, and measure the transmission spectrum using a similar arrangement, the spectrum of the source being known or measured separately—a *single-beam* instrument. Alternatively, one can make a continuous comparison of the spectral distribution of the radiation transmitted by the specimen with that of the incident radiation using a *double-beam* instrument. Between the source and the detector of a double-beam instrument the radiation from a monochromator is passed alternately through the specimen and through a variable aperture (or a variable density filter) as the spectrum is scanned. If the reduction in intensity is different in the two radiation paths, there is an a.c. component in the output from the detector. A servo-mechanism can then operate the variable aperture or filter until this a.c. signal falls to zero, the setting of the aperture or filter indicating the absorption in the specimen. This can be recorded as the wavelength from the monochromator is varied and the absorption spectrum recorded automatically. If the specimen is a liquid, one does, of course, have to make an allowance for the absorption of the liquid cell—in the double beam instrument, for example, an empty cell can be inserted in the comparison beam. (The problem is more difficult than it appears to be at first sight since the liquid specimen not only absorbs radiation but also changes the losses caused by reflection at the walls of the cell.)

An advantage of a double beam system is that one eliminates the effect of slow variations in source intensity and detector sensitivity, but recent developments have improved the stability of both the source and the detector and its associated amplifying and recording equipment to such an extent that adequate stability can now often be obtained with a single beam instrument.

18–12. The Radiometric Properties of High Resolution Spectroscopic Instruments

Expressed in general terms, spectrometers can be divided into two broad classes: single channel instruments, which employ a single detector at any one time, and multichannel instruments which employ an array of detectors simultaneously.

The single channel class can be further subdivided: in what is essentially a *monochromator* the detector records information from each resolution element in the spectrum separately—the spectrum is scanned in some way and the intensity distribution is measured directly; in what is known as a *multiplex system* information from all resolution elements is recorded simultaneously as a chosen parameter is varied, and the information from which the spectrum is obtained must be decoded in some way. Instruments that are essentially monochromators include the classical prism and grating spectrometers and the Fabry-Perot interferometer. The only common multiplex system is the Fourier transform spectrometer in which the intensity is measured as the path difference in the interferometer is varied, and the spectral distribution is calculated by taking the Fourier transform of the interferogram; the Mock interferometer can be included in this category.

It can be said at once that the advantage of a multiplex system, which is known as the *multiplex principle* or *multiplex gain*, or sometimes as *Fellgett's advantage* (since it was Fellgett who in 1951 first discussed the principle explicitly in this context and used it successfully), is that for each setting of the interferometer all the emergent light—containing all the wavelengths in the spectrum being studied—falls on the detector, and this gives a considerable gain in signal strength. In absorption spectroradiometry it can happen that with the Fourier transform spectrometer the energy passing through the sample is too great!

The only commonly used multichannel spectroscopic instrument is the spectrograph, which records a wide spectral range with a single exposure—the resolution elements of the photographic emulsion constitute the array of detectors. To examine small spectral ranges with high resolution the classical prism or grating spectrograph can be combined with, or replaced by, a Fabry-Perot interferometer giving spatially displayed fringes. These systems are not well suited to spectroradiometry because it is difficult to obtain a reliable relationship between the intensity of the radiation falling on the photographic emulsion and the resulting optical density after development. In consequence, the discussion that follows is concerned with the relative merits of various single channel instruments. In practice the relative radiometric efficiencies of the various classes of spectrometers depend upon the special circumstances in which they are used and upon the spectral region being investigated. A full account of the theory is far beyond the scope of the present book and the discussion to be given here can do no more than indicate some general results.

The chromatic resolving power of any spectrometer (which is taken to include the optical system, radiation detector, and any amplifying and recording equipment employed) is determined by its *instrumental profile* (or *instrumental* or *scanning function*), i.e. the output obtained when the input is a strictly monochromatic disturbance. Although it has not always been pointed out explicitly in the earlier discussions, it will be seen that for any form of spectrometer the instrumental profile gives the *apparent* spectral distribution of a source that is, in fact, strictly monochromatic. (The physical impossibility of such a source does not ruin this concept.) This instrumental profile is analogous to the spread function in image-forming systems. Each in its own context defines the output corresponding to what is regarded in that context as an input of elementary form, and by a summation process gives the output when the form of the input is complex.

In most of the discussions in earlier chapters ideal and unrealistic conditions have been assumed to exist. For example, in discussing the simple case of the resolving power of a prism or grating spectrometer it was assumed that the entrance slit had infinitesimal width. Furthermore, by discussing the intensity at any point in the spectrum it was implied that this could in fact be observed! However, to do this one would need to scan the spectrum with a detector whose resolution was unlimited. Thus the implication was that if the spectrum were recorded by using a detector such as a photo-cell, this would be located behind an exit slit of infinitesimal width. Similar considerations apply to all the spectroscopic instruments that have been discussed. As a further illustration, consider the Fabry-Perot spectrometer. Here the intensity in the fringe pattern was discussed and, although it was not essential to make unrealistic assumptions about the geometry of the source, one would need a scanning aperture of infinitesimal diameter in order to record the intensity. Obviously, in these and in all similar arrangements the energy reaching the detector would, in fact, be infinitesimal; they are theoretically ideal arrangements from which one must depart in practice.

The instrumental profile of a classical prism or grating spectrometer with infinitesimal slits has the form $\left(\dfrac{\sin \alpha}{\alpha}\right)^2$, which describes the intensity distribution in the diffraction image of the entrance slit, the value of α being related to the aperture of the prism or grating (see §11–5). The instrumental profile of the idealised Fabry-Perot spectrometer employing an infinitesimal exit aperture has the form of the Airy formula (see §9–3). Both SISAM and the Fourier transform spectrometer have idealised profiles of the form $\left(\dfrac{\sin\alpha}{\alpha}\right)$ (see §§12–12 and 8–11). (As has already been pointed out, the instrumental profiles of some instruments can be modified with advantage by the technique of apodization.)

When finite apertures are employed the instrumental profiles are broadened and there are consequent losses in resolution. For example, the slits in a prism or grating spectrometer must, in practice, have finite widths and the extent to which the slits can be opened depends upon the acceptable compromise between the associated loss of resolution and the gain in energy reaching the detector. One must remember that manufacturing imperfections such as imperfect gratings and lack of flatness of interferometer mirrors cause a further broadening of the instrumental profile and a further loss of resolution.

In paraxial geometrical optical terms, the ability of an optical system to transmit energy is determined by a combination of the sizes of the field stop and the pupil in the same optical space; it is measured by the product of the area of one and the solid angle subtended at its centre by the other. This is the three-dimensional equivalent of the Helmhotz-Lagrange invariant or the Sine Relation—see §§2–4 and 18–5. (It is assumed that the system is in air, and absorption and similar losses are, of course, neglected.) With the usual chaos associated with terminology, this product is referred to as the *Throughput, Luminosity, Light Grasp, Acceptance*, or *étendue*. This is mentioned so that the long-suffering student will realise that various authors using these different terms are probably referring to the same thing! *Throughput* will be used here; one is tempted to follow Jacquinot who, in his classic review (1960), used Luminosity, but unfortunately this term is widely used in a different photometric context (see §17–6).

The field stop and pupil referred to above correspond to the slit and the prism or grating aperture for the classical spectrometer, and to the focal diaphragm and the mirrors for the Fabry-Perot interferometer. It is easy to see why the profile will be broadened and the resolving power reduced when the throughput of a prism or grating spectrometer is increased by widening the slits: the geometrical image of the entrance slit becomes finite and the finite exit slit obviously transmits a broader range of wavelengths. For the Fabry-Perot etalon an increase in throughput that is achieved by opening the focal diaphragm causes a loss of resolving power because light that reaches the detector has passed through the mirrors with a range of directions, corresponding to a range of path differences.

It will be appreciated that when comparing the photometric properties of different instruments it is pointless to discuss the throughput T that can be achieved without considering the penalty paid in terms of lost resolving power R. The product TR can be used as a measure of overall performance. For most devices the decrease in R caused by an increase in T is such that TR is constant, i.e. TR is characteristic of the particular class of device. It can be shown that the classical Fabry-Perot instruments have a much larger value of TR than classical prism or grating spectrometers. Expressed briefly (but loosely): for a just-acceptable loss of resolution the focal diaphragm in a Fabry-Perot spectrometer subtends a larger solid angle than does the slit in the equivalent prism or grating spectrometer. A multislit or grille spectrometer has greatly increased throughput with a value of TR comparable with that of a classical Fabry-Perot. The field widened (spherical) Fabry-Perot gives a gain in throughput compared with the classical form because the path difference is much less sensitive to variations in the direction of the light passing through the system and hence one can use a larger focal diaphragm: in practice this gain becomes larger for larger mirror separations i.e. for higher resolving powers. When comparing the plane and spherical Fabry-Perot instruments it should be noted that there is no theoretical limit to the mirror diameters with the classical system whereas one has to limit the apertures of the mirrors of a spherical Fabry-Perot. However, in practice the gain from field widening is such that overall there is a considerable gain in throughput.

It has been shown by Jacquinot that the increased throughput of Fabry-Perot instruments compared with classical spectrometers which employ slits arises from the circular symmetry which the interferometer has about the axis. Hence a similar advantage is obtained with SISAM and also with the Fourier transform spectrometer (although it is lost if the latter employs a lamellar grating since this requires a slit).

It must be pointed out that the throughput of, say, a Fabry-Perot spectro meter can be reduced considerably if one uses a pre-monochromator consisting of a classical spectrometer with slits; on the other hand the loss is relatively small if, as is often the case, one can use wide slits in the pre-monochromator. Another point worth noting is that it is becoming more common to employ as a pre-monochromator a narrow pass-band interference filter—which is itself a thin etalon.

When comparing the properties of interference spectrometers and classical grating spectrometers employing slits, it is of interest to note that even if only a relatively low resolving power is required with a grating system it is photometrically advantageous to employ the largest possible grating (i.e. one capable of giving a resolving power which is as large as possible and in excess of what is actually required) and to increase the slit width as far as is consistent with the resolving power required for the problem in hand. In this way both the aperture and field are increased so that the throughput is increased. In contrast, it can be shown that one should use a Fabry-Perot that gives just the resolving power required.

At this stage in the discussion a general point should be reiterated: electronic amplifiers are available which can amplify the output from any detector but these, of course, will amplify the noise as well as the signal. Obviously it is the signal-to-noise ratio from the detector that is important, and not merely the signal. Increasing the throughput of the spectrometer increases the signal and if the noise is largely detector noise which is independent of the signal, this will increase the signal-to-noise ratio. This is often the case, particularly in the infra-red. If, on the other hand, photon noise is a significant factor (which often happens with the highly sensitive detectors employed in the ultra-violet and visible regions) a gain in throughput does not yield a corresponding gain in signal-to-noise ratio. This is of particular importance in selective modulator systems such as SISAM where unmodulated light can reach the detector. In such a situation the noise level can be increased until the system loses the advantage of its high throughput.

The Fourier transform spectrometer is a multiplex system, and if the system is detector-noise limited i.e. the noise is independent of the signal this gives a very large gain in the signal-to-noise ratio. It is assumed that if the detector receives radiation for a time t, the signal integrates in direct proportion to t, whereas the noise integrates in proportion to $\sqrt{t}$. Suppose, then, that a total time τ is to be used to record the spectrum. With each of the spectrometers discussed above the spectrum is scanned so that if m spectral elements are to be measured the exposure time for each is τ/m. Thus the signal-to-noise ratio is proportional to $(\tau/m) \div \sqrt{\tau/m} = \sqrt{\tau/m}$. With the Fourier transform spectrometer every spectral element is exposed for the full time τ so that the signal-to-noise ratio is proportional to $\tau \div \sqrt{\tau} = \sqrt{\tau}$. That is, there is a gain in the signal-to-noise ratio of a factor $\sqrt{m}$; this is the multiplex gain. To give some idea of the importance of this factor it can be mentioned that in recent laboratory work several workers have multiplexed about a million spectral elements. In addition to the requirement that the system should be detector-noise limited there is the further requirement that the detector should have a large dynamic range—it must remain linear over a wide range of powers if it is to yield a sufficiently accurate interferogram. In this respect the Fourier transform spectrometer is more demanding than a scanning system.

The original motivation for the development of the modern Fourier transform spectrometer was the desire to investigate the infra-red, where the radiation level is very low so that with traditional spectrometers the signal-to-noise problem is severe. Fortunately, with the detectors employed the noise is independent of the signal strength so that by multiplexing one attacks the problem in the ideal way, i.e. by increasing the signal without increasing the noise.

In addition, if a Michelson interferometer is employed, the Fourier transform spectrometer has the same circular symmetry and therefore the same high throughput as the Fabry-Perot interferometer. The throughput can be increased still further by employing a *field-widened interferometer*. As with the spherical Fabry-Perot, the aim is to cause the path difference to vary less rapidly with the direction in which the light enters the system, so that a larger solid angle can be accepted. Mertz produced a field widened interferometer by inserting a plane-parallel slab of glass between beam splitter and mirror. This increases the optical path for a beam entering the system within a given solid angle i.e. increases the permitted solid angle for a given optical path and hence increases the throughput for a given resolving power.

The multiplex gain and high throughput coupled with virtually unlimited resolving power and high precision measurements of intensity and wavelength, are such that there is a wide range of problems that only the Fourier transform spectrometer can solve in practice—although other instruments could solve them in principle. For example, Jacquinot (1969) has remarked that the infra-red planetary spectra obtained by P. and J. Connes could have been obtained with a good grating spectrometer in a total time of a few thousand years! Another example of the power of the method is the measurement of rare earth emission spectra in the near infra-red by Guelachvili and Maillard (1971). The results obtained by multiplexing about a million elements would have required a large number of separate Fabry-Perot or SISAM recordings to be made after a preliminary low resolution examination of the entire spectrum. However, there are situations in which the full potentialities of the Fourier transform method may not be realized. In the far infra-red one possibility is that with a Michelson system the beam splitter might limit the width of the spectral band that can be transmitted; obviously if fewer spectral elements are recorded simultaneously the multiplex gain is reduced. The lamellar grating system would overcome this by enabling a broader spectral band to be transmitted but one would then lose the high throughput of the Michelson system. Another possibility is that the advantage of the high throughput of the Michelson system might be lost because the physical dimensions of the detector render it incapable of receiving the whole of the output. The same difficulty can arise with a grille spectrometer if the grilles are large. At the other end of the spectrum—in the ultra-violet—the multiplex gain is more likely to be reduced because the predominant noise is photon noise. Both the beam splitters and the detectors are being improved, so that the range of application of the Fourier transform method is constantly being extended. The advantages of the Fourier transform spectrometer are most marked when one wishes to examine with high resolution a broad and complicated weak spectrum and can use a detector that can receive the entire energy flux from the interferometer and for which the noise is independent of the signal; it has virtually displaced all other systems in infra-red astrophysics. A further advantage of the Fourier transform method is that it reduces the effect of stray light since, in effect, the system gives frequency modulated coding. Another point worth noting is that very small efficient two-beam interferometers of moderate resolution have been produced by Mertz and others and these are well suited to observation from, for example, satellites and space probes. Recent developments include the cooling of the complete interferometer to liquid helium temperatures in order to investigate very weak infra-red spectra that might be partially masked by thermal radiation from the equipment itself. However, rapid advances are being made in the development of high sensitivity infra-red detectors such as very low temperature ($<1°K$) and super-conducting bolometers. If this improvement in detectors reduces the detector noise until it no longer predominates, there is nothing to be gained from multiplexing, and although very low temperature operation complicates the experimental technique it has been suggested by Richards (1971) that the Fourier transform method may not, in the long term, be the permanent solution for infra-red spectroscopy; it does, of course, retain the advantages of high throughput, the ability to analyse very broad spectra, and the maintenance of very high resolving power and very high precision absolute wavelength measurement over a wide range of wavelengths. As far as ultimate resolving power is concerned, it has been pointed out by P. Connes (1971) that the stage has now been reached where the resolving power that can be employed in a particular investigation is no longer limited by the spectrometer that could be built, but by the intensity of the source under investigation.

18–13. Directional Properties of Sources

As has already been pointed out the luminous intensity of a source is the flux emitted per unit solid angle in the direction considered. In general, sources are not uniform, and it is frequently necessary to examine the variations of luminous

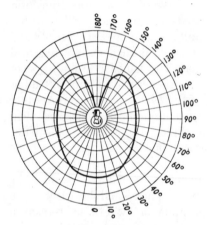

intensity with direction. If the source can be used in any orientation, the ordinary visual optical bench can be used to measure the intensity for various positions of the source. Actually very few sources are unaffected by orientation (a vacuum filament lamp is one of the few) and if a visual photometer bench is used, it is necessary to employ a system of mirrors to collect light radiated in directions out of the horizontal plane.

Fig. 18–16. Polar curve for a gas-filled filament lamp.

If a physical receiver such as a photocell is employed, it can be moved round the source and this is frequently a more convenient method. It should be remembered, however, that it is necessary for the photo-cell to be positioned sufficiently far from the source for the latter to be regarded as a point source.

Using the above method, it is possible to measure the luminous intensity of a source in any direction. The results of a number of such measurements on a source are most conveniently shown on a polar diagram in which the source is origin and the length of the radius vector in any direction represents the luminous intensity in that direction. Fig. 18–16 shows a diagram of this type for the vertical section through a gas-filled electric lamp. Sometimes the length of a radius vector at a given angle to the vertical represents the mean luminous intensity for all azimuths measured from a vertical plane.

If the source does not radiate approximately symmetrically about an axis it is better to construct a so-called *isocandela diagram*. A sphere is imagined surrounding the source and on this sphere are drawn loci of points from which the source appears to have the same luminous intensity. This diagram is then projected on to a plane by one of the projections used in ordinary map making. An equal area projection is desirable so that by comparing two diagrams one may obtain an approximate idea of the relative values of the total fluxes emitted by the two sources.

18–14. The Integrating Sphere

It is possible to find an approximate value of the total flux radiated from a source by finding the mean spherical candle power from measurements of the luminous intensity in a number of directions. It is much easier to use an *integrating sphere*, the theory of which will now be given. In Fig. 18–17 ds_1 and ds_2 are elementary areas of the surface of a sphere, centre C, radius r, the inner surface of the sphere being a uniformly diffuse reflector. Suppose there is a source of light inside the sphere so that the luminance of ds_1 is L. Then it follows from equation (18–7) that the flux $d\Phi$ received by ds_2 from ds_1 is $Lds_1ds_2\cos^2\theta/d^2$. Now $\cos\theta = d/2r$

$$\therefore \quad d\Phi = \frac{Lds_1ds_2}{4r^2} . \qquad \cdots \cdots \quad (18\text{--}20)$$

Hence, for a given element ds_1, the flux received by ds_2 is independent of its position on the sphere, i.e. the flux reflected from each part of the sphere is equally distributed over the other parts, or the flux received by an element *after reflection at other points* is everywhere the same. For a given sphere it can therefore depend only upon the total flux radiated by the enclosed source. The exact relation between the two is easily found as follows. Suppose that the total flux radiated by the enclosed source is Φ. The illumination at any point on the surface of the sphere is due to direct and reflected light. Let ϱ be the reflection factor of the surface of the sphere, and let L be the luminance of any point due to light received directly from the source. Then an element ds emits a total flux πLds and must have received a total flux $\pi Lds/\varrho$ directly from the source.

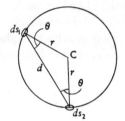

Fig. 18–17.
Integrating sphere.

$$\therefore \quad \frac{\pi}{\varrho}\int Lds = \Phi, \qquad \cdots \cdots \quad (18\text{--}21)$$

where the summation is taken over the whole surface of the sphere. Now the total flux received by an element ds' due to a single reflection from every point of the sphere is, from equation (18–20), given by $(ds'/4r^2)\int Lds$ or, from (18–21), $ds'\Phi\varrho/4\pi r^2$. This flux will give the element ds' a luminance $\varrho^2\Phi/4\pi^2r^2$. Thus the luminance of any point of the sphere due to the direct light and light which has undergone one previous reflection is $L+\varrho^2\Phi/4\pi^2r^2$. Consequently, the flux received by any element of area ds' due to light which has undergone the first two reflections is

$$\frac{ds'}{4r^2}\int\left(L+\frac{\varrho^2\Phi}{4\pi^2r^2}\right)ds = \frac{ds'}{4r^2}\int Lds + \frac{ds'\Phi\varrho^2}{16\pi^2r^4}\int ds$$

$$= \frac{ds'\Phi}{4\pi r^2}(\varrho+\varrho^2).$$

Similarly, when further reflections are considered, it may be shown that the total flux, due to reflected light, which reaches any elementary area ds' is

$$\frac{ds'\Phi}{4\pi r^2}(\varrho+\varrho^2+\varrho^3+\varrho^4+\ldots) = \frac{ds'\Phi}{4\pi r^2}\cdot\frac{\varrho}{1-\varrho}. \qquad \cdots \quad (18\text{--}22)$$

This is independent of the position and directional properties of the source. It assumes that the source causes no obstruction of the reflected light and that the surface of the sphere is a uniformly diffuse reflector. In order to measure the illumination at any point due to reflected light it is necessary to introduce a screen to obstruct the direct light and this will also obstruct the reflected light. Nevertheless an integrating sphere which is large compared with the source of light does provide an accurate method of comparing the total fluxes radiated by two sources. In practice a uniformly diffusing translucent window is built into the side of the sphere and a screen positioned to prevent light from the source falling directly on the window (see Fig. 18–18). The luminance of the window viewed from outside the sphere is then measured. It will be seen that this is directly proportional to the total flux emitted by the source inside the sphere. The luminance or radiance of the window can be measured by means of a photo-cell placed immediately outside it or one can adapt a photometer head such as the Lummer-Brodhun as indicated in Fig. 18–18. If the luminance of the window is high, the luminous intensity can be measured in the usual way, using an ordinary photometer bench.

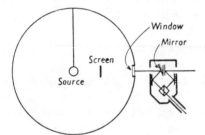

Fig. 18–18. Visual photometry with an integrating sphere.

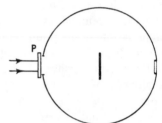

Fig. 18–19. Measurement of transmission factor.

18–15. Transmission and Reflection Factors

The specular reflection factor or the transmission factor of a non-scattering specimen can be measured in an obvious manner simply by inserting the specimen between the source and a physical detector or between the source and a visual photometer head. The reflection or transmission factor can be obtained in either radiometric or photometric terms. For example, the straightforward use of a visual photometer head will yield the integrated visible reflection or transmission factor. If one wishes to investigate a small region of a specimen, it is usual to employ a physical detector and to insert a suitable lens system in order to ensure that the detector receives only radiation that was incident on the region of interest. The arrangement is called a *microphotometer* or *microradiometer*. For transmission measurements—for example, where one wishes to measure the absorption (density) of a filter or a developed photographic plate or film—one speaks of a *densitometer* or *microdensitometer*.

If one wishes to make reflection or transmission measurements with a specimen that introduces convergence, divergence, or scattering of the radiation, the above method cannot be used in its simplest form. (In fact this applies to photographic emulsions since these do scatter some radiation, but the effect is often ignored and comparative measurements taken with the distance between the emulsion and the detector held constant.) In general the difficulty can be overcome by using an integrating sphere. For example, Fig. 18–19 shows the arrangement for measuring the transmission factor of a translucent plate P. The cross section of the incident beam must be less than the entrance aperture in the sphere. The transmission factor is the ratio of the illuminations (irradiances) of the window with and without the plate in position. A similar method can be used to find the transmission factor of an optical instrument. This is placed at P and the measurements made as before. Care must be taken that interposing the instrument does not cause the light to be obstructed by the edge of the sphere aperture. For example, when measuring the transmission factor of a small terrestrial telescope the eyepiece should be placed nearer the sphere, so that the beam which emerges has a smaller diameter than that which enters.

18–16. Illumination Photometers

The measurement of the illumination on a surface is frequently of considerable practical importance. For example, it is necessary to know whether there is adequate illumination on a desk or in a street at night. In many cases it is necessary for the apparatus to be portable. One of the simplest methods employs a barrier layer photo-cell provided with the usual correcting filter. This is simply placed face upwards on the surface where the illumination is required. The only auxiliary apparatus required is a micro-ammeter, and this can be calibrated to read the illumination directly. The chief disadvantage of this method is that the output for a given illumination falls off rapidly when the light is incident at an angle exceeding about 50°. This is mainly due to the directional properties of the cell itself and is made worse by the presence of the filter whose transmission decreases with increased obliquity. Obviously other physical detectors can be used in a similar manner.

An alternative approach to the problem of measuring the illumination on a surface is to place a diffuse test surface of known reflectivity on the surface concerned. This can then be viewed visually and its brightness matched to the brightness of a comparison field which is usually produced in a portable instrument by means of a lamp run from a battery under controlled conditions. There are numerous optical systems that employ this principle. An advantage of the method is that the brightness of a diffuse surface is independent of the distance and direction from which it is viewed so that the test surface can be viewed from any convenient position.

It is interesting to note that the illumination on a horizontal surface in England at mid-day in summer may reach about 10^4 lumens per sq. ft. The recommended minimum illuminations for a number of surfaces are (in lumens per sq. ft.):

Library table	15
Drawing-board	30
Hospital operating-table	300

For street lighting the optimum height of the lamps is important since if they are high the illumination will tend to be more uniform but rather weak, and if they are low it will be patchy.

18–17. Optical Pyrometers

An optical pyrometer measures the temperature of a black body, or a "grey" body which radiates a temperature-dependent continuous spectrum, by observing the radiance or the luminance and colour of the radiator. With a visual pyrometer one can obtain an accuracy of a few degrees over a range extending from about 800°K to about 2300°K (the precision being within 1° over most of the range). Another order of magnitude of accuracy can be obtained with a physical instrument using a photo-multiplier. Usually the method is based on the temperature dependence of the radiance in a narrow spectral region; to obtain an absolute temperature measurement one needs to know the emissivity of the radiator—a true black body radiator can obviously be measured with maximum accuracy.

In the *disappearing filament pyrometer* a filament is viewed against the radiator as background (often in conjunction with a filter). The filament is heated electrically until it radiates and is no longer seen because it matches the background in brightness and colour. The instrument can be calibrated by using a black body at known temperatures.

In the *total radiation pyrometer* an image of the radiator is formed and the radiation of all wavelengths heats a thermocouple which measures the irradiance of the image.

CHAPTER XIX

THE ELECTROMAGNETIC THEORY OF LIGHT

19–1. Vector Notation

IN this chapter it will be assumed that the reader is familiar with the laws of electromagnetism and with the fundamentals of vector calculus. With regard to the latter, it is possible for those who are not familiar with the subject to look upon the notation as a form of shorthand. Thus if ϕ is a scalar point function, grad ϕ is a vector whose x, y, and z components are $\partial\phi/\partial x$, $\partial\phi/\partial y$, and $\partial\phi/\partial z$, respectively. If one has a vector $\mathbf{E}$ whose components are E_x, E_y, and E_z, then div $\mathbf{E}$ is the scalar quantity $(\partial E_x/\partial x + \partial E_y/\partial y + \partial E_z/\partial z)$, and curl $\mathbf{E}$ is the vector whose components are $(\partial E_z/\partial y - \partial E_y/\partial z)$, $(\partial E_x/\partial z - \partial E_z/\partial x)$, and $(\partial E_y/\partial x - \partial E_x/\partial y)$. The symbol ∇^2 applied to the scalar ϕ means $(\partial^2\phi/\partial x^2 + \partial^2\phi/\partial y^2 + \partial^2\phi/\partial z^2)$, and $\nabla^2\mathbf{E}$ is the vector whose components are $\nabla^2 E_x$, $\nabla^2 E_y$, and $\nabla^2 E_z$. If $\mathbf{E}$ and $\mathbf{H}$ are two vectors inclined at angle θ, the *scalar product* is the scalar quantity $EH \cos\theta$ where E and H are the magnitudes of $\mathbf{E}$ and $\mathbf{H}$. The *vector product* is a vector whose magnitude is $EH \sin\theta$, and whose direction is normal to the plane containing $\mathbf{E}$ and $\mathbf{H}$ and such that, to an observer looking along the direction of the vector product, the rotation from $\mathbf{E}$ to $\mathbf{H}$ is clockwise. The scalar product is written $\mathbf{E.H}$ and the vector product $\mathbf{E} \times \mathbf{H}$. It will be seen that $\mathbf{E} \times \mathbf{H} = -\mathbf{H} \times \mathbf{E}$. As indicated above, vectors are printed in heavy type and the magnitudes in italics. Finally it should be pointed out that the right-handed cartesian co-ordinates are employed. That is to say, when looking along the z-axis, a rotation from the x-axis to the y-axis is clockwise. A number of results in vector analysis will be used in what follows, and these will be quoted when they are required.

19–2. Maxwell's Equations. The Displacement Current

The fundamental equations of the electromagnetic field are:

$$\text{div } \mathbf{D} = 4\pi\varrho, \qquad \ldots \ldots \quad (19\text{–}1)$$

$$\text{div } \mathbf{B} = 0, \qquad \ldots \ldots \quad (19\text{–}2)$$

$$\text{curl } \mathbf{H} = 4\pi\mathbf{j} + \frac{1}{c}\frac{\partial \mathbf{D}}{\partial t}, \qquad \ldots \ldots \quad (19\text{–}3)$$

$$\text{curl } \mathbf{E} = -\frac{1}{c}\frac{\partial \mathbf{B}}{\partial t}, \qquad \ldots \ldots \quad (19\text{–}4)$$

where $\mathbf{D}$ is the electric displacement, ϱ the charge density, $\mathbf{B}$ the magnetic induction, $\mathbf{H}$ the magnetic intensity, $\mathbf{j}$ the current density, and $\mathbf{E}$ the electric intensity. $\mathbf{D}$, $\mathbf{E}$, and ϱ, are measured in e.s.u. and $\mathbf{B}$, $\mathbf{H}$, and $\mathbf{j}$, in e.m.u. The magnitude of the constant c is the ratio of the e.m.u. to the e.s.u. of charge. Equation (19–1) is an expression of Gauss's flux law and (19–2) expresses the fact that isolated magnetic poles do not exist. Equation (19–4) is Maxwell's generalization of Neumann's law of electromagnetic induction. A far more important contribution by Maxwell was the inclusion of the second term on the right-hand side of (19–3). In the absence of this term, the equation follows directly from Ampere's law concerning the work done in taking a magnetic pole round a path linking a current and, in that form, is applicable to steady and slowly varying conditions. Now for any vector V div curl $V \equiv 0$ so that, if curl $H = 4\pi j$, it follows that div $j = 0$. But the equation of continuity of charge is, from (19–1) (using e.s.u. throughout),

$$\operatorname{div} j = -\frac{\partial \varrho}{\partial t} = -\frac{\partial}{\partial t}\left(\frac{1}{4\pi}\operatorname{div} D\right), \quad \cdots \quad (19\text{–}5)$$

i.e. div j is zero for the steady state only.

From (19–5) one may write

$$\operatorname{div}\left(j + \frac{1}{4\pi}\frac{\partial D}{\partial t}\right) = 0, \quad \cdots \quad (19\text{–}6)$$

which reduces to div $j = 0$ in the steady state. Equation (19–6) suggests that, in general, j should be replaced by $[j + (1/4\pi)(\partial D/\partial t)]$ so that, in the non-steady state, the current density is effectively $[j + (1/4\pi)(\partial D/\partial t)]$. j is, of course, the conduction current density and $(1/4\pi)(\partial D/\partial t)$ is called the *displacement current* density. Now the displacement D is related to the electric intensity E and the polarization P by the equation

$$D = E + 4\pi P.$$

Therefore the displacement current may be written

$$\frac{1}{4\pi}\frac{\partial D}{\partial t} = \frac{\partial P}{\partial t} + \frac{1}{4\pi}\frac{\partial E}{\partial t}. \quad \cdots \quad (19\text{–}7)$$

One can interpret the so-called *polarization current*, $\partial P/\partial t$, as follows: when a dielectric is polarized by an applied electric field, positive and negative charges are separated. While this separation is taking place the charges are moving and constitute an electric current. The last term of (19–7) is present even when there is no material medium; it is the displacement current in a vacuum. This was called the *aether displacement current* since it was assumed that a polarizable medium, the aether, pervaded all space and was the medium for the transmission of light disturbances. Probably it is better to regard (19–3) (omitting the term $4\pi j$) simply as an analogue of the more familiar equation (19–4). That is to say, just as a change in magnetic flux induces an electric field, a change in electric flux induces a magnetic field. It is found that it is not necessary to consider the displacement current unless one has rapidly varying fields, i.e. high frequencies.

19–3. Propagation of an Electromagnetic Disturbance in Isotropic Dielectrics

For a non-conducting medium the conduction current j is zero. If the medium is isotropic, the electric displacement at any point is in the same direction as the electric intensity, and the magnetic induction is in the same direction as the magnetic intensity, i.e. one may write

$$D = KE, \quad \cdots \cdots \quad (19\text{–}8)$$

$$B = \mu H, \quad \cdots \cdots \quad (19\text{–}9)$$

where K and μ are scalar quantities known respectively as the dielectric constant and the permeability. The medium is assumed to be homogeneous so that K and μ are not functions of position; it is assumed that they are independent of E and H. It is also assumed that there is no volume distribution of charge— $\rho=0$. With these assumptions Maxwell's equations reduce to

$$\text{div } \mathbf{E}=0, \quad \ldots \ldots \ldots \quad (19\text{--}10)$$

$$\text{div } \mathbf{H}=0, \quad \ldots \ldots \ldots \quad (19\text{--}11)$$

$$\text{curl } \mathbf{H}=\frac{K}{c}\frac{\partial \mathbf{E}}{\partial t}, \quad \ldots \ldots \quad (19\text{--}12)$$

$$\text{curl } \mathbf{E}=-\frac{\mu}{c}\frac{\partial \mathbf{H}}{\partial t}. \quad \ldots \ldots \quad (19\text{--}13)$$

From (19–13) one has

$$\text{curl curl } \mathbf{E}=-\frac{\mu}{c}\text{curl }\frac{\partial \mathbf{H}}{\partial t}=-\frac{\mu}{c}\frac{\partial}{\partial t}(\text{curl } \mathbf{H}).$$

Now it can be shown that for any vector $\mathbf{V}$ curl curl $\mathbf{V} \equiv \text{grad div } \mathbf{V} - \nabla^2\mathbf{V}$ so that, using (19–12), one has

$$\text{grad div } \mathbf{E} - \nabla^2\mathbf{E}=-\frac{\mu K}{c^2}\frac{\partial^2 \mathbf{E}}{\partial t^2},$$

and since div E=0 one has

$$\nabla^2\mathbf{E}=\frac{\mu K}{c^2}\frac{\partial^2 \mathbf{E}}{\partial t^2}. \quad \ldots \ldots \quad (19\text{--}14)$$

Similarly, E can be eliminated to obtain

$$\nabla^2\mathbf{H}=\frac{\mu K}{c^2}\frac{\partial^2 \mathbf{H}}{\partial t^2}. \quad \ldots \ldots \quad (19\text{--}15)$$

These equations have the form of the general equation of wave motion and state that variations of E and H are propagated with a speed $c/\sqrt{\mu K}$. It is important to notice that these equations are linear so that the sum of a number of solutions is also a solution. Consequently, the so-called *electromagnetic waves* interfere according to the principle of superposition.

19-4. Plane Waves

The case of a plane wave will now be considered, the values of E and H being constant in wavefronts which are perpendicular to the x-axis. Thus the components of E and H are constant in these planes and their partial derivatives with respect to y and z are zero. Now equation (19–10) is

$$\text{div } \mathbf{E}=\frac{\partial E_x}{\partial x}+\frac{\partial E_y}{\partial y}+\frac{\partial E_z}{\partial z}=0,$$

$$\therefore \quad \frac{\partial E_x}{\partial x}=0.$$

Similarly, from (19–11),

$$\frac{\partial H_x}{\partial x} = 0.$$

In addition, (19–12) and (19–13) show that

$$\frac{\partial E_x}{\partial t} = 0 \quad \text{and} \quad \frac{\partial H_x}{\partial t} = 0.$$

Now there may be a constant field superposed on the propagated disturbances, but, as far as the latter are concerned, these equations show that E_x and H_x are zero. The existence of any constant field will be ignored since it is of no interest in the present discussions. Thus at any point both **E** and **H** lie in a plane perpendicular to the x-axis, i.e. in the wavefront. In addition, from (19–14) one has

$$\left. \begin{array}{c} \dfrac{\partial^2 E_y}{\partial x^2} = \dfrac{\mu K}{c^2} \dfrac{\partial^2 E_y}{\partial t^2} \\[3mm] \text{and} \quad \dfrac{\partial^2 E_z}{\partial x^2} = \dfrac{\mu K}{c^2} \dfrac{\partial^2 E_z}{\partial t^2} \end{array} \right\} \quad \dotsb \quad (19\text{--}16)$$

with similar expressions for H_y and H_z. These equations show that the disturbances are propagated along the x-axis, i.e. in the direction of the normal to the wavefront. It follows that **E** and **H** are always perpendicular to the direction of propagation, i.e. one has transverse waves.

It will now be assumed that the variations of **E** are simple harmonic and that **E** is parallel to the y-axis. Then $E_z=0$ and from (19–16) one may write, for a wave travelling in the positive x-direction,

$$E_y = E_{y0} e^{ik(x-vt)}, \quad \dotsb \quad (19\text{--}17)$$

where

$$v^2 = \frac{c^2}{\mu K}$$

A wave in which the variations of **E** are confined to one plane containing the direction of propagation (in this case the xy plane) is known as a *plane-polarized wave*. From (19–12) one has, considering the z components,

$$\frac{\partial H_y}{\partial x} - \frac{\partial H_x}{\partial y} = \frac{K}{c} \frac{\partial E_z}{\partial t} = 0.$$

It has already been stated that $\partial H_x/\partial y$ is zero,

$$\therefore \quad \frac{\partial H_y}{\partial x} = 0.$$

Hence the variations of **H** are confined to the z component. That is to say, **E** and **H** are perpendicular and the direction of propagation, **E**, and **H**, are oriented as in a right-handed co-ordinate system.

From (19–13), considering z components, one has

$$\frac{\partial E_y}{\partial x} - \frac{\partial E_x}{\partial y} = -\frac{\mu}{c} \frac{\partial H_z}{\partial t}.$$

Now $\partial E_x/\partial y = 0$ and E_y is given by (19–17). Hence

$$\frac{\partial H_z}{\partial t} = -\frac{c}{\mu} ikE_{y0}\, e^{ik\left(x-\frac{c}{\sqrt{\mu K}}t\right)},$$

$$\therefore\quad H_z = \sqrt{\frac{K}{\mu}}\, E_{y0}\, e^{ik(x-vt)},$$

i.e. $\qquad\qquad H_z = \sqrt{\frac{K}{\mu}}\, E_y, \quad\ldots\quad\ldots\quad\ldots$ (19–18)

the constants of integration being omitted since one is not interested in a constant component of the field. Since $\sqrt{K/\mu}$ is a real and positive constant, (19–18) shows that the variations of **H** and **E** are in phase. The properties of an electromagnetic wave depend to a great extent upon the frequency. By comparing the properties predicted by theory with those observed experimentally, many waves are found to be electromagnetic in character. The several divisions into which electromagnetic waves fall are indicated approximately in Fig. 19–1.

Equation (19–17) shows that $c/\sqrt{\mu K}$ is the phase velocity. In a vacuum μ and K are unity and the phase velocity is c. The refractive index of a material medium is defined as the ratio of the phase velocity in a vacuum to that in the medium.

$$\therefore\quad n = \sqrt{\mu K}. \quad\ldots\quad (19\text{–}19)$$

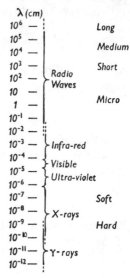

Fig. 19–1. The electromagnetic spectrum.

It will be seen that if μ and K are independent of frequency there is no dispersion. It is found that this is approximately true for wavelengths longer than about 1 cm., but for shorter wavelengths (higher frequencies) the refractive indices of many substances differ widely from those given by (19–19) using values of μ and K appropriate to slowly varying fields. For optical frequencies one may put $\mu=1$ for transparent media so that $n=\sqrt{K}$. Using the low frequency value of K, one has $\sqrt{K}=1.405$ for paraffin, which agrees fairly well with the observed refractive index of 1.422. On the other hand, for water $\sqrt{K}=9$ whereas n is about 1.33. For water, and for many other substances, the high value of the static dielectric constant is associated with a contribution to the electric polarization caused by the orientation of polar molecules when an electric field is applied. This is ineffective at optical frequencies, and the refractive index is correspondingly reduced.

19–5. The Poynting Vector

In the presence of a steady electromagnetic field one may say that the energy per unit volume of a homogeneous dielectric is

$$\frac{1}{8\pi}\,(\mathbf{H}.\mathbf{B}+\mathbf{E}.\mathbf{D}).$$

Consider now the volume V which is enclosed by the closed surface S. As the electromagnetic field changes, the rate of increase of energy in V is given by

$$\frac{\partial W}{\partial t} = \frac{\partial}{\partial t} \left\{ \iiint \frac{1}{8\pi} (\mathbf{H}.\mathbf{B} + \mathbf{E}.\mathbf{D}) dV \right\}$$

and, for an isotropic medium, one has $\mathbf{D} = K\mathbf{E}$ and $\mathbf{B} = \mu\mathbf{H}$ so that $\mathbf{H}.\mathbf{B} = \mu\mathbf{H}^2$ and $\mathbf{E}.\mathbf{D} = K\mathbf{E}^2$. Thus one has

$$\frac{\partial W}{\partial t} = \iiint \frac{1}{4\pi} \left(\mathbf{H}.\frac{\partial \mathbf{B}}{\partial t} + \mathbf{E}.\frac{\partial \mathbf{D}}{\partial t} \right) dV.$$

When the medium is anisotropic, K and μ are tensors (see § 19–7), but $\partial W/\partial t$ is still given by the above expression.

Using (19–3) and (19–4) one has (since $\mathbf{j} = 0$ for a dielectric)

$$\frac{\partial W}{\partial t} = -\iiint \frac{c}{4\pi} (\mathbf{H}.\text{curl } \mathbf{E} - \mathbf{E}.\text{curl } \mathbf{H}) dV.$$

Now for any two vectors $\mathbf{a}_1$ and $\mathbf{a}_2$ it can be shown that $\mathbf{a}_1.\text{curl } \mathbf{a}_2 - \mathbf{a}_2.\text{curl } \mathbf{a}_1 \equiv \text{div} (\mathbf{a}_2 \times \mathbf{a}_1)$,

$$\therefore \quad \frac{\partial W}{\partial t} = -\iiint \frac{c}{4\pi} \text{div}(\mathbf{E} \times \mathbf{H}) dV.$$

Gauss's theorem states that if the surface S encloses the volume V, $\iiint \text{div } \mathbf{a} dV = -\iint \mathbf{a}.d\mathbf{S}$ if the vector $d\mathbf{S}$ has the direction of the inward normal to the surface element $d\mathbf{S}$.

$$\therefore \quad \frac{\partial W}{\partial t} = \iint \frac{c}{4\pi} (\mathbf{E} \times \mathbf{H}).d\mathbf{S}. \quad \ldots \ldots \quad (19\text{--}20)$$

The right-hand side is the integral over the surface S of the normal component of the vector

$$\mathbf{\Pi} = \frac{c}{4\pi} (\mathbf{E} \times \mathbf{H}). \quad \ldots \ldots \quad (19\text{--}21)$$

Now the integral gives the rate at which energy enters the enclosed volume. Hence (assuming conservation of energy) the vector $\mathbf{\Pi}$ can be interpreted as giving the direction of the energy flow at any point and the amount of energy per second crossing unit area perpendicular to the direction of flow. $\mathbf{\Pi}$ is known as *Poynting's vector*. It should be noted that (19–20) indicates that the *integral* on the right-hand side of equation (19–20) gives the *total* energy flow and it has not been proved that $\mathbf{\Pi}$ gives the energy flow across an isolated element of area. However, this interpretation of $\mathbf{\Pi}$ is a very useful one. Now the vector $\mathbf{E} \times \mathbf{H}$ is normal to the plane containing $\mathbf{E}$ and $\mathbf{H}$, so that in an electromagnetic disturbance the energy travels instantaneously in this direction. It will be seen that for a plane wave the energy obviously flows in the direction of propagation. $\mathbf{E}$ and $\mathbf{H}$ are oscillatory and change direction, but $\mathbf{\Pi}$ is unidirectional. However, its magnitude varies and is a maximum when $\mathbf{E}$ and $\mathbf{H}$ are at their maxima and is zero when $\mathbf{E}$ and $\mathbf{H}$ are zero.

For a plane wave in an isotropic medium as discussed in the previous section, $\mathbf{\Pi}$ has magnitude

$$E_y H_z = \frac{c}{4\pi} \sqrt{\frac{K}{\mu}} E_{y0}^2 e^{2ik(x-vt)}.$$

This has mean value

$$\frac{c}{8\pi}\sqrt{\frac{K}{\mu}}E_{v0}{}^2.$$

Now the energy density is $(1/8\pi)(\mu H^2 + KE^2)$ and from equations (19–17) and (19–18) it will be seen that this has mean value $(K/8\pi)E_{v0}{}^2$. Hence the average speed of flow of the energy is given by

$$u = \frac{c}{8\pi}\sqrt{\frac{K}{\mu}}E_{v0}{}^2 \div \frac{K}{8\pi}E_{v0}{}^2,$$

$$\therefore \quad u = \frac{c}{\sqrt{\mu K}}. \qquad \cdots \cdots \quad (19\text{–}22)$$

Thus the average speed of flow of the energy is equal to the phase velocity or speed of propagation of the wave.

It is important to bear in mind that this discussion applies to a monochromatic wave.

19-6. Hertz's Experiments

It has been seen that Maxwell's equations lead to the result that an electromagnetic disturbance can be propagated as a wave motion and that the velocity

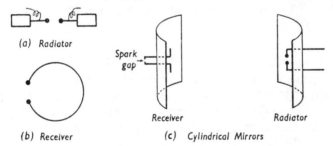

(a) Radiator

(b) Receiver

Spark
gap

Receiver

(c) Cylindrical Mirrors

Radiator

Fig. 19-2. Apparatus used by Hertz.

is $c/\sqrt{\mu K}$. Maxwell's *Treatise on Electricity and Magnetism* was published in 1873, and in 1887 Hertz discovered waves which were undoubtedly electromagnetic and whose properties resembled those of light. It was later shown that the speed of these waves is equal to the speed of light.

Hertz generated electromagnetic waves by means of an oscillatory discharge across a spark gap. The arrangement is illustrated in Fig. 19–2(a). The spark gap is between two brass knobs connected to rectangular brass plates each of about 40 sq. cm. area. These form a condenser of low capacity, and, when the knobs are connected to the secondary of an induction coil, an oscillatory discharge passes across the gap. This obviously generates an electromagnetic disturbance which, according to the above theory, will be propagated through the air. To detect these waves Hertz used a very simple resonance circuit consisting of a circle of wire broken by a spark gap, as shown in Fig. 19–2(b). The gap between the knobs is regulated by means of a micrometer screw. The diameter of the circle is chosen so that the resonant frequency of this simple circuit is equal to the frequency of the oscillatory discharge across the gap of the source. If the magnetic flux threading the circuit changes, an e.m.f. is induced, and when this is sufficiently large, a spark passes across the gap. Thus Hertz was able to detect the magnetic component of the wave. The maximum length of the gap across which sparks will pass is a measure of the component of the magnetic intensity normal to the plane of the circle. This was found to be a maximum in the direction at right angles to the line joining the knobs of the source, and zero in the direction of propagation and in the direction parallel to the spark gap of the source. This indicates that the waves are transverse and plane polarized. One of the most convincing demonstrations of the existence of a wave motion is to produce interference, and this was done by reflecting the waves from a large metal screen on which they were incident normally. The existence of the resulting stationary waves was shown since the receiver sparked at the antinodes but not at the nodes. Hertz also demonstrated the reflection of the waves from large cylindrical parabolic zinc mirrors, a much stronger spark being obtained when the source was placed at the focus of one and the receiver at the focus of the other. In this case the receiver consisted of two pieces of thick wire placed along the focal line of the receiving mirror and connected to a spark gap behind the mirror. A spark was produced in the receiver when the axes of the mirrors were parallel [Fig. 19–2(c)] but not when they were perpendicular, again indicating that the waves are plane polarized. By interposing a large prism of pitch between the source and receiver, refraction of the waves was demonstrated.

19–7. Electrical Polarization in Anisotropic Media

Many crystals are not isotropic and the above theory cannot describe the propagation of light in such substances. In an anisotropic medium the displacement is not, in general, in the same direction as the electric intensity. That is, the dielectric constant is not a simple scalar constant. The components of $\mathbf{D}$ are related to the components of $\mathbf{E}$ by the equations

$$\left.\begin{aligned} D_x &= K_{xx}E_x + K_{xy}E_y + K_{xz}E_z, \\ D_y &= K_{yx}E_x + K_{yy}E_y + K_{yz}E_z, \\ D_z &= K_{zx}E_x + K_{zy}E_y + K_{zz}E_z, \end{aligned}\right\} \qquad \ldots \quad (19\text{--}23)$$

where the coefficients are scalar constants if the material is homogeneous.

One writes

$$\mathbf{D} = \boldsymbol{\Phi}\mathbf{E}, \qquad \ldots \ldots \quad (19\text{--}24)$$

where $\boldsymbol{\Phi}$ is a tensor whose components are K_{xx}, etc. It will be seen that, in general, multiplication by a tensor changes the direction as well as the magnitude of a vector. As with isotropic media, it is assumed that the permeability is unity at optical frequencies, so that only the electrical anisotropy is considered. Since the coefficients of $\boldsymbol{\Phi}$ vary with frequency, only monochromatic disturbances will be considered in the present discussion.

Maxwell's equations (19–1) to (19–4) remain valid and the energy density is $(1/8\pi)(\mathbf{E}.\mathbf{D}+\mathbf{H}^2)$. The energy flux vector is still $(c/4\pi)(\mathbf{E}\times\mathbf{H})$ and it can be shown that it follows that the tensor $\boldsymbol{\Phi}$ is symmetrical. That is, $K_{xy}=K_{yx}$, $K_{yz}=K_{zy}$, and $K_{xz}=K_{zx}$. That is to say, $\boldsymbol{\Phi}$ has only six independent coefficients.

The connection between $\mathbf{D}$ and $\mathbf{E}$ can then be described by referring to the so-called *tensor ellipsoid* which represents the properties of $\boldsymbol{\Phi}$. The scalar product of $\mathbf{E}$ and $\mathbf{D}$ is given by

$$\mathbf{E}.\mathbf{D}=\mathbf{E}.(\boldsymbol{\Phi}\mathbf{E})=K_{xx}E_x^2+K_{yy}E_y^2+K_{zz}E_z^2$$
$$+2K_{xy}E_xE_y+2K_{yz}E_yE_z+2K_{zx}E_zE_x=p \text{ (say).}$$

If one represents $\mathbf{E}$ in magnitude and direction by the radius vector $\mathbf{r}$ from a fixed origin, $E_x=x$, $E_y=y$, $E_z=z$, and the surface $p=1$ is an ellipsoid known as the tensor ellipsoid since its shape characterizes the tensor $\boldsymbol{\Phi}$. In this particular case it is usually referred to as the *Fresnel Ellipsoid*. It is easily shown that for any value of $\mathbf{E}$ the direction of $\mathbf{D}$ is that of the normal to the surface at the end point of $\mathbf{r}$, and that the magnitude of $\mathbf{D}$ is the reciprocal of the projection of $\mathbf{r}$ on the normal.

If the ellipsoid is referred to its principal axes, the coefficients E_{xy}, E_{yz}, and E_{zx}, vanish and the equation to the ellipsoid becomes

$$K_x x^2+K_y y^2+K_z z^2=1 \quad \ldots \ldots \quad (19\text{--}25)$$

(only single suffixes being necessary). The semi-axes are

$$\frac{1}{\sqrt{K_x}}, \quad \frac{1}{\sqrt{K_y}}, \quad \frac{1}{\sqrt{K_z}}.$$

It will be seen that, of the six independent coefficients of $\boldsymbol{\Phi}$, three define the shape of the Fresnel ellipsoid, so that the other three define its orientation in the medium (i.e. locate the axes of the ellipsoid with respect to the crystal structure). In what follows it will be assumed that the principal axes of the Fresnel ellipsoid are always chosen as the co-ordinate axes.

K_x, K_y, and K_z, are called the *principal dielectric constants* and one now has

$$D_x=K_xE_x, \quad D_y=K_yE_y, \quad D_z=K_zE_z.$$

One defines the *principal velocities* of light by the relations

$$v_x=\frac{c}{\sqrt{K_x}}, \quad v_y=\frac{c}{\sqrt{K_y}}, \quad v_z=\frac{c}{\sqrt{K_z}},$$

so that the axes of the Fresnel ellipsoid are proportional to the principal velocities. These are the velocities in the three principal directions. It is important to remember that they are not the components of the velocity if a disturbance is propagated in another direction.

Since one has $E_x=(1/K_x)D_x$, $E_y=(1/K_y)D_y$, and $E_z=(1/K_z)D_z$, the principal axes of the Fresnel ellipsoid are also the principal axes of the tensor ellipsoid which gives $\mathbf{E}$ as a function of $\mathbf{D}$. One writes $\mathbf{E}=\boldsymbol{\Phi}^{-1}\mathbf{D}$, and the equation to the ellipsoid corresponding to $\boldsymbol{\Phi}^{-1}$ is

$$\frac{x^2}{K_x}+\frac{y^2}{K_y}+\frac{z^2}{K_z}=1. \quad \ldots \ldots \quad (19\text{--}26)$$

Thus its semi-axes are equal to the square roots of the dielectric constants, i.e. to the so-called *principal indices of refraction*, n_x, n_y, and n_z. It is called the *index ellipsoid*.

19–8. Plane Waves in Anisotropic Media

As before, it is assumed that there is no volume charge and no conduction current, and that the permeability is unity. Equations (19–1) to (19–4) then give

$$\operatorname{div} \mathbf{D} = 0 \quad \ldots \ldots \ldots \quad (19\text{--}27)$$

$$\operatorname{div} \mathbf{H} = 0 \quad \ldots \ldots \ldots \quad (19\text{--}28)$$

$$\operatorname{curl} \mathbf{H} = \frac{1}{c} \frac{\partial \mathbf{D}}{\partial t} \quad \ldots \ldots \quad (19\text{--}29)$$

$$\operatorname{curl} \mathbf{E} = \frac{1}{c} \frac{\partial \mathbf{H}}{\partial t} \cdot \quad \ldots \ldots \quad (19\text{--}30)$$

It should be noted that div $\mathbf{E}$ is not now zero.

In §19–4 it was possible, without loss of generality, to consider a plane wave propagated in the direction of a co-ordinate axis. This is no longer possible since it is assumed that the co-ordinate axes are the principal axes of the Fresnel ellipsoid for the medium. Assume, therefore, that a plane wave advances with phase velocity v along the direction of $\mathbf{n}$, the unit vector normal to the planes of equal phase. For a simple harmonic solution one may write

$$\mathbf{E} = \mathbf{E}_0 e^{ik(vt - \mathbf{r}.\mathbf{n})}, \quad \ldots \ldots \quad (19\text{--}31)$$

where $\mathbf{r}$ is the radius vector from the origin.

From (19–30) one has

$$\frac{\partial \mathbf{H}}{\partial t} = -c \operatorname{curl} \mathbf{E}_0 e^{ik(vt - \mathbf{r}.\mathbf{n})}. \quad \ldots \ldots \quad (19\text{--}32)$$

Now it can be shown that if S and $\mathbf{V}$ represent a scalar and a vector field respectively, curl $S\mathbf{V} = S$ curl $\mathbf{V} - \mathbf{V} \times$ grad S.

$$\therefore \quad \frac{\partial \mathbf{H}}{\partial t} = c\mathbf{E}_0 \times \operatorname{grad} e^{ik(vt - \mathbf{r}.\mathbf{n})}$$

since curl $\mathbf{E}_0 = 0$.

Now $\qquad$ grad $e^{ik(vt - \mathbf{r}.\mathbf{n})} = -ike^{ik(vt - \mathbf{r}.\mathbf{n})}$ grad $\mathbf{r}.\mathbf{n}$

and if cos α, cos β, and cos γ, are the direction cosines of $\mathbf{n}$,

$$\operatorname{grad} \mathbf{r}.\mathbf{n} = \left(\mathbf{i} \frac{\partial}{\partial x} + \mathbf{j} \frac{\partial}{\partial y} + \mathbf{k} \frac{\partial}{\partial z} \right) \left(x \cos \alpha + y \cos \beta + z \cos \gamma \right) = \mathbf{n},$$

$$\therefore \quad \frac{\partial \mathbf{H}}{\partial t} = -cike^{ik(vt - \mathbf{r}.\mathbf{n})} \mathbf{E}_0 \times \mathbf{n},$$

$$\therefore \quad \mathbf{H} = \frac{c}{v} e^{ik(vt - \mathbf{r}.\mathbf{n})} (\mathbf{n} \times \mathbf{E}_0) = \mathbf{H}_0 e^{ik(vt - \mathbf{r}.\mathbf{n})} \text{ (say).} \quad (19\text{--}33)$$

This shows that $\mathbf{H}$ is normal to the plane of $\mathbf{E}$ and $\mathbf{n}$.

From (19–29) one has

$$\frac{\partial \mathbf{D}}{\partial t} = c \operatorname{curl} \mathbf{H} = c \operatorname{curl} \mathbf{H}_0 e^{ik(vt - \mathbf{r}.\mathbf{n})}.$$

This is the analogue of (19-32) so that, proceeding as above, one has

$$\frac{\partial \mathbf{D}}{\partial t} = -cike^{ik(vt-\mathbf{r.n})}(\mathbf{n} \times \mathbf{H}_0).$$

Integrating, and substituting for $\mathbf{H}_0$,

$$\mathbf{D} = -\frac{c^2}{v^2}\mathbf{n} \times (\mathbf{n} \times \mathbf{E}). \quad \ldots \quad (19\text{-}34)$$

This shows that $\mathbf{D}$ is normal to $\mathbf{n}$ and to $\mathbf{H}$. Thus both $\mathbf{H}$ and $\mathbf{D}$ are normal to $\mathbf{n}$, i.e. lie in the wavefront. Hence with respect to $\mathbf{H}$ and $\mathbf{D}$ one has transverse waves. Since $\mathbf{H}$ is normal to $\mathbf{D}$, $\mathbf{E}$, and $\mathbf{n}$, these three must be coplanar. Now if $\mathbf{V}_1$, $\mathbf{V}_2$, and $\mathbf{V}_3$, are three vectors, it can be shown that $\mathbf{V}_1 \times (\mathbf{V}_2 \times \mathbf{V}_3) = \mathbf{V}_2(\mathbf{V}_1.\mathbf{V}_3) - \mathbf{V}_3(\mathbf{V}_1.\mathbf{V}_2)$. Hence from (19-34)

$$\mathbf{D} = \frac{c^2}{v^2}[\mathbf{E} - \mathbf{n}(\mathbf{n}.\mathbf{E})], \quad (19\text{-}35)$$

which again shows that $\mathbf{D}$, $\mathbf{E}$, and $\mathbf{n}$, are coplanar.

Now the Poynting vector

$$\mathbf{\Pi} = \frac{c}{4\pi}\mathbf{E} \times \mathbf{H}$$

is normal to the plane of $\mathbf{E}$ and $\mathbf{H}$ and therefore $\mathbf{D}$, $\mathbf{E}$, $\mathbf{n}$, and $\mathbf{\Pi}$, are coplanar. Also, since $\mathbf{n}$ is normal to $\mathbf{D}$ and $\mathbf{\Pi}$ is normal to $\mathbf{E}$, the angle between $\mathbf{\Pi}$ and $\mathbf{n}$ is equal to the angle between $\mathbf{E}$

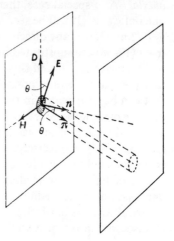

Fig. 19-3. Orientation of ray and wave normal.

and $\mathbf{D}$. The direction of $\mathbf{\Pi}$ is the direction of energy flow, i.e. the direction of the ray. One may say that a given element of wave travels in this direction (see Fig. 19-3). Its velocity—the ray velocity V—is obviously $v/\cos \theta$.

Equation (19-35) gives

$$\left. D_x = \frac{c^2}{v^2}(E_x - (\mathbf{n}.\mathbf{E})\cos \alpha) \right\} \quad \ldots \ldots \quad (19\text{-}36)$$

with similar expressions for D_y and D_z.

Now $E_x = D_x/K_x = D_x v_x^2/c^2$ and E_y and E_z can similarly be expressed in terms of the other principal velocities v_y and v_z.

Making these substitutions in (19-36), one obtains

$$\left. \begin{aligned} D_x &= \frac{c^2(\mathbf{n}.\mathbf{E}) \cos \alpha}{v_x^2 - v^2}, \\ D_y &= \frac{c^2(\mathbf{n}.\mathbf{E}) \cos \beta}{v_y^2 - v^2}, \\ D_z &= \frac{c^2(\mathbf{n}.\mathbf{E}) \cos \gamma}{v_z^2 - v^2}. \end{aligned} \right\} \quad \ldots \ldots \quad (19\text{-}37)$$

Now $\mathbf{D}$ is normal to $\mathbf{n}$

$$\therefore \quad \mathbf{D.n} = D_x \cos \alpha + D_y \cos \beta + D_z \cos \gamma = 0.$$

Using (19–37), this gives

$$\frac{\cos^2 \alpha}{v_x{}^2 - v^2} + \frac{\cos^2 \beta}{v_y{}^2 - v^2} + \frac{\cos^2 \gamma}{v_z{}^2 - v^2} = 0. \quad \ldots \quad (19\text{–}38)$$

It will be seen that, in general, the phase velocity v can have two values in any given direction. However, for most anisotropic media, it is found that there are two directions for which (19–38) has only one solution, i.e. gives only one phase velocity. These directions are called the *optic axes*, and the medium is said to be *biaxial*. As a special case, there are also *uniaxial* crystals, in which there is only one optic axis. It will be seen that the optic axis is a *direction* in the medium and not a line. If, for any direction other than an optic axis, the values of v obtained from (19–38) are substituted into (19–37), the corresponding directions of $\mathbf{D}$ can be obtained. It is found that they are at right angles. It is important to notice that these are the only possible directions $\mathbf{D}$. That is to say, the disturbances that can be propagated in the medium are limited to those in which $\mathbf{D}$ has one of these directions. Since, in general, there are two possible directions for $\mathbf{D}$ for a given direction of the wave normal, there are also two possible ray directions (see Fig. 19–3). Conversely, for a given direction of the ray there are, in general, two possible directions for the wave normal. Consequently, if one has a point source of light in the medium, the rays spread out from the point in straight lines (assuming the medium is homogeneous), but there is not a single expanding spherical wave as in isotropic media. The form of the expanding wavefront will be discussed in the next section.

19–9. Wave Surface Expanding from a Point Source

In a homogeneous medium the rays from a point source are directed radially outwards. After a time t the disturbances from the source are located at distances along the rays determined by the ray velocities in the appropriate directions. To find the surface of constant phase (the wavefront) it is therefore necessary to plot the ray velocity V (not the wave or phase velocity) on a polar diagram.

It was shown in § 19–8 that $\mathbf{D}$, $\mathbf{E}$, and $\mathbf{\Pi}$, lie in the same plane. Thus one may write

$$\mathbf{D} + l\mathbf{E} + m\mathbf{\pi} = 0 \quad \ldots \ldots \quad (19\text{–}39)$$

where l and m are scalars and $\mathbf{\pi}$ is the unit vector. Taking the scalar product with $\mathbf{\pi}$, one has

$$\mathbf{\pi.D} + m = 0 \quad \ldots \ldots \ldots \quad (19\text{–}40)$$

since $\mathbf{\pi.\pi} = 1$, and $\mathbf{\pi.E} = 0$ because $\mathbf{\pi}$ is normal to $\mathbf{E}$.

$$\therefore \quad \mathbf{D} + l\mathbf{E} - (\mathbf{\pi.D})\mathbf{\pi} = 0.$$

Taking the scalar product with $\mathbf{n}$, one has $\mathbf{n.D} = 0$ because $\mathbf{n}$ is normal to $\mathbf{D}$, and this leaves

$$l(\mathbf{n.E}) = (\mathbf{\pi.D})(\mathbf{n.\pi})$$

or
$$l = \frac{(\mathbf{\pi.D})(\mathbf{n.\pi})}{(\mathbf{n.E})}. \quad \ldots \ldots \quad (19\text{–}41)$$

Also, taking the scalar product of (19–35) with $\boldsymbol{\pi}$ and remembering that $\boldsymbol{\pi}.\mathbf{E}=0$ one has

$$\boldsymbol{\pi}.\mathbf{D}=-\frac{c^2}{v^2}(\boldsymbol{\pi}.\mathbf{n})(\mathbf{n}.\mathbf{E}). \quad \ldots \ldots \quad (19\text{–}42)$$

(19–41) and (19–42) give

$$l=-\frac{c^2}{v^2}(\mathbf{n}.\boldsymbol{\pi})^2=-\frac{c^2}{v^2}\cos^2\theta=-\frac{c^2}{V^2}, \quad \ldots \quad (19\text{–}43)$$

since $\mathbf{n}$ and $\boldsymbol{\pi}$ are unit vectors inclined at angle θ. Substituting for l and m from (19–40) and (19–43), (19–39) gives

$$\mathbf{D}-\frac{c^2}{V^2}\mathbf{E}-(\boldsymbol{\pi}.\mathbf{D})\boldsymbol{\pi}=0,$$

or

$$\mathbf{E}=\frac{V^2}{c^2}[\mathbf{D}-\boldsymbol{\pi}(\boldsymbol{\pi}.\mathbf{D})]. \quad \ldots \ldots \quad (19\text{–}44)$$

On comparing this with (19–35), it will be seen that $\mathbf{D}$, $\mathbf{E}$, $\mathbf{n}$, and v^2/c^2, are simply replaced by $\mathbf{E}$, $\mathbf{D}$, $\boldsymbol{\pi}$, and c^2/V^2. Proceeding in a similar manner, one has, if ξ, η, and ζ, are the direction cosines of $\boldsymbol{\pi}$, i.e. of the ray,

$$\left. E_x=\frac{V^2}{c^2}(D_x-(\boldsymbol{\pi}.\mathbf{D})\cos\xi) \right\} \quad \ldots \ldots \quad (19\text{–}45)$$

with similar expressions for E_y and E_z.

And since

$$D_x=K_xE_x=\frac{c^2E_x}{v_x^2}, \text{ etc.}$$

one has

$$\left. \begin{aligned} E_x&=\frac{v_x^2V^2(\boldsymbol{\pi}.\mathbf{D})\cos\xi}{c^2(V^2-v_x^2)}, \\ E_y&=\frac{v_y^2V^2(\boldsymbol{\pi}.\mathbf{D})\cos\eta}{c^2(V^2-v_y^2)}, \\ E_z&=\frac{v_z^2V^2(\boldsymbol{\pi}.\mathbf{D})\cos\zeta}{c^2(V^2-v_z^2)}. \end{aligned} \right\} \quad \ldots \ldots \quad (19\text{–}46)$$

Now $\mathbf{E}$ is normal to $\boldsymbol{\pi}$ so that

$$\mathbf{E}.\boldsymbol{\pi}=E_x\cos\xi+E_y\cos\eta+E_z\cos\zeta=0$$

and, using (19–46), this gives

$$\frac{v_x^2\cos^2\xi}{v_x^2-V^2}+\frac{v_y^2\cos^2\eta}{v_y^2-V^2}+\frac{v_z^2\cos^2\zeta}{v_z^2-V^2}=0, \quad \ldots \quad (19\text{–}47)$$

which gives the ray velocity V for any direction of the ray. One can now find the shape of the wavefront by representing V by the length of the radius vector drawn in direction (ξ, η, ζ). Thus, putting $V=r$, $\cos\xi=x/r$, $\cos\eta=y/r$, and $\cos\zeta=z/r$, one has, for the equation to the wavefront,

$$v_x^2(v_y^2-r^2)(v_z^2-r^2)x^2+v_y^2(v_z^2-r^2)(v_x^2-r^2)y^2+v_z^2(v_x^2-r^2)(v_y^2-r^2)z^2=0,$$

i.e.

$$(v_x^2x^2+v_y^2y^2+v_z^2z^2)r^2-v_x^2(v_y^2+v_z^2)x^2$$
$$-v_y^2(v_z^2+v_x^2)y^2-v_z^2(v_x^2+v_y^2)z^2+v_x^2v_y^2v_z^2=0. \quad (19\text{–}48)$$

This surface, which is a wavefront, is frequently referred to as the *wave surface*; since it is obtained from the ray velocity, it is sometimes called the *ray* (or *ray velocity*) *surface*. To obtain an idea of the shape of the surface, consider the sections in the co-ordinate planes. It will be assumed that $v_x > v_y > v_z$. The intersection with the xz plane ($y=0$) is given by

$$(r^2 - v_y^2)(x^2 v_x^2 + z^2 v_z^2 - v_x^2 v_z^2) = 0,$$

i.e. the circle

$$r^2 = x^2 + z^2 = v_y^2$$

and the ellipse

$$\left. \frac{x^2}{v_z^2} + \frac{z^2}{v_x^2} = 1. \right\} \qquad \cdots \cdots \quad (19\text{--}49)$$

Since $v_x > v_y > v_z$, the two curves intersect in four points A,B,C,D [Fig. 19–4(a)]. The directions AOC and DOB are the *ray axes* or directions in which there is a

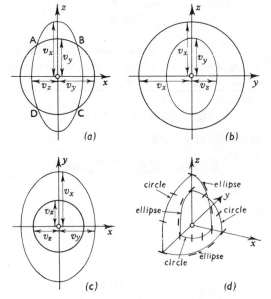

Fig. 19–4. Wave surface for a biaxial medium.

single ray velocity; they are not the *optic axes*. The latter are the directions in which there is only one phase velocity. This will be discussed further in § 22–9. It will be seen from the symmetrical nature of (19–48) that the intersection with each of the planes $x=0$ and $z=0$ is also a circle and an ellipse. Thus in the yz plane, one has

$$y^2 + z^2 = v_x^2$$

and

$$\frac{y^2}{v_z^2} + \frac{z^2}{v_y^2} = 1,$$

i.e. a circle entirely enclosing an ellipse [Fig. 19–4(b)]. In the xy plane one has

$$x^2 + y^2 = v_z^2$$

and

$$\frac{x^2}{v_y^2} + \frac{y^2}{v_x^2} = 1,$$

i.e. an ellipse entirely enclosing a circle [Fig. 19–4(c)].

The direction of **D** can be deduced from the fact that, at any point, **D** is in the tangent plane of the wave and is coplanar with the ray and the wave normal. On the elliptic sections it is therefore in the plane of the diagram, and is consequently normal to this plane on the circular sections. The shape of the wave surface and the directions of **D** are illustrated in Fig. 19–4(d), which shows quadrant sections of the wave surface. In all these diagrams the differences between the velocities v_x, v_y, and v_z, are very much greater than those encountered in practice.

19–10. Uniaxial Media

If the difference between v_x and v_y is reduced, the ray axes (and the optic axes) approach the z-axis, and when $v_x=v_y=v_{xy}$ (say) there is only one optic axis. Since $v_{xy}>v_z$ it will be seen from Fig. 19–4 that the sections in both the xz and

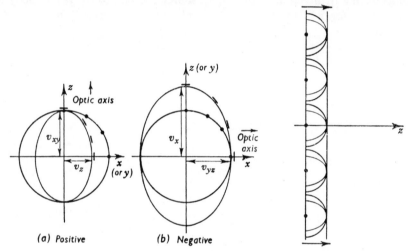

Fig. 19–5. Wave surface for uniaxial media.

Fig. 19–6. Propagation of a plane wave along the optic axis.

yz planes consist of a circle which touches an inscribed ellipse on the z-axis. The section in the xy plane consists of two concentric circles. Thus the complete wave surface consists of a sphere which touches an inscribed prolate spheroid on the axis of the latter. It is generated by rotating Fig. 19–5(a) about the z-axis. One is said to have a *positive uniaxial medium*. On the other hand, if the difference between v_y and v_z is reduced until $v_y=v_z=v_{yz}$ (keeping $v_x>v_{yz}$), one obtains a so-called *negative uniaxial medium* and the x-axis becomes the optic axis. The sections in both the xz and xy planes consist of an ellipse which touches an inscribed sphere on the x-axis [Fig. 19–5(b)]. The section in the yz plane consists of concentric circles, and the complete wave surface is generated by rotating Fig. 19–5(b) about the x-axis. That is, one has an oblate spheroid and an inscribed sphere.

The fact that for a uniaxial medium the optic axis coincides with the ray axis is easily shown as follows. Suppose that for a positive uniaxial medium one has a plane wave whose normal lies along the ray axis. Then according to Huygens' principle each element of a wavefront acts as a source of secondary waves and the two waves from each source spread out as shown in Fig. 19–6; the plane envelope of the elliptic sections obviously coincides with that of the circular sections. That is, the wavefront associated with the former has the same wave velocity as that associated with the latter. Further examples of the application of Huygens' principle to anisotropic media are discussed in Chapter XXII.

One should point out, in conclusion, that the discussion in §§ 19–9 and 19–10 is not entirely satisfactory. The aim was to find the shape of the wavefront expanding from a point source in an anisotropic medium. What is derived is the ray velocity surface, describing the variation with direction of the ray velocity, but it is, in fact, the ray velocity appropriate to a plane wave. That is to say, it has been assumed, without proof, that the ray velocity with which a disturbance spreads out in any direction from a point source is equal to the ray velocity in that direction for a plane wave. It is a plausible assumption, but it is an assumption.

CHAPTER XX

ABSORPTION, EMISSION, DISPERSION, AND SCATTERING

WHEN a beam of light is propagated in a material medium its velocity is less than its velocity in a vacuum and its intensity gradually decreases as it progresses through the medium. The velocity of light in a material medium varies with the wavelength, and this variation is referred to as *dispersion*. The reduction in the intensity of a beam traversing a medium is partly due to the fact that some light is *scattered*, and partly due to *absorption*. In true absorption the light disappears, the energy being converted into heat. The phenomena of scattering, absorption, and dispersion, are inseparably connected as will be seen from the discussions which follow.

20–1. General and Selective Absorption

All media show some absorption. Those which are referred to as being transparent simply have very small absorptions. Some media absorb all wavelengths more or less equally; they are said to show *general absorption*. Others show *selective absorption*—they absorb some wavelengths very much more strongly than others. Lamp black shows very general absorption, whereas dyes and pigments are obvious examples of materials showing selective absorption. Metals show very strong general absorption although many also show selective absorption; the associated selective reflection accounts for their colour. For example, a thick film of gold appears opaque, but a thin film appears blue-green by transmission, showing that although all wavelengths are strongly absorbed blue-green light is absorbed rather less strongly than the rest. Again, it is found that thin films of silver transmit wavelengths in the region of 3160 angstroms. Glasses and other optical materials which are transparent throughout the visible region show very strong absorption in the infra-red and ultra-violet regions. For example, crown glass is transparent from $\lambda 0{\cdot}35\mu$ to $\lambda 2\mu$, quartz from $0{\cdot}19\mu$ to 4μ (and again beyond 40μ except for an absorption band at 78μ), fluorite from $0{\cdot}125\mu$ to $9{\cdot}5\mu$, and rock salt from $0{\cdot}175\mu$ to 17μ. Air absorbs wavelengths below $0{\cdot}185\mu$ so that spectroscopes intended for use in this region must be evacuated.

It is fairly obvious how selective absorption influences the colour of a substance viewed by transmission. The colour of an opaque body depends upon selective absorption in a less obvious manner. For most coloured objects such as paints or flowers, the light penetrates a short distance below the surface and is reflected, refracted, diffracted, and scattered, within the body before emerging minus the selectively absorbed wavelengths. Such a body is said to have *pigment* or *body colour*; it appears to have the same colour by transmission if a sufficiently thin layer is employed. If the surface of such a body is highly polished, the specularly reflected light is almost white. If, however, the selective absorption of a substance is extremely strong, it shows true *selective reflection* and the colour of the specularly reflected light is found to be complementary to the colour of a thin film viewed by transmission. The wavelengths that are selectively absorbed are also selectively reflected to give the *surface colour*. Thus gold shows surface colour in addition to the high absorption and consequent high reflectivity for all wavelengths which characterizes metals. Powerful aniline dyes illustrate the way in which extremely strong selective absorption is associated with selective reflection. Thus a dye which absorbs green light appears purple by transmission when in solution, and green by reflection when in the solid state.

The *absorption spectrum* of a substance can be observed by passing white light through it and examining the transmitted light with a spectroscope. In general, light will be removed from the transmitted beam by scattering as well as absorption, but the scattering is reduced if the substance is in a homogeneous state such as a single crystal, a liquid, or an amorphous solid. Dielectrics show pronounced

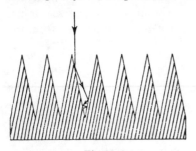

Fig. 20-1.
Penetration of light into a surface.

selective absorption. For solids and liquids this usually takes the form of *absorption bands* spreading over continuous bands of wavelengths. Speaking generally, it may be said that most insulators are more or less transparent to γ-rays and X-rays; show very strong absorption in the far ultra-violet (sometimes spreading up to or through the visible region); become transparent in or near the visible region; show strong absorption bands in the infra-red; and become almost completely transparent in the region of radio waves.

The absorption spectra of all gases at ordinary pressures show narrow absorption lines. Sometimes regions of continuous absorption can also be observed. The lines in the absorption spectra correspond to lines in the emission spectra, and if the gas consists of diatomic or polyatomic molecules, the sharp lines form the structure of the bands characteristic of molecular spectra. The absorption of an insulator in an absorption band can be comparable *in magnitude* to the absorption of a metal. It is completely different in origin.

Under certain circumstances, a body composed of a highly absorbing and therefore highly reflecting material such as a metal is almost completely non-reflecting. For example, chemically deposited silver appears black. The effect is due to the surface structure of the body. If the surface has the form represented by Fig. 20-1, the light is multiply reflected down the interstices between the needle-like structures and gradually becomes absorbed. A surface of platinum black has a spongy texture and light penetrates into the surface in this way and becomes almost completely absorbed. Near grazing incidence the light will not usually enter the interstices to any great extent. For example, a smoked glass plate has high specular reflectivity near grazing incidence.

20-2. Lambert's Law

Lambert's law states that equal paths in the same absorbing medium absorb equal fractions of the light that enters them. Suppose, then, that in traversing a path of length dx in a medium the intensity is reduced from I to $I-dI$. Then Lambert's law states that dI/I is the same for all elementary paths of length dx. One can put

$$\frac{dI}{I} = -K \cdot dx,$$

where K is a constant and is known as the *absorption coefficient*.

$$\therefore \quad \log I = -Kx + C,$$

where C is a constant. If $I = I_0$ at $x = 0$, $C = \log I_0$ and one has

$$I = I_0 e^{-Kx}. \quad \cdots \quad \cdots \quad (20\text{--}1)$$

This expresses Lambert's law.

It is sometimes more convenient to specify the absorption of a medium by means of the *extinction coefficient* $\varkappa$. This is defined by saying that if a beam of light traverses a path whose length is equal to the wavelength of the light in the medium (λ_m), the intensity is multiplied by $e^{-4\pi\varkappa}$. Thus after traversing a path x one has

$$I = I_0 e^{-(4\pi/\lambda_m)\varkappa x}. \quad \cdots \quad \cdots \quad (20\text{--}2)$$

On comparing (20–2) and (20–1) it will be seen that $\varkappa = K\lambda_m/4\pi$. In terms of the amplitudes one has

$$A = A_0 e^{-(2\pi/\lambda_m)\varkappa x}. \quad \cdots \quad \cdots \quad (20\text{--}3)$$

Since for a given frequency the wavelength is different in different media, the factor $e^{-4\pi\varkappa}$ corresponds to different path lengths in different media. This difficulty can be overcome by referring to the absorption in a path whose length is equal to the wavelength of the light in a vacuum. The factor is then $e^{-4\pi n\varkappa x}$, and in terms of amplitudes one has

$$A = A_0 e^{-(2\pi/\lambda)n\varkappa x},$$

λ being the wavelength in a vacuum. $n\varkappa$ is sometimes called the *absorption index*.*

If one has a solution of an absorbing substance, the absorption coefficient of the solution will depend upon the concentration. Beer's law states that it is directly proportional to the concentration. This will be true only if the absorbing power of a molecule is not influenced by the proximity of its neighbours. Thus, whilst Lambert's law always holds, Beer's law sometimes doesn't.

When attempting to determine the absorption coefficient, it must be remembered that in (20–1) I_0 is the intensity of the light that enters the body; since some light is reflected at the first surface, it is not the intensity of the incident light. Again, owing to reflection at the second surface, I is nôt the emergent intensity for a body of total thickness x. The losses due to reflection can be eliminated by measuring the emergent intensity for two different thicknesses. Let I_0 be the incident intensity and I_1 and I_2 the emergent intensities for path lengths x_1 and x_2 respectively. Let A_1 and A_2 be the corresponding absorbed fractions of the light that enters the medium, and let R be the fraction of the incident energy that

---

* There seems to be no general agreement regarding the names to be given to the various constants that can be used to specify the absorption of a medium.

is reflected at each surface. The intensity of the light actually entering the medium is $I_0(1-R)$, the intensity incident on the second surface after traversing a path of length x_1 is $I_0(1-R)(1-A_1)$, and the emergent intensity is $I_1=I_0(1-R)^2(1-A_1)$. Similarly, the emergent intensity for a thickness x_2 is $I_2=I_0(1-R)^2(1-A_2)$. Now $A_1=1-e^{-Kx_1}$, and $A_2=1-e^{-Kx_2}$,

$$\therefore \quad \frac{I_2}{I_1}=e^{-K(x_2-x_1)} \quad \quad \ldots \ldots \ldots \quad (20\text{-}4)$$

and K can be found by measuring I_1 and I_2.

20–3. Dichromatism

Suppose one has a medium which very strongly absorbs the yellow and the blue regions of the spectrum, weakly absorbs the green region, and is almost transparent to red light. A fairly small thickness of the medium will be sufficient to absorb the yellow and blue regions of the spectrum, and when viewed by transmission with incident white light, the emergent red and green light cause it to appear yellow (see § 17–9). If a thick layer of the medium is viewed by transmission, the green light will be absent and it will appear red. Thus the colour of the emergent light changes as the thickness is increased. This phenomenon is referred to as *dichromatism*. Certain mixtures of dyes show the effect.

20–4. The Scattering of Light by Small Particles

The scattering of light by a particle that is large compared with the wavelength simply consists of a mixture of diffraction and diffuse reflection and refraction. The scattering from a cloud of such particles is more or less independent of the wavelength unless the particles themselves are large and selectively absorbing. The path of a beam of sunlight entering a room is rendered visible by the white light scattered by the dust particles in the air. A far more interesting phenomenon is the scattering of light from particles that are small compared with the wavelength. Tobacco smoke consists of a cloud of such particles and its characteristic blue colour is caused by the dependence of the scattering upon the wavelength. The scattering of light by very small particles was investigated experimentally by Tyndall and is sometimes referred to as the *Tyndall effect*. The theory was studied by Rayleigh, so that it is frequently referred to as *Rayleigh scattering*. It is found that the scattered intensity is proportional to the square of the volume of the particle and proportional to $1/\lambda^4$. Thus the scattering for light at the blue end of the spectrum is about ten times as great as that for light at the red end. This explains the blue appearance of tobacco smoke. As will be seen below, it also explains the blue colour of the sky.

It will be seen from above that as the particle size is reduced the intensity of the scattered light decreases rapidly. For particles of atomic and molecular dimensions the scattering is very small, but it is still observable. Thus if a strong beam of white light is passed through a pure liquid which is completely free from suspended particles, it is possible to observe the bluish light which is scattered laterally by the molecules themselves. The effect is less marked with a gas since there are fewer molecules per c.c. However, Rayleigh has shown that almost all the light seen in a clear sky is due to this *molecular scattering* by the air molecules. Here one has a very large number of particles so that the total intensity of the scattered light is easily observable. If it were not for the scattering of the molecules of air, a cloudless sky would appear completely black except in the direction of the sun.

The red colour of the sun itself at sunset can be explained in a similar manner. Fig. 20–2 shows that in this case the direct sunlight traverses an extra long path through the lowest and densest parts of the atmosphere. Since more blue than red light is scattered out of the direct sunlight, the sun appears red. The effect is enhanced by the presence of dust or mist particles. In a fine mist the same effect explains why one can obtain clearer photographs of distant objects by using light of longer wavelength, infra-red photographs being clearer than those taken with

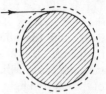

Fig. 20–2. Sunlight at grazing incidence.

visible light. In a thick fog the particles are usually not small compared with the wavelength and there is little advantage in using red or infra-red light.

20–5. The Quantum Theory of Absorption, Emission, and Scattering. The Laser

Absorption and molecular scattering are processes which are concerned with the interaction between light and matter, and it is only in terms of the quantum theory that one can give an adequate account of them. The present book is chiefly concerned with classical theory, and only a brief description of the phenomena can be given.

When discussing the interchange of energy between light and matter, it is found to be necessary to assume that the light consists of discrete packets of energy. These are referred to as *photons* and, if v is the frequency of the light in terms of the wave theory, each photon has energy hv where h is *Planck's constant*. Thus in some respects light behaves as a wave motion and in others it behaves as if it consisted of particles. These wave and particle aspects are not now associated with rival theories of light, but are complementary. Modern quantum mechanics includes a complete theory of light and shows when and to what extent light behaves as particles, and when and to what extent it behaves as waves. The theory is beyond the scope of the present book.

Rayleigh scattering results from an *elastic collision* between a photon and an atom or molecule. Here the photon is scattered without loss or gain in energy, so that there is no change of frequency. This is sometimes referred to as *coherent scattering* since the scattered light is coherent with the incident light.

Absorption also results from a collision between a photon and an atom. If the energy of the incident photon is exactly equal to an excitation energy of the atom, the photon is absorbed. The fact that the photon must have exactly the right energy accounts for the narrow absorption lines observed for gases at low pressures. For gases at high pressures, and for liquids and solids, the relative closeness of the molecules causes the excitation energies of different molecules to be slightly different. Hence the medium can absorb photons over a finite range of energies (i.e. over a finite range of frequencies). This accounts for the broader absorption lines. A number of close lines can merge to give a broad absorption band. For media of this type there is considerable interaction between molecules, and the energy of the excited molecules is quickly dissipated.

This gives true absorption. For gases at low pressures intermolecular collisions are relatively rare and the excited atom is undisturbed for long enough to be able to reradiate and return to its unexcited state. In doing so, the frequency of the emitted radiation is exactly the same as that of the light originally absorbed. This is referred to as *resonance radiation* for a reason which becomes obvious in the next section. The light is not simply radiated in the direction of the incident light, and the effect is of pronounced scattering of light of that particular frequency. It is also possible for an excited atom to return to its unexcited state via some intermediate state. In this case the energy (and therefore the frequency) of the emitted light must be lower than that of the light originally absorbed. This is known as *Stokes' Law* and the phenomenon is referred to as fluorescence; it is exhibited by solids, liquids, and gases.

In the phenomena described above, the incident photon is completely absorbed. Another possibility is that the incident photon has more than enough energy to excite the atom. In this case it is possible for the excess energy to be carried away as a photon of smaller energy (lower frequency). It is also possible for a photon to collide with an excited atom which falls to a lower state and gives some energy to the photon so that the frequency is increased. It will be seen that in this *inelastic collision* the frequency of the scattered light can be either lower than, or higher than, that of the incident light. This is known as the *Raman Effect*. It is sometimes referred to as *incoherent scattering* since the scattered light is not coherent with the incident light. Since the probability of this type of collision occurring is rather small, the effect is best observed in gases at high pressure, or in liquids or solids; it has become much easier to observe with the advent of powerful laser light sources.

In the ordinary absorption process a photon raises an atom from a lower to a higher state, and in spontaneous emission the photon is emitted as the atom falls from the higher to the lower state. If, however, an identical photon strikes the atom while it is in the higher state, it can trigger the transition so that a second photon is emitted. It will be seen that as a result of this *stimulated emission* process the total light intensity will increase if there is what is known as a *population inversion* where there are more atoms in the higher state than there are in the lower—there can be said to be negative absorption. This is the principle of the *laser* (*l*ight *a*mplification by the *s*timulated *e*mission of *r*adiation).

The population inversion can be achieved as in the ruby laser by "optical pumping", where light from a separate source raises the atoms concerned to the higher energy state to await the arrival of the stimulating photon. Alternatively, the inversion can be achieved as a result of inelastic collisions between the atoms in, for example, a mixture of gases: in the helium-neon gas laser helium atoms (which are excited to a metastable state in an ordinary gas discharge tube arrangement) collide with, and excite, neon atoms. In practice the medium in which the stimulated emission occurs is contained in a Fabry-Perot etalon so that light of the appropriate wavelength is reflected to and fro. If the amplification due to stimulated emission is sufficient to balance the loss of light from the system, one has the positive feed-back characteristic of an oscillator—a laser source is, in fact, an optical oscillator. If the gain is sufficient, the emission can be maintained while some of the light emerges through one of the etalon mirrors.

The crucial factor is that in stimulated emission the light emitted is coherent and in phase with the incident light, and since the losses from the etalon are smallest for light travelling parallel to the axis, the light emerging from a laser is predominantly in the form of a highly coherent, highly collimated, beam. Since the amplification is larger if the light path is longer, lasers tend to use long etalons; plane mirrors would then be difficult to adjust for parallelism. In consequence, lasers often use confocal spherical etalons (§9–9) or etalons consisting of one spherical and one plane mirror; each of these is easier to align. It is usual for the gas tube to have end windows inclined at the polarizing (or Brewster) angle (see §6–17). This gives maximum transmission for light with **D** in the plane of incidence and, in fact, it is only this component whose intensity builds up strongly; in consequence, the ultimate output is plane polarized.

Einstein (1917) first drew attention to the existence of the phenomenon of stimulated emission but it was not until 1958 that Schawlow and Townes showed that an optical laser source was feasible. The first operative laser was the ruby laser produced in 1960 by Maiman, and this was followed in 1961 by the helium-neon gas laser operated by Javan, Bennett, and Herriott. The ruby laser was a pulsed laser which was pumped by an electronic flash tube, and produced a high-power pulse of short duration; the gas laser operated continuously at very much lower power. More recently, Patel (1964) developed the carbon dioxide laser, which is capable of operating continuously with much higher power than the helium-neon laser. Another development, which could have far-reaching consequences in absorption spectroscopy, is the continuously tunable *dye laser*. Sorokin and Lankard (1966) discovered that one could obtain laser emission from luminescent organic molecules. Certain dyes are characterised by a broad fluorescence spectrum and Soffer and McFarland (1967) achieved wavelength selection to give a continuously tunable laser by replacing one of the etalon mirrors with a plane diffraction grating in a Littrow mount; the wavelength that satisfies the normal Littrow condition is the wavelength for which one effectively has an etalon which gives the required positive feedback.

Lasers are capable of producing extremely high intensities, giving a wide range of new phenomena associated with the interaction between light and matter and between different photons. The theory of lasers and this associated field of *non-linear optics* is a subject in itself and is far outside the scope of the present book, which deals with the classical theory of the propagation of light.

20–6. The Electromagnetic Theory of Absorption and Scattering

Although a completely satisfactory theory of absorption and scattering can only be given in terms of the processes outlined in the previous section, the simpler classical picture is adequate for many purposes. It is assumed that the molecules in a dielectric may be represented as *dipoles* or *bound charges*, i.e. charges that are bound to their centres of equilibrium by an elastic force. For an uncharged medium there must be equal numbers of positive and negative charges. When an electromagnetic wave impinges on a bound charge it is caused to oscillate and therefore to radiate. If the frequency of the wave is not equal to the natural frequency of the bound charge, one has forced oscillation of small amplitude, and the radiation from the charge is therefore weak and of the same frequency as that of the wave. This corresponds to molecular scattering. If the frequency of the wave is equal to the natural frequency of the bound charge, one has resonance, and a much larger amount of energy from the wave goes into the vibrating charge. In a solid, liquid, or gas at high pressure, there is strong inter-molecular action and friction type forces cause heavy damping of the dipole with the result that its energy is quickly dissipated. This corresponds to true absorption. In a gas at low pressure the motion of the charge or dipole is relatively undamped and it radiates strongly. This is resonance radiation.

It is possible to extend this theory to give a qualitative explanation of fluorescence and the Raman effect by assuming that the charges are bound by forces which do not obey Hooke's law.

The absorption of an electromagnetic wave by a conducting medium is easily explained by assuming that a conductor has a large number of *free electrons*. When the wave arrives, its energy goes into causing the charges to move. The moving charges constitute currents and the usual dissipation of energy by these currents explains the absorption of the radiant energy. It will be seen that although the absorption in an absorption band of an insulator may be comparable in magnitude with the absorption of a metal, the two processes are rather different.

The electromagnetic theory gives a simple explanation of the distribution of the scattered light in different directions. Suppose, for example, that plane polarized light with the electric vector parallel to the y-axis is travelling along the x-axis and is incident on a small particle at the origin O (Fig. 20–3). Since the incident light has no component of E perpendicular to the y-axis, there can be no scattered light along this axis. Light can obviously be scattered in all directions in the xz plane and will be plane polarized with E parallel to that of the incident light. For light scattered in the direction OP the incident light has component $E \cos \theta$ perpendicular to the direction of propagation. Thus the intensity of the light scattered in a direction making angle θ with the xz plane is proportional to $\cos^2 \theta$. If the incident light is unpolarized, it has equal com-

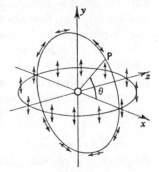

Fig. 20–3.
Polarization of scattered light.

ponents of E in all directions perpendicular to the x-axis. The light scattered along the x-axis is unpolarized, and that scattered in any direction in the yz plane is plane polarized with E transverse and in that plane.

In some cases of the molecular scattering of polarized incident light, the scattered light is found to be only partially polarized. This depolarization has been explained by assuming that the molecules are anisotropic, so that the moment of the induced dipole is not in the direction of the electric vector of the incident light.

20-7. Normal Dispersion. Cauchy's Formula

It is not surprising that the earliest measurements of dispersion were for visible light and that the materials whose dispersions were investigated first were those which are transparent throughout the visible region. Thus the earliest measurements of dispersion were for a range of wavelengths in or near which the medium has no absorption band. It was found that, although the absolute value of the refractive index is different for different substances, its variation with wavelength is always of approximately the same form. Fig. 20–4 shows the dispersion curves in the visible region for a number of colourless transparent substances. It will be seen that in each case both n and $dn/d\lambda$ decrease in magnitude as λ increases. It should be noted, however, that the curves are *not exactly* the same shape. Although there are many exceptions, it is useful to note that, in general, denser materials have higher refractive indices than less dense media.

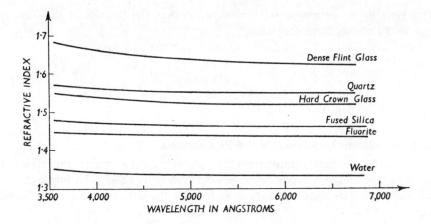

Fig. 20–4. Normal dispersion.

Dispersion of the form described above was later referred to as *normal dispersion*, to distinguish it from the *anomalous dispersion* which was observed in the region of an absorption band. It was found that the refractive index increases with increasing wavelength as one passes through an absorption band. It must be pointed out at once that there is nothing abnormal about anomalous dispersion. All substances exhibit normal dispersion for wavelengths that are far removed from absorption bands, and anomalous dispersion in the region of a band.

It was shown by Cauchy in 1836 that in a region of normal dispersion the variation of refractive index with wavelength for a given medium can be represented by an expression of the form

$$n = A + \frac{B}{\lambda^2} + \frac{C}{\lambda^4}, \quad \cdots \cdots \quad (20\text{–}5)$$

where A, B, and C, are constants characteristic of the medium concerned, and λ is usually the wavelength in a vacuum. The constants can be determined experimentally by measuring n for three values of λ. For many purposes it is sufficiently accurate to include the first two terms only. Then

$$n = A + \frac{B}{\lambda^2} \qquad \cdots \qquad \cdots \qquad (20\text{-}6)$$

and

$$\frac{dn}{d\lambda} = -\frac{2B}{\lambda^3}. \qquad \cdots \qquad \cdots \qquad (20\text{-}7)$$

Since A and B are positive, these equations express the fact that n decreases as λ increases, and that the slope of the dispersion curve is negative, its magnitude decreasing as λ increases.

The following alternative expression for the dispersion of glasses has been given by Hartmann:

$$n = n_0 + \frac{A}{(\lambda - \lambda_0)^{1.2}}. \qquad \cdots \qquad \cdots \qquad (20\text{-}8)$$

This gives better agreement with observation than the formula of the Cauchy type that also has three constants.

For a gas, Gladstone and Dale showed experimentally that

$$\frac{n-1}{\varrho} = \text{constant}, \qquad \cdots \qquad \cdots \qquad (20\text{-}9)$$

where ϱ is the density of the gas.

20-8. Anomalous Dispersion. Sellmeier's Formula

For a substance that is transparent throughout the visible region, the refractive index in that region is represented very accurately by (20-5). If, however, the refractive index measurements are extended into the infra-red region, it is found

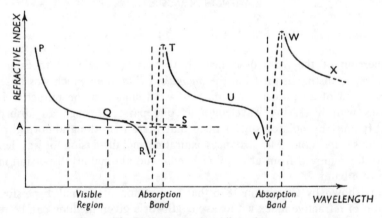

Fig. 20-5. Anomalous dispersion.

that, as one approaches wavelengths which the medium absorbs, the refractive index begins to decrease much more rapidly than is indicated by Cauchy's formula. In Fig. 20-5 the curve PQR represents the dispersion curve actually obtained and the broken line QS is that indicated by Cauchy's formula. Measurements beyond R become difficult owing to the absorption of the medium, but for wavelengths beyond the absorption band measurements can again be made quite easily; they yield a curve of the form TU. If there is a second absorption band, the curve continues along UV and, beyond this band, along WX. A more complete investigation shows that the curves PQR, TUV, and WX, are connected up as indicated to form a continuous curve. Since the light is absorbed very strongly, measurements within a band must be made with very thin layers of the substance. It will be seen that as one passes through an absorption band n increases as λ increases. As already stated, this is referred to as anomalous dispersion.

In order to obtain a satisfactory dispersion formula, it is necessary to consider the mechanism by which the light is absorbed by the medium. Sellmeier (1871) assumed that the medium contains elastically bound particles which have a natural frequency of vibration, and that if light of this frequency is passed through the medium, the particles resonate and energy is absorbed. For light of other frequencies the particles execute forced oscillations, the amplitudes increasing as the frequency of the light approaches the resonant frequency.

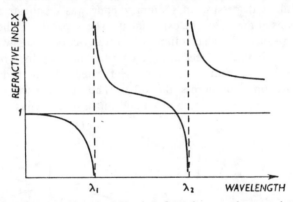

Fig. 20–6. Dispersion curve given by Sellmeier's formula, assuming two absorption bands.

These vibrations cause a change in the velocity of the light in a manner which will be discussed briefly later (see § 20–14). If all the elastically bound particles in the medium have a natural frequency corresponding to light of wavelength λ_0 in a vacuum, Sellmeier's theory gives

$$n^2 = 1 + \frac{A\lambda^2}{\lambda^2 - \lambda_0^2}, \qquad \cdots \cdots \quad (20\text{--}10)$$

where A is a constant. This would correspond to a medium having only one absorption band. The existence of several bands corresponds to the existence of particles with several different natural frequencies. This gives

$$n^2 = 1 + \sum \frac{A_j\lambda^2}{\lambda^2 - \lambda_j^2}, \qquad \cdots \cdots \quad (20\text{--}11)$$

where A_j is proportional to the number of particles whose natural frequency corresponds to λ_j. Fig. 20–6 shows the form of the dispersion curve given by Sellmeier's formula, assuming there are two absorption bands.

Although Sellmeier's equation represents a considerable improvement over Cauchy's formula, it has an obvious important defect: it indicates that n is infinite and discontinuous at λ_1, λ_2, etc. Thus Sellmeier's formula breaks down in the region of an absorption band. In regions remote from an absorption band Sellmeier's formula gives better agreement with observation than a formula of the Cauchy type having the same number of constants. It will be seen below that Sellmeier's equation can be derived from the more exact results obtained from electromagnetic theory. It is easily shown that Sellmeier's equation reduces to Cauchy's, when certain approximations are made. If there is one absorption band and this is a considerable distance to the short wavelength side of the region for which the dispersion is required, one can write $\lambda^2 \gg \lambda_0^2$ in (20–10) and obtain

$$n^2 = 1 + A \left(1 - \frac{\lambda_0^2}{\lambda^2} \right)^{-1} = 1 + A + \frac{A\lambda_0^2}{\lambda^2} + \frac{A\lambda_0^4}{\lambda^4},$$

which is of the same form as Cauchy's equation.

20–9. Experimental Demonstration of Anomalous Dispersion

The anomalous dispersion of sodium vapour was demonstrated by R. W. Wood by means of the following experiment. The apparatus is shown schematically in Fig. 20–7(a). T is a steel tube with a water-cooled glass window at each end. Pieces of sodium are placed along the bottom of this tube, which is partially evacuated and heated from the bottom as indicated. As the sodium is heated, vapour diffuses upwards across the tube. Thus the density of the vapour gradually increases from top to bottom and this is equivalent to a prism of the

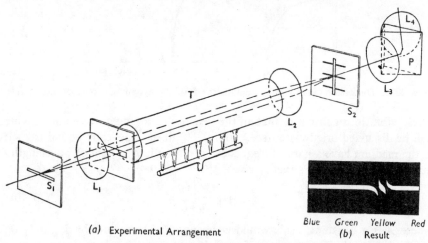

(a) Experimental Arrangement

Blue   Green   Yellow   Red
(b) Result

Fig. 20–7. The anomalous dispersion of sodium vapour.

vapour whose thickness increases downwards, the refracting edge being perpendicular to the axis of the tube. White light from the horizontal slit S_1 is collimated by the lens L_1 and is brought to a focus in the plane of the vertical slit S_2 by means of the lens L_2. S_2, L_3, P, and L_4, represent a simple prism spectroscope, the refracting edge of P being vertical. When the tube is cold it contains a homogeneous gas, and a white image of S_1 is produced across the spectroscope slit S_2, giving a continuous spectrum along a horizontal strip of the focal plane of the spectroscope. When the sodium is vaporized, the prismatic action of the tube of vapour causes the images of S_1 to be dispersed along the length of

S_2. For wavelengths remote from the D lines the refractive index of the vapour is approximately unity, and the horizontal continuous spectrum in the spectroscope is unaffected. As the D lines are approached from the short wavelength side, the refractive index of the vapour becomes less than one, and the light is deviated upwards by the vapour. The inversion introduced by the spectroscope lenses causes a downward displacement of the narrow continuous spectrum. Thus displacements upwards and downwards of the spectrum correspond to increases and decreases of the refractive index of the vapour, and the narrow spectrum gives a qualitative representation of a dispersion curve (such as Fig. 20–5) for the two neighbouring absorption lines D_1 and D_2 at 5896 and 5890 angstroms. The result is represented by Fig. 20–7(b). If the density of the vapour in the tube is high, the absorption lines broaden and it is not possible to distinguish the part of the curve that lies between them. A diaphragm immediately after lens L_1 is used to select a section of the tube across which the density gradient is uniform.

20–10. Propagation of Light in an Absorbing Medium

It follows from §19–4 that, for a plane polarized plane wave travelling with velocity v in the positive x-direction in a non-absorbing medium, one can write

$$E_y = E_{v0}e^{ik(vt-x)} = E_{v0}e^{i\omega(t-\frac{x}{v})} = E_{v0}e^{i\omega(t-\frac{n}{c}x)}, \quad . \quad . \quad (20\text{–}12)$$

where n is the refractive index of the medium. Suppose now that E_{v0} is the amplitude at $x=0$ in an absorbing medium whose extinction coefficient is $\varkappa$. Then it follows from (20–3) that the disturbance at x is represented by

$$E_y = E_{v0}e^{-k\varkappa x}e^{i\omega\left(t-\frac{n}{c}x\right)}$$

$$= E_{v0}e^{i\omega\left(t-\frac{n(1-i\varkappa)}{c}x\right)},$$

since $k=\omega/v=n\omega/c$. This can be written

$$E_y = E_{v0}e^{i\omega\left(t-\frac{\mathbf{n}}{c}x\right)}, \quad . \quad . \quad . \quad . \quad (20\text{–}13)$$

where

$$\mathbf{n} = n(1-i\varkappa). \quad . \quad . \quad . \quad . \quad (20\text{–}14)$$

Owing to the formal similarity between (20–12) and (20–13), $\mathbf{n}$ is often called the *complex index*. This is not entirely satisfactory because the real part of $\mathbf{n}$ is only indirectly related to the angles of refraction with absorbing media.

It will be seen in the next section that when an electromagnetic wave passes through an absorbing dielectric the ratio of the polarization to the electric intensity is complex [see equation (20–19)]. This means that the dielectric constant is complex. It is easily shown that a complex dielectric constant implies absorption and that $\sqrt{K}$ becomes $\mathbf{n}$ instead of n. Thus the components of the field vectors satisfy equations of the form of (19–16) where K is complex. For example, for a plane polarized plane wave travelling in the positive x-direction one has, setting $\mu=1$,

$$E_y = E_{v0}e^{i(\omega t - k'x)},$$

where

$$k' = (\omega/c)\sqrt{K} = (\omega/c)(\alpha - i\beta) \text{ (say)}.$$

$$\therefore E_y = E_{v0}e^{-(k\beta/n)x}e^{i\omega\left(t-\frac{\alpha}{c}x\right)}.$$

The first term obviously represents absorption and it will be seen that $\beta = n\varkappa$ and $\alpha = n$. That is to say, $K = \mathbf{n}^2$.

20-11. The Theory of Dispersion in Dielectrics

The electromagnetic theory of Maxwell is based on the assumption that the light is propagated in a medium that is continuous and has a certain dielectric constant and permeability. It follows from that theory that the refractive index of a dielectric is also constant, so that the phenomenon of dispersion is excluded from the theory. It has already been pointed out that the dielectric constant is, in fact, not a constant but that it depends upon the frequency. This agrees with the well known fact that the refractive index depends upon the frequency.

Now it is known that matter consists of electrically charged particles, and it should be possible to explain the electrical properties of matter (such as the way in which electromagnetic waves are propagated) directly in terms of the electronic structure. That is to say, it ought not to be necessary to resort to empirical relations such as that implied by a constant or variable dielectric constant. The first attempt to derive the electrical properties of matter from the electron theory was made by H. A. Lorentz. In this theory it is impossible to take into account the actual contribution of each electron or dipole of which the medium is composed, but the average effect of a large number can be found. Thus in discussing the polarization **P**, one is concerned with its average value over a volume which is sufficiently large to contain many dipoles although the linear dimensions of the volume are considerably smaller than the wavelength of the light.

Let **E** be the electric vector of an incident light wave and let **P** be the resulting polarization. Then one has

$$\mathbf{D} = K\mathbf{E} = \mathbf{E} + 4\pi\mathbf{P}$$

and, as shown in §20–10, one has, in general,

$$4\pi \frac{\mathbf{P}}{\mathbf{E}} = K - 1 = \mathbf{n}^2 - 1. \qquad \ldots \ldots \quad (20\text{–}15)$$

It can be shown that the total electric intensity acting on a charge at any point within a polarized isotropic medium is $\mathbf{E} + (4\pi/3)\mathbf{P}$. (Actually this is the average value for charges contained within a small volume surrounding the point in question.) Suppose that the bound charges are electrons, each of which has mass m and charge e. If the medium is isotropic and the magnetic force on an electron associated with its own motion is neglected (there being no external magnetic field), the displacement of the electron x is along the same straight line as **E** and **P**. Suppose the electron is subject to an elastic restoring force which is proportional to the displacement from the equilibrium position, and that the free period is $2\pi/\omega_0$. In addition, there will be a damping force, and the equation of motion for an "average electron" is therefore of the form

$$\ddot{x} + r\dot{x} + \omega_0^2 x = \frac{1}{m}\left(E + \frac{4\pi}{3}P\right)e, \qquad \ldots \quad (20\text{–}16)$$

where r is the resisting force per unit mass per unit velocity, the right-hand side being the applied force per unit mass.

Now the polarization is equal to the dipole moment per unit volume, so that if there are N electrons per unit volume one has

$$P = Nex. \qquad \ldots \ldots \ldots \quad (20\text{–}17)$$

Using (20–16), this gives

$$\ddot{P}+r\dot{P}+\omega_0{}^2P=\frac{Ne^2}{m}\left(E+\frac{4\pi}{3}P\right). \quad \ldots \quad (20\text{–}18)$$

It is now assumed that P has the same frequency as E but that there may be a phase difference between them. Thus if

$$E=E_0e^{i\omega t}$$

one has

$$P=zE_0e^{i\omega t},$$

where z is a constant which is, in general, complex. Hence from (20–18)

$$z(-\omega^2+ir\omega+\omega_0{}^2)=\frac{Ne^2}{m}\left(1+\frac{4\pi}{3}z\right).$$

Thus from (20–15) one has

$$n^2-1=4\pi\frac{P}{E}=4\pi z=\frac{4\pi Ne^2}{m}\cdot\frac{1}{\omega_0{}^2-\omega^2-\dfrac{4\pi}{3}\cdot\dfrac{Ne^2}{m}+ir\omega}, \quad (20\text{–}19)$$

or

$$\frac{n^2-1}{n^2+2}=\frac{4\pi z}{4\pi z+3}=\frac{4\pi Ne^2}{3m}\cdot\frac{1}{\omega_0{}^2-\omega^2+ir\omega}. \quad \ldots \quad (20\text{–}20)$$

Now in deriving these results it was assumed that the medium contained a distribution of oscillators all of which had the same natural frequency. This would correspond to the existence of only one absorption line. To generalize the result, suppose that there are N molecules per unit volume and that there are f_1 electrons of type 1, f_2 of type 2, etc. Then (20–17) becomes

$$P=Ne\sum_j f_j x_j, \quad \ldots \quad \ldots \quad (20\text{–}21)$$

and (20–20) becomes

$$\frac{n^2-1}{n^2+2}=\frac{4\pi}{3}\cdot\frac{Ne^2}{m}\sum_j\frac{f_j}{\omega_j{}^2-\omega^2+ir_j\omega}, \quad \ldots \quad (20\text{–}22)$$

and (20–19) can be written

$$n^2-1=4\pi\frac{Ne^2}{m}\sum_j\frac{f_j}{\omega_j'^2-\omega^2+ir_j\omega}, \quad \ldots \quad (20\text{–}23)$$

where

$$\omega_j'^2=\omega_j{}^2-\frac{4\pi}{3}\cdot\frac{Ne^2}{m}f_j. \quad \ldots \quad \ldots \quad (20\text{–}24)$$

For stationary or slowly varying fields one can put $\omega=0$ so that (20–23) gives

$$n^2-1=K-1=4\pi\frac{Ne^2}{m}\sum_j\frac{f_j}{\omega_j'^2}. \quad \ldots \quad (20\text{–}25)$$

As introduced above, the f's are integers, but it is found that in order to obtain agreement between the observed and calculated values of the dielectric constant, non-integral values must be substituted into (20–25). It is then supposed that each molecule contains only one oscillator of each type, the f's being the strengths of the various oscillators. In the classical theory the ω_j's correspond to the natural frequencies of oscillators that are distributed throughout the medium. In quantum mechanics these oscillators are associated with the transitions that an atom can undergo. With each transition there is associated an oscillator strength defined as the number of classical oscillators per atom which would absorb the same amount of radiation from a parallel beam of light. Although it involves a different interpretation of the quantities that occur, the quantum theory of dispersion yields results which are similar to those given by classical theory. A detailed discussion of the quantum theory is beyond the scope of the present book.

20-12. Regions Remote from Absorption Bands

If $\omega_j'^2 - \omega^2 \gg r\omega$, the absorption may be neglected and n is real. (20-23) gives

$$n^2 - 1 = \frac{4\pi N e^2}{m} \sum_j \frac{f_j}{\omega_j'^2 - \omega^2}. \qquad \ldots \quad (20\text{-}26)$$

Now the wavelength in a vacuum is given by $\lambda = 2\pi c/\omega$, and if $\lambda_j' = 2\pi c/\omega'$ one has

$$n^2 = 1 + \frac{Ne^2}{\pi c^2 m} \sum_j \frac{f_j \lambda^2 \lambda_j'^2}{\lambda^2 - \lambda_j'^2} \qquad \ldots \ldots \quad (20\text{-}27)$$

or, putting

$$A_j = \frac{Ne^2}{\pi c^2 m} f_j \lambda_j'^2,$$

$$n^2 = 1 + \sum_j \frac{A_j \lambda^2}{\lambda^2 - \lambda_j^2}, \qquad \ldots \ldots \quad (20\text{-}28)$$

which is Sellmeier's formula.

Since the density ϱ of a substance is proportional to N, (20-22) shows that for light of a given frequency

$$\frac{n^2 - 1}{(n^2 + 2)\varrho} = \text{constant.} \qquad \ldots \ldots \quad (20\text{-}29)$$

This is known as the Lorenz-Lorentz law. It is confirmed experimentally for a wide range of substances, even through a change of state such as the transition from liquid to vapour. For low frequencies it follows that the dielectric constant is related to the density by the equation

$$\frac{K - 1}{(K + 2)\varrho} = \text{constant.} \qquad \ldots \ldots \quad (20\text{-}30)$$

Equation (20-29) gives

$$\frac{(n-1)}{\varrho} \cdot \frac{(n+1)}{(n^2 + 2)} = \text{constant.}$$

Now for a gas the refractive index does not differ widely from unity, so that the second factor is almost constant. Hence for a gas one has

$$\frac{n - 1}{\varrho} = \text{constant,} \qquad \ldots \ldots \quad (20\text{-}31)$$

which is the law of Gladstone and Dale.

The quantity $(n^2 - 1)/(n^2 + 2)\varrho$ has been termed the *specific refraction* and the product of this and the molecular weight M of a substance is known as the *molecular (or molar) refraction*. If N_0 is Avogadro's number one has

$$N = \frac{N_0 \varrho}{M}$$

so that (20–22) gives, for the molecular refraction,

$$[n] = \frac{(n^2-1)M}{(n^2+2)\varrho} = \frac{4\pi N_0 e^2}{3m} \sum_j \frac{f_j}{\omega_j{}^2 - \omega^2} . \qquad . \quad . \quad (20\text{–}32)$$

This suggests that in a mixture of substances which do not interact the molar refractions are additive, a result which is confirmed experimentally. When there is chemical reaction between two or more substances, it is sometimes possible to add the molar refractions plus some extra terms which take account of the electrons which are associated with the chemical bond.

20–13. Regions Close to Absorption Bands

For simplicity it is assumed initially that there is only one absorption band, so that (20–23) becomes

$$\mathbf{n}^2 = n^2(1-i\varkappa)^2 = 1 + \frac{4\pi N e^2}{m} \frac{f}{\omega_0{}'^2 - \omega^2 + ir\omega} . \quad . \quad . \quad (20\text{–}33)$$

On equating the real parts, one obtains

$$n^2 - n^2\varkappa^2 = 1 + \frac{4\pi N e^2}{m} \frac{(\omega_0{}'^2 - \omega^2)f}{(\omega_0{}'^2 - \omega^2)^2 + r^2\omega^2} , \quad . \quad . \quad (20\text{–}34)$$

and on equating the imaginary parts,

$$n^2\varkappa = \frac{2\pi N e^2}{m} \frac{r\omega f}{(\omega_0{}'^2 - \omega^2)^2 + r^2\omega^2} . \qquad . \quad . \quad (20\text{–}35)$$

If one has a weak absorption band, $\varkappa$ is small and $n^2\varkappa^2 \ll n^2$. Hence, putting $\omega_0' = 2\pi c/\lambda_0$ and $\omega = 2\pi c/\lambda$, where c and λ refer to the light in a vacuum, (20–34) gives

$$n^2 = 1 + \frac{4\pi N e^2}{m} \cdot \frac{\lambda_0{}^2\lambda^2(\lambda^2 - \lambda_0{}^2)f}{4\pi^2 c^2(\lambda^2 - \lambda_0{}^2)^2 + r^2\lambda_0{}^4\lambda^2} . \quad . \quad . \quad (20\text{–}36)$$

Now in the region of the absorption band $\lambda \doteqdot \lambda_0$ and one can write $\lambda_0{}^2\lambda^2 \doteqdot \lambda_0{}^4$, $\lambda_0{}^4\lambda^2 \doteqdot \lambda_0{}^6$, and $(\lambda + \lambda_0) \doteqdot 2\lambda_0$. Hence, since n does not differ widely from unity (20–36) is approximately of the form

$$n = 1 + \frac{A(\lambda - \lambda_0)}{B(\lambda - \lambda_0)^2 + C} . \qquad . \quad . \quad . \quad . \quad (20\text{–}37)$$

Also, since $n \doteqdot 1$ (20–35) gives, for the absorption index,

$$n\varkappa = \frac{N e^2}{cm} \frac{\lambda_0{}^4\lambda^3 r f}{4\pi^2 c^2(\lambda^2 - \lambda_0{}^2)^2 + r^2\lambda_0{}^4\lambda^2} , \qquad . \quad . \quad (20\text{–}38)$$

which is approximately of the form

$$n\varkappa = \frac{P}{Q(\lambda - \lambda_0)^2 + R} \qquad . \quad . \quad . \quad . \quad . \quad (20\text{–}39)$$

Fig. 20–8(a) shows the variation of n and $n\varkappa$ in the region of a single weak absorption band. This is typical of the results found experimentally for gases, the refractive index never being very different from unity. It can be shown that for a strong absorption band the variation of n takes the form shown in Fig. 20–8(b). It will be seen that n does not approach unity beyond the absorption band, so that if there is a second band at a longer wavelength, there will be a displacement of the curve of the form already shown in Fig. 20–5. The above figures show the anomalous dispersion (increase of n with λ) which is observed as one passes through an absorption band. In a complete theory it would be possible to calculate the values of the constants in the above equations from the atomic structure.

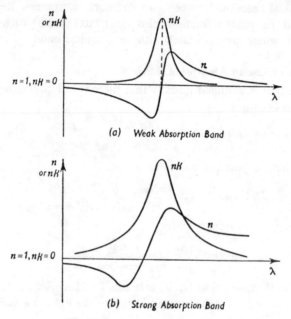

Fig. 20–8. Anomalous dispersion.

20–14. Connection Between Dispersion and Molecular Scattering

It has already been remarked that in molecular scattering the scattered light is coherent with the incident light. When the incident light impinges on a "molecular oscillator", the latter executes forced oscillations and sends out secondary waves in all directions. Referring to Fig. 20–9(a), it will be seen that

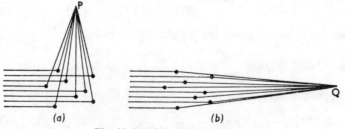

Fig. 20–9. Molecular scattering.

for a random distribution of scattering centres there will be a random variation of optical path for light scattered laterally from different oscillators to the point P. Hence the resultant intensity at P will be small. On the other hand, the light that is scattered more or less straight ahead arrives almost in phase at Q [Fig. 20–9(b)]. The disturbance at Q due to the scattered light can then interfere with the primary wave. The amplitude of the scattered disturbance is small and the resultant has an amplitude and phase which is not very different from that of the primary wave. Now if the phase of the resultant at Q is ahead of that of the primary wave, it means that the effect of the scattered disturbances is an increase in the phase velocity of the light—the disturbance at Q has a given phase earlier than it would have had in the absence of scattering. That is to say, the scattered light affects the refractive index. Now the phase and amplitude of the scattered light depends upon the frequency of the incident beam, and it can be shown that the effect of the scattered light on the refractive index is precisely that represented by the dispersion formulae given above.

20–15. The Propagation of Light in Metals

Initially, a metal is considered as a continuous medium with large conductivity, the dielectric constant being not very different from unity. Maxwell's equation (19–3) gives

$$\text{curl } \mathbf{H} = 4\pi \mathbf{j} + \frac{1}{c} \frac{\partial \mathbf{D}}{\partial t}.$$

If the medium is isotropic, $\mathbf{D} = K\mathbf{E}$, and if σ is the conductivity

$$\mathbf{j} = \frac{1}{c} \sigma \mathbf{E},$$

where $\mathbf{j}$ is in e.m.u. and σ and $\mathbf{E}$ in e.s.u. Hence Maxwell's equation becomes

$$\text{curl } \mathbf{H} = \left(\frac{4\pi}{c} \sigma + \frac{K}{c} \frac{\partial}{\partial t} \right) \mathbf{E}.$$

It has already been seen that for any vector $\mathbf{V}$ curl curl $\mathbf{V} = \text{grad (div } \mathbf{V}) - \nabla^2 \mathbf{V}$. Now one may put $\mu = 1$ for optical frequencies so that equation (19–4) becomes curl $\mathbf{E} = -(1/c)\partial \mathbf{H}/\partial t$, and if there is no volume charge (19–1) gives div $\mathbf{E} = 0$.

Hence
$$\nabla^2 \mathbf{E} = \frac{1}{c} \frac{\partial}{\partial t} \left(\frac{4\pi}{c} \sigma + \frac{K}{c} \frac{\partial}{\partial t} \right) \mathbf{E}$$

$$= \frac{K}{c^2} \frac{\partial^2 \mathbf{E}}{\partial t^2} + \frac{4\pi\sigma}{c^2} \frac{\partial \mathbf{E}}{\partial t}. \quad \ldots \ldots \quad (20\text{–}40)$$

This is satisfied by
$$E_y = E_{y0} e^{i\omega\left(t - \frac{\mathbf{n}}{c}x\right)}, \quad \ldots \ldots \quad (20\text{–}41)$$

where
$$\mathbf{n}^2 = K - i \frac{4\pi\sigma}{\omega}. \quad \ldots \ldots \quad (20\text{–}42)$$

(20–41) is of the same form as (20–13), which applies to any absorbing medium whose extinction coefficient is given by

$$\mathbf{n}^2 = n^2 (1 - i\varkappa)^2. \quad \ldots \ldots \quad (20\text{–}43)$$

Equating the real parts and the imaginary parts in (20–42) and (20–43), one obtains

$$n^2(1-\varkappa^2)=K \qquad \cdots \cdots \qquad (20\text{--}44)$$

and

$$n^2\varkappa=\frac{2\pi\sigma}{\omega}=\sigma T=\frac{\sigma\lambda}{c}, \qquad \cdots \cdots \qquad (20\text{--}45)$$

where T is the period of the light and λ its wavelength in a vacuum. On eliminating $\varkappa$ one has

$$n^2=\frac{K+\sqrt{K^2+4\sigma^2T^2}}{2}. \qquad \cdots \cdots \qquad (20\text{--}46)$$

Except for electromagnetic waves of very low frequency, the observed values of n and $\varkappa$ do not agree with the values calculated from these equations. The above theory is obviously concerned with the general absorption exhibited by metals, and does not deal with the selective absorption which they also exhibit. The fact that the experimental results agree fairly well with the above theory for long wavelengths indicates that the selectively absorbed frequencies are confined to the higher (e.g. optical) frequencies. As for dielectrics, it is necessary, for a more complete theory, to consider the structure of the medium.

20–16. The Dispersion of Metals

It is assumed that metals possess free electrons which when in motion give the conduction current, and that they also have bound charges as possessed by dielectrics. Suppose, then, that there are N atoms per unit volume and f_e free electrons per atom. When an electric field E is applied, each free electron is given an acceleration Ee/m. Hence a drift is superposed on the random motion of the free electrons which is assumed to exist in the absence of a field. If the free electrons move under the influence of the field for time t, they attain a velocity of Eet/m, but collisions between the free electrons and the atoms quickly destroy this drift and the collisions increase the random (thermal) motion. It is in this way that energy is dissipated. At any instant the average velocity of the electrons depends on the average time between collisions. If the average velocity is $Ee\tau/m$, τ is the *relaxation time* and the current is $(Nf_e e)(Ee\tau/m)$ $=Nf_e e^2\tau E/m$. Hence the conductivity is given by

$$\sigma=\frac{\text{current}}{\text{field}}=\frac{Nf_e e^2\tau}{m}. \qquad \cdots \cdots \qquad (20\text{--}47)$$

If the field E is due to an incident electromagnetic wave of period T, one would expect (20–47) to hold only if $T\gg\tau$.

The effect of the free electrons can be included in the dispersion formula simply by assuming that they correspond to electrons whose natural frequency is zero. This gives one extra term in equation (20–23). It is also assumed that the field within the metal is E and not $E+(4\pi/3)P$, so that ω_j' reduces to ω_j and (20–23) gives

$$n^2-1=\frac{4\pi Ne^2}{m}\left[\frac{f_e}{-\omega^2+ir_e\omega}+\sum\frac{f_j}{\omega_j^2-\omega^2+ir_j\omega}\right]. \qquad (20\text{--}48)$$

Suppose the frequency is low so that $\omega \ll r_e$. If the effect of the bound electrons can be neglected, one then has

$$\mathbf{n}^2 - 1 = \frac{4\pi Ne^2 f_e}{imr_e\omega}. \qquad \qquad \ldots \quad (20\text{–}49)$$

Also, if the effect of the bound electrons is neglected, one has $K=1$. (20–42) gives

$$\mathbf{n}^2 = 1 - i\frac{4\pi\sigma}{\omega}. \qquad \ldots \ldots \ldots (20\text{–}50)$$

(20–49) and (20–50) give

$$r_e = \frac{Ne^2 f_e}{m\sigma} \qquad \ldots \ldots \ldots (20\text{–}51)$$

or, from (20–47),

$$r_e = \frac{1}{\tau}. \qquad \ldots \ldots \ldots (20\text{–}52)$$

Suppose now that the frequency is sufficiently high to write $r_e \ll \omega$. This corresponds to a very small dissipation of energy by the free electrons associated with the fact that the time between collisions is much larger than the period of the wave. If the effect of the bound electrons can again be neglected, (20–48) gives

$$\mathbf{n}^2 = 1 - \frac{4\pi Ne^2 f_e}{m\omega^2}. \qquad \ldots \ldots (20\text{–}53)$$

This equation indicates that there is a transition wavelength for which $\mathbf{n}=0$. For shorter wavelengths $\mathbf{n}$ is real so that there is no absorption and n is less than unity. For longer wavelengths $\mathbf{n}$ is purely imaginary so that no periodic wave can be propagated in the metal. Thus when the wave strikes the metal the refracted disturbance dies away exponentially and there is almost total reflection for all angles of incidence. The reflection coefficient for normal incidence changes suddenly from almost zero to almost unity at the transition wavelength. Although a transition of this extreme nature does not occur in practice, a very sudden change has been observed with some metals.

In general it is found that if one neglects the terms corresponding to the bound electrons, the absorption given by (20–48) is smaller than the observed value. It is also found that the variation of $\varkappa$ with temperature does not correspond to the temperature coefficient of σ. It must be concluded that the absorption of many metals is associated with bound electrons as well as free electrons, i.e. that the metals absorb selectively as well as generally. This is obvious anyway if they are markedly coloured.

The experimental results give a qualitative agreement with the theory given above but it is only in terms of the quantum theory that one can give a satisfactory account of the optical properties of metals. In particular, the above theory is not able to predict the natural frequencies of the bound electrons. In the infra-red region the effect of the bound electrons may be neglected, but even here the above theory gives only a qualitative agreement with observation.

CHAPTER XXI

REFLECTION AND REFRACTION

21–1. Boundary Conditions

WHEN an electromagnetic wave strikes the boundary between two media, there are reflected and refracted disturbances. The total electromagnetic field in the first medium therefore includes both the incident and the reflected waves. Now it is known from electromagnetic theory that at any boundary separating two media the tangential components of $\mathbf{E}$ and $\mathbf{H}$ and the normal components of $\mathbf{D}$ and $\mathbf{B}$ are continuous. These boundary conditions determine the reflected and refracted disturbances. At optical frequencies all materials can be regarded as non-magnetic so that $\mu=1$ and $\mathbf{B}=\mathbf{H}$. For isotropic media $\mathbf{D}=K\mathbf{E}$. Thus at an interface E_t, H_t, KE_n, and H_n, are continuous.

It will be assumed that the plane boundary separating the two media occupies the xy plane. Thus if $\mathbf{E}_1$ and $\mathbf{H}_1$, $\mathbf{E}_2$ and $\mathbf{H}_2$, and $\mathbf{E}_3$ and $\mathbf{H}_3$, represent the incident, reflected, and refracted disturbances respectively, the boundary conditions give

$$E_{1x}+E_{2x}=E_{3x}, \qquad \cdots \cdots \quad (21\text{–}1)$$

$$E_{1y}+E_{2y}=E_{3y}, \qquad \cdots \cdots \quad (21\text{–}2)$$

$$H_{1x}+H_{2x}=H_{3x}, \qquad \cdots \cdots \quad (21\text{–}3)$$

$$H_{1y}+H_{2y}=H_{3y}, \qquad \cdots \cdots \quad (21\text{–}4)$$

$$KE_{1z}+KE_{2z}=K'E_{3z}, \qquad \cdots \cdots \quad (21\text{–}5)$$

$$H_{1z}+H_{2z}=H_{3z}, \qquad \cdots \cdots \quad (21\text{–}6)$$

where K and K' are the dielectric constants for the first and second media respectively. These conditions supply only four new independent equations since the last two conditions can be deduced from the first four using Maxwell's equations.

21–2. Reflection and Refraction by Transparent Isotropic Dielectrics

It will be assumed that the incident disturbance is a monochromatic plane polarized plane wave. The boundary between media whose dielectric constants are K and K' is taken as the xy plane and the incident ray is assumed to be in the xz plane, the z-axis being directed into the first medium. The angles of incidence, reflection, and refraction, are denoted by I_1, I_2, and I_3, respectively. One can then write, for the y components of the electric vectors,

$$E_{1y}=A_{1y}e^{i\{\omega_1 t-k_1(l_1 x+m_1 y+n_1 z)\}}, \qquad \cdots \cdots \quad (21\text{–}7)$$

$$E_{2y}=A_{2y}e^{i\{\omega_1 t-k_2(l_2 x+m_2 y+n_2 z)\}}, \qquad \cdots \cdots \quad (21\text{–}8)$$

$$E_{3y}=A_{3y}e^{i\{\omega_1 t-k_3(l_3 x+m_3 y+n_3 z)\}}, \qquad \cdots \cdots \quad (21\text{–}9)$$

where A_1, A_2, and A_3, are used instead of E_{10}, E_{20}, and E_{30}, for the amplitudes at the surface in order to reduce the number of suffixes. (With absorbing media the E's would require suffixes $z=0$.) If A_1 is assumed to be real and positive, a real and negative value of A_2 or A_3 will indicate a phase change of π at the boundary, and a complex value will indicate a phase change other than π. Since the incident ray is in the xz plane as defined above,

$$l_1=\sin I_1, \qquad m_1=0, \qquad n_1=-\cos I_1. \quad . \quad . \quad (21\text{--}10)$$

Now at the boundary (that is, at $z=0$) the condition (21–2) must be satisfied for all values of x, y, and t. Hence one must have the same coefficients of x, y, and t, in the exponentials of (21–7), (21–8), and (21–9). On equating the coefficients of t one has

$$\omega_1=\omega_2=\omega_3, \quad . \quad . \quad . \quad . \quad . \quad . \quad (21\text{--}11)$$

which shows that the incident, reflected, and refracted disturbances have the same frequency. The numerical suffixes can therefore be omitted. Since the incident and reflected disturbances are in the same medium (and therefore have the same velocity) it also follows that

$$k_1=k_2=k \text{ (say)}. \quad . \quad . \quad . \quad . \quad (21\text{--}12)$$

Again, if the velocities in the first and second media are c and c' respectively, one has $\omega=kc=k'c'$ where k' is now written for k_3.

$$\therefore \quad \frac{k'}{k}=\frac{c}{c'}=\frac{n'}{n}, \quad . \quad . \quad . \quad . \quad (21\text{--}13)$$

where n and n' are the refractive indices of the media. On equating coefficients of y in (21–7), (21–8), and (21–9), one has, in view of (21–10),

$$m_2=m_3=0. \quad . \quad . \quad . \quad . \quad . \quad (21\text{--}14)$$

This confirms one of the classical laws of reflection and refraction: the reflected and refracted rays lie in the plane of incidence. One can therefore represent the rays on a plane figure (Fig. 21–1) and write

$$\left.\begin{array}{l} l_2=\sin I_2 \quad l_3=\sin I_3, \\ n_2=\cos I_2 \quad n_3=-\cos I_3. \end{array}\right\} \quad . \quad . \quad . \quad (21\text{--}15)$$

Equating coefficients of x in (21–7), (21–8), and (21–9), gives

$$\sin I_1=\sin I_2=\sin I \text{ (say)} \quad . \quad . \quad . \quad (21\text{--}16)$$

and

$$k \sin I_1=k' \sin I_3=k' \sin I' \text{ (say)} \quad . \quad . \quad (21\text{--}17)$$

or

$$n \sin I=n' \sin I', \quad . \quad . \quad . \quad . \quad (21\text{--}18)$$

which is Snell's law.

The above results would follow from any self-consistent wave theory and do not depend on the particular boundary conditions given in §21–1. Whatever boundary conditions one has, the relationship between E_1, E_2, and E_3, must be independent of x, y, and t, and the above results must follow. A far more important test of the theory is made when it is used to calculate the intensities and states of polarization of the reflected and refracted disturbances.

Whatever the direction of the electric vector of the incident beam, it can be resolved into components in and perpendicular to the plane of incidence. Since the angle of incidence is specified, the disturbance is completely determined by specifying these *two* components. Hence these are used instead of the *three* components parallel to the co-ordinate axes. The latter must, however, be introduced in order to apply the boundary conditions. The components of E_1 parallel to and perpendicular to the plane of incidence are denoted by E_{1l} and E_{1r} respectively (the initial letters being unsatisfactory suffixes) and the components of E_2 and E_3 are similarly represented. Obviously E_r is parallel to the y-axis in each case but it is also necessary to specify the directions in which E_r and E_l are positive. The obvious choice for the perpendicular component is to

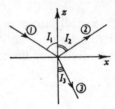

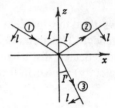

Fig. 21-1. Reflection and refraction at a plane surface.

Fig. 21-2. Definition of the l-directions.

make it positive when in the positive direction of y so that $E_{1r} \equiv E_{1y}$, $E_{2r} \equiv E_{2y}$, and $E_{3r} \equiv E_{3y}$. The directions in which E_{1l}, E_{2l}, and E_{3l}, are positive are shown in Fig. 21-2. Thus for the incident disturbance one has

$$\left.\begin{aligned} E_{1l} &= A_{1l} e^{i\{\omega t - k(x \, \sin I - z \, \cos I)\}}, \\ E_{1r} &= A_{1r} e^{i\{\omega t - k(x \, \sin I - z \, \cos I)\}}, \end{aligned}\right\} \quad \ldots \quad (21\text{--}19)$$

with similar expressions for E_{2l}, E_{2r}, E_{3l}, and E_{3r}. The boundary condition (21-2) gives, at $z=0$,

$$E_{1r} + E_{2r} = E_{3r} \quad \cdots \cdots \quad (21\text{--}20)$$

and (21-3) gives

$$-H_{1l} \cos I + H_{2l} \cos I = -H_{3l} \cos I'. \quad \cdots \quad (21\text{--}21)$$

Now it follows from equation (19-18) (page 479) that $H_{1l} = -\sqrt{K} E_{1r}$, $H_{2l} = -\sqrt{K} E_{2r}$, and $H_{3l} = -\sqrt{K'} E_{3r}$, where K and K' are the dielectric constants of the first and second media, the negative signs being due to the orientation of the positive l-directions. Hence (21-21) gives

$$(E_{1r} - E_{2r}) \, n \cos I = E_{3r} \, n' \cos I' \quad \cdots \quad (21\text{--}22)$$

since $n = \sqrt{K}$ and $n' = \sqrt{K'}$.

The boundary condition (21-1) gives

$$(E_{1l} - E_{2l}) \cos I = E_{3l} \cos I' \quad \cdots \cdots \quad (21\text{--}23)$$

and (21-4) gives

$$H_{1r} + H_{2r} = H_{3r}.$$

But $H_{1r}=\sqrt{K}E_{1l}$, $H_{2r}=\sqrt{K}E_{2l}$, and $H_{3r}=\sqrt{K'}E_{3l}$, so that one has

$$(E_{1l}+E_{2l})n=E_{3l}n'. \qquad \ldots \ldots \quad (21\text{–}24)$$

Now in the expressions (21–19) for $\mathbf{E}_1$, and in the corresponding expressions for $\mathbf{E}_2$ and $\mathbf{E}_3$, one has an exponential factor and all these factors are equal at $z=0$. Hence in (21–20), (21–22), (21–23), and (21–24), one may replace E's by A's to give relationships between the amplitudes.

Division of (21–23) by (21–24) gives

$$\frac{(A_{1l}-A_{2l})\cos I}{(A_{1l}+A_{2l})n}=\frac{\cos I'}{n'}$$

or, using Snell's law,

$$(A_{1l}-A_{2l})\sin I \cos I=(A_{1l}+A_{2l})\sin I' \cos I'.$$

$$\therefore \quad A_{2l}=A_{1l}\frac{\sin 2I-\sin 2I'}{\sin 2I+\sin 2I'}, \qquad \ldots \ldots \quad (21\text{–}25)$$

or

$$A_{2l}=A_{1l}\frac{\tan (I-I')}{\tan (I+I')}. \qquad \ldots \ldots \quad (21\text{–}26)$$

Division of (21–22) by (21–20) gives

$$\frac{(A_{1r}-A_{2r})n \cos I}{(A_{1r}+A_{2r})}=n' \cos I'$$

or, using Snell's law,

$$A_{2r}\left(\frac{\sin I'}{\cos I'}+\frac{\sin I}{\cos I}\right)=A_{1r}\left(\frac{\sin I'}{\cos I'}-\frac{\sin I}{\cos I}\right),$$

$$\therefore \quad A_{2r} (\sin I' \cos I+\cos I' \sin I)=A_{1r} (\sin I' \cos I-\cos I' \sin I)$$

or

$$A_{2r}=-A_{1r}\frac{\sin (I-I')}{\sin (I+I')}. \qquad \ldots \ldots \quad (21\text{–}27)$$

From (21–25) and (21–23)

$$A_{3l}\cos I'=A_{1l}\cos I\left(\frac{2 \sin 2I'}{\sin 2I+\sin 2I'}\right),$$

$$\therefore \quad A_{3l}=A_{1l}\frac{2 \sin I' \cos I}{\sin (I+I') \cos (I-I')}. \qquad \ldots \quad (21\text{–}28)$$

Finally, from (21–20) and (21–27),

$$A_{3r}=A_{1r}\frac{\sin (I+I')-\sin (I-I')}{\sin (I+I')}$$

or

$$A_{3r}=A_{1r}\frac{2 \sin I' \cos I}{\sin (I+I')}. \qquad \ldots \ldots \quad (21\text{–}29)$$

(21–26), (21–27), (21–28), and (21–29), are known as *Fresnel's equations*; they were first derived by Fresnel from the elastic solid theory.

21-3. Normal Incidence

Fresnel's equations as given above are indeterminate at normal incidence but (21-20), (21-22), (21-23), and (21-24) are easily solved directly. Thus, putting $\cos I = \cos I' = 1$, (21-23) and (21-24) give

$$(A_{1i}+A_{2i})n=(A_{1i}-A_{2i})n',$$

$$\therefore \quad A_{2i}=A_{1i}\frac{n'-n}{n'+n}. \quad \cdots \quad \cdots \quad (21\text{-}30)$$

(21-22) and (21-20) give

$$(A_{1r}-A_{2r})n=(A_{1r}+A_{2r})n',$$

$$\therefore \quad A_{2r}=-A_{1r}\frac{n'-n}{n'+n}. \quad \cdots \quad \cdots \quad (21\text{-}31)$$

(21-30) and (21-23) give

$$A_{3i}=A_{1i}\frac{2n}{n'+n}. \quad \cdots \quad \cdots \quad (21\text{-}32)$$

(21-31) and (21-20) give

$$A_{3r}=A_{1r}\frac{2n}{n'+n}. \quad \cdots \quad \cdots \quad (21\text{-}33)$$

For normal incidence the plane of incidence is indeterminate and at first sight it may be surprising that (21-31) contains a negative sign and (21-30) does not. The explanation is simply that, as defined above, the positive direction of l is reversed on reflection at normal incidence whereas the direction of the y-axis (i.e. of r) is fixed in space.

21-4. Discussion of Results

The equations deduced in the previous two sections give the magnitude and direction of the electric vector in the reflected and refracted disturbances. These are sufficient to specify the disturbances completely since in any wave $\mathbf{E}$ and $\mathbf{H}$ are not independent. It is more useful to specify $\mathbf{E}$ than $\mathbf{H}$ since it is found that it is the magnitude of $\mathbf{E}$ which determines the observed intensity (see § 21-10).

Whatever the state of polarization of the incident light, $\mathbf{E}$ can be resolved into components in and perpendicular to the plane of incidence as indicated in Chapter VI. Fresnel's equations then give the components of the reflected and refracted disturbances, and therefore give the intensities and states of polarization of these waves. For example, equation (21-26) shows that A_{2i} is always zero when $I+I'=90°$, i.e. when the reflected and refracted rays are at right angles. Hence, whatever the state of polarization of the incident light, the reflected disturbance has no component in the plane of incidence—it is plane polarized with $\mathbf{E}$ perpendicular to that plane. This corresponds to Brewster's law as given in § 6-17 if the vibrations associated with the light in Chapter VI correspond to the variations of $\mathbf{E}$.

To find the intensity and state of polarization of the reflected and refracted rays, it is necessary to know the amplitudes and phases of the components in and perpendicular to the plane of incidence. If $n'>n$ the expressions for A_2 and A_3 are real. Hence the phases are either 0 or π with respect to that of the incident light and are indicated simply by the signs of the amplitudes. Fig. 21–3 shows the amplitudes for the reflected and refracted disturbances as fractions of the incident amplitudes in the corresponding planes for $n'/n=1\cdot5$. That is, the figure gives the ratios A_{2r}/A_{1r}, A_{2l}/A_{1l}, etc., and corresponds to light incident from air to glass.

The reflection coefficients are the ratios of the intensities of the reflected and incident disturbances and are simply obtained by squaring the ordinates for the A_{2l} and A_{2r} curves. They are shown in Fig. 21–4.

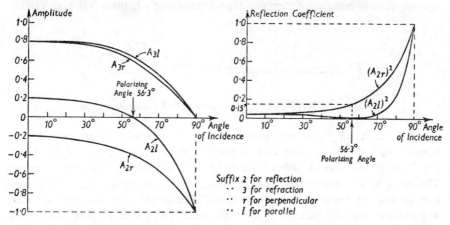

Fig. 21–3. Reflected and refracted amplitudes for $n'/n=1\cdot5$.

Fig. 21–4. Reflected intensities.

Suffix 2 for reflection
    ·· 3 for refraction
    ·· r for perpendicular
    ·· l for parallel

It must be remembered that the intensity is defined as the energy in unit time crossing unit area perpendicular to the direction of propagation. It is proportional to the product of the refractive index and the square of the amplitude (see § 19–5) and, since the incident and reflected beams are in the same medium, the ratio of the squares of the amplitudes is equal to the ratio of the intensities. It is also the ratio of the energy fluxes since the two beams have the same cross-section area. On the other hand, when a beam is refracted the cross-section area is increased by a factor (cos I'/cos I) so that the total energy in the refracted beam is proportional to the intensity multiplied by (cos I'/cos I). It should be noted that it is the total energy fluxes in the reflected and refracted beams which are complementary—not the intensities nor the squares of the amplitudes (see Fig. 21–3).

Fig. 21–4 shows that the reflecting power at normal incidence is about 4%, and that when it falls to zero at the polarizing angle (56·3°) for light having **E** in the plane of incidence, it is about 15% for light having **E** perpendicular to the plane of incidence. All the incident light is reflected at grazing incidence.

Fig. 21–3 shows that for all angles of incidence there is a phase change of π on reflection for light whose vibration direction is perpendicular to the plane of incidence. On the other hand, E_{2l} is in phase with E_{1l} for angles of incidence less than the polarizing angle and is π out of phase for larger angles of incidence.

In interpreting this result one must be careful to take into account the definition of the positive l-direction. This is shown in Fig. 21–2. For light near grazing incidence the positive directions of l for the incident and reflected disturbances are roughly parallel as in Fig. 21–5(a), and the phase change of π means that, to an observer somewhere along the x-axis, the direction of E_l is suddenly reversed on reflection. For light near normal incidence, the positive directions of l for the incident and reflected rays are roughly in opposite directions [Fig. 21–5(b)], and, for an observer somewhere along the z-axis, a zero phase change for E_l really means a reversal of its direction on reflection. Thus there is a sense in which one always has a phase change of π on reflection (see also §21–3). This is important in certain interference experiments (see Chapters VII and VIII).

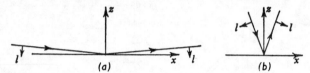

Fig. 21–5. The l-directions near grazing and normal incidence.

If the incident light is unpolarized, it can be resolved into incoherent components of equal amplitude in and perpendicular to the plane of incidence. Fig. 21–4 shows that at normal incidence the reflection coefficients are equal. This must be so because the plane of incidence is indeterminate and there is no real distinction between the components. Hence the reflected and refracted disturbances are unpolarized. The totally reflected light at grazing incidence is also unpolarized. For all other angles of incidence the reflection coefficients for the components are different. For light incident at the polarizing angle the reflected light is plane polarized and the refracted light partially plane polarized. For other angles of incidence both the reflected and refracted beams are partially plane polarized. The *degree of polarization* is defined as $(A^2_{max}-A^2_{min})/(A^2_{max}+A^2_{min})$ where A_{max} and A_{min} are the maximum and minimum amplitude components of the electric vector. It can be calculated from Fresnel's formulae for the reflected and refracted beams.

If the incident light is plane polarized one can again discuss the reflection and refraction of the separate components. For the reflected light Fig. 21–3 shows that, except for normal or grazing incidence, $|A_{2r}/A_{2l}|$ is always greater than $|A_{1r}/A_{1l}|$. This means that the plane of the electric vector is inclined to the plane of incidence at a larger angle in the reflected than in the incident light. For the refracted light $|A_{3r}/A_{3l}|<|A_{1r}/A_{1l}|$ so that the plane of vibration is rotated towards the plane of incidence.

Generally speaking, the results predicted by the theory are found to be in very close agreement with the experimental results. (Methods of determining the state of polarization of a beam of light are discussed in Chapters XXII and XXIII.) In practice, light reflected at the polarizing angle is found to be elliptically polarized, although the ellipse is a very thin one. Measurements on freshly formed liquid surfaces show that the ellipse is made more slender when the surfaces are cleaner. The residual ellipticity is associated with the arrangement of the molecules in the surface film.

21–5. Total Internal Reflection

Except for the reference to Brewster's law, the discussion of the previous section was concerned only with the case $n'>n$. When $n'<n$ one has the well known phenomenon of total internal reflection for angles of incidence greater than the critical angle.

For angles of incidence less than the critical angle Fig. 21–6 shows the amplitudes of the reflected components as fractions of the incident amplitudes for $n/n'=1.5$. Fig. 21–7 shows the reflection coefficients. At normal incidence the reflection coefficients for both components are 4%—the same as for light incident on the same boundary from the other side. The polarizing angle (33·7°) is simply the complement of that for light incident from the less dense side of the boundary, and is also such that the ratio of their sines is 1·5. That is, if the angle of incidence

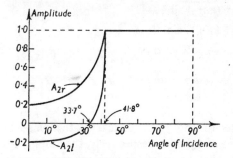

Fig. 21–6. Internally reflected amplitudes for $n/n'=1.5$.

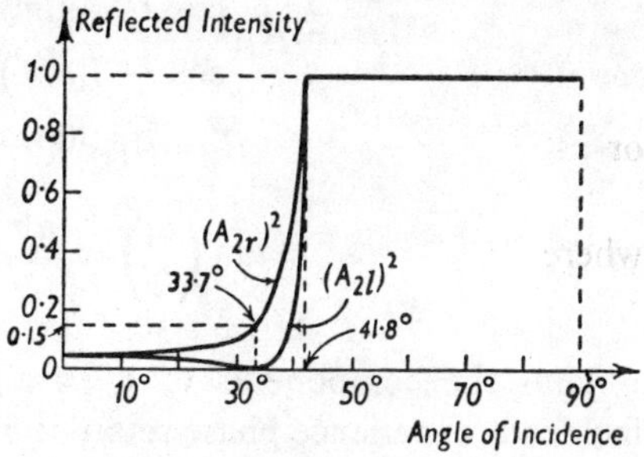

Fig. 21–7. Reflected intensities for $n/n'=1.5$.

for light in one direction is the polarizing angle and the directions of the incident and refracted rays are reversed, the new angle of incidence is the polarizing angle for light in the reverse direction. This follows at once from the original statement of Brewster's law.

For angles of incidence greater than the critical angle (41·8°), the sine of the angle of refraction as given by Snell's law is greater than unity. One has

$$\sin I' = \left(\frac{n}{n'}\right) \sin I.$$

$$\therefore \quad \cos I' = \sqrt{1-(n/n')^2 \sin^2 I}$$

$$= i\sqrt{(n/n')^2 \sin^2 I - 1}. \quad \ldots \ldots \quad (21\text{–}34)$$

Using this value of $\cos I'$, equation (21–27) gives, for the *reflected* light,

$$A_{2r} = -A_{1r} \frac{i \sin I \sqrt{(n/n')^2 \sin^2 I-1} - \cos I\,(n/n')\sin I}{i \sin I \sqrt{(n/n')^2 \sin^2 I-1} + \cos I\,(n/n')\sin I}$$

$$= A_{1r} \frac{(n/n') \cos I - i\sqrt{(n/n')^2 \sin^2 I-1}}{(n/n') \cos I + i\sqrt{(n/n')^2 \sin^2 I-1}}.$$

Now $\dfrac{a-ib}{a+ib}=e^{-i\delta}$, where $\tan\dfrac{\delta}{2}=\dfrac{b}{a}$. Hence one may write

$$A_{2r}=A_{1r}e^{-i\delta_r} \qquad \qquad (21\text{--}35)$$

or

$$E_{2r}=A_{1r}e^{i\{\omega t-k(x\,\sin I+z\,\cos I)-\delta_r\}}, \qquad (21\text{--}36)$$

where

$$\tan\left(\frac{\delta_r}{2}\right)=\frac{\sqrt{(n/n')^2\sin^2 I-1}}{(n/n')\cos I}. \qquad (21\text{--}37)$$

Similarly, from (21-25),

$$A_{2l}=A_{1l}\,\frac{\sin I\cos I-i(n/n')\sin I\sqrt{(n/n')^2\sin^2 I-1}}{\sin I\cos I+i(n/n')\sin I\sqrt{(n/n')^2\sin^2 I-1}}$$

$$=A_{1l}\,\frac{\cos I-i(n/n')\sqrt{(n/n')^2\sin^2 I-1}}{\cos I+i(n/n')\sqrt{(n/n')^2\sin^2 I-1}}$$

or

$$E_{2l}=A_{1l}e^{i\{\omega t-k(x\,\sin I+z\,\cos I)-\delta_l\}}, \qquad (21\text{--}38)$$

where

$$\tan\left(\frac{\delta_l}{2}\right)=\frac{(n/n')\sqrt{(n/n')^2\sin^2 I-1}}{\cos I}. \qquad (21\text{--}39)$$

Thus the components of **E** parallel to and perpendicular to the plane of incidence experience phase retardations of δ_l and δ_r respectively, the amplitudes of the reflected components being equal to those of the incident components. For example, if the incident light is plane polarized, the incident components are in phase and the totally reflected components are out of phase. Hence the reflected light is elliptically polarized, the phase difference (δ) between the components being given by

$$\tan\left(\frac{\delta}{2}\right)=\tan\left(\frac{\delta_l}{2}-\frac{\delta_r}{2}\right)$$

$$=\frac{\tan(\delta_l/2)-\tan(\delta_r/2)}{1+\tan(\delta_l/2)\tan(\delta_r/2)}$$

$$=\frac{\dfrac{[(n/n')^2-1]\sqrt{(n/n')^2\sin^2 I-1}}{(n/n')\cos I}}{\dfrac{(n/n')\cos^2 I+(n/n')[(n/n')^2\sin^2 I-1]}{(n/n')\cos^2 I}}$$

$$=\frac{\cos I\sqrt{(n/n')^2\sin^2 I-1}}{(n/n')\sin^2 I}. \qquad (21\text{--}40)$$

Fig. 21-8 shows the phase changes for the reflected components of internally reflected light for $n/n'=1{\cdot}5$ (corresponding to internal reflection at a glass-air surface). For angles of incidence less than the critical angle they are 0 or π as indicated by Fig. 21-6, and for total reflection they are given by (21-37) and (21-39). The figure also shows the phase difference δ between the reflected components when the incident components are in phase.

Suppose one has plane polarized light incident on a glass-air surface and that the plane of **E** is inclined at 45° to the plane of incidence. The components parallel to and perpendicular to the plane of incidence are then equal. Using (21–40) it is found that for $n/n'=1\cdot5$ the phase difference between the reflected components is 45° when the angle of incidence is either 53°15′ or 50°14′ (see Fig. 21–8). The reflected light is, of course, elliptically polarized. If there is now a second internal reflection at one of these angles a further 45° phase difference is introduced, so that a total phase difference of 90° is introduced between the two perpendicular components. Since the components have equal amplitudes the light is circularly polarized. *Fresnel's rhomb* produces circularly polarized

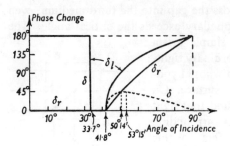

Fig. 21–8. Phase changes for internal reflection ($n/n'=1\cdot5$).

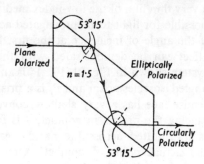

Fig. 21–9. Fresnel's rhomb.

light in this way (Fig. 21–9). The example illustrated employs glass of index 1·5 and has angles of incidence of 53°15′. Light is incident normally on the first face, the plane of vibration being at 45° to the plane of the diagram.

21–6. Disturbance in the Second Medium

When one has total internal reflection the phase difference between the incident and reflected disturbances is not π. That is to say, there is a resultant field in the first medium at the boundary. It follows that in order to satisfy the boundary conditions there must also be a disturbance in the second medium. The component perpendicular to the plane of incidence is represented by

$$E_{3r}=A_{3r}e^{i\{\omega t-k'(x\ \sin\ I'-z\ \cos\ I')\}}$$

(see 21–19) where, from (21–20) and (21–35),

$$A_{3r}=A_{1r}(1+e^{-i\delta_r})$$
$$=A_{1r}e^{-i(\delta_r/2)}(e^{+i(\delta_r/2)}+e^{-i(\delta_r/2)})$$
$$=2A_{1r}\cos\left(\frac{\delta_r}{2}\right)e^{-i(\delta_r/2)}.$$

Substituting for $k'\sin I'$ from (21–17) and for $\cos I'$ from (21–34), one has

$$E_{3r}=2A_{1r}\cos\left(\frac{\delta_r}{2}\right)e^{-i(\delta_r/2)}e^{i\{\omega t-kx\ \sin\ I+k'\ zi\sqrt{(n/n')^2\ \sin^2\ I-1}\}}$$

$$=2A_{1r}\cos\left(\frac{\delta_r}{2}\right)e^{-k'z\sqrt{(n/n')^2\ \sin^2\ I-1}}e^{i\{\omega t-kx\ \sin\ I-(\delta_r/2)\}}. \quad (21\text{--}41)$$

Thus E_{3r} is not periodic in z; the expression does not represent a wave whose direction of propagation has a component in the z-direction. The amplitude of E_{3r} decreases exponentially with increasing distance beyond the boundary and becomes negligibly small after a distance of a few wavelengths. It follows that, although there is a finite disturbance beyond the boundary, no energy is conveyed away from the surface and none is absorbed since the media are transparent. It is all reflected as indicated by the equations given above for the reflected wave.

It has been shown both theoretically and experimentally that if one has only a very thin film of the low-index medium followed by a high-index medium, it is possible for light to be propagated across the gap into the third medium, even if the angle of incidence at the first boundary exceeds the critical angle. The effect can be demonstrated with the simple prism system shown in Fig. 21–10. P_1 is an ordinary right-angled isosceles prism and P_2 is a prism whose hypotenuse face has a very shallow convex surface. On placing the prisms in contact, it is found that light is transmitted across the narrow gap surrounding the actual point of contact. One refers to this as *frustrated total reflection*. The transmitted light gets weaker as the thickness of the film increases, being negligible if the air gap exceeds a few wavelengths. If both hypotenuse faces are plane and

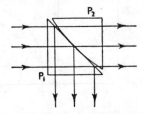

Fig. 21–10. Demonstration of frustrated total reflection.

can be kept accurately parallel and brought very close together, one can obtain uniform transmitted and reflected beams of any desired relative intensity. Since there is no absorption one has a very efficient beam splitter. A convenient method of spacing the prisms is to coat one with a low-index solid film of the necessary thickness and to place the other in contact with it. Non-absorbing, semi-reflecting films of this type have been used in the production of interference filters (see § 9–17).

21–7. Reflection by an Absorbing Medium

It was shown in § 20–10 that the expression representing a disturbance in an absorbing medium is formally the same as for a transparent medium, the refractive index being complex. The results of §§ 21–2 and 21–3 can therefore be used for light incident on an absorbing medium. It will be assumed that the first medium is air (or, more strictly, a vacuum) so that $n=1$. Then, if $\mathbf{n}$ is the complex index of the second medium, one can write

$$E_{1r} = A_{1r}e^{i\omega\{t-\frac{1}{c}(x \sin I - z \cos I)\}}, \quad \cdots \quad (21\text{–}42)$$

$$E_{2r} = A_{2r}e^{i\omega\{t-\frac{1}{c}(x \sin I + z \cos I)\}}, \quad \cdots \quad (21\text{–}43)$$

$$E_{3r} = A_{3r}e^{i\omega\{t-\frac{1}{c}(x \sin I' - z \cos I')\}}. \quad \cdots \quad (21\text{–}44)$$

In order to satisfy the boundary condition (21–20) one must have

$$\sin I' = \frac{\sin I}{\mathbf{n}} \quad \cdots \quad (21\text{–}45)$$

and, except for normal incidence, I' is complex. A_{2r} and A_{3r} are also complex, and are given by the expressions in §§ 21-2 and 21-3 where I' is as given above. One has

$$\cos I' = \frac{\sqrt{\mathbf{n}^2 - \sin^2 I}}{\mathbf{n}} \qquad \ldots \ldots (21\text{-}46)$$

and

$$\tan I' = \frac{\sin I}{\sqrt{\mathbf{n}^2 - \sin^2 I}}. \qquad \ldots \ldots (21\text{-}47)$$

The reflection coefficient or reflecting power (R) is the ratio of the reflected to the incident intensity and, for normal incidence, it follows at once from equation (21-31) on putting $n=1$, $n'=\mathbf{n}$. A_{2r} is complex and the reflected intensity is the product of A_{2r} with its conjugate so that

$$R = \frac{A_{2r}A_{2r}^*}{A^2_{1r}} = \frac{\mathbf{n}-1}{\mathbf{n}+1} \cdot \frac{\mathbf{n}^*-1}{\mathbf{n}^*+1}, \qquad \ldots \ldots (21\text{-}48)$$

$$\therefore \quad R = \frac{(n-1)^2 + n^2 \varkappa^2}{(n+1)^2 + n^2 \varkappa^2}. \qquad \ldots \ldots (21\text{-}49)$$

For oblique incidence R depends, of course, on the polarization of the incident beam. R_r and R_l can be deduced from the expressions given in § 21-2 but the calculations are rather tedious. It can be shown that when $n^2 + n^2\varkappa^2 \gg 1$ one has

$$R_r = \frac{(n-\cos I)^2 + n^2\varkappa^2}{(n+\cos I)^2 + n^2\varkappa^2}, \qquad \ldots \ldots (21\text{-}50)$$

$$R_l = \frac{\left(n - \dfrac{1}{\cos I}\right)^2 + n^2\varkappa^2}{\left(n + \dfrac{1}{\cos I}\right)^2 + n^2\varkappa^2}. \qquad \ldots \ldots (21\text{-}51)$$

Fig. 21-11 shows, qualitatively, the form of the variation of R_r and R_l with angle of incidence. Equations (21-49), (21-50), and (21-51), show that a high reflecting power is associated with strong absorption. It will be seen that when the absorption is very strong the reflecting power approaches unity. It has already been pointed out in § 20-1 that a body selectively absorbs those wavelengths which it selectively reflects, and that metals reflect and absorb strongly over a wide range of wavelengths. It was also pointed out that it is only when the selective absorption is very strong that the selective reflection determines the colour (surface colour). For less power-ful absorption even the maximum reflection coefficient is relatively small and the colour observed is body colour.

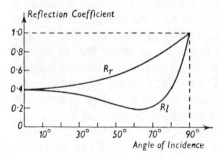

Fig. 21-11. Reflection coefficients for an absorbing medium.

The phenomenon of selective reflection has been used as a method of isolating the wavelengths concerned. For example, in their famous determinations of emissivities Hagen and Rubens made use of the fact that after three reflections by fluorite crystals the thermal radiation is reduced to a narrow band of wavelengths in the region of $25\cdot5\mu$. Radiations isolated in this way are referred to as the *residual rays* or *reststrahlen*.

The propagation of light in metals was discussed in § 20–15, the metal being regarded simply as a continuous medium of finite electrical conductivity. It follows from equations (20–45) and (20–46) that, if $K \ll \sigma T$, $n = \sqrt{\sigma T}$ and $\varkappa = 1$. If $\sigma T \gg 1$ (21–49) gives, for the reflecting power at normal incidence,

$$R = 1 - \frac{2}{\sqrt{\sigma T}} = 1 - \sqrt{\frac{c}{\sigma \lambda}}. \qquad \ldots \ldots \quad (21\text{–}52)$$

Since the reflecting powers of metals are very high, direct measurements of R do not give a sufficiently accurate method of investigating the variations of R from metal to metal, nor of the variation of R with temperature. In consequence, it is better to measure the emissivity E which, according to Kirchhoff's law, is given by

$$E = 1 - R.$$

Experiments by Hagen and Rubens and subsequent workers have given results in fairly good agreement with (21–52) for wavelengths greater than about 4μ. The results show a variation of R with temperature which agrees fairly well with that indicated by the variation of the conductivity.

For wavelengths in the visible region the experimental values of R differ widely from those given by (21–52). In this region metals show selective reflection and absorption so that they must have bound as well as free electrons. (See §§ 20–15 and 20–16.)

21–8. The Reflected Disturbance

For all metals the reflection coefficients for disturbances (or components) whose electric vector is in or perpendicular to the plane of incidence are of the form illustrated in Fig. 21–11. The curves show a resemblance to those for transparent media but there is one important difference: for metals R falls to a minimum for a particular angle of incidence but does not fall to zero as it does for dielectrics. Both components undergo a change of phase on reflection and the difference (δ) between these phase changes varies from 180° at normal incidence to zero at grazing incidence. Whereas for transparent media δ changes from 180° to zero at the polarizing angle, the value of δ for a metal changes gradually. If, as usual, $n^2(1 + \varkappa^2) \gg 1$, δ is almost exactly 90° when R_l is a minimum. δ is exactly 90° at the so-called *principal angle of incidence* (I_p).

It will be seen that if plane polarized light is incident on a metal the reflected components are, in general, unequal and out of phase—one has elliptically polarized reflected light. In particular, suppose plane polarized light is incident at the principal angle of incidence and that its electric vector is at 45° to the plane of incidence. The reflected components will be unequal and will be 90° out of phase. Hence the reflected light will be elliptically polarized and one of the axes of the ellipse will be in the plane of incidence. Since the incident light has equal components in and perpendicular to the plane of incidence, the ratio of the axes is simply $|A_{2l}|/|A_{2r}|$ (i.e. $\sqrt{R_l/R_r}$). In general, to analyse elliptically polarized light the phase difference is compensated to give plane polarized light, and the plane of vibration of this light is determined by means of an analyser (see §22–16). In the particular case under discussion here the compensation can be effected by a quarter-wave plate and the angle between the plane of incidence and the plane perpendicular to the plane of vibration of the resulting plane polarized light is known as the *principal azimuth* (ψ_p).* In Fig. 21–12, AB represents the vibration direction of the plane polarized light and it will be seen that $\tan \psi_p$ is the ratio of the axes of the ellipse, i.e. $\tan \psi_p = \sqrt{R_l/R_r}$.

If circularly polarized light (from a Fresnel rhomb, for example) is incident at the principal angle of incidence, the 90° phase difference between the components in and perpendicular to the plane of incidence becomes zero (or 180°)

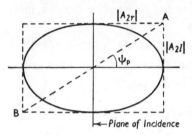

Fig. 21–12. The principal azimuth.

after reflection. Hence the reflected light is plane polarized and the angle between the plane of incidence and the plane perpendicular to the electric vector is the principal azimuth.

If the principal angle of incidence and the principal azimuth are known, n and $\varkappa$ can be calculated as shown below, and hence the normal reflecting power can be found from (21–49). This can then be compared with the experimental value.

To find n and $\varkappa$ from I_p and ψ_p one proceeds as follows. If a is the ratio of the amplitudes of the reflected components and δ is their phase difference one has, in general,

$$ae^{i\delta} = \frac{A_{2l}}{A_{2r}}. \quad \ldots \ldots \ldots (21\text{–}53)$$

If the incident light has its electric vector at 45° to the plane of incidence $A_{1l} = A_{1r}$, and if the angle of incidence is I_p, $\delta = 90°$ and $e^{i\delta} = i$. Hence, using (21–26) and (21–27), one has

$$ia = -\frac{\cos (I_p + I')}{\cos (I_p - I')}$$

$$= \frac{\tan I_p \tan I' - 1}{\tan I_p \tan I' + 1}.$$

$$\therefore \quad \frac{1 + ia}{1 - ia} = \tan I_p \tan I'$$

or, using (21–47) and assuming that $n^2 + n^2\varkappa^2 \gg \sin^2 I_p$, one has

$$\frac{1 + ia}{1 - ia} = \frac{\tan I_p \sin I_p}{n}.$$

* The reason for this rather clumsy definition is that the term *plane of polarization* was at one time used for the plane perpendicular to the plane of the electric vector and ψ_p was naturally defined as the angle between the plane of polarization and the plane of incidence. It is convenient to retain this angle rather than to use the complementary angle between the plane of incidence and the plane of the electric vector itself.

As has already been shown (Fig. 21–12),

$$a = \tan \psi_p.$$

$$\therefore \quad \tan I_p \sin I_p = n \frac{1 + i \tan \psi_p}{1 - i \tan \psi_p}$$

$$= n \frac{\cos \psi_p + i \sin \psi_p}{\cos \psi_p - i \sin \psi_p}$$

$$= n e^{2i\psi_p}$$

$$= n(1 - i\varkappa)(\cos 2\psi_p + i \sin 2\psi_p).$$

$$\therefore \quad \tan I_p \sin I_p = (n \cos 2\psi_p + n\varkappa \sin 2\psi_p)$$

$$+ i (n \sin 2\psi_p - n\varkappa \cos 2\psi_p). \quad . \quad (21\text{–}54)$$

Now the left-hand side of (21–54) is real so that one may equate the imaginary part of the right-hand side to zero. This gives

$$\varkappa = \tan 2\psi_p. \quad . \quad . \quad . \quad . \quad . \quad (21\text{–}55)$$

On equating the real parts of (21–54) and using this value of $\varkappa$, one has

$$\tan I_p \sin I_p = n \cos 2\psi_p + n \sin 2\psi_p \tan 2\psi_p$$

$$= \frac{n}{\cos 2\psi_p} (\cos^2 2\psi_p + \sin^2 2\psi_p),$$

or $\qquad\qquad n = \tan I_p \sin I_p \cos 2\psi_p. \quad . \quad . \quad . \quad (21\text{–}56)$

As pointed out above, the values of n and $\varkappa$ given by these equations can be used to calculate the normal reflecting power. The values of R calculated in this way from the observed values of I_p and ψ_p give fairly good agreement with the values of R observed directly. An important difficulty encountered in this work is that the properties of a metal surface depend upon the method used to polish it since this affects the structure of the surface films. It is also very important to ensure that the surface is clean. By using thin films and thin prisms, direct measurements of n and $\varkappa$ have been made but here one encounters the additional difficulty that the properties of thin films differ from those of the metal in bulk. The results of these investigations will not be discussed in detail but it is interesting to note that for some metals n is less than unity.

21–9. The Refracted Disturbance

Fig. 21–13 represents the refraction of a plane wave at a surface separating an absorbing medium and air (or more strictly, a vacuum). When the edge B of the wave reaches the boundary the edge A has already travelled into the medium. Hence the amplitude decreases across the wavefront from B to A. It can be shown that, in a direction normal to the wavefront, a wave of this type is propagated with a wave velocity which depends upon the rate at which the amplitude decreases across the wave. That is, the wave velocity depends on the angle

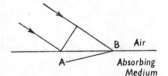

Fig. 21–13. Refraction into an absorbing medium.

between the surface of constant phase (the wavefront) and the surfaces of constant amplitude (which are parallel to the refracting surface). Hence the wave velocity in the medium depends upon the angle of incidence. It differs in magnitude and direction from the ray velocity, and the propagation of the wave resembles that of an extraordinary wave in a crystal (see Chapter XXII). Snell's law is not obeyed, and the ratio of the sines of I' and I is not equal to the real part of $\mathbf{n}$.

To show that the velocity of the refracted wave depends upon the angle of incidence one proceeds as follows. According to equations (21–44), (21–45), and (21–46), the perpendicular component of the refracted disturbance is given by

$$E_{3r}=A_{3r}e^{i\omega\left\{t-\frac{\mathbf{n}}{c}(x\ \sin\ I'-z\ \cos\ I')\right\}},$$

where $\mathbf{n} \sin I'=\sin I$

and $\mathbf{n} \cos I'=\sqrt{\mathbf{n}^2-\sin^2 I}=a-ib$ (say).

$$\therefore \quad E_{3r}=A_{3r}e^{\frac{\omega bz}{c}}e^{i\omega\left\{t-\frac{x\ \sin\ I-az}{c}\right\}}.$$

Hence the direction cosines of the normal to the wavefront are

$$\frac{\sin I}{\sqrt{a^2+\sin^2 I}}, \quad 0, \quad \frac{-a}{\sqrt{a^2+\sin^2 I}}$$

and the wave velocity is

$$\frac{c}{\sqrt{a^2+\sin^2 I}}$$

and obviously depends on the angle of incidence.

21–10. Stationary Light Waves

It was shown in §6–14 that when one has wave trains of equal velocity, amplitude, and frequency, travelling in opposite directions, a system of standing

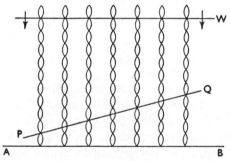

Fig. 21–14. Standing waves—normal incidence.

waves is set up. Now when a beam of light is incident normally on a plane reflecting surface, the incident and reflected disturbances can interfere in this way to give stationary waves. The existence of these stationary light waves was first demonstrated by Wiener in 1890. If the reflecting surface is only partially reflecting, the superposed wave trains have unequal amplitudes and, superposed on the stationary waves, there is a progressive component towards the surface. The present discussion is concerned only with the stationary waves.

The arrangement used by Wiener is illustrated in Fig. 21–14. W represents a plane wave incident on the plane reflecting surface AB. Whatever its direction in the incident light, the electric vector is reversed on reflection if AB is the surface of a dielectric (see § 21–4). This is approximately true if AB is the surface of a highly reflecting metal. In consequence, one has a node at the surface as far as the electric vector is concerned. It can be shown that for the magnetic intensity there is an antinode at the reflecting surface. The figure shows the formation of the nodes and antinodes for the electric intensity. Wiener showed the existence of the nodes and antinodes by placing a very thin photographic film in an inclined position such as PQ. It was found that the film showed a series of fringes corresponding to planes of maximum light intensity. The spacing of the fringes across PQ is obviously increased if the angle between PQ and AB is decreased and this makes it easier to resolve the fringes. It is found

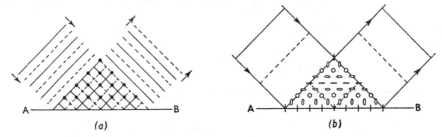

Fig. 21–15. Stationary waves with light incident at 45°.

that the blackening of the photographic film is a maximum at the electric antinodes as opposed to the magnetic ones, which indicates that the "effective light disturbance" is determined by the magnitude of the electric intensity associated with the wave.

Drude and Nernst repeated Wiener's experiment with a fluorescent film in place of the photographic film and obtained similar results. They also found that no fluorescence was observed in a very thin film of fluorescent material placed on a silver reflecting surface. This also shows that there is no fluorescence where the total electric intensity is zero.

Fig. 21–15 shows another experiment due to Wiener. In (a) plane polarized light is incident on a reflecting surface AB at an angle of incidence of 45°, the electric vector being perpendicular to the plane of incidence. Although the incident and reflected beams are perpendicular, their electric vectors are parallel and there will be destructive interference in planes parallel to AB. Hence fringes can be observed with an inclined photographic film as in Wiener's first experiment. In the figure crests and troughs (in both the incident and reflected waves) are represented by full and broken lines respectively. A trough is reflected as a crest and a crest as a trough. Destructive interference occurs where an incident crest coincides with a reflected trough and vice versa. If the electric vector is in the plane of incidence as in (b), the vectors for the incident and reflected beams are at right angles. In this case there is no destructive interference, the resultants being circles, ellipses, or straight lines, according to the phase difference as indicated in the figure. No fringes are observed with this arrangement whatever form of detector—photographic emulsion, fluorescent screen, or photo-electric cell—is used to examine the light distribution. It is important to notice that the effect of the superposition of the magnetic vectors in the first experiment is represented by the second figure, and that for the second experiment it is represented by the first figure. That is to say, observable fringes are caused by a variation of the amplitude of the electric intensity but not by a variation of the amplitude of the magnetic intensity.

It will be seen that the experiments on standing waves suggest that it is the magnitude of the electric vector rather than that of the magnetic vector which determines the intensity of the observed light. It is often convenient to refer to the oscillatory electric intensity simply as the light vibration.

21–11. Thin Dielectric Films and Multilayers

The use of thin films and multilayers for anti-reflection coatings, high reflection coatings, and interference filters, has increased very greatly in recent years as a result of improved production techniques; an introduction to the basic theory is given here using the matrix method.

Consider first a single dielectric film separating what are formally taken to be two semi-infinite media. In practice, this often becomes a thin film on a thick substrate in air, but to maintain generality three non-unit refractive indices will be retained. The theoretical approach can be illustrated and a number of important results obtained by a discussion that is simplified by considering the case of normal incidence. Thus in Fig. 21–16 monochromatic light is incident normally from left to right on the thin film of thickness d_1' and index n_1' on the semi-infinite substrate of index n_2'; the index of the initial medium is n_1.

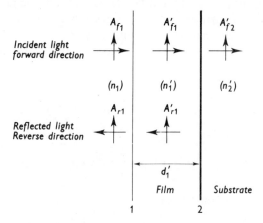

Fig. 21–16. Single film.

As in geometrical optics, the primed symbols refer to quantities to the right of a surface and the numerical suffix indicates the surface, numbering from the left. Since the light is incident normally, there is no distinction between the electric and magnetic field components parallel to and perpendicular to the plane of incidence, so that the suffixes r and l used earlier in this chapter become superfluous. Suffixes f and r will now be used throughout for fields corresponding to light travelling in the forward (left to right) and reverse (right to left) directions respectively. (To use i for incidence direction might lead to confusion since light in the "reverse" direction is obviously *incident* on a surface from the right!) As before, A will be used to denote the *amplitude* of the electric vector $\mathbf{E}$. Thus, for light travelling in the forward direction the amplitudes of the electric field vectors in the three media are A_{f1}, A'_{f1}, A'_{f2} and for light travelling in the reverse direction one has A_{r1} and A'_{r1}; obviously there is no light travelling from right to left in the substrate. The disturbance in the forward direction at surface 1 has a phase lead ahead of that at surface 2 and the disturbance in the reverse direction at surface 1 has a phase lag behind that at surface 2. Let the phase lag corresponding to propagation across the film be δ. Hence the boundary conditions, which state that the electric and magnetic fields are continuous at each surface, give, for surface 1, for the electric field

$$A_{f1} + A_{r1} = A_{f1}' + A_{r1}', \qquad \cdots \quad (21\text{--}57)$$

and for the magnetic field (since $H = nE$)

$$n_1 A_{f1} - n_1 A_{r1} = n_1' A_{f1}' - n_1' A_{r1}'. \qquad \cdots \quad (21\text{--}58)$$

Similarly, for surface 2 one has

$$A_{f1}' e^{-i\delta} + A_{r1}' e^{i\delta} = A_{f2}' \qquad \cdots \quad (21\text{--}59)$$

and

$$n_1' A_{f1}' e^{-i\delta} - n_1' A_{r1}' e^{i\delta} = n_2' A_{f2}'. \qquad \cdots \quad (21\text{--}60)$$

The negative signs in (21–58) and (21–60) arise from the fact that, for light travelling in either the forward or reverse direction, the ray direction, $\mathbf{E}$, and $\mathbf{H}$, are oriented as in a right-handed co-ordinate system.

If one takes (21–57), (21–58), (21–59), and (21–60), and eliminates A_{f1}' and A_{r1}', one obtains

$$1 + \frac{A_{r1}}{A_{f1}} = \left(\cos \delta + i \frac{n_2'}{n_1'} \sin \delta \right) \frac{A_{f2}'}{A_{f1}} \qquad \cdots \quad (21\text{--}61)$$

and

$$n_1 - n_1 \frac{A_{r1}}{A_{f1}} = \left(i n_1' \sin \delta + n_2' \cos \delta \right) \frac{A_{f2}'}{A_{f1}} \qquad \cdots \quad (21\text{--}62)$$

Using matrix notation, one writes

$$\begin{bmatrix} 1 \\ n_1 \end{bmatrix} + \begin{bmatrix} 1 \\ -n_1 \end{bmatrix} \frac{A_{r1}}{A_{f1}} = \begin{bmatrix} \cos \delta & i/n_1' \sin \delta \\ i n_1' \sin \delta & \cos \delta \end{bmatrix} \begin{bmatrix} 1 \\ n_2' \end{bmatrix} \frac{A_{f2}'}{A_{f1}}, \quad (21\text{--}63)$$

or

$$\begin{bmatrix} 1 \\ n_1 \end{bmatrix} + \begin{bmatrix} 1 \\ -n_1 \end{bmatrix} r = \mathbf{M} \begin{bmatrix} 1 \\ n_2' \end{bmatrix} t, \qquad \cdots \quad (21\text{--}64)$$

where $r(= A_{r1}/A_{f1})$ and $t(= A_{f2}'/A_{f1})$ are the reflection and transmission coefficients. The matrix $\mathbf{M}$, given by

$$\mathbf{M} = \begin{bmatrix} \cos\delta & i/n_1{}'\sin\delta \\ in_1{}'\sin\delta & \cos\delta \end{bmatrix}, \quad \ldots \quad (21\text{–}65)$$

is called the *transfer matrix* or *characteristic matrix* of the film.

The two simultaneous equations represented by (21–64) can be solved to give

$$r = \frac{(n_1 - n_2{}')n_1{}'\cos\delta + i(n_1 n_2{}' - n_1{}'^2)\sin\delta}{(n_1 + n_2{}')n_1{}'\cos\delta + i(n_1 n_2{}' + n_1{}'^2)\sin\delta}. \quad \ldots \quad (21\text{–}66)$$

If one has a film of optical thickness $\lambda/4$, $\delta = \pi/2$ and (21–66) gives at once

$$r = 0 \quad \text{if} \quad n_1{}'^2 = n_1 n_2{}'. \quad \ldots \ldots \quad (21\text{–}67)$$

That is to say, if the initial medium is air, a single quarter-wave film whose refractive index is the square root of that of the substrate gives a perfect anti-reflection film for the wavelength concerned. Some points of practical importance are discussed in §8–5. It also follows from (21–66) that by using a high index quarter-wave film one can obtain increased reflectivity.

It is interesting to note that if one has a single $\lambda/2$ film $(\delta = \pi)$, r is $(n_1 - n_2{}')/(n_1 + n_2{}')$, i.e. is independent of the refractive index of the film and is the same as it would be with no film.

Suppose, now, that one has a multilayer film consisting of N layers, the light being incident initially on film 1. With the notation used here the refractive indices of the films are $n_1{}', n_2{}', \ldots n_N{}'$. Let the transfer matrices of the films, as defined by (21–65), be $\mathbf{M}_1, \mathbf{M}_2, \mathbf{M}_3, \ldots \mathbf{M}_N$. $\mathbf{M}_{j+1}$ relates the fields to the left and right of film $j+1$ and $\mathbf{M}_j$ relates the fields to the left and right of film j. Since the field is continuous across the interface between films j and $j+1$, $\mathbf{M}_j\mathbf{M}_{j+1}$ relates the fields before film j and after $j+1$. It follows that for the complete multilayer film one has

$$\begin{bmatrix} 1 \\ n_1 \end{bmatrix} + \begin{bmatrix} 1 \\ -n_1 \end{bmatrix} r = \mathbf{M} \begin{bmatrix} 1 \\ n'_{N+1} \end{bmatrix} t, \quad \ldots \quad (21\text{–}68)$$

where n_1 and n'_{N+1} are the refractive indices of the initial medium and the substrate respectively, and r and t are the reflection and transmission coefficients of the multilayer. $\mathbf{M}$ is the transfer matrix of the multilayer and is given by

$$\mathbf{M} = \mathbf{M}_1\mathbf{M}_2\mathbf{M}_3 \ldots \mathbf{M}_N. \quad \ldots \ldots \quad (21\text{–}69)$$

Following the usual notation, put

$$\mathbf{M} = \begin{bmatrix} M_{11} & M_{12} \\ M_{21} & M_{22} \end{bmatrix}. \quad \ldots \ldots \quad (21\text{–}70)$$

The two simultaneous equations represented by (21–68) can be solved to give

$$r = \frac{M_{11}n_1 + M_{12}n_1 n'_{N+1} - M_{21} - M_{22}n'_{N+1}}{M_{11}n_1 + M_{12}n_1 n'_{N+1} + M_{21} + M_{22}n'_{N+1}} \quad \ldots \quad (21\text{–}71)$$

and

$$t = \frac{2n_1}{M_{11}n_1 + M_{12}n_1 n'_{N+1} + M_{21} + M_{22}n'_{N+1}}. \quad \ldots \quad (21\text{–}72)$$

It will be appreciated that to explore the properties of multilayers will, in general, involve extensive computation. However, one particularly interesting case, known as the *quarter-wave stack* can be examined here.

Suppose, then, that one has a multilayer consisting of $2N$ quarter-wave films whose indices alternate between n_A and n_B and suppose the initial and final media are air (i.e. $n_1 = n'_{N+1} = 1$). Since $\delta = \pi/2$ for each film, one has

$$\mathbf{M}_1\mathbf{M}_2 = \begin{bmatrix} 0 & i/n_A \\ in_A & 0 \end{bmatrix} \begin{bmatrix} 0 & i/n_B \\ in_B & 0 \end{bmatrix} = \begin{bmatrix} -n_B/n_A & 0 \\ 0 & -n_A/n_B \end{bmatrix}$$

and

$$\mathbf{M}_1\mathbf{M}_2 \dots \mathbf{M}_{2N} = (\mathbf{M}_1\mathbf{M}_2)^N = \begin{bmatrix} (-n_B/n_A)^N & 0 \\ 0 & (-n_A/n_B)^N \end{bmatrix}$$

(21–71) then gives

$$r = \frac{(-n_B/n_A)^N - (-n_A/n_B)^N}{(-n_B/n_A)^N + (-n_A/n_B)^N} = \frac{(-n_B/n_A)^{2N} - 1}{(-n_B/n_A)^{2N} + 1}. \quad . \quad (21\text{–}73)$$

Equation (21–73) shows that whatever are the values of n_A and n_B, provided they are different, $|r|^2 \to 1$ as $N \to \infty$. This points the way to the use of multilayers to produce very high reflectivities. In practice, dielectric multilayers are widely used in place of metal films for Fabry-Perot interferometers and interference filters (see §9–17).

THE PROPAGATION OF LIGHT IN ANISOTROPIC MEDIA AND THE ANALYSIS OF POLARIZED LIGHT

ALTHOUGH some results from Chapter XIX are used, it is not essential to have a knowledge of the electromagnetic theory of light before reading the present chapter. The chief result assumed is the nature of the wave surface in a crystal. The nature of polarized light is discussed in simple terms in Chapter VI. The vibration direction introduced there is the same as the direction of the electric displacement as indicated by electromagnetic theory.

22-1. Waves in Anisotropic Media

The path of a wave through a medium can be found by applying Huygens' Principle. In its simple form, which can be used when there is no diffraction, this principle states that each point on a wavefront acts as a source of secondary waves and that these are effective only on their forward envelope. It has already been seen that for isotropic media the disturbance from a point source spreads out as a spherical wave so that, from each point on a wavefront, one has spherical secondary waves. The rays (lines of energy flow) are then the wave normals and obey the classical laws of refraction. Huygens' principle can be applied in the same way in order to find the path of a disturbance in an anisotropic medium. When light enters such a medium it is found that there are, in general, two refracted paths. The medium is said to be *doubly refracting* or *birefringent*.

For a uniaxial medium, it was shown in § 19–10 that the disturbance from a point source can spread out as two waves—a sphere and a spheroid—and that each is plane polarized. It will be seen that there can be a disturbance propagated by means of spherical secondary waves as in isotropic media. One speaks of *ordinary* rays and waves since the rays are wave normals and obey the classical laws of refraction when a disturbance enters the anisotropic medium from one which is isotropic. The disturbances that are propagated by the spheroidal secondary waves behave in a rather different manner and one speaks of *extraordinary* waves and rays. In general, the rays are no longer wave normals and do not obey Snell's law of refraction. The refracted extraordinary ray is not, in general, in the plane of incidence.

For biaxial media the situation is different. As shown in § 19–9, there is no spherical sheet in the wave surface expanding from a point source. Consequently there are no spherical secondary waves and therefore, in general, no ordinary rays or waves. Fig. 19–4 showed that even with a biaxial crystal the wave surface has one circular section in each of three mutually perpendicular planes. If light is incident on the crystal from an isotropic medium such that one of these planes is the plane of incidence, one of the refracted disturbances is propagated by waves having a circular section and, in that plane, $\sin I'/\sin I$ is constant. That is, one has an ordinary ray.

The directions of the electric displacement **D** in the polarized disturbances are as given in §§ 19–9 and 19–10. As has already been pointed out, **D** and **H** are not independent when the direction of propagation is specified, so that it is sufficient to specify only **D**. **D** is chosen and is often referred to as the light vibration because, as indicated by Wiener's experiment, it is more directly related to the light intensity as ordinarily observed (see §21–10).

22–2. Double Refraction in Uniaxial Crystals

Fig. 22–1 shows a plane wave incident normally on the plane interface AB between an isotropic medium, such as air, and a negative uniaxial crystal. An example of the latter is calcite or iceland spar—naturally occurring forms of calcium carbonate. The figure shows a section through the crystal which is normal to the crystal surface and parallel to the optic axis. (A section containing

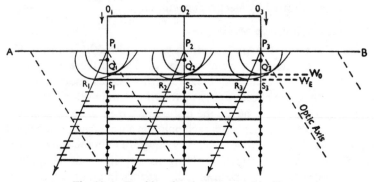

Fig. 22–1. Double refraction at normal incidence.

the optic axis is called a *principal section*.) The direction of the optic axis (which is a *direction* and not a single line) is indicated by the broken lines. It should be noted that in all these figures the eccentricity of the elliptic section of the wave surface is very much exaggerated. When the wave strikes the interface, each point such as P_1, P_2, and P_3, acts as a source of secondary waves. In the crystal the wave from each of these points takes the form of a sphere inscribed within an oblate spheroid. These touch on the axis of rotation of the latter and this is parallel to the optic axis. For the circular wave section the direction of **D** is perpendicular to the plane of the figure; for the elliptic wave section it is in the plane of the figure. If the incident wave has components of **D** in both these directions, both waves spread out from each point and the two sets of secondary waves give two forward envelopes. These are the ordinary and extraordinary waves W_O and W_E. If the incident wave is plane polarized with the direction of **D** either in or perpendicular to the principal section, there will be only one wave in the crystal—that for which the direction of **D** is the same as that in the incident wave. The other refracted wave will be absent since, in the incident wave, there is no component of **D** in the appropriate direction. If the incident wave is polarized in any other way, or is unpolarized, it will have components in and perpendicular to the principal section, and there will be double refraction. Thus one may say that for each ray such as O_1P_1 incident on the interface there will be two refracted rays—the ordinary ray P_1Q_1 for which **D** is perpendicular to the principal section, and the extraordinary ray P_1R_1 for which **D** is in the principal section. It will be seen that the ordinary ray is normal to the wave and enters the crystal without deviation as would a ray entering an isotropic medium at normal incidence. The extraordinary ray is refracted in the principal section as shown. The ray velocities are proportional to P_1Q_1 and P_1R_1, and the wave velocities are measured along the wave normals and are proportional to P_1Q_1 and P_1S_1. Thus the velocities of the extraordinary ray and wave are greater than those of the ordinary ray and wave. The reverse is true for positive uniaxial crystals in which the spherical sheet of the wave surface encloses the spheroidal sheet. Except for disturbances propagated in directions close to the optic axis,

quartz behaves as a positive uniaxial crystal. The properties of quartz will be discussed later.

Two special cases of normal incidence should be mentioned. These are illustrated for negative crystals by Figs 22–2 and 22–3. In the first, light is incident normally on a crystal face cut perpendicular to the optic axis, i.e. the light is travelling along the axis. It will be seen that the two ray velocities are equal and

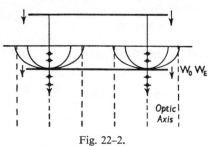

Fig. 22–2.
Crystal cut perpendicular to the axis.

equal to the two wave velocities. There is no double refraction and the O- and E-waves advance together. Figs 22–3(a) and (b) show light incident normally on a crystal face cut parallel to the optic axis. (a) shows a principal section and (b) a section at right angles. It will be seen that, again, neither the O- nor the E-rays are refracted but the waves advance with different speeds. The direction of **D** for each ray is as shown. These follow from the discussions of §§ 19–9 and 19–10 but may alternatively be described, by referring to the so-called *principal planes*, as follows. The principal section that contains the ordinary ray is sometimes called the principal plane of the ordinary ray. The principal plane of the E-ray is similarly defined, i.e. it is the plane containing the E-ray and the optic axis. The direction of **D** for the O-ray is always perpendicular to the principal

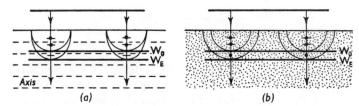

Fig. 22–3. Crystal cut parallel to the axis.

plane of the O-ray, whereas for the E-ray **D** is *in* the corresponding principal plane. Since the ordinary wave surface is spherical, the ordinary ray and wave velocities are equal and the same in all directions. This velocity (v_0) is one of the two *principal velocities* for the uniaxial medium (see Chapter XIX). The corresponding principal refractive index, known as the *ordinary index*, is therefore given by

$$n_0 = \frac{\text{velocity of light in a vacuum}}{\text{ordinary wave velocity}}.$$

Along the optic axis the E-wave and ray velocities are equal and equal to the ordinary velocity (see Fig. 22–2). As one moves away from the axis, the velocity of the E-wave becomes less than that of the E-ray. For positive crystals both become smaller than the ordinary velocity, and for negative crystals both become larger than v_0 (see Fig. 22–1). The E-wave and ray velocities again become equal in directions at right angles to the optic axis, reaching a maximum value for negative crystals (Fig. 22–3) and a minimum for positive crystals. Thus for positive crystals the second principal refractive index (the *extraordinary index*) is given by

$$n_E = \frac{\text{velocity of light in a vacuum}}{\text{minimum velocity of E-wave}}$$

and for negative crystals,

$$n_E = \frac{\text{velocity of light in a vacuum}}{\text{maximum velocity of E-wave}}.$$

For sodium light (λ 5893) one has, for calcite at 18°C., $n_0 = 1·65836$, $n_E = 1·48641$ and for quartz $n_0 = 1·54425$, $n_E = 1·55336$. (For fused quartz, which is isotropic, one has $n = 1·45845$.)

.22–3. Plane Waves at Oblique Incidence

Fig. 22–4 shows parallel rays (i.e. a plane wave) incident obliquely on the plane surface of a negative uniaxial crystal. It is assumed that the plane of

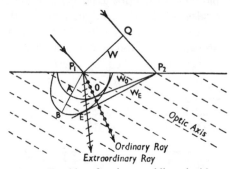

Fig. 22–4. Double refraction at oblique incidence.

incidence is a principal section, so that the optic axis lies in the plane of the figure; its direction is indicated by the broken lines. To find the refracted path of the wave one proceeds in the same way as for isotropic media (see § 1–7). The wave W reaches P_1 before it reaches P_2, and, during the time taken for the dis- turbance to travel from Q to P_2 (assumed, for convenience, to be in air), the secondary waves expand from P_1 into the crystal. One draws the circular section of the O-wave surface with radius $P_1A = QP_2/n_0$, the major axis of the E-wave surface being $P_1B = QP_2/n_E$. The refracted wavefronts W_0 and W_E are then found by drawing, from P_2, a tangent to each sheet of the wave surface. In terms of rays, the ray incident at P_1 is doubly refracted along P_1O (the O-ray) and P_1E (the E-ray), the directions of **D** being as indicated. As before, both refracted rays are present if the incident ray has a component of **D** in the appropriate directions. The special cases of a crystal cut with its surface (i) parallel to and (ii) perpendicular to the optic axis are shown in Figs 22–5 and 22–6 respectively.

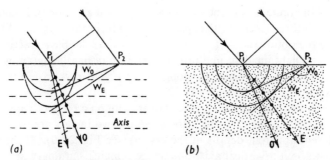

Fig. 22-5. Double refraction with crystal cut parallel to the axis.

In the examples illustrated the optic axis is either in or perpendicular to the plane of incidence, and in each case the E-ray lies in the plane of incidence. When the plane of incidence is not a principal section the extraordinary ray does not lie in the plane of incidence. It will be seen, from above, that Snell's law does not apply to the E-ray.

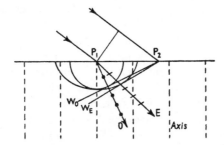

Fig. 22-6. Double refraction with crystal cut perpendicular to the axis.

22–4. Observation of Double Refraction in Calcite

The fundamental crystal form of calcite is a rhombohedron as shown in Fig. 22–7. Each rhombic face has angles of approximately 78° and 102°. Large

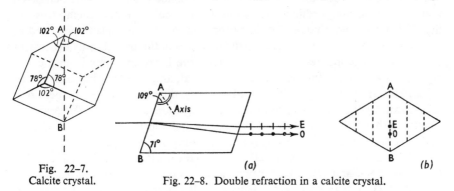

Fig. 22-7.
Calcite crystal.

Fig. 22-8. Double refraction in a calcite crystal.

crystals may be reduced by cleavage to parallelopipeds or rhombohedra containing these angles. It will be seen that there are two diametrically opposite corners (A and B) which are bounded by three obtuse (102°) angles. The direction of the optic axis is such that, passing through A or B, it makes equal angles with all three faces. In the rhombohedral form the diagonal AB is in the direction of the optic axis. Fig. 22–8(a) illustrates the double refraction of a ray of unpolarized light incident in a principal section (a parallelogram with angles of about 109° and 71°). Looking along the emergent rays from the right, the principal sections are as indicated in 22–8(b).

Suppose that a calcite crystal is placed over a dot on a piece of paper, and that the dot is viewed normally through the crystal. Two images are seen and these are at different apparent depths in the crystal (see Fig. 22–9). If the crystal is now rotated about a vertical axis, one image remains stationary and the other

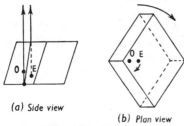

(a) Side view

(b) Plan view

Fig. 22–9.
Demonstration of double image.

moves round it. The former is, of course, the ordinary, and the latter the extra-ordinary image.

An interesting experiment with two equal calcite crystals is illustrated by Fig. 22–10. A ray of unpolarized light is incident normally on the first crystal and is doubly refracted. Two plane polarized rays emerge from the crystal with directions of **D** as indicated. If the second crystal is similarly orientated, these beams are further separated, but each passes through the second crystal as a single ray [see Fig. 22–10(a)]. This is because a ray incident on a crystal can give rise to two refracted rays only if it has a component of **D** in *each* of the perpendicular directions which correspond to the O- and the E-rays. If the second crystal is rotated 180° to the position shown in (b), the two rays are reunited as shown and the emergent ray shows all the properties of unpolarized light. If the second crystal is rotated 90° so that the principal sections of the two crystals are at right angles, the direction of **D** for the ordinary ray emerging from the first crystal is exactly the direction of **D** for the extraordinary ray in the second crystal. It is therefore singly refracted as an E-ray. Similarly, the ray which was the E-ray in the first crystal becomes the O-ray in the second. It should be remembered that there is nothing ordinary or extraordinary about the rays themselves—in the air between the crystals they are simply rays which are plane polarized at right angles. They behave as E or O disturbances in the crystals only because of the directional properties of the latter. Thus the E-ray emerging from the first crystal passes straight through the second crystal, and the O-ray becomes the E-ray and is refracted. The positions and vibration directions of the emergent rays are shown by the end-on view in (c). The rays are denoted by OE and EO, signifying that one is the O-ray in the first crystal and the E-ray in the second, and the other is the E-ray in the first and the O-ray in the second.

If the angle between the principal planes of the crystals is between 0° and 90° (or between 90° and 180°), each ray incident on the second crystal has a component of **D** both in and perpendicular to the principal section. Each ray is therefore doubly refracted as an O- and an E-ray. Figs 22–10(d) and (e) show the positions of the four emergent rays when the principal sections of the crystals are inclined at 45° and 135°. In general, the relative intensities of the four rays emerging from the second crystal are determined by the components of **D** in and perpendicular to the principal section of the second crystal. Suppose the angle between the principal sections is θ. Then, for the E-ray from the first crystal, **D** is in the principal section of that crystal; it has components proportional to $\cos \theta$ and $\sin \theta$ respectively in and perpendicular to the principal section of the second crystal. The intensity of the emergent ray EE is therefore proportional to $\cos^2 \theta$ (Law of Malus—see § 6–16), and that of the ray EO is proportional to $\sin^2 \theta$. Similarly, the intensities of the rays OO and OE are also proportional to $\cos^2 \theta$ and $\sin^2 \theta$ respectively. Thus as θ is increased from 0 to 90° the intensities of the rays OO and EE gradually decrease from a maximum down to zero,

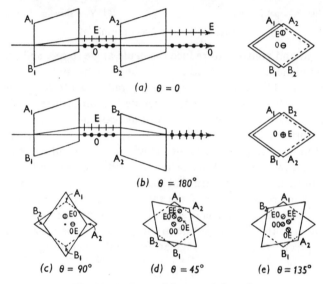

(a) $\theta = 0$

(b) $\theta = 180°$

(c) $\theta = 90°$ (d) $\theta = 45°$ (e) $\theta = 135°$

Fig. 22–10. Two calcite crystals in series.

while the intensities of the rays OE and EO gradually increase from zero to a maximum. When θ is 45° or 135° [Figs 22–10(d) and (e)] all four emergent rays have the same intensity. It has been assumed here that unpolarized light is incident on the first crystal. If the original beam is polarized, the O- and E-rays from the first crystal may have unequal intensities and the relative intensities of the final emergent rays will be changed accordingly.

22–5. The Nicol Prism

It is frequently desirable to be able to produce a single plane polarized beam of light. Obviously this can be done by double refraction if one of the refracted beams can be suitably removed. Now the O- and E-rays in a calcite crystal travel with different speeds, the speed of the E-ray varying with direction. In the Nicol prism (Fig. 22–11) a calcite crystal is cut across a diagonal as shown and the halves cemented together with canada balsam or some similar material. The refractive index of the cement is such that with the chosen E-ray direction in the calcite the E-ray is transmitted at the calcite-cement interface while the O-ray is totally reflected and absorbed by the black coating which is applied to the edges of the prism. The details of the construction of a Nicol are as follows. A crystal is chosen whose long edges are about three times as long as its short edges. It was pointed out in § 22–4 that a principal section is a parallelogram with angles of approximately 109° and 71°. The end faces of the crystal

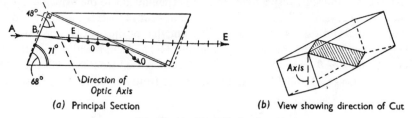

(a) Principal Section   (b) View showing direction of Cut

Fig. 22–11. The Nicol prism.

are now ground and polished to reduce the acute angle to 68°, and the crystal is cut across the plane that is perpendicular to the principal section and to the end faces (see Fig. 22–11). If light is incident parallel to the long edges of the prism, the O-ray is totally reflected at the calcite-balsam surface as shown in the figure. A clockwise rotation of the rays will increase the angles of incidence at the cemented surface, and when that of the E-ray exceeds the critical angle, both rays will be totally reflected. An anti-clockwise rotation will eventually reduce the angle of incidence of the O-ray below the critical angle, and both rays will then be transmitted. It will be seen that the O- and E-rays will be separated for only a small range of directions of the ray entering the crystal. In the Nicol prism the end faces are ground as described above in order to make the ray AB (incident parallel to the long edges) lie in the middle of this range of about 24°. It will be appreciated that this angular field will depend upon the cement employed.

22–6. Two Nicols in Series

Suppose that the plane polarized ray emerging from one Nicol prism enters another, the long edges of the prisms being parallel. If the principal sections of

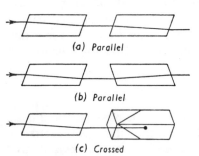

(a) Parallel

(b) Parallel

(c) Crossed

Fig. 22–12. Two Nicols in series.

the prisms are parallel [Fig. 22–12(a) or (b)], the ray will pass through the second prism unaffected (apart from the normal reflection and absorption losses). If the principal sections are perpendicular [Fig. 22–12(c)], the prisms are said to be *crossed*, and the ray which was the E-ray in the first prism becomes the O-ray in the second crystal and is totally reflected. That is, no light is transmitted by a pair of crossed Nicols. Frequently one employs two Nicols in series, with some optical system between them. The purpose of the first Nicol—the *polarizer*— is to plane-polarize the light. The system between the Nicols may change the state of polarization. For example, it may rotate the plane of vibration. By rotating the second Nicol about an axis parallel to the long edges, one can then find the new plane of vibration—it is perpendicular to the principal section of the second Nicol when the intensity of the emergent beam falls to zero. The second Nicol, which is used to examine the state of polarization of the light emerging from the system, is referred to as the *analyser*. More complicated arrangements involving the use of a polarizer and an analyser will be described later. With only a fixed polarizer and a rotating analyser, the component of the vibrations (i.e. of **D**) which can be transmitted by the analyser is proportional to $\cos \theta$ where θ is the angle between the principal sections of the Nicols. The intensity of the light emerging from the analyser is therefore proportional to $\cos^2 \theta$ and is independent of the orientation of the polarizer if the original light is unpolarized. If the original incident beam contains plane or elliptically polarized light, the intensity of the beam emerging from the polarizer will depend upon the orientation of the latter, and the intensity of the final emergent beam will not depend simply upon the angle between the principal sections of the two crystals.

22–7. Other Polarizing Prisms

In the previous section, only a single incident ray was discussed and it was assumed that this was incident parallel to the long edges of the polarizer. It was stated in §22–5 that a Nicol has an angular field of about 24°; but if two Nicols are placed in series, it is found that the whole field of view does not go

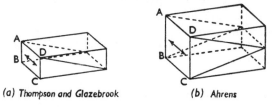

(a) *Thompson and Glazebrook*   (b) *Ahrens*

Fig. 22–13. Square-ended prisms.

uniformly dark when the Nicols are crossed. As the analyser is rotated, a dark shadow crosses the field of view. The reason is that the light emerging from a Nicol is not uniformly plane polarized. Another disadvantage of the simple Nicol prism is that, since the E-ray passes obliquely through a parallel-sided plate, the emergent ray is displaced laterally. This latter defect can be overcome by making the end faces perpendicular to the long edges. A prism of this type is often said to be square ended but this does not, of course, mean that the cross-section is a square. The square ended Nicol has the additional advantage of a smaller loss of light by reflection at the end face since less light is reflected at normal incidence.

In order to obtain more uniformly plane polarized light, the square ended prism shown in Fig. 22–13(a) was used by S. P. Thompson and by Glazebrook; it is rather longer than an ordinary Nicol of the same width. Other prisms of this type have been cut with the optic axis parallel to AB instead of AD. A prism which has been widely used is that due to Ahrens [Fig. 22–13(b)]. This has three calcite components cemented with balsam and obviously can be regarded as a combination of Thompson prisms. It has a field of about 26° and, for a given width, has a length only about half that of a Nicol. Ahrens prisms have also been cut with the optic axis parallel to AB instead of AD.

Nicol prisms and others employing Canada balsam cannot be used in the ultra-violet because the balsam absorbs in that region. Foucault devised a prism of the Nicol type but with an air gap and this had the advantage of shortening the prism. However, the angular field was small (about 8°) and the intensity of the E-ray was reduced by reflections. Square ended Foucault prisms were also employed, and that due to Glan bears the same relation to the

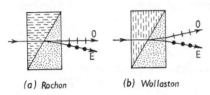

(a) Rochon   (b) Wollaston

Fig. 22-14. Quartz double image prisms.

Thompson prism as does the Foucault to the Nicol. All have the disadvantage of multiple reflections in the air film.

Most of the prisms used in the ultra-violet have been of the double image type. Fig. 22-14 shows those due to Rochon and Wollaston. Each of these employs a pair of quartz prisms, the directions of the optic axes being as shown. In each case the plane polarized components are separated by double refraction at the interface. In the Rochon prism the O-ray is transmitted without deviation for all wavelengths. In the Wollaston prism the ray labelled O is, in fact, the extra-ordinary ray in the first prism, so that both rays are deviated at the interface. With this arrangement a greater separation is obtained. It should be noted that with the Rochon prism the light must pass through in the direction indicated. If the right-hand component is traversed first the plane of vibration will be rotated by the left-hand component and this rotation depends upon wavelength (see Chapter XXIII).

22–8. Polarization by Selective Absorption (Dichroism). Sheet-type Polarizers

The prisms described above provide methods of obtaining linearly, i.e. plane, polarized light by double refraction, one of the components being removed by reflection or by mechanical obstruction. Now, some materials show different absorptions for the O- and E- rays. This effect (discovered in 1815 by Biot) is known as *dichroism* and, if it is very marked, provides a means of producing light that is almost completely plane polarized. One of the best-known naturally occurring dichroic materials is the mineral tourmaline. In this uniaxial crystal the O-ray is much more strongly absorbed than the E-ray, and a single crystal cut parallel to the axis can be used as a polarizer or analyser. To a first approximation, the intensity of the light emerging from the analyser is again proportional to $\cos^2 \theta$ (see Fig. 22-15). Unfortunately, neither the absorption of the O-ray nor the transmission of the E-ray is complete, and the absorptions depend upon

Fig. 22–15.
Two tourmaline plates in series.

the wavelength. The search then began for dichroic polarizers that were more efficient, available with larger apertures, and less expensive. There was little hope that a solution would be found in the form of a naturally-occurring material, and over a very long period there were many unsuccessful attempts to solve the problems by growing large single dichroic crystals in the laboratory.

The breakthrough in the production of dichroic polarizers was made by Land (c. 1930) who succeeded in producing the first *sheet-type* polarizer. Instead of producing a single dichroic crystal, Land obtained a similar effect by aligning millions of dichroic micro-crystals. The scattering of light that one would expect to obtain with such an array of separate crystals was reduced to a very low level by using crystals whose thickness was small compared with the wavelength of light and by embedding these micro-crystals in a sheet of a transparent medium of approximately the same refractive index; thus the sheet behaved like a single crystal. Micro-crystals of herapathite, which tend to be needle-shaped and strongly absorbing for light whose vibration direction is parallel to the long axis, were contained in a sheet of cellulose acetate. When this sheet was stretched unidirectionally the micro-crystals became oriented with their axes approximately parallel, so that the complete sheet behaved as a dichroic material with its major absorption axis parallel to the stretch direction. This material, which is now known as J-sheet, was mass-produced by the Polaroid Corporation, U.S.A.; it was the first form of what is often referred to loosely as "polaroid".

Although the light scatter in J-sheet is small, it is not zero, and a major advance was made by Land (c. 1938) by the introduction of H-sheet, which is a molecular as opposed to a micro-crystalline polarizer; that is to say, it depends for its action on aligned long thin dichroic molecules. Molecular polarizers are now the most commonly used sheet-type polarizers. H-sheet, which consists largely of polymeric molecules that have a preferred direction, and is stained with a material that causes the sheet to be dichroic, can be produced as follows: A thin sheet of polyvinyl alcohol is heated, stretched unidirectionally, and then mounted on a support sheet of cellulose acetate butyrate. The PVA face is then brought into contact with H-ink, a solution which is rich in iodine, so that iodine is absorbed. The sheet is then dried and stored as a roll that can be up to 18 inches wide and of almost unlimited length. Subsequently the required polarizer is cut from the roll and mounted between cover plates of glass or plastic.

The precise properties of H-sheet depend upon the amount of iodine taken up, and several forms are available. The properties of a linear polarizer can be defined by the *principal transmittances*, k_1 and k_2, which give, respectively, the maximum and minimum values of the ratio of the transmitted intensity to the incident intensity for linearly polarized incident light. Ideally, of course, one would have $k_1 = 1$ and $k_2 = 0$ for all wavelengths, the ratio being infinite. In practice, a ratio of about 10^5 can now be achieved over the greater part of the visible region, the higher ratios being accompanied by lower values of both k_1 and k_2: the penalty for a high ratio and excellent extinction is relatively poor transmission for the "transparent direction".

Another molecular polarizer in common use is K-sheet (Land and Rogers c. 1939). This has greater resistance to heat and humidity. Whereas H-sheet is produced by adding atoms to PVA sheet, K-sheet is produced by subtraction: the PVA sheet is heated in the presence of a catalyst (HCl is an example) and dehydration occurs, which causes a small proportion of the PVA to be converted into polyvinylene, which is the dichroic molecule. Alignment is again achieved by stretching, and the sheet is mounted on a supporting material as before and subsequently mounted between cover plates.

The performance of H-sheet tends to fall off at the blue end of the visible spectrum, crossed sheets of some forms giving noticeable transmission; crossed K-sheets have appreciable transmission at the extreme red end of the spectrum. For both H- and K-sheets the direction of maximum transmission is perpendicular to the direction in which the sheet is stretched.

In the visible region sheet-type polarizers have now displaced polarizing prisms such as Nicols in most applications; in the U.V. birefringent polarizers are still used where high performance is required.

22–9. Ray and Wave Velocity for Biaxial Crystals

The occurrence of double refraction in a biaxial crystal can be demonstrated by using Huygens' principle as for uniaxial media. Fig. 22–16 shows the special

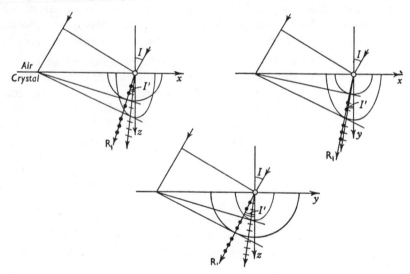

Fig. 22–16. Oblique incidence on a biaxial crystal.

cases where the plane of incidence contains two of the principal axes (see Chapter XIX). In each plane, one of the sections of the wave surface is circular and, as pointed out in §22–1, one has sin I'/sin $I=$constant for one of the refracted rays (OR$_1$).

As defined in Chapter XIX, the phase or wave velocity is the velocity with which a plane wave advances in the direction of the wave normal, and an optic axis is a direction in which there is only one possible phase velocity. In Fig. 22–17, OB is a direction of single ray velocity but it is not an optic axis. Fig. 22–18 shows that a plane wave parallel to T_1T_2 is propagated with the same phase (wave normal) velocity by means of secondary waves having either the elliptic or the circular sections. That is to say, one of the optic axes is the direction OT_2 which is normal to T_1T_2, the tangent common to the ellipse and the circle. The other optic axis is the direction OT_4. Let the direction of the optic axis OT_2 be defined by the angle ϕ which it makes with the x-axis. The equations to the circular and elliptic sections of the wave surface are, respectively,

$$x^2+z^2=v_y^2$$

and

$$\frac{x^2}{v_z^2}+\frac{z^2}{v_x^2}=1.$$

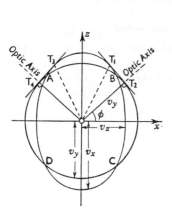

Fig. 22–17. The optic axes.

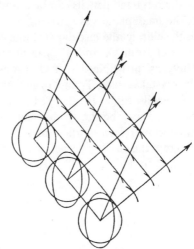

Fig. 22–18.
Propagation of plane waves along an axis.

Now T_2 is the point $(v_y \cos \phi, v_y \sin \phi)$ so that the equation to the tangent to the circle at T_2 is

$$x \cos \phi+z \sin \phi=v_y.$$

If this touches the ellipse one has*

$$v_z^2 \cos^2 \phi+v_x^2 \sin^2 \phi=v_y^2,$$

$$\therefore \quad v_z^2 \cos^2 \phi+v_x^2-v_x^2 \cos^2 \phi=v_y^2$$

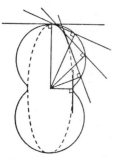

Fig. 22–19.
Connection between
wave surface and
wave velocity surface.

or $$\cos \phi=\sqrt{\frac{v_x^2-v_y^2}{v_x^2-v_z^2}}. \quad . \quad . \quad . \quad (22\text{-}1)$$

This defines the direction of the optic axis OT_2. The angle between the optic axes is usually taken to be acute. That is to say, it is 2ϕ if $(v_x^2-v_y^2)>(v_y^2-v_z^2)$ and $2(\pi/2-\phi)$ if $(v_x^2-v_y^2)<(v_y^2-v_z^2)$; the former inequality defines a negative, and the latter a positive biaxial crystal.

If, in general, one draws a tangent to the elliptic section of the wave surface, the perpendicular distance from O to that tangent represents the wave velocity in that direction. If one does this for each point on the

* The straight line $lx+my=n$ touches the ellipse $\dfrac{x^2}{a^2}+\dfrac{y^2}{b^2}=1$ if $a^2l^2+b^2m^2=n^2$.

ellipse, one obtains a polar diagram of the wave velocity. This is the *wave velocity surface* (sometimes called the *normal surface*). It will be seen from Fig. 22–19 that, conversely, the wave surface (or ray velocity surface) is the envelope of the plane wavefronts that advance in all directions through the origin.

22–10. Internal Conical Refraction

Suppose a narrow, collimated pencil of unpolarized light is incident normally on a biaxial crystal cut perpendicular to the optic axis [Fig. 22–20(*a*)]. The paths of the refracted pencils can be found as follows. Take the point O in the middle of the small plane wave striking the crystal. The line T_1T_2 is the tangent common to the elliptic and circular sections of the wave surface for a disturbance spreading out from O, and T_1T_2 is parallel to the crystal surface. It follows from Huygens' principle that if O is moved over the element of plane wave striking the crystal surface, T_1 and T_2 generate the refracted elements of the wave. Hence the pencil of light is refracted as shown. Actually this is an over-simplification because it can be shown that the tangent plane T_1T_2 actually touches the three-dimensional wave surface in a circle of which T_1T_2 is the diameter. (In the diagram the circle is perpendicular to the paper.) Thus while disturbances travel from O to T_1 and to T_2 they also travel to each point on the circumference of

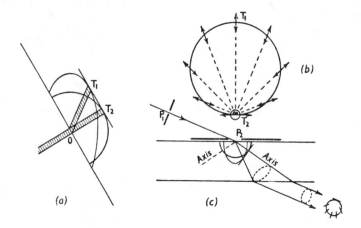

Fig. 22–20. Internal conical refraction.

this circle; as O generates the incident plane wave element, each point on the circumference of the circle generates a similar wave element. Consequently, light spreads into the crystal from O towards each point on the circumference of the circle whose diameter is T_1T_2. That is to say, a hollow cone of light is propagated in the crystal. For each of the generating rays of this cone, the direction of **D** is coplanar with the ray and the wave normal. Thus, viewing the oncoming light, the direction of **D** is as represented by Fig. 22–20(*b*).

If the second surface of the crystal is parallel to the first, a hollow cylinder of light will emerge. The phenomenon, known as *internal conical refraction*, was predicted by Sir William Hamilton and verified by Lloyd using a plate of arragonite ($CaCO_3$) cut with its faces equally inclined to the two optic axes. That is, oblique incidence was used as in Fig. 22–20(c). If pinhole P_1 is kept stationary and P_2 is moved about on the crystal surface, there will, in general, be two refracted rays. When the correct position for P_2 is found, the refracted light spreads out into a hollow cone as indicated. In practice, only an infinitesimal amount of light is incident in *precisely* the correct direction, so that the cone of light predicted by the theory is not seen. Rays in closely neighbouring directions of incidence give light close to the hollow cone referred to and the cone itself is observed as a *dark* cone, i.e. the emerging hollow cylinder has "walls" of finite thickness and the end-on view shows a fine dark ring dividing inner and outer rings of light.

22–11. External Conical Refraction

At the point B [Fig. 22–21(a)], which on the three-dimensional wave surface has the appearance of a dimple, one can draw an infinite number of tangent planes to the wave surface. Hence, for a narrow pencil of rays travelling along the ray axis, there are an infinite number of possible directions of the wave

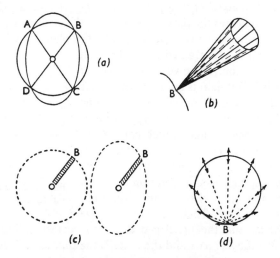

Fig. 22–21. External conical refraction.

normal. These are indicated in (b). Pencils of rays along OB with the two wave normals that lie in the plane of Fig. (a) are of the form shown in (c). Suppose a pencil of rays along OB in the crystal contains plane wave elements with every possible orientation of wave normal. Suppose, further, that the surface of the crystal is at B. There are then an infinite number of emergent rays, and each corresponds to the ray OB and one of the possible associated wave normals in the crystal. The emergent rays constitute a hollow cone of light spreading out from B. For each of the wave normals that can be associated with the ray OB, there is only one possible vibration direction—it is in the plane defined by the

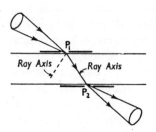

Fig. 22–22. Observation of external conical refraction.

ray and the wave normal. This determines the vibration direction of each of the emergent rays. Looking towards the oncoming cone of rays, the vibration directions are as indicated by (*d*). Conversely, a hollow cone of suitably polarized rays incident at B from outside the crystal will unite to give a single ray BO inside.

Experimentally, the phenomenon of *external conical refraction* can be observed as follows. A convergent cone of unpolarized light is incident on the crystal through the pinhole P_1 (Fig. 22–22). Those rays in the incident cone that have the appropriate vibration directions combine into a single ray, and others give other refracted rays. If the position of P_2 is adjusted until P_1P_2 is the ray axis, the single ray P_1P_2 gives an emergent cone of rays, and all other rays in the crystal are obstructed.

22–12. Passage of Light Through Crystal Plates

Suppose a plane polarized beam of light is incident normally on the surface of a plane parallel uniaxial crystal plate. Fig. 22–23 shows a principal section of the crystal and indicates the passage of the light through the plate. In general, there will be two emergent rays, one (O) with the plane of vibration perpendicular to the principal section, and the other (E) with **D** in the principal section. If, however, the plane of vibration of the incident light is in or perpendicular to the principal section, there will be only one plane polarized emergent beam. The two directions of vibration of the incident light which correspond to a single plane polarized emergent beam are sometimes referred to as the *privileged directions* of the crystal. One corresponds to the vibration direction of the O-ray, and the other is the vibration direction of the E-ray. These two rays have

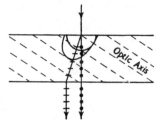

Fig. 22–23. Passage of light through a crystal plate.

different velocities, and the privileged direction corresponding to **D** for the faster ray is called the *fast axis* and the other is the *slow axis*. These correspond to the O- and E-rays respectively for a positive crystal and vice versa for a negative crystal. If the incident light is unpolarized, there will always be two plane polarized emergent rays, and their intensities will not change as the crystal is rotated about the incident ray as axis.

If the crystal plate is cut perpendicular to the optic axis, the state of polarization of the emergent beam is always the same as that of the incident beam, and there are no privileged directions.

If the crystal plate is cut parallel to the optic axis, the O- and E-rays travel along the same line but with different speeds. Fig. 22–24 shows a collimated incident beam of light plane polarized with **D** neither in nor perpendicular to the principal section of a negative crystal. In this case the O and E disturbances will emerge from the plate with different phases and, since they are coherent and along the same path, the perpendicular vibrations are superposed to give plane, circularly, or elliptically polarized light (see below). If the incident light is unpolarized, it can be resolved into *incoherent* plane polarized components with any two mutually perpendicular vibration directions. That is, incoherent O and E disturbances are propagated through the crystal and emerge as unpolarized light. (One can, in fact, be said to have unpolarized light at all points through the plate.)

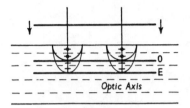

Fig. 22–24. Plate cut parallel to the axis.

Consider, again, the case of a plane polarized incident beam. With the negative crystal shown in the figure, the privileged directions of the plate will be along and perpendicular to the optic axis, the former being the fast and the latter the slow axis. If the vibration direction of the incident light is not in one of the privileged directions, it will have components in both these directions. For example, if the vibration direction of the incident beam makes angle θ with the optic axis, the amplitudes of the O- and E-waves will be $A \sin \theta$ and $A \cos \theta$ respectively, where A is the amplitude of the incident light. Fig. 22–25 shows the end-on view looking against the oncoming light. Suppose the thickness of the crystal plate is d and let n_O and n_E be the ordinary and extraordinary refractive indices. Then the optical path through the crystal is $n_O d$ for the O-ray and $n_E d$ for the E-ray. That is to say, the O- and E-rays pass through optical paths which differ by $(n_O - n_E)d$. Now the vibrations of the O and E disturbances commence in phase since they are components of a linear vibration. Consequently, they emerge from the plate with phase difference given by

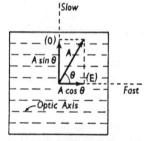

Fig. 22–25. Resolution along fast and slow axes.

$$\delta = \frac{2\pi}{\lambda}(n_O - n_E)d, \quad . \quad . \quad (22\text{-}2)$$

where λ is the wavelength in air.

When the coherent O and E disturbances emerge from the crystal, their perpendicular linear simple harmonic vibrations are superposed. Imagine now that the thickness of the crystal can be increased gradually. When $\delta = 0$ (or $2p\pi$, where p is an integer) the resultant is a linear vibration identical to that of the incident light. As δ increases from zero, the resultant becomes an elliptic motion, and, as δ continues to increase, the eccentricity changes and the axes rotate as indicated in Fig. 22–26. When δ reaches π [or $(2p+1)\pi$] the resultant is again linear so that the emergent light is plane polarized, although it does not have the same vibration direction as the incident light. Further increase of δ gives the elliptic resultants shown until one returns to the plane polarization of the incident light when $\delta = 2\pi$. Fig. 22–26 can be looked upon as giving the state of polarization at different distances through a thick crystal as the light traverses it. It should be noted that between $\delta = 0$ and $\delta = \pi$ the elliptic motion is anti-clockwise, and between $\delta = \pi$ and $\delta = 2\pi$ it is clockwise. Obviously, the reverse is true if the vibration direction of the incident light is inclined in the opposite direction, or if one looks along the rays from the opposite direction. It should be noted also that if, in the above example, the vibration direction of the incident

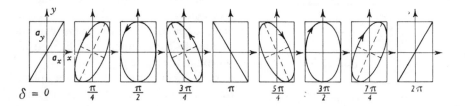

Fig. 22–26. State of polarization of light passing through a crystal plate. (y lags.)

light had been inclined to the optic axis at $\theta=45°$, one would have obtained circularly polarized emergent light when $\delta=\pi/2$ [or $(2p+\frac{1}{2})\pi$] and $\delta=3\pi/2$ [or $(2p+3/2)\pi$].

If the light emerging from the crystal plate is passed into an analyser (such as a Nicol), only the component parallel to the principal section of the analyser will be transmitted. That is to say, after compounding the O and E disturbances from the plate the resultant must be resolved into a new pair of perpendicular directions—that for which the analyser is opaque, and that for which it is transparent. This can be done by resolving the O and E disturbances from the plate in the relevant directions. Thus to find the disturbance transmitted by the analyser, one finds the components of the O and E vibrations in the direction for which the analyser is transparent and, allowing for the phase difference between them, the two components are superposed (see also § 22–17).

Before proceeding further with a discussion of the effect of crystal plates on polarized light, it should be pointed out that elliptically polarized light can be resolved into any two perpendicularly polarized plane components. If it is resolved into plane polarized components whose vibration directions are parallel to the axes of the ellipse, the components are $\pi/2$ out of phase, but if it is resolved into any other perpendicular directions, the components are not $\pi/2$ out of phase. That is to say, when elliptically polarized light enters a crystal plate cut parallel to the optic axis, the phase difference between the O and E components depends on the orientation of the ellipse. The relation between the orientation of the axes of the ellipse and the phase difference between its components has already been given [equation (6–57)].

If unpolarized light enters the crystal the components are incoherent and the emergent light is also unpolarized.

22–13. Quarter- and Half-wave Plates

A crystal plate that introduces a phase difference of $\pi/2$ (a quarter of a cycle) between the O and E disturbances is called a *quarter-wave plate*. One also has half-wave plates, which introduce a phase difference of π, and so on.

From the discussion of the previous section it will be seen that for a full-wave plate the state of polarization of the emergent light is always identical to that of the incident light.

With a half-wave plate the general state of polarization of the emergent light is the same as that of the incident light, but usually with a different orientation. For example, suppose that the incident light is plane polarized and that the vibration direction makes angle θ with the optic axis. The plate introduces a phase retardation of π in the O-ray relative to the E-ray so that, on emergence, the vector representing the O vibration (Fig. 22–25) is reversed. Thus the resultant is in the direction $-\theta$. That is to say, the plane of vibration is rotated 2θ. There is, of course, no rotation if the plane of vibration of the incident light is in one of the privileged directions, i.e. if $\theta=0$ or $\pi/2$. Reference to Fig. 22–26 will also illustrate the fact that, unless the axes are in the privileged directions, an elliptically polarized incident beam gives a similarly polarized emergent beam, the axes of the ellipse being rotated.

Consider now a quarter-wave plate. When the incident light is plane polarized one has, in general, elliptically polarized emergent light, the axes of the ellipse being parallel to the privileged directions. There are two special cases: if the incident light has its vibration direction parallel to, or perpendicular to, the optic axis, the emergent light is identically polarized; and if the vibration direction is initially at 45° to the optic axis, the emergent light is circularly polarized.

If, with a quarter-wave plate, the incident light is circularly polarized, or elliptically polarized with axes parallel to the privileged directions of the plate, the incident disturbance can be resolved into plane polarized O and E components which are $\pi/2$ out of phase. The plate then removes this phase difference so that plane polarized light emerges. With elliptically polarized incident light, the O and E components are not, in general, $\pi/2$ out of phase, so that they do not emerge in phase. That is, one has elliptically polarized emergent light.

Quarter-wave plates are frequently made of quartz or mica. Actually mica is biaxial but there is a clear colourless form for which the angle between the optic axes is vanishingly small. Mica splits easily into thin sheets and by trial and error one can select a sheet of the thickness necessary to give a quarter-wave plate. It is not always easy to cleave a mica plate for which the optic axis is parallel to the surface, but with quartz—which has to be cut and polished—it is usual to produce a plate in which the axis is so orientated. The principal refractive indices of quartz for sodium light are 1·55336 and 1·54425. Hence, since $\delta=(2\pi/\lambda)(n_E-n_O)d$, one has $d=0\cdot0164$ mm. when $\delta=\pi/2$. One can, of course, use more robust plates giving $\delta=5\pi/2$, $9\pi/2$, etc. Such plates give identical effects with polarized light but it is easy to distinguish between them with a micrometer screw gauge.

A sheet or plate of birefringent material, which introduces a phase difference between the O- and E-disturbances, is often called a *linear retarder*. Since a quarter-wave plate can convert plane polarized light into circularly polarized light, one can produce a *sheet-type circular polarizer* which will produce circularly polarized light from unpolarized light by combining a sheet-type linear polarizer such as H-sheet (see §22–8) with a $\pi/2$ sheet-type retarder with its axis at 45°; stretched polyvinyl alcohol has been used for the linear retarder. As a generalization, one obtains an *elliptical polarizer* if the angle between the linear polarizer and the $\pi/2$ linear retarder is not 45° or if the retarder does not retard by $\pi/2$. It will be seen that it is essential for the light to pass through a circular or elliptical polarizer in the correct direction. If unpolarized light enters in the wrong direction, the retarder has no effect and the linear polarizer simply gives linearly (plane) polarized light.

22–14. Use of a Quarter-wave Plate in the Analysis of Polarized Light

A theorem due to Stokes states that there are only seven possible states of polarization of a beam of light: unpolarized, plane polarized, elliptically polarized, circularly polarized, and three mixtures—each of unpolarized and one of the other three types. On passing the light through an analyser which can rotate about the direction of the light as axis, one can detect the presence of elliptically or plane polarized light since this will cause the intensity of the transmitted light to vary with the orientation of the analyser. If there is no variation with orientation, the original light is unpolarized, circularly polarized, or a mixture of the two. By combining an analyser (such as a Nicol) and a quarter-wave plate, it is possible to make a complete systematic qualitative analysis of the light. One proceeds as follows:

I. Insert analyser and rotate it. There are three possibilities:
   (*a*) No light is transmitted for one orientation of the analyser.
   (*b*) The intensity of the transmitted light is independent of the orientation of the analyser.
   (*c*) The intensity varies with orientation but never falls to zero.
   (*a*) indicates that the light is completely plane polarized; (*b*) indicates that the light is unpolarized, circularly polarized, or a mixture of the two; (*c*) indicates elliptically polarized light, a mixture of elliptically polarized and unpolarized light, or a mixture of plane polarized and unpolarized light.

II. To investigate I (*b*) further, pass the light through a quarter-wave plate followed by an analyser, and rotate the analyser. There are the same three possibilities: (*a*), (*b*), and (*c*). (*a*) indicates that the original beam is circularly polarized; (*b*) indicates that it is unpolarized; (*c*) indicates that it is a mixture of circularly polarized and unpolarized light.

III. To investigate I (*c*) further, one again uses a quarter-wave plate and an analyser but, now, the plate and analyser are rotated independently. There are only two possibilities:
   (*a*) The light is extinguished for one setting.
   (*b*) The intensity of the light varies as the components are rotated, but never falls to zero.
   In this case (*a*) indicates that the light is elliptically polarized and that the axes of the ellipse are parallel to the privileged directions of the plate when the light is extinguished; (*b*) indicates that the light is a mixture of elliptically polarized and unpolarized light, or a mixture of plane polarized

and unpolarized light. For the former, when the intensity is a minimum the privileged directions of the plate are parallel to the axes of the ellipse, and the position of the analyser gives the ratio of the axes (see §22–16). A mixture of plane polarized and unpolarized light emerges as a special case of the last mixture and is indicated by the fact that the analyser is parallel to one of the privileged directions of the plate when the intensity is a minimum. That is to say, one of the axes of the ellipse contracts to zero.

The student should satisfy himself that he understands why the effects described lead to the conclusions stated. Note that, after passage through a quarter-wave plate, circularly polarized light becomes plane polarized; plane polarized light can become elliptically, circularly, or plane polarized, according to the orientation of the plate; elliptically polarized light emerges either plane or elliptically polarized; and unpolarized light remains unpolarized.

22–15. The Babinet and Soleil Compensators

It is convenient, for the production and analysis of elliptically polarized light, to have a crystal plate of variable thickness. Such plates are known as compensators. Fig. 22–27 shows the Babinet and Soleil compensators. In the former one has two quartz wedges of small angle, the optic axes being as shown. Light enters from above and, in contrast with the Wollaston prism, the wedge angles are sufficiently small for one to be able to neglect the separation of

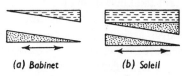

(a) Babinet   (b) Soleil

Fig. 22–27. Compensators.

the O- and E-rays. The upper wedge is fixed and, by means of a micrometer screw, the lower can be moved as indicated. On entering the first wedge, the O-ray travels faster than the E-ray (quartz being a positive crystal). Since the optic axis of the second wedge is perpendicular to that of the first, the O-ray for the first wedge becomes the E-ray for the second and vice versa. That is to say, the rays exchange speeds on passing into the second wedge. By moving the point of incidence along the upper wedge, or by moving the lower wedge, one can cause the paths in the wedges to be different. In this way one can control the phase difference between the emergent rays so that one has, in effect, a plate of variable thickness.

The fact that the effective thickness of the plate depends on the point of incidence is sometimes a disadvantage. This difficulty is overcome in the Soleil compensator which consists of two thin wedges and a plane-parallel plate cut as shown. The bottom wedge can be moved as indicated. The combined effect of the wedges is that of a plane-parallel plate of variable thickness and this, combined with the top plane-parallel plate, gives, in effect, a thin plate of variable thickness but whose thickness is the same at all points across it.

22–16. Analysis of Elliptically Polarized Light using a Babinet Compensator

It is first necessary to calibrate the compensator. To do this, it is placed between a polarizer and a crossed analyser so that the plane polarized light from the polarizer has its plane of vibration at about 45° to the axes of the wedges [see Fig. 22–28(a)]. At points where the effective thickness of the compensator is $2p\pi$ (p integral) the emergent light is plane polarized in the same plane as the

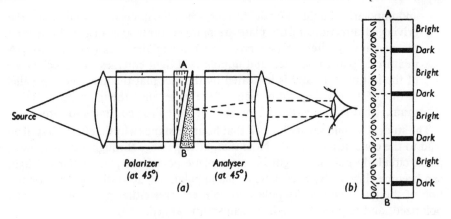

Fig. 22–28. Analysis of elliptically polarized light.

incident light and is removed by the analyser. Now the compensator has constant thickness along lines parallel to the edges of the wedges so that, across the aperture of the compensator, one has a series of dark fringes as in (b). As one moves across the aperture from A to B, the effective thickness of the compensator varies and the polarization of the emergent light varies as shown in (b). There is, in general, a component of **D** in the plane of the analyser so that some light is transmitted. If the original plane of vibration is exactly at 45° to the axes of the wedges, one has maxima midway between the minima since the plate gives a retardation of $(2p+1)\pi$ and the emergent light is plane polarized perpendicular to the incident light, i.e. in the transparent direction of the analyser. If the original plane of vibration is not exactly at 45°, the emergent light is not exactly parallel to the transparent direction of the analyser when the retardation is $(2p+1)\pi$ and the contrast in the fringes is reduced. As the wedge of the compensator is moved, the position of zero effective thickness moves across the wedge, i.e. the fringes move across the field of view. The movement of the screw (θ) necessary to produce a shift of one fringe spacing is that which changes the phase difference by 2π. The compensator is then adjusted until a minimum falls on the cross-wires. The phase difference at that point in the field of view is then $2p\pi$. The micrometer screw is then turned through $\theta/4$. This makes the phase difference at the cross-wires $(2p+\frac{1}{2})\pi$. That is, one has a quarter-wave plate in that part of the field of view. The polarizer is then removed and the elliptically polarized light to be measured is passed into the compensator and analyser. These are rotated independently until the bands are as clear as possible and a dark band is brought on to the cross-wires. This corresponds to plane polarized light leaving the quarter-wave plate so that the axes of the ellipse must be parallel to the optic axes of the compensator wedges (the privileged directions).

The $\pi/2$ phase difference between the components of the ellipse is removed by the compensator to give the plane polarized emergent light, and the plane of vibration of the latter is determined by the ratio of the axes of the ellipse (see Fig. 22–29). It is perpendicular to the transparent direction of the analyser so

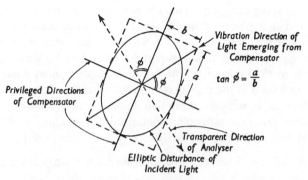

Fig. 22–29. Analysis of elliptically polarized light.

that the ratio of the axes of the ellipse is given by the tangent of the angle that the analyser makes with the principal directions of the compensator.

In the procedure described above the compensator is simply used as a quarter-wave plate. It has already been pointed out in § 22–12 that elliptically polarized light can be resolved into linear components in *any* two perpendicular directions. Suppose, then, that one passes elliptically polarized light into a calibrated compensator. Whatever the orientations of the ellipse and the compensator axes, the ellipse can be resolved into linear components parallel to the axes of the compensator. One can then adjust the compensator wedges until the phase difference between the linear components is compensated and plane polarized light emerges. An analyser will indicate when the setting of the compensator is correct since it will give extinction when suitably orientated. The orientation of the analyser will indicate the ratio of the components parallel to the compensator axes and the compensator will indicate the phase difference between them. (6–57) then gives the orientation of the ellipse, but its eccentricity (although calculable) is not indicated directly.

22–17. Passage of Collimated White Light through a Crystal Plate

In the previous section it has been assumed that the light is monochromatic. In general, a crystal plate introduces a phase difference which depends upon the wavelength. Frequently the path difference $(n_E - n_O)d$ remains approximately constant throughout the visible spectrum so that the phase difference, $(2\pi/\lambda)(n_E - n_O)d$, varies by a factor of about 2. Thus if plane polarized white light is passed through a crystal cut parallel to the optic axis, the polarization of the emergent light will vary with wavelength. If this light is passed through an analyser, the transmitted intensity will vary with wavelength and the emergent beam will be coloured. In particular, if the plate is replaced by a Babinet compensator the fringes will be coloured. (The central fringe, corresponding to zero effective thickness, will be achromatic, e.g. will be black when the polarizer and analyser are crossed.) Again, a quarter-wave plate will give the nominal $\pi/2$ phase change for one wavelength only—usually for sodium light. If plane polarized white light enters a half-wave plate, only the sodium light will emerge plane polarized and an analyser set to extinguish this will transmit some light of other wavelengths. The emergent light has the characteristic reddish-violet colour which is usually called the *tint of passage* (see also § 23–8).

Consider, now, the case of a crystal cut parallel to the axis and placed between a polarizer and an analyser. Suppose the transparent directions of the polarizer and the analyser make angles θ and ϕ with the optic axis of the crystal. The incident plane polarized light has amplitudes $a \cos \theta$ and $a \sin \theta$ parallel to and perpendicular to the optic axis. The former has component $a \cos \theta \cos \phi$ and the latter $a \sin \theta \sin \phi$ in the transparent direction of the analyser. There will be a phase difference, δ, between these two components which are coherent since they are derived from the same plane polarized beam. Consequently, the resultant intensity transmitted by the analyser is given [from (6–9)] by

$$I = a^2 \cos^2 \theta \cos^2 \phi + a^2 \sin^2 \theta \sin^2 \phi + 2a^2 \sin \theta \cos \theta \sin \phi \cos \phi \cos \delta$$

$$= a^2 \cos^2 \theta \cos^2 \phi + a^2 \sin^2 \theta \sin^2 \phi + 2a^2 \sin \theta \cos \theta \sin \phi \cos \phi \left(1 - 2 \sin^2 \frac{\delta}{2} \right)$$

$$= a^2 \cos^2 (\theta - \phi) - a^2 \sin 2\theta \sin 2\phi \sin^2 \frac{\delta}{2}. \quad . \quad . \quad . \quad . \quad . \quad . \quad . \quad (22\text{-}3)$$

The first term does not contain δ and therefore does not depend upon the crystal thickness or the wavelength. It depends solely upon the angle $(\theta - \phi)$ between the polarizer and the analyser, being a maximum when they are parallel and zero when they are crossed. The second term, involving δ, varies with wavelength, but is zero when either the polarizer or analyser is parallel to one of the privileged directions of the crystal. The first term represents the light which would be transmitted by the analyser in the absence of the crystal and shows that the transmitted light would be white. When summed for all wavelengths, the second term represents coloured light added to or subtracted from the white light. Now the sign of this term changes when either the polarizer or the analyser is rotated 90°, keeping the rest of the system fixed. Hence such a rotation changes the colour of the emergent beam to the complementary colour. For example, the colour of the emergent beam with crossed Nicols is complementary to that obtained when the analyser is rotated to make them parallel. If the analyser is replaced by a simple calcite crystal one will obtain two emergent beams—polarized at right angles and of complementary colours.

As a further example of the passage of white light through a crystal, suppose a thick plate is placed between a polarizer and a crossed analyser with the optic axis of the plate inclined at 45° to the plane of vibration of the light incident upon it. The emergent light will be plane, elliptically, or circularly polarized, depending upon the phase shift introduced by the plate, i.e. depending upon the wavelength. The analyser will remove those wavelengths for which the plate gives a phase shift that is a multiple of 2π and, if the light from the analyser enters a spectroscope, the spectrum will show a series of dark fringes. If the crystal plate is replaced by a Soleil compensator, one can adjust the equivalent thickness and control the fringe spacing.

When a very thick crystal plate is placed between crossed Nicols with white light, the wavelengths removed are very numerous and, since they are distributed throughout the spectrum, the emergent light still appears white. With thin plates relatively few wavelengths are removed and, in general, the emergent light is coloured. Thin pieces of many anisotropic substances such as mica, when chosen at random, show considerable variations of thickness and, between crossed Nicols in white light, show many variations of colour. The general effect is, of course, not restricted to uniaxial crystals cut and orientated in the particular way discussed above, although the discussion is restricted to the passage of approximately parallel light through the crystal.

22–18. Passage of Convergent Plane Polarized Pencils through Crystal Plates

(a) Experimental Arrangement

The light transmitted by a crystal plate depends upon the angle of incidence of the light and the orientation of the crystal axis. For a crystal cut in a given manner one can investigate the effect of varying the angle of incidence by mounting the crystal as shown in Fig. 22–30. The screen AB is in the focal plane of the lens L. P is the crystal plate and an analyser (not shown in the figure) is placed between P and L or between L and AB. It will be seen that rays which come to a focus at a point on AB must have been parallel when they passed through the crystal. That is to say, each point on AB corresponds to a given *direction* through the crystal and not to a given local region of it. Consequently, the variation of intensity across AB indicates the way in which the state of polarization of the transmitted light varies with angle of incidence on the crystal. It is necessary for all rays incident on the crystal in a given direction to have the same path through the crystal, so that the plate must be accurately plane parallel. For visual observation, and

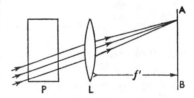

Fig. 22–30. Passage of an oblique pencil through a crystal plate.

to obtain the necessary angular aperture, it is convenient to use a polarizing microscope with the Amici-Bertrand lens in position (see §22–19). The system is then known as a *conoscope*. It will be seen that, owing to the effect of the Amici-Bertrand lens, each direction in the field of view corresponds to a single direction through the crystal and not to a single point on its surface.

(b) Uniaxial Crystal Cut Perpendicular to the Optic Axis

For any collimated pencil of coherent plane polarized rays incident obliquely on the crystal plate, there will, in general, be two emergent pencils. These will be coherent and plane polarized at right angles. If the crystal is fairly thin and the pencil of rays fairly broad, the two pencils will not be separated on emergence and one can superpose the emergent disturbances in the usual way. The crystal will introduce a phase difference between the perpendicularly polarized disturbances. This phase difference will depend upon the optical path difference through the crystal and, for a given crystal thickness and given wavelength, it will depend only upon the angle between the incident ray and the optic axis. Consequently, a movement of the crystal along the axis of the system does not affect the transmitted light. That is to say, it makes no difference to the observed effects if the crystal is located beyond the focus of the principal rays of the collimated pencils. Many students find it easier to visualize this arrangement with divergent pencils. Thus in Fig. 22–31 suppose the principal rays of the pencils diverge from the point P on the axis of the system. Consider the pencils whose principal rays are incident at points on the circumference of the circle

ABCD, centred on the axis. These pencils are equally inclined to the axis of the system and come to a focus at points on the circumference of a circle in the image plane. They also make equal angles with the optic axis of the crystal, so that equal phase differences are introduced between their emergent components. In particular, if the analyser and polarizer are crossed and the phase difference introduced by the crystal is a multiple of 2π, the resultant emergent disturbances

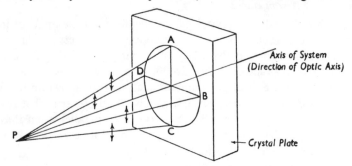

Fig. 22–31. Diverging cone incident on a crystal plate.

will be obstructed and there will be a dark ring in the field of view. For principal rays inclined to the axis at certain other angles, the phase differences between emergent components will be other multiples of 2π. That is, there will be a series of concentric dark rings in the image plane or field of view and, between these dark rings, there will be maxima. The radii of the bright and dark rings depend upon the wavelength and coloured rings will be observed when white light is used. However, since there is symmetry about the axis of the system, the colour is constant round any circle centred on the axis. Such a line is called an *isochromatic line*. Suppose, now, that the plane of vibration of the incident light

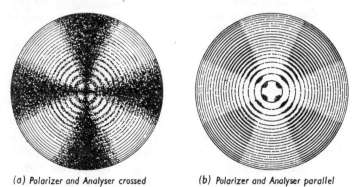

(a) Polarizer and Analyser crossed   (b) Polarizer and Analyser parallel

Fig. 22–32. Rings and brushes for a uniaxial crystal cut perpendicular to the axis.

is parallel to AC. Pencils whose principal rays are incident at points along AC have their plane of vibration in the plane of incidence, and those whose principal. rays are incident along BD have their plane of vibration perpendicular to the plane of incidence. The former have no component of **D** in the direction corresponding to the O-ray, and the latter give no E-ray in the crystal; for these pencils the vibration is in a privileged direction. Consequently, the emergent rays will be polarized in the same plane as the incident rays and will be obstructed by the crossed analyser. In the field of view this gives two perpendicular diameters of zero intensity. Since these dark lines will be the same for all wavelengths, they are called the *achromatic* lines or *isogyres*. The isochromatic maxima are, of course, interrupted by the achromatic lines, and the intensities of the former build up slowly as one moves away from the latter. The isochromatic lines are usually called the *rings* and the achromatic lines are called the *brushes* since they are less sharp. When the analyser and polarizer are parallel, the complementary patterns are observed. Fig. 22–32 shows the rings and brushes for a uniaxial crystal cut perpendicular to the axis and placed between crossed and parallel Nicols.

(c) Isochromatic Surfaces

With a crystal plate in converging (or diverging) pencils it is necessary to know the way in which the phase difference between the O and E disturbances depends

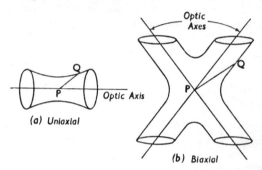

Fig. 22–33. Isochromatic surfaces.

upon the direction of the light. Suppose that one takes a point P in an anisotropic medium and that one has plane waves passing through P such that the O- and E-wave normals are along PQ [Fig. 22–33(a) or (b)]. If the disturbances are in phase at P the phase difference, δ, at Q will depend on the length and direction of PQ. Taking P as origin, let Q move in such a way that the length PQ always corresponds to a given phase difference. The locus of Q will be a surface of constant phase difference for waves from P, and by taking a number of different values of δ a family of surfaces is obtained. They are known as *isochromatic surfaces* (sometimes as Bertin's surfaces). Fig. 22–33 shows typical isochromatic surfaces for uniaxial and biaxial media.

As was pointed out above, a movement of the crystal along the axis of the system does not affect the pattern observed with a conoscope. It is now convenient to take the intersection point of the principal rays (the point P in Fig. 22–31) to be on the first surface of the crystal. Taking this as the origin for the isochromatic surfaces (P in Fig. 22–33), the wave normal directions in the crystal that correspond to a given phase difference on emergence are found by joining P to points on the curve of intersection of the appropriate isochromatic surface with the second surface of the crystal plate. For a uniaxial crystal cut perpendicular to the optic axis as described above, this intersection is a circle for each isochromatic surface. In this case the observed isochromatic lines have the same form as the intersections of the isochromatic surface with the crystal surface but, in general, this is not so because, for crystals cut with other orientations of the optic axis, wave normals in different azimuths are refracted differently on emergence. That is to say, those which are equally inclined to the axis of the system within the crystal are not equally inclined to the axis after emergence. However, the observed isochromatic lines do bear a superficial resemblance to the intersections of the isochromatic surfaces with the crystal face, and these intersections can be used to give a qualitative explanation of the form of the isochromatic lines. It should be noted that for a given incident wave normal there are, in general, two wave normals in the crystal. Hence the phase difference required is not that between waves whose normals are parallel. This effect also can be neglected in an elementary treatment of the subject.

(d) Uniaxial Crystal Cut Parallel to the Optic Axis

By considering the intersections of the isochromatic surfaces with the crystal surface as described above one can explain, in general terms, the hyperbolic form of the isochromatic lines obtained when a uniaxial crystal is cut parallel to the optic axis (see Fig. 22–34).

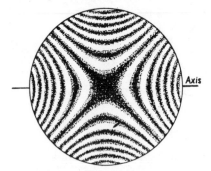
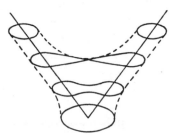

Fig. 22–34. Isochromatic lines for a uni-axial crystal cut parallel to the axis.

Fig. 22–35. Sections across an isochro-matic of a biaxial crystal.

(e) Biaxial Crystal Cut Perpendicular to the Direction Bisecting the Optic Axes

The shape of an isochromatic surface is shown in Fig. 22–33(b). If, at various distances from P, one takes sections normal to the direction bisecting the optic axes, one obtains the curves shown in Fig. 22–35. These, then, are the forms of

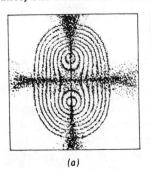
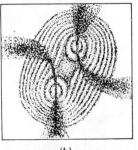
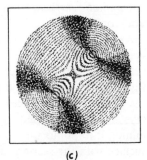

(a)   (b)   (c)

Fig. 22–36. Isochromatic lines for a biaxial crystal cut perpendicular to the bisector of the angle between the axes.

the intersections of the second crystal surface with various isochromatic surfaces if one takes the origin P to be on the first surface of the crystal plate. This gives a qualitative explanation of the isochromatic lines actually observed (see Fig. 22–36). The shape of the brushes, or achromatic lines, depends upon the orientation of the polarizer and analyser to the planes of the optic axes. In (a) the plane containing the axes is parallel to the principal section of the polarizer; in (b) the crystal has been rotated slightly from this position; and (c) shows the pattern for a crystal at an angle of 45°. In each case the polarizer and analyser are crossed and the crystal is cut perpendicular to the bisector of the angle between the axes.

22–19. The Polarizing Microscope

Fig. 22–37(a) shows, schematically, the arrangement for straightforward examination of specimens between a polarizer and analyser. In the past, Ahrens

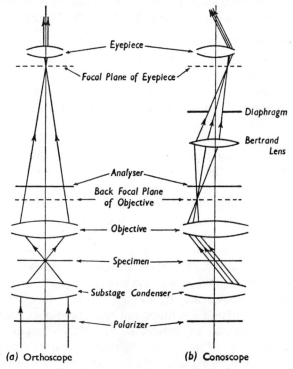

(a) Orthoscope   (b) Conoscope

Fig. 22–37. The polarizing microscope.

prisms were in common use, and since it was necessary for collimated light to traverse the analyser, the prism was placed between two extra lenses. Sheets of Polaroid are now used for the polarizer and analyser and the extra lenses are no longer required. It is usually convenient for the specimen to be carried on a revolving circular stage or on a stage which enables it to be tilted as well as rotated. When it is desired to examine a crystal plate in converging pencils of parallel rays, an additional *Amici-Bertrand lens* is interposed as in (b). This converts the system into a *conoscope*. It will be seen that each point in the field of view corresponds to a given *direction* through the crystal. In effect, the Bertrand lens and the eyepiece constitute a system used to examine the pattern in the back focal plane of the objective. This corresponds to the pattern in the plane AB of the lens L in Fig. 22–30 (§22–18). When the Bertrand lens is in use, a small aperture is positioned conjugate to the specimen as shown in the figure. In consequence, only a small area of the specimen is used and a plane-parallel element can be selected.

In order to analyse the light transmitted by the specimen, a quartz compensator or a quarter-wave plate can be interposed between the objective and the analyser. It is also possible in some instruments to locate a quartz wedge in the focal plane of the eyepiece. The advantage of this method is that the graduations on the wedge, which indicate its thickness, are in focus with either of the arrangements shown in Fig. 22–37. In addition, the eyepiece is often fitted with a rotating analyser; obviously this can be done less clumsily with Polaroid than with any calcite prism analyser. With the conoscopic system it is sometimes convenient to mount a quartz wedge in the back focal plane of the objective.

22–20. Jones Vectors and Matrices[1]

The Jones vectors and matrices provide a method of specifying the intensity and the state of polarization of a beam of completely polarized light and of specifying the properties of a system of polarizers and retarders. Suppose one has a collimated beam of completely polarized, nominally monochromatic, light travelling in the positive z-direction. As has already been explained (§§6–18 and 22–12), whatever the state of polarization, the (transverse) light disturbance of a completely polarized beam can be resolved into x and y components that are coherent. In any z-plane the amplitudes and phases of these components can be specified by the corresponding complex amplitudes. In the most general example of an ideal system of polarizers and retarders each component of the incident light gives a contribution to each component of the emergent light, and since the contributions are still coherent they simply interfere to give the total x and y components and hence the intensity and state of polarization of the emergent light. Suppose, then, that the system is such that an incident x component of unit amplitude and zero phase gives emergent x and y components whose complex amplitudes are $J_{xx} = T_{xx}e^{-i\delta_{xx}}$, $J_{yx} = T_{yx}e^{-i\delta_{yx}}$ and that an incident y component of unit amplitude and zero phase gives emergent x and y components whose complex amplitudes are $J_{xy} = T_{xy}e^{-i\delta_{xy}}$ and $J_{yy} = T_{yy}e^{-i\delta_{yy}}$. If the x and y components of the incident light have complex amplitudes ψ_x and ψ_y, the complex amplitudes of the x and y components of the emergent light are given by

$$\left. \begin{array}{l} \psi_x' = J_{xx}\psi_x + J_{xy}\psi_y \\ \psi_y' = J_{yx}\psi_x + J_{yy}\psi_y \end{array} \right\} \qquad \ldots \ldots \quad (22\text{–}4)$$

[1] The literature is chaotic on the subject of notation and sign convention in this field. Whatever symbols and sign convention are used there will be agreement with some authorities and disagreement with others. If one uses the notation of the originators of the various techniques to be discussed, unavoidable ambiguities arise. It has been decided to attempt to aid the memory of the reader during his passage through the following pages by using **J** for the Jones matrix, **C** for the Coherency matrix, S for the Stokes parameters and **M** for the Mueller matrix. The reader should also note that in this book a complex amplitude is written as $\psi = ae^{-i\delta}$ where δ denotes a phase lag.

or, in matrix form,

$$\begin{bmatrix} \psi_x' \\ \psi_y' \end{bmatrix} = \begin{bmatrix} J_{xx} & J_{xy} \\ J_{yx} & J_{yy} \end{bmatrix} \begin{bmatrix} \psi_x \\ \psi_y \end{bmatrix} \qquad \ldots \ldots \quad (22\text{–}5)$$

One writes $$\Psi' = J\Psi \qquad \ldots \ldots \ldots \quad (22\text{–}6)$$

The column vectors Ψ, Ψ', which specify the incident and emergent disturbances, are, in effect, the *Jones vectors* (Jones, 1941) and the matrix J corresponds to the *Jones matrix* of the system. Often one is not interested in absolute intensities and phases but simply in the intensity of the emergent light relative to the incident light and in the states of polarization of the incident and emergent beams. That is, the Jones vector of the incident light can be simplified by taking the intensity to be unity and the phase of one component to be zero. Similarly the matrix can be simplified by taking one of the phase changes to be zero. This simplifying procedure is illustrated in the examples given below.

If light passes through a succession of n polarizing and retarding components in the order 1 to n the Jones matrices of the components being $J_1 \ldots \ldots J_n$, one has

$$\Psi_1' = J_1\Psi_1, \ \Psi_2' = J_2\Psi_2 \equiv J_2\Psi_1' = J_2J_1\Psi_1, \quad \text{etc.}$$

i.e. $$\Psi_n' = J\Psi_1 \qquad \ldots \ldots \ldots \ldots \quad (22\text{–}7)$$

where the overall Jones matrix is given by

$$J = J_n J_{n-1} \ldots \ldots J_2 J_1 \qquad \ldots \ldots \quad (22\text{–}8)$$

and must, of course, be evaluated by taking the products in this order.

A situation may arise where one needs to know the effect of a system on incident beams with various states of polarization in various orientations. The Jones matrix for the complete system is then calculated from equation (22–8) using a particular pair of axes (x, y) and the x and y components of the various incident beams then found in turn before applying equation (22–7) (remembering that this gives the x and y components of the emergent beams). Thus the matrix for the system needs to be calculated once only.

It should be noted that if two coherent beams of polarized light are added, the Jones vector of the resultant is simply the sum of the separate vectors since for both x and y components the complex amplitude of the sum is the sum of the complex amplitudes.

Examples
1. Consider the Jones vectors for (*i*) plane polarized light with vibration direction (*a*) horizontal, (*b*) vertical, (*c*) at $+45°$ to horizontal and (*ii*) for clockwise circularly polarized light.
    (*i*) (*a*) $a_y = 0$. Normalize by setting $a_x = 1$. The phase can be set equal to zero. Hence one obtains

$$\begin{bmatrix} 1 \\ 0 \end{bmatrix}$$

(b) $a_x = 0$, $a_y = 1$, and one can set $\delta_y = 0$ to give

$$\begin{bmatrix} 0 \\ 1 \end{bmatrix}$$

(c) $a_x = a_y$, and the disturbances are in phase. Hence set $\delta_x = \delta_y = 0$. One could set $a_x = a_y = 1$ to obtain $\begin{bmatrix} 1 \\ 1 \end{bmatrix}$. If one normalizes for unit intensity ($I = a_x^2 + a_y^2$) one has $a_x = a_y = 1/\sqrt{2}$ and the vector becomes

$$\frac{1}{\sqrt{2}} \begin{bmatrix} 1 \\ 1 \end{bmatrix}$$

(ii) $a_x = a_y$. The y disturbance leads by $\pi/2$. Hence if one sets $\delta_x = 0$, $\delta_y = -\pi/2$ (remembering that $\psi = ae^{-i\delta}$ where δ is a phase *lag*). One can write

$$\begin{bmatrix} 1 \\ e^{i\pi/2} \end{bmatrix} = \begin{bmatrix} 1 \\ i \end{bmatrix} \text{ or, normalizing for unit intensity, } \frac{1}{\sqrt{2}} \begin{bmatrix} 1 \\ i \end{bmatrix}.$$

Alternatively, one can set $\delta_y = 0$, giving $\delta_x = +\pi/2$ and, for the normalized vector, $\frac{1}{\sqrt{2}} \begin{bmatrix} -i \\ 1 \end{bmatrix}$.

2. Consider the Jones matrix for a quarter-wave plate with its fast axis vertical.

Whatever the state of polarization of the incident light $a_x' = a_x$ and is independent of a_y, whilst $a_y' = a_y$ and is independent of a_x. Hence $T_{xx} = T_{yy} = 1$ and $T_{xy} = T_{yx} = 0$.

The plate will introduce a phase *lag* of $\pi/2$ of the x component relative to the y component. One can take $\delta_{yy} = 0$ and $\delta_{xx} = +\pi/2$ giving $J_{xx} = e^{-i\pi/2} = -i$ (or $\delta_{xx} = 0$ and $\delta_{yy} = -\pi/2$ giving $J_{yy} = +i$)

$$\therefore \quad \mathbf{J} = \begin{bmatrix} -i & 0 \\ 0 & 1 \end{bmatrix} \text{ or } \begin{bmatrix} 1 & 0 \\ 0 & +i \end{bmatrix}.$$

These can be written in the symmetrical form

$$\begin{bmatrix} e^{-i\pi/4} & 0 \\ 0 & e^{+i\pi/4} \end{bmatrix}$$

3. Consider the Jones matrix for a horizontal linear polarizer.

This introduces no changes of phase. I.e. each δ is zero. $a_x' = a_x$ and $a_y' = 0$.

$$\therefore \quad \mathbf{J} = \begin{bmatrix} 1 & 0 \\ 0 & 0 \end{bmatrix}.$$

4. Consider a Linear Polarizer at $+45°$ to the horizontal.

Each δ is zero. a_x has component $a_x/\sqrt{2}$ in transparent azimuth of polarizer and this has component $(a_x/\sqrt{2})/\sqrt{2} = a_x/2$ in x-direction. Thus $T_{xx} = \frac{1}{2}$. Also a_y has component $a_y/\sqrt{2}$ in transparent azimuth of polarizer and this has component $a_y/2$ in x-direction. Thus $T_{yx} = \frac{1}{2}$. Similarly $T_{yy} = T_{xy} = \frac{1}{2}$

$$\therefore \quad \mathbf{J} = \begin{bmatrix} \frac{1}{2} & \frac{1}{2} \\ \frac{1}{2} & \frac{1}{2} \end{bmatrix} \text{ or } \frac{1}{2}\begin{bmatrix} 1 & 1 \\ 1 & 1 \end{bmatrix}.$$

5. Suppose plane polarized light at $+45°$ to x axis enters a $\lambda/4$ plate whose fast axis is vertical.

For the emergent light one has

$$\Psi' = \begin{bmatrix} -i & 0 \\ 0 & 1 \end{bmatrix} \frac{1}{\sqrt{2}} \begin{bmatrix} 1 \\ 1 \end{bmatrix} = \frac{1}{\sqrt{2}} \begin{bmatrix} -i \\ 1 \end{bmatrix}$$

which, as shown in 1 (*ii*) above, is clockwise circularly polarized light.

22–21. The Coherency Matrix

It has been pointed out that a truly monochromatic beam of light (which cannot, in fact, exist!) would necessarily be completely polarized and that two monochromatic beams must be coherent. The term quasi-monochromatic has been used to describe polychromatic light in which the spread of frequencies is sufficiently small for a single frequency to be associated with it for many purposes whilst the non-monochromaticity permits two beams which nominally have the same frequency to be incoherent. It follows that quasi-monochromatic light can be completely or partially polarized or unpolarized.

Suppose, then, that one has a collimated beam of quasi-monochromatic light travelling in the positive z-direction and that in some z-plane the complex amplitudes of the x and y components of the light disturbance are $\psi_x = a_x e^{-i\delta_x}$ and $\psi_y = a_y e^{-i\delta_y}$ (see footnote on p. 562).

If the light is completely unpolarized orthogonal components have equal intensities but are completely incoherent. This means that the phase difference $\delta(=\delta_y - \delta_x)$ varies rapidly and randomly so that the time average of δ over the "exposure time" required by a detector is zero. For light of a given frequency one simply needs to state the total intensity recorded by the detector; it is $a_x^2 + a_y^2 = 2a_x^2 = 2a_y^2$.

If the light is completely polarized orthogonal components are coherent so that δ does not vary with time and, in general, the components have different amplitudes. Hence, in order to give an adequate description of the light one must specify the state of polarization in addition to the intensity and this can be done by specifying the relative amplitudes and relative phases of the components; the intensity remains equal to $a_x^2 + a_y^2$.

If the light is partially polarized it can be looked upon as light whose components are partially coherent, so that in addition to specifying the intensities of the components one must specify the nature of the correlation that exists between their sinusoidal variations; the total intensity is still $a_x^2 + a_y^2$. An alternative point of view is to look upon partially polarized light as a mixture of unpolarized light and polarized light where one needs to specify the relative intensities of the unpolarized and the polarized fractions and the state of polarization of the polarized fraction.

One way in which a beam of partially polarized quasi-monochromatic light can be described using the first point of view is by means of the *Coherency matrix* introduced by Wolf (1959). This is defined by

$$C = \begin{bmatrix} C_{xx} & C_{xy} \\ C_{yx} & C_{yy} \end{bmatrix} = \begin{bmatrix} \langle a_x^2 \rangle & \langle a_x a_y e^{i\delta} \rangle \\ \langle a_x a_y e^{-i\delta} \rangle & \langle a_y^2 \rangle \end{bmatrix} \quad . \quad . \quad (22\text{--}9)$$

where $\langle \ldots \rangle$ denotes the time average over a typical detector exposure time. (In fact Wolf's δ is the negative of δ used here.) Obviously C_{xx} and C_{yy} specify the intensities of the components (so that $C_{xx} + C_{yy}$ gives the total intensity of the light) and the other two elements specify the correlation between them (see equation (6–19)). If the two components are completely coherent δ is constant and the elements do not vary with time; if the components are incoherent the time averages are zero; and if the components are partially coherent the time averages take intermediate values. It will be seen that $C_{xy} = C_{yx}{}^*$ and one has, in effect, four independent scalar parameters: C_{xx}, C_{yy}, and the modulus and argument of, say, C_{xy}. Wolf puts

$$\mu_{xy} = |\mu_{xy}| e^{i\beta_{xy}} = \frac{C_{xy}}{\sqrt{C_{xx}} \sqrt{C_{yy}}} \qquad \ldots \ldots \quad (22\text{--}10)$$

where μ_{xy} corresponds to the *complex degree of coherence*. $|\mu_{xy}|$ (which can be shown to be $\leqslant 1$) is a measure of the *degree of coherence* and β_{xy} is referred to as the *effective phase difference* between the components.

Equation (22–9) is

$$C = \begin{bmatrix} \langle \psi_x \psi_x{}^* \rangle & \langle \psi_x \psi_y{}^* \rangle \\ \langle \psi_y \psi_x{}^* \rangle & \langle \psi_y \psi_y{}^* \rangle \end{bmatrix} \qquad \ldots \ldots \quad (22\text{--}11)$$

$$= \left\langle \begin{bmatrix} \psi_x \\ \psi_y \end{bmatrix} \times [\psi_x{}^* \psi_y{}^*] \right\rangle \qquad \ldots \ldots \quad (22\text{--}12)$$

where $\times$ denotes the *Kronecker* (or *Direct*) *product* of the two element column and row vectors. The column vector is simply the Jones vector Ψ and the row vector is its *Hermitian Conjugate*, which is denoted by $\Psi^\dagger$. I.e. one writes

$$C = \langle \Psi \times \Psi^\dagger \rangle. \qquad \ldots \ldots \quad (22\text{--}13)$$

Suppose, now, that a collimated beam of quasi-monochromatic partially polarized light passes through a system of polarizers and retarders. Let C and C' be the coherency matrices of the incident and emergent beams. It was shown in § 22–20 that one has

$$\Psi' = J\Psi. \qquad \ldots \ldots \quad (22\text{--}14)$$

Hence one has

$$C' = \langle \Psi' \times \Psi'^\dagger \rangle = \langle J\Psi \times \Psi^\dagger J^\dagger \rangle \qquad \ldots \quad (22\text{--}15)$$

i.e.

$$C' = JCJ^\dagger \qquad \ldots \ldots \quad (22\text{--}16)$$

and this relates the coherency matrices of the incident and emergent beams.

It was pointed out in § 22–20 that if one adds two coherent beams of completely polarized light one adds the Jones vectors. It can be shown that if one adds two *incoherent* beams of partially polarized light one adds the coherency matrices to obtain the coherency matrix of the sum.

Example

Consider plane polarized light at $45°$ entering a $\lambda/4$ plate whose fast axis is vertical.

For the incident light $a_x = a_y$ and $\delta = 0$. Hence **C** can be written $\begin{bmatrix} 1 & 1 \\ 1 & 1 \end{bmatrix}$.

(22–16) then gives, for the emergent light,

$$\mathbf{C'} = \begin{bmatrix} -i & 0 \\ 0 & 1 \end{bmatrix} \begin{bmatrix} 1 & 1 \\ 1 & 1 \end{bmatrix} \begin{bmatrix} +i & 0 \\ 0 & 1 \end{bmatrix}$$

$$= \begin{bmatrix} -i & 0 \\ 0 & 1 \end{bmatrix} \begin{bmatrix} +i & 1 \\ +i & 1 \end{bmatrix}$$

$$= \begin{bmatrix} 1 & -i \\ +i & 1 \end{bmatrix}$$

which, from (22–9), corresponds to light which has the x and y components equal but with the y component lagging by $-\pi/2$ i.e. clockwise circularly polarized light—in agreement with the result using the Jones method (§ 22–20, example 5).

22–22. Stokes Vectors and Mueller Matrices

When a beam of partially polarized light is specified by means of the coherency matrix two real quantities (C_{xx} and C_{yy}) and one complex quantity (C_{xy} or C_{yx}) are involved. This corresponds to the use of four scalar parameters. The properties of an optical system are then specified by a 2×2 matrix whose elements, in general, are complex. As an alternative to the coherency matrix, partially polarized quasi-monochromatic light can be specified by a four element column vector, known as the *Stokes vector*, the elements of which are real. The corresponding system matrix is then the 4×4 *Mueller matrix*, each element of which is also real since it transforms one set of real parameters into another. The four *Stokes parameters* (Stokes, 1852) and the Stokes vector can be defined as follows:

$$\begin{aligned} S_0 &= \langle a_x^2 \rangle + \langle a_y^2 \rangle \\ S_1 &= \langle a_x^2 \rangle - \langle a_y^2 \rangle \\ S_2 &= 2\langle a_x a_y \cos \delta \rangle \\ S_3 &= -2\langle a_x a_y \sin \delta \rangle \end{aligned} \qquad \mathbf{S} = \begin{bmatrix} S_0 \\ S_1 \\ S_2 \\ S_3 \end{bmatrix}. \qquad \ldots \quad (22\text{–}17)$$

(The negative sign is inserted in S_3 in order to conform to the majority view of this parameter.)

From equations (22–9) it follows that the Stokes parameters and the elements of the coherency matrix are related as follows:

$$\begin{aligned} S_0 &= C_{xx} + C_{yy} & C_{xx} &= \tfrac{1}{2}(S_0 + S_1) \\ S_1 &= C_{xx} - C_{yy} & C_{yy} &= \tfrac{1}{2}(S_0 - S_1) \\ S_2 &= C_{xy} + C_{yx} & C_{xy} &= \tfrac{1}{2}(S_2 - iS_3) \\ S_3 &= i(C_{xy} - C_{yx}) & C_{yx} &= \tfrac{1}{2}(S_2 + iS_3) \end{aligned} \qquad \ldots \quad (22\text{–}18)$$

Stokes vectors, like coherency matrices, are additive for incoherent beams of partially polarized quasi-monochromatic light.

Taking the alternative point of view of partially polarized light referred to in § 22–21, let A be the amplitude of the unpolarized fraction, let A_x and A_y be the amplitudes of the components of the polarized fraction, and let δ be the phase difference between these components. The Stokes parameters are then given by

$$
\begin{aligned}
S_0 &= A_x{}^2 + A_y{}^2 + A^2 \\
S_1 &= A_x{}^2 - A_y{}^2 \\
S_2 &= 2A_xA_y \cos \delta \\
S_3 &= -2A_xA_y \sin \delta
\end{aligned}
\qquad \cdots \quad (22\text{–}19)
$$

These expressions are equivalent to those given in (22–17). In each case S_0 gives the total intensity and S_1 gives the difference between the intensities of the components. In (22–17) the incoherent fraction of the light would give zero contributions to the time averages in S_2 and S_3, leaving the values given by (22–19).

For completely polarized light one has

$$
S_0{}^2 = S_1{}^2 + S_2{}^2 + S_3{}^2. \qquad \cdots \quad (22\text{–}20)
$$

For partially polarized light $x^2 + y^2 + z^2 = 1$

$$
S_0{}^2 > S_1{}^2 + S_2{}^2 + S_3{}^2. \qquad \cdots \quad (22\text{–}21)
$$

and the degree of polarization is given by

$$
P = \frac{\text{Intensity of polarized fraction}}{\text{Total Intensity}}
$$

$$
= \frac{A_x{}^2 + A_y{}^2}{A_x{}^2 + A_y{}^2 + A^2} = \frac{\sqrt{S_1{}^2 + S_2{}^2 + S_3{}^2}}{S_0}. \qquad \cdots \quad (22\text{–}22)
$$

Soleillet (1929) pointed out that if light is passed through an optical system the Stokes parameters of the emergent light are linear functions of those of the incident light. Hence in the most general situation that can be imagined one would have

$$
S_0' = M_{11}S_0 + M_{12}S_1 + M_{13}S_2 + M_{14}S_3
$$

with similar equations for S_1', S_2', S_3'. That is to say,

$$
\mathbf{S'} = \mathbf{MS} \qquad \cdots \quad (22\text{–}23)
$$

where $\mathbf{M}$ is a 4×4 matrix; it is called the *Mueller matrix* of the system. If the light passes through n systems in succession (in the order $1 \ldots n$) one has $\mathbf{S_1'} = \mathbf{M_1 S_1}$, $\mathbf{S_2'} = \mathbf{M_2 S_2} \equiv \mathbf{M_2 S_1'} = \mathbf{M_2 M_1 S_1} \ldots$ and so on. That is,

$$
\mathbf{S_n'} = \mathbf{M S_1} \qquad \cdots \quad (22\text{–}24)
$$

where $\mathbf{M} = \mathbf{M_n M_{n-1} \ldots M_2 M_1}$ $\cdots$ (22–25)

$\mathbf{M}$ is the Mueller matrix of the complete system and need be calculated once only.

The Jones vectors and the Stokes vectors for the various forms of polarized light, and the Jones and the Mueller matrices for the most commonly used polarizing and phase changing devices, have been calculated once for all and are tabulated in, for example, Shurcliff's "Polarized Light" (Harvard/Oxford, 1962).

For any beam of light, the elements of the coherency matrix and of the Stokes vector can be determined by means of simple experiments employing polarizers and retarders; in fact the experiments provide a means of defining the parameters operationally. Consider, for example, the Stokes parameters. As was pointed out, it follows from (22–19) that S_0 gives the total intensity of the light and S_1 gives the difference between intensities of the x and y components. That is, S_1 gives the difference between the intensities transmitted by ideal linear polarizers that transmit horizontally and vertically polarized light respectively. It can be shown that S_2 is given by the difference between the intensity transmitted by a linear polarizer in the azimuth $+45°$ and one in the azimuth $-45°$, and S_3 is given by the difference between the intensity transmitted by an ideal circular polarizer that transmits clockwise circularly polarized light and one that transmits anticlockwise circularly polarized light. The proofs provide a useful exercise to help one to clarify one's thinking. Consider the disturbances in the azimuth $+45°$. The components A_x and A_y give amplitudes $A_x/\sqrt{2}$ and $A_y/\sqrt{2}$ and the latter has a phase lag of δ. Hence the intensity in the $45°$ azimuth is

$$I_{45} = \tfrac{1}{2}A_x{}^2 + \tfrac{1}{2}A_y{}^2 + A_x A_y \cos \delta.$$

In the azimuth $-45°$ the amplitudes are $A_x/\sqrt{2}$ and $A_y/\sqrt{2}$ and the phase difference is $\pi + \delta$. Hence

$$I_{-45} = \tfrac{1}{2}A_x{}^2 + \tfrac{1}{2}A_y{}^2 + A_x A_y \cos(\pi + \delta).$$

Since $\cos(\pi + \delta) = -\cos \delta$ one has at once

$$I_{45} - I_{-45} = 2A_x A_y \cos \delta = S_2.$$

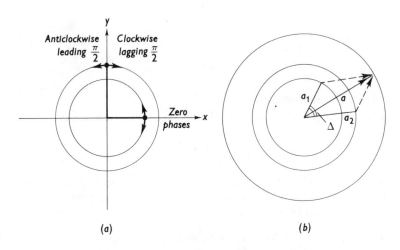

(a)　　　　　　　　　　　(b)

Fig. 22–38. Combination of circular motions.

Now a linear S.H.M. of amplitude A can be regarded as the superposition of clockwise and anticlockwise circular motions, each of radius $\frac{1}{2}A$ (see §23–3). If one takes the linearly polarized component A_x to be the superposition of clockwise and anticlockwise circularly polarized components, each having zero phase, then a component A_y *in phase* with A_x would be the superposition of clockwise and anticlockwise circular components whose phases were $+\pi/2$ and $-\pi/2$ (see Fig. 22–38 (a)). In fact A_y has a phase lag of δ compared with A_x so that the corresponding circular components have phases $\delta \pm \pi/2$. It will be seen from Fig. 22–38(b) that two circular motions whose radii are a_1 and a_2 and which are in the same direction but with a phase difference Δ combine to give a circular motion whose radius a is given by

$$a^2 = a_1{}^2 + a_2{}^2 + 2a_1 a_2 \cos \Delta$$

Remembering that this represents an intensity $2a^2$, one has for the clockwise and anticlockwise circularly polarized resultants

$$I_C = 2 \left\{ \tfrac{1}{4}A_x{}^2 + \tfrac{1}{4}A_y{}^2 + \tfrac{1}{2}A_x A_y \cos(\delta + \pi/2) \right\}$$

$$I_A = 2 \left\{ \tfrac{1}{4}A_x{}^2 + \tfrac{1}{4}A_y{}^2 + \tfrac{1}{2}A_x A_y \cos(\delta - \pi/2) \right\}$$

Since $\cos(\delta + \pi/2) = -\sin \delta$ and $\cos(\delta - \pi/2) = +\sin \delta$, one has

$$I_C - I_A = -2A_x A_y \sin \delta = S_3.$$

Examples of Stokes Vectors

1. For completely plane polarized light at $+45°$ one has

$$\begin{bmatrix} 1 \\ 0 \\ 1 \\ 0 \end{bmatrix}$$

2. For clockwise circularly polarized light one has

$$\begin{bmatrix} 1 \\ 0 \\ 0 \\ 1 \end{bmatrix}$$

22–23. The Poincaré Sphere

For completely polarized light the Stokes parameters are related by equation 22–20:

$$S_0{}^2 = S_1{}^2 + S_2{}^2 + S_3{}^2.$$

If one takes Cartesian axes S_1, S_2, S_3, this equation defines a sphere of radius S_0. The sphere is known as the *Poincaré sphere* (Poincaré, 1892) and conventionally is taken to have unit radius. Thus, for the *normalized* Stokes parameters, one has

$$S_1{}^2 + S_2{}^2 + S_3{}^2 = 1 \quad \ldots \ldots \ldots \quad (22\text{–}26)$$

It will be seen that each point on the surface of the sphere corresponds to a unique set of values of the Stokes parameters and hence to a particular state of polarization. The sphere is illustrated in Fig. 22–39.

At the north and south poles one is on the S_3 axis so that $S_1 = S_2 = 0$. It follows from equations (22–19) that $A_x = A_y$ and $\delta = +\pi/2$. If $\delta = +\pi/2$ the y component lags behind the x component and the light is anticlockwise circularly polarized (see Fig. 22–26). This corresponds to $S_3 = -1$: the south pole. Similarly the north pole corresponds to clockwise circularly polarized light.

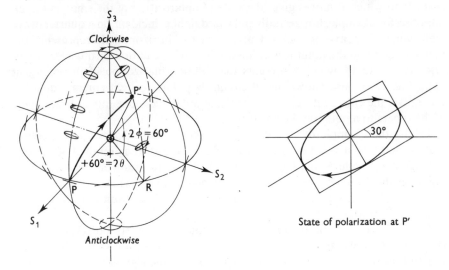

Fig. 22–39. The Poincaré sphere.

For points on the equator $S_3 = 0$ and $\delta = 0$ or π, corresponding to plane polarized light. For $\delta = 0$ the azimuth is between 0 and $+\pi/2$ and corresponds to a positive value of S_2; for $\delta = \pi$ the azimuth is between 0 and $-\pi/2$ and corresponds to a negative value of S_2. When $S_1 = 0$, $A_x{}^2 = A_y{}^2$, or $A_x = \pm A_y$ and the azimuth is $\pm 45°$: $+45°$ for $S_2 = +1$ and $-45°$ for $S_2 = -1$. If $S_2 = 0$, $S_1{}^2 = 1$ and $S_1 = \pm 1$. If $S_1 = +1$, $A_y = 0$ and one has horizontal plane polarized light; if $S_1 = -1$, $A_x = 0$ and one has vertical plane polarized light. Other equatorial points correspond to plane polarized light in intermediate azimuths.

It can be shown that as one moves down a line of longitude one passes from circularly polarized light at the pole, through elliptically polarized states of increasing eccentricity, to plane polarized on the equator, the major axes of the ellipses having the same orientation; if 2θ is the longitude measured from the $S_1 S_3$ plane, the major axes are inclined to the horizontal at angle θ. Also, it can be shown that points around a line of latitude represent various ellipses which have the same eccentricity; if 2ϕ is the latitude measured from the $S_1 S_2$ (equatorial) plane, the ratio of the axes of each ellipse is $\tan \phi$.

It can be shown that θ and ϕ are related to the normalized Stokes parameters as follows:

$$S_1 = \cos 2\phi \cos 2\theta$$
$$S_2 = \cos 2\phi \sin 2\theta \quad \quad \ldots \ldots \quad (22\text{–}27)$$
$$S_3 = \sin 2\phi.$$

The Poincaré sphere can be used to find the state of polarization of the emergent light when light with a given initial state of polarization is passed through any form of retarder or combination of retarders. Suppose the point P on the sphere represents the state of polarization of the incident light and the point R represents the form of polarization that has the greater speed in the retarder. One rotates the sphere clockwise (as viewed in the direction RO) about OR as axis through an angle equal to the phase change introduced. The point P′ to which P moves gives the state of polarization of the emergent light. Suppose, for example, horizontally polarized light is incident on a quarter-wave plate whose fast axis is inclined at 30° to the horizontal. P represents the incident light and the point R is at longitude $+60°$. Rotation through 90° about OR takes P to P′, which represents the elliptic polarization of the emergent light. Fig. 22–39 also shows that if circularly polarized light enters a quarter-wave plate the emergent light must be plane polarized—one moves from a pole of the sphere to a point on the equator.

Taking the example discussed earlier in terms of the Jones and coherency matrix methods, it will be seen that if light polarized in azimuth $+45°$ passes through a $\lambda/4$ plate whose fast axis is vertical, one rotates from the equatorial point on the S_2 axis ($S_2 = 1$) to the north pole ($S_3 = 1$), which corresponds to clockwise circularly polarized emergent light and is the result obtained by the other two methods.

The passage of light through a succession of retarders corresponds to a succession of movements around the surface of the Poincaré sphere. It is worth noting that there is no need to stop and interpret the significance of the intermediate points if one needs to know only the state of polarization of the final emergent light.

22–24. The Kerr Electro-optical Effect

When placed in an electric field, many substances behave like a uniaxial crystal (usually positive) whose optic axis is parallel to the lines of force. The effect was discovered by Kerr, in 1895, using glass. If n_1 and n_2 are the refractive indices for light whose planes of vibration are respectively parallel to and perpendicular to the field, it is found that

$$n_1 - n_2 = k\lambda E^2, \quad \ldots \ldots \quad (22\text{–}28)$$

where E is the electric intensity and λ is the wavelength. k is known as Kerr's constant and has been determined for a wide range of solids, liquids, and gases. If E is in e.s.u. one has, for sodium light, $k = 2·5 \times 10^{-11}$ for carbon dioxide at N.T.P., $4·7 \times 10^{-7}$ for water, $2·2 \times 10^{-5}$ for nitrobenzene, and $2·9 \times 10^{-9} - 1·5 \times 10^{-8}$ for various glasses, λ being measured in cm.

The effect can be investigated by employing the substance concerned as the dielectric in a parallel plate condenser, this so-called *Kerr Cell* being placed between a polarizer and an analyser.

The theory of the Kerr effect is beyond the scope of this book. It has been assumed that an applied electric field induces an electric moment in non-polar molecules and changes the moment of polar molecules. There is also a reorientation of the molecules which causes the medium as a whole to become anisotropic.

This explains the delay that occurs between the application of the field and the appearance of the maximum effect. The delay can be several seconds in the case of some solids but is very small ($\sim 10^{-11}$ sec.) for non-polar liquids with small molecules.

The Kerr effect must not be confused with the *electric double refraction* observed in the region of absorption lines. It is found that in the presence of an electric field the absorption frequencies are dependent upon the polarization of the light. Thus for frequencies near an absorption line the velocity of light will depend on the state of polarization, i.e. the medium is birefringent.

22–25. The Voigt and Cotton-Mouton Effects

The *Cotton-Mouton effect* (discovered in 1905) is the magnetic analogue of the Kerr electro-optical effect. It is found that in the presence of a transverse magnetic field many liquids became doubly refracting. As with the Kerr effect, the difference between the refractive indices is proportional to the square of the field strength and, in a similar manner, it is attributed to a re-orientation of molecules in the presence of the field. In the analogue of equation (22–28) the constant has rather smaller values, e.g. $-1 \cdot 1 \times 10^{-14}$ for water. A very much smaller effect is the *magnetic double refraction* or *Voigt effect* (discovered 1902) which is analogous to the electric double refraction referred to in the previous section. In a similar manner it is attributed to the effect of the field on the absorption frequencies which become dependent on the polarization of the light. Electric and magnetic double refraction are related, in similar ways, to the Stark and Zeeman effects respectively.

22–26. The Photo-elastic Effect

It was found by Brewster in 1816 that a transparent isotropic medium becomes optically anisotropic when a mechanical stress is applied. If a body under stress is placed between a polarizer and a crossed analyser, the birefringence of the body will cause some light to be transmitted by the analyser. When investigating the effect, the observing system must be focused on the specimen itself, and the variations of intensity (and of colour if white light is used) will depend on the local variations of birefringence throughout the specimen. It is found that the privileged directions of $\mathbf{D}$ are along the directions of the principal stresses. If n_P and n_Q are the refractive indices for light whose vibration directions are parallel to the principal stresses P and Q respectively, then one has

$$n_P - n_Q \propto (Q - P).$$

In many engineering problems it is very useful to build a model of the structure concerned in some transparent material; when the structure is loaded the nature of the stresses can be observed. For example, *isochromatic lines* are the loci of points in the model at which $(P - Q)$ has the same value.

In the manufacture of many optical instruments, care must be taken to avoid the setting up of stresses in the glass. In the production of the glass itself the annealing process must be controlled with great care and, in the instrument, attention must be given to the way in which the components are held in position. At every stage the glass is viewed between a polarizer and crossed analyser in a *polariscope*.

CHAPTER XXIII

OPTICAL ACTIVITY

23–1. Optically Active Crystals

IF a beam of light enters a plane-parallel plate cut from an ordinary uniaxial crystal, there are, in general, two emergent beams, and these are plane polarized at right angles. If, however, one has light incident normally on a plate cut perpendicular to the optic axis, the O and E disturbances are propagated with equal velocities and the phase difference between them remains constant. That is to say, the state of polarization of the emergent light is identical with that of the incident light. With some crystals, of which quartz is a well known example, it is found that if plane polarized light is incident normally on a plate cut perpendicular to the axis, the emergent beam is also plane polarized but its plane of vibration is different from that of the incident beam. The effect was discovered in 1811 by Arago. It is found that the rotation of the plane of vibration is proportional to the thickness of the crystal plate, and does not depend upon the orientation of the plane of vibration of the incident light. Crystals that rotate the plane of vibration in this way are said to be *optically active*. Looking against the oncoming light, crystals that rotate the plane of vibration in a clockwise direction are said to be *right-handed* or *dextro-rotatory*; those that rotate it in an anticlockwise direction are *left-handed* or *laevo-rotatory*. Quartz exists in both forms. The rotation produced by a plate 1 mm. in thickness is called the *specific rotation*. For sodium light the specific rotation is 21·7° for quartz, 3·67° for sodium chlorate, and 325° for cinnabar. The effect is found also in biaxial crystals.

23–2. Rotatory Dispersion

The specific rotation depends on the wavelength of the light and was found by Biot to be approximately proportional to $(1/\lambda^2)$. More accurate investigations showed that rotatory dispersion follows a law similar to that governing ordinary dispersion of the refractive index. For example, throughout most of the visible spectrum the specific rotation of quartz is given by an equation of the Cauchy type:

$$\alpha = A + \frac{B}{\lambda^2}. \qquad \ldots \ldots \ldots \quad (23\text{–}1)$$

Near absorption lines one has anomalous rotatory dispersion comparable with anomalous dispersion of the refractive index.

Suppose a collimated beam of white light is incident normally on a quartz plate cut perpendicular to the axis and placed between crossed Nicols. Wavelengths for which the plate gives a rotation of a multiple of π will be removed by the analyser, and, if the transmitted light is passed into a spectroscope, the spectrum will show a number of dark bands. The intensity will reach a maximum for wavelengths that have been rotated an odd multiple of $\pi/2$, and for intermediate wavelengths the analyser will transmit the component that is parallel to the transparent vibration direction. If the quartz plate is very thin, the rotation is small and does not vary widely throughout the visible region. Consequently, the light transmitted by the analyser appears white or grey. If the quartz plate is very thick, the transmitted light is white because the many obstructed wavelengths are distributed throughout the spectrum. For a plate of moderate thickness, however, one or two spectral regions are obstructed and the transmitted light appears coloured. If the analyser is replaced by an ordinary calcite prism, there are two emergent beams. One contains wavelengths which a crossed analyser would pass, and the other those which the analyser would obstruct. Consequently, the two beams have complementary colours. If they are made to overlap, a white beam is produced—another convenient way of demonstrating the combination of complementary colours.

23–3. Fresnel's Theory

For a non-active crystal, only plane polarized disturbances can be propagated in directions other than the optic axis. For a point source in such a medium the wave surface is in two sheets, and these touch to give a common diameter parallel to the optic axis. This suggests that light in any state of polarization can be propagated along the optic axis. Fresnel assumed that in an optically active crystal only circularly polarized disturbances can be propagated in the direction of what would otherwise be the optic axis, and that the velocity of propagation is different for different directions of rotation. It is easy to show that a plane polarized disturbance can be resolved into two circularly polarized components with opposite directions of rotation. If these travel with different speeds, the phase difference between them is different at different distances through the crystal and this corresponds to a rotation of the plane of the resultant.

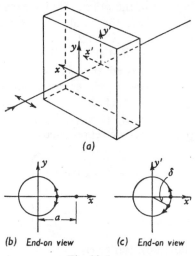

(a)

(b) *End-on view*   (c) *End-on view*

Fig. 23–1.
Fresnel's theory of optical activity.

Suppose, for example, that plane polarized light is incident normally on an optically active crystal plate cut perpendicular to the axis. Let the first crystal face be in the xy-plane [see Fig. 23–1(a)], and let the plane polarized incident disturbance be represented by

$$x = a \cos \omega t. \qquad \cdots \qquad (23\text{–}2)$$

This can be resolved into the two circular motions represented by

$$\left. \begin{array}{l} x = \dfrac{a}{2} \cos \omega t, \\[2mm] y = \dfrac{a}{2} \sin \omega t \end{array} \right\} \qquad \cdots \qquad (23\text{–}3)$$

$$x = \frac{a}{2} \cos \omega t,$$

and
$$\left. \begin{array}{c} \\ \\ \end{array} \right\} \quad \cdots \cdots \quad (23\text{-}4)$$

$$y = -\frac{a}{2} \sin \omega t.$$

The first pair of equations represents an anticlockwise, and the second a clockwise, circular motion. Both start from the point $x = a/2$ at time $t = 0$, and reference to the equations and to Fig. 23–1(b) shows that on superposition they give the linear motion represented by (23–2). It will be seen that two oppositely directed circular motions of equal frequency combine to give a linear motion along the diameter joining the points where the rotating radius vectors cross.

Now, according to Fresnel, the two circular components are propagated through the plate with different speeds. (The two circles are, of course, traversed at the same speed since they correspond to disturbances of the same frequency.) Consequently, when they emerge from the plate there is a phase difference δ between them. Suppose, for example, that the clockwise component advances in front of the other. Taking axes x' and y' (parallel to x and y) in the second crystal face, the result of this phase difference is that, if the anticlockwise rotation is at $x' = a/2$ at $t = 0$, the clockwise rotation has advanced through an angle δ beyond the x'-axis [Fig. 23–1(c)]. That is to say, the rotating radius vectors crossed at $-\delta/2$, and this gives the direction of the linear resultant disturbance which emerges—the plane of vibration has rotated clockwise through an angle $\delta/2$.

Analytically (since only relative phases matter) the emergent circular components can be represented by

$$\left. \begin{array}{l} x' = \dfrac{a}{2} \cos \omega t, \\[2mm] y' = \dfrac{a}{2} \sin \omega t \end{array} \right\} \quad \text{and} \quad \left\{ \begin{array}{l} x' = \dfrac{a}{2} \cos (\omega t + \delta), \\[2mm] y' = -\dfrac{a}{2} \sin (\omega t + \delta). \end{array} \right.$$

Superposition gives

$$x' = \frac{a}{2} [\cos \omega t + \cos (\omega t + \delta)],$$

$$y' = \frac{a}{2} [\sin \omega t - \sin (\omega t + \delta)].$$

$$\therefore \quad \left. \begin{array}{l} x' = a \cos \left(\dfrac{\delta}{2} \right) \cos \left(\omega t + \dfrac{\delta}{2} \right), \\[3mm] y' = -a \sin \left(\dfrac{\delta}{2} \right) \cos \left(\omega t + \dfrac{\delta}{2} \right). \end{array} \right\} \quad \cdots \quad (23\text{-}5)$$

Since these two perpendicular linear components are in phase the resultant is along the line inclined to the x' axis at angle $-\delta/2$.

If n_O and n_A are the refractive indices of the crystal in the direction of the optic axis for clockwise and anticlockwise circularly polarized light, and d is the thickness of the crystal plate, the phase difference δ is given by

$$\delta = \frac{2\pi d}{\lambda}(n_A - n_C), \qquad \ldots \ldots \ldots \text{(23–6)}$$

i.e.   rotation of plane of vibration $= (\pi d/\lambda)(n_A - n_C)$.   . (23–7)

It will be seen that in a right-handed (dextro-rotatory) crystal, the clockwise circularly polarized component travels more quickly than the other, whereas in a left-handed crystal it is the slower component. The existence of these circularly polarized components was demonstrated by Fresnel using a train of left- and right-handed quartz prisms as indicated by Fig. 23–2. For the clockwise component the refractive index of the second prism is higher than that of the first,

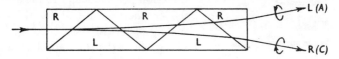

Fig. 23–2. Fresnel's prism system for separating circular components of plane polarized or unpolarized light.

but for the anticlockwise component it is lower. Consequently, there is double refraction at this surface and the circularly polarized components are separated slightly. At the interface between the second and third prisms the right-handed component is entering a less dense medium and the left-handed component is entering a denser medium. The obliquity of the surface causes the angular separation between the rays to be increased. This continues throughout the train of prisms until two approximately circularly polarized beams emerge. (As will be seen below, the disturbances propagated in a direction which is slightly inclined to the optic axis are, in fact, elliptically polarized.) If the incident light is unpolarized each linear component splits up in this way.

23-4. Wave Surface for Quartz

For a non-active uniaxial crystal the wave surface consists of a sphere and either an inscribed prolate spheroid or an escribed oblate spheroid. These two sheets of the wave surface touch at two points and the line joining the points of contact gives the direction of the optic axis. The radius vector from the centre to any point on one of the sheets of the wave surface is proportional to the ray velocity in that direction and corresponds to one of the two possible plane polarized disturbances. If the crystal is optically active there are two possible velocities in the direction corresponding to the optic axis, so that the two sheets of the wave surface do not touch. There is another important difference: in the direction of the axis the radius vector to each sheet of the wavefront represents a possible velocity of circularly, not linearly, polarized light. In directions perpendicular to the axis the possible velocities now correspond to *approximately* plane polarized rays. For quartz the two sheets of the wave surface resemble those of a positive uniaxial crystal except that they do not touch. What would normally be the spherical sheet is distorted outwards by a small amount in the region of the axis, and the other sheet is distorted inwards. The wave surface takes the form indicated in Fig. 23–3. It is important to realize that the separa-

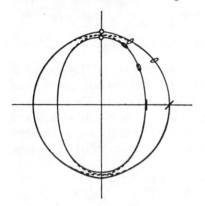

tion of the two sheets along the axis is very much smaller than that in the perpendicular direction and that both are exaggerated in the figure. For example, for sodium light travelling perpendicular to the optic axis one has (at 18°C.) $n_E = 1·55336$, $n_O = 1·54425$, the latter being also the mean of n_C and n_A for light along the axis. It follows that the separation of the sheets of the wave surface in the direction perpendicular to the axis is about $0·006r$ where r is the radius of the approximately spherical sheet. For sodium light along the axis, the specific rotation is found to be 21·7° per mm. Using equation

Fig. 23–3. Wave surface for quartz.

(23–7) this gives, for right-handed quartz $(n_A - n_C) \doteq 0·00007$ so that the separation of the sheets of the wave surface is about $0·000045r$.

As stated above, the wave surface gives the velocities of circularly polarized light along the axis, and of approximately plane polarized light in the perpendicular directions. They are the only types of disturbance which can be propagated without change in those directions. It was shown by Airy that in intermediate directions the wave surface gives the velocities of elliptically polarized disturbances. Thus any disturbance in these directions is resolved into elliptically polarized components. The major axes of the ellipses are in the vibration directions of the plane polarized disturbances for an inactive crystal. As one moves away from the direction of the axis the eccentricities of the ellipses increase rapidly and in most directions one can look upon the wave surface as indicating the velocities of the two possible plane polarized disturbances appropriate to an inactive crystal. That is to say, it is only for directions close to the optic axis that the properties of an active crystal differ markedly from those of an inactive one. For example, at 10° away from the optic axis the ratio of the axes of the ellipse is 7·8.

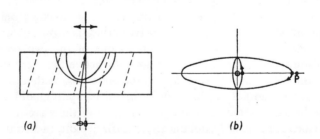

Fig. 23–4. Resolution of plane polarized light into elliptically polarized components.

Fig. 23–4(a) illustrates the double refraction of a ray incident normally on a quartz plate cut almost perpendicular to the axis. Both the refracted rays are elliptically polarized; they lie very close together and the eccentricities of the ellipses are the same. The relative intensities of the two refracted rays are determined by the state of polarization of the incident light, and if the latter is unpolarized the refracted rays will have equal intensities. If the incident light is plane polarized with **D** in a principal section as illustrated, the major axis of one ellipse will be the same as the minor axis of the other. For example, in Fig. 23–4(*b*) the two elliptic motions illustrated are the components of the linear motion of amplitude OP. For other states of polarization of the incident light the relative intensities and phases of the elliptic components will be different.

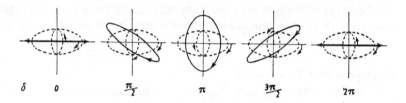

$$\delta \qquad 0 \qquad \frac{\pi}{2} \qquad \pi \qquad \frac{3\pi}{2} \qquad 2\pi$$

Fig. 23–5. Recombination of elliptic components.

Again, for the example illustrated, the two refracted paths are not widely different and, for a moderately thin crystal plate and a beam of finite width, the emergent disturbances can be assumed to be superposed. There will be a certain phase difference between the elliptic components shown in (*b*) and, in general, the resultant will be elliptically polarized. Fig. 23–5 shows the state of polarization of the resultant for different values of the phase difference δ.

When a quartz plate cut perpendicular to the axis is placed between a polarizer and an analyser and illuminated by convergent pencils as in § 22–18, the patterns observed are very similar to those obtained with a non-active crystal. The chief difference is that for light which has passed parallel, or nearly parallel, to the optic axis the crystal gives a rotation of the plane of vibration. The result is that, in general, the centre of the pattern with a crossed analyser is bright instead of dark.

23–5. The Cornu Prism

If it is desired to use a uniaxial crystal in a prism spectrograph, the prism is cut with the optic axis along the path of the light in the prism [see Fig. 23–6(*a*)]. However, with an optically active crystal such as quartz, this arrangement still gives a doubling of each spectrum line since for each incident ray, one has two circularly polarized emergent rays corresponding to slightly different refractive indices. To overcome this difficulty Cornu used a composite prism employing both left- and right-handed quartz as indicated in (*b*). When used at minimum

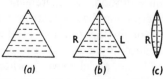

Fig. 23–6. Crystal prisms and lens.

deviation, this prism gives a single emergent beam since the phase difference introduced in one half is removed in the other.

If a Littrow system is required, one of the components of the Cornu prism can be used alone, the surface AB being coated to give the plane mirror. The clockwise circularly polarized vibrations approaching the reflecting surface become the anticlockwise components after reflection, and the anticlockwise become the clockwise. It will be seen that the phase difference resulting from the passage through the prism in one direction is removed by the passage in the reverse direction.

One can also use quartz lenses with a construction similar to that of the Cornu prism [see Fig. 23-6(c)]. The composite lens consists of two plano-convex components, one of left- and the other of right-handed quartz.

Fused quartz components are sometimes used since these are not birefringent. It is, however, difficult to produce large components free from inhomogeneities.

23–6. Optical Activity in Liquids

It was found by Biot that many liquids and solutions are optically active, i.e. rotate the plane of vibration when plane polarized light is passed through them. The optical activity of most crystals is associated with the absence of a centre of symmetry in the fundamental crystal form; this causes the velocity of circularly polarized light to depend upon the direction of rotation. Now a liquid is iso-tropic in the sense that the molecules are randomly orientated throughout it and, consequently, the optical activity is attributed to the asymmetrical structure of the molecules themselves. One might say that there is a preferential direction of rotation associated with each molecule. If, at first sight, it seems that the random orientation of the molecules in a liquid would cause this to be ineffective in a finite volume, it is worth noting that, no matter how the screw is positioned, the rotation associated with the advance of a right-handed screw is always the same. Thus some optical activity is caused by the structural dissymmetry of crystals and some by molecular dissymmetry. The latter remains when the substance is in the fused state or is in solution but the former disappears when the crystal is dissolved or fused. When a substance shows both kinds of dissym-metry the activity associated with the crystal structure predominates.

It was shown above that the rotation produced by an active crystal is propor-tional to the thickness of crystal traversed. This means that it is proportional to the number of fundamental rotating elements traversed. In the same way it is found that the rotation produced by a solution of an active substance in an inactive solvent is proportional to the length of path in the solution and to the concentration. Measurement of the rotation is now a widely used method of determining the concentration of sugar solutions.

Usually the rotation produced by a liquid is considerably less than that produced by a crystal. For example, 10 cm. of turpentine rotates sodium light $-37°$ (i.e. anticlockwise), whereas an equal thickness of quartz would rotate it $\pm 2172°$. For this reason the standard path length for a liquid is taken as 10 cm. instead of 1 mm. For an active liquid, or for a solution of an active substance such as sugar in an inactive solvent, one defines the *specific rotation* or *rotatory power* (α) as the rotation produced by a 10 cm. column of liquid containing 1 gm. of active substance per c.c. of solution. Thus for a solution containing *m* grams per c.c. the rotation for a path length *l* is given by

$$\theta = \frac{\alpha ml}{10},$$

i.e.

$$\alpha = \frac{10\theta}{ml} \quad \cdot \quad \cdot \quad \cdot \quad \cdot \quad \cdot \quad \cdot \quad \cdot \quad (23\text{-}8)$$

and for a pure liquid *m* is simply the density.

The product of the specific rotation and the molecular weight of the active substance is known as the *molecular rotation*.

Careful investigation has shown that the rotation is not *exactly* proportional to the concentration, and in accurate work a suitable correction must be made. It is found also that the rotation depends upon the temperature and, to a slight extent, upon the solvent employed.

Liquids show rotatory dispersion similar to that of a crystal. In particular, the rotatory dispersions of sugars are very similar to the rotatory dispersion of quartz; the importance of this fact will be seen in § 23-9.

23-7. Polarimeters

Instruments that measure the rotation of the plane of vibration produced by a solution are known as *polarimeters*; those that are intended for measurements on sugar solutions in particular are called *saccharimeters*. If an optically active solution is placed between crossed Nicols (polarizer and analyser), some light is transmitted by the analyser. Obviously the rotation of the solution for a given wavelength can be measured by rotating the analyser or the polarizer until there is again no transmitted light (see § 22-6). Until recently, most saccharimeters have employed a fixed polarizer and analyser, the rotation of the solution being balanced out by a quartz compensator. As will be seen in § 23-9, this enabled white light to be used. This was an advantage because high intensity mono-chromatic sources were not available or were inconvenient for routine use such as that for which saccharimeters are required. Instruments designed as sacchari-meters are usually provided with a scale which indicates sugar concentration directly.

The simplest form of polarimeter contains only a polarizer and an analyser. For example, Biot used a reflecting plate of glass as the polarizer and employed a calcite crystal as an analyser, and Nörrenberg used glass plates for both polarizer and analyser. The first polarimeter to employ Nicols was that due to Mitscherlich (1844). The rotation produced by the optically active medium was measured by rotating the analyser until the transmitted light was extinguished. This gave the vibration direction of the light entering the analyser but did not indicate whether the plane of vibration was rotated clockwise or anticlockwise in order to reach that orientation. In addition, one could not tell whether the plane of vibration had rotated through (say) θ or $p\pi + \theta$. These ambiguities, which always arise in polarimetry, can be removed by reducing the thickness of active medium traversed and observing the new position of the analyser for extinction.

23–8. Half-shadow End-point Indicators

With a simple polarizer-analyser system the intensity of the field of view is very insensitive to small rotations of the analyser from the position of extinction. Consequently, the analyser cannot be set with great accuracy. Suppose that the polarizer gives plane polarized light but that one has slightly different planes of vibration in the two halves of the aperture. For example, suppose that, looking towards the oncoming beam of light, the planes of vibration are as indicated by Fig. 23–7. Suppose, now, that the analyser is rotated until its principal section is parallel to AB. Then the light in each half of the aperture has a small component of **D** in the transparent direction of the analyser, and the two halves of the aperture will appear equally and weakly illuminated. If the analyser is rotated through an angle $\theta/2$, the intensity of one side falls to zero while that

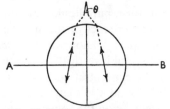

Fig. 23–7. The half-shadow principle: cross-section of beams showing vibration directions.

of the other increases. The position of the analyser that gives equal brightness in the two halves of the aperture can be found more precisely than the position that gives extinction in one half. In a polarimeter employing this method one simply adjusts the analyser for equality of illumination in the two halves of the field. The adjustment is repeated with the optically active medium between polarizer and analyser and the angle turned through by the analyser gives the rotation produced by the medium under test. It is interesting to note that the two halves of the field are also of equal brightness when the principal section of the analyser is perpendicular to AB. However, the brightnesses are then near their maxima and a small rotation of the analyser produces only a small change in the *relative* brightness.

The angle θ is known as the *shadow angle* so that the *half-shadow angle*, $\theta/2$, is the angle through which the analyser must be turned from the equality-of-brightness position in order to extinguish one half of the field. A number of half-shadow devices will now be described.

The Biquartz. The biquartz, devised by Soleil, was originally intended for use with white light in a saccharimeter but, as will be seen below, it can be used also with monochromatic light to give a half-shadow end-point device of the type described above. The biquartz is a quartz disc cut perpendicular to the axis, one half of the disc being of laevo- and the other of dextro-quartz. The thickness of the disc (3·75 mm.) is such that, for the mean yellow light in the region of maximum luminosity with a normal white source, the plane of vibration undergoes a rotation of $\pi/2$—clockwise in one half of the disc and anticlockwise in the other. Thus for yellow light the biquartz consists of two quarter-wave plates, where the term is used in a rather different sense from that employed in §22–13. The biquartz is placed immediately after an ordinary polarizer whose principal section PP′ (Fig. 23–8) is parallel to the diameter separating the two halves of the disc. With white light the rotatory dispersion of the quartz causes the

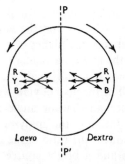

Fig. 23–8.　The biquartz.

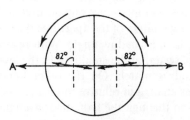

Fig. 23–9.　The biquartz as mono-chromatic half-shadow device.

vibration directions of red, yellow, and blue light to be as indicated in the figure. If the principal section of the analyser is parallel to that of the polarizer, the yellow light is extinguished, but in each half of the aperture some red and blue light is transmitted. The two halves of the field are equally illuminated with a reddish-violet *sensitive* or *transition tint*—the *tint of passage*. A small rotation of the analyser causes one half of the field to become distinctly red and the other to become blue. The setting of the analyser for a colour match can be done very accurately. The application of the biquartz in a compensating saccharimeter with white light is discussed in §23–9. When sodium light alone is used the biquartz becomes an ordinary half-shadow device. Each half gives a rotation of 82° so that the vibration directions are as shown in Fig. 23–9. Obviously, the two halves of the field of view appear equally bright if the principal section of the analyser is normal to AB. It will be seen that the half-shadow angle is 8°. In the figure the dividing line of the biquartz is parallel to the principal section of the polarizer, but this is not essential. The two halves of the field will always be matched when the polarizer and analyser are parallel.

The Cornu-Jellett Prism. A half-shadow device was constructed from a Nicol

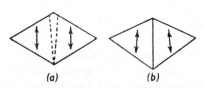

(a)　　　　　*(b)*

Fig. 23–10.　The Cornu-Jellett prism.

by Jellett in 1860 and modified by Cornu in 1870. In the later form a thin wedge was removed from a Nicol as indicated by Fig. 23–10(*a*) and the remaining parts brought together as in (*b*). The vibration directions of the transmitted rays are then as indicated, and the mode of action is as described above. One can use such a prism as the polarizer or as the analyser in a polarimeter employing monochromatic light. The half-shadow angle is half the angle of the wedge which is removed; usually it was $2\frac{1}{2}°$. Obviously a similar effect can be produced by using two pieces of sheet-type dichroic linear polarizer.

Laurent Half-wave Plate. It was shown in §22–13 that if plane polarized light enters a half-wave plate (cut parallel to the optic axis) with the vibration direction inclined at angle θ to the axis, the vibration direction of the emergent light is inclined at angle $-\theta$—the plane of vibration is rotated through angle 2θ. The discussion in §22–13 referred to a negative crystal but applies equally well to a positive crystal such as quartz. In the Laurent polarimeter, a half-wave plate of quartz is placed after, and across one half of, the aperture of the polarizer.

By adjusting the angle between the axes of the polarizer and the half-wave plate one can introduce any desired inclination between the planes of vibration in the two halves of the aperture; this was, in fact, the first polarimeter to have a variable half-shadow angle. With regard to the choice of half-shadow angle, it will be seen that the smaller it is, the less bright will be the field of view when the two halves match. On the other hand, a smaller half-shadow angle gives a greater change in relative intensities for a given rotation of the analyser since a rotation through the half-shadow angle always reduces the ratio of the intensities from unity to zero. With transparent media and a bright source a half-shadow angle of $2\frac{1}{2}°$ has been widely used for visual work; for absorbing media it may be desirable to increase it to between $5°$ and $10°$ in order to secure a sufficiently bright field of view.

The Lippich Polarizer. The chief disadvantage of the Laurent system is that it can be used only with monochromatic light of the wavelength for which the plate is a half-wave plate. A more satisfactory system was devised by Lippich (1885) and this has formed the basis of many high precision polarimeters and saccharimeters. After the polarizer (a prism of the Glazebrook-Thompson type)

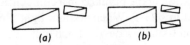

(a)   (b)

Fig. 23-11. Lippich polarizers.

a smaller prism is placed across half the aperture [Fig. 23-11(a)]. The observer sights on the edge of the smaller prism and this prism must be tilted as in the figure until there is no overlapping of the halves of the field nor any separation between them. The principal section of the smaller prism is inclined at angle θ to that of the larger. Consequently, disturbances emerging from the smaller prism have their vibration direction inclined at angle θ to that of disturbances in the other half of the field. The intensity of the light emerging from the small prism is reduced by a factor $\cos^2 \theta$ since the prisms are partially crossed, and the intensity is further reduced by the extra absorption and partial reflections. In consequence, the principal section of the analyser is not exactly perpendicular to the bisector of the shadow angle θ when the halves of the field are matched. Obviously the shadow angle can be varied simply by rotating one of the prisms —usually the larger.

Lippich also produced a triple-field polarizer by employing two small prisms, each occupying one-third of the aperture of the large prism [see Fig. 23-11(b)]. The principal sections of the small prisms are parallel and are inclined at angle θ to that of the large prism, the centre of the aperture of the large prism being left clear. The analyser is rotated until all three parts of the field are matched. This can be done only if the source and eye are lined up symmetrically with respect to the two small prisms. With the double-field polarizer it is possible to change the relative intensities of the two halves of the field by an error of alignment, so that an incorrect setting of the analyser is made.

Since convenient high-intensity monochromatic sources have become readily available, ordinary polarimeters (frequently of the Lippich double- or triple-field type) have been calibrated for use as routine saccharimeters. With instruments of this type an accuracy of 0·005° has been obtained in the measurement of the rotation. Modern sheet-type dichroic polarizers can now be used instead of polarizing prisms.

In recent years photo-electric polarimeters have been developed. In principle, the two halves of the field of a visual instrument are presented in succession to a photo-electric detector, and the end-point is a zero A.C. component in the output.

23–9. Compensating Saccharimeters

As mentioned in § 23–7, many saccharimeters have employed a compensator to balance out the rotation introduced by the solution. Fig. 23–12 shows two

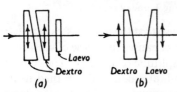

Fig. 23–12. Quartz compensators.

forms of quartz wedge compensator. In (a) the two wedges of dextro-quartz and the plate of laevo-quartz are cut perpendicular to the optic axis. The wedge angles are usually about 3°. By moving the wedges as indicated the path in dextro-quartz can be varied and can be made smaller than, or greater than, the thickness of the fixed plate. That is, the combination can produce a net rotation in either direction. In (b) one wedge is of dextro- and the other of laevo-quartz. Again the net rotation can be in either direction. With both compensators the net rotation is the same for all parts of the aperture.

The compensator can be mounted immediately in front of the analyser. With no solution present the analyser or polarizer is adjusted until the parts of the field are matched (using one of the end-point indicators described in the previous section). The solution is then interposed and the rotation produced by the solution is balanced out by the compensator by adjusting the latter until the parts of the field again match. The movement of the compensator wedges is indicated on a linear scale which can be calibrated to indicate either rotation in degrees or sugar concentration.

Fortunately the rotatory dispersion of quartz in the visible region is almost identical with that of cane sugar. Consequently, if the compensator balances the rotation of the sugar solution for one wavelength, it does so for all, and one may therefore employ white light. For example, a biquartz can be used and the compensator adjusted to bring back the sensitive tint in both halves of the field. It is easy to see that the rotatory dispersion of the substance under test precludes the use of white light in an ordinary polarimeter. Some compensating saccharimeters have used Lippich polarizers in order to obtain a variable half-shadow angle; although such a system can use white light, it cannot employ the sensitive tint.

It was pointed out above that compensating saccharimeters were devised mainly in order to allow white light to be used, but this point has now lost much of its importance since convenient high intensity (quasi-) monochromatic sources are available and physical detectors are being used instead of visual observation. However, it may be an advantage to avoid a rotating analyser involving a circular scale, and Gates (1958) has described a recording photo-electric compensating saccharimeter that uses the Faraday effect (see §23–11) to provide the compensating rotation. The essential elements of the system are shown in Fig. 23–13. The modulator consists of a solenoid wound round a glass

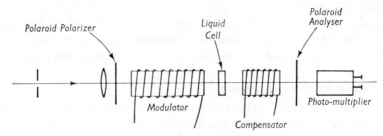

Fig. 23–13. Recording saccharimeter using Faraday rotation.

core. A 50 cycle a.c. is passed through the solenoid so that the light passes through an alternating magnetic field and the Faraday effect causes the plane of vibration to oscillate about the central position with an amplitude of about $3°$. The central position is rotated by the solution under test and, at balance, is returned to its original position by the d.c. field of the Faraday compensator. The analyser and polarizer are crossed, so that the intensity emerging from the analyser would be zero if there were no modulator. With the modulator operating, the intensity reaches a maximum at each extremity of the oscillating plane of vibration, so that it varies with a frequency double that of the applied a.c. If the compensation is incorrect the plane of vibration does not oscillate symmetrically about the extinction position and the intensity maxima are unequal. The varying intensity then has a component of the same frequency as the applied a.c., the phase of this component depending on which maximum is the larger and therefore indicating whether the compensating rotation is too large or too small, and the amplitude indicating the extent of the imbalance. The output from the photo-multiplier is then passed through a filter to remove the double frequency component and then to a phase-sensitive rectifier keyed to the modulation signal. The output then indicates the direction and magnitude of the imbalance and is used to control the current in the compensating solenoid. The current in the compensator is proportional to the rotation in the sample and is recorded.

23–10. Circular Dichroism. The Cotton Effect

The phenomenon of dichroism (selective absorption of plane polarized light) was described in §22–8, and it is to be expected that some substances will show unequal absorption of left- and right-handed circularly polarized light. The effect was first observed by Haidinger in 1847 in certain quartz crystals, and was observed in solutions by Cotton in 1895. Cotton suggested the term *circular dichroism* and this phenomenon, together with the anomalous rotatory dispersion in the region of absorption which accompanies it, is now referred to as the *Cotton effect*.

23–11. The Faraday Effect

In 1845 Faraday discovered that if glass is placed in a strong magnetic field it rotates the plane of polarization when plane polarized light is passed in a direction in which the field has a component. This was one of the earliest indications of a connection between light and electromagnetism. The effect has since been observed in many solids, liquids, and gases, and it is found that, if H is the magnetic intensity parallel to the direction of propagation of the light, the rotation of the plane of polarization for light of a given wavelength is given by

$$\theta = VHl,$$

where l is the length of the path in the field and V is a constant known as Verdet's constant. V is different for different substances, a positive value indicating that the rotation is in the clockwise direction *when looking along the direction of the field.* That is to say, θ is in the direction of rotation of a current which would give a field in the direction of H. It is important to notice that with magnetic rotation the direction of rotation does not depend on the direction of propagation of the light. Consequently, if light is reflected to and fro through the substance concerned, the rotation continues to increase. With *natural* rotation, on the other hand, the rotation changes direction when the direction of the light is reversed. (It was seen that a positive crystal always rotates the plane of vibration clockwise when looking *towards the oncoming light.*)

For sodium light the value of V in minutes of arc per oersted per cm. is 0·013 for water, 0·043 for carbon bisulphide, and about 0·08 for flint glass. For solutions one defines the *specific magnetic rotation* as the ratio of the rotation to the rotation produced by the same path length in water divided by the strength of the solution in grams of solute per c.c. of solution. The molecular rotation is the specific magnetic rotation multiplied by the ratio of the molecular weight of the solute to that of water.

Magnetic rotation can be explained by assuming that in the presence of a magnetic field the velocity of circularly polarized light depends on the direction of rotation. This variation of velocity is associated with the fact that the absorption frequencies, and therefore the dispersion of the medium, depend on the polarization of the light. (See also § 22–21.) Away from an absorption band Verdet's constant is approximately proportional to $(1/\lambda^2)$.

Very strong effects are observed with thin transparent films of ferro-magnetic materials. For example, for iron the effect is many thousand times larger than for water.

CHAPTER XXIV

THE VELOCITY OF LIGHT

24-1. Introduction

IN view of the enormous speed of light, it is not surprising that it was in the field of astronomy that it was first shown that the speed is, in fact, finite rather than infinite; it was only when astronomical distances were involved that the time of transit could be measured. The first calculation of the speed of light was made by Römer in 1676 from the observed times of the eclipses of a satellite of the planet Jupiter (see § 24–2). Bradley's observations in 1727 on the aberration of light led to a value of the speed of light which was in reasonable agreement with Römer's result (see § 24–3). It was not until 1849 that Fizeau succeeded in making the first determination of the time of transit of a beam of light over a terrestrial distance (§ 24–4). This was followed by Foucault's rotating mirror experiment in 1860 (§ 24–4). The rotating mirror method was used in a modified form by Michelson in 1879 and in later experiments by Michelson and by Newcomb.

Meanwhile, Maxwell's electromagnetic theory was published (1873) and Hertz's experiments (1887) demonstrated the existence of the waves predicted by the theory. The agreement between the predicted properties of electro-magnetic waves and the observed properties of light quickly confirmed beyond all reasonable doubt that light waves were electromagnetic waves. Now the theory indicates that the velocity of electromagnetic waves in free space is independent of the frequency and is equal to the ratio of the e.m.u. to the e.s.u. of charge. Thus the speed of light can be determined by measuring the speed of radio waves or by measuring the ratio of the units. These methods are capable of about the same accuracy as the modern optical methods. Recent experiments have included direct measurement of the speed of the high frequency waves used in radar (§24–7), measurement of the resonant frequency of a cavity resonator (§24–7), measurements with a micro-wave interferometer (§24–7), and direct measurement of the speed of light using a shutter involving a Kerr cell or a piezo-electric crystal (§§24–5, 24–8).

It is only in terms of the theory of relativity that one can give a full discussion of the velocity of light in moving media and of the velocity relative to an observer in motion. These subjects are referred to briefly in this chapter but the theory of relativity is beyond the scope of the present book.

24-2. Römer's Method

At a time when the earth was in that part of its orbit which is nearest to Jupiter, Römer observed the interval between successive eclipses of Jupiter's innermost satellite. He then calculated when the hundredth eclipse would occur but found that the observed time was about 15 minutes later. Römer pointed out that the earth had moved farther away from Jupiter and that the delay could be explained by assuming that the light took that much longer to travel from

Jupiter to the earth. The increase in distance between the earth and Jupiter could be deduced from astronomical observations so that the velocity of light could be calculated. Römer pointed out that if his explanation was correct the observed times of the eclipses would be delayed more and more as the distance between the earth and Jupiter increased, and that the delays would decrease as the planets approached one another. The delay would become zero when they returned to their original positions. That is to say, the observed interval between eclipses should show a periodic variation. This was confirmed experimentally and Römer obtained the value of $3 \cdot 10 \times 10^{10}$ cm. sec.$^{-1}$ for the speed of light. More recent observations have given about $2 \cdot 98 \times 10^{10}$ cm. sec.$^{-1}$.

24–3. The Aberration of Light

Owing to parallax, there is an apparent movement of nearer stars relative to more distant ones as the earth moves round its orbit. Bradley discovered that, in addition, every star appears to be displaced in the direction in which the earth is moving, and that this displacement is independent of the distance of the star from the earth. The necessary change in the direction of the observing telescope can be likened to the change in tilt which one must give to one's umbrella when, after standing still, one commences to walk in the rain. The effect is known as the *aberration of light*. Suppose AB (Fig. 24–1) is the axis of an observing telescope which is moving from left to right with speed v. Suppose light from a distant star is inclined at angle θ to the horizontal and enters the telescope at A. If the telescope points in the *apparent* direction of the star, the light must travel

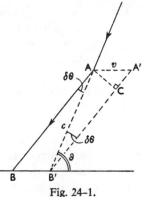

Fig. 24–1.
The aberration of light.

along the axis of the telescope. While it is doing so, the telescope axis moves to the position A′B′. That is, the light travels from the original position of A to the displaced position of B (i.e. to B′). Thus AA′ ∝ v and AB′ ∝ c. In the figure the magnitude of the angle $\delta\theta$ has been very much exaggerated—since $v \ll c$, $\delta\theta$ is always small. Therefore $\angle$ABB′ $\doteqdot \theta$ and $\triangle$ABB′ gives sin $\delta\theta$/sin θ = BB′/AB′ = v/c. Hence one can write

$$\delta\theta = \frac{v}{c} \sin \theta. \qquad \cdots \cdots \cdots \quad (24\text{–}1)$$

By measuring the magnitude of the aberration Bradley obtained about $3 \cdot 08 \times 10^{10}$ cm. sec.$^{-1}$ for the velocity of light. More recent measurements have given 299,714 km. sec.$^{-1}$. The above is a simple non-relativistic account of the phenomenon of aberration; the theory of relativity is beyond the scope of the present discussions and a complete account of the aberration of light cannot be given here.

24–4. Early Terrestrial Methods

Fizeau's experiment (1849) was the first successful attempt to find the time taken for a beam of light to traverse a terrestrial distance. A beam of light was interrupted by a rotating toothed wheel and directed over a path of 1.7266×10^6 cm. before returning to the wheel. If the transit time of the light is equal to the time taken for a tooth to move into the position initially occupied by a space, the returning beam is obstructed. Fizeau observed successive eclipses as the speed of the wheel was increased and found the transit time to be 5.5×10^{-5} sec., giving $c = 3.14 \times 10^{10}$ cm. sec.$^{-1}$. Similar methods were used by Cornu (1874), by Young and Forbes (1881), and by Perrotin and Prim (1903).

The rotating mirror method used by Foucault in 1860 was originally suggested in 1834 by Wheatstone and was used subsequently by several other workers. A light path of up to 20 metres was used and the light was reflected on its outward and return journeys by a rotating mirror. The small rotation of the mirror during the transit time of the light caused a deviation of the returning beam from its original path; the transit time was found by measuring this deviation and the speed of rotation of the mirror. Foucault's result was 2.98×10^{10} cm. sec.$^{-1}$.

Foucault's rotating mirror method was also used by Cornu, by Michelson (1879, 1882, 1927), and by Newcomb (1882). Newcomb introduced the idea of employing a many-sided mirror; for example, he used a highly polished square steel prism (so that light was transmitted and collected four times per revolution) and obtained a brighter image. Michelson and Pearson subsequently used 8-, 12-, and 16-sided mirrors. In the later determinations attempts were made to find the refractive index of the air and to calculate the speed of light in a vacuum.

Probably the most accurate series of experiments using the rotating mirror method was that made in Michelson's Mount Wilson determination (1926). An octagonal mirror block was used and its speed of rotation was adjusted so that the transit time was approximately one-eighth of the time for one revolution. Thus on the return journey the light was reflected by the mirror face *next* to the one used on the outward journey; the image displacement was made small and was measured with a micrometer eyepiece. According to Dorsey, the light path (about 22 miles in each direction) was determined to an accuracy of about 1 in 200,000, which was better than the accuracy of the transit time and of the estimated refractive index. The result, as given by Michelson, was 299,796 km. sec.$^{-1}$ for the speed in a vacuum. Birge (1941) has pointed out that the correction to a vacuum was made incorrectly (see §24–6) and that the result should have been 2 km. sec.$^{-1}$ higher. This result is closer than any other pre-war result to the currently accepted value.

In order to minimise the refractive index correction Michelson planned a rotating mirror determination in which the light path was in air at low pressure contained within a sealed iron pipe about a mile long. The experiment was eventually completed by Pearson and Pease after Michelson's death in 1931. By reflecting the light to and fro, a total path of nearly 13 km. was obtained and the air pressure in the pipe varied between 5.5 and 0.5 mm. of mercury. A correction was made for the residual air and the result for the speed of light in a vacuum was given as $299,774 \pm 11$ km. sec.$^{-1}$.

24–5. The Use of a Kerr Cell Optical Shutter

By placing a Kerr cell (see §22–24) between crossed Nicols or crossed polaroids, one can construct a very convenient optical shutter. The arrangement

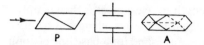

Fig. 24–2. Kerr cell optical shutter.

is shown in Fig. 24–2. The light is plane polarized by the polarizer P and passes through the Kerr cell to the analyser A. (Nitrobenzene is a suitable liquid for use in the cell since its Kerr constant is relatively large.) When there is no p.d. across the cell no light emerges from the analyser. If a p.d. is applied, the liquid in the cell acts as a uniaxial crystal whose axis is parallel to the lines of force. It follows that if the lines of force are not parallel to or perpendicular to the plane of vibration of the incident light, the latter is resolved into O and E components. These components travel with different speeds, so that a phase difference is introduced between them. Hence, in general, elliptically polarized light emerges from the cell and some light is transmitted by the analyser. Usually the plane of vibration of the light entering the Kerr cell is at 45° to the field. The phase difference δ between the O and E components emerging from the cell is proportional to the square of the applied p.d. (see § 22–24), and the intensity of the light transmitted by the analyser is given by $I = I_0 \sin^2 (\delta/2)$ [see equation (22–3)]. Actually it is usual to superpose a steady p.d. across the plates in order to work on the more linear part of the cell characteristic (see Fig. 24–3). The intensity of the light transmitted by the analyser then varies about a mean value with a frequency equal to that of the alternating components of the applied field.

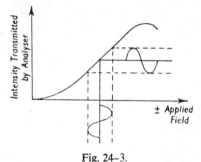

Fig. 24–3.
Characteristic of a Kerr cell shutter.

A Kerr cell operated as an optical shutter was first used in a determination of the velocity of light by Karolus and Mittelstaedt in 1925. In principle the method resembles that of Fizeau, the toothed wheel being replaced by the Kerr cell shutter. The advantage of using the Kerr cell is that the frequency with which the light beam can be modulated is increased enormously and much shorter light paths can be employed. A light path of over three hundred metres was used and the final result for the velocity of light in a vacuum was $299{,}778 \pm 10$ km. sec.$^{-1}$. Further experiments employing Kerr cells were carried out by Hüttel (1940) and by Anderson (1937 and 1940).

Anderson used a Kerr cell to modulate a light beam at a frequency of 19·2 Mc/s. This beam was amplitude-divided so that one component was used to provide a reference beam and the other was reflected in turn (a) from the nearer end and (b) from the farther end of the accurately measured base line before being recombined with the reference beam and received by the detector. The reference beam was used to compare the phases of the modulation waves of the beams returning from the nearer end and from the farther end of the base line—

if the modulation waves of the superposed beams are in phase one has a large signal at that frequency whereas if they are out of phase one has a minimum. The principle of the length measurement is then simply that for the modulation waves a path length equal to the wavelength introduces a phase lag of 2π. In this way the transit time was found for a base line of about 172 metres. Anderson produced his first result in 1937 and his final result in 1940 as indicated in the table given below.

These experiments can be looked upon as marking the end of an era, and a careful analysis of the position was given in 1941 by Birge. The situation was again reviewed in 1944 in a very extensive critical review by Dorsey.

24–6. The Position in 1941

In 1941 Birge gave a critical review of the various determinations of the velocity of light and other physical constants. He pointed out that up to 1940 all workers had converted their results for the velocity in air to those for the velocity in a vacuum by using the wave refractive index instead of the group refractive index. The refractive index as determined by Snell's law of refraction gives the ratio of the wave velocities in the media concerned, but in all the experiments on the velocity of light the light beam is modulated so that the disturbances are propagated with the group velocity (see §6–10). Thus one should convert the results from air to vacuum by means of the group index. Anderson did this in his second determination. The small correction needed in the evacuated tube experiment can be neglected, but the results given by other workers require revision.

One of the difficulties in deciding the "best value" for the velocity of light is that of estimating the relative accuracy of the various experiments. For example, the ranges of uncertainty quoted by various workers do not always have the same significance. Birge analysed the various methods and calculated or estimated the so-called probable error for each.

In addition to the ordinary optical methods described above, there are two indirect determinations of c which should be mentioned:

(1) In 1906 and 1907 Rosa and Dorsey carried out a very accurate determination of the ratio of the electromagnetic and electrostatic units. As has already been pointed out, this gives a value for the velocity of electromagnetic waves in a vacuum. The value of c given by the results of Rosa and Dorsey could not be decided until later, when the ratio of the international ohm to the absolute ohm had been found with sufficient accuracy.

(2) In 1923 Mercier measured the velocity of propagation of electromagnetic waves guided by parallel wires. It is necessary to make certain corrections to obtain a value for the propagation of waves in the free air and a further correction is required to find the velocity in a vacuum.

The results obtained by Mercier and by Rosa and Dorsey were analysed by Birge and the final results and estimated accuracy are given in the table below. Birge considered that none of the results obtained before 1906 should be included when calculating the most probable value of c. The results obtained between 1906 and 1941 are as follows:

Workers	Date	Method	Result Originally Announced (km.sec.$^{-1}$)	Corrected Result (km.sec.$^{-1}$)	Probable Error
Rosa and Dorsey	1906	Ratio of Units	299,710	299,784	10
Mercier	1923	Waves on Wires	299,700	299,782	30
Michelson	1927	Rotating Mirror	299,796	299,798	15
Mittelstaedt	1928	Kerr cell	299,778	299,786	10
Michelson, Pease, and Pearson	1933	Rotating Mirror	299,774	299,774	4
Anderson	1937	Kerr cell	299,764	299,771	10
Hüttel	1937	Kerr cell	299,768	299,771	10
Anderson	1940	Kerr cell	299,776	299,776	6

The weighted mean given by Birge was $299{,}776 \pm 4$ km. sec.$^{-1}$.

Mention has already been made of the very extensive critical review given in 1944 by Dorsey. He pointed out that many of the earlier reports did not give sufficient details to enable a reader to estimate the reliability of the results. After a careful analysis of the various reports he suggested a new weighted mean of $299{,}773 \pm 10$ km. sec.$^{-1}$. Dorsey pointed out that the search for systematic errors does not usually seem to have been sufficiently thorough.

Later reviewers have made further adjustments to the "corrected" results obtainable from the early experiments but it now seems that all the pre-war determinations must be regarded as merely of historic interest.

24–7. Use of Radar, Cavity Resonators, and Micro-wave Interferometry and Spectroscopy

When radar came into use during the war, it was necessary to assume a value for the velocity of electromagnetic waves in order to be able to calculate the distance of the target. The accuracy of the radar method was checked by using targets at known distances. The above value was assumed for the velocity of electromagnetic waves in a vacuum, and corrections were made for the refractive index of the air and for the effect of radiation reflected or scattered from the ground. It was then found that the distances given by the radar method were always too small by about 1 part in 20,000. This could be put right by assuming a correspondingly larger value for the velocity of the waves but an error of this magnitude came as a surprise in view of the results quoted in §24–6. It is obvious that by employing known distances the radar apparatus can be used to determine the velocity of the radio waves, and the result obtained in 1949 by Aslakson for the velocity in a vacuum was $299{,}792 \pm 2{\cdot}4$ km. sec.$^{-1}$. A subsequent determination in 1951 gave $299{,}794{\cdot}2 \pm 1{\cdot}9$ km. sec.$^{-1}$.

Meanwhile, Essen and Gordon-Smith (1947) had made a very accurate measurement of the resonant frequency of a cavity resonator. Since micro-waves travel along a metal tube or wave guide with little attenuation, a closed guide constitutes an electrical resonator. Essen and Gordon-Smith used a resonator which consisted of an accurately made cylindrical cavity in a block of copper. Copper was used because it is desirable for the walls of the cavity to have a high conductivity. The dimensions of the cavity were accurately measured interferometrically, the cavity was evacuated, and the high frequency fed to a probe. The frequency was then adjusted until resonance occurred, a second probe being connected to a detector. If the dimensions of the cavity and the resonant frequency are known, the velocity of the waves can obviously be found. Although the waves in a guide do not travel with the free-space velocity, the latter can be calculated; the value obtained was $299,792 \pm 3$ km. sec.$^{-1}$. Further experiments with cavity resonators were conducted in 1950 by Essen in England, and by Hansen and Bol in America. Essen's result was $299,792 \cdot 5 \pm 1$ km. sec.$^{-1}$, the maximum possible error being given as ± 3. The result announced by Bol was $299,789 \cdot 3 \pm 0 \cdot 3$ km. sec.$^{-1}$, the systematic error being estimated at not more than $+0 \cdot 5$. However, Dayhoff (1952) considered that if a proper correction is made for the skin effect Bol's result becomes $299,794 \cdot 3$ km. sec.$^{-1}$.

In 1951 Froome used a micro-wave analogue of the Michelson interferometer in order to make an accurate determination of a wavelength in the region of $1 \cdot 25$ cm. The fixed arm and the beam divider were enclosed in a wave guide, and the variable path was in free space. The intensity of the beam traversing the fixed path was varied to give good contrast between the maxima and minima. These were detected by a superheterodyne receiver as the path difference was changed. The frequency (24,000 Mc/s) was measured accurately by comparison with a high harmonic of a quartz oscillator, and the velocity was calculated from the product of the frequency and wavelength. The result obtained for the velocity in air was converted to the result for a vacuum by measuring the atmospheric temperature, pressure, and humidity. The result was $299,792 \cdot 6 \pm 0 \cdot 7$ km. sec.$^{-1}$. More recently, this method has been repeated by Simkin, Lukin, Sikora, and Strelenskii (1967) using a higher frequency (37 Gc/s.— λ 8 mm.). For this determination the microwave refractive index was determined with a cavity resonator refractometer. The result was $299,792 \cdot 56 \pm 0 \cdot 11$ km. sec.$^{-1}$.

It should be noted that with this microwave analogue of the Michelson interferometer diffraction effects are much greater than with the optical interferometer. This is because the wavelength is very much longer. Since the source is only a few wavelengths in diameter one has almost true spherical waves instead of the plane waves one gets with the Twyman-Green modification. In a second series of experiments Froome (1954) used a new form of interferometer. This had two transmitting horns facing one another and two receiving horns placed back to back between them. The outputs from the receivers were superposed and the phase difference between them varied by moving the receiving horns together along the line joining the transmitters. The wavelength was measured by finding the positions for interference minima. The result obtained from the second determination was $299,792 \cdot 7 \pm 0 \cdot 3$ km. sec.$^{-1}$. Important sources of error were the effect of diffraction and random scattering by fixed objects in the laboratory; in fact Froome's 1954 results were obtained from experiments that were really performed in order to study the diffraction effects with an interferometer that was an experimental prototype. In his final series of experiments Froome (1958) used a high precision interferometer; doubled the range over which the receiving horns could be moved to 2 metres; employed a higher frequency (72,000 Mc/s—

wavelength about 4 mm.) and improved the location of the equipment to give a tenfold reduction of the diffraction and scattering errors; and used a cavity resonator to measure the refractive index of the air in the neighbourhood of the interferometer. As a result of these improvements Froome was able to make what is probably the most accurate determination of c so far achieved, the largest remaining uncertainty then arising from the use of the length standards in measuring the displacement of the receiving horns. Froome's final value was 299,792·50 ± 0·10 km.sec.$^{-1}$.

The advantage of employing micro-waves in an interferometer is that it is possible to measure both the wavelength and the frequency; in optical interference experiments only the wavelength can be determined. However, it is possible to combine wavelength measurements made optically with frequency measurements made electrically in the following manner. Energy levels associated with the rotation-vibration bands in infra-red spectra are also responsible for the micro-wave spectra, and measurements in the infra-red region in terms of wave-length can therefore be combined with micro-wave measurements in terms of frequency to give a value for c.

The method, first suggested by Douglas, involves the calculation of the rotational constant B'' of the ground state of a linear molecule; I.R. spectro-scopy yields a value in wave-number units and microwave spectroscopy yields a value in frequency units, the ratio giving c. In the first determination of c by this method Rank, Vander Sluis, and Ruth (1952) measured the wavelengths of the absorption lines of the 004–000 and 103–000 bands of HCN in the "photographic infra-red" (0.8μ). They crossed a Fabry-Perot etalon with a plane grating spectrograph and combined their results with the microwave measurements of Nethercot, Klein, and Townes, to get $c=299,776$ km. sec.$^{-1}$. The original estimate of the accuracy was ±6 but subsequently it was pointed out that defects in the grating could have contributed a systematic error as large as 14 km. sec.$^{-1}$. Rank, Shearer, and Wiggins (1954) used the HCN 002–000 band at 1.53μ. They employed an etalon in conjunction with a grating and used a grating value for the wavelength of an absorption line in order to calibrate the interferometer, this calibration being the least reliable part of the experiment. Combination with the microwave measurements of Nethercot and Klein gave $c=299,789.8 \pm 3.0$ km. sec.$^{-1}$. A better value was derived from these measure-ments by Rank, Bennett, and Bennett (1955) by obtaining a more accurate value of the wavelength of the line used to calibrate the interferometer. They were able to use the interferometer to calibrate this line against the Hg^{198} green line, thus avoiding the grating calibration. The new value for c was 299,791·9 ± 2·2 km. sec.$^{-1}$. The measurements on the HCN 002–000 band were repeated by Rank, Guenther, Shearer, and Wiggins (1957) to get $c=299,793·2 \pm 1·8$ km. sec.$^{-1}$ Rank, Guenther, Saksena, Shearer, and Wiggins (1957) proceeded to measure interferometrically 33 lines of the 2–0 band of CO at 2.3μ, and combined their results with the microwave measurements of Cowan and Gordy to get $c=299,793·7 \pm 0·7$ km. sec.$^{-1}$. Plyler, Blaine, and Conner (1955) had made some rather less accurate measurements of the CO band using a grating and had taken the microwave results of Gilliam, Johnson, and Gordy, to get $c=299,792 \pm 6$ km. sec.$^{-1}$ (the Cowan and Gordy results give 299,793·4 ± 5·8 km. sec.$^{-1}$).

The accuracy assumed by Rank *et al* (1957) was based on the assumption that the microwave measurements made a negligible contribution to the errors but some later measurements by Jones and Gordy (1964) throw some doubt on this; these later results decrease the 1957 value for c by 2·8 km. sec.$^{-1}$ and put it well below the values obtained in all other recent determinations; one feels inclined to suspect the 1964 microwave result.

A new infra-red technique was used by Rank, Eastman, Rao, and Wiggins (1962). They used a double-pass echelle spectrograph with a replica echelle of high quality to measure the 1–0, 2–0, 3–0, bands of HCl^{35} and the 1–0, 2–0, bands of DCl^{35} in the range 2–5μ. The microwave measurements of Gordy and Cowan give $c=299,793\cdot1$ $\pm0\cdot65$ km. sec.$^{-1}$ from the DCl^{35} bands and the 1964 microwave measurements by Gordy and Jones yield 299,792·8 $\pm0\cdot4$ km. sec.$^{-1}$ from the HCl^{35} bands.

Florman (1955) has made the lowest frequency (173 Mc/s) high precision determination of the velocity of light. He measured the wavelength by measuring the phase difference between disturbances received at two different distances from the R.F. transmitter, the distance between the receivers having been measured by the usual surveying methods. The frequency was measured and the speed calculated to be 299,795·1 $\pm3\cdot1$ km. sec.$^{-1}$.

In 1956 and 1957 Wadley used equipment known as a Tellurometer to determine c. This is equivalent to Bergstrand's Geodimeter (see §24–8) but employs a microwave carrier frequency. The Tellurometer, like the Geodimeter, is really designed to measure distances, assuming a value for the speed of light.

24–8. Recent Optical Experiments and Summary

In view of the disagreement between the results given by the experiments on radio waves and those given by the pre-war optical methods, it is obvious that further optical experiments were called for. High precision optical determinations have been made by Bergstrand, by Schöldström, by Edge, and by Karolus and Helmberger.

The principle of Bergstrand's method is as follows. By using a Kerr cell the light from the source S [Fig. 24–4(a)] is modulated by the high frequency crystal-controlled oscillator O. The light is reflected from the mirror M to the photo-tube P which gets its operating voltage from O. Thus the sensitivity of the tube varies with the same frequency as the intensity of the light. Suppose, for example, that the maxima of the source intensity and photo-tube sensitivity occur simultaneously. It will be seen that an intensity maximum will arrive at P at a time of high or low sensitivity, depending on the distance $D=$ SMP. That is, the photo-electric current will vary with D, the variation being of the form represented by curve (1) in Fig. 24–4(b). If one had a similar apparatus in which the maxima of the source intensity coincided with the minima of the photo-tube sensitivity, the current would follow curve (2). If these two currents are passed in opposite directions through a measuring instrument, the difference curve (3) is recorded. The advantage of an arrangement of this type is that one has sharply defined values of D where the current is zero, instead of rather ill-defined values where the current is either a maximum or a minimum. It is impracticable to arrange two identical sets of apparatus as suggested above, but it is possible to

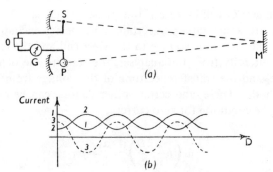

Fig. 24–4.   Principle of Bergstrand's method.

make one apparatus do the work of each in turn. This can be achieved by periodically reversing the phase of the light modulation relative to that of the detector, the period being long compared with that of the modulation, but short compared with the response time of the measuring instrument. The phase of the light modulation is reversed by reversing the direction of the bias voltage that is applied to the Kerr cell. Ideally, the bias voltage would have a square wave form; this was approximated to by cutting off the top of a sine wave.

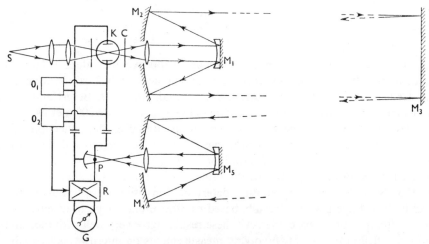

Fig. 24–5.   Bergstrand's method.

The experimental arrangement is illustrated by Fig. 24–5. S is the source (a 30-watt filament lamp); K C is the Kerr cell arrangement; O_1 is the high frequency modulation oscillator (2000 volts, 8 Mc/s); O_2 is the low frequency bias oscillator (5000 volts, 50 c/s); P is the photo-tube; M_2 and M_4 are concave spherical mirrors (apertures 46 cm., focal lengths 75 cm.); M_3 is plane, and M_1 and M_5 are plano-concave lenses silvered on their rear surfaces. The spherical aberration of M_2 and M_4 is balanced by the lens mirrors M_1 and M_5, and the chromatic aberration of the latter does not matter since only yellow light is used. If $2D$ is the total light path, the phase lag between source and receiver for the modulation waves is $\omega(2D/c)$ where $\omega/2\pi$ is the modulation frequency. The current in the meter G will be zero if this phase change is $\pi/2-p\pi$ (p integral). Actually there are some rather uncertain time delays including the interval between the application of a voltage to the Kerr cell and the emergence of the corresponding intensity from the analyser, the transit time of the electrons in the photo-tube, and the transference time of the voltage from the Kerr plates to the photo-anode. These, and certain other factors, can be taken account of by writing, for the condition for zero current,

$$\omega\left(t-\frac{2D}{c}\right)=\frac{\pi}{2}-p\pi,$$

t being the total time uncertainty.

Hence
$$D=K+\frac{2p-1}{8}\lambda,$$

where λ ($=2\pi c/\omega$) is the modulation wavelength and K is a constant. K can be eliminated by taking two values of D. If D_N is the difference one has

$$D_N=\frac{N}{4}\lambda,$$

where N is an integer. N can be found by using approximate values of λ and D_N since it must be integral, so that if D_N is known accurately λ can be found. If the frequency is known c can then be calculated. In effect, Bergstrand made adjustments for values of D of approximately 90 metres and 7000 metres. With M_3 fixed at 7000 metres the frequency was adjusted to give zero deflection. The light was then reflected from a movable mirror at 90 metres and the mirror adjusted for zero deflection. The mirror at 90 metres was convex so that the image formed at the receiver was the same as that for a plane mirror at 7 km. In 1955 Schöldström made another determination using Bergstrand's method and in 1956 Edge gave two results based on measurements by Mackenzie over two ordinance survey base lines. All these results agree very well with the values obtained from experiments involving measurements on microwaves (see table, p. 600). Apparatus similar to that used in Bergstrand's method (known as a *Geodimeter*) is now employed to measure distances, assuming a value for c (see below).

In 1950 Houstoun reported the results obtained with a new type of optical shutter. Stationary ultrasonic waves were set up in a piezo-electric quartz oscillator. When a high frequency p.d. is applied across a quartz crystal, standing ultrasonic waves are set up so that one has variations of density across the crystal, the magnitude of the spatial variations varying periodically with time. In consequence, the crystal behaves as an intermittent diffraction grating and the intensity of, say, the first order maximum varies periodically. In this way Houstoun was able to obtain a beam of light which was modulated with a very high frequency (about 100 Mc/s). The light in the first order maximum passed to a mirror and back, and returned towards the source if it arrived back when the grating was effective. Thus the crystal replaced the toothed wheel in an experiment which resembled that of Fizeau. The accuracy of Houstoun's initial results is less than that of Bergstrand's results, but Houstoun claims that the accuracy of his method can be improved considerably. The result obtained was 299,782 ± 9 km. sec.$^{-1}$ but was originally announced as 299,775.

A rather similar light modulator was used by Karolus and Helmberger (1966) in a far more accurate determination. An ultrasonic standing wave system in a liquid is produced by quartz crystal oscillators immersed in the liquid. Collimated monochromatic light from a mercury lamp is passed through the resulting diffraction grating and focused in the plane of a slit which transmits only the zero order beam. Since the ultrasonic grating diffracts a varying intensity into the non-zero orders, the light transmitted by the slit exhibits a shallow modulation (at about 19 Mc/s). The general layout of the experiment is shown schematically in Fig. 24–6. L is the source, U the ultrasonic grating cell, and D a detector system. The modulated beam of light from the slit S is amplitude-divided at A. One beam provides the detector (photomultiplier) with a reference signal whose phase can be varied by moving the mirror assembly R.

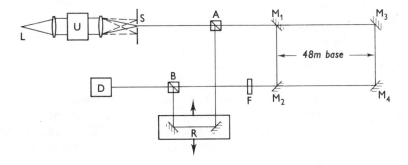

Fig. 24–6. Layout used by Karolus and Helmberger.

The other beam passes to the base line of 48 metres, and is returned from the nearer end by M_1 and M_2 or from the farther end by M_3 and M_4. The two beams are received by the detector system D. If the modulations of the beam that returns from the base are in phase with those of the beam that has travelled via R, the output from D is a maximum; the output is a minimum if the beams arrive in anti-phase. To obtain high sensitivity of modulation phase setting the intensities of the beams must be equal, and this is achieved by means of the adjustable neutral filter F. Movements of R of less than 0·1 mm. could be detected, which gives the precision with which the base-line end-points can be determined. The error of the refractive index correction was estimated to be $\pm 1 \times 10^{-7}$. Following the correction of an error in the base length the final result for c was 299,792·44 $\pm 0·2$ km. sec.$^{-1}$.

Subsequently, Karolus and Helmberger made another determination, in which the light source was a He–Ne gas laser, and announced (1967) a result of 299,792·5 $\pm 0·15$ km. sec.$^{-1}$.

Date	Workers	Method	Result (km. sec.$^{-1}$)
1947	Essen and Gordon-Smith	Cavity Resonator	299,792 ± 4
1949	Aslakson	Radar	299,792·4 $\pm 2·4$
1949	Bergstrand	Geodimeter	299·793 ± 2
1950	Essen	Cavity Resonator	299,792·5 ± 1
1950	Bol	Cavity Resonator	299,794 ± 1
1950–1	Bergstrand	Geodimeter	299,793·1 $\pm 0·2$
1951	Aslakson	Radar	299,794·2 $\pm 1·9$
1951	Froome	Microwave Interferometer	299,792·6 $\pm 0·7$
1954	Froome	Microwave Interferometer	299,792·7 $\pm 0·3$
1955	Florman	R.F.	299,795·1 $\pm 3·1$
1955	Rank et al	Band Spectra	299,791·9 $\pm 2·2$
1955	Schöldström	Geodimeter	299,792·4 $\pm 0·4$
1956	Wadley	Tellurometer	299,792·9 $\pm 2·0$ 299,792·7 $\pm 2·0$
1956	Edge	Geodimeter	299,792·4 $\pm 0·5$ 299,792·2 $\pm 0·4$
1957	Wadley	Tellurometer	299,792·6 $\pm 1·2$
1957	Rank et al	Band Spectra	299,793·2 $\pm 1·8$
1957	Rank et al	Band Spectra	299,793·7 $\pm 0·7$
1958	Froome	Microwave Interferometer	299,792·50 $\pm 0·10$
1962	Rank et al	Band Spectra	299,793·1 $\pm 0·65$
1964	Rank et al	Band Spectra	299,792·8 $\pm 0·4$
1966–7	Karolus and Helmberger	Ultrasonic Modulator	299,792·44 $\pm 0·2$
1967	Karolus and Helmberger	Ultrasonic Modulator	299,792·5 $\pm 0·15$
1967	Simkin et al	Microwave Interferometer	299,792·56 $\pm 0·11$

The table gives a summary of the best post-war determinations of c and excludes certain results that do not seem to constitute a useful advance. Some of the values given in the table differ from those announced originally owing to the subsequent discovery of sources of error. Some of the "possible errors" also differ from those announced originally; as far as possible those given in the table represent estimates (not always by the original workers) of combinations of random and possible systematic errors. When comparing these results with the earlier ones it must be remembered that the limits given by Birge are simply the estimated probable errors* and exclude systematic errors. In 1957 the XIIth General Assembly of International Scientific Radio Union recommended the adoption of the value 299,792·5 $\pm 0·4$ km. sec.$^{-1}$ and this value was accepted by the International Union for Geodesy and Geophysics.

The use of the agreed value of c to form the basis of the measurement of distance is now a well established practice. Froome and Essen consider that it is likely that by using a sub-mm. interferometer c can be determined to an accuracy of 2 in 10^8; they then draw attention to the possibilities of defining c by international agreement and then using frequency measurement to derive the standard of length.

One of the attractive features of the microwave interferometer method of determining c is that it is possible to measure both the frequency and the wave-

---

* For experimental results exhibiting only random errors the probable error defines the range for which the probability that the true result is included is $\frac{1}{2}$

length and calculate c, so that there is no need for an inconveniently long base line. Work is currently (1972) in progress at the National Physical Laboratory on a new method that is based on the determination of both frequency and wavelength of radiation in the infra-red. The method is made possible largely as a result of techniques being developed in what might be called laser technology. The final details are not yet settled but the present intention is to measure the frequency and wavelength of radiation emitted by a CO_2 laser. There are several steps in the proposed frequency measurement: (1) Measure the beat frequency between the CO_2 line at 32·1 THz (λ 9·3 μ) and the 3rd harmonic of H_2O laser radiation at 10·7 THz (28 μ) ($3 \times 10·7 = 32·1$). (2) Measure the beat frequency between the H_2O line and the 12th harmonic of HCN laser radiation at 891 GHz (337 μ) ($12 \times 891 = 10692$). (3) Measure the beat frequency between the HCN line and the 9th harmonic of a klystron at 99 GHz (3 mm.) ($9 \times 99 = 891$). (4) Compare the klystron with a harmonic of a quartz crystal oscillator whose frequency can be derived directly from the caesium atomic clock. This chain of measurements gives the frequency of the CO_2 line in terms of the atomic clock. It is proposed to measure the wavelength of the CO_2 line by mixing this radiation with light from a He–Ne laser (0·633 μ) to give a difference frequency sideband (0·679 μ) comparable with the He–Ne line, the final comparison being between the He–Ne line and the current standard krypton wavelength (0·605 μ), using a Fabry-Perot interferometer. The proposed procedure should give c with a precision limited almost entirely by the reproducibility of the present length standard, which is about 1 part in 10^8; this is roughly 100 times better than the present precision and would add weight to the proposal to base the length standard on c. (1974: $c = 299\,792\,459 \pm 0·8$ ms^{-1}).

24–9. The Velocity of Light in Material Media

The experiments described in the preceding sections were devised in order to measure the velocity of light in air or a vacuum. Foucault (1850) was the first to compare the velocity of light in a medium denser than air with that in air. At the time this was regarded as an important experiment which would decide between the rival theories of light. Newton's corpuscular theory explained refraction by assuming that light travelled faster in the medium with the larger refractive index (defined from Snell's law in the form $n \sin I = n' \sin I'$), whereas the wave theory indicated that the velocity decreased as the refractive index increased.

Foucault used the rotating mirror method with light paths (a) in air and (b) in water. His results, which were only qualitative, indicated a lower speed in water. Michelson repeated the experiment in 1885 and obtained quantitative results. He obtained 1·33 for the ratio of the velocity of light in air to the velocity in water, and 1·758 for ratio of the velocity in air to the velocity in carbon disulphide. Now the refractive index calculated from Snell's law gives the wave velocity, and from this the group velocity can be calculated if the dispersion is known. The calculated ratio of the group velocities in air and water is not widely different from the calculated ratio of the wave velocities and, within the limits of experimental error, agrees with Michelson's result. On the other hand, the ratio of the group velocities for air and carbon disulphide is found to be about 1·745 for λ5800 angstroms whereas the ratio of the wave velocities is 1·64. There is some doubt about the mean wavelength of the light used by Michelson but the results quoted above indicate that it is the group velocities which are measured in these experiments. It is rather surprising that until 1941 this fact was not taken into account in the results given by various observers for the velocity of light in a vacuum (see § 24–6). Michelson's experiments also indicate that the velocity of red light in CS_2 is slightly greater than the velocity of blue light, which agrees with the observed dispersion of the refractive index.

Michelson's result for the velocity of light in carbon disulphide was confirmed by Gouy in 1886 using the rotating mirror method. Further experiments by Gutton in 1912 gave results for carbon disulphide and for monobromnaphtha-lene which were also in agreement with the calculated group velocities.

In a more accurate series of experiments in 1944 using the piezo-electric light modulator already described, Houstoun measured the group velocity in water for the red, green, and blue light transmitted by a set of filters. The group indices obtained from these results were in good agreement with the calculated values. The results for red, green, and blue light were as follows (calculated values of the group and phase indices are shown in brackets): red, 1·3485 (1·3482, 1·333); green, 1·3570 (1·3572, 1·336); blue, 1·3668 (1·3655, 1·338).

In 1954 Bergstrand measured the group velocity of light in glass and in calcite. The method employs the piezo-electric light modulator and the results confirm the theoretical expression for the group velocity. For calcite, measurements in various directions confirm the shapes of the wave surfaces.

24–10. The Velocity of Light in a Moving Medium

Fig. 24–7 shows the apparatus used by Fizeau in 1859 in order to find the effect of the motion of the medium on the velocity of light. In principle, the

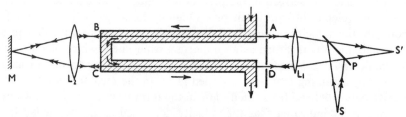

Fig. 24–7. Fizeau's determination of the speed of light in a moving medium.

arrangement resembles the Rayleigh refractometer. Light from the slit source S is collimated by lens L_1 and passes through the water circulating system to M, and back. It will be seen that a double slit fringe system will be formed at S'. One beam passes along AB and returns along CD, while the other passes along DC and returns along BA. Consequently, if water circulates as shown, one beam always travels in the same direction as the water, and the other travels in the opposite direction. Thus if the velocity of the light in the moving water is different in the two directions, a phase difference will be introduced between the beams and there will be a displacement of the fringes at S'. The experiment was repeated by Michelson with a modified form of the apparatus. The results obtained by Fizeau and Michelson do not correspond to a Newtonian addition of the velocities of light and medium but are consistent with the relation

$$u'=u+v\left(1-\frac{1}{n^2}\right)=\frac{c}{n}+v\left(1-\frac{1}{n^2}\right)$$

where v is the speed of the water. This result was first explained by assuming that light was propagated with a fixed velocity relative to an aether which was dragged along by the water, but the results can be explained in terms of relativity theory without any special assumptions.

24–11. The Michelson-Morley Experiment

At one time it was thought that light is propagated with a fixed speed relative to an all-pervading aether with respect to which the earth can be either stationary or in motion. If the earth is moving relative to the aether the speed of light with respect to the earth is then different in different directions. The Michelson-Morley experiment (1881–7) was devised in order to detect this motion of the earth through the aether. In principle the apparatus is simply a Michelson interferometer. Fig. 24–8 (a) shows the light paths on the assumption that the

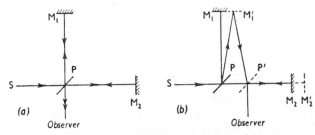

Fig. 24–8.   The Michelson-Morley experiment.

earth and apparatus are at rest in the aether and (b) shows the paths on the assumption that the apparatus is moving from left to right through the aether with uniform speed. In (a) $PM_1=PM_2=l$ so that the beams traverse equal paths and localized fringes are observed in the plane of M_1. In (b) M_1 and M_2 are the positions of the mirrors in the aether when the light leaves P. M_1 moves to M_1' while the light travels from P to the mirror, and P moves to P' while the light travels to M_1' and back. M_2 is at M_2' when reflection occurs. Let the velocity of the light relative to the aether be c and let the velocity of the apparatus through the aether be v. Let t_1 be the time taken for the light to travel from P to M_2', and t_2 the time from M_2' to P'. Then $M_2M_2'=vt_1$, $PM_2'=l+vt_1$, and $M_2'P'=l-vt_2$. Also, since these distances are traversed at speed c, one has

$$ct_1=l+vt_1$$

and

$$ct_2=l-vt_2.$$

Hence

$$t_1=\frac{l}{c-v}$$

and

$$t_2=\frac{l}{c+v}.$$

Now in the total time (t_1+t_2) the path length through the aether traversed by the light is given by

$$c(t_1+t_2)=c\left(\frac{l}{c-v}+\frac{l}{c+v}\right)=\frac{2lc^2}{c^2-v^2}\fallingdotseq 2l\left(1+\frac{v^2}{c^2}\right),$$

where powers of v/c higher than the second have been neglected since it can be assumed that v/c is small. For the other beam the path length is $PM_1'P'=2PM_1'$,

where $PM_1'=\sqrt{l^2+(M_1M_1')^2}.$

Since M_1 travels a distance M_1M_1' while light travels a distance PM_1', one has

$$M_1M_1'=PM_1'\frac{v}{c},$$

$$\therefore\quad (PM_1')^2=l^2+(PM_1')^2\frac{v^2}{c^2}$$

or $$(PM_1')^2=\frac{l^2}{1-\frac{v^2}{c^2}},$$

$$\therefore\quad \text{Path length}=2PM_1'=\frac{2l}{\left(1-\frac{v^2}{c^2}\right)^{\frac{1}{2}}}\fallingdotseq 2l\left(1+\frac{v^2}{2c^2}\right).$$

Hence the beams traverse paths which differ by lv^2/c^2.

It will be seen that if the interferometer is rotated through 90° the path lengths of the beams are interchanged and there will be a displacement of the fringes if there is any relative motion of the apparatus through the aether. Michelson and Morley made the light paths very large by reflecting the light to and fro across the apparatus, but no displacement of the fringes was observed as the apparatus was rotated in the laboratory.

The null result of the Michelson-Morley experiment could be explained by assuming that at the time of the experiment the earth was stationary with respect to the aether. However, the null result was always obtained when the experiment was repeated, and it seems very unlikely that the earth is the one body in the universe which is permanently at rest in the aether. FitzGerald and Lorentz showed that one can explain the null result by assuming that when an arm of the interferometer lies in the direction of the aether-earth velocity, it contracts to a length $l\sqrt{1-v^2/c^2}$. In a sense this contraction is included in Einstein's theory of relativity but in that theory absolute rest and absolute velocity are meaningless and one cannot speak of an aether at absolute rest with respect to which the velocity of light must be measured and with respect to which anything may be said to have absolute motion.

APPENDIX

SOME SUGGESTIONS FOR FURTHER READING

I. BOOKS

An indication of the field covered is given when this is not obvious from the title.

ANDERS, H., *Thin Films in Optics* (Focal Press).

ARENS, H. (Trans. K. H. Ruddock), *Colour Measurement* (Focal Press). Primarily for those interested in colour photography.

BAKER, B. B., and COPSON, E. T., *The Mathematical Theory of Huygens' Principle* (O.U.P.).

BARNES, K. R., *The Optical Transfer Function* (Hilger).

BAUER, G., *Measurement of Optical Radiations* (Focal Press). Covers all spectral regions.

BENNETT, A. H., JUPNIK, H., OSTERBERG, H., and RICHARDS, O. W., *Phase Microscopy* (Wiley). Includes a general discussion of diffraction theory of the microscope.

BERAN, M. J., and PARRENT, G. B., *Theory of Partial Coherence* (Prentice-Hall).

BORN, M., and WOLF, E., *Principles of Optics* (Pergamon). Covers a field similar to that covered by the present book but at a more advanced level.

BOUMA, P. J. (Ed. W. de Groot, A. A. Kruithof, J. L. Ouweltjes), *Physical Aspects of Colour* (MacMillan).

BOUSQUET, P. (Trans. K. M. Greenland), *Spectroscopy and its Instrumentation* (Hilger). Mostly instrumentation.

BOUTRY, G. A. (Trans. R. Auerbach), *Instrumental Optics* (Hilger). Covers paraxial and elementary aberration theory of image-forming and spectroscopic instruments, and includes vision and visual instruments.

BOUWERS, A., *Achievements in Optics* (Elsevier). Includes reflecting systems, aberrant images, and phase contrast.

BROUWER, W., *Matrix Methods in Optical Instrument Design* (W. A. Benjamin). Includes paraxial and primary aberration theory of telescopes, microscopes, cameras, and spectroscopes.

BROWN, E. B., *Modern Optics* (Reinhold). Geometrical and physical optics and applications to optical instruments and devices. Includes discussion of laser sources.

BUCHDAHL, H. A., *An Introduction to Hamilton Optics* (C.U.P.). Very advanced.

BUCHDAHL, H. A., *Optical Aberration Coefficients* (O.U.P.). Very advanced.

CAGNET, M., FRANÇON, M., and THRIERR, J. C., *Atlas of Optical Phenomena* (Springer-Verlag). Photographs of interference and diffraction patterns.

CANDLER, C., *Modern Interferometers* (Hilger). Gives basic theory of the use of interferometers in metrology and spectroscopy.

CAULFIELD, H. J., and SUN LU., *Applications of Holography* (Wiley-Interscience).

CLARKE, D., and GRAINGER, J. F., *Polarized Light and Optical Measurement* (Pergamon). Matrix methods. Ellipsometry.

COLLIER, R. J., BURCKHARDT, C. B., and LIN, L. H., *Optical Holography* (Academic Press). Covers relevant general optical theory, including Fourier optics as well as theory, experimental methods, and applications of holography.

COOK, A. H., *Interference of Electromagnetic Waves* (O.U.P.). Fundamental concepts and theory plus applications and instrumentation; sophisticated treatment.

COX, A., *A System of Optical Design* (Focal Press). Mainly on aberration theory and lens design, but includes chapters on image formation.

DE VELIS, J. B., and REYNOLDS, G. O., *Theory and Applications of Holography* (Addison-Wesley).

DITCHBURN, R. W., *Light* (Blackie). Coverage similar to present book with extension into quantum optics. Emphasis on Fourier methods.

EAGLESFIELD, C.C., *Laser Light: Fundamentals and Optical Communications* (St. Martin's Press).

FOWLES, G. R., *Introduction to Modern Optics* (Holt, Rinehart, and Winston). Essential principles of physical optics, including coherence, and extension into quantum optics and principles of lasers.

FRANÇON, M. (Trans. Barbara Jeffrey and ed. J. H. Sanders), *Diffraction— Coherence in Optics* (Pergamon). On coherence, diffraction, image formation.

FRANÇON, M., *Modern Applications of Physical Optics* (Interscience). A brief review, mainly on applications of interferometry and on image formation.

FRANÇON, M., *Optical Interferometry* (Academic Press). Extensive account of coherence, interference and applications of interferometry.

FRANÇON, M., *Progress in Microscopy* (Pergamon). Includes a general discussion of image formation in a microscope.

FROOME, K. D., and ESSEN, L., *The Velocity of Light and Radio Waves* (Academic Press). Comprehensive and up to date.

GARBUNY, M., *Optical Physics* (Academic Press). Deals with the interaction between light and matter—emission, absorption, and propagation in material media.

GOLDWASSER, E. L., *Optics, Waves, Atoms, and Nuclei: An introduction* (W. A. Benjamin). Includes classical and quantum optics.

GOODMAN, J. W., *Introduction to Fourier Optics* (McGraw-Hill). Includes very useful discussion of Kirchhoff diffraction theory as well as extensive discussion of Fourier methods, including spatial filtering.

GRAHAM, C. H. (Ed.), *Vision and Visual Perception* (Wiley).

GUILD, J., *Diffraction Gratings as Measuring Scales* (O.U.P.). A practical guide to the metrological use of moire fringes.

GUILD, J., *The Interference Systems of Crossed Diffraction Gratings* (O.U.P.). A full discussion of the theory of moire fringes.

HAVELOCK, T., *The Propagation of Disturbances in Dispersing Media* (C.U.P.). Gives a full discussion of group velocity.

HEAVENS, O. S., *Optical Properties of Thin Solid Films* (Butterworth).

HERZBERGER, M., *Modern Geometrical Optics* (Interscience). An advanced book covering paraxial and primary and higher order aberration theory.

HOPKINS, H. H., *The Wave Theory of Aberrations* (O.U.P.).

JAMES, J. F., and STERNBERG, R. S., *The Design of Optical Spectrometers* (Chapman and Hall).

JENKINS, F. A., and WHITE, H. E., *Fundamentals of Optics* (McGraw-Hill). Coverage similar to present book.

JUDD, D. B., and WYSZECKI, G., *Colour in Business, Science, and Industry* (Wiley).

KAPANY, N. S., *Fiber Optics: Principles and Applications* (Academic Press).

KERKER, M., *The Scattering of Light and Other Electromagnetic Radiation* (Academic Press). Very extensive bibliography.

KIMMITT, M. F., *Far-infrared Techniques* (Pion). Sources, detectors, spectroscopic systems.

KINGSLAKE, R. (Ed.), *Applied Optics and Optical Engineering* (Academic Press). 5 vols. Vol. 1 includes photometry, light sources, optical materials survey of basic geometrical and physical optics. Vol. II includes vision, photography, and photo-electric and thermal detectors. Vol. III discusses optical components, including design, production, and testing, and Vols. IV and V deal with the main kinds of optical instruments.

KLEIN, M. V., *Optics* (Wiley). Covers field similar to that of present book but somewhat more mathematical approach and including more extensive discussion of Fourier methods and correlation functions.

KLINE, M. K., and KAY, I. W., *Electromagnetic Theory and Geometrical Optics* (Interscience). An advanced book developing geometrical optics and diffraction theory from Maxwell's equations, including propagation in anisotropic media.

KOCK, W. E., *Lasers and Holography* (Doubleday).

KRUG, W., RIENITZ, J., and SCHULZ, G. (Trans, J. Home Dickson), *Contributions to Interference Microscopy* (Hilger).

LE GRAND, Y. (Trans. R. W. G. Hunt and F. R. W. Hunt), *Light, Colour, and Vision* (Chapman and Hall).

LEVI, L., *Applied Optics: A Guide to Modern Optical System Design* (Wiley). Includes radiometry, colorimetry, interference, diffraction, communication theory in optics, quantum optics, sources, including luminescence and lasers. Optical systems.

LINFOOT, E. H., *Fourier Methods in Optical Image Evaluation* (Focal Press).

LINFOOT, E. H., *Recent Advances in Optics* (O.U.P.). Covers coherence and image formation, Foucault test, Schmidt and other aspheric systems.

LIPSON, S. G., and LIPSON, H., *Optical Physics* (C.U.P.). Concise account of principles of geometrical and physical optics, including elementary quantum optics.

LUNEBURG, R. K., *Mathematical Theory of Optics* (University of California Press). An advanced book developing both the geometrical and diffraction theories of optical instruments from Maxwell's equations, including geometrical and diffraction theories of aberrations.

MACLEOD, H. A., *Thin Film Optical Filters* (Hilger). Full account of theory, production, and applications of thin films.

MARION, J. B., *Classical Electromagnetic Radiation* (Academic Press). Includes basic electromagnetic theory, reflection, refraction, dispersion, interference, and diffraction.

MARTIN, A. E., *Infrared Instrumentation and Techniques* (Elsevier).

MARTIN, L. C., *Geometrical Optics* (Pitman). Includes basic geometrical optics, paraxial theory of instruments, introduction to aberrations, photometry.

MARTIN, L. C., and WELFORD, W. T., *Technical Optics* (Pitman). Vol. I includes paraxial theory, aberration, image formation, vision; Vol. II includes more detail discussion of telescopes, microscopes, and cameras, photometry of instruments, testing of instruments, and aspherics.

MARTIN, L. C., *The Theory of the Microscope* (Blackie). On the theory of image formation with special reference to the microscope.

MERTZ, L., *Transformations in Optics* (Wiley). Includes Fourier transform spectroscopy, and holography.

MEYER, C. F., *Diffraction of Light, X-rays, and Material Particles* (Chicago U.P.). Easy to read, mainly on light, including gratings.

NUSSBAUM, A., *Geometrical Optics: An Introduction* (Addison-Wesley). Uses matrix method and includes introduction to computer methods for paraxial optics.

O'NEILL, E. L., *Introduction to Statistical Optics* (Addison-Wesley). Deals with diffraction and image formation (including effect of aberrations), partial coherence, and partial polarization.

PAPAULIS, A., *Systems and Transforms with Applications in Optics* (McGraw-Hill).

ROGERS, G. L., *Handbook of Gas Laser Experiments* (Butterworth).

ROSSI, B., *Optics* (Addison-Wesley). Coverage similar to present book, with less on optical instruments but extension into quantum optics.

SANDERS, J. H., *Velocity of Light* (Pergamon). Includes reprints of the original papers by Michelson, Pease, and Pearson; by Essen and Gordon-Smith; and by Bergstrand.

SHULMAN, A. R., *Optical Data Processing* (Wiley). Includes extensive review of Fourier methods; holographic techniques.

SHURCLIFF, W. A., *Polarized Light* (Harvard U.P. and O.U.P.).

SHURCLIFF, W. A., and BALLARD, S. S., *Polarized Light* (van Nostrand).

SMITH, F. G., and THOMSON, J. H., *Optics* (Wiley). Concise account of principles of geometrical and physical optics, including modern developments.

SMITH, H. M., *Principles of Holography* (Interscience).

SMITH, R. A., JONES, F. E., and CHASMAR, R. P., *The Detection and Measurement of Infra-Red Radiation* (Clarendon Press).

SMITH, W. J., *Modern Optical Engineering* (McGraw-Hill). Design of optical systems, paraxial theory and aberrations, vision and photometry, image evaluation.

SOMMERFELD, A., (Trans. O. Laporte and P. A. Moldauer), *Optics (Lectures in Theoretical Physics, Vol. IV)* (Academic Press). Particularly useful for diffraction; also covers reflection and refraction, dispersion, crystal optics.

STEEL, W. H., *Interferometry* (C.U.P.). Extensive discussion of theory of coherence, interference, and applications of interferometry.

STOKES, A. R., *The Theory of the Optical Properties of Inhomogeneous Materials* (Spon).

STONE, J. M., *Radiation and Optics* (McGraw-Hill). Gives systematic treatment of electromagnetic theory, interference, diffraction, image formation, crystal optics. Emphasis on Fourier methods.

STROKE, G. W., *An Introduction to Coherent Optics and Holography* (Academic Press).

STRONG, J., *Concepts of Classical Optics* (Freeman). Ranges over whole field of geometrical and physical optics.

TOLANSKY, S., *An Introduction to Interferometry* (Longmans).

TOLANSKY, S., *Microstructures of Surfaces using Interferometers* (Arnold)

TOLANSKY, S., *Multiple Beam Interferometry of Surfaces and Films* (O.U.P.).

TOLANSKY, S., *Surface Microtopography* (Longmans).

TROUP, C. J., *Optical Coherence Theory—Recent Developments* (Methuen).

VALYUS, N. A., *Stereoscopy* (Focal Press).

VAN DE HULST, H. C., *Light Scattering by Small Particles* (Wiley).

VAŠÍČEK, A., *Optics of Thin Films* (North-Holland).

WALSH, J. W., *Photometry* (Constable).

WEBB, R. H., *Elementary Wave Optics* (Academic Press). Interference, diffraction, polarization; uses mechanical analogies and phasors.

WELFORD, W. T., *Geometrical Optics* (North-Holland). Deals with paraxial and elementary aberration theory of optical instruments.

WRIGHT, W. D., *The Measurement of Colour* (Hilger).

WRIGHT, W. D., *Photometry and the Eye* (Hatton Press).

WYSZECKI, G., and STILES, W. S., *Color Science: Concepts and Methods, Qualitative Data and Formulas* (Wiley).

YOUNG, H. D., *Fundamentals of Optics and Modern Physics* (McGraw-Hill). Includes classical and quantum optics.

II. REVIEW ARTICLES

The series *Progress in Optics* edited by E. Wolf and published by North-Holland contains review articles ranging over the whole field of theoretical and experimental optics; the list of reviews given below includes those that have appeared in the first nine volumes of this series in the fields covered in the present book.

ABELÈS, F., "Methods for Determining Optical Parameters of Thin Films (*Progress in Optics*, **II**, p. 249, 1963).

AMMANN, E. O., "Synthesis of Optical Birefringent Networks" (*Progress in Optics*, **IX**, p. 125, 1971). Deals with birefringent filters.

ARMSTRONG, J. A., and SMITH, A. W., "Experimental Studies of Intensity Fluctuations in Lasers" (*Progress in Optics*, **VI**, p. 213, 1967).

BARAKAT, R., "The Intensity Distribution and Total Illumination of Aberration-Free Diffraction Images" (*Progress in Optics*, **I**, p. 67, 1961).

BELL, E. E., "Optical Constants and their Measurement" (*Encyclopedia of Physics*, **XXV/2a**, p. 1, Springer-Verlag 1967).

BERGSTRAND, E., "Determination of the Velocity of Light" (*Encyclopedia of Physics*, **XXIV**, p. 1, Springer-Verlag 1956). Gives classified list of references.

BILLMEYER, F. W., JR, "Current American Practice in Color Measurement" (*Applied Optics*, **8**, p. 737, 1969). Useful bibliography.

BLOOM, A. L., "Gas Lasers and their Application to Precise Length Measurement" (*Progress in Optics*, **IX**, p. 3, 1971).

BOUWKAMP, C. J., "Diffraction Theory (*Reports on Progress in Physics*, **XVII**, p. 35, 1954).

BRADDICK, H. J. J., "Photoelectric Photometry" (*Reports on Progress in Physics*, **XXIII**, p. 154, 1960).

BRADLEY, D. J., "Applications of high resolution spectroscopic techniques to space research and laser physics" (*Optica Acta*, **15**, p. 431, 1968).

BREMMER, H., "Propagation of Electromagnetic Waves" (*Encyclopedia of Physics*, **XVI**, p. 423, Springer-Verlag 1958).

BRYNGDAHL, O., "Applications of Shearing Interferometry" (*Progress in Optics*, **IV**, p. 37, 1965).

BURCH, J. M., "The Metrological Application of Diffraction Gratings" (*Progress in Optics*, **II**, p. 73, 1963).

DELANO, E., and PEGIS, R. J., "Methods of Synthesis for Dielectric Multilayer Filters" (*Progress in Optics*, **VII**, p. 69, 1969).

DORSEY, N. E., "The Velocity of Light" (*Trans. Phil. Soc. Am.*, **34**, (Part I), 1944).

ENNOS, A. E., "Holography and its Application" (*Contemporary Physics*, **3**, p. 153, 1967).

FIORENTINI, A., "Dynamic Characteristics of Visual Processes" (*Progress in Optics*, **I**, p. 253, 1961).

FRANÇON, M., and MALLICK, S., "Measurement of the Second Order Degree of Coherence" (*Progress in Optics*, **VI**, p. 73, 1967).

FRY, G. A., "The Optical Performance of the Human Eye" (*Progress in Optics*, **VIII**, p. 53, 1970).

FRYER, P. A., "Vibration Analysis by Holography" (*Reports on Progress in Physics*, **XXXIII**, p. 487, 1970).

GABOR, D., "Light and Information" (*Progress in Optics*, **I**, p. 109, 1961).

GABOR, D., "Progress in Holography" (*Reports on Progress in Physics*, **XXXII**, p. 395, 1969).

GAMO, H., "Matrix Treatment of Partial Coherence" (*Progress in Optics*, **III**, p. 187, 1964).

GEBBIE, H. A., "Fourier Transform versus Grating Spectroscopy" (*Applied Optics*, **8**, p. 501, 1969).

GEBBIE, H. A., and TWISS, R. Q., "Two-beam Interferometric Spectroscopy" (*Reports on Progress in Physics*, **XXIX**, p. 729, 1966).

GNIADEK, K., and PETYKIEWICZ, J., "Application of Optical Methods in the Diffraction Theory of Elastic Waves" (*Progress in Optics*, **IX**, p. 283, 1971). Discusses Kirchhoff and Rubinowicz theories.

GOODMAN, J. W., "Synthetic-aperture Optics" (*Progress in Optics*, **VIII**, p. 3, 1970). Review of techniques for achieving with small apertures resolution corresponding to a large aperture.

HARTSHORN, L., and SAXTON, J. A., "The Dispersion and Absorption of Electromagnetic Waves" (*Encyclopedia of Physics*, **XVI**, p. 640, Springer-Verlag 1958).

HEAVENS, O. S., "Optical Properties of Thin Films" (*Reports on Progress in Physics*, **XXIII**, p. 1, 1960).

HERRIOT, D. R., "Some Applications of Lasers to Interferometry" (*Progress in Optics*, **VI**, p. 173, 1967).

HOPKINS, H. H., "Application of Frequency Response Techniques in Optics" (*Proc. Phys. Soc.*, **79**, p. 889, 1962).

JACOBSSON, R., "Light Reflection from Films of Continuously Varying Refractive Index" (*Progress in Optics*, **V**, p. 249, 1966).

JACQUINOT, P., "Interferometry and Grating Spectroscopy: an Introductory Survey" (*Applied Optics*, **8**, p. 497, 1969).

JACQUINOT, P., "New Developments in Interference Spectroscopy" (*Reports on Progress in Physics*, **XXIII**, p. 267, 1960).

JACQUINOT, P., and ROIZEN-DOSSIER, B., "Apodisation" (*Progress in Optics*, **III**, p. 29, 1964). Appendices on classical, Fourier transform, SISAM, and Girard Grille spectrometers.

KNEUBUHL, F., "Diffraction Grating Spectroscopy" (*Applied Optics*, **8**, p. 505, 1969). Extensive bibliography, including reviews, on grating and Fourier transform spectroscopy, with summary of properties of I.R. detectors, filters, etc.

KOTTLER, F., "Diffraction at a Black Screen" (*Progress in Optics*, **IV**, p. 281, 1965).

KOTTLER, F., "Diffraction at a Black Screen, Part II, Electromagnetic Theory" (*Progress in Optics*, **VI**, p. 333, 1967).

KUHN, H., "New Techniques in Optical Interferometry" (*Reports on Progress in Physics*, **XIV**, p. 64, 1951).

LATTA, J. N., "A Classified Bibliography on Holography and Related Fields" (*J. Soc. Motion Pict. Tel. Engrs. (U.S.A.)*, **77**, p. 540, 1968). 655 references.

LEITH, E. N., and UPATNIEKS, J., "Recent Advances in Holography" (*Progress in Optics*, **VI**, p. 3, 1967).

LEVI, L., "On Image Evaluation and Enhancement" (*Optica Acta*, **17**, p. 59, 1970).

LEVI, L., "Vision In Communication" (*Progress in Optics*, **VIII**, p. 345, 1970).

LIPSON, H., and TAYLOR, C. A., "X-ray Crystal-Structure Determination as a Branch of Physical Optics" (*Progress in Optics*, **V**, p. 289, 1966).

LISSBERGER, P. H., "Optical Applications of Dielectric Thin Films" (*Reports on Progress in Physics*, **XXXIII**, p. 197, 1970).

LOEWENSTEIN, E. V., "The History and Current Status of Fourier Transform Spectroscopy" (*Applied Optics*, **5**, p. 845, 1966). Includes general classification of spectrometers.

MANDEL, L., "Fluctuations of Light Beams" (*Progress in Optics*, **II**, p. 181, 1963).

MANDEL, L., and WOLF, E., "Coherence Properties of Optical Fields" (*Reviews of Modern Physics*, **37**, p. 231, 1965). Includes both classical and quantum theories; extensive bibliography.

MEYER-ARENDT, J. R., "Radiometry and Photometry: Units and Conversion Factors" (*Applied Optics*, **7**, p. 2081, 1968).

MIKAELIAN, A. L., and TER-MIKAELIAN, M. L., "Quasi-classical Theory of Laser Radiation" (*Progress in Optics*, **VII**, p. 233, 1969).

MIYAMOTO, K., "Wave Optics and Geometrical Optics in Optical Design" (*Progress in Optics*, **I**, p. 31, 1961).

MURATA, K., "Instruments for the Measuring of Optical Transfer Functions" (*Progress in Optics*, **V**, p. 201, 1966).

MUSSET, A., and THELEN, A., "Multilayer Antireflection Coatings" (*Progress in Optics*, **VIII**, p. 203, 1970).

OOUE, S., "The Photographic Image" (*Progress in Optics*, **VII**, p. 301, 1969).

PALIK, E. D., "A Brief Survey of Magneto-optics" (*Applied Optics*, **6**, p. 597, 1967).

PALIK, E. D., and HENVIS, B. W., "A Bibliography of Magneto-optics of Solids" (*Applied Optics*, **6**, p. 603, 1967). 1238 references.

PALMER, E. W., and VERRILL, J. F., "Diffraction Gratings" (*Contemporary Physics*, **9**, p. 257, 1968). Production and uses.

POLSTER, H. D., et al. "New Developments in Interferometry (*Applied Optics*, **8**, p. 521, 1969).

POTTS, W. J., JR., and SMITH, A. L., "Optimizing the Operating Parameters of Infrared Spectrometers" (*Applied Optics*, **6**, p. 257, 1967).

RAMACHANDRAN, G. N., and RAMASESHAN, S., "Crystal Optics" (*Encyclopedia of Physics*, **XXV/I**, p. 1, Springer-Verlag 1961).

RANK, D. H., "The Band Spectrum Method of Measuring the Velocity of Light" (*J. Molecular Spec.* (U.S.A.), **17**, 1965).

RATCLIFFE, J. A., "Some Aspects of Diffraction Theory and Their Applications to the Ionosphere" (*Reports on Progress in Physics*, **XIX**, p. 188, 1956).

RISKEN, H., "Statistical Properties of Laser Light" (*Progress in Optics*, **VIII**, p. 241, 1970).

ROSENHAUER, K., and ROSENBRUCH, K.-J., "The Measurement of the Optical Transfer Function of Lenses" (*Reports on Progress in Physics*, **XXX**, p. 1, 1967).

ROUARD, P., and BOUSQUET, P., "Optical Constants of Thin Films" (*Progress in Optics*, **IV**, p. 145, 1965).

RUBINOWICZ, A., "The Miyamoto-Wolf Diffraction Wave" (*Progress in Optics*, **IV**, p. 199, 1965).

SINCLAIR, R. S., and WRIGHT, W. D., "Color Measurement in Europe" (*Applied Optics*, **8**, p. 751, 1969). Useful bibliography.

STEEL, W. H., "Two-Beam Interferometry" (*Progress in Optics*, **V**, p. 147, 1966).

STROKE, G. W., "Diffraction Gratings" (*Encyclopedia of Physics*, **XXIX**, p. 426, Springer-Verlag 1967). Comprehensive and detailed account of production, testing, and use of gratings with general theory of plane and concave gratings and grating instruments, including multislit and SISAM spectrometers. Extensive bibliography.

STROKE, G. W., "Ruling, Testing and Use of Optical Gratings for High Resolution Spectroscopy" (*Progress in Optics*, **II**, p. 1, 1963).

THETFORD, A., "The Basic Theory of Multilayers" (*Opt. and Laser Tech.* (G.B.), **3**, p. 131, 1971).

THOMPSON, B. J., "Image Formation with Partially Coherent Light" (*Progress in Optics*, **VII**, p. 171, 1969).

TOUSEY, R., "Highlights of Twenty Years of Optical Space Research" (*Applied Optics*, **6**, p. 2044, 1967). Includes discussion of problems of instrumentation. Extensive bibliography.

TSUJIUCHI, J., "Correction of Optical Images by Compensation of Aberrations and by Spatial Frequency Filtering" (*Progress in Optics*, **II**, p. 131, 1963).

URBACH, J. C., and MEIER, R. W., "Properties and Limitations of Hologram Recording Materials" (*Applied Optics*, **8**, p. 2269, 1969).

VANASSE, G. A., and SAKAI, H., "Fourier Spectroscopy" (*Progress in Optics*, **VI**, p. 261, 1967).

VANDER LUGT, "A review of optical data processing techniques" (*Optica Acta*, **15**, p. 1, 1968).

VAN HEEL, A. C. S., (Ed.), "Advanced Optical Techniques" (North-Holland 1967). A collection of reviews: "Precision Measurements", by J. B. Saunders; "Isotropic and Anisotropic Media. Applications of Anisotropic Materials to Interferometry", by M. Françon; "Principles of Instrumental Methods in Spectroscopy", by A. Girard and P. Jacquinot; "Interferometry: Some Modern Techniques", by K. M. Baird; "Optics of Thin Films", by

F. Abelès; "The Theory of Coherence and its Applications", by H. H. Hopkins; "Fiber Optics", by R. Draugard and R. J. Potler; "Lasers", by H. G. Friei, and A. L. Schawlow; "Geometrical Optics" and "Design of Optical Instruments", by W. Brouwer and A. Walther; "Measurement of Aberrations and O.T.F.", by K. Rosenhauer.

VAN VLIET, K. M., "Noise Limitations in Solid State Photodetectors" (*Applied Optics*, **6**, p. 1145, 1967).

WELFORD, W. T., "Aberration Theory of Gratings and Grating Mountings" (*Progress in Optics*, **IV**, p. 241, 1965).

WELFORD, W. T., "Optical Calculations and Optical Instruments, an Introduction" (*Encyclopedia of Physics*, **XXIX**, p. 1, Springer-Verlag 1967). Includes introduction to automatic computer methods.

WOLF, E., "The Diffraction Theory of Aberrations" (*Reports on Progress in Physics*, **XIV**, p. 95, 1951).

WORMSER, E. M., "Sensing the Invisible World" (*Applied Optics*, **7**, p. 1667, 1968). Reviews techniques for translating from I.R. to visible images.

WYNNE, C. G., "New Lens Systems" (*Reports on Progress in Physics*, **XIX**, p. 298, 1956).

YAMAJI, K., "Design of Zoom Lenses" (*Progress in Optics*, **VI**, p. 107, 1967).

YAMAMOTO, Y., "Coherence Theory of Source-size Compensation in Interference Microscopy" (*Progress in Optics*, **VIII**, p. 297, 1970).

YOUNG, M., "Holography and Coherent Imaging" (*Am. J. Phys.*, **37**, p. 304, 1969). Review at undergraduate level.

MISCELLANEOUS EXAMPLES

QUESTIONS* set at London, Oxford, and Cambridge examinations are marked as follows:

London: M.Sc., B.Sc. Special Parts I and II, and B.Sc. General Part II—LM, LS I, LS II, and LG II.

Oxford: Honour Moderations in Natural Science, and Finals—OM and OF.

Cambridge: Preliminary Examination in Natural Science, Parts I and II, and Tripos, Parts I and II—CP I, CP II, CT I, and CT II.

CHAPTERS I–V

1. Explain "Fermat's Principle of Stationary Time", illustrating your account by using the principle to solve some problems in geometrical optics.

   (CT I)

2. An object is placed 1 metre in front of a screen. Using a certain thin converging lens, it is found that there are two positions of the lens between the object and the screen which give sharp images on the screen. The images have lengths 4 cm. and 0·25 cm. Find the length of the object and the focal length of the lens.

3. A thin plano-convex lens of refractive index 1·64 has a power of 0·1 cm.$^{-1}$ in air. Find the focal lengths, and the positions of the principal and nodal points (a) when the lens has water (index 1·33) on the plane side and air on the curved side, (b) when the lens has water on the curved side and air on the plane side.

4. A converging meniscus lens has radii of curvature 10 cm. and 20 cm., and has an axial thickness of 5 cm.; it is made of glass of refractive index 1·64 and is in air. Find the focal length of the lens and the positions of the nodal points.

5. Find the cardinal points for a hemispherical glass lens of radius 10 cm. and refractive index 1·64 in air.

6. A converging meniscus lens has radii of curvature 20 cm. and 30 cm. and axial thickness 6 cm. it is made of glass of refractive index 1·52 and has water (index 1·33) on its concave side and air on its convex side. If the (real) object plane is 120 cm. from the convex side, find the position of the image plane and find the magnification for this pair of conjugate planes.

7. Verify the results given on page 426 for the positions of the cardinal points of the schematic eye specified on page 425.

8. State briefly what is meant by the principal planes, nodal planes, and principal foci of a lens system. Make the necessary calculations and show

* *Acknowledgments.* We are indebted to the following for permission to reproduce questions from past examination papers: The University of Cambridge; the University of London; the University of Oxford.

by a clear drawing the paths of (i) a ray incident parallel to the axis, (ii) a ray incident through the first principal point of a system of which the first component is a thin converging lens, focal length $7\frac{1}{2}$ cm., and the second, distant 6 cm., a thin diverging lens, focal length 2 cm. Derive any formulae employed.   (LS I)

9. Define the principal points of a lens system and derive an expression for the focal length of a combination of two thin lenses in air in terms of their focal lengths and distance apart. How would you determine the focal length of such a combination experimentally?

A thin converging lens of focal length 20 cm. forms an image 4 mm. high of a distant object. Determine the nature and focal length of a second lens which, when placed 5 cm. behind the first lens, produces an image 1 cm. high of the same object on a screen in the focal plane of the combination.   (LS I)

10. With the aid of a diagram, explain the method of formation of a real image 0·75 cm. long of an extremely distant object subtending an angle of 0·015 radians at the point of observation, by a coaxial lens system comprising one converging and one diverging lens, of focal lengths 10 and 5 cm. respectively.

Deduce the positions of the principal and focal planes of the lens system and represent them on a diagram drawn very approximately to scale.
                                                                                                         (LS I)

11. Define the cardinal points of a lens system. Light from a distant object is incident on a system consisting of a thin converging lens of focal length 12 cm. placed 6 cm. in front of a thin diverging lens of focal length 24 cm. Explain, with the aid of a diagram, why the position of the image is unaffected if the lens system is slightly rotated about an axis perpendicular to its principal axis and 2 cm. in front of the converging lens.

At what distance from the converging lens do conjugate object and image planes coincide?   (LS I)

12. Define the terms *principal points*, *nodal points* and *principal foci* as applied to a coaxial lens system and illustrate the use of these points in the construction giving the position of the image of a point object on the axis of the system. An object is located 100 cm. in front of a thin converging lens of focal length 20 cm. behind which there is a diverging lens of the same focal length at a distance of 10 cm. Show that the lens system can be rotated through a small angle about a point midway between the two principal points without causing displacement of the image.   (LS I)

13. Define the terms entrance pupil, exit pupil, and chief ray. A thin converging lens of focal length 5 cm. and diameter 6 cm. has a 2·5 cm. diameter stop situated 2 cm. in front of it. A line object 2·1 cm. high is located with its lower end on the lens axis 10 cm. in front of the lens. Find the position and size of the exit pupil and draw a diagram approximately to scale of the marginal rays and the chief ray from the upper end of the object.

                                                                                                         (LS I)

14. Discuss the relative merits of Ramsden and Huygens eyepieces. The lenses in a Huygens eyepiece have focal lengths of 2 cm. and 4 cm. What is the distance between the lenses and where are the cardinal points of the eyepiece?

15. Find the positions of the cardinal points in an eyepiece consisting of two thin converging lenses each of focal length 4 cm. and separated by a distance of 3 cm.

   The eyepiece is used in an astronomical telescope in normal adjustment. The diameter of the objective is 5 cm. and its focal length is 60 cm. Find the position and diameter of the exit pupil. (LG II)

16. A thin lens of crown glass is to be combined with a thin lens of flint glass to form a thin doublet that has the same focal length for a given pair of wavelengths. Deduce expressions for the powers of the crown and flint components in terms of the power of the doublet and the dispersive powers of the glasses.

   Using the glasses specified below, find the focal lengths of the crown and flint components necessary to give a diverging doublet of focal length 10 cm., achromatized for the C and F lines.

   Crown $n_d=1\cdot51690$, $n_C=1\cdot51429$, $n_F=1\cdot52282$

   Flint $n_d=1\cdot62258$, $n_C=1\cdot61749$, $n_F=1\cdot63476$

17. Describe how the refractive index of a liquid may be measured in the visible region as a function of wavelength, and estimate the accuracy you could achieve. Explain why liquid prisms are sometimes used in spectrometers. (OM)

18. Discuss the relative merits of critical-angle and non-critical-angle methods in the precision refractometry of liquids.

   Describe in detail a method of investigating the variation, with concentration, of the refractive index of a *dilute* solution. (LS I)

19. A Pulfrich refractometer is provided with standard prisms having refractive indices of $1\cdot6200$ and $1\cdot7500$. What ranges of refractive indices can be measured with these prisms using a refracting angle of (a) 90°, (b) 60°?

20. A prism spectrometer (Fig. 5–5) is adjusted in the usual way employing monochromatic light, and the slit is made fairly narrow. Light from the collimator is directed on to a 60° prism at almost grazing incidence and is refracted through the prism. The slit is observed through the telescope and the prism is rotated continuously so that the angle of incidence is decreased. Describe what happens to the slit image.

CHAPTERS VI–IX

1. Explain carefully what is meant by group velocity, giving examples from various types of wave motion. Derive an expression for the relation between group velocity and phase velocity.

   The refractive index of a certain glass is given by the formula $\mu=A+(B/\lambda_0^2)$ where λ_0 is the wavelength *in vacuo*. Show that the ratio of group velocity to phase velocity is

$$(A\lambda_0^2+B)/(A\lambda_0^2+3B).$$

(CT I)

2. In Young's double slit experiment the separation of the slits is $1\cdot9$ mm. and the fringe spacing is $0\cdot31$ mm. at a distance of 1 metre from the slits. Find the wavelength of the light. What is the visibility of the fringes if (a) the slits give equal intensities, (b) the intensity from one slit is four times that from the other.

3. Explain the resultant intensity when two beams of light are superimposed (a) when they are coherent, and (b) when they are incoherent. Apply this to a discussion of the interference fringes obtained in the Lloyd's mirror arrangement.

In a Lloyd's mirror experiment calculate the ratio of the intensities at the interference maxima and minima if the mirror reflects only 75% of the light incident upon it. (OM)

4. Explain the principles involved in the production of interference fringes by the method of division of wave-front. What form would such fringes take when two images of a point source are produced on the axis by a split convex lens, the two halves being displaced along the axis relative to one another?

In an experiment with Fresnel's biprism, fringes for light of wavelength 5×10^{-5} cm. are observed 0·2 mm. apart at a distance of 175 cm. from the prism. The prism is made of glass of refractive index 1·5 and is 25 cm. from the illuminated slit.

Calculate the angle at the vertex of the biprism. (OM)

5. Discuss the interference effects observed when a wave-front of mono-chromatic light is divided into two parts of equal width. Illustrate your answer by considering the character and location of the fringes for various kinds of interferometer, and in each case state what happens when white light is substituted for monochromatic light.

A concave mirror is cut along a diameter and the two halves rejoined at an angle of 40′ with each other so as to increase the concavity. A point source of monochromatic light, wavelength 6×10^{-5} cm., is placed at the focal distance from the mirror and close to its axis. Draw a sketch diagram of the region in which fringes are to be expected. Describe the form of the fringes and calculate their width. (CT I)

6. Fringes of equal thickness are observed in a thin glass wedge of refractive index 1·52. The fringe spacing is 1 mm. and the vacuum wavelength of the light is 5893 Å. Find the angle of the wedge in seconds of arc.

7. Fringes of equal inclination are observed in a Michelson interferometer. As one of the mirrors is moved back 1 mm. 3663 fringes move out from the centre of the pattern. Calculate the wavelength of the light. What will be observed if one then changes to white light?

8. A Michelson interferometer is adjusted so that fringes can be observed with white light, and the source is changed to a quasi-monochromatic source. Describe the appearance of the fringes. Fringes are photographed using black and white film with a quasi-monochromatic source emitting (a) blue light (b) green light. How can one distinguish between the photographs?

Using quasi-monochromatic light, what is observed if

1. One mirror is moved very slowly in the direction of its normal.
2. One mirror is rotated so that to the observer it appears to be rotated about a fringe as axis.
3. One mirror, although smooth, is not flat.
4. One mirror is rotated about an axis which lies in the plane of the mirror but to the observer appears to be perpendicular to the fringes.

9. Describe the use of an interferometer as applied *either* to the accurate measurement of refractive index *or* to the examination of surfaces.

A Michelson interferometer is set up using parallel light from a point source and interference fringes are observed arising from imperfections in a plane glass plate which is inserted normal to one of the beams. The interferometer mirrors are then removed and the fringes formed by multiple reflection within the plate itself are observed. If the same number, m, of fringes occur between the points P_1, P_2 on the plate in each of these experiments, show that the thickness of the plate is the same at P_1 and P_2 but that the mean refractive index differs by $m\lambda/2D$, where D is the thickness of the plate and λ is the wavelength of the light. (OF)

10. Explain the formation of Newton's rings and of the rings seen in a Fabry-Perot interferometer. How would you obtain photographs of these two ring systems whose general appearance was very nearly the same? Would there remain any geometrical difference in your photographs?

The Newton's rings are formed between a plane glass plate and a spherical glass surface of radius 100 cm., and the photograph is taken with magnification 5. The photograph of the Fabry-Perot rings is taken on a photographic plate in the focal plane of a telescope of focal length 50 cm. When the two photographs are superimposed a few rings near the centre are found to coincide. What is the separation of the reflecting surfaces in the Fabry-Perot interferometer? (CT I)

11. Describe a refractometer suitable for the determination of the refractive index of a gas and explain how it could be used to investigate the variation of the refractive index of air with density.

A refractometer with tubes 1 metre long initially contains air at atmospheric pressure at 20°C. When the pressure in one tube is reduced by 20 cm. of Hg, a shift of 130 fringes of light of wavelength 5460 Å is observed. Calculate the refractive index of air at S.T.P. for this wavelength. (LS I)

12. Explain the formation of fringes in a Michelson interferometer illuminated by (*a*) monochromatic light, (*b*) white light. If one mirror is slowly moved back with velocity v calculate directly the number of monochromatic fringes passing a point in the field of view in one second. Show that the same result is obtained by a treatment which considers the beats formed between the incident light and the reflected light whose frequency has been modified by the Doppler effect at the moving mirror. (LS I)

13. Describe the formation of the fringes in the Jamin interferometer. Explain how the instrument is used for refractive index measurements. What would be the effect upon the fringe pattern if the reflectivity of the silvering on one plate were to fall to half that on the other? (LS I)

14. Describe how interference fringes are formed in the Fabry-Perot interferometer. Show that if there is no loss at the reflecting surfaces the maximum intensity in the interference pattern viewed by transmission is the same as the intensity in the absence of the interferometer.

What are the possible values of the distance between the plates of a Fabry-Perot interferometer which gives fringes in which the maxima due to the sodium D_1 line coincide with the minima due to the sodium D_2 line?

Outline the method of use of the interferometer to measure the relative intensities of the sodium D lines.

(Wavelength of D_1 line = 5896 Å, of D_2 line = 5890 Å.) (OF)

15. How small must one make the plate separation of a Fabry-Perot interferometer in order to obtain a free spectral range of 0·05 Å at a wavelength of 5000 Å? What is the smallest resolvable wavelength difference if the reflectivity of the plates is 95%?

16. Explain the formation of Brewster's fringes and describe briefly their use in interferometry.

Two different Fabry-Perot etalons are placed one behind the other. White light passes normally through both. When the shorter of the two etalons is turned through an angle of 0·01571 radian, Brewster's fringes are seen centrally in the field of view. If the length of the inclined etalon is 0·50027 cm. and that of the other nearly four times as great, calculate the length of the latter. (LS I)

17. The upper half of the slit of a spectroscope is illiminated with a mercury lamp and the lower half of the slit is illuminated with white light through a thin Fabry-Perot etalon. If there are 20 fringes in the continuous spectrum between the blue and green mercury lines, what is the thickness of the etalon? (Wavelengths of mercury lines: 5461 Å and 4358 Å.)

CHAPTERS X–XIII

1. Parallel monochromatic light is incident normally on a slit, passes through a converging lens near the slit, and is brought to a focus on a screen. Obtain an expression for the intensity distribution on the screen. Point out (a) the inclinations to the axis of the system for which minima occur, (b) the conditions under which maxima occur.

   If the incident light is *white*, and the slit has a width of 0·080 cm. and the lens has a focal length of 80 cm., find the wavelengths in the visible region which are missing on the screen at a distance of 0·30 cm. from the axis of the system. (LG II)

2. Explain what is meant by the chromatic resolving power, $\lambda/\delta\lambda$, of an optical instrument.

   Deduce an expression for the resolving power of a prism of material for which the refractive index n is given by $n = a + b/\lambda^2$, where a and b are constants. Calculate a value for the minimum length of side of an equilateral prism, of material for which $b = 9·84 \times 10^{-11}$ cm², which will resolve the D lines of sodium (wavelengths 5890, 5896 Å). (LS I)

3. Describe and give the theory of Michelson's stellar interferometer. For the star Betelgeuse the mirror separation for the first disappearance of the fringes was 3 metres. If the distance of the star from the earth is $1·7 \times 10^{15}$ km., calculate its diameter, assuming the effective wavelength of the emitted light to be 5750 Å. (LS I)

4. What is meant by Fraunhofer diffraction? Under what conditions may it be observed?

   A reflecting astronomical telescope focuses light from a field of stars on to a photographic plate, using a concave mirror with radius 60 cm. and focal length 700 cm. Describe and explain the photographs which may be obtained when the front of the telescope tube is covered with the following screens:

(*a*) An opaque cover with two circular apertures 20 cm. in diameter and 100 cm. apart.

(*b*) A transmission grating in which the amplitude of transmitted light varies sinusoidally across the grating with a periodicity of 0·1 cm.

(*c*) A square mesh wire net, with wires 0·1 cm. thick spaced apart by 2 cm.

(*d*) A mosaic screen complementary to the wire mesh in (*c*).   (CT I)

5. Explain Babinet's principle of complementary screens and describe how it can be used to determine the size of small particles. A lunar corona sub-tends an angle of 5° at the eye of an observer. Assuming this to be due to diffraction by water droplets in the atmosphere, calculate their diameter. Take the effective wavelength of the light to be 5000 Å.   (LS I)

6. Sodium light is incident normally on a plane transmission grating having 3000 lines per cm. Find the direction of the first order for the D lines and the width of the grating necessary to resolve them. (Wavelengths of D lines: 5890 Å and 5896 Å.)

7. Light is incident at 70° on a plane reflection grating which has 6000 lines per cm. ruled over a width of 3 cm. What is the maximum resolving power available at a wavelength of 5000 Å? Find the corresponding angle of diffraction and the smallest resolvable difference in wavelength.

8. Derive, with full explanation, an expression for the intensity of light diffracted in the nth order by a plane grating consisting of parallel opaque strips of width a separated by transparent spaces of width b, when light of intensity I and wavelength λ falls normally on the grating. (Assume a and b large compared with λ.)

   The following angles of diffraction, and estimates of intensity, were obtained with such a grating used at normal incidence with a wavelength of 5000 Å.

0° 17′ very strong	1° 9′ weak	2° 18′ weak
0° 34′ strong	1° 26′ medium	2° 35′ weak
0° 51′ weak	1° 43′ weak	3° 9′ weak

Obtain from these such information as you can about the values of a and b.
      (LS I)

9. Explain the use of amplitude phase diagrams in diffraction problems.

   A diffraction grating consists of two types of slit, A and B, in a regular alternating sequence ABABAB. . . . The slits are equally spaced and of the same width, but the amplitude diffracted by B is twice that diffracted by A. Compare the diffraction pattern from this grating with that from an ordinary grating consisting only of A slits spaced at a distance equal to that between adjacent A and B slits in the first grating.   (CP I)

10. Obtain an expression for the intensity of a beam diffracted from a plane grating when monochromatic light is incident normally upon it. If the widths of the grating bars and spaces are respectively $2·5 \times 10^{-4}$ and $1·5 \times 10^{-4}$ cm., compare the intensities of the 3rd and 4th orders of spectra for light of wavelength 5000 Å.   (LS I)

11. A plane transmission grating having 5000 lines per cm. is mounted in the usual way on an ordinary table spectrometer (Fig. 5–5), the plane of the grating being perpendicular to the axis of the collimator. A sodium lamp is used as the source, the spectrometer slit is very narrow, and the clearly resolved D lines are viewed in the first order. What would be observed if:

    (a) The width of the spectrometer slit were slowly increased from zero.

    (b) A slit were placed immediately after the grating and the slit width slowly increased from zero, the slit being parallel to the grating rulings.

    (c) The grating were moved in its own plane in a direction perpendicular to the rulings.

    (d) The grating were rotated about its central ruling as axis.

    (e) The grating were rotated about the axis of the collimator.

12. A collimated beam of light of wavelength $5 \cdot 46 \times 10^{-5}$ cm. is incident normally on a thin opaque screen containing a circular aperture of diameter 5 mm. Find the positions of the axial points of maximum and minimum intensity.

13. Explain the action of a zone plate and show that the object and image distances obey the ordinary thin lens formula.

    A zone plate is found to give a series of images of a point source on its axis. If the strongest and second strongest images are at distances of 30 cm. and 6 cm. respectively from the zone plate, both on the same side remote from the source, calculate the distance of the source from the zone plate, the principal focal length and the radius of the first zone.

    (Wavelength of light 5×10^{-5} cm.)　　　　　　　　　　　　　(CP I)

14. Describe the construction and action of a zone plate. Show that for points on its axis it is equivalent to a lens with chromatic aberration.

    A zone plate is illuminated by parallel white light. A lens of crown glass is to be placed in contact with it so that the resulting image is free from chromatic aberration. Design a suitable lens and find the position of the image, if the zone plate has a 'principal' focal length of 100 cm. for red light of wavelength 8×10^{-5} cm.

    (Refractive index of crown glass $= 1 \cdot 500$ for wavelength 8×10^{-5} and $1 \cdot 535$ for wavelength 4×10^{-5} cm.)　　　　　　　　　　(CT I)

15. Light of wavelength 5893 Å emerges from a narrow slit and is diffracted by a thin wire parallel to the slit. Approximately evenly spaced fringes are observed within the geometrical shadow of the wire. Find the diameter of the wire if the bright fringes are $0 \cdot 1$ mm. apart at a distance of 20 cm. from the wire.

CHAPTERS XIV–XVI

1. The largest useful magnifying power (M.P.) of a microscope is about 600 times the numerical aperture (N.A.). Justify this statement and calculate the M.P. of an eyepiece to be used with an objective of focal length 3 mm. and N.A. $0 \cdot 80$, in order to give the recommended overall M.P. of 480. (Take $\lambda = 5000$ Å, the optical tube length as 16 cm., the least distance of distinct vision as 25 cm., and the angular resolution of the eye as 2′.)　　　(LS I)

2. A compound microscope is focused so that the final image is at infinity. The instrument uses an objective of focal length 4 mm. and N.A. 0·72, working at a tube length of 160 mm., and employing light of wavelength 0·59 μ. The observer can resolve 0·1 mm. at the near point (D_v=25 cm.), and is employing the minimum necessary magnifying power. Find the focal length of the eyepiece and the diameter of the exit pupil. (Use eq. 14–14.)

3. Outline the use of Fourier methods in the diffraction theory of image formation. To what extent does the diffraction pattern of a diffraction pattern reproduce the original diffracting object?

Discuss the meaning of resolving power in the case of (*a*) a microscope objective, and (*b*) a spectrometer. (CT II)

4. Discuss the Rayleigh limit of resolution as applied to the resolving power, the longitudinal range of focusing without impairing image sharpness, and the tolerable aberrations of optical systems.

An astronomical telescope has a well-corrected objective of diameter 30 cm. and focal length 450 cm. Calculate (*a*) the minimum magnifying power of the eyepiece and its focal length if the optimum resolution is to be achieved, (*b*) the precision with which this eyepiece must be focused in order to achieve the optimum resolution, and (*c*) the resolving power of the instrument under these conditions. Assume the resolving power of the eye is 1·5 minutes of arc. (CT II)

5. Deduce from first principles the nature of the monochromatic aberrations that an axially symmetrical system may exhibit.

A wide aperture lens is to form a geometrically perfect image of a small object lying near to and perpendicular to the optical axis. Show that the sine condition should be satisfied for all zones of the aperture, and state which aberrations are thereby corrected. Give an example of an optical instrument in which you would expect this form of correction to be employed. Why are the other aberrations less important in this case? (LM)

6. What determines the resolving power of optical instruments?

A telescope has an objective with a diameter of 5 cm. and a focal length of 1·5 m. If the eye can just distinguish points with an angular separation of 1′, estimate the maximum value of the focal length of the eyepiece which may be employed in order that the resolution should be limited by the telescope rather than by the eye. The wavelength of light may be taken to be 5×10^{-5} cm.

An opaque screen having a pair of narrow parallel slits of separation d is placed over the objective of a telescope, the diameter of the objective being greater than d. The telescope is used to view a distant slit which is of width W and parallel to the other slits. Explain what is seen as W is increased from a very small value. What practical use has been made of the effect? (OF)

7. Discuss the defects that may occur in photographic objectives and the methods by which they are corrected.

The radius of the entrance pupil of a camera objective is 1 cm. With an object at 4 metres the magnification is 1/100. If the permissible radius of confusion on the plate is 0·05 mm., what is the depth of the field? (LS I)

8. With each of two lenses it is found that with the reference sphere centred at the paraxial focus the primary marginal spherical wavefront aberration is one wavelength. The image-forming cones of rays have semi-angles of (a) 2° and (b) 10°. Find the longitudinal spherical ray aberration for each lens. ($\lambda = 5000$ Å.)

9. A point object is imaged by a small concave mirror of radius of curvature 10 cm. The object is 30 cm. from the mirror and is 10° off axis. Find the separation of the focal lines.

10. A thin converging crown glass lens ($n_d = 1\cdot51690$, $V = 60\cdot6$) has a focal length of 10 cm. If the aperture stop is at the lens and the object plane is at infinity, find the radius of curvature of the Petzval surface and the separation of the focal lines for an object 5° off axis.

    The above lens is replaced by a thin converging achromatic doublet of the same focal length, employing the same crown glass in conjunction with a dense flint ($n_d = 1\cdot62258$, $V = 36\cdot0$). Find the new Petzval radius and the new astigmatic difference of focus.

11. (a) In making a thin converging achromatic doublet the diverging component was accidentally made from the same glass as the converging component. What difference would you expect this to make to the paraxial properties of the doublet?

    (b) The objective of a simple prism spectrometer (Fig. 5–5) consists of a thin cemented achromatic doublet. Accidentally, this doublet is mounted the wrong way round in the telescope tube. What difference will this make to the paraxial properties of the system? Will it make any difference to the extra-paraxial properties?

    (c) The components of a telescope objective which consists of a thin achromatic doublet are moved a short distance apart. How will this affect the quality of the axial image?

CHAPTERS XVII and XVIII

1. A colour C is matched by a mixture of certain monochromatic red, green and blue matching stimuli with luminous fluxes in the proportions $1 : 10 : 0\cdot1$. The luminous fluxes of the matching stimuli necessary to match the standard white are in the proportions $1 : 4\cdot6 : 0\cdot06$. Find the unit trichromatic equation for the colour C.

2. A plane screen receives light from an infinite, uniformly diffuse, plane surface of luminance L lying parallel to it. Find the illumination on the screen.

3. Show how the brightness of a perfectly diffusing surface is related to its intensity of illumination.

    A short, opaque cylinder, with a perfectly diffusing surface is illuminated with parallel light from a direction perpendicular to the axis of the cylinder. It is viewed from a considerable distance in a direction perpendicular to the axis of the cylinder and making an angle ϕ with the incident light. Find an expression for the apparent brightness of different parts of the illuminated surface. Calculate the total luminous flux per unit solid angle diffused in the direction such that $\phi = \pi/2$ if the incident flux of F lumens cm.$^{-2}$ and the radius of the cylinder is equal to its length l. (OM)

4. The aperture settings on a photographic lens are labelled $f/2\cdot8$, $f/4$, $f/5\cdot6$, $f/8$, $f/11$, $f/16$. Why is this a convenient choice?

5. Discuss the brightness of the image of an extended source when viewed through a telescope.

   With the object of increasing the intensity of spectrum lines by improving the light gathering power of a standard spectrometer, an instrument is built in which the collimator lens has a focal length one quarter that of the telescope lens, the diameters of the two lenses being the same.

   Discuss the efficiency of such an optical system. (LS I)

6. A uniform point source is placed at the centre of a sphere of radius 100 cm., the inside surface of the sphere being uniformly diffusing with a reflection factor $0\cdot8$. An identical source is placed at the centre of a similar sphere of radius 150 cm. and reflection factor $0\cdot9$. Compare the illuminations on the surfaces of the spheres.

CHAPTERS XIX–XXI

1. Assuming Maxwell's equations of the electromagnetic field, establish relations between (*a*) the directions, (*b*) the amplitudes of the electric and magnetic fields in a progressive plane wave in a homogeneous isotropic dielectric.

   The root-mean-square value of the displacement current density in a uniform plane-polarized sinusoidal electromagnetic wave in free space is 10^{-9} amp cm.$^{-2}$. If the frequency is 10^8 sec^{-1}, obtain values for the amplitudes of the electric and magnetic fields in the wave. (LS I)

2. Prove by means of Maxwell's equations that in a plane electromagnetic wave in an isotropic dielectric medium the vectors **E** and **H** of the electric and magnetic field strength are normal to each other and normal to the direction of the wave propagation. Derive expressions for the wave velocity and the electromagnetic energy flow.

   Calculate the r.m.s. electric field strength in the sun's radiation at the surface of the earth from the known value of the radiation intensity ($0\cdot132$ joule sec^{-1} cm.$^{-2}$). (LS I)

3. Discuss the concept of Displacement Current. Deduce the conductivity of a medium of dielectric constant 5 when the magnitude of the conduction and displacement currents are equal at a frequency of 10^6 sec^{-1}.

   What is the attenuation per cm. of the amplitude of a plane wave of the above frequency when propagated in such a medium? (Assume $\mu=1$.) (LS I)

4. Obtain an expression for the ratio of the amplitudes of the electric and magnetic vectors in a plane polarized light wave in a dielectric.

   Find the reflected intensity when the light is incident normally at a plane surface of separation between air and diamond of refractive index $2\cdot42$. (LS I)

5. Unpolarized light is incident at $45°$ on a glass of refractive index $1\cdot52$. The reflected light is passed through a Nicol. Find the ratio of the maximum and minimum intensities transmitted by the Nicol as it is rotated.

6. Circularly polarized light is to be produced by means of two successive internal reflections in a Fresnel rhomb. If the refractive index is $1\cdot52$, find the appropriate angles of incidence.

7. Circularly polarized monochromatic light is incident on the surface of a metal. For an angle of incidence of $72° 30'$ it is found that the reflected light is extinguished by a Nicol whose principal section is inclined to the plane of incidence at an angle of $43° 30'$. Calculate the refractive index, extinction coefficient, and reflection coefficient at normal incidence.

CHAPTERS XXII and XXIII

1. A plane-parallel quartz plate 0·5 mm. thick is cut with its faces parallel to the optic axis and placed between crossed Nicols with its axis at 45° to the principal sections of the Nicols. If the difference between the principal refractive indices for quartz is 0·009 and is independent of wavelength, find the wavelengths in the visible region that will not be transmitted through the system.

2. A plate of uniaxial crystal cut with its faces parallel to the axis is viewed between crossed Nicols in a parallel beam of monochromatic light. Explain the effects observed when the crystal is rotated.
   A thin quartz wedge, of angle 1/50th radian, cut with the optic axis at right angles to the edge, is placed between crossed Nicol prisms such that the optic axis bisects the angle between the vibration directions of polarizer and analyser. Calculate the separation between the fringes seen on the wedge for a wavelength of 6000 Å given that the ordinary and extraordinary refractive indices of quartz for this wavelength differ by 0·009.   (LS I)

3. What is meant by elliptically polarized light? Show that plane and circularly polarized light are special cases of elliptically polarized light.
   Describe the propagation of light in a uniaxial doubly refracting crystal. How may a plate cut from such a crystal be used to convert plane polarized light into circularly polarized light? Calculate the thickness of such a plate made from calcite, for which the principal refractive indices are $n_o = 1·658$ and $n_e = 1·486$ at the wavelength of the light used, 5890 Å.   (OF)

4. Elliptically polarized light is passed through a quarter-wave plate and then through a Nicol. Extinction occurs when the optic axis of the plate and the principal section of the Nicol are inclined to the horizontal at angles of 30° and 60° respectively, the angles being measured in the same direction. Find the orientation of the ellipse and the ratio of its axes.

5. Explain how some form of polarizing prism works.
   A quartz quarter-wave plate which is designed for light of wavelength 5893 Å is used in conjunction with a Nicol prism to produce circularly polarized light. Show that if there is also present some light of wavelength 5460 Å, this latter will be elliptically polarized with axes at 45° to the axes of the plate. Find the ratio of the major and minor axes of the ellipse.
   At 5893 Å the ordinary and extraordinary indices of refraction of quartz are 1·544 and 1·553 respectively, and at 5460 Å they are 1·546 and 1·555.   (OF)

6. What is a *quarter-wave plate* and how is it used? In a Laurent polarimeter, where a half-wave plate is used to divide the field, the plane of vibration of the incident plane polarized light makes an angle of 5° with the optic axis of the half-wave plate and the analysing Nicol is set in turn in the two

positions giving equal brightness in the two parts of the field of view. What is the ratio of the intensities observed in these two positions? (LG II)

7. Discuss the phenomenon of the rotation of the plane of polarization of light by optically active materials.

The specific rotation of quartz at 5086 Å is 29·73 deg./mm. Calculate the difference in the refractive indices. (LS I)

8. A plane parallel quartz plate of thickness 2 cm. is cut with its faces perpendicular to the optic axis and placed between crossed Nicols. Find the wavelengths in the visible region that will not be transmitted through the system. (The specific rotation of quartz in degrees per mm. is given by a formula of the Cauchy type where $A = -2$ and $B = 8·2 \times 10^8$, λ being in angstroms.)

9. Plane polarized monochromatic light is passed through a 20 cm. tube containing an aqueous solution of sucrose and then along the axis of a glass rod of length 25 cm. around which is wound a uniform solenoid containing 5000 turns of wire. It is found that with a current of 0·96 amp. round the solenoid the Faraday rotation in the glass exactly balances the rotation in the solution. Find the concentration of the sucrose solution. (Verdet constant for the glass = 0·08 minutes oersted^{-1} cm.$^{-1}$, specific rotation of sucrose solution (1 gm. per cc.) = 6·6 deg. per cm.)

10. The plane of vibration of a beam of plane-polarized monochromatic light is rotated 90° (a) by passing it through an optically active uniaxial crystal in a direction parallel to the optic axis, (b) by passing it through a non-active uniaxial crystal in a direction perpendicular to the axis. In each case the crystal has the minimum thickness suitable for the purpose. What is the state of polarization of the light half-way through each crystal plate? Explain how you arrived at your conclusions. Suppose that in each case the plane of vibration of the incident light is rotated. Describe and explain what happens to the state of polarization (a) of each emergent beam and (b) of the light midway through each crystal.

11. Light passes normally through two identical quarter-wave plates placed in series (a) with their optic axes parallel (b) with their optic axes perpendicular. The incident light is plane polarized with its plane of vibration rotated 30° clockwise from the fast axis of the first plate. For each of the arrangements (a) and (b), what is the state of polarization of the emergent light? What is the state of polarization after the first plate?

OUTLINES OF SOLUTIONS, WITH COMMENTS

Unless otherwise stated a number in the form (2–53) refers to an equation in the main text.

1. See §1–9.

2. See §1–16. $h = \sqrt{(h_1'h_2')} = \sqrt{(4 \times \frac{1}{4})} = 1$ cm. $|l'|/|l| = 4$ and $|l'| + |l| = 100$ cm. Hence $|l'| = 80$ cm and $|l| = 20$ cm. Inserting $l' = +80$ cm and $l = -20$ cm into the equation for a thin lens in air gives $f' = +16$ cm.

3. Note that (2–53) and (2–56) apply only to a thick lens *in air*. In general, it is probably best to use (2–35), (2–36), (2–37). In air, the power of the convex surface $= 0.1$ cm^{-1}, giving $r = 6.4$ cm.

   Take the convex surface to the left.
   (a) $F_1 = 0.1$ cm^{-1}, $F_2 = 0$, $F = 0.1$ cm^{-1}. P and P′ at lens since $d = 0$. $f = -10.0$ cm, $f' = +13.3$ cm. (2–10) gives N and N′ 3.3 cm outside the plane surface.
   Using the matrix method: The refraction matrices are, from (2–81),

   $$\mathbf{R}_1 = \begin{bmatrix} 1 & 0.1 \\ 0 & 1 \end{bmatrix} \quad \text{and} \quad \mathbf{R}_2 = \begin{bmatrix} 1 & 0 \\ 0 & 1 \end{bmatrix}$$

   and the translation matrix is, from (2–82),

   $$_1\mathbf{T}_2 = \begin{bmatrix} 1 & 0 \\ 0 & 1 \end{bmatrix}$$

   Hence system matrix is, from (2–85),

   $$\mathbf{M} = \mathbf{R}_2 \,_1\mathbf{T}_2\, \mathbf{R}_1 = \begin{bmatrix} 1 & 0.1 \\ 0 & 1 \end{bmatrix}.$$

   (2–101) and (2–102) give $P_1P = P_2'P' = 0$, and (2–104) and (2–105) give $(l_1)_N = (l_2')_{N'} = +3.3$ cm. (2–106) and (2–107) give $f = -10$ cm and $f' = +13.3$ cm.
   (b) One still has $r_1 = 6.4$ cm, but $F_1 = 0.048438$. Again $F_2 = 0$, so that $F = F_1$. $f = -27.458$ cm, $f' = +20.645$ cm. Again $P_1P = P_2'P' = 0$. N and N′ coincide 6.813 cm. to left of convex surface.

   The system matrix becomes $\begin{bmatrix} 1 & 0.048438 \\ 0 & 1 \end{bmatrix}$, whence $P_1P = P_2'P' = 0$, $(l_1)_N = (l_2')_N' = -6.812$ cm, $f = 27.458$ cm, and $f' = +20.645$ cm.

4. Take the convex surface to the left. The powers of the surfaces are $F_1 = +0.064$ cm.$^{-1}$, $F_2 = -0.032$ cm.$^{-1}$.
   $d = 3.0488$ cm. (2–37) gives $F = +0.038244$ cm.$^{-1}$ whence $f' = -f = +26.148$ cm. Since lens in air $P_1N = P_1P$ and $P_2'N' = P_2'P'$. (2–36) and (2–35) give $P_1N = -2.551$ cm, $P_2'N' = -5.102$ cm. Using the matrix method, the system matrix is

$$\mathbf{M} = \begin{bmatrix} 1 & -0.032 \\ 0 & 1 \end{bmatrix} \begin{bmatrix} 1 & 0 \\ -3.0488 & 1 \end{bmatrix} \begin{bmatrix} 1 & 0.064 \\ 0 & 1 \end{bmatrix}$$

$$= \begin{bmatrix} 1.09756 & 0.038244 \\ -3.0488 & 0.80488 \end{bmatrix}$$

(2–108) gives $f' = 26.148$ cm, and (2–104) and (2–103) give $(l_1)_N = -2.551$ cm, $(l_2')_N' = -5.102$ cm.

5. Take the convex surface to the left. (2–37) gives $F = F_1 = 0.064$ cm^{-1}. $\bar{d} = 6.0976$ cm. (2–36) and (2–35) give $P_1P = 0$, $P_2'P' = -6.0976$ cm. N is at P, N' at P'. $f' = 1/F = +15.625$ cm $= -f$, whence $P_1F = 15.625$ cm and $P_2'F' = 9.528$ cm. Since one is not asked to find the focal length explicitly, P_1F and $P_2'F'$ could have been obtained directly from (2–40) and (2–39).

Using the matrix method, the system matrix is

$$\mathbf{M} = \begin{bmatrix} 1 & 0 \\ 0 & 1 \end{bmatrix} \begin{bmatrix} 1 & 0 \\ -6.0976 & 1 \end{bmatrix} \begin{bmatrix} 1 & 0.064 \\ 0 & 1 \end{bmatrix}$$

$$= \begin{bmatrix} 1 & 0.064 \\ -6.0976 & 0.60976 \end{bmatrix}$$

and the positions of $P(=N)$, $P'(=N')$, F, F', follow from (2–101), (2–102), (2–105), (2–107).

6. Take the convex face to the left. Powers of surfaces: $F_1 = +0.0260$ cm^{-1}, $F_2 = -0.0063333$ cm^{-1}. $\bar{d} = 3.9474$ cm.

Method (a) (2–35), (2–36), (2–37) yield $F = +0.0203167$ cm^{-1}, $f = -49.221$ cm, $f' = +65.463$ cm, $P_1P = -1.230$ cm, $P_2'P' = -6.719$ cm, whence, from (2–10), $P_1N = +15.012$ cm, $P_2'N' = +9.523$ cm. It is noteworthy that if one wishes to locate all the cardinal points to an accuracy of, say, 0.001 cm as has been done here, one requires a different number of significant figures at different stages in the calculation.

In (2–16) $z = -135.012$ cm, whence $z' = +95.549$ cm. Hence image plane is 105.07 cm. from concave face and $m = z'/z = -0.7077$.

Method (b) The refraction matrices (see (2–81)) are

$$\mathbf{R}_1 = \begin{bmatrix} 1 & 0.026 \\ 0 & 1 \end{bmatrix} \quad \text{and} \quad \mathbf{R}_2 = \begin{bmatrix} 1 & -0.0063333 \\ 0 & 1 \end{bmatrix}$$

and the translation matrix (see (2–82)) is

$$_1\mathbf{T}_2 = \begin{bmatrix} 1 & 0 \\ -3.9474 & 1 \end{bmatrix}.$$

The system matrix (see (2–85) and (2–87)) is

$$\mathbf{M} = \mathbf{R}_2\,_1\mathbf{T}_2\,\mathbf{R}_1 = \begin{bmatrix} 1.02500 & 0.0203167 \\ -3.9474 & 0.89737 \end{bmatrix}.$$

Check that M has determinant value of unity.

From (2–95), since $\bar{l}_1 = -120$, $l_k' = 105.07$ and (2–96) gives $m = -0.7077$. Note that one is not, in fact, asked to find the cardinal planes and if one uses the matrix method it is not necessary to find them in order to find the position of the image plane and the magnification.

7. One can take an incident ray parallel to the axis at, say, $y_1 = 1$ and apply (2–62) and (2–63) until y_3 and u_3' are found and (2–64) and (2–65) can be applied to give P' and f'. A ray in the opposite direction is required to find P and f.

This is a problem that is probably best solved by the matrix method. The powers of the individual refracting surfaces in cm^{-1} are $+0.4308$, $+0.077$, $+0.1283$, and the reduced separations are 0.2694 and 0.2548. Using (2–81), (2–82), (2–86), the system matrix is

$$
\mathbf{M} = \begin{bmatrix} 1 & 0.1283 \\ 0 & 1 \end{bmatrix} \begin{bmatrix} 1 & 0 \\ -0.2548 & 1 \end{bmatrix} \begin{bmatrix} 1 & 0.077 \\ 0 & 1 \end{bmatrix} \begin{bmatrix} 1 & 0 \\ -0.2694 & 1 \end{bmatrix} \begin{bmatrix} 1 & 0.4308 \\ 0 & 1 \end{bmatrix}
$$
$$
= \begin{bmatrix} 0.91268 & 0.5959 \\ -0.5190 & 0.7570 \end{bmatrix}.
$$

Check that $\mathbf{M}$ has determinant value unity. This is an especially useful check on a fairly long calculation. The results given on page 426 then follow at once from equations (2–101) to (2–109).

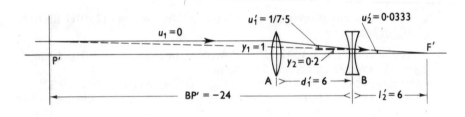

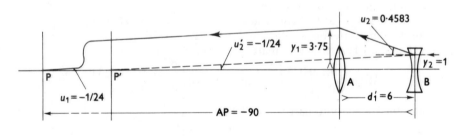

Fig. 8

8. (i) Probably the simplest procedure is to set $y_1 = 1$ and use (2–62) [with $n = n' = 1$ for each lens] and (2–63). One has $F_1 = +1/7.5$, $F_2 = -1/2$, $\bar{d} = d = 6$, which gives $u_1' = 1/7.5$, $y_2 = 1 - 6/7.5 = 0.2000$, $u_2' = 1/7.5 - \frac{1}{2} \times 0.2$ $= +0.0333$. The path of the ray is now found; it cuts the axis at $l_2' = y_2/u_2' = 6$. (ii) (2–64) locates P' (BP' = -24), and the required ray can be traced in reverse by taking a ray directed towards P' from the right of B. One can set $y_2 = 1$, whence $u_2' = -1/24$. (2–62) gives $u_2 = 0.4583$ and (2–63) gives $y_1 = 3.75$. $u_1 = u_2'$ since the principal points are also the nodal points and hence the emergent ray is found; it cuts the axis at P where $AP = y_1/u_1 = -90$.

9. For an image point 0·4 cm from the axis in the focal plane of a lens with $f' = 20$ cm, the incident collimated beam from the left is inclined to the axis at angle $0·4/20 = 0·02$. If the same incident beam gives an image point 1 cm from the axis, the focal length of the lens *combination* must be $1/0·02 = 50$ cm. The focal length of the required lens then follows from (2–60). One obtains $f_2' = -25$, i.e. the lens is diverging.

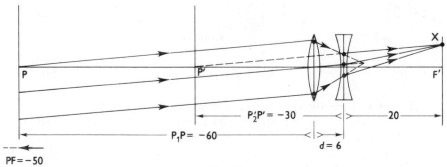

Fig. 10

10. Take object to left of system and converging lens to left of diverging lens. The separation of the lenses must be found. Focal length of complete system is $f' = 0·75/0·015 = +50$ cm. (2–60) yields $d = 6$ cm. (2–35) and (2–36) or (2–61) give $P_1P = -60$ cm, $P_2'P' = -30$ cm. Alternatively, use matrix method. System matrix has elements 2·2, 0·02, -6, 0·4. For image point X in figure incident rays are parallel to P'X and between lenses converge to point in focal plane of first lens. Hence paths of image-forming rays obtained. For convenience in diagram vertical distances have been increased by factor ×6 relative to horizontal distances.

11. Take object to left of system. Any of the usual methods yield positions of cardinal points. Note: N at P, N' at P'. Rotation referred to is about second nodal point. Hence situation corresponds to Fig. 4–10. One also finds $f' = +16$ cm and $PP' = 2$ cm. Hence image and object planes coincide if $l' = l - 2$, whence $l = 6·74$ cm or $-4·74$ cm, corresponding, respectively, to real object 8·74 cm to the left, or virtual object 2·74 cm to right of first (converging) lens.

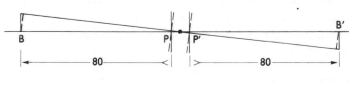

Fig. 12

12. Take object to left of system. Any of usual methods yield positions of cardinal points. $P_1P = P_2'P' = -20$ cm and $f' = +40$ cm. N at P, N' at P'. Hence object is 80 cm $(2f)$ in front of P and image is 80 cm $(2f')$ behind P'. Since magnification is -1, B and B' remain conjugate when system rotated. The figure illustrates the symmetry of the system.

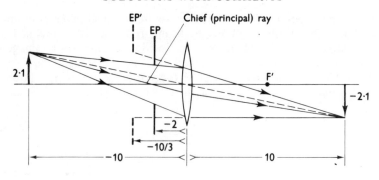

Fig. 13

13. The figure shows the system. The thin lens equation (2–49) where $n=n'=1$ gives exit pupil 10/3 cm from lens and diameter of exit pupil $= 2.5 \times (10/3)/2 = 25/6$. Note that $l=2f$. Hence $l'=2f'$ and $m=-1$.

14. See §§3–11 and 3–12. $d=3$ cm. System similar to Fig. 3–12(a) with $AP=4$ cm, $BP'=-2$ cm, $f'=P'F'=-PF=8/3$ cm. Hence $AF=4/3$ cm, $BF'=2/3$ cm.

15. Normal methods give $f'=3.2$ cm. $P_1P=-P_2'P'=2.4$ cm. (N at P, N' at P'). $P_2'F'=-P_1F=0.8$ cm. Telescope similar to Fig. 3–13(b). O.G. is entrance pupil and is imaged by eyepiece. $l=-63.2$ cm, giving $l'=+3.37$ cm, whence exit pupil is $3.37-2.4=0.97$ cm beyond eye lens. Exit pupil diameter $= 5 \times l'/l = 0.267$ cm.

16. Dispersive power $\omega=1/V$ and (3–11) give $F_1=F\omega_2/(\omega_2-\omega_1)$ and $F_2=-F\omega_1/(\omega_2-\omega_1)$. Using (3–8), crown has $V=60.60$. flint has $V=36.05$. (3–11) yields f' (crown) $=-4.05$ cm (diverging), f' (flint) $=+6.81$ cm (converging). (N.B. *diverging* doublet)

17. See §§5–8, 5–9, 5–12, 5–13, 5–14, 5–15. Note the importance of temperature control.

18. See §5–14. See §§5–8, 5–12, 5–14, 8–13, 11–14. For dilute solutions the method must be capable of measuring *small changes* of refractive index.

19. (a) 1.2746 to 1.6200 and 1.4362 to 1.7500 ($\sqrt{n^2-1}$ to n) (b) 1 to 1.6200 and 1 to 1.7500 (1 to n). Upper limit for any critical angle method must be n. With 90° prism occurrence of total reflection at face of emergence sets lower limit; with 60° prism no such lower limit.

20. Two main effects. (i) Image moves sideways, corresponding to reduction in angle of deviation; reaches extreme position, corresponding to minimum deviation (when angle of emergence equals angle of incidence); reverses direction of movement, corresponding to increase in angle of deviation. (ii) Image gets broader. This follows from Helmholtz equation: width of beam entering prism increases and width of emergent beam decreases; prism effectively a telescopic system (§2–11). Alternatively, consider complete optical system: for beam that passes through prism angle of divergence from collimator slit increases whilst angle of convergence to image at focus of telescope O.G. decreases. Since actual slit width constant, slit image width increases (Helmholtz—§2–4).

CHAPTERS VI–IX

1. If one uses the approach given in §6–10, (6–43) emerges as the most simple solution to the first part of the question. Although this can be used for the second part of the question, fairly extensive manipulation is involved. This is because c and λ in (6–43) are the phase velocity and the wavelength *in the medium*. The corresponding values in a vacuum are $c_o = \mu c$ and $\lambda_o = \mu\lambda$.

$$\frac{v}{c} = 1 - \frac{\lambda}{c}\frac{dc}{d\lambda} = 1 - \frac{\lambda_o}{c_o}\frac{dc}{d\lambda}$$

$$\frac{dc}{d\lambda} = c_o\frac{d}{d\lambda}\left(\frac{1}{\mu}\right) = -\frac{c_o}{\mu^2}\frac{d\mu}{d\lambda} = -\frac{c_o}{\mu^2}\frac{d\mu}{d\lambda_o}\frac{d\lambda_o}{d\lambda}$$

$$\frac{d\mu}{d\lambda_o} = -\frac{2B}{\lambda_o^{\,3}}$$

Using $\lambda_o = \left(A + \dfrac{B}{\lambda_o^{\,2}}\right)\lambda$, one obtains, after some manipulation

$$\frac{d\lambda_o}{d\lambda} = \frac{(A\lambda_o^{\,2} + B)^2}{A\lambda_o^{\,4} + 3B\lambda_o^{\,2}}$$

Whence

$$\frac{v}{c} = \frac{A\lambda_o^{\,2} + B}{A\lambda_o^{\,2} + 3B}$$

If one writes a simple harmonic wave in the form $a\cos(\omega t - kx)$, (6–44) emerges easily. This is not an explicit answer to the first part of the question but yields (6–46) and (6–47), which do relate c and v explicitly. It is, in fact, (6–47) which provides the simplest solution to the second part of the question. One has

$$\frac{c}{v} = 1 - \frac{c}{c_o}\lambda_o\frac{d\mu}{d\lambda_o}$$

$$= 1 - \frac{\lambda_o}{\mu}\frac{d\mu}{d\lambda_o}$$

$$= 1 + \frac{\lambda_o}{1 + B/\lambda_o^{\,2}}\frac{2B}{\lambda_o^{\,3}}$$

$$= \frac{A\lambda_o^{\,2} + 3B}{A\lambda_o^{\,2} + B}$$

2. In (7–2) adjacent fringes have values of p that differ by 1. I.e. fringe spacing $= \lambda D/d$.

$$\therefore \quad \lambda = \frac{0{\cdot}31 \times 1{\cdot}9}{1000} = 5{\cdot}89 \times 10^{-4} \text{ mm.}$$

Fringe visibility is defined by equation (8–6):

$$V = \frac{I_{max} - I_{min}}{I_{max} + I_{min}}$$

(a) If the slits give equal intensities $I_{min} = 0$ and $V = 1$.
(b) If $I_1 = 4I_2$, the amplitudes are related by $a_1 = 2a_2$.
In the fringe system one then has $a_{max} = 3a_2$, $a_{min} = a_2$ or $I_{max} = 9a_2^{\,2}$ and $I_{min} = a_2^{\,2}$, whence $V = 0{\cdot}8$.

3. See §§6–1 to 6–4. With a narrow source slit the disturbances are coherent but they have slightly different amplitudes because one beam loses energy when it is reflected.

The intensities of the interfering disturbances can be written as a^2 and $\frac{3}{4}a^2$.

Hence the amplitudes are a and $\frac{1}{2}a\sqrt{3}$.

At a maximum, amplitude $a_{max} = a + \frac{1}{2}a\sqrt{3}$.

At a minimum, amplitude $a_{min} = a - \frac{1}{2}a\sqrt{3}$.

Whence $\qquad \dfrac{I_{max}}{I_{min}} = \dfrac{a_{max}{}^2}{a_{min}{}^2} = 194.$

4. See §§7–1 and 7–7. From §7–4, fringe spacing $= x = \lambda(D_1 + D_2)/d$ (see Fig. 7–3). One has $D_1 = 25$ cm, $D_2 = 175$ cm, $x = 0.02$ cm, $\lambda = 5 \times 10^{-5}$ cm. This gives $d = 0.5$ cm. The refracting angle α of each half of the biprism is then given by $0.5 = 25 \times 2 \times 0.5 \times \alpha$, whence $\alpha = 0.02$ radian $= 1°9'$.

$$\therefore \quad \text{Vertex angle} = 180° - 2\,\alpha = 177°42'.$$

5. See §§7–4 and 7–5, assuming complete wavefront utilized. Fringes not localized. With white light fringes observable only in vicinity of zero order.

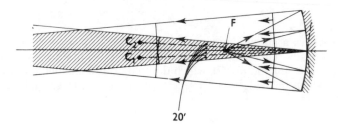

Fig. 5

Each mirror has a point source in its focal plane and hence gives plane waves. Angle between plane waves $= 80'$.

Linear, non-localized $\cos^2$ fringes in region of overlap of wavefronts (shaded). Fringes parallel to axis of rotation of one half of mirror relative to other, i.e. perpendicular to plane of figure. Fringe spacing x given by

$$\frac{\lambda}{x} = 80 \times \frac{\pi}{60 \times 180} \quad \text{whence } x = 2.58 \times 10^{-3} \text{ cm.}$$

6. If the wedge angle $= \theta$ radian, and fringe spacing is x, $x\theta = \lambda/2$ where λ is the wavelength in glass. In terms of the vacuum wavelength λ_o, $x\theta = \lambda_o/2n$. One obtains $\theta = 40''$.

7. If d is the separation of M_1 and M_2' (Fig. 8–9), o.p.d $= 2d$ at centre of pattern. If o.p.d. changes by $p\lambda$, p fringes emerge i.e. $2d = p\lambda$, whence

$$\lambda = \frac{2 \times 1}{3663} = 5.46 \times 10^{-4} \text{ mm.}$$

If one changes to white light the field of view will simply be white with no fringes visible. One can say *either* that the constituent wavelengths produce fringes that have various diameters and are intermingled to such an extent that they are not discernible *or* that the effective coherence length is virtually zero and very much smaller than $2d$ so that no interference effects are observable.

8. Referring throughout to Fig. 8–9 (page 156), initial adjustment requires M_1 and M_2' to form what is, in effect, a wedge-shaped film of air with M_2' intersecting M_1 within the field of view. With quasi-monochromatic: equidistant linear $\cos^2$ fringes parallel to line of intersection of M_1 and M_2'. Green light gives more widely spaced fringes; λ is larger and one must move farther to increase the wedge thickness by $\lambda/2$ and pass from one fringe to next.

   1. Fringe spacing unchanged. All fringes move in same direction as line of intersection of M_1 and M_2'.
   2. Assume rotation is about line of intersection of M_1 and M_2'. If angle between M_1 and M_2' decreased, fringes broadened; if angle increased, fringes become more widely spaced. If rotation not about line of intersection (i.e. zero order fringe), then this line also displaced and associated displacement of fringes superimposed.
   3. Fringes not straight—they are lines of constant wedge thickness.
   4. Fringes remain linear and equidistant but rotate to remain parallel to line of intersection of M_1 and M_2'. Fringe spacing increases because wedge angle increases.

9. See §§8–11 (end), 8–13, 9–10 (end), 11–14 for refractive index or §§8–6, 8–12, 9–15, 9–16 for surfaces. The system referred to is really a Twyman-

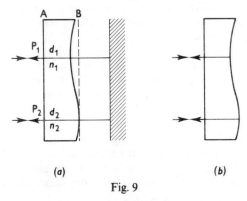

(a)   (b)

Fig. 9

Green interferometer. With mirrors in place (Fig. a): optical paths between planes A and B are $n_1 d_1 + (d_2 - d_1)$ and $n_2 d_2$. These are traversed twice. Hence

$$\text{o.p.d.} = 2[(n_2 d_2 - n_1 d_1) - (d_2 - d_1)] = m\lambda.$$

With mirrors removed (Fig. b): optical paths with plate are $2n_1 d_1$ and $2n_2 d_2$.
Hence o.p.d. $= 2(n_2 d_2 - n_1 d_1) = m\lambda.$

$$\therefore \quad d_2 - d_1 = 0. \qquad\qquad \text{i.e. } d_1 = d_2 = D$$

$$2D(n_2 - n_1) = m\lambda$$

$$\text{or} \qquad n_2 - n_1 = m\lambda/2D.$$

10. It was shown in §§8–1 and 8–4 respectively that for fringes of equal inclination and for Newton's rings, (ring diameter)^2 varies approximately linearly with ring number, counting from centre of pattern. With perfect contact at centre Newton's rings when viewed, in the usual way, by reflection have a central dark fringe. For Fabry-Perot discussion of §8–1 is applicable if one puts $n=1$ and omits the π phase change at one surface; i.e. $2d \cos \theta = p\lambda$ for bright fringes. d can be adjusted to make $2d=(p+\frac{1}{2})\lambda$, thus making central fringe dark, and can also be adjusted to change diameters of rings. To photograph Newton's rings camera must be focused on plane of thin film; for Fabry-Perot camera must be focused at infinity. Rings on photographs can be matched in size by appropriate choice of focal lengths of lens systems used and also by convenient choice of Fabry-Perot plate separation, provided $2d=(p+\frac{1}{2})\lambda$. Intensity distributions in rings could be made similar by silvering surfaces used for Newton's rings and viewing by transmission. These multiple beam Newton's rings would show the surface imperfections; exceptionally good quality surfaces would be required to produce fringes as smooth as Fabry-Perot fringes.

For both sets of fringes the result (ring riameter)$^2 \propto$ (ring number) is obtained as a consequence of an approximation that is valid near centre of pattern where it is assumed the two sets of fringes are matched in size. For Fabry-Perot approximation involved is $\cos \theta = 1 - \frac{1}{2}\theta^2$. In fact $\cos \theta > 1 - \frac{1}{2}\theta^2$ so that in outer part of pattern actual o.p.d. is larger than approximate value. In other words, assumed (approximate) o.p.d. actually corresponds to a direction for which $\cos \theta$ is smaller i.e. θ is larger than supposed. Now a particular fringe always corresponds to a particular o.p.d. so that for any given fringe the corresponding o.p.d. is found at a larger value of θ than is indicated by the approximate theory, i.e. the fringes have slightly larger diameters than is indicated by the approximate theory. On the other hand for Newton's rings actual film thickness is larger than approximate value and displacement of any given ring is such that o.p.d. is reduced, i.e. actual ring diameters are smaller than those indicated by approximate theory. Thus if diameters of two sets of rings matched in the central part of field, Fabry-Perot rings have slightly larger diameters than Newton's rings in outer part of field.

Moving from centre of Newton's rings, next fringe occurs where $2d=\lambda$ i.e. $r^2=\lambda R$. For Fabry-Perot o.p.d. at centre $=2d$, and at next fringe o.p.d. $= 2d(1-\frac{1}{2}\theta^2)=2d-\lambda$, i.e. $\theta^2 d=\lambda$. For Newton's ring of radius r, radius on the photograph $= 5r$, whilst for Fabry-Perot ring of angular radius θ radius in the focal plane of 50 cm lens is 50θ.

$$\therefore \quad 50\theta = 5r$$
$$\text{or} \quad 2500\theta^2 = 25r^2$$
$$\text{i.e.} \quad 2500\lambda/d = 25\,\lambda R \qquad \text{where } R=100$$
$$\therefore \quad d=1 \text{ cm.}$$

11. See §§8–13 and 11–14. Investigate variation of n with pressure at constant temperature.

$$\text{o.p.d.} = 130 \times 5 \cdot 46 \times 10^{-5} = \delta n \times 100$$
$$\therefore \quad \delta n = 7 \cdot 098 \times 10^{-5}$$

Law of Gladstone and Dale: $(n-1)/\rho = \text{constant}$ ($\rho = \text{density}$).
For a perfect gas $\qquad p/\rho T = \text{constant}$

$$\therefore \quad \frac{p}{(n-1)T} = \text{constant}$$

If n is the index at 76 cm Hg and 20°C one has

$$\frac{76}{(n-1)} = \frac{56}{(n-\delta n-1)}$$

Inserting the above value of δn yields $(n-1) = 0\cdot0002697$. n_o at S.T.P. is given by $(n_o-1)273 = (n-1)293$, whence $n = 1\cdot0002895$.

12. See §§8–7, 8–8. Mirror displacement of $\lambda/2$ corresponds to 1 fringe
∴ Mirror displacement of v corresponds to $2v/\lambda$ fringes.
∴ Mirror velocity v corresponds to $2v/\lambda$ fringes per second.
If mirror moves with speed v, virtual image of source behind mirror moves with speed $2v$ away from observer. If f is frequency of the light from stationary source, frequency of light from receding source obtained by substituting $-2v$ for v in (6–53).

$$f' = f\sqrt{\frac{1-2v/c}{1+2v/c}} = f\sqrt{\frac{(1-2v/c)^2}{(1+2v/c)(1-2v/c)}}$$

or

$$f' = \frac{f(1-2v/c)}{\sqrt{(1-4v^2/c^2)}}$$

Since $v \ll c$ one can take $f' = f(1-2v/c)$.
Light from stationary mirror has frequency f so that one observes beat frequency $(f - f')$ where

$$f - f' = \frac{2vf}{c} = \frac{2v}{\lambda}$$

13. See §8–13.
Intensities of two beams can be written a^2 and $\tfrac{1}{2}a^2$

∴ amplitudes are a and $a/\sqrt{2}$.

At a maximum, amplitude $= a + a/\sqrt{2} = 1\cdot7071a$.

At a minimum, amplitude $= a - a/\sqrt{2} = 0\cdot2929a$.
Intensities are $I_{max} = 2\cdot9148a^2$ and $I_{min} = 0\cdot08580a^2$ instead of $4a^2$ and zero respectively.

Fringe visibility falls from unity to $\dfrac{I_{max} - I_{min}}{I_{max} + I_{min}} = 0\cdot943$.

14. See §9–1. For a maximum for wavelength λ_1 $2d \cos \theta = p\lambda_1$.
For a minimum for $\lambda_2(<\lambda_1)$ $2d \cos \theta = (p+k+\tfrac{1}{2})\lambda_2$
λ_1 maxima and λ_2 minima coincide if

$$p\lambda_1 = (p+k+\tfrac{1}{2})\lambda_2$$

i.e. $\quad 2d \cos \theta = \left(\dfrac{2d \cos \theta}{\lambda_1} + k + \tfrac{1}{2}\right)\lambda_2$

If condition satisfied on axis $\cos\theta = 1$: fringes then out of step but one does not necessarily have a maximum or a minimum on the axis. One obtains

$$d = (k + \tfrac{1}{2}) \frac{\lambda_1 \lambda_2}{2(\lambda_1 - \lambda_2)}$$

Inserting wavelengths of D lines gives

$$d = (2k + 1) \times 0{\cdot}01447 \text{ cm.}$$

i.e. an odd multiple of $0{\cdot}01447$ cm.
Intensity measurements best made on axis—using central spot scanning (see §9–8).

15. Equation (9–11) gives

$$d = \frac{\lambda^2}{2\varDelta\lambda} = \frac{(5 \times 10^{-5})^2}{2 \times 0{\cdot}05 \times 10^{-8}} = 2{\cdot}5 \text{ cm.}$$

Using Taylor criterion,

$$\frac{\lambda}{\delta\lambda} = \frac{\pi d \sqrt{F}}{\lambda} \quad \text{or} \quad \delta\lambda = \frac{\lambda^2}{\pi d \sqrt{F}} \quad \text{where} \quad F = \frac{4R}{(1-R)^2}$$

$R = 0{\cdot}95$ gives $F = 1520$ or $\sqrt{F} = 38{\cdot}99$

$$\therefore \quad \delta\lambda = \frac{(5 \times 10^{-5})^2}{\pi \times 2{\cdot}5 \times 38{\cdot}99} = 8{\cdot}16 \times 10^{-12} \text{ cm.}$$

The precision of the Taylor (or any other) criterion for resolution is such that there is little sense in specifying $\delta\lambda$ with a precision better than about 10%. I.e. $\delta\lambda \sim 0{\cdot}0008$ Å.

16. See §§9–7 and 9–10. Note that although in absolute measure it is the thinner of the two etalons that is rotated, it is the one whose thickness is slightly *greater* than is required to make the ratio of the thicknesses N_1/N_2 (see §9–7)—in this case d_1 is slightly greater than $\tfrac{1}{4}d_2$.

One has $\quad d_2 = 4 \times 0{\cdot}50027 \times \cos 0{\cdot}01571$
$\qquad\qquad = 4 \times 0{\cdot}50027(1 - \tfrac{1}{2}0{\cdot}01571^2)$
$\qquad\qquad = 2{\cdot}00108 - 0{\cdot}00025 = 2{\cdot}00083$ cm.

17. See §9–13. If p is order for green line, order for blue is $p + 20$.
$\quad\therefore \quad 2d = p \times 5{\cdot}461 \times 10^{-5} = (p + 20) \times 4{\cdot}358 \times 10^{-5}$ cm.
$\quad\therefore \quad p(5{\cdot}461 - 4{\cdot}358) = 20 \times 4{\cdot}358$

$$d = \frac{1}{2} \times \frac{20 \times 4{\cdot}358}{1{\cdot}103} \times 5{\cdot}461 \times 10^{-5} = 2{\cdot}16 \times 10^{-3} \text{ cm.}$$

CHAPTERS X–XIII

1. See §§11–1 to 11–3. (a) Minima in directions θ given by $a \sin\theta = p\lambda$ (p integral) (b) Maxima when $\tan\alpha = \alpha$, where $\alpha = (\pi a \sin\theta)/\lambda$. In Fig. 11–1(a) $a = 0{\cdot}080$ cm, $f_2' = 80$ cm, $h'(S'P') = 0{\cdot}30$ cm. Hence θ is small ($= 0{\cdot}30/80$) and the condition for a minimum becomes $a\theta = p\lambda$ (p integral). Wavelengths

that satisfy this equation are missing. Set $\lambda = 8 \times 10^{-5}$ cm (extreme red) to find minimum value of p. One has

$$p = \frac{0 \cdot 080 \times 0 \cdot 30}{80 \times 8 \times 10^{-5}} = 3 \cdot 7$$

so that values of $p > 3 \cdot 7$ give wavelengths in the visible region. Hence missing wavelengths given by

$$\lambda = \frac{0 \cdot 080 \times 0 \cdot 30}{80 \times p} \text{ with } p > 3 \cdot 7$$

$p = 4$ gives $\lambda = 7 \cdot 5 \times 10^{-5}$ cm.
$p = 5$ gives $\lambda = 6 \times 10^{-5}$ cm.
$p = 6$ gives $\lambda = 5 \times 10^{-5}$ cm.
$p = 7$ gives $\lambda = 4 \cdot 286 \times 10^{-5}$ cm.

Since $p = 8$ gives λ in U.V. these four wavelengths are the ones in the visible region that are missing.

2. See §§ 11–5 and 9–3. Equation (11–6) gives, for isosceles prism at minimum deviation,

$$\frac{\lambda}{\delta\lambda} = t\frac{dn}{d\lambda} = -\frac{2bt}{\lambda^3}$$

R.P. required $= 5893/6$. Hence value of t required is

$$t = \frac{5893}{6} \times \frac{(5 \cdot 893 \times 10^{-5})^3}{2 \times 9 \cdot 84 \times 10^{-11}} = 1 \cdot 02 \text{ cm.}$$

3. See § 11–13. Angular diameter of star $= \alpha = 1 \cdot 22 \times 5 \cdot 75 \times 10^{-5}/300$. Linear diameter $= \alpha \times 1 \cdot 7 \times 10^{15}$ km $= 3 \cdot 98 \times 10^8$ km.

4. See § 10–9. (a) For each star each aperture alone would give Airy pattern. Diameter of each Airy disc $= 1 \cdot 22\lambda/u'$ where $u' = 10/700$. Airy disc diameter $- 85 \cdot 4\lambda$.

Over area corresponding to Airy pattern get double slit fringes with fringe spacing λ/θ where $\theta = 100/700$. $\therefore$ Fringe spacing $= 7\lambda$.

(b) The grating contains only one spatial frequency so that for star on axis maxima occur at only one angle on each side of the normal (see §§ 11–15 and 14–11). Angle $= \lambda/d$. Hence star gives two images with angular separation $2\lambda/d$ or linear separation in focal plane $2f'\lambda/d = 1 \cdot 4 \times 10^4\lambda$. Similar effect for other stars.

(c) Star on axis gives plane wave incident on mesh. Mesh is two-dimensional grating—periodicity in two perpendicular directions. Each periodicity gives series of maxima, and combined result is square array of maxima. Since separation of wires much larger than thickness, central maxima much stronger than others. In directions parallel to wires separation of maxima $= f'\lambda/d$ where d is grating interval $= 2$ cm. $\therefore$ separation $= 350\lambda$. Dispersion occurs in directions parallel to both sets of wires. Hence general effect of radial spectra. Similar effect for each star.

(d) Similar arrays. Same periodicity but much weaker central maxima because grating spaces now small compared with grating interval.

5. See §§ 10–12, 11–9. For droplet of diameter a, angular radius of bright ring is $1 \cdot 64\lambda/a$ (§ 11–9). Here angular diameter $= 5°$.

$$\therefore \quad 2\cdot5 \times \frac{\pi}{180} = \frac{1\cdot64 \times 5 \times 10^{-4}}{a}$$

whence $\qquad a = 0\cdot0188$ mm.

6. First order at θ where $d \sin \theta = \lambda$ (12–1). Using mean wavelength, $\sin \theta = 5\cdot893 \times 10^{-5} \times 3000 = 0\cdot17679$, whence $\theta = 10° \ 11'$. From §12–6 lines just resolved when

$$\frac{\lambda}{\delta\lambda} = \frac{Nd\,(\sin\theta_1 - \sin I)}{\lambda}$$

whence $\qquad (Nd) = 3\cdot27$ mm.

7. See §12–6. Max R.P. when max θ used, i.e. when highest possible order used. $d\,(\sin 70° + \sin\theta) = p\lambda$. Put $\theta = 90°$ to find maximum p.

$$p = \frac{1}{6000}\,\frac{(\sin 70° + 1)}{5 \times 10^{-5}} = 6\cdot4.$$

Hence highest usable order is 6. Total number of lines $= N = 18,000$. $\therefore$ Maximum R.P. given by

$$\frac{\lambda}{\delta\lambda} = Np = 1\cdot08 \times 10^5.$$

When $\lambda = 5000$ Å, $\delta\lambda = 0\cdot046$ Å. Angle of diffraction for the 6th order $= 59°21'$.

8. See §12–8. With such a grating 1st order cannot be absent. Hence $0° \ 17'$ corresponds to 1st order. Since angles small, expected positions for various orders would be (approximately)

order	1	2	3	4	5	6	7	8	9	10	11
angle	0° 17'	0° 34'	0° 51'	1° 8'	1° 25'	1° 42'	1° 59'	2° 16'	2° 33'	2° 50'	3° 7'

7th and 10th orders unobserved. Hence envelope has zeros in these regions. Also, since 5th stronger than 4th, envelope has a zero before 5th. From §12–8 envelope has zeros corresponding to orders $3\frac{1}{2}$, 7, $10\frac{1}{2}$ if $b = 2\frac{1}{2}a$. But 11th seen and 10th not seen; also 3rd and 4th both classed as weak although 3rd is in central envelope maximum. Hence probable that envelope has zeros corresponding to orders a little less than $3\frac{1}{2}$, 7, $10\frac{1}{2}$ so that b is probably slightly less $2\frac{1}{2}a$. Grating interval $(a+b)$ calculable from any order but accuracy best from highest order. $3° \ 9'$ for 11th gives $(a+b)\sin 3° \ 9' = 11 \times 5 \times 10^{-4}$ or $(a+b) = 0\cdot100$ mm.

9. See §10–10, 11–2, 12–1, 13–1, 13–4. For both simple grating and mixed grating disturbances from all slits in phase when $d(\sin\theta \pm \sin I) = p\lambda$. I.e. principal maxima of both gratings in same positions. A and B slits give diffraction patterns of same shape. Hence relative amplitudes of principal maxima same for both gratings. If each A slit gives amplitude a at a given principal maximum, each B gives $2a$ so that total amplitude for $2N$ slits is $2N \times a$ (Intensity $4N^2a^2$) if all A, and amplitude $= N \times a + N \times 2a = 3Na$ (Intensity $9N^2a^2$) if mixed. Hence intensities of principal maxima for mixed grating are 9/4 times intensities for simple grating.

Simple grating has $2N-1$ minima between adjacent principal maxima. At position of Nth minimum phase difference between disturbances from adjacent slits is π and disturbances from alternate slits are in phase. Hence for mixed grating disturbances from all B slits in phase and disturbances from all A slits in phase but those from B are π out of phase with those from A. B slits give total amplitude $2Na$, A slits give Na and resultant amplitude is Na and intensity is $N^2 a^2 = 1/9$th intensity of principal maximum (ignoring effect of diffraction envelope).

10. See §§12–7 and 12–8. For 3rd order $d \sin \theta_3 = 3\lambda$ and for 4th $d \sin \theta_4 = 4\lambda$. Putting $d = 4 \times 10^{-4}$ cm yields $\sin \theta_3 = 0.375$, $\sin \theta_4 = 0.500$. Ratio of intensities is

$$\frac{I_3}{I_4} = \left(\frac{\sin \alpha_3}{\alpha_3}\right)^2 \Big/ \left(\frac{\sin \alpha_4}{\alpha_4}\right)^2 \quad \text{where} \quad \alpha = \frac{\pi a \sin \theta}{\lambda}.$$

Inserting θ_3 and θ_4 from above and putting $a = 1.5 \times 10^{-4}$ cm yields $I_3/I_4 = 0.26$.

11. (a) Slit images broaden and eventually lines unresolved.
(b) Initially one very broad maximum. As slit widened maximum narrows until divides into two as D lines become resolved. Then resolved lines become narrower. When lines just resolved

$$\frac{\lambda}{\delta\lambda} = \frac{Nd \sin \theta}{\lambda} \quad (\S 12\text{–}6).$$

But for first order $d \sin \theta = \lambda$. Hence lines become resolved for slit width $(Nd) = \lambda d/\delta\lambda = 1.96$ mm.
(c) Pattern unaffected until edge of grating crosses aperture. (The deviation in § 12–7 is unaffected by a translation of the grating.)
(d) Angle of deviation decreases to a minimum when $\theta = -I$ (§ 12–3).
(e) Lines displaced vertically.

12. Maxima when aperture transmits odd number of $\frac{1}{2}$-period annular zones; minima when even number transmitted. For axial point P distance b from aperture, odd number of zones transmitted when distance from P to edge of aperture $= b + (p + \frac{1}{2})\lambda$ (p integral). I.e. for maxima $[b + (p + \frac{1}{2})\lambda]^2 = b^2 + 2 \cdot 5^2$. (See Fig. 13–4.) For minima $[b + p\lambda]^2 = b^2 + 2 \cdot 5^2$. These relations yield: for minima $b = 5.7/p$ metres, for maxima $b = 5.7/(p + \frac{1}{2})$ metres.

13. See § 13–3. Principal focal length $= f_1' = R^2/\lambda$ where $R =$ radius of central zone. Secondary focal length $= f_2' = \frac{1}{3}R^2/\lambda$. If source is to left of zone plate, using usual sign convention

$$\frac{1}{30} - \frac{1}{l} = \frac{1}{f_1'} = \frac{\lambda}{R^2} \quad \text{and} \quad \frac{1}{6} - \frac{1}{l} = \frac{1}{f_2'} = \frac{3\lambda}{R^2}$$

$$\therefore \quad \frac{3(l-30)}{30l} = \frac{l-6}{6l}, \quad \text{whence} \quad l = -30 \text{ cm.}$$

i.e. source is 30 cm from zone plate.

$$\frac{1}{30} + \frac{1}{30} = \frac{1}{f_1'}, \quad \text{whence} \quad f_1' = 15 \text{ cm.}$$

$R^2 = f_1'\lambda$, whence $R = 2.74 \times 10^{-2}$ cm.

14. See § 13–3. Since f' depends on λ, image distance depends on λ i.e. there is chromatic aberration. Since $f' \propto 1/\lambda$ zone plate corresponds to converging

lens with negative chromatic aberration. To correct this aberration need lens with positive aberration, i.e. converging lens. Contact with flat zone plate best if lens plano-convex. Radius of convex surface $=r$. Use suffixes R for red, V for violet, Z for zone plate, L for lens. For power of combination

$$F_R = F_{ZR} + F_{LR} = F_V = F_{ZV} + F_{LV}$$

$$F_{ZR} = 0.010. \quad \therefore \quad F_{ZV} = 0.005. \quad F_{LR} = (n_R - 1)/r. \quad F_{LV} = (n_V - 1)/r$$

$$\therefore \quad 0.010 + 0.500/r = 0.005 + 0.535/r$$

$$\text{whence} \quad r = 7 \text{ cm.}$$

$$F_{RV} = 0.010 + 0.500/7 = 0.0814$$

or $f_{RV}' = 1/0.0814 = 12.3$ cm $=$ image distance.

15. Fringes formed by edge waves from edges of wire (§ 13–9). Equivalent to two slit sources of separation $d =$ diameter of wire. Fringe spacing $= x = \lambda D/d$ (Equation 7–2).

$$\therefore \quad x = 5.893 \times 10^{-5} \times 20/0.01 = 0.118 \text{ cm.}$$

CHAPTERS XIV–XVI

1. See §§ 14–8 and 14–12. Actual value obtained for M depends on expression used for limit of resolution of O.G., on assumed limit of resolution of eye, and on wavelength. 2′ actually corresponds to 0.015 cm at 25 cm. Taking limit of resolution of O.G. to be $0.5\lambda/(\text{N.A.})$ and $\lambda = 5000$ Å, equation 14–17 becomes

$$M = \frac{0.015 \times (\text{N.A.})}{0.5 \times 5 \times 10^{-5}} = 600(\text{N.A.})$$

(14–16) then gives, for power of eyepiece

$$\left(\frac{D_v}{f_E'}\right) = \frac{480 \times 0.3}{16} = 9.$$

2. See § 14–8. The numerical values in this question were chosen to be realistic and also to simplify the arithmetic if one takes the limit of resolution of an O.G. to be $0.61\lambda/(\text{N.A.})$.
(14–16) and (14–17) give

$$f_E' = \frac{0.61g\,D_v\lambda}{0.01 f_o'(\text{N.A.})},$$

giving $f_E' = 5$ cm.

Semi-angle of cone of light leaving O.G. is small and given by Sine Relation (14–4): $u' = (\text{N.A.})/m$ where m is magnification of O.G. $= g/f_o'$. This gives $u' = 0.018$. This is also semi-angle of cone of light entering eyepiece. Hence diameter of exit pupil $= 2 \times 0.018 \times f_E' = 0.18$ cm.

3. For first part see §§ 14–11, 14–16. Second part draws attention to the completely different contexts in which the term *resolving power* is often used. (In this book the term *limit of resolution* has been used in the image formation context.) For an image-forming system such as a microscope one is concerned with the detail in the object that can be reproduced in the image;

this can be expressed as a distance in the object plane or as a spatial frequency. For a spectroscopic instrument one is concerned with *chromatic resolving power*, i.e. the smallest resolvable difference (in wavelength, wavenumber, or frequency) between neighbouring spectral lines. (See §§ 11–6 and **14–5**, 6, 7, 11, 12, 17 for limit of resolution and §§ **11–5**, **8–11**, **9–3**, 8, 9; **12–6**, 17; **18–12** for chromatic resolving power of various kinds of spectrometer.

4. See §§ 9–3, 11–5, 11–6, 14–5, 14–9, 15–14.

Angular limit of resolution of O.G. $= 1.22\lambda/30$ radian.

Angular limit of resolution of eye $= 1.5\pi/(60 \times 180)$ radian.

$$\text{Necessary Magnifying power} = M = \frac{1.5 \times \pi \times 30}{60 \times 180 \times 1.22 \times \lambda} = \frac{f_o'}{f_E'}$$

$$\therefore \quad f_E' = \frac{450 \times 60 \times 180 \times 1.22 \times \lambda}{1.5 \times \pi \times 30}.$$

Taking $\lambda = 5 \times 10^{-5}$ cm yields $f_E' = 2.1$ cm.

Eyepiece power $= D_v/f_E' = 11.9 \sim 12$, taking $D_v = 25$ cm.

(Note that since criteria for resolution are not precise it is pointless to give results with high precision.)

The limit of resolution of the O.G. is then utilised. It is

$$\frac{1.22 \times 5 \times 10^{-5}}{30} \times \frac{180 \times 60 \times 60}{\pi} = 0.42 \text{ sec.}$$

$$\text{Depth of focus} = \pm \frac{\lambda}{8 \sin^2\left(\dfrac{u'}{2}\right)} \qquad \text{where} \quad u' = \frac{15}{450} = \frac{1}{30}$$

$$= \pm 0.0225 \text{ cm.} \sim 1/5 \text{th mm.}$$

5. See §§ 15–5; 14–3, 15–11. If there is no spherical aberration and the Sine Condition is satisfied for all zones of the aperture, the system is well corrected for a small field. A microscope O.G. is a wide aperture lens having this state of correction. The other aberrations are less important since they depend on powers of the field higher than the first.

6. The context implies that *resolving power* here means *limit of resolution*. See §§ 11–6; 14–5, 6, 7, 11, 12, 17; 15–14.

Angular limit of resolution of O.G. $= 1.22\lambda/a = 1.22 \times 5 \times 10^{-5}/5 = 1.22 \times 10^{-5}$ radian.

Angular limit of resolution of eye $= 1' = 2.91 \times 10^{-4}$ radian.

Maximum value of f_E' is that which gives magnifying power $= M = f_o'/f_E' =$ resolution of eye $\div$ resolution of O.G.

$$\therefore \quad f_E' = 150 \times 1.22 \times 10^{-5}/(2.91 \times 10^{-4}) = 6.29 \sim 6.3 \text{ cm.}$$

Last part: see §§ 11–12, 11–13. Assume incoherent monochromatic source immediately behind distant slit. Initially see $\cos^2$ fringes. As W increased mean brightness level rises but fringe visibility decreases and reaches zero when W subtends λ/d. As W increased farther, fringes reappear with positions of maxima and minima interchanged. Visibility passes through a maximum and again falls to zero when W subtends $2\lambda/d$. Progressive increase of W gives fringe visibility rising and falling, being zero when W subtends $p\lambda/d$; positions of bright and dark fringes interchanged with each re-appearance. The visibility varies as $(\sin \alpha)/\alpha$ where $\alpha = \pi d\theta/\lambda$ where θ is angle subtended by W. Used (with distant slit replaced by star) in Michelson stellar interferometer.

7. See §16–11. From Sine Relation (14–4)—angular magnification is 100. Since $u = 1/400$, $\sin u' = 1/4$. The reliability of a geometrical discussion such as this is such that one may make the approximation $\tan u' = 1/4$. Depth of focus is then $\delta l' = \pm 4 \times 0.005 = \pm 0.02$ cm. Try using (2–11). This gives depth of field as δl conjugate to $\delta l'$ where $m^2 \delta l = \delta l'$ or $\delta l = \pm 200$ cm. This is too large to justify (2–11). Hence find actual object distances for image positions 0.02 cm on each side of original. Take object to left. Put $l = -400$ and $l'/l = -1/100$; $l' = +4$. This gives $f' = 400/101$ cm. If $l' = +4.02$ cm, $l = -2.7$ metres and if $l' = +3.98$ cm, $l = -8$ metres. ∴ Depth of field is from 2.7 metres to 8 metres. Note magnitude of error using approximate method.

8. $2° = 0.03491$ radian. $10° = 0.17453$ radian.
One can write $W' = Cu'^4$ and $\delta l' = 4Cu'^2$ (§15–6)

$$\therefore \quad \delta l' = 4W'/u'^2$$

(a) $\quad \delta l' = \dfrac{4 \times 5 \times 10^{-4}}{(0.03491)^2} = 1.64$ mm.

(b) $\quad \delta l' = \dfrac{4 \times 5 \times 10^{-4}}{(0.17453)^2} = 0.066$ mm.

9. From (16–13)

$$\frac{1}{30} + \frac{1}{s'} = \frac{2 \cos 10°}{10}, \text{ whence } s' = 6.1118 \text{ cm.}$$

From (16–15)

$$\frac{1}{30} + \frac{1}{t'} = \frac{2}{10 \cos 10°}, \text{ whence } t' = 5.8888 \text{ cm.}$$

$$s' - t' = 0.223 \text{ cm.}$$

(Note the rather unsatisfactory way in which a small distance is obtained as the difference between two relatively large distances.)

10. (16–8) gives, for single thin lens,

$$\frac{1}{R'} - \frac{1}{R} = \frac{1}{nf'}$$

Here $R = \infty$. ∴ $R' = nf' = 15.17$ cm.
$\delta f' = f' \sin^2 u_{pr}' = 10 \sin^2 5° = 0.0760$ cm.
(16–12) gives, for the doublet,

$$P = \frac{1}{R'} = \frac{0.1}{24.6} \left[\frac{60.6}{1.5169} - \frac{36.0}{1.62258} \right]$$

whence $R' = 13.85$.

For a thin lens at the stop the astigmatic difference of focus depends only upon the focal length and the field, i.e. is the same for the doublet as for the single lens.

11. (a) Chromatic correction lost. Power of complete lens increased because power of diverging component smaller if crown used instead of flint with same surface curvatures.

(b) Paraxial properties unaffected. Spherical aberration and coma correction lost. Astigmatism, field curvature and distortion unchanged (§16–5—stop at lens).

(c) Chromatic correction lost—at second lens incidence height of ray in axial pencil changed. Angles of incidence at second lens changed and spherical aberration correction lost.

CHAPTERS XVII and XVIII

1. One has $L'_C(C) \equiv 1(R) + 10(G) + 0.1(B)$
   and $\quad L_W'(W) \equiv 1(R) + 4.6(G) + 0.06(B)$
   $\therefore$ Tristimulus coefficients for C are

$$\frac{1}{1} = 1, \quad \frac{10}{4.6} = 2.174, \quad \frac{0.1}{0.06} = 1.667$$

Sum of tristimulus coefficients $= 4.841$.

Unit trichromatic coefficients are tristimulus coefficients $\div 4.841$, so that unit trichromatic equation is

$$1.0(C) \equiv 0.207(R) + 0.449(G) + 0.344(B).$$

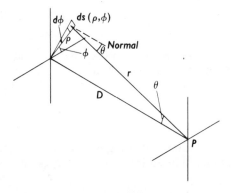

Fig. 2

2. Perpendicular separation of planes $= D$. Equation (18–7) gives: illumination at P due to $ds = dE = \dfrac{L ds \cos^2 \theta}{r^2}$. From the figure $ds = \varrho\,d\varrho\,d\phi$ and $\varrho = D \tan \theta$.

Hence $d\varrho = D \sec^2 \theta\, d\theta$ and $ds \cos^2 \theta = D^2 \tan \theta\, d\theta d\phi$. But $D = r \cos \theta$

$$\therefore \quad E = \iint dE = \int_{\phi=0}^{2\pi} \int_{\theta=0}^{\pi/2} L \sin \theta \cos \theta\, d\phi\, d\theta.$$
$$= \pi L.$$

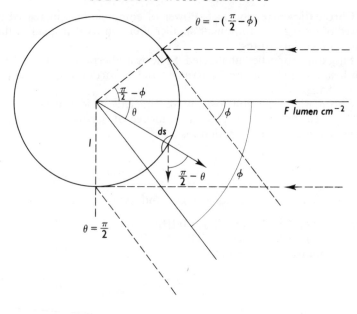

Fig. 3

3. First part: See §18–3. Second part: Apparent brightness $\propto$ luminance L at point observed. Luminous flux incident on element ds at θ (see Fig.) = $F\,ds\cos\theta$ = Total flux leaving $= \pi L\,ds$.

$$\therefore \quad \text{Luminance of } ds = L = \frac{F\,ds\cos\theta}{\pi\,ds} = \frac{F\cos\theta}{\pi}$$

This gives apparent brightness between $\theta = \pi/2$ and $\theta = -(\pi/2-\phi)$ (No light reaches points beyond $\theta = \pi/2$, and $\theta = -(\pi/2-\phi)$ defines edge of cylinder as seen by observer.) Luminous intensity of ds in direction $\phi = \pi/2$

is $L\,ds\cos\left(\dfrac{\pi}{2}-\theta\right) = \dfrac{F\,ds\sin\theta\cos\theta}{\pi}$ = flux diffused into unit solid angle

by ds. When $\phi = \pi/2$ illuminated region of cylinder extends from $\theta = 0$ to $\theta = \pi/2$. $\therefore$ Total flux diffused into unit solid angle at $\phi = \pi/2$ is

$$\int_{\theta=0}^{\pi/2} \int_{l=0}^{l} \frac{F\sin\theta\cos\theta\,ds}{\pi} \text{ where } ds = (l\,d\theta)dl.$$

This gives $F\,l^2/2\pi$.

4. Squares of denominators increase by factor of approximately 2 at each step. Hence area of entrance pupil and illumination of image reduced by factor of approximately 2.

5. First part: See §18–6. Second part: (a) If slit wide: no improvement because more light spread over proportionally greater image area. (b) If slit narrow enough for image width to be determined by diffraction: gain inversely proportional to focal length of collimator.

6. If luminous intensity of source is I, total flux emitted $=4\pi I$. From (18–22), illumination on surface of sphere of radius r due to reflected light $=$ $\dfrac{4\pi I}{4\pi r^2}\cdot\dfrac{\rho}{1-\rho}$. Illumination due to direct light $=I/r^2$.

$$\therefore\quad \text{Total illumination} = \frac{I}{r^2}\left(\frac{\rho}{1-\rho}+1\right) = \frac{I}{r^2}\left(\frac{1}{1-\rho}\right).$$

$$\therefore\quad \frac{E_1}{E_2} = \frac{150^2\times 0\cdot 1}{100^2\times 0\cdot 2} = 1\cdot 125$$

CHAPTERS XIX–XXI

1. First part: See §19–4. Second part: If E is in e.s.u., displacement current density in e.m.u. is $\dfrac{1}{4\pi c}\dfrac{\partial E}{\partial t}$. For plane polarized plane wave of frequency v

$$E=E_o\sin\frac{2\pi v}{c}(ct-x).$$

$$\therefore\quad \left(\frac{\partial E}{\partial t}\right)_{max} = 2\pi v E_o$$

$$\therefore\quad \text{r.m.s. displacement current density} = \frac{v E_o}{2c\sqrt{2}}$$

10^{-9} amp cm$^{-2}=10^{-10}$ e.m.u. cm^{-2}. $v=10^8$ sec.$^{-1}$

$$\therefore\quad 10^{-10} = \frac{10^8\,E_o}{3\times 10^{10}\times 2\sqrt{2}}$$

whence $E_o=6\sqrt{2}\times 10^{-8}=8\cdot 485\times 10^{-8}$ e.s.u.

$$H_o \text{ in e.m.u.} = E_o \text{ in e.s.u.}$$

2. First part: See §§19–4, 19–5. Second part: $K=\mu=1$. Mean value of magnitude of Poynting vector $=\dfrac{c}{8\pi}E_o^2$ erg sec.$^{-1}$ cm.$^{-2}$

$$\frac{c}{8\pi}E_o^2 = 0\cdot 132\times 10^7$$

whence $E_o=0\cdot 0333$ e.s.u.: r.m.s. value $=0\cdot 0235$ e.s.u.

3. See §19–2 for concept of displacement current.

$$D=KE=KE_o\sin\frac{2\pi v}{c}(ct-x).$$

Displacement current max. (e.s.u.) $=\dfrac{1}{4\pi}\left(\dfrac{\partial D}{\partial t}\right)_{max} = \dfrac{KE_o v}{2}$

Conduction current max. (e.s.u.)$=\sigma E_o$.

Equate to get conductivity $=\sigma=\dfrac{Kv}{2}=\dfrac{5\times 10^6}{2}=2\cdot 5\times 10^6$ e.s.u.

(20–3) gave $A=A_o\exp-(2\pi/\lambda)n\kappa x$ where λ is the vacuum wavelength and from (20–44) and (20–46) one has $2n^2\kappa^2=\sqrt{(K^2+4\sigma^2 T^2)}-K$, where

$T = 1/v$. This yields $n\kappa = 1 \cdot 018$ which gives $A = A_o \exp(-0 \cdot 00021)$. But $\exp(x) = 1 + x$ if x is small. Hence amplitude decreases at rate of $0 \cdot 021 \%$ per cm.

4. First part: See §19–4. Second part: (21–30) gives ratio of amplitudes.

$$\frac{\text{reflected intensity}}{\text{incident intensity}} = \left(\frac{n' - n}{n' + n}\right)^2 = \left(\frac{1 \cdot 42}{3 \cdot 42}\right)^2 = 0 \cdot 172$$

5. Angle of refraction, I', given by $\sin I' = (\sin I)/n$: $I' = 27°43'$. For component parallel to plane of incidence, equation (21–26) gives $A_{2l} = A_1 \tan 17°17'/\tan 72°43' = 0 \cdot 0968 \, A_1$. For component perpendicular to plane of incidence, (21–27) gives $A_{2r} = A_1 \sin 17°17'/\sin 72°43' = 0 \cdot 311 \, A_1$.

$$\frac{I_r}{I_l} = \left(\frac{A_r}{A_l}\right)^2 = \left(\frac{0 \cdot 311}{0 \cdot 0968}\right)^2 = 10 \cdot 3.$$

6. (21–40) yields the possible values of I at each surface. Put $\tan \frac{1}{2}\delta = T$ and $n/n' = n$ and manipulate (21–40) to obtain a quadratic in $\sin^2 I$ yielding

$$\sin^2 I = \frac{(n^2 + 1) \pm \sqrt{(n^2 + 1) - 4(T^2 + 1)n^2}}{2(T^2 + 1)n^2}$$

One requires $\delta = 45°$. Inserting $T = \tan 22\frac{1}{2}°$ and $n = 1 \cdot 52$, one obtains

$$I = 55°\,27' \quad \text{or} \quad 47°\,33'.$$

7. See §21–8. Here $I_p = 72°\,30'$ and $\psi_p = 43°\,30'$. (21–55) gives $\kappa = \tan 87° = 19 \cdot 1$. (21–56) gives $n = \tan 72°\,30' \sin 72°\,30' \cos 87° = 0 \cdot 158$. (21–49) then gives $R = 0 \cdot 94$.

CHAPTERS XXII and XXIII

1. When o.p.d. between O and E disturbances is an integral number of wavelengths the polarization of the light emerging from the quartz plate is the same as that of the incident light and there is extinction. The condition is o.p.d. $= d\delta n = p\lambda$ where $d = 0 \cdot 5$ mm and $\delta n = 0 \cdot 009$. Take long wavelength end of visible region to be $\lambda = 8000 \, \text{Å} = 8 \times 10^{-4}$ mm. If this is inserted into $d \, \delta n = p\lambda$ one obtains $p = 5 \cdot 6$. In fact p must be an integer. Hence minimum value of p yielding λ in visible region is 6. Calculate values of λ for $p \geq 6$. The resulting wavelengths in the visible region are 7500 Å, 6429 Å, 5625 Å, 5000 Å, 4500 Å, 4091 Å. $p > 11$ yields wavelengths beyond the visible region.

2. First part: See §22–17, equation (22–3). For plate between crossed Nicols $(\theta - \phi) = \pi/2$ and first term is zero. Second term is $a^2[\cos 2(\theta + \phi) - \cos 2(\theta - \phi)] \sin^2 (\delta/2) = a^2[\cos 2(\theta + \phi) + 1] \sin^2 (\delta/2) = a^2(1 - \cos 4\theta) \sin^2 (\delta/2) = 2a^2 \sin^2 2\theta \sin^2 (\delta/2)$. As plate rotated intensity is zero at $\theta = 0$ and $\pi/2$ (plane polarized light from first Nicol simply passes straight through crystal, and second Nicol obviously gives extinction). Between $\theta = 0$ and $\pi/2$ intensity $\propto \sin^2 2\theta$, i.e. passes through a maximum at $\theta = \pi/4$.

Second part: Emergent light same as incident light (and hence extinction) when quartz introduces phase difference $p \times 2\pi$. I.e. dark fringes at wedge thicknesses that satisfy o.p.d. $= d\,\delta n = p\lambda$. For adjacent fringes wedge thicknesses differ by $d_1 - d_2 = \lambda/\delta n$. If fringe separation $= x$, $(d_1 - d_2)/x = 1/50$. Hence $x = 50\,\lambda/\delta n = 50 \times 6 \times 10^{-5}/0\cdot009 = \frac{1}{3}$ cm.

3. First part: See §6–18. Second part: See §§22–2, 22–12. Cut crystal parallel to axis to obtain $\lambda/4$ plate and pass light normally into plate polarized at 45° to axis. Plate thickness required $= d = \lambda/4(n_o - n_E) = 5\cdot89 \times 10^{-4}/4 \times 0\cdot172 = 8\cdot56 \times 10^{-4}$ mm. Plates with thicknesses $5d, 9d, 13d, \ldots$ give same emergent light; thicknesses $3d, 7d, 11d, \ldots$ give circularly polarized light with opposite direction of rotation.

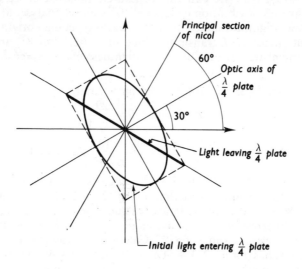

Fig. 4

4. To get extinction $\lambda/4$ plate converts elliptically into plane polarized. Hence axes of ellipse must be parallel to axes of plate (§22–12) and resulting plane of vibration perpendicular to principal section of Nicol. Angle between plane of vibration and optic axis $= 60°$. ∴ ratio of axes of ellipse $= \tan 60° = 1\cdot732$, and minor axis at 30° to horizontal.

5. First part: See §§22–5, 22–7. Second part: Since intention was to produce circularly polarized, principal section of Nicol at 45° to axis of plate. For λ 5893 Å, thickness of $\lambda/4$ plate $= d = \lambda/4(n_E - n_o)$. For λ 5460 Å this introduces phase difference $\delta = (2\pi/\lambda)d(n_E - n_o)$. $(n_E - n_o)$ is same in both cases

$$\therefore \quad \delta = \frac{2\pi}{5460} \times \frac{5893}{4} \text{ radian} = 97\cdot14°.$$

Emergent light of λ 5460 Å has perpendicular components whose amplitudes are equal and whose phase difference $= \delta$. (6–57) shows that, *whatever the value of δ, $\theta = 45°$ when $a_x = a_y$.* Alternatively, from first principles, if x and y are axes of plate (6–56), with $a_x = a_y = a$, gives ellipse for emergent light. Change to new axes X and Y at $45°$ to x and y. $y = (Y - X)/\sqrt{2}$ and $x = (Y + X)/\sqrt{2}$. This yields

$$\frac{X^2}{\left(\dfrac{a^2 \sin^2 \delta}{1 + \cos \delta}\right)} + \frac{Y^2}{\left(\dfrac{a^2 \sin^2 \delta}{1 - \cos \delta}\right)} = 1$$

This is equation of ellipse referred to its own axes, which proves that emergent light has axes at $45°$ to axes of plate.

$$\text{Ratio of axes} = \left(\frac{1 - \cos \delta}{1 + \cos \delta}\right)^{1/2}$$
$$= \tan (\delta/2) = \tan 48 \cdot 57° = 1 \cdot 13.$$

6. For $\lambda/4$ and $\lambda/2$ plate see §22–13. For Laurent polarimeter see §23–8. After $\lambda/2$ plate planes of vibration inclined at $10°$. Can set analyser at $85°$ to each and get low brightness match or at $5°$ to each to get high brightness match. If incident intensity $= I$, transmitted intensities are $I \cos^2 85°$ and $I \cos^2 5°$. Ratio of intensities for high and low brightness matches $= \cos^2 5°/\cos^2 85° = 131$.

7. First part: See §§23–3, 23–4. Second part: rotation $= \theta = \pi \, d \, \delta n/\lambda$ radian.

$$\text{For } d = 1 \text{ mm} \quad \theta = \frac{29 \cdot 73 \times \pi}{180}$$

$$\therefore \quad \delta n = \frac{29 \cdot 73 \times \pi}{180} \times \frac{5 \cdot 086 \times 10^{-4}}{\pi \times 1} = 8 \cdot 4 \times 10^{-5}.$$

8. Rotation $= \theta = (-2 + 8 \cdot 2 \times 10^8/\lambda^2) \times 20$ degrees (λ in Å). If $\theta = p \times 180°$ (p integral) there will be extinction. Put $(-2 + 8 \cdot 2 \times 10^8/\lambda^2) \times 20 = p \times 180°$, insert $\lambda = 8000$ Å (extreme red), and calculate p. One obtains $p = 1 \cdot 2$. This means $p = 1$ corresponds to a wavelength beyond red end of visible region, and smallest relevant value of p is 2. Calculate wavelengths in visible region for $p \geqslant 2$. One has (for $p = 2, 3, 4, 5$) $\lambda = 6400$ Å, 5320 Å, 4650 Å, 4180 Å. These are the wavelengths in the visible region that are not transmitted. ($p = 6$ gives λ beyond violet end of visible region.)

9. Rotation for solution in min $= 6 \cdot 6 \times 20m \times 60$ where m is concentration in grams per cc.
Faraday rotation in min $= 0 \cdot 08 \times 25 \times H$ where H is field in oersted.

$$H = \frac{4\pi \times 5000}{25} \times \frac{0 \cdot 96}{10} \text{ oersted.}$$

$$6 \cdot 6 \times 20 \times 60 \times m = \frac{0 \cdot 08 \times 25 \times 4\pi \times 5000 \times 0 \cdot 96}{25 \times 10}$$

$$m = \frac{0 \cdot 08 \times 25 \times 4\pi \times 5000 \times 0 \cdot 96}{25 \times 10 \times 6 \cdot 6 \times 20 \times 60} = 0 \cdot 061 \text{ gram per cc.}$$

10. Travelling parallel to axis in active crystal: remains plane polarized but plane of vibration rotates through angle $\propto$ distance travelled. Travelling perpendicular to axis in non-active crystal: state of polarization changes (§22–12).
I Active crystal. Half-way through, rotation simply $45°$. As plane of vibration of incident rotated get identical rotation at every point along path.
II Inactive crystal. Plane polarized emergent light rotated $90°$ corresponds

to $\lambda/2$ plate with axis at 45° to vibration direction of incident light (Fig. 22–26). Incident then resolved into equal components in phase. Half-way through corresponds to $\lambda/4$ plate, components have phase difference $\pi/2$, and resultant is circularly polarized. As incident plane of vibration rotated emergent remains plane polarized with plane of vibration at 2θ to incident where θ is inclination to optic axis. Half-way through light passes though elliptical states with axes parallel to axes of plate and ratio $=\tan\theta$, i.e. circularly polarized when $\theta=\pm45°$ and plane when $\theta=0$, or 90°.

11. Take fast axis of first plate to be horizontal (Fig. 22–26). If (b) phase change for combination zero and emergent same as incident. If (a) combination is $\lambda/2$ plate and emergent light plane polarized with plane of vibration at 30° to horizontal in direction opposite to incident light. After first plate light elliptically polarized, clockwise, major axis horizontal, ratio of axes $=$ $\tan 60° = 1·732$.

INDEX

Abbe, E., 326, 406, 410, 418
  compensator, 94
  condenser, 411
  principle for resolution, 338–40
  refractometer, 93–4
  theory of the microscope, 334–8, 355
Abelès, F., 608, 611
Aberration, chromatic, 52, **61–5**, **402–4**, 409–410, 419–21
  of the entrance pupil, 452
  function, 379–82
  high order, 409, 410, 412, 414
  of light, 589
  longitudinal chromatic, 61–5, 402–4, 409–410
  ray, 376–7, 382–7, 392
  relation, ray to wavefront, 377, 382–3
  spherical, **382–3**, 392–4, 397–8, **399–400**, **404–5**, 408–16, **419–20**; of principal rays, 419
  tolerance for, 390–1
  transverse chromatic, 61, **64–5**, 410, 418
  wavefront, 376–87, 419–20
  zonal, 382, 409, 410
Aberrations, balancing of, 389, 404–5, 409, 411–12
  classification of, 382
  of a concave grating, 284, 288
  dependence on stop position, 393–4
  of a doublet, 402–6
  of the eye, 426, 432
  factors influencing, 392–3
  of a mirror, 406–7, 419–20
  monochromatic, Chapters XV and XVI; chromatic variations of, 410
  occurrence and control of, Chapter XVI
  primary, 382–7
  of a single surface, 395–9
  in spectrometers, 98, 285–6, 288
  of a thin lens, 399–402
Absent orders, 247–9, 274–5
Absolute photometry (radiometry), 455–8
Absolute wavelength measurement, 166, 301, 468
Absorbing medium, 503, 507, 509–10
  dispersion by, 507–8, 510–11
  reflection by, 511, **522–6**
  refraction into, 526–7
Absorption, 491–8
  bands, 492, 495, 498, 500–2, 505–8
  Beer's law of, 493
  coefficient, 493
  by conductors, 491, **509–11**, 524
  e.m. theory of, 497–8
  by gases, 492, 495, 498
  general, **491–2**, 509–11, 524
  index, **493**, 507
  by insulators, 492, 495, 497–8, 500, 522–4, 526–7
  Lambert's law of, 492–3
  lines, 492, 495
  by liquids, 492, 495, 498, 506
  measurement, 472–3
  by metals, 491, **509–11**, 524
  quantum theory of, 495–6

selective, **491–2**, 494–6, 510–11, 523–4
  of polarized light, 542–4
  by solids, 491–2, 495–6
  spectrophotometer, 301–2, 463–4
  spectrum, 492
Acceptance of spectrometer, 466
Accommodation, 55, **425**
  amplitude of, 426
Achromatic doublet, **62–4**, 66, **402–6**, 408, 412
  Chevalier, 412
  Fraunhofer, 408
  Gauss, 408, 415
  Grubb, 412
  monochromatic aberrations, 404–6, 412
  secondary spectrum, 402–4
  Steinheil, 408
Achromatic fringes, 145
  with Lloyd's mirror, 144
Achromatic lines, 559, 561
Achromatic triplet, 404, 408
Achromatism with separated lenses, 64–5
Achromatism of visual instruments, 404
  of photographic lenses, 404
Achromatized Ramsden eyepiece, 418–19
Acoustic grating, 276, 598–9, 602
Active fibres, 375
Activity, optical, Chapter XXIII
  of crystals, 574
  Fresnel's theory, 575–7
  of liquids, 580–1
  of quartz, 577–9
Acuity, contour, 433
  vernier, 433
  visual, 432–3
Adaptation, 428–9
Addition of disturbances, 101–5
Additive colour mixture, 433–8
Aerial survey lens, 417
Aether, 1, 603
  displacement current, 476
  drag, 603
After-image, 429
Ahrens prism, **541**, 561
Air, dispersion of, 200, 201
  refractive index of, 179, 199
Airy disc, **244–5**, 321, 326, 333
Airy pattern, **241–5**, 321, 326, 327, 331, 333, 338, 355
Airy, G. B., on wave surface of quartz, 578
Alignment by auto-collimator, 91–2
All-positive diagram, 9
Aluminium reflecting films, 177
Amici-Bertrand lens, 557, **561**
Amici objectives, 409
  prisms, **86**, 94
  reflecting microscope, 70
Ammann, E. O., 609
α-mono-bromnaphthalene, 96
Ampere's law, 476
Amplitude, 2
  complex, **104**, 221, 238, 243, 249, 257, 315
Amplitude spectroscopy, 166
Amplitudes, vector addition of, 103
Analogue computer, optical, 260

Analyser, **131**, **541**, 542, 581–5
Analysis, Fourier, 114–18
  of Gaussian profile, 119–20
  of sharply limited wave train, 118–20
Analysis of polarized light, 552–5
Analytic signal, 108
Anastigmat, 393, 414
Anders, H., 605
Anderson, W. C., 591–2, 593
Angle-gauge interferometer, 173
Angle of incidence, principal, 524
Angle of minimum deviation, grating, 266
  prism, 82–3, 88
Angle, polarizing, 132–3, 517, 518, 519
Angle of prism, 82, 88
Angular dispersion, gratings, 267–9, 276, 286
  prism, 84–5
Angular limit of resolution of telescope, 327–328
Angular magnification, **20**, 56
Anisotropic media, 6, 482–90, Chapters XXII
  and XXIII
  biaxial, 486–9, 544–6
  electric polarization, 482–3
  optic axis, **486**, 488–9, 534–7, 544–5
  phase velocity, 484–6, 535–6, 544–6
  plane waves in, **484–6**, 489, 534–7, 544–6
  ray axis, **488**, 489–90, 544–6
  ray velocity, 485, 486–8, 535, 544–6
  uniaxial, 486, 489–90, 533–9
  wave surface, 486–9, 533–7, 544–6
  wave velocity, 484–6, 535–6, 544–6
Annular zones, 216–20, 306–11
Anomalous dispersion, 499, 500–3, 507–8
  rotatory, 574
  sodium vapour, 502–3
Anterior chamber, 424
Antinode, **128**, 528
Anti-reflecting films, **154–5**, 529
Anti-Stokes lines, 496
Aperture, camera, 52–3, 451–2
  circular, 220, 241–5, 307–8
  double slit, 229–30, 246–51, 253–4
  function, 258, 273, 336
  numerical, 328–31, 335, 338–40, 408–10
  rectangular, 227–9, 232–41, 257–9, 311–16
  318–19
Aperture stop, 52
  effect of position on aberrations, 393–4
Aperture of a telescope, 328
Aplanat, 393, 394, 405, 406
Aplanatic achromatic doublet, 405, 406
Aplanatic points, 397, 398, 400
Aplanatic refraction, 397–9
  in microscope objective, 409, 410
Apochromatic correction, 404
  in microscope objective, 409–10, 411
Apodization, 164, 259, 291, 333, 359
Apostilb, 457
Apparent luminance, 452–3
Apparent luminous intensity, 453
Apparent size of an object, 54–5
Aqueous humour, 424
Arago, D. F., 220, 574
Arago-Fresnel laws, 133–4
Arens, H., 605
Argument, 104
Armstrong, J. A., 609
Arragonite, 547
Aslakson, C. I., **593**, 600

Astigmatic correction, 389, 412
Astigmatic focal lines, 386
Astigmatism, 382, **385–6**, 389, 393
  of a concave grating, 284, 288, 407
  of a concave mirror, 406–7
  of a doublet, 405
  of the eye, 427
  high order, 414
  of a single surface, 397, 398
  of a thin lens, 401–2
Astronomical telescope, 37–8, **55**, 69, 420–1
  range of, 453–4
Astrophysics, I.R., 469
Asymmetric mode of operation of interfero-
  meter, 166
Atomic beam source, 202
Auerbach, R., 605
Auto-collimation, 77, 78, 87, 91–2, 97
Auto-collimator, 91–2
Auto-correlation function, 109, 162
Automatic lathe, 305
Automatic lens design, 422–3
Axial magnification, 22
Axis, fast, 548
Axis of an optical system, 17
Axis, optic, **486**, 488–9, 534–7, 544–5
  ray, **488**, 489–90, 544–5
  slow, 548
  visual, 426
Azimuth angle, 380
Azimuth, principal, 525–6

Babinet compensator, 553–5
Babinet's principle, **230–1**, 245
Backwave, 219, 224
Baird, K. M., 201, 611
Baker, B. B., 605
Baker, J. G., 421
Ballard, S. S., 608
Band spectra, 595–6, 600
Banded spectrum, 203–4
Barakat, R., 609
Barnes, K. R., 605
Barrel distortion, 387
Barrell, H., and Sears, J. E., determination of
  Metre, 196–210
Barrier layer cell, 461
Batarchukova, N. R., 201
Bates, W. J., 172
Bauer, G., 605
Bausch, C. L., 289, 296
Beam splitter, 164–5, 469, 522
Beats with two lasers, 112
Beck binocular microscope, 443
Beck dark-ground illuminator, 346
Beer's law, 493
Bell, E. E., 609
Bellingham and Stanley refractometer, 93
Bennett, A. H., 605
Bennett, H. E., 595
Bennett, J. M., 595
Bennett, W. R., 497
Benoît, J. R., and Michelson, A. A., determin-
  ation of Metre, 174, 197, 201
Benoît, J. R., Fabry, C., and Perot, A., deter-
  mination of Metre, 197, 200–1
Benzene, 96
Beran, M. J., 605
Bergstrand, E., 596–9, 600, 602, 609
Bertin's surfaces, 559

Bertrand lens, 557, 561–2
Best focus, with astigmatism, 386, 389
  with spherical aberration, 383
Beutler, H. G., 282
Biaxial crystals, 486–9, 544–8, 559–61, 574
  double refraction in, 533, 544
  external conical refraction, 547–8
  internal conical refraction, 546–7
  normal surface, 546
  optic axes, 486, 488, 544–6, 559–61
  phase velocity, 484–6, 488, 544–6
  plane waves in, 544–6
  ray axes, 488, 544–6, 547
  ray velocity, 485–8, 544–6
  wave surface, 486–9, 544–6
  wave velocity surface, 546
Bibliography, 605
Billet split lens, 143
Billmeyer, F. W., 609
Binocular microscope, 443–4
Binocular telescope, 442
  prismatic, 68–9, 442
Binocular vision, 439–41
Biot, J. B., 542, 574, 580, 581
Biplate, 143
Biprism, 140–2
Biquartz, 582–3, 585
Birefringence, 533–7, 544
  in calcite, 483–5
Birge, R. T., 590, **592**, 593, 600
Black body radiator, 576–8
Blaine, L. R., 595
Blaze angle, 277
Blazed gratings, 276–8, 295, 300
Blind spot, 425
Blindness, colour, 438
  night, 428
Bloom, A. L., 609
Blooming, 155
Blue colour of sky, 494–5
Body colour, **491**, 523
Bol, K., 594, 600
Bolometer, 462
  superconducting, 302
Booth, L. B., 416, 417
Born, M., 605
Bouma, P. J., 605
Bound charges, 497–8, 504–5, 510–11, 524
Boundary conditions, **225**, **512–14**, 521–2
Boundary waves, **231**, 317–18, 319, 345
Bousquet, P., 605, 611
Boutry, G. A., 605
Bouwers, A., 421, 605
Bouwkamp, C. J., 609
Braddick, H. J. J., 609
Bradley, D. J., 194, 609
Bradley, J., 588, 589
Bragg, W. L., 303–4
  equation, 303
Bravais points, 38
Breadth of interference fringes, 176–8, 180–1
  of spectral lines, 118–21, 160–1, 163
Bremmer, H., 609
Brewster, D., 410, 573
Brewster window, 497
Brewster's fringes, **189–92**, 197, 200
Brewster's law, 132–3, 516, 519
Brightness, 430–1, 445–6, 448
  of diffuse radiator, 449
  of image, 450–4

Broadening of spectral lines, 118–21, 160–1
Brodhun-Lummer photometer, 458–60
Bronchoscope, fibre, 374
Brouwer, W., 605, 612
Brown, D., 172
Brown, E. B., 605
Bruce, C. F., 195
Brushes and rings, 558–61
Bryngdahl, O., 609
Buchdahl, H. A., 605
Burch, J. M., 609
Burckhardt, C. B., 605
Butler and Edser fringes, 203–4, 209, 213

C.I.E. system, 437
Cadmium wavelength standard, 197, 199–202
Cagnet, M., 605
Calcite, double refraction in, 537–9
  fundamental crystal form of, 537
  group velocity in, 602
  refractive indices of, 536
Calcium fluoride, 96
Camera lens, 57, 411–18
  achromatic doublet, 412–13
  aperture, 52–3, 451–2
  chromatic correction of, 414
  image formation, 354–9
  image illumination, 451–2
  inverted telephoto, 417
  meniscus, 411–12
  Petzval type, 416
  symmetrical type, 413–14
  telephoto, 57, 58, 417–18
  triplet type, 415–16
  wide-angle type, 416–17
  zoom, 421–2
Camera, Schmidt, 419–20
Candela, 456
Candle, international, 455
Candle power, 446
  mean spherical, 446
Candler, C., 605
Carbon bisulphide, 96, 587, 601–2
Carbon dioxide, 572
Carbon dioxide laser, 111, 497, 601
Carbon monoxide, 595
Cardinal points, 21
  of combination of two systems, 26–30, 40–1
  of complicated systems, 39, 41, 43–6
  location of, 39, 76–8
  numerical example, 47–51
  of single refracting surface, 30–1
  from system matrix, 43–6
  of telescopic system, 25
  of thick lens, 33–6, 40–1
  of thin lens, 32–3
  of two separated thin lenses, 36–8
Cassegrain telescope, 69, 420–1
Catadioptric systems, 346, 349, 419–21
Cauchy's dispersion formula, 498–500, 502
Caulfield, H. J., 605
Cavity resonator, 588, **593–4**, 600
Celor lens, 414, 417
Central coma for thin lens, 400–1
Centre, optical, 34–5
Centre spot scanning, 192–5
Chamberlain, J. E., 166
Characteristic matrix, 531
Charge density, 475

Charges, bound, 497–8, 504–5, 510–11, 524
   free, 498, 510–11, 524
Chasmar, R. P., 608
Chief ray, 53
Choroid, 425
Chromatic aberration, 52, 61–5, 402–4, 409,
      420–1
   correction of, 62–5, 402–4, 409–10, 420–1
   longitudinal, 61–5
   secondary spectrum, 63, **403**
   of thin lens, 62, 402
   transverse, 61, 64–5, 410, 418
Chromatic difference of focal length, 62
Chromatic difference of magnification, 61,
      64–5
Chromatic resolving power, 181, 240
   comparison of instruments, 298–301
   of an echelon 279, 299, 301
   of Fabry-Perot instruments, **180–4**, 193–4,
      196, 299, 301
   of Fourier transform spectrometer, 163–4,
      300, 301
   of grating, 124, **269–70**, 286, 294, 298, 299,
      300
   of Lummer plate, 205, 299
   of prism, 125, 239–40
Chromatic variation of monochromatic aber-
      ration, 410
Chromaticity diagram, 437
Ciliary muscle, 425
Cine-anastigmat lens, 414
Ciné projector, 58
Cinnabar, 574
Circular aperture, 220
   Fraunhofer pattern, 241–5
   Fresnel pattern, 307–8
Circular dichroism, 586
Circular division of wavefront, 216–20, 306–
      311
Circular obstacle, 220, 307–8
Circularly polarized light, **135**, 521, 550, 551,
      552, 553, 563–4, 567, 569–70, 571
   as component of plane polarized, 570,
      575–7, 578
   production of, 135, 521, 551
   resolution into plane components, 135, 551,
      552
Cittert, van, –Zernike, theorem, 253
Clad fibres, 371–2
Clarke, D., 605
Classification of interference and diffraction
      phenomena, 137
Classification of diffraction patterns, 235–7
Cleavage of calcite, 537
   of mica, 551
Coated surfaces, 154–5, 177, 210
Coding, with fibre bundles, 374
Coefficients, trichromatic, 436
Coherence, 105–9
   degree of, **110**, 253, 254, 566
   with a laser, 111–13, 496–7
   lateral, **108**, 109–10, 251–4
   length, **107**, 108, 119–21
   longitudinal, **108**, 109, 119–21
   mutual, 108
   partial, **106–10**, 136, 251–4, 565–6
   patch, 253
   self, 109
   spatial, **108**, 109–10, 251–4
   temporal, **108**, 109, 119–21

   time, **107**, 108, 119–21
   transverse, **108**, 109–10, 251–4
Coherency matrix, 562, 565–7, 572
Coherent background, 340, 342, 360–1, 362
Coherent disturbances, 106, 107, 566
   addition of, 101–5
Coherent scattering, 495
Coincidence methods, object-image, 71–3, 75,
      78
Coincidences, method of, 160, 184–5, 269
Collier, R. J., 605
Collimator, 76, 87–8, 96, 407
   of spectrometer, adjustment of, 87–8
Collimation, auto-, 77, 78, 87, 91–2, 97
Collimator, auto-, 91–2
Collinear transformation, 17, 30
Collisions, elastic, 495
   inelastic, 496
Colorimeter, 438
Colour, 433–9
   blindness, 438
   body, **491**, 523
   complementary, 435, 437, 575
   desaturated, 434
   equation, 436–7
   filter, interference, 211–13
   measurement, 435–8
   mixture, additive, 433–8; subtractive, 439
   of opaque bodies, **438–9**, **491–2**, 523
   pigment, 438–9, 491–2, 523
   rendering properties of sources, 439
   saturated, 434, 437
   of sky, 494–5
   of sunset, 495
   surface, **492**, 523
   of thin films, 153
   of tobacco smoke, 494
   by transmission, 491, 492
   vision, 433–8
Colours, interference, 153
Coma, 382, **384–5**, 407, 409, 411–13, 419
   dependence on stop position, 393–4
   and the Sine Condition, 388
   of a single refracting surface, 397–8
   of a thin doublet, 404–5, 406
   of a thin lens, 400–1
Combination of disturbances, 101–10
   of prisms, 85–6
   of refracting surfaces, 11–12, 31–6, 39–41
   of thin lenses in contact, 14, 40, 62–3, 71,
      402–6
   of two separated thin lenses, 36–8, 64–6, 72
   of two systems, 26–30, 40–1; in air, 30
Common path interferometer, 172
Communications using lasers, 113
Comparator, Dowell's, 173
   Kösters', 172–3
   N.P.L., 172, 173, 201–2
Compensating eyepiece, 410, 418
Compensating plate in interferometer, 156,
      171
Compensating saccharimeter, 585–6
Compensator, Babinet, 553–5
   Jamin, 170
   Rayleigh, 255–6
   Soleil, 553
Complementary colour, 435, 437, 575
Complementary diffracting screens, **230**, 245
Complex amplitude, 104, 221, 238, 243, 249,
      257, 315

Complex degree of coherence, **110**, 566
Complex quantities, use in wave theory, 103–5
Complex refractive index, **503**, 504–5, 507, 509–10, 522–3, 525–7
Compound Fabry-Perot interferometer, 203, 211–12
Compound microscope, 60–1, 67–8, 330
    as a diverging system, 27
Computer, optical analogue, 260
Computerised lens design, 422–3
Concave grating, 281–5, 286–8, 299–300
    astigmatism of, 284, 288, 407
Concave mirror, 15–16, 419–20
    astigmatism of, 406–7
    correction of aberrations of, 419–20
    determination of radius of, 72, 74–6
    testing of, 369
Concentric system, 420
Condenser, projector, 58–9
    substage, **67**, 333–4, 339, **345–7**, **411**
Conductors, dispersion of, 510–11
    e.m. waves in, 498, 509–11
    reflection by, 511, 524–6
Cones, **428**, 429, 431, 432, 435
Confocal etalon, 195–6, 497
Conical fibres, 372–3
Conical refraction, external, 547–8
    internal, 546–7
Conjugate distance parameter, 399
Conjugate distances, reference to any conjugate planes, 24
    reference to focal planes, 18–19
    reference to nodal planes, 23–4
    reference to principal planes, 22–3, 46–7
Conjugate, Hermitian, 566
Conjugate planes and points, 17–18, 43–5
    location of, 18–19, 22–4, 43–5, 71–3
Conjugate rays, 17–18
    location of, 19
Conner, W. S., 595
Connes, J., 163, 166, 469
Connes, P., 166, 195, 290, 469
Conoscope, 557, 559, 561–2
Conservation of energy, 102, 140, 480
Constant deviation spectrograph, 97
Constringence, 62
Constructive interference, 102
Contact lens, 427
Continuity of charge, equation of, 430
Continuous laser, 111, 497
Continuous spectrum, 83, 121, 160, 163, 241
Contour acuity, 433
Contrast transfer factor, 356–7
Convex mirror, determination of radius, 72, 74–5, 76
Cook, A. H., 605
Co-ordinates, optical, 258, 356
Co-ordinates, trichromatic, 436
Copson, E. T., 605
Cornea, 424
Cornu, M. A., 590
Cornu-Jellett polarizing prism, 583
Cornu prism, 579–80
Cornu's spiral, 229, **313–14**, 315–20
Correlation function, auto- (self-), 109, 162
    cross-, 108
Cos^2 fringes, **140**, **157**, 247–8, 250
Cos^4 law of illumination, 452
Cosine law of emission, 448
    of illumination, 447

Cosine transform, 118, 162
Cotton effect, 586
Cotton-Mouton effect, 573
Cowan, M., 595
Cox, A., 606
Crawford-Stiles effect, **433**, 449, 452, 453, 455, 457
Critical angle refractometry, 89–91, 92–4
Critical illumination, 333–4
Cross-correlation function, 108
Crossed Nicols, 540
Crown glass, 63, 403, 406, 491, 499
Cryolite, 155, 211, 212
Crystal plate in collimated beam, 548–50, 555–7
    in convergent light, 557–61
Crystallography, X-ray, 303
Crystals, active, 574–9
    biaxial, 486–9, 544–8, 559–61, 574
    double refraction by, 533–9, 544
    e.m. waves in, 484–90
    electric polarization in, 482–3
    optic axes of, **486**, 488–9, 534–7, 544–6
    refractive indices of, 483, 535–6
    uniaxial, 486, 489–90, 533–9
    wave surface, **486–9**, 533–7, 544–6
Crystalline lens, 424
Curl of a vector, 475
Current aether displacement, 476
    conduction, 475, 476, 509–10
    displacement, 476
    polarization, 476
Curvature of field, 382, **386–7**, 389, 393, 411, 415
    of a concave mirror, 419
    of a doublet, 405–6
    of a single surface, 395–6
    of a thin lens, 402
Curvature of prism spectrum line, 83–4
Curvature of a surface, 12
Cylindrical lens, 232, 427
    wave, 232
    wave envelope, **138**, 140, 142, 143, **232**, 238–9, 315–16
Cystoscope, fibre, 374
Czerny, M., 98, 285

Dagor-lens, 414
Dallmeyer, J. H., 413
Dallon lens, 417
Dale and Gladstone law, 500, 506
Damping, natural, of atomic oscillator, 161, 498, 504
Dark adaptation, 428–9
Dark-ground illuminator, 345–7
Dark-ground microscopy, 338, **340–7**, 369
Dayhoff, 594
DCL, 596
De Velis, J. B., 606
Debye, P., 304
Degree of coherence, 110, 252–3, 566
Degree of polarization, 518
Delano, E., 609
Delouis, H., 163
Dense barium crown glass, 406
Density, optical, 472
Depolarization by scattering, 498
Depth of field, 332–3
Depth of focus, 331–3
Derivation of formulae, 10

Desaturated colours, 434
Design, automatic lens, 422–3
Destructive interference, 102–3
Detector, absolute, 457–8
Detector, infra-red, 301–2, 469
Detector noise, 462, 467–8, 469
Detectors for physical photometry, 461–2, 469
Deviation, constant, spectroscope, 97
Deviation method of tracing rays, 38–9
Deviation method for two systems, 40–1
Deviation, minimum, by a prism, 82–3, 88
Deviation of a ray by a lens, 38, 40, 415
  by a prism, 82–3
  by a single refracting surface, 5, 392–3
  by a train of prisms, 85–6
Dextro-rotatory substances, 574
Dichroism, 542–4
  circular, 586
Dichromatism, 494
Dickson, J. Home, 607
Dielectric constant, 477, 479, 482–3, 503–6, 509–10
  constants, principal, 483
Dielectric ellipsoid, 483
Dielectric films, 154–5, 210–13, 529–32
Dielectrics, dispersion of, 498–503, 504–9
  e.m. waves in, 476–90, 503–8
  reflection by, 512–26
Differential shearing microscope, 350–1
Diffracting screens, complementary, **230**, 245
Diffraction, 137, Chapters X–XIII
  by a circular aperture, 220, 241–5, 308
  by a circular obstacle, 220, 307–8
Diffraction, classification of patterns, 225–7
  distinction from interference, 137
Diffraction by a double slit, 229–30, 246–51, 253–4, 320
Diffraction, edge waves, 231, 317–18, 319, 345
Diffraction formula, Kirchhoff's, **224–5**, 237, 243, 249, 314
Diffraction fringes, 230, 236, 244, 247, 263, 308, 316, 318, 319
Diffraction grating, Chapter XII
  absent orders, 274–5
  arbitrarily shaped elements, 274
  blazed, 276–8, 295, 300
  concave, 281–5, 299–300; astigmatism of, 284, 288, 407; mountings for, 286–8
  dispersion of, 267–8, 276, 279, 286
  echelle, 278, 288–90, 299
  echelette, 278, 295
  echelon, 276–81, 300–1
  effect on a pulse, 124
  equation, 262, 265–6, 279, 282, 302–3
  Fraunhofer, 261, 274
  ghosts, 297–8
  groove form, 274–9, 295–7
  holographic, 297 ,
  immersion, 270, 285
  infra-red, 278; 296
  intensity distribution for narrow slits, 271–3; for wide slits, 273–6
  maxima positions, 262–6, 272, 274–5, 279, 282–4, 302–3
  minima positions, 262–5, 272, 274–5, 280
  minimum deviation, 266
  missing orders, 274–5
  mountings for, 280, 285–91
  narrow apertures, 261–73
  oblique incidence, 265–6, 268, 270, 298, 299

order, 262, 264, 279, 299, 300
photometric efficiency of, 286, 288, 291, 298, 299, 300, 466, 467, 469
plane, **261–81**, 285–6, 288–91, 299, 300, 304–5
principal maxima, 262–76, 279–80, 282–4, 302–3
production of, 274, 276–9, 294–7
radial, 305
reflection, 266, 269, 270, 276–92
replicas, 274, 296, 297
resolving power, **269–70**, 276, 278, 279, 294, 295, 298, 299, 300
ruled, 274, 276, 294–7
secondary maxima, 264, 272–3, 275
sharpness of principal maxima, 262–5, 269–70, 298
sine wave, 355–6
spectral range, **269**, 279, 287, 288–9, 299
spectrum, dispersion in, **267–8**, 276, 279, 287; formation of, 266–7; ghosts in, 297–8; missing orders in, 274–5
square wave, 114, 273–6, 334–8
three-dimensional, 303–4
transmission, 361–78, 296
two-dimensional, 302
ultrasonic, 276, 598–9, 600, 602
wavelength measurement, 268, 288
wide slits, 273–6
Diffraction gratings, crossed, 304–5
Diffraction gratings in series, 285–6, 289–90
Diffraction by irregular obstacle, 220
  by opaque strip, 319–20
Diffraction in optical instruments, 226, 239–41, 257–60, 326–48, 353–9
Diffraction pattern, far-field, 228
Diffraction patterns, classification of, 225–7
  Fraunhofer, Chapter XI
  Fresnel, Chapter XIII
  with white light, 249
Diffraction by rectangular aperture, 232–9, 311–16
  by single slit, 227–9, 238–9, 311–16
  by straight edge, 316–8
Diffraction theory of the microscope, 328–31, 333–48, 355
Diffraction, X-ray, 303–4
Diffractometer, optical, 304
Diffusing surface, 448–9
  brightness of, 449
  flux radiated by, 449–50
Dioptre, 12
Dipole, 497, 504
Dipping refractometer, 94
Direct product, 566
Direct vision spectroscope, 86
Disc, Airy, **244–5**, 321, 326, 333
Disc of least confusion, 383, 386
Disc, opaque, diffraction by, 220, 307–8
Dispersion, 61–2, 122–3, 491, 498–511
  of air, 200, 201
  anomalous, 499, 500–3, 507–8; of sodium vapour, 502–3
  Cauchy formula, 499–500, 502
  of Fabry-Perot, 184–5, 194
  of a gas, 500, 506, 508
  Gladstone and Dale formula, 500, 506
  of a grating, 267–8, 276, 279, 286, 287, 289–90
Hartmann formula, 500

interferometric measurement, 166
of a lens, 61–2
Lorenz-Lorentz formula, 506
mean, 404
and molecular scattering, 508–9
normal, 498–500, 506–7
partial, 404
of a prism, 84–5
of a prism train, 85–6
in a region of absorption, 500–1, 507–8
relative partial, 404
rotatory, crystals, 574; liquids, 581, 585;
    quartz, 574–5, 581, 585
Sellmeier formula, 501, 506
theory for dielectrics, 503–9
theory for metals, 510–11
Dispersive power, 62, 63, 86
Displacement current, 476
Displacements, interferometric measurement
    of, 156, 173–4
Distinct vision, least distance of, 58, 59,
    329–30, **426**
Distortion, **387**, 388–9
dependence on stop position, 393, 394
effect on image illumination, 452
of a single surface, 397, 398
of a thin lens, 402
Ditchburn, R. W., 606
Divergence of a vector, 475
Dominant wavelength, 434
Doppler effect, **128–30**, 161
Dorsey, N. E., 592, 593, 609
Double beam spectrophotometer, 463–4
Double-Gauss lenses, 415
Double image prism, Rochon, 542
Wollaston, 542
Double mirror, Fresnel, 142
Double refraction, 533–9, 544
in calcite, 537–9
electric, 573
magnetic, 573
Double slit, Fraunhofer pattern, 229–30,
    246–51, 253–7
with double source, 249–50
with wide source, 250–1, 253–4
Double slit, Fresnel pattern, 229–30, 320
Young's experiment, 138–40
Double stars, **130**, 249–50, 253
Doublet, achromatic, **62–4**, 66, 402–6, 407–8,
    412–13
Chevalier, 412–13
Fraunhofer, 408
Gauss, 408
Grubb, 412
monochromatic aberrations, **404–6**, 412
secondary spectrum, 403–4
Steinheil, 408
Douglas, A. E., 595
Dowell, J. H., 173
Draugard, R., 612
Drew, R. L., 172
Drude, P., 528
Duplication, image, 350–1
axial, 351–2
Dye laser, 497
Dynamic scanning of fibres, 374
Dyson interference microscope, 349–50

Eagle mounting, 287–8
Eaglesfield, C. C., 606

Eastman, D. P., 596
Ebert grating mounting, 285–6
Echelle, 278, 288–90, 299, 300
mountings for, 288–90
Echelette, 278, 295
Echelon, 276–81, 300–1
Eclipse, 588–9
Edge, R. C. A., 598, 600
Edge, straight, diffraction at, 316–18
Edge waves, **231**, 317–18, 319, 345
Edlén, B., 201
Edser and Butler fringes, 203–4, 209, 213
Efficiency, relative luminous, 430–1, 433
Einstein, A., 497, 604
Elastic collisions, 495
Elastic solid theory of light, 100, 135
Electric displacement, 475
Electric double refraction, 573
Electromagnetic spectrum, 479
Electromagnetic theory, Chapters XIX–XXI
of absorption and scattering, 497–8
of dispersion, 503–11
of reflection and refraction, Chapter XXI
Electromagnetic waves, in absorbing media,
    503, 509–10
in anisotropic media, 484–90
energy flux, 480–1, 482, 485
Hertz's experiments, 481–2
in isotropic dielectrics, 476–81
in metals, 509–11
velocity of, 477–9, 483–90, 592–3
Electron microscope, 329, 361–2
scanning, 353
Electron probe micro-analyser, 353
Electrons, bound, 497–8, 504–5, 510–11, 524
free, 498, 510–11, 524
Ellipsoid, Fresnel, 483, 484
Index (or dielectric), 483
Tensor, 483
Elliptically polarized light, **135–6**, 521,
    549–55, 571–2
analysis of, 552–5
as component of plane, 578–9
production of, 135–6, 520–1, 524, 549–52
resolution into plane components, 135,
    549–51, 555
Emissivity, 524
Emittance, luminous or radiant, 450
Empty magnification, 329
End gauge testing, 155–6, 168, 172–3, 198–9
Endoscope, fibre, 374
Energy, conservation of, 102, 140, 480
Energy in e.m. field, 479–81, 482, 485
Energy flux, 2, 445–6, 480–1, 482, 485
Energy of S.H. oscillator, 2
Engelhard, E., 201
Ennos, A. E., 609
Entrance pupil, 53
Entrance window, 54
Episcope, 59
Equivalent focal length, 28
Equivalent lens, 28, 35
Erecting eyepiece, 68
Erecting prisms, 68–9
Eriometer, Young's, 245–6
Errors in gratings, 297–8
Esselbach, 203
Essen, L., **593–4**, 600, 606
Etalon, Fabry-Perot, 179–204, 211, 299, 300,
    301

all-glass, 204, 205
centre spot scanning, 192–5
confocal, 195–6, 497
dispersion, 184–5, 194
field widened, 196, 467
as filter, 211
finesse, 181–3, 194, 299
in metrology, 197–202
photometric efficiency, 193, 195–6, 299, 466–7
as refractometer, 179, 200
resolving power, 180–4, 193–4, 196, 299, 300, 301, 465
spectral range, 183–4, 194–5, 299
spherical, 195–6, 467
Etalons in series, 189–92, 203, 212
Etendue of a spectrometer, 466
Evacuated pipe experiment, 590
Exact fractions, method of, 185–9, 197
Exit pupil, 53
Exit window, 54
Exitance, luminous or radiant, 450
Exposure, photographic, 451
    meter, 462
External conical refraction, 547–8
Extinction coefficient, 493, 503, 507–8, 509–10
Extraordinary index, 536
Extraordinary rays and waves, 533–9
Eye, 424
    aberrations of, 426, 427, 432
    clearance, 56
    lens, 65
    lens of, 424
    movements, 433
    optical system of, 425–6
    paraxial properties of, 54–5
    point, 56
    pupil, 424, 428
    refractive errors, 427
    resolving power, 428, 432–3
    schematic, 425–6
    spectral sensitivity, 430–1
    structure of, 424–5
Eyepiece, 55, 60–1, 64, 65–7, 418–19
    achromatized Ramsden, 418–19
    compensating, 410, **418**
    erecting, 68
    Gauss, 75, **87**
    Huygens, 37, **65–8**, 418
    Kellner, 418–19
    magnification, **61**, 330
    micrometer, 67
    orthoscopic, 418
    power of, 61, 330
    Ramsden, **65–8**, 418–19
    scale, 67

Fabry, C., Perot, A., and Benoît, J. R., determination of Metre, 197, 201
Fabry-Perot etalon, 179–204, 211, 299, 300, 301
    all glass, 204, 205
    centre spot scanning, 192–5
    confocal, 195–6, 497
    dispersion, 184–5, 194
    field widened, 196, 467
    as filter, 211
    finesse, 181–3, 194, 299
    in metrology, 197–202
    photometric efficiency, 193, 195–6, 299, 466–7
    as refractometer, 179, 200
    resolving power, 180–4, 193–4, 196, 299, 300, 301, 465
    spectral range, 183–4, 194–5, 299
    spherical, 195–6, 467
Fabry-Perot fringes, 175–9, 213
Fabry-Perot instruments in series, 189–92, 203, 467
Fabry-Perot interferometer, 179–204
    oscillating, 192–5
    See also Fabry-Perot etalon
False detail in a microscope, 338
Fans, sagittal, 386
    reflection of, 406–7
Fans, tangential, 385–6
    reflection of, 406–7
Far point, 426
Faraday compensator, 586
Faraday effect, 587
Far-field pattern, 228
Fast axis, 548
Fastie, W. G., 285
FECO fringes, 208–10, 214
Feed-back in lasers, 496
Fellgett, P., 166, 464
Fermat's principle, 6–7, 321–2
Fibre optics, 370–5
Finesse, **181–3**, 194, 299
Field curvature, 382, **386–7**, 389–90, 393, 411
    of a concave mirror, 419
    of a doublet, **405–6**, 412
    of a single surface, 395–6
    of a thin lens, **402**, 415
Field flattener, 416
Field lens, 66–7, 397, 416, 419
Field stop, 53
Field widened spectrometer, 196, 467, 468
Figuring of surfaces, 167–8
Films, dielectric, 154–5, 210–13, 529–32
    highly reflecting dielectric, 2, 10–12
    highly reflecting metal, 177
    multilayer, 155, 210–12, 529–32
    non-reflecting, 154–5, 210, 531
    quarter-wave, 154–5, 210–2, 531–2
    thin, 150–3, 522
    wedge, 150–1, 158
Filter, frustrated total reflection, **213**, 522
    interference, 210–13, 467, 532
    spatial frequency, 353–4
Finite wavefront, 257–60
Finite wave trains, 105–9, **118–21**, 161
Fiorentini, A., 609
FitzGerald-Lorentz contraction, 604
Fizeau fringes, 150–3, 155–6
    multiple beam, 205–8
Fizeau, H., on velocity of light, 588, 589–90
    in a moving medium, 602–3
Flat field microscope objective, 411
Flicker photometer, 460–1
Flicker, sensitivity to, 433
Flint glass, 63, 403, 404, 406, 499, 587
Florman, E. F., 590
Fluorite, 96, 409, 410, 421, 491, 499, 524
Fluorescence, 496
Flux, energy, 2, 445–6, 480–1, 482, 485
    luminous, 431, 445–6; unit of, 456
    radiated from a source, 446, 447–50
    radiated from a uniform diffuser, 448–50
Flying spot microscope, 352–3

F-number, 451
Focal length, 18
  back, 30
  chromatic difference of, 62
  of combination of two systems, 28, 41; in air, 30
  of combination of two thin lenses, 36–7, 40
  of complicated systems, 39, 46
  equivalent, 28
  first, 18
  front, 30
  measurement, auto-collimation method, 72, 77; goniometer method, 80; magnification method, 79
  second, 18
  of a single surface, 10, 31
  of a spherical lens, 35
  from system matrix, 43–6
  of a thick lens, 33, 41
  of a thin lens, 12, 32–3
Focal lines, 386
  of a concave mirror, 406–7
  of a thin lens, 401–2
Focal plane, 18, 44
  from system matrix, 43–6
Focon, 373
Focus, best, with spherical aberration, 383
  with astigmatism, 386
Focus, choice of, effect on aberrations, 377–9, 381
Focus, principal, 18
Foot-candle, 457
Foot-lambert, 457
Forbes, G., 590
Formulae, derivation of, 10
Foucault knife-edge test, 75, 77, **369–70**
Foucault prism, 542
Foucault, L., on velocity of light, 588, 590
  in material medium, 601
Fourier, J. B. J., 114
Fourier analysis, 114–18
  of finite wavefront, 257–9
  of Gaussian profile, 119–20
  of sharply limited wave train, 118–20
  of top hat function, 259
Fourier integral, 116–18
Fourier methods in image theory, 335–8, 353–9
Fourier series, 114–16; in diffraction theory, 273, 336
Fourier theorem, 114
Fourier transform, complex, 117; cosine, 118, 162; in diffraction theory, 258–60, 273, 336–7, 353–4; spectroscopy, 161–6, 299, 300, 301, 464, 466, 468–9
Fovea centralis, **425**, 428, 432, 433
Fowles, G. R., 606
Fractions, method of exact, 185–9, 197
Françon, M., 350, 605, 606, 609, 611
Fraunhofer, I. von, 226
Fraunhofer diffraction gratings, 261, 274
Fraunhofer diffraction pattern, 226, Chapter XI
  circular aperture, 241–5
  double slit, 229–39, 246–9
  as Fourier transform, 258–9
  as image, 226, 233, 237, 241, 321
  in microscope, 334–8
  rectangular aperture, 227–9, 232–8, 257–9
  single slit, 227–9, 238, 257–9

Fraunhofer doublet, 408
Free electrons, 498, 510–11, 524
French, 410
Frequency, 3
  distribution, 118–22
  response, spatial, 356–9, 390–1
  spatial, 117, 258–9, 336, 337, 353–9; filtering, 353–4
Fresnel, A. J., 215
Fresnel-Arago laws, 133–4
Fresnel biprism, 140–2
Fresnel diffraction pattern, 226, Chapter XIII
  circular aperture, 220, **308**
  circular obstacle, 220, **307–8**
  double slit, 229–30, **320**
  opaque strip, 319–20
  rectangular apertures, 227–9, **311–16**, 318–19
  single slit, 227–9, 311–16, 318–19
  straight edge, 316–18
Fresnel double mirror, 142
Fresnel equations for refraction, 515
  at normal incidence, 516
Fresnel half-period zones, plane wave, 306–7
  spherical wave, 216–20, 307–8, 312
Fresnel-Huygens principle, 215–19
  defects, 219
  successes, 220
Fresnel integrals, 313, 315
Fresnel on optical activity, 575–7
Fresnel prism train, 577
Fresnel rhomb, 521
Friei, H. G., 612
Fringes, achromatic, 145; with Lloyd's mirror, 144–5
  biprism, 141
  Brewster, **189–92**, 197, 200
  cos², **140**, **157**, 247–8, 250
  diffraction, 230, 236, 244, 247, 263, 308, 316, 318, 319
  disappearance of, 249–51, 253–4
  displacement by inserting plate, 145–6
  Edser-Butler, 203–4, 213
  of equal chromatic order, 208–10, 214
  of equal inclination, 147–9, 157, 213
    multiple beam, 175–8, 204–5
  of equal thickness, 150–3, 155–6, 158, 213
    multiple beam, 205–8
  Fabry-Perot, 175–9, 213
  FECO, 208–10, 214
  Fizeau, 150–3, 155–6, 213
    multiple beam, 205–8
  Haidinger, **148**, 152, 175–8
  half-value width of, 180–1
  localized, **151–3**, 155–6, 158
    multiple beam, 205–8
  Michelson, 157–61, 213
  Newton's rings, 152–4, 213
  order of, **140**, 152, 157, 180–1, 184
  types of, 213–4
  visibility of, **159–61**, 177, 249–50, 251, 253–4
  wedge, 150–1, 158, 205–8
  with white light, 122, 140, **144–5**, **158**, 166, 174, **191**, 249
Froome, K. D., 594–5, 600, 606
Fry, G. A., 609
Fryer, P. A., 609
Frustrated total reflection, 522
  filter, 212–13

Full radiator, 445–6
Funnel, light, 372–3
Fused silica, 499, 536, 580
Fused thoria, 455

Gabor, D., 359–62, 609
Galilean telescope, 37, 55–6
Garbuny, M., 606
Gas laser, 111–13, 211, 497
Gases, dispersion of, 500, 506, 508
    refractive index of, 500, 506, 508; measurement, 170, 179, 200, 255–7
Gates, J. W. C., 586
Gauges, interferometric tests, 155–6, 168, 172–3, 198–9
Gauss, K. F., 52
    doublet, 408
    eyepiece, 75, **87**
    flux law, 475*
    lens, double-, 415
    theorem, 480
Gaussian constants, 43
Gaussian image, definition, 43, 376
Gaussian spectral line profile, 119–20, 160, 163
Gaussian wave train, 119–20
Gebbie, H. A., 166, 298, 609
Gehrcke-Lummer interferometer, 204–5, 299, 300
General absorption, **491–2**, 509–10, 523
General equation of wave motion, 100, 221, 477
Geodimeter, 598, 600
Ghosts in spectra, 165, 297–9
Gibbs, J. E., 166
Girard, A., 291–2, 611
Gladstone and Dale law, 500, 506
Glan prism, 542
Glass, 500, 573, 603
    crowns, 63, 403, 406, 491, 499
    flints, 63, 403, 404, 406, 499, 587
Glasses, choice for doublets, 404–6
Glazebrook prism, 541
Gniadek, K., 609
Golay, M. J. E., 291
    cell, 462
Goldwasser, E. L., 606
Goniometer, Searle, 80
Goodman, J. W., 606, 609
Gordon-Smith, A. L., 593, 600
Gordy, W., 595, 596
Gouy, L. G., 602
Gradient of a vector, 475
Graham, C. H., 606
Grainger, J. F., 605
Graphical method of adding disturbances, 103
Grasp, light, 466
Grass in grating spectra, 298
Grating, diffraction, Chapter XII
    absent orders, 274–5
    arbitrarily shaped elements, 274
    blazed, 276–8, 295, 300
    concave, 281–5, 299–300; astigmatism of, 284, 288, 407; mountings for, 286–8
    dispersion of, 267–8, 276, 279, 286
    echelle, 278, 288–90, 299
    echelette, 278, 295
    echelon, 276–81, 300–1
    effect on a pulse, 124
    equation, 262, 265–6, 279, 282, 302–3

Fraunhofer, 261, 274
    ghosts, 297–8
    groove form, 274–9, 285–7
    holographic, 297 ,
    immersion, 270, 285
    infra-red, 278, 296
    intensity distribution for narrow slits, 271–3; for wide slits, 273–6
    maxima positions, 262–6, 272, 274–5, 279, 282–4, 302–3
    minima positions, 262–5, 272, 274–5, 280
    minimum deviation, 266
    missing orders, 274–5
    mountings for, 280, 285–91
    narrow apertures, 261–73
    oblique incidence, 265–6, 268, 270, 298, 299
    order, 262, 264, 279, 299, 300
    photometric efficiency of, 286, 288, 291, 298, 299, 300, 466, 467, 469
    plane, **261–81**, 285–6, 288–91, 299, 300, 304–5
    principal maxima, 262–76, 279–80, 282–4, 302–3
    production of, 274, 276–9, 294–7
    radial, 305
    reflection, 266, 269, 270, 276–92
    replicas, 274, 296, 297
    resolving power, **269–70**, 276, 278, 279, 294, 295, 298, 299, 300
    ruled, 274, 276, 294–7
    secondary maxima, 264, 272–3, 275
    sharpness of principal maxima, 262–5, 269–70, 298
    sine wave, 355–6
    spectral range, **269**, 279, 287, 288–9, 299
    spectrum, dispersion in, **267–8**
        formation of, 266–7; ghosts in, 297–8; missing orders in, 274–5
    square wave, 114, 273–6, 334–8
    three-dimensional, 303–4
    transmission, 261–78, 296
    two-dimensional, 302
    ultrasonic, 276, 298–9, 600, 602
    wavelength measurement, 268, 288
    wide slits, 273–6
Grating, lamellar, 165, 467
Grating monochromator, 286, 300
Grating spectrograph, 285–90, 299, 300
Grating spectrometer, 285–94, 298, 299, 300, 464–7, 469
Gratings, crossed, 304–5
Gratings in series, 285–6, 289–90
Green and Twyman interferometer, 161, **166–8**, 171, 173, 290, 296, 298, 348, 594
Green's theorem, **221**, 223, 225
Greenland, K. M., 605
Greenough microscope, 443
Gregorian telescope, 70, 421
Grey, D. S., 421
Grille spectrometer, 291–2, 299, 467
Grimaldi, F. M., 137
Groot, W. de., 605
Groove form in grating, 274–8, 295–7
Group of waves, 121, 122
Group velocity, **122–4**, 125, 592, 601–2
Guelachvili, G., 469
Guenther, A. H., 595
Guild, J., 606
    flicker photometer, 461
Gullstrand, A., 425

Gutton, C., 602

Hagen, E., 524
Haidinger, W., 586
  fringes, **148**, 152, 175–8
Half-period lunes, 314
Half-period zones on plane wave, 306–7
  on spherical wave, 216–20, 307–8
Half-shadow angle, 582–5
  variable, 584
Half-shadow devices, 582–5
Half-value width, 160, 180–1
Half-wave plate, 550–1, 583–4
Half-width, 160
Hamilton, W. R., 546
Hansen, W. W., 594, 600
Harcourt-Vernon pentane lamp, 455
Harrison, G. R., 278, 289–90, 295, 296
Hart, K. H., 201
Hartmann dispersion formula, 500
Hartshorn, L., 609
Havelock, T., 606
HC1, 596
HCN, 595
Heavens, O. S., 606, 610
Heffner lamp, 455
Helium, 63, 404
Helium-neon laser, 111, 211, 496, 497
Helmberger, M., 599, 600
Helmholtz, H. von, 438
  equation, **20**, 323, 466; for infinite conjugate, 24–5; paraxial form, 20, 31–2, 45
  formula for Huygens' principle, 222
  invariant 20–1, 53
Henvis, B. W., 611
Herapathite, 543
Hermitian conjugate, 566
Herriott, D. R., 497, 610
Herschelian telescope, 70, 421
Herschel's condition, 324–6
Hertz's experiments, **481–2**, 588
Herzberger, M., 606
High resolution spectroscopic instruments compared, 298–302, 464–70
Highly reflecting dielectric film, 210–12, 529–232
  metal films, 177, 207–8
Hilger-Chance refractometer, 93, **94–5**, 96
Hill, R. M., 195
H-ink, 543
Hoegh, E. von, 414, 417
Hologram, 360
Holography, 359–68
Hopkins, H. H., 109, 370, 422, 606, 610, 612
Hopkins, R. E., 373
Houstoun, R. A., **598–9**, 602
H-sheet polaroid, 543, 544
Hue, 433
Hull, A. W., 304
Hulst, H. C. Van de, 608
Hulthén, E., 285
Humour, aqueous and vitreous, 424
Hunt, F. R. W., 607
Hunt, R. W. G., 607
Hüttel, A., **591**, 593
Huygens, C., 215
Huygens' eyepiece, 37, **65–8**, 418
Huygens' Principle, 3–5, 215–24
  biaxial media, 533, 544–5
  Fresnel theory, 215–20

Helmholtz formula, 222
  Kirchhoff theory, 221–4
  uniaxial media, 489–90, 533–7
Hydrogen, 63, 404
Hyperfine structure, 185
Hypergon lens, 417
Hypermetropia, 427

Illuminance, 446
Illumination, 446–7
  cos^4 law, 452
  cosine law, 447
  critical, 333–4, 346
  of images, 450–2
  Köhler, **334**, 338, 346, 347
  necessary, 473
  photometer, 473
  retinal, 449
  units of, 456–7
Image assessment, 354–9, 390–1
Image brightness, 450–5
Image duplication, 350–1
Image formation, aberration theory, 376–421
  by a camera lens, 57–8, 354–9, 451–2
  coherent object, 334–45, 353–4
  by the eye, 54–5, 424–33
  Fourier methods, 333–8, 353–9
  geometrical theory, 7–70
  by holograph, 359–67
  incoherent object, 326–31, 354–9
  by a microscope, 59–61, 67–8, 328–31, 333–53
  paraxial theory, 11–81
  photometric theory, 450–4
  physical theory, 226, 241, 321–67
  by rays, 7–70
  by reconstructed wavefronts, 359–67
  by a telescope, 55–7, 66–7, 327–8, 452–4
Image as Fraunhofer pattern, **226**, 233, 241, 257–60, 321, 335
Image illumination, 450–2
Image luminance, 450–2
  apparent, 452–3
Image luminous intensity, apparent, 453–4
Image position specification, 19, 22–4, 399
Image storage and retrieval, 368
Imagery by reconstructed wavefronts, 359–67
Imaizumi, M., 201
Immersion objective, 410
Immersion refractometer, 94
Inaccessible object, measurement, 14
Incidence, principal angle of, 524–6
Inclination factor, **216**, 217–19, **224**, 225, 227, 236, 307, 311
Incoherent disturbances, 106, 566, 568
Incoherent scattering, 496
Index ellipsoid, 483
Index of refraction, **3**, 5, 9–10, 18, 31, **321**, 479, 498–502, 503–11, 526
  of biaxial crystals, 483
  for circular components, 576, 578
  complex, **503**, 504–5, 509, 522–3, 525–7, 571
  and dielectric constant, 479, 503
  extraordinary, 536
  of gases, 500, 506, 508; measurement, 170, 179, 200, 255–7
  of a lens, 72–3
  of a liquid, 506; measurement, 73, 89–95, 170, 255–7

of a metal, 509–11, 524, 526
ordinary, 535
of a solid, measurement, 88–96
of uniaxial crystals, 535–6
Indices, principal refractive, 483, 535–6
Inelastic collision, 496
Infinite wave trains, 105, 107, 118, 136
Infra-red astrophysics, 469
Infra-red detector, 301–2, 462, 468, 469
Infra-red photography, 495
Infra-red, prisms for, 96–7
Infra-red spectroscopy, 96–7, 164–6, 192, 278, 291, 292–4, 301–2, 467–9
    determination of c, 595–6, 600, 601
Instrumental profile (function) **164**, 181, 193, **239**, 269, 291, 294, 465
Integrating sphere, 471–2
Intensity, **2**, 480, 517
    from complex amplitude, 104–5
    luminous, 446; apparent, 453; unit of, 455–456
    mutual, 110
    of polarized light, 136
    radiant, 446
    of resultant, 102–5, 108, 110
Interference, 102
    colours, 153
    conditions for, 105–9
    constructive, 102
    destructive, 102–3
    distinction from diffraction, 137
    with division of amplitude, 137, 147–214
    with division of wavefront, 137–46
    filter, 210–13, 467, 532
Interference fringes, achromatic, 145; with Lloyd's mirror, 144–5
    biprism, 141
    Brewster, **189–92**, 197, 200
    $\cos^2$, **140**, **157**, 247–8, 250
    disappearance of, 249–51, 253–4
    displacement by inserting plate, 145–6
    Edser-Butler, 203–4, 213
    of equal chromatic order, 208–10, 214
    of equal inclination, 147–9, 157, 213
        multiple beam, 175–8, 204–5
    of equal thickness, 150–3, 155–6, 158, 213
        multiple beam, 205–8
    Fabry-Perot, 175–9, 213
    FECO, 208–10, 214
    Fizeau, 150–3, 155–6, 213
        multiple beam, 205–8
    Haidinger, **148**, 152, 175–8
    half-value width of, 180–1
    localized, **151–3**, 155–6, 158
        multiple beam, 205–8
    Michelson, 157–61, 213
    Newton's rings, 152–4, 213
    order of, **140**, 152, 157, 180–1, 184
    types of, 213–14
    visibility of, **159–61**, 177, 249–50, 251, 253–4
    wedge, 150–1, 158, 205–8
    with white light, 122, 140, **144–5**, **158**, 174, **191**, 249
Interference microscope, 341, 348–52
Interference in plane-parallel plate, 147–9, 175–6, 204–5
Interference of polarized light, 133–6
Interference spectroscopy, 160–6, 180–9, 192–196, 203–4, 298–301, 464–9

Interference in thick plates, 147–9
    in thin films, 150–6, 159
    of two beams, 137–74
    with two lasers, 112
Interferometer, common path, 172
Interferometer, Dowell, 173
    Fabry-Perot, 179–204; adjustment of, 179; all glass, 204, 205; centre spot scanning, 192–5; confocal, 195–6, 497; field widened, 196, 467; as filter, 211; finesse, 181–3, 194, 299; in metrology, 197–202; Mock, 292–4, 464; oscillating, 192–5; photometric efficiency, 193, 195–6, 299, 466–7; as refractometer, 179, 200; resolving power, 180–4, 193–4, 196, 299, 300, 301, 465; spectral range, 183–4, 194–5, 299; spherical, 195–6, 467; two in series, 189–192, 203, 467
    Gauge measuring, 155–6, 168, 172–3, 198–9
    Jamin, 170
    Kösters, 172–3
    Lummer-Gehrcke, 204–5, 299, 300; resolving power of, 205, 299; spectral range of, 205
    Mach-Zehnder, **170–1**, 172, 348
    Michelson, **156–8**, 161, 213, 290, 349, 594–5; adjustment of, 158–9, 167; application to metrology, 173–4, 197, 201; to structure of spectral lines, 159–61, 290; field widened, 468
    Michelson stellar, 253–4
    microwave, 594–5, 600
    Mock, 292–4, 464
    N.P.L., 172–3
    Rayleigh, 255–7
    Twyman and Green, 161, **166–8**, 171, 173, 290, 296, 298, 348, 594
    wavefront shearing, **171–2**, 348, 350–1
Interferometry, Chapters VII–IX
    holographic, 367
Internal conical refraction, 546–7
Internal reflection, total, 519–21
    disturbance in second medium, 521–2
International candle, 455
Inter-ocular distance, 440, 442
Inverted telephoto lens, 417
Invariant, Helmholtz, (Lagrange), 20–1, 53
Inverse square law, 2, 447
Inversion, population, 496
Iris, 54, **424**, 428
Iron, 587
Irradiance, 447
    of images, 450–2
Irregular obstacle, diffraction by, 220
Irregular solid, refractive index of, 95–6
Isocandela diagram, 470
Isochromatic line, 558–61, 573
    surface, 559–60
Isogyre, 559
Isoplanatism, 358–9
Isotope lamp, 201–2

Jacobsson, R., 610
Jacquinot, P., 166, 466, 467, 468, 610, 611
James, J. F., 606
Jamin, compensator, 170
    interferometer, 170
Javan, A., 497
Jellett-Cornu prism, 583
Jenkins, F. A., 606

Johnson, C. M., 595
Johnson, P. D., 288
Jones, F. E., 608
Jones, G., 596
Jones, R. C., 563
Jones matrix, 562–5, 569
Jones vector, 562–5, 567, 569
Judd, D. B., 606
Jupiter, 588–9
Jupnik, H., 605

Kapany, N. S., 370, 372, 373, 606
Karolus, A., 591, 596, 599, 600
Kartashev, A. I., 201
Kay, I. W., 607
Kellner eyepiece, 418–19
Kerker, M., 606
Kerr cell, 572, **590–1**, 596–8
Kerr constant, 572
Kerr effect, 572–3
Khintchine-Wiener theorem, 162
Kidger, M. J., 423
Kimmit, M. F., 606
Kinematograph projector, 58–9
Kingslake, R., 606
Kirchhoff, G., 220
Kirchhoff diffraction integral, 224–5; application to circular aperture, 243; to double slit, 249; to Fresnel patterns, 314; to rectangular aperture, 237
Kirchhoff formulation of Huygens' principle, 221–4
Kirchhoff law of emission and absorption, 524
Klein, J. A., 595
Klein, M. V., 607
Kline, M. K., 607
Kneubühl, F., 610
Knife-edge method for radius of mirror, 75–6
Knife-edge test, 369–70
Koch, J., 201
Kock, W. E., 607
Koester, C. J., 374
Köhler illumination, **334**, 338, 346, 347
Kösters, W., and Lampe, P., determination of Metre, 197, 201
Kösters interferometer, 172–3
Kottler, F., 610
Kuhn, H., 170, 610
Kronecker product, 566
Krug, W., 607
Kruithof, A. A., 605
Krypton wavelength standard, 201–2
K-sheet, 543–4

Laevo-rotatory substances, 574, 577
Lagrange's equation, 20, 466
Lambert, 457
Lambert radiator, 448
Lambert's cosine law of emission, 448
Lambert's law of absorption, 493
Lamellar grating, 163, 467
Lamp, photometric standard, 455
  wavelength standard, 201–2
Lampe, P., and Kösters, W., determination of Metre, 197, 201
Land, E. H., 543
Lankard, J., 497
Landscape lens, 412
Laporte, O., 608
Laser, 101, 111–13, 496–7

carbon dioxide, 111, 497
cavity, 1,11, 196, 211, 496–7
coherence properties, 111–13
  in communications, 113
continuous, 111–13, 496–7
dye, 497
fibre, 375
gas, 111–13, 211, 496–7
helium-neon, 111, 211, 496, 497
high power, 111, 113
  in holography, 363
liquid, 111
modes, 111
probe, 375
pulsed, 111, 113, 497
ruby, 111, 496, 497
solid state, 111
spectral line width, 111, 112, 113, 160–1
tunable, 301, 497
Lateral coherence, **108**, 109–10, 251–4
Lateral magnification, 18
  relation to angular, 20, 56–7
Latta, J. N., 610
Laue, M. von, 304
Laurent polarimeter, 583–4
Least distance of distinct vision, 58, 59, 329–30, **426**
Lee, H. W., 415, 418
Le Grand, Y., 607
Leith, E. N., 362, 363, 365, 368, 610
Leitz interference microscope, 348
Left- and right-handed optical activity, 574, 577
Lens, assessment, 354–9, 390–1
  cardinal points location, 76–8
  contact, 427
  crystalline, of eye, 424
  design, automatic, 422–3
  equivalent, 28, 35
  eye, 65
  field, **66–7**, 397, 416, 419
  focal length measurement, 72, 79–80
  frequency response, 356
  optical transfer function, 356
  perfect, 17, 52, 321, 323–4, 376; impossibility of, 323–4, 376
  performance, 354–9, 390–1
  radii of curvature measurement, 72–5
  spherical, 35
  testing, 155, 167–8, 171–2, 359
  thick, in air, 33–6; cardinal points of, 33–6, 41; focal length of, 33, 41; numerical example, 47–51; optical centre of, 34–5; power of, 33, 41
  thin, 11–12; aberrations of, 399–402; in air, 11–12, 32–3; astigmatism of, 401–2; cardinal points of, 32–3; coma of, 400–1; distortion of, 402; equivalent, 28, 35; field curvature of, 402; neutralization test for power, 71; power of, 12, 32; refractive index measurement, 72–3; smallest separation of object and real image, 12–13; spherical aberration of, 399–400
  Zoom, 421–2
  See also Camera, Eyepiece, Microscope, Telescope
Lenses, thin, in contact, 14, 40
  achromatic combination of, 62–3, 403–4
Lenses, choice between imperfect, 357

Lenses, two separated, 36–8
  achromatic combination, 64–5
Levi, L., 607, 610
Light adaptation, 429
Light flint glass, 406
Light grasp, 466
Light, monochromatic, 105, 107, 118, 136
Light pulse, 121
  passage through a grating, 124
  passage through a prism, 125
Light, quasi-monochromatic, 107, 108, 109–110, 118–21
Light vibration, effective, 131, **529**, 533–4
Lin, L. H., 605
Limit of resolution, 326–7, 337, 338–40, 356–7
  with circular aperture, 326–7
  of eye, 329, 330, 428, 432–3
  of microscope, 328–9, 338–40
  Rayleigh criterion, 326–7
  with rectangular aperture, 241
  of stellar interferometer, 253–4
  of telescope, 327–8
Line profile, spectral, 107, 112, 119–21, 160–1, 163
Line spectrum, 83, 121, 239, 266
Line width, spectral, 107, 112, 119–21, 160–1, 162–3
Linearly (plane) polarized light, 1, 131, 135, 478, 533, 563, 570, 571
Linfoot, E. H., 607
Linnik, W., 348
Lipson, H., 607, 610
Lipson, S. G., 607
Lippich polarizer, 584–5
Liquid, optical activity of, 580–1
  measurement of, 581–6
Liquid, refractive index measurement, 73, 89–95, 170, 255–7
Lissberger, P. H., 610
Lister objective, 408–9
Lithium fluoride, 96
Littrow mountings for gratings, 268, 269, 270, 285, 286, 287, 289, 299
Littrow prism spectrograph, 97–8
  quartz, 580
Lloyd, H., on conical refraction, 546
Lloyd's mirror, 142
  achromatic fringes, 144–5
Localised fringes, **151–3**, 158–9
  applications, 155–6, 173–4, 205–10
  multiple beam, 205–8
  with white light, 158, 208–10
Loewenstein, E. V., 610
Long sight, 427
Longitudinal chromatic aberration, 61–5, 402–4, 409–10
Longitudinal coherence, **108**, 109, 119–21
Longitudinal magnification, 22
Longitudinal shift of focus, 378, 381
Lorentz, H. A., 504
Lorentz-FitzGerald contraction, 604
Lorenz-Lorentz law, 506
Lukin, I. V., 594
Lumen, 456
Luminance, 447–8
  of an image, 450–1; apparent, 452–3
  of a uniform diffuser, 448–9
  unit of, 457
Luminosity, **430–1**, 436

of a spectrometer, 466–7
Luminous efficiency, relative, 430–1, 433
Luminous emittance, 450
Luminous exitance, 450
Luminous flux, 431, 436, **445–6**
  unit of, 456
Luminous intensity, 446
  apparent, of image, 453
  units of, 456
Lummer-Brodhun photometer heads, 458–60
Lummer-Gehrcke interferometer, 204–5, 299, 300
  resolving power of, 205, 299
  spectral range of, 205
Lunar division of spherical wavefront, 312–14
Luneburg, R. K., 607
Lux, 456
Lyman ghosts, 298

Mach-Zehder interferometer, **170–1**, 172, 348
Mackenzie, I. C. C., 598
Macleod, H. A., 607
Magnesium fluoride, 155
Magnetic double refraction, 573
Magnetic rotation, specific, 587
Magnification, angular, 20, 56
  axial, 22
  chromatic difference of, 61, 64–5
  of a compound microscope, 61
  empty, 329
  eyepiece, **61**, 330–1
  lateral, 18–20
  longitudinal, 22
  method for focal length, 79–80
  necessary in microscope, 329–31
  normal, 453
  objective, 61
  primary, 61
  of a simple microscope, 59–60
  telephoto, 57
  of a telescope, 26, 56–7
  transverse, 18, 20, 45
Magnifier, simple, 59–60
Magnifying power, chromatic difference of, 64–5
  of a compound microscope, 61
  measurement of, 81
  necessary in microscope, 329–31
  necessary in telescope, 331
  of a simple microscope, 59–60
  of a telescope, 56–7
Magyar, G., 112
Maillard, J. P., 469
Maiman, T. H., 497
Maksutov, D. D., 421
Mallick, S., 609
Malus' experiment, 130–2
  law, **131**, 539
Mandel, L., 112, 610
Marion, J. B., 607
Martin, A. E., 607
Martin, L. C., 607
Match, colour, 434–8
Matching method in photometry, 458–60, 463
Matrix, characteristic (film), 531
  coherency, 562, 565–7, 572
  Jones, 562–5, 569
  Methods, paraxial optics, 41–7
    polarization, 562–70
    thin films, 529–32

Mueller, 562, 567–8
refraction, 42
system, 42
transfer (film), 531
translation, 42
Matt surface, 448–50
Maxwell, J. C., 17, 438, 481, 588
Maxwellian view, 58, 148, 150, 166, **454–5**
Maxwell's equations, 475, 477, 484
McCubbin, T. K., 165
McFarland, B. B., 497
Mean dispersion, 404
Mean spherical candle power, 446
Mechanical scanning Fabry-Perot, 194–5
Meier, R. W., 611
Meniscus camera lens, 411–12
Meniscus, concentric, 420
Mercier, J., 592, 593
Mercury isotope lamp, 202
Meridian plane, 380
Merit function of a lens, 423
Merton, T., 296–7
Mertz, L., 166, 292, 468, 469, 607
Metal, absorption in, 509–11
dispersion of, 510–11
propagation of light in, 509–11
reflection by, 511, 524, 526
Metre-candle, 456
Metre, definition, 200–2
determination, 174, 196–202
Meyer, C. F., 607
Meyer-Arendt, J. R., 610
Mica, 551
Michelson, A. A., and Benoît, J. R., determination of the Metre, 174, 197, 201
Michelson, on echelon, 277
Michelson interferometer, **156–8**, 161, 213, 290, 349, 594–5
adjustment of, 158–9, 167
field widened, 468
Michelson-Morley experiment, 603–4
Michelson, on spectral lines, 159–61
Michelson stellar interferometer, 253–4
Michelson, on velocity of light, 590, 593, 601–3, 608
Micro-analyser, electron probe, 353
Microdensitometer, 472
Micrometer eyepiece, 67
Microphotometer (microradiometer), 472
Microscope, Abbe theory, 334–40, 355
binocular, 443–4
compound, 37, **60–1**, 67, 329–31
condenser, sub-stage, **67**, 333–4, 339–40, **346, 411**
dark-ground, 338, **340–7**, 369
diffraction theory, 328–31, 333–48, 355
electron, 329, 361–2
scanning, 353
flying spot, 352–3
geometrical theory, 59–61, 66–8
Greenough, 443
image luminance, 453
image theory, 328–31, 333–48, 355
interference, 341, 348–52
limit of resolution, 328–9, 338–40
magnifying power, 59–61, 329–31; determination, 81
measuring, 67–8
necessary magnifying power, 329–31
normal magnification, 453

numerical aperture, 328–31, 338–40, 408
objective, **60**, 61, 67, 70, 328–9, 333–40, 347–8, 408–11, 421; apochromatic, 409, 410, 411; catadioptic, 421; flat field, 411; fluorite, 409, 410; immersion, 411; reflecting, 70, 421; ultra-violet, 329, 353, 421
optical system of, 67, 333–4
phase-contrast, 338, 340–5, 347–9
photoelectric, 202
polarizing, 557, **561–2**
Rayleigh theory, 338
simple, 59–60
stereoscopic, 443–4
ultra-violet, 329, 353, 421
Microscopy, experimental arrangements, 345–353
Microscopy by wavefront reconstruction (holography), 361–2
Microwave interferometer, 594–5, 600
spectroscopy for c, 595–6, 600
Mikaelian, A. L., 610
Mile pipe experiment, 590, 593
Minimum deviation, grating, 266
prism, 82–3, 88–9
Minimum separation, object-image, 12–13
Mirror, concave, astigmatism of, 406–7
correction of aberrations, 419–21
testing of, 369
Mirror, Lloyd's, 142
Mirror, spherical, measurement of radius, 72, 74–6
Mirror systems, 69–70, 98–9, 346–7, 419–21
Missing orders, 247–9, 274–5
Mitscherlich polarimeter, 581
Mittelstaedt, O., 591, 593
Mixture, additive colour, 433–8
of pigments, 438–9
subtractive colour, 439
Mock interferometry, 292–4, 464
Mode, laser, 111
Modulation noise, 463
Modulation transfer factor, 356–7, 390
Modulus of complex quantity, 104
Moire fringes, 166, 304–5
Moldauer, P. A., 608
Molecular (molar) refraction, 507
Molecular rotation, 581
Molecular scattering, 494–5, 498, 508–9
Monochromatic light, 105, 107, 118, 136, 159
source of, 105–13, 121, **201–2**, 496–7
Monochromator, 97–8, 193, 285, 286, 288, 291, 300, 301, 467
Morley-Michelson experiment, 603–4
Mount Wilson experiment, 590, 593
Mounting for echelon, 280
for ruled grating, 285–92
Mouton-Cotton effect, 573
Muellar matrix, 562, 567–8
Multilayer films, 155, 210–12, 529–32
Multiple beam fringes of equal inclination, 175–8, 204–5
of equal thickness (Fizeau), 205–8
Multiple fibres, 373–4
Multiplex spectrometers, 464, 468
Multislit spectrometer, **291–2**, 299, 300, 467, 469
Murata, K., 610
Muscle, ciliary, 425
Musset, A., 610

Mutual coherence, 108
Mutual intensity, 110
Myopia, 427

N.A., 328–31, 338–40, 408
N.P.L. interferometer, 172–3
Natural damping, **161**, 495, 504
Near point, 329–31, 426
Necessary magnifying power of a microscope, 329–31
    of a telescope, 331
Negative uniaxial media, 489, 534–7, 548–50
Negative biaxial media, 545
Neighbouring rays, paths along, 321–2
Neon-helium laser, 111, 211, 496, 497
Nernst, W., 528
Nerve, optic, 425
Nethercot, A. H., 595
Neuhaus, H., 285
Neumann's law, 475
Neutralization test for lens power, 71
Newcomb, S., 590
Newton, I., 6, 62, 601
Newtonian telescope, 70, 421
Newton's equation for conjugate distances, 19
Newton's rings, 76, **152 4**, 157, 213
Nicol prism, 539–40
    square ended, 541
Nicols in series, 540–1
Night blindness, 428
Nit, 457
Nitrobenzene, 96, 572, 591
Nodal planes, definition, 21
Nodal points, definition, 21
    experimental location, 77–8
    of eye, 426
    as reference points, 23
    of a single surface, 31
    from system matrix, 43–6
    of a thick lens, 34
    of a thin lens, 33
Nodal slide, 78
Node, 128, 528
Noise in radiometry, 462–3
    in spectrometers, 467–8, 469
Nomarski, G., 351
Non-linear optics, 101, 113, 497
Non-reflecting films, 154–5, 531
Non-uniform wavefront, 258–60
Normal dispersion, 498–500, 506–7
Normal magnification, 453
Normal spectrum, 268, 287
Normal surface, 546
Normalization, 238, 244
Nörrenberg polarimeter, 581
Notation – preface,
    in electromagnetism, Chapter XIX
    in geometrical optics, 8–10
    in photometry (radiometry), 445
    in polarization, 562
Numerical aperture, 328–31, 338–40, 408
Nussbaum, A., 607

Object-image, coincidence methods, 71–3, 78
    smallest separation, 12–13
Object, measurement if inaccessible, 14
Object position, specification of, 19, 22–4, 30, 399
Objective magnification, 61
Objective, design of small, 407–8

Objective, microscope, **60**, 61, 67, 70, 328–9, 333–40, 347–8, **408–11**, 421
    apochromatic, 409, 410, 411
    catadioptric, 421
    flat field, 411
    fluorite, 409, 410
    immersion, 411
    limit of resolution, 328–9, 338–40
    reflecting, 70, 421
    ultra-violet, 329, 353, 421
Objective, photographic, 57, 411–18
    achromatic doublet, 412–13
    chromatic correction of, 404
    image illumination, 451–2
    inverted telephoto, 417
    meniscus, 411–12
    Petzval type, 416
    symmetrical type, 413–14
    telephoto, 57, 58, 417–18
    triplet type, 415–16
    wide-angle type, 416–17
    zoom, 421–2
Objective, telescope, 55, 66, 70, 405, 408–9, 420–1
    limit of resolution of, 327–8
Obliquity factor, **216**, 217–19, **224**, 225, 227, 236, 307, 311
Obstacle, circular, 220, 307–8
    irregular, 220
Observer, standard, 431, 461
Onaka, R., 288
O'Neill, E. L., 607
Ooue, S., 610
Opaque object, microscopic examination of, 346–7, 348–9
Opaque strip, Fresnel pattern, 319–20
Opic lens, 415
Optic axis, **486**, 488–9, 534–7, 544–5
Optic nerve, 425
Optical activity, Chapter XXIII
    of crystals, 574
    Fresnel's theory, 575–6
    of liquids, 580–1
    of quartz, 577–9
Optical centre of thick lens, 34–5
Optical constants of metals, 524, 526
Optical coordinates, 258, 356
Optical path, 3, 6–7, 102
Optical paths along neighbouring rays, 321–2
Optical pumping, 496
Optical pyrometer, 474
Optical shutter, 590–1, 598–9
Optical systems, of eye, 425
    of a microscope, 66–7, 333–4, 367
    of a telescope, 66–7
    perfect, 17, 52, 321, 323–4, 376; impossibility of, 23–4, 376
Optical transfer function, 354–9, 390
Optical tube length, **61**, 330, 410
Optimisation in lens design, 422
Order of interference, **140**, 156, 180–1
Orders, coincidence of, 184, 269
    missing, 247–9, 274–5
    overlapping of, 183–4, 268–9, 278, 279–80, 289
    separation of, 181, 183–4, 279–80, 289
Ordinary index, 535
Ordinary ray, 533–9, 544–5
Ordinary wave, 533–7, 544–5
Oscillating Fabry-Perot systems, 193–5

Oscillator (bound charge), 497, 504–5, 510–11
Oscillator (laser), 496
Oscillator strength, 505–6
Osterberg, H., 605
Ouweltjes, J. L., 605

Palik, E. D., 611
Palmer, E. W., 611
Papaulis, A., 607
Paraffin, 479
Paraxial constants, determination of, Chapter IV
Paraxial imagery as a collinear transformation, 17, 30
Paraxial rays, 11
    refraction equation, 41
    tracing by deviation method, 38–9
    tracing by matrix method, 41–2
    transfer (translation) equation, 41
Paraxial region, definition, 11
    imagery in, 11–16, 17, 30–41, 43–6
Paraxial theory, general, Chapter II
    of particular instruments, Chapter III
Parrent, G. B., 605
Parseval theorem, 260
Partial coherence, 105–10, 251–4, 565–6
Partial dispersion, 404
    relative, 404
Partially polarized light, 133, 136, 518, 565–6, 567–9
Particle size analysis, 353
Particles, light as, 1
Paschen grating mounting, 286–7
Patel, C. K. N., 497
Path, optical, 3, 6–7, 102
Paths along neighbouring rays, 321–2
Pattern function, 258–60, 273, 336
Patterns, types of diffraction, 225–30
Pearson, F., 590, 593, 608
Pease, F. G., 590, 593, 608
Pegis, R. J., 609
Penetration in total reflection, 521–2
Pentac lens, 416
Perfect optical system, 17, 52, 321, 323–4, 376
    impossiblity of, 323–4, 376
Period, 2
Periodic error in grating, 297
Peripheral vision, 426, 428, 432, 441
Permeability, 477
Perot, A., Fabry, C., and Benoît, J. R., determination of Metre, 197, 201
Perot-Fabry etalon, 179–204, 211, 299, 300, 301
    all glass, 204, 205
    centre spot scanning, 192–5
    confocal, 195–6, 497
    dispersion, 184–5, 194
    field widened, 196, 467
    as filter, 211
    finesse, 181–3, 194, 299
    in metrology, 197–202
    photometric efficiency, 193, 195–6, 299, 466–7
    as refractometer, 179, 200
    resolving power, 180–4, 193–4, 196, 299, 300, 301, 465
    spectral range, 183–4, 194–5, 299
    spherical, 195–6, 467
Perot-Fabry fringes, 175–9, 213
Perot-Fabry instruments in series, 189–92,
203, 467
Perot-Fabry interferometer, 179–204
    oscillating, 192–5
    See also Perot-Fabry etalon,
Perrotin, H., 590
Perry, D. L., 211
Persistence of vision, 429
Perspective, 57–8, 440
    centre of, 47
Petykiewicz, J., 609
Petzval, J., 416
Petzval field curvature, 386–7, 389–90, 393, 411
    of a concave mirror, 419
    of a doublet, 405–6, 412
    of a single surface, 395–6
    of a thin lens, 402, 415
Petzval lens, 416
Petzval sum, 396
Phase, 2
    change, on reflection, 142, 517–18, 524–5
        on passing through a focus, 144
    coherence factor, 110, 252–4
    constant, 101
Phase-contrast microscope, 338, 340–5, 347–8, 352
Phase-contrast test, 369
Phase difference, 2, 3, 102
Phase object, 340–5, 352
Phase plate, 343, 344–5, 347, 352
Phase reversal zone plate, 309
Phase velocity, 2, 122–3, 477–9
    in anisotropic media, 483–6, 535–6, 544, 6
Phot, 457
Photo-conductive cell, 462
Photo-elastic effect, 573
Photo-electric cell, 461, 462
Photo-electric effect, 1
Photo-electric microscope, 202
Photo-electric polarimeter, 585, 586
Photo-emissive cell, 462
Photographic lens, 57, 411–18
    achromatic doublet, 412–13
    aperture, 52–3, 451–2
    chromatic correction of, 404
    image formation, 354–9
    image illumination, 451–2
    inverse telephoto, 417
    meniscus, 411–12
    Petzval type, 416
    symmetrical type, 413–14
    telephoto, 57, 58, 417–18
    triplet type, 415–16
    wide-angle type, 416–17
    zoom, 421–2
Photometer bench, 458–60
    Guild flicker, 461
    illumination, 473
    Lummer-Brodhum, 458–60
Photometric brightness (luminance), 448
Photometric properties of image forming systems, 450–4
    of spectrometers, 163, 164, 166, 184, 193, 195, 196, 288, 291–2, 294, 298, 464–9
Photometric quantities, definitions, 445–8
    units, 455–8
Photometry, Chapter XVIII
    absolute, 455–6, 457–8
    flicker, 460–1
    heterochromatic, 460–3

physical, 461–4
visual, 458–61
    soundness of, 431
Photo-multiplier, 462
Photon, 1, 495–6
Photon noise, 462, 469
Photopic vision, **428**, 429–33
Photo-visual achromatism, 404
Photo-voltaic cell, 461
Piazzi-Smyth, C., 416
Piezo-electric light modulator, 598–9, 602
Pigment (body) colour, 491
Pigments, 438–9
Pile of plates, 133
Pin-cushion distortion, 387
Planckian radiator, 455–6
Planck's constant, 495
Plane e.m. waves, 477–9, 480–1
    in absorbing media, 503
    in anisotropic media, 484–6
Plane of polarization, 131
Plane polarized light, 1, 131, 533–4, 563–5, 567, 570, 571
    e.m. theory, 478–9; anisotropic media, 484–90
    interference of, 133–5
    production by double refraction, 534–42; by reflection, 130–3, 516; by selective absorption, 542–4
    resolution into circular components, 570, 575–7; into elliptic components, 578–9; into plane components, 131–5, 534–41, 556, 557
Plane of vibration, 131, 529, 533, 534
Plane waves, 3, 100, 477–9, 480–1
    in absorbing media, 503
    in anisotropic media, 484–6
    in biaxial crystals, 544–6
    diffraction of, 226, 232–3, 234–8, 246–9, 257–60, 261–76
    division into half-period zones, 306–7
    division into strips, **228–9**, 232–3, 234–5, 237, 241–2, 246–7, 273
    interference between, 150–1, 166–8, 528
    passage through large aperture, 4
    reflection at plane surface, 4–5
    refraction at plane surface, 5
    in uniaxial crystals, 490, 534–7
Plettig, V., 98
Plyler, E. K., 595
Pneumatic detector, 462
Poincaré sphere, 570–2
Point source, non-uniform, 446, 447
    uniform, 446–7
Poisson, S. D., 220
Polar diagram of source, 470
Polarimeter, 581–5
    Biot, 542, 581
    Faraday compensating, 586
    half-shadow, 582–5
    Laurent, 583–4
    Lippich, 584–5
    Mitscherlich, 581
    Nörrenberg, 581
    photo-electric, 585, 586
Polariscope, 573
Polarization, degree of, 568
Polarization, electric, 476, 504
    in anisotropic media, 482–3
Polarization, degree of, 518

plane of, 131; rotation by reflection and refraction, 518
Polarization, rotatory, Chapter XXIII
    Fresnel's theory, 575–7
    of crystals, 574
    of liquids, 580–1
    of quartz, 577–9
Polarization of scattered light, 498
Polarized light, 131, 133–6, 562–72
    circularly, **135**, 521, 550, 551, 552, 553, 563–4, 565, 567, 569–70, 571; as component of plane, 570, **575–7**, 578; production of, 135, 521, 551; resolution into plane components, 135, 551, 552
    elliptically, **135–6**, 521, 541–55, 571–2; analysis of, 552–5; as component of plane, 578–9; production of, 135–6, 520–1, 524, 549–52; resolution into plane components, 135, 549–51, 555
    interference of, 133–6, 562–70
    matrix methods, 562–70
    partially, 132, 136, 518, 552, 565–6, 567–8
    passage through crystal plates (retarders), 548–61, 563, 564, 566, 568, 572
    plane (linear), 1, 131, 533–4, 563–5, 567, 570, 571; e.m. theory, 478–9, 480–1, anisotropic media, 484–90; production by double refraction, 534–42, by reflection, 130–3, 516, by selective absorption, 542–4; resolution into circular components, 570, 575–7, into elliptic components, 578–9, into plane components, 131–5, 534–41, 556, 557
    specification of, 135–6, 562–3, 565–8, 570–2
Polarizer, **131**, 539–44, 564, 581–2, 583–4,
    Cornu-Jellett, 583
    Lippich, 584–5
    micro-crystalline, 543
    molecular, 543–4
    sheet-type (dichroic), 543–4, 551–2
Polarizers and retarders in series, 562–3, 566, 568, 572
Polarizing angle, **132**, 517, 518, 519
Polarizing microscope, 557, 561–2
Polarizing prism, Ahrens, 541–2, 561
    Cornu-Jellett, 583
    Foucault, 542
    Glan, 542
    Glazebrook, 541
    Nicol, 539–40; square ended, 541–2
    Rochon, 542
    Thompson, 541
    Wollaston, 542
Polaroid, 543
Polster, H. D., 611
Polychromatic source, 108
Population inversion, 496
Positive parameter, 399
Positive biaxial crystal, 545
    uniaxial crystal, 489, 536, 548, 577
Posterior chamber, 424
Potassium bromide, 96–7
Potler, R. J., 612
Potts, W. J., 611
Power, 23
    back vertex, 30
    of a combination of systems, 29, 41; of two separated thin lenses, 36; of two thin lenses in contact, 14, 40

dispersive, **62**, 63, 86
of an eyepiece, **61**, 330–1
front vertex, 30
of a single surface, 10, 31
of a spherical lens, 35
from system matrix, 46
of a thick lens, 33, 41
of a thin lens, 12, 32; neutralization test, 71
Poynting vector, 480–1
in anisotropic media, 485–6
Presbyopia, 427
Pressure broadening, 161
Pressure scanning, 193–4
Prim, 590
Primary aberrations, 382
astigmatism, 382, **385–6**, 389–90, 393, 401–2
coma, 382, **384–5**, 388, 393, 400–1, 405
distortion, 382, **387**, 393
field curvature, 382, **386–7**, 389, 393, **395–6**
spherical, **382–3**, 393, 399–400, 404–5, 419–20
Primary image, 56, 61, 66, 67, 68
magnification, 61, 68, 331
Primary (tangential) plane, 380
Principal angle of incidence, 524–6
Principal azimuth, 525–6
Principal dielectric constants, 483
Principal focus, 18
Principal indices of refraction, 483
Principal maxima, 262–76, 279–80, 282–4, 302–3
Principal planes of E-ray and O-ray, 535
Principal planes, 18
of a combination of systems, 28–9, 41; in air, 30; of two separated thin lenses, 36–7
of complicated systems, 39, 43–6
experimental location of, 77
of a single surface, 30–1
from system matrix, 43–6
of a thick lens, 33–6, 41
of a thin lens, 32, 33
Principal points, 18
See also Principal planes,
Principal ray, 53
spherical aberration of, 389, 419
Principal refractive indices, 483
Principal section of a crystal, **534**, 538
of a prism, 82
Principal transmittance (polarizer), 543
Principal velocities, 483, 535
Principle of superposition, 101, 477
Prism, action on a pulse, 125
Ahrens, **541–2**, 561
angle of, 82, 88–9
chromatic resolving power, 239–40
constant deviation, 97
Cornu, 579–80
Cornu-Jellett, 583
curvature of spectral line, 84
dispersion of, 84–5
formation of a spectrum, 83–4
Foucault, 542
Glan, 542
Glazebrook, 541
infra-red, 96–7
minimum deviation, 82–3, 88–9
Nicol, 539–40; square ended, 541–2
principal section, 82

refracting angle, 82, 88–9; edge, 82; face, 82
refraction through, 82–3
resolving power, 239–40
Rochon, 542
spectrograph, 96–9
spectrometer, 86–7, 97, 298–9, 300, 465; adjustment of, 87–8
thin, 85–6
Thompson, 541
ultra-violet, 96
Wollaston, 542
Prismatic binocular telescope, 69, 442
Prisms, Amici system of, 86, 94
combinations of, 86
erecting, 68–9
Privileged directions, **548**, 551, 556, 558
Profile, instrumental, **164**, 181, 193, **239**, 269, 291, 294, 465
Profile, spectral line, 107, 112, 119–21, 160–1, 163
Progressive error in a grating, 297
Progressive waves, 1–2
Projector, kinematograph, 58–9
lantern slide, 58
Projection Ektar lens, 416
Projection lens, 58, 416
Propagation of light, in absorbing media, 503, 507–8
in crystals, 482–90, 533–9, 544–61, 574–9
general equation of wave motion, 100
Huygens' principle, 3–4, 215–24
in isotropic dielectrics, 476–81, 504–9
in metals, 509–12
Propagation, rectilinear, 1, 4
Protar lens, 414
Pseudoscopic image, 366
Pulfrich refractometer, 93
Pulse, 121
passage through a grating, 124
passage through a prism, 125
Pulsed laser, 111, 113, 497
Pumping, optical, 496
Pupil, entrance, 53; aberration of, 452
exit, 53
of eye, 424, 428
Pupil function, 359
Purkinje effect, 431, 462
Pyrometer, 474

Quantum theory of absorption, 495–6
of dispersion, 506
of scattering, 495–6
Quarter-wave films, 154–5, 210–12, 531–2
stack, 210, 212, 532
Quarter-wave limit, Rayleigh, 331–2, 390
Quarter-wave plate, 550–3, 564–5, 567, 582
Quartz, 96–7, 491, 535, 551, 585
circularly polarized light in, 577
compensators, 553, 585
dispersion of, 499
elliptically polarized light in, 578–9
optical activity of, 577–9
plane polarized light in, 535, 577
refractive indices of, 536, 578
rotatory dispersion of, 574, 585
spectrograph, 96, 579–80
wave surface for, 577–8
Quasi-monochromatic light, **107–10**, 118–21, 160, 251–3
polarization of, 565–6

source, 109–13, 251–3

Radar, 593, 600
Radial grating, 305
Radiance, 447
Radiant emittance, 450
Radiant energy, 445, 455
Radiant exitance, 450
Radiant intensity, 446
Radiant power, 455
Radiation pyrometer, 474
Radiator, black body, 455–6
Radiator, Lambert, 448–50
Radiometric properties of image forming systems, 450–2
  of spectrometers, 163, 164, 166, 184, 193, 195, 196, 288, 291–2, 294, 298, **464–9**
Radiometric quantities, definitions, 445–8
Radiometric units, 455
Radiometry, Chapter XVIII
  absolute, 458
Radius of curvature, accurate measurement, 74–6
  long, measurement, 75–6
  short, measurement, 75
  simple methods of measurement, 71–2
Ratio-recording, 463
Ramachandran, G. N., 611
Raman effect, 496
Ramaseshan, S., 611
Ramsden eyepiece, 65–8, 418–19
  achromatized, 418
Random errors in gratings, 297
Range, spectral, 183
  echelle, 288–9, 299
  echelon, 279–80
  Fabry-Perot system, 183–4, 194–5, 299
  grating, 269, 279, 287, 288–9, 299
  Lummer plate, 205
Range, stereoscopic, 442
Range of a telescope, 453–4
Rank, D. H., 595–6, 600, 611
Rao, B. S., 596
Rapid rectilinear lens, 413
Ratcliffe, J. A., 611
Ray, 1, 5–7
  aberration, 61–5, 376–7, 382–99, 401–7, 420
  axis, **488**, 489–90, 544–5, 547–8
  chief, 53
  conjugate, location of, 19
  deviation on refraction, 38, 392
  direction in a crystal, 485–6, 534–7, 544–6
  extraordinary, 533–7
  ordinary, 533–7
  paraxial, 11; deviation method for tracing, 38–9; matrix method for tracing, 41–3; refraction equation, 41; transfer equation, 41
  path, 6–9
  paths, neighbouring, 321–2
  principal, 53; aberration of, 389, 419
  surface, **488**–9, 533–7, 544–5
  velocity in a crystal, 485–8, 535–6, 544
  velocity surface, **488**–9, 533–7, 544–6
  as wave normal, 5–6
Rayleigh, Lord, on blazed gratings, 276
Rayleigh criterion for resolution, 181, 239–40, 269, 326–7
Rayleigh on dispersion of prism, 84–5

Rayleigh on line structure, 161, 163
Rayleigh-Parseval theorem, 260
Rayleigh quarter-wave limit, 331–2, 390
Rayleigh refractometer, 255–7
Rayleigh scattering, 494–5, 498
Rayleigh theory of microscope, 338
Real image and object, minimum separation, 13
Real-time holographic interferometry, 367
Reconstruction, wavefront, 359–67
Rectangular aperture, Fraunhofer pattern, 227–9, **232–8**, 257–9
  Fresnel pattern, 227–9, **311–16**, 318–19
  limit of resolution with, 241
Rectilinear propagation, 1, 4, 5, 215, 220
Reduced quantities, 29, 41
Reference surface for aberration, 377–8
Reflecting microscope, 70, 421
Reflecting power, 517, 519–20, 523–4
Reflecting systems, 69–70, 419–21
Reflecting telescopes, 70, 420–1
Reflection, Chapter XXI
  by absorbing media, 522–6
  coefficient, 517, 519–20, 523–4
  by dielectrics, 130–3, 512–21
  diffuse, 448
  echelon, 277, 278–81, 301
  factor, measurement, 472
  grating, 266, 269, 270, 276–92
  by metals, 524, 526
  phase change, 142, 517–8, 524–5
  of plane wave at plane surface, 4–5
  polarization by, 130–3, 516
  selective, **491–2**, 511, **523–4**
  as special case of refraction, 16
  at a spherical surface, 15–16
  total internal, 519–22
Reflector, diffuse, 448
Refracting angle of prism, 82, 88–9
Refracting edge, 82
Refracting face, 82
Refracting surface, spherical, 10, 11, 14–15, 30–2, 38
  aberrations of, 392, **397–9**
  cardinal points of, 31
  conjugate distance relation, 10
  deviation of a ray, 38, 392
  notation, 8–10
Refraction, Chapter XXI
  into an absorbing medium, 526–7
  by a dielectric, 512–18
  double, 533–7, 544; by calcite, 537–9
  external conical, 547–8
  internal conical, 546–7
  matrix, 41
  molecular (molar), 507
  of plane wave at plane surface, 5
  Snell's law, 5, 7, 321, 513
  specific, 506
  at spherical surface, cardinal points, 31; conjugate distance relation, 10, 14–15; deviation of a ray, 38, 392; notation, 8–10
  of a spherical wave at a spherical surface, 14–15
  through a prism, 82–3
Refractive errors of eye, 427
Refractive index, **3**, 5, 9–10, 18, 31, **321**, 479, 498–502, 503–11, 526
  of biaxial crystals, 483
  for circular components, 576, 578

complex, **503**, 504–5, 509, 511, 522–3, 525–7
and dielectric constant, 479, 503
extraordinary, 536
of gases, 500, 506, 508; measurement, 170, 179, 200, 255–7
of a lens, 72–3
of a liquid, 506; measurement, 73, 89–95, 170, 255–7
of a metal, 509–11, 524, 526
ordinary, 535
of a solid, measurement, 88–96
of uniaxial crystals, 535–6
Refractive indices, principal, 483, 535–6
Refractometer, Abbe, 93–4
dipping, 94
fibre, 374–5
Hilger-Chance, 93, **94–5**
Jamin, 168–70
Pulfrich, 93
Rayleigh, 255–7
Relative luminous efficiency, 430–1, 433
Relative partial dispersion, 404
Relativity, 129, 603, 604
Relaxation time, 501
Replicas of gratings, 274, 295, 296, 297
Residual rays, 524
Resolution criterion, Rayleigh, 181, **239–40**, 269, **326–7**
Taylor, 181–2, 184, 194
Resolution, limit of, 326–7, 337, 338–40, 356–7
with circular aperture, 326–7
of eye, 329, 330, 428, 432–3
of microscope, 328–9, 338–40
with rectangular aperture, 241
of stellar interferometer, 253–4
of telescope, 327–8
Resolution, requirement for maximum, 298–9, 301
Resolution into spatial frequencies, 336–7, 355–9
Resolution, spurious, 357
Resolving limit, 181
of echelon, 279, 301
of Fabry-Perot instrument, 182–3, 299
of grating, **270**, 276, 299
Resolving power, see Resolution, limit of
Resolving power, chromatic, 181, 240
comparison of instruments, 298–301
of echelon, 279, 299, 301
of Fabry-Perot instruments, **180–4**, 193–4, 196, 299, 301
of Fourier transform spectrometer, 163–4
of grating, 124, **269–70**, 286, 294, 298, 299, 300
of Lummer plate, 205, 299
of prism, 125, 239–40
Resonance radiation, 496
Resonator, cavity, 593–4, 600
Response, spatial frequency, 356–9, 390–1
Reststrahlen, 524
Retarded potential, 223
Retarder, birefringent, 551
Retarders in series, 562–3, 566, 568, 572
Retina, 54, 424–5, 426, 428–33, 434
Retinal illumination, 449
Retinal image, 54, 449, 453
Reynolds, G. O., 606
Rhodopsin, 428

Richards, O. W., 605
Richards, P. L., 302, **469**
Rienitz, J., 607
Right- and left-handed optical activity, 574, 577
Ring, J., 293
Rings and brushes, 559, 561
Risken, H., 611
Rochon, A. M. de, 542
Rock salt, 96, 491
Rods, **428**, 429, 431
Rogers, G. L., 607
Rogers, H. G., 543
Rohr, Mvon, 416
Roizen-Dossier, B., 610
Romanova, M. F., 201
Römer, O., 588–7
Roosinov, M., 417
Rosa, E. B., 592, 593
Rosenbruch, K.,-J., 611
Rosenhauer, K., 611, 612
Ross wide-angle lens, 417
Rossi, B., 607
Rotation, molecular, 581
magnetic, 587
Rotation of plane of vibration on reflection, 518
on refraction, 518
Rotation, specific, of liquid, 580–1
magnetic, 587
of solid, 574
Rotation-vibration bands, 595–6
Rotatory dispersion of liquids, 581, 585
of quartz, 574–5, 585
of solids, 574–5
of sugar, 585
Rotatory polarization, Chapter XXIII
Fresnel theory, 575–7
measurement of rotation, 581–6
Rotatory power, 580–1
Rouard, P., 611
Rowland circle, 281–2, 284, 286–8
Rowland ghosts, 297
Rowland gratings, 295
Rowland mounting, 286
Rubens, H., 524
Rubinowicz, A., 231, 611
Ruby laser, 111, 496, 497
Ruddock, K. H., 605
Rudolph, P., 414, 415
Ruling of gratings, 294–7
Ruth, R. P., 595

Saccharimeter, 581, 585–6
Sagittal fans, 386
reflection of, 406–7
Sagittal focus, 385–7, 401, 406, 407
Saggital plane, 380
St. Venant's Principle, 225
Sakai, H., 611
Saksena, A. H., 595
Sanders, J. H., 607
Satellites in grating spectra, 298
Satellites, interferometers for, 469
Saturated colour, 434, 437
Saunders, J. B., 611
Saxton, J. A., 609
Scalar product of vectors, 475
Scanning function, **164**, 181, 193, **239**, 269, 291, 294, 465

Scanning microscope, 352–3
Scattered light, depolarization of, 498
　　polarization of, 498
Scattering, coherent, 495
　　e.m. theory of, 497–8
　　incoherent, 496
　　molecular, 494–5, 508–9
　　quantum theory of, 495–6
　　Rayleigh, 494–5
Schade, W., 416
Schawlow, A. L., 497, 612
Schematic eye, 425–6
Scherrer, P., 304
Schlieren systems, 369–70
Schmidt camera, 419–20
Schöldström, R., 596, 600
Schott, O., 406
Schultz, G., 607
Scintillation noise, 463
Sclera, 424
Scoptopic vision, **428–9**, 431, 433, 434
Searle, G. F. C., 80
Sears, J. E., and Barrell, H., determination of
　　Metre, 196–202
　　index and dispersion of air, 200
Secondary maxima, **264**, 265, 267, 272–3
Secondary (sagittal) plane, 380
Secondary spectrum, 63, **403–4**
Secondary sources and waves, 3–4, 215–16,
　　219, 224
　　in biaxial media, 533, 544–5
　　in uniaxial media, 533–7
Seidel aberrations, 382
Selby, M. J., 293
Selective absorption, **491–2**, 510–11, 523–4
Selective reflection, **491–2**, 511, **523–4**
Self-coherence, 109
Sellmeier dispersion formula, 501, 506
Sensitive tint, 555, 583, 585
Sensitivity of eye, 428–9
　　spectral, 430–1
Separated thin lenses, 36–8
　　achromatism of, 64–5
Seya, M., 288
Shadow angle, 582
Shadows, 1, 4
Shape parameter, 399
Sharply limited wave train, 118–20
Shearer, J. N., 595
Shearing interferometer, wavefront, 171–2
Shearing microscope, 341, 350–1
Sheet-type polarizer, 543–4, 551–2
Sheet-type retarder, 551–2
Shift of focus effect on aberration, 377–9,
　　381
Short sight, 427
Shulman, A. R., 608
Shurcliff, W. A., 569, 608
Sign convention, geometrical optics, 9
　　polarized light, 562
Signal, analytic, 108
Signal-to-noise ratio, 462, 567–8
Sikora, S. V., 594
Silica, fused, 499, 536, 580
Silver, 177
Silver chloride, 96
Simkin, G. S., 594
Simple harmonic motions, addition of, 101–5
　　at right angles, 134–5
Simple harmonic waves, **2–3**, 100, 102–3, 105,

　　118
electromagnetic, **479–80**, 481–2, 484
Simple microscope, 59–60
Sinc function, 119
Sinclair, R. S., 611
Sine Condition, **324**, 326, 388, 397–9, 409
Sine Relation, **323**, 328, 450, 466
Sine wave grating, 355–6
Single order position for echelon, 280
Single slit, Fraunhofer pattern, 227–9, 232–8,
　　257–60
　　Fresnel pattern, 227–9, 311–16, 318–19
SISAM, **290–1**, 299, 300, 466, 467, 468, 469
Size of an object, apparent, 54–5
Size, particle, 353
Sky, blue colour of, 494–5
Slip gauge testing, 155–6, 168, 172–3
Slit, double, Fraunhofer pattern, 229–30,
　　246–51; with double source, 249–50;
　　with wide source, 250–4
　　Fresnel pattern, 227–9, 311–16, 318–19
Slit source in diffraction experiments, 232,
　　238–9, 311, 315–16
　　Fresnel pattern, 227–9, 311–6, 318–9
Slit source in diffraction experiments, 232,
　　238–9, 311, 315–6
Slow axis, 548
Slussareff, G., 452
Smith, A. L., 611
Smith, A. W., 609
Smith, F. G., 608
Smith, F. H., 351–2
Smith, H. M., 608
Smith, R. A., 608
Smith, W. J., 608
Snell's law, 5, 7, 321, 513
Snitzer, E., 375
Sodium, 404
　　anomalous dispersion of, 502–3
Sodium chloride, 96
Soffer, B. H., 497
Soleil biquartz, 582, 585
Soleil compensator, **553**, 556, 585
Soleillet, P., 568
Solid state laser, 111
Sommerfeld, A., 231, 608
Sorokin, P. P., 497
Sound waves, 100
Source, directional properties of, 470
　　double, 249–50, 253
　　monochromatic, 105–13, 121, **201–2**, 496–7
　　photometric standard, 455–6
　　point, non-uniform, 446, 447; uniform,
　　446–7
　　polychromatic, 108
　　quasi-monochromatic, 109–13, 251–3
　　slit or line, 232, 238–9, 311, 315–16
　　thermal, 101, 105
　　wavelength standard, 200–2
　　wide, 140, 239, 250–4
Space probe interferometer, 469
Spaces, left- and right-hand, 8–9
Spatial coherence, 108, 109–10, 251–4
Spatial filtering, 353–4
Spatial frequency, 117, 258–9, 336, 337, 353–9
　　filtering, 353–4
　　response, 356–9, 390–1
Specific magnetic rotation, 587
Specific refraction, 506
Specific rotation, liquids, 580–1

solids, 574
Spectacles, 427
Spectral distribution, 161–3, 435
Spectral line profile, 107, 112, 119–21, 160–1, 163
  structure, 121, 159–61, 162–3, 185, 201
  width, 107, 112, 119–21, 160–1, 162–3
Spectral locus, 437
Spectral range, 183
  echelle, 288–9, 290
  echelon, 279–81
  Fabry-Perot system, 183–4, 194–5, 299
  grating, 269, 279, 287, 288–9, 299
  Lummer plate, 205
Spectral sensitivity of eye, 317–18
Spectrograph, 98–9, 464–5
  echelle, 288–90, 300
  echelon, 280–1, 300–1
  Fabry-Perot, 192–3, 301
  grating, 285–90, 299–300
  Lummer plate, 300
  prism, 96–8, 579–80
Spectrometer, spectrophotometer, spectro-radiometer, 99, 298–9, 461–4
  absorption, 301–2, 463–4
  comparison of various classes, 281, 298–302, 464–70
  Fabry-Perot, **184–9**, **192–5**, 299, 300–1, 464–9
  Fourier transform, **161–6**, 299, 300–1, 464, 467, **468–9**
  grating, **266–70**, **285–94**, 298–300, 464–7, 469
  Michelson, 159–61
  Mock interferometer, **292–4**, 464
  multislit or grille, **291–2**, 299, 300, 467, 469
  photometric properties, 164, 165, 166, 184, 193, 195–6, 286, 288, 291–2, 294, 298, **464–9**
  prism, 83–4, **86–7**, **97–8**, 298–9, 300, 465; adjustment, 87–8
  SISAM, **290–1**, 299, 300, 466, 467, 468, 469
Spectroscope, 86–7
  constant deviation, 97
  direct vision, 86
  Edser and Butler calibration, 203–4
Spectroscopy, amplitude, 166
  high resolution, 160, 163–6, 183–6, 193–5, 278–91, 288–91, 295–6, 298–302, 464–70
  infra-red, 96–7, 164–5, 166, 192, 278, 291, 292–4, 301–2, 467, 468–9
  microwave for c, 595–6, 600
  sub-mm, 301
  ultra-violet, 96, 177, 288, 300, 301, 468
  of weak sources, 291, 301, 468, 469
Spectrum, absorption, 492
Spectrum, continuous, 83, 121, 160, 163, 241
  dispersion, grating, 267–9; prism, 84–5, 268
  equal energy, 434, 435
  formation by a grating, 267; prism, 83–4
  line, 83, **121**, 266
  normal, 268, 287
  visible, 430, 433–4
Speculum metal, 70
Speed of light, 477, 479
  adopted value, 600
  determination of, 588–603; aberration method, 589; band spectra method, 595–6; cavity resonator method, 593–4; correction to a vacuum, 592; eclipse

method, 588–9; I.R. frequency measurement, 601; Kerr cell methods, 590–2, 596–8; Lecher wires method, 592; microwave interferometer, 594–5; mile pipe experiment, 590; Mount Wilson experiment, 590; Piezo-electric shutter, **598–9**, 602; radar method, 593; R.F. method, 596; ratio of units method, 592; rotating mirror methods, 590, 601–3; summary of results, 592–3, 600; toothed wheel method, 589–90; U.S. grating shutter method, 599
Sphere, integrating, 271–2
Sphere, Poincaré, 570–2
Spherical aberration, **382–3**, 392, 411, 416
  chromatic variation of, 410
  of a concave mirror, 419
  connection between ray and wavefront, 382–3
  correction in concentric system, 420; in microscope objective, 408–10; in Protar lens, 414; in Schmidt camera, 420
  of a doublet, **404–5**, 408, 412
  high order, 409
  primary, **382–3**, 393, 399–400, 420
  of principal rays, 419
  secondary, 409, 412, 414
  of a single surface, 397, 398
  of sub-stage condensers, 409
  of a thin lens, 399–400
  zonal, 409, 410
Spherical Fabry-Perot, 195–6, 467
Spherical lens, 35
Spherical mirror, 15–16, 70, 406–7, 419–21
  astigmatism of, 406–7
  spherical aberration of, 420
Spherical refracting surface, 10, 11, 14–15, 30–2, 38
  aberrations of, 392, **397–9**
  cardinal points of, 31
  conjugate distance relation, 10
  deviation of a ray, 38, 392
  notation, 8–10
Spherical surfaces, combination of, 31, 39
  forming thick lens, 33–6, 40–1
  forming thin lens, 11–12, 32–3
Spherical wave, 2, 100
  annular division of, 216–20, 307–8
  converging, 6, 219
  diffraction of, 242–5, 307–20
  lunar division of, 312–14
  passage through a large aperture, 4
  propagation of, 215–20, 223–4
  refraction of, 14–15
Spherometer, 74
Spiral, Cornu, 229, **313–14**, 315–20
Split lens, Billet, 143
Spot diagram, 423
Spread function, 358–9, 465
Spurious maxima in interferogram, 193, 203
Spurious resolution, 338, 357
Square wave grating, 114, 273–6, 334–8
Stack, quarter wave, 210, 212, 532
Standard observer, 431, 461
Standard source, photometric, 455–6
  wavelength, 201–2
Standard wavelength, 200–2
Standing waves, 128, 527–9
Star, double, **130**, 253
Stark effect, 573

Stationary waves, 128, 527–9
Steel, W. H., 608, 611
Steinheil doublet, 408
Stellar interferometer, 253–4
Stereoscope, 440
Stereoscopic effect, 439–44
  film, 441
  microscope, 443–4
  range, 442
  telescope, 442
Sternberg, R. S., 606
Stilb, 457
Stiles, W. S., 608
Stiles-Crawford effect, **433**, 449, 452, 453, 455, 457
Stiles factor, 433, 452, 453
Stimulated emission, 496
Stokes, A. R., 608
Stokes, G. G., 552
Stokes' Law, 496
Stokes parameters, 562, 567–72
  normalized, 570
Stokes vectors, 567–70
Stone, J. M., 608
Stop, aperture, 53
  field, 53
Stop position, effect on aberrations, 393
  coma, 394
  distortion, 394
  vignetting, 54
Straight edge, Fresnel pattern, 316–18
Strehl intensity, 391
Strelenskii, V. E., 594
Strip division of a wavefront, **228–9**, **232–3**, 234–5, 236, **241–2**, 246–7, 273
Stroboscope, 429
Stroke, G. W., 286, 290, 295, 298, 365, 608, 611
Strong, J., 165, 166, 295, 608
Structure of spectral lines, 121, 159–61, 162–3, 185, 201
Strutt, J. W., see Rayleigh, Lord.
Sub-mm spectroscopy, 301
Substage condenser, 67, 333–4, 339–40, **345–6**, **411**
Substitution method in photometry, 463
Sugar, optical activity of, 585
Sun Lu, 605
Sunset, colour of, 495
Superposition of circularly polarized components, 575–7
  of elliptically polarized components, 578–9
  of plane polarized components, 133–6, 538, 549–50
Superposition, principle of, **101**, 477
Superposition of S.H. disturbances, 101–5
  graphical method, 103
Superposition of S.H.M.s at right angles, 134–5
Surface colour, **492**, 523
Surface, spherical refracting, see Spherical refracting surface
Surface topography, 206–10
Swan cube, 459
Symmetrical camera lenses, 413–15
Symmetrical position for echelon, 280
Synthesis, Fourier, 114–18, 258–60
  of image, 335–8, 353–9
System matrix, 41–6
Systems in air, combination of, 30
Systems of refracting surfaces, 31, 39, 41–6

Talbot's bands, 125–7
Tangent condition, 388–9
Tangential fans, 385–6
  reflection of, 406–7
Tangential focus, 385–7, 401, 406, 407
Tangential plane, 380
Taylor, C. A., 610
Taylor, H. D., 415
Taylor criterion for resolution, 181, 183, 194
Telecentric system, 53
Telephoto lens, 38, 57, 58, 417–18
  inverted, 417
Telephoto magnification, 57
Telescope, astronomical, 38, **55–6**, 66–7, 420–1, 453–4; range of, 453–4
  binocular, 69, 442
  brightness of image, 452–3
  Cassegrainian, 70, 420–1
  catadioptic 420–1
  Calilean 38, 55–6
  Gregorian, 70, 421
  Herschelian, 70, 421
  image brightness, 452–3
  limit of resolution, 327–8
  magnification, 25–6, 56–7
  magnifying power, 56–7, 452, 453
    determination of, 81
    necessary, 331
  Newtonian, 70, 421
  objective, **55**, 66, 70, 405, **407–8**; limit of resolution, 327–8
  prismatic, 68–9, 442
  stereoscopic, 442
  terrestrial, 58–9, 421
Telescopic systems, **25–6**, 28, 44, 56–7, 326
Tellurometer, 596, 600
Temporal coherence, **108**, 109, 119–21
Tensor ellipsoid, 483
Ter-mikaelian, M. L., 610
Terminology, in geometrical optics, 8–10
  in photometry and radiometry, 445
  in polarization, 562
Tessar lens, 415–16
Test plate, 155
Thallium bromide-iodide, 97
Thelen, A., 610
Thermal source, 101, 105
Thermopile, 461
Thetford, A., 611
Thick lens, 33–6, 40–1
  in air, 33–6
  cardinal points of, 33–6, 41
  focal length of, 33, 41
  numerical example, 47, 51
  optical centre of, 34–5
  power of, 33, 41
  spherical, 35
Thin achromatic doublet, **62–4**, 66, **402–6**, 408, 412
  monochromatic aberrations of, 404–6, 412
  secondary spectrum of, 402–4
Thin films, 150–3, 522
  highly reflecting, 177, 210–2, 532
  interference in, 150–5, 203–4, 205–13, 522, 529–32
  matrix methods, 529–32
  multilayer, 155, 210–12, 529–32
  multiple reflections in, 205–8, 529–32
  non-reflecting, 154–5, 210, 531
Thin lens, 11–12

aberrations of, 399–402
  in air, 11–12, 32–3
  astigmatism of, 401–2
  cardinal points of, 32–3
  coma of, 400–1
  distortion of, 402
  equivalent, 28, 35
  field curvature of, 402
  neutralization test for power, 71
  power of, 12, 32
  refractive index measurement, 72–3
  smallest separation, object-image, 12–13
  spherical aberration of, 399–400
Thin lenses in contact, 14, 40
  achromatic combination of, 62–3, 403–4
Thin lenses, two separated in air, 36–8
  achromatic combination of, 64–5
Thin prisms, 85–6
Third order aberrations, 382
Thompson, B. J., 611
Thompson, S. W., 278, 296
Thompson prism, 541
Thomson, J. H., 608
Thoria, fused, 455
Three-dimensional grating, 303–4
Threshold of visibility, 316
Thrier, J. C., 605
Throughput, **466**, 467, 468, 469
Time-average holographic interferometry, 367
Time-lapse interferometry, 367
Tint of passage (sensitive tint), 555, **583**, 585
Toepler, A., 369–70
Tolansky, S., 181, 194, 207, 209, 608
Tolerance for aberration, 390–1
Tolerance, coherence, 253
Tolles, R. B., 410
Top hat function, 259
Topogon lens, 417
Topography, surface, 205–10
Toric lens, 427
Total internal reflection, 519–22
Total radiator, 455–6
Total reflection, fustrated filter, 212–13, 522
Total reflection refractometry, 89–91, 92–4
Tourmaline, 542
Tousey, R., 611
Townes, C. H., 497, 595
Transfer equation for paraxial rays, 41
Transfer factor, contrast (modulation), 356–7
Transfer function, optical, 354–9, 390
Transfer (translation) matrix, 42–3
Transfer matrix for thin film, 531
Transform, Fourier, 117–18
  in analysis of waveforms, 118–20
  cosine, 118
  in diffraction theory, 258–60, 273, 336–7, 353–4
  spectroscopy, 161–6, 299, 300, 301, 464, 466, 468–9
Transition tint, 555, **583**, 585
Transition wavelength, 511
Translation equation for paraxial rays, 41
Translucent surface, 448
Transmission echelon, 277, 300
Transmission factor, 450–2
  measurement, 472
Transmission grating, 261–76, 296
Transmittance of polarizer, 544
Transverse chromatic aberration, 61, 64–5, 410, 418

Transverse coherence, **108**, 109–10, 251–4
Transverse magnification, 18, 45
  relation to angular, 20, 56–7
Transverse shift of focus, 378–9, 381
Transverse waves, 1, 131–2, 478, 485
Traub, W., 201
Trichromatic coefficients, 436
  coordinates, 436
  equation, 436
  system of colour specification, 435–8
  units, 436
Triplet, Cooke, 415
Triplet thin lens, 404, 408
Tristimulus colorimeter, 438
Troland, 457
Troup, C. J., 608
Tsujiuchi, J., 611
Tube length, optical, **61**, 330, 410
Tunable laser, 301, 497
Turner, A. F., 285
Turpentine, 580
Twiss, R. Q., 609
Two-dimensional grating, 302
Two systems, combination of, 26–30, 40–1
Two thin lenses in contact, 14, 40, 71
  achromatic combination of, 62–3, 403–4
Two thin lenses, separated, 36–8
  achromatic combination of, 64–5
Twyman and Green interferometer, 161, **166–8**, 171, 173, 290, 296, 298, 348, 594
Tyndall effect, 494

Ultra-microscope, 346
Ultrasonic grating, 276, 598–9
Ultra-violet microscope, 329, 353, 421
Ultra-violet, prisms for, 96
Ultra-violet spectroscopy, 96, 177, 192, 288, 300, 301, 468
Uniaxial crystals, 486, 489–90, 533–9
  double refraction in, 533–9
  optic axis of, 486, 489, 534–7
  plane waves in, 489–90, 534–7
  ray velocity in, 486–9, 535–6
  refractive indices of, 535–6
  wave surface in, 489, 533–7
Uniform point source, 446–7
Uniformly diffusing surface, 448–50
  brightness of, 449
  flux radiated by, 449–50
Unit planes, 18
  See also Principal planes
Unit trichromatic equation, 436
Units, preface
  in colorimetry, 436
  in electromagnetism, 475
  M.K.S. preface,
  in photometry, 446, 455–7
  ratio of emu to esu, 588, 592
  trichromatic, 436
Unpolarized light, 131, 136
  resolution into circular components, 577;
    into elliptic components, 579; into plane
    components, 131–3, 136, 549
Upatnieks, J., 362, 363, 365, 368, 610
Urbach, J. C., 611

V number, **62**, 63, 403, 404, 405–6
Vacuum-grown crystals, 96
Valyus, N. A., 608
Van Cittert-Zernike theorem, 253

Van de Hulst, H. C., 608
Van Heel, A. C. S., 370, 371, 611
Van Vliet, K. M., 612
Vanasse, G. A., 165, 166, 611
Vander Lugt, A., 611
Vander Sluis, K. L., 595
Vašíček, A., 608
Vector addition of disturbances, 103
Vector, Jones, 562–5, 567, 569
Vector notation, 475
Vector polygon, 103
  for diffraction grating, 262, 264–5
  for Fraunhofer circular aperture pattern, 241–2
  for Fraunhofer single slit pattern, 234–5
  for Fresnel pattern, 312–14
  for multiple beam interference, 177–8
Vector polygons for Fraunhofer and Fresnel patterns compared, 228–9
Vector, Poynting, 480
  for anisotropic media, 485, 486
Vector product, definition, 475
Velocities, principal, 483, 535
Velocity of circularly polarized light, 575–7
  of e.m.waves, 477, 483–90, 588, 592
Velocity, group, 122–4, 592, 601–2
Velocity of light, 477, 479
  adopted value, 600
  determination of, 588–603; aberration method, 589; band spectra method, 595–6; cavity resonator method, 593–4; correction to a vacuum, 592; eclipse method, 588–9; I.R. frequency measurement, 601; Kerr cell methods, 590–2, 596–8; Lecher wires method, 592; microwave interferometer method, 594–5; mile pipe experiment, 590; Mount Wilson experiment, 590; Piezo-electric shutter, 598–9, 602; radar method, 593; R.F. method, 596; ratio of units method, 592; rotating mirror methods, 590, 601–3; summary of results, 592–3, 600; toothed wheel method, 589–90; U.S. grating shutter method, 599
Velocity of progressive waves, 1–2
Velocity, phase, 2, 477–9
  in anisotropic media, 483–6, 535–6, 544–6
Velocity potential, 100
Velocity, ray, in anisotropic media, 485–8, 535–6, 544–6
Velocity, wave, 2, 122–4, 477–9
  in anisotropic media, 483–6, 535–6, 544–6
Verdet's constant, 587
Vernier acuity, 433
Vernon-Harcourt lamp, 455
Verrill, J. F., 611
Vertex power, 30
Vibration curve for Fraunhofer single slit pattern, 234–5
  for Fresnel pattern, 312–14
Vibration curves for Fraunhofer and Fresnel patterns compared, 228–9
Vibration direction, effective for light, 131, 529, 533–4
Vibration, plane of, 131, 529, 533, 534
Viewing distance, 57–8
Vignetting, 54, 419
Virtual spaces in optical systems, 8–9
Visibility, 430 •
Visibility of fringes, 159–61, 177, 249–50, 251,

253–4
Visible spectrum, 430, 433–4
Vision, Chapter XVII,
  binocular, 439–44
  colour, 433–8
  persistence of, 429
  photopic, 428, 429, 432–5
  scotopic, 428, 429, 431, 433
  stereoscopic, 439–44
Visual acuity, 432–3
Visual axis of eye, 426
Visual instruments, achromatism of, 404
  image brightness, 452–4
Visual photometry, 458–61
Visual purple, 428–9
Vitreous humour, 424
Voigt effect, 573
Von Rohr, M., 416

Wadley, T. L., 596
Wadsworth mounting, 284, 288
Wahrlich, G. V., 201
Walsh, J. W., 608
Walther, A., 612
Watanabe, N., 201
Water, 479, 499, 572, 573, 587, 601, 602
Watson interference objective, 348–9
Wave, cylindrical, 232
Wave equation, 100, 221, 477
Wave group, 119–24
Wave of finite width, 257–60
Wave motion, general equation of, 100, 221, 477
Wave, plane, 3, 100, 477–9, 480–1
  in absorbing media, 503
  in anisotropic media, 484–6
  in biaxial crystals, 544–6
  diffraction of, 226, 232–3, 234–8, 246–9, 257–60, 261–76
  division into half-period zones, 306–7; into strips, 228–9, 232–3, 234–5, 237, 241–2, 246–7, 273
  passage through a large aperture, 4
  reflection at a plane surface, 4–5
  refraction at a plane surface, 5
  in uniaxial crystals, 490, 534–7
Wave, spherical, 2, 100
  annular division of, 216–20, 307–8
  converging, 6, 219
  diffraction of, 242–5, 307–20
  lunar division of, 312–14
  passage through a large aperture, 4
  propagation of, 215–20, 223–4
  refraction of 14–15
Wave surface for active crystals, 577–8
  for biaxial crystals, 486–9, 544–6
  for uniaxial crystals, 489, 533–7
Wave theory of light, Chapter VI
Wave train of finite length, 105–8, 118–21, 259
  passage through a grating, 124; through a prism, 125
Wave train, infinite, 105, 118, 136
Wave train from a laser, 111–13, 160–1, 497
Wave velocity, 2, 122–4, 477–9
  in anisotropic media, 483–6, 535–6, 544–6
Wave velocity surface, 546
Wavefront, 2–4
Wavefront aberration, 376–87, 420
Wavefront of finite width, 257–60
Wavefront, non uniform, 258–60

Wavefront reconstruction, 359–67
Wavefront shearing interferometer, 171–2, 341, 350–1
Wave-guide, optical, 370
Wavelength, 2, 3
  dominant, 434
  in material medium, 3
  measurement, absolute, 166, 301, 468
  measurement by exact fractions, 185–9
  standard, 200–2
Wavelengths, comparison by coincidences, 160, 184–5, 268
Waves, boundary (edge), **231**, 317–18, 319, 345
  electromagnetic in absorbing media, 503, 509–10; in anisotropic media, 484–90; energy flux, 480–1, 482, 485; Hertz's experiments, 481–2; in isotropic dielectrics, 476–81; in metals, 509–11; velocity of, 477–9, 483–90, 492–3; progressive, 1–2
  secondary, 3–4, 215–16, 219, 224
  simple harmonic, **2–3**, 100, 102–3, 105, 118
  stationary, 128, 527–9
  transverse, 1, 131–2, 478, 485
Webb, R. H., 608
Wedge film, 150–1
Welford, W. T., 607, 608, 612
Wheatley, G. A., 170
Wheatstone, C., 440, 590
White, H. E., 606
White light, 121–2
  equal energy, 434, 435, 439
White light, fringes with, 122, 140, **144–5**, **158**, 166, 174, **191**
Wide-angle lenses, 416–17
Wide screen cinema, 441
Wide source, effect on fringes, 140, 239, 250–1, 253–4
Width of fringes, 176–8, 180–1
Width of principal maxima, 262–4, 265

Width of spectral lines, 107, 112, 119–21, 160–1, 162–3
Wiener, O., 527, 528
Wiener-Khintchine theorem, 162
Wiggins, T. A., 595, 596
Williams, W. E., 278
Windows, entrance and exit, 54
Wolf, E., 565, 566, 605, 610
Wollaston, W. H., 92
Wollaston prism, 351, 542
Wood, R. W., 278, 295, 502
Wormser, E. M., 612
Wright, W. D., 608, 611
Wynne, C. G., 415, 423, 612
Wyszecki, G., 606, 608

X-ray diffraction, 303–4
X-ray microscope, 353

Yamaji, K., 612
Yamanoto, Y., 612
Yard, determination of, 200, 202
Young, H. D., 608
Young, J., 590
Young, M., 612
Young, T., 137, 438
Young's edge waves, 231
Young's eriometer, 245–6
Young's interference experiments, 137–8
  with slits, **138–40**, 248

Zeeman effect, 573
Zehnder-Mach interferometer, **170–1**, 172, 348
Zernike, F., 341, 343
Zernike-van Cittert theorem, 253
Zinc sulphide, 211
Zonal aberration, 382, 409, 410
Zone plate, 309–11
Zones, Fresnel half-period, 216–19, 306–7
Zoom lens, 421–2